Climate change, environmental impact and the limited natural resources urge scientific research and novel technical solutions. The monograph series Green Energy and Technology serves as a publishing platform for scientific and technological approaches to "green"—i.e. environmentally friendly and sustainable—technologies. While a focus lies on energy and power supply, it also covers "green" solutions in industrial engineering and engineering design. Green Energy and Technology addresses researchers, advanced students, technical consultants as well as decision makers in industries and politics. Hence, the level of presentation spans from instructional to highly technical.

Indexed in Scopus.

Indexed in Ei Compendex.

More information about this series at http://www.springer.com/series/8059

Ashwani K. Gupta · Ashoke De ·
Suresh K. Aggarwal · Abhijit Kushari ·
Akshai K. Runchal
Editors

Advances in Energy and Combustion

Safety and sustainability

Springer

Editors
Ashwani K. Gupta
Department of Mechanical Engineering
University of Maryland
College Park, MD, USA

Ashoke De
Department of Aerospace Engineering
Indian Institute of Technology Kanpur
Kanpur, Uttar Pradesh, India

Suresh K. Aggarwal
Department of Mechanical and Industrial Engineering
University of Illinois at Chicago
Chicago, IL, USA

Abhijit Kushari
Department of Aerospace Engineering
Indian Institute of Technology
Kanpur, Uttar Pradesh, India

Akshai K. Runchal
Analytic and Computational Research, Inc.
Los Angeles, CA, USA

ISSN 1865-3529 ISSN 1865-3537 (electronic)
Green Energy and Technology
ISBN 978-981-16-2647-0 ISBN 978-981-16-2648-7 (eBook)
https://doi.org/10.1007/978-981-16-2648-7

This Springer imprint is published by the registered company Springer Nature Singapore Pte Ltd.
The registered company address is: 152 Beach Road, #21-01/04 Gateway East, Singapore 189721, Singapore

Preface

The amount of energy used worldwide is significant, with more than 500 quads of energy per year. Most of the propulsion and power systems use liquid fuels representing 19 million barrels of liquid fuel per day alone in the USA and about 5 times more this amount worldwide. The development of environmentally benign and sustainable energy sources that drive the society and industry continues to be an important fundamental challenge for engineers and scientists. The known available fossil fuel resources are rapidly depleting, and as a result, there is an increased emphasis on developing renewable and alternative energy sources in a sustainable manner. In the near-term, fossil fuels are expected to remain the main driver of our economy and infrastructure as they offer dispatchability and high energy density. We have seen continued improvements in fossil fuel utilization for different technologies over the past century. Current technologies using fossil fuels for energy are also recognized as contributing to global warming, degrading of the environment, and a burden on the economy with extensive cleanup costs. Greenhouse gas emissions are known to be significant contributor to climate change, rising global temperatures, and lasting impact on the protective ozone layer. The consequent increase in global average temperatures, melting of glaciers, and sea-level rise are now serious threats to the human society's very core of existence. Simultaneously, our requirements for energy needs continue to grow due to the considerable growth of the human population in developing countries and increasing living standards in the emerging economies. This means that the average per capita energy use will only increase substantially in the coming decades, with fossil fuels to remain important major sources of energy for all industrial and power applications through direct combustion of fuels. Society urgently needs safe, sustainable, transportable, and more efficient energy sources for various industrial and domestic needs at affordable costs.

The world population has reached over 7.8 billion and is projected to increase by more than 25% to 9.9 billion by 2050. The worldwide energy supply in 2019 was about 170,000 TWh, or 580 quads (quadrillions BTU), and is expected to increase to 815 quads by 2040. Per IEA, if we continue along the current path, without any additional policy changes, the energy demand will rise by 1.3% per year up to 2040. If the policy intentions and targets stated by major energy user nations are implemented, this energy demand rate could decrease to 1% per year. While considerable progress is being made to shift to renewable energy sources, fossil fuels are expected to provide 75% of the world's energy in 2040. While the environment is already under stress, either of these scenarios places a significant additional burden on the environment with a somewhat slower rise in emissions but still far short of sustainability and safety. Therefore, we urgently need a sustainable development scenario that maps out a way to meet energy demand goals, requiring rapid and widespread changes across all energy sectors. The current and foreseeable clean energy technologies are not likely to provide the answer. We need new and innovative technologies with improved efficiency to meet the Paris Agreement's goals by controlling the rise in global temperatures to well below 2 °C and pursuing efforts to limit it to 1.5 °C and meet global objectives cost-effective energy access and cleaner air. The spectrum of the world's energy needs requires multiple solutions. Reduction in emissions (in particular CO_2 and particulates) must be achieved from all kinds of gas, liquid, and solid fuel use. The dispatchable technologies provide efficient and cost-effective energy solutions in all energy-using sectors.

An international forum to discuss the challenges outlined above was started in 2004, focusing on combustion, power, and propulsion technologies using different kinds of fuels available to us. Over the years, it has evolved into annual international workshops on Sustainable Energy, Power and Propulsion (ISEPP), which provide a broad framework of international collaborations and share ideas about research and developments related to cleaner and efficient utilization of energy sources and to develop sustainable alternative and renewable energy technologies. The 11th workshop in this series was held from February 24 to 26, 2020, at the Indian Institute of Technology, Kanpur, India. The workshop's objectives were to promote international collaborations and disseminate knowledge related to safe and sustainable energy initiatives and technology development. The workshop attracted over 100 participants from different academic and research organizations and representatives from governments, industry, and academia. Many world-renowned scientists and researchers from Japan, Malaysia, Thailand, Saudi Arabia, the UK, USA, and India presented their latest research and development findings and shared ideas, informal talks and informal technical exchanges. The TEQIP (Ministry of Human Resource Development, India), the Defense Research and Development Organization (DRDO), India, The National Science Foundation (NSF), USA, and the Office of Naval Research (ONR) Global sponsored the workshop. It was co-sponsored by industries, several leading research organizations, and universities, including Begell House Publishers, Springer/Springer Nature Publishers, Siemens, TSI, Inc., ACRi Los Angeles, NIT Kurukshetra, IIT Kanpur, University of Maryland, College Park, and the University of Illinois at Chicago. The editors wish to express their sincere thanks to the faculty, staff, and students at IIT Kanpur for their outstanding support and warm hospitality to all participants at this international workshop.

This research monograph provides a wealth of knowledge on the latest research and development presented by the eminent scientists and engineers at the workshop. The broad theme of the monograph is energy, combustion, power, safety, and sustainability. The chapters are grouped into six sections: Propulsion Systems, Fuels,

Flames, and Reacting Systems, Combustion Systems, Fires, and Sustainable Energy Technologies. These chapters deal with prototype systems using state-of-the-art diagnostic and analytic techniques. Each chapter is written by a renowned international authority in the field of study. Several chapters deal with specific design aspects. In contrast, the others deal with novel concepts using alternative energy sources for improved performance of next generation of advanced power and propulsion systems, with due considerations to the emission of various trace pollutants, such as polycyclic aromatic hydrocarbons (PAH), dioxins, furans, NOx, unburnt hydrocarbons, carbon monoxide, carbon, and particulates. Thc contributions of cutting-edge research and development technologies should be of paramount interest to researchers and engineers working on safe and sustainable energy technologies in academia and industry. We hope the book proves to be an asset to those working in combustion and clean energy.

College Park, USA	Ashwani K. Gupta
Kanpur, India	Ashoke De
Chicago, USA	Suresh K. Aggarwal
Kanpur, India	Abhijit Kushari
Los Angeles, USA	Akshai K. Runchal

Introduction

This book is evolved from the International Workshop on Energy, Environment and Multiphase Flows held at IIT Kanpur during February 24–26, 2020, with the sponsorship from the Office of Naval Research Global (ONR Global), US National Science Foundation (NSF), TEQUIP (India), Defense Research and Development Organization (India), Begell House Publishers, Springer/Springer Nature Publishers, Siemens, and TSI Inc. The workshop was co-sponsored by IIT Kanpur, University of Maryland College Park, USA, the University of Illinois at Chicago, USA, and Analytic and Computational Research Inc. (ACRI) Los Angeles, USA. This workshop represented the 11th International Workshop on the Broad Theme of Energy, Power, and Propulsion, with the first workshop in 2004. Over the years, the workshop has grown into a world-class forum for sustainable fuels, energy, power, and propulsion, focusing on clean energy production and utilization. This volume is focused on fuels, combustion, energy, safety, and sustainability representing many areas of engineering importance in energy, power, and propulsion. The Chief Guest at the workshop was Dr. Satish Kumar, Director of National Institute of Technology, Kurukshetra, Haryana, India, who provided a wealth of information from his rich, extensive prior experiences in various leadership positions at Defense Research and Development Organization (DRDO), India. Renowned international authorities delivered some 28 invited lectures in their specific expertise on the various topics covered at the workshop. Twenty-four topics on advances in energy and combustion, emphasizing safety and sustainability from the selected authorities, were delivered. Subsequently, they contributed a chapter on their topic, which are included in this book. This book is aimed at researchers working in energy conversion, waste, pollution control, modeling, diagnostics, fluid mechanics, heat transfer, biofuels production, and energy sustainability and safety. This book should also benefit the design and development engineers of power plant equipment and managers from industry and government with the latest scientific developments reported on safe and sustainable use of fossil fuels, biofuels, and low-grade fuels in specific applications.

Ashwani K. Gupta
Ashoke De
Suresh K. Aggarwal
Abhijit Kushari
Akshai K. Runchal

Contents

Flames and Reacting Systems

Combustion Systems

Fires

Sustainable Energy Technologies

About the Editors

Dr. Ashwani K. Gupta is a Distinguished University Professor at the University of Maryland (UMD), USA. He obtained his Ph.D. and D.Sc. from the University of Sheffield, UK. He was awarded D.Sc. from Southampton University, UK, and Honorary Doctorates from the University of Wisconsin–Milwaukee, USA, King Mongkut's University of Technology North Bangkok, Thailand, and University of Derby, UK. He received the President Kirwan Research Award and College of Engineering Research Award at UMD. He has received several honors and awards from AIAA and ASME, and has co-authored over 800 papers, three books, five patents, and edited 18 books in the areas of combustion, swirl flows, high temperature air combustion (HiTAC), distributed combustion, wastes to energy, acid gas treatment, fuel reforming, and air pollution. He is Honorary Fellow of ASME and RAeS (UK), and Fellow of AIAA, SAE, and AAAS, and Member of European Academy of Sciences and Arts (EASA).

Dr. Ashoke De is a Professor in the Department of Aerospace Engineering, Indian Institute of Technology (IIT) Kanpur. He received his MS from IIT Kanpur, and Ph.D. from Louisiana State University, USA. Before joining IIT Kanpur, he has worked as post-doctoral scholar at Technical University of Delft (TU-Delft), Netherlands and as Research Engineer in GE Global research at Bangalore, India. He is the recipient of DAAD Fellowship (2016), IEI-Young Engineer's Award-2014 (Aerospace Engineering), Early Career Research award-2015 (DST, India), Expertise Award-2010 (GE Global research), and P. K. Kelkar Young Research Fellowship (2015–2018, IITK). He is AvHumboldt Fellow, Fellow of West Bengal Academy of Science and Technology, Associate Editor of *International Journal of Energy for a Clean Environment* (2019–present), and currently member of APS, AIAA (senior member), ASME, SIAM, FMFP, ISHMT and Combustion Institute. He leads large scale initiatives in the modelling of turbulent flows at IIT Kanpur and authored over 130 papers, and four edited books. His current research interests include multiphase modeling, hybrid RANS/LES model development, supersonic flows and Fluid-Structure interactions (FSI).

Dr. Suresh K. Aggarwal received his Ph.D. from Georgia Institute of Technology, Atlanta, USA. He was a member of the research staff at Princeton University, New Jersey, USA and a Senior Research Engineer at CMU. He then joined the University of Illinois at Chicago, USA and was promoted to Professor in 1995. He served as Director of Graduate Studies, and was also a Visiting Scientist at ANL, Visiting Professor at Ecole Centrale-Paris, France and Guest Professor at Jiangsu University, China. His research interests include Combustion, Multiphase Reacting Flows, Emissions, Clean Energy, and Fire Suppression. He has authored over 340 publications and graduated 17 Ph.D. and 46 MS students. He is Fellow of ASME, AAAS and Associate Fellow of AIAA.

Dr. Abhijit Kushari is Professor in the Department of Aerospace Engineering, Indian Institute of Technology (IIT) Kanpur, India. Dr. Kushari received his Ph.D. from Georgia Institute of Technology, Atlanta, Georgia, in 2000 and joined IIT Kanpur as an Assistant Professor in 2001. He became a Professor in 2014. His research interests are Aerospace Propulsion, Gas Turbine engines, Turbo-machinery, Liquid Atomization Flow Control and Combustion Dynamics. He has authored more than 130 technical papers. He has supervised six Ph.D. and 50 M.Tech. theses in Aerospace Propulsion fields.

Dr. Akshai K. Runchal is an expert in Computational Fluid Dynamics (CFD) and numerical simulation of flow, heat and mass transport processes in engineering and environmental sciences. In late 1960s under the guidance of Prof. D. B. Spalding at Imperial College, London, Dr. Runchal, and Dr. Wolfshtein, developed the Finite Volume Method (FVM) for CFD applications. His experience spans a range of problems including those related to design, production, operation and environmental impact of industrial and urban projects. One of his specific areas of expertise has been analysis and management of natural and man-made disaster scenarios. Dr. Runchal received his Ph.D. from Imperial College, London (UK) in 1969 and a Bachelor of Engineering with Honors from Punjab Engineering College (Chandigarh, India) in 1964. He has taught and guided research students at UCLA, Cal Tech, CSU (Northridge), Imperial College (London), and IIT (Kanpur). He is the author or co-author of six books and over 200 technical publications. He has delivered keynote and invited talks at more than 100 conferences and seminars in the USA, the UK, France, Germany, India, Australia, Belgium, Sweden, Croatia, Czechoslovakia and Slovenia.

Propulsion Systems

Injector Dynamics and Pressure Gain in Rotating Detonation Engines

K. Kailasanath

Abstract Rotating detonation engines are an example of a pressure gain combustion device. By definition, detonation is a pressure gain combustion process, and hence, a combustor based on detonations will always produce pressure gain. However, this does not always translate into pressure gain for an engine based on detonations since there are other components involved in an engine. Numerical simulations of an idealized "engine," consisting of a simple inlet (with injectors), combustor and exhaust nozzle is used to illustrate this issue. The inlet injector area ratio is shown to be a key parameter. A poor choice of this parameter could result in inlet losses that are larger than the pressure gain in the combustor, resulting in an overall system that does not produce pressure gain. Some other choices will produce overall pressure gain but may induce unwanted pressure oscillations. Hence, a suitable design of the inlet-combustor interface is shown to be a critical factor in the overall development of rotating detonation engines.

Keywords Detonation engines · Pressure gain · Injectors

1 Background

Typical combustion engines in current use for propulsion or power use flames to convert the fuel into products and release the chemical energy. In theory, the conversion process takes place under constant pressure conditions, while in practice, there is some pressure loss. Detonations provide an alternate form of energy conversion. Detonations are a pressure gain combustion process and hence result in higher thermodynamic efficiency, which can result in better fuel efficiency and reduced emissions.

There is world-wide interest in exploring and developing detonation engines [1]. An overview of detonation engines as well as recent developments has been discussed elsewhere [2]. For the sake of completeness and to motivate the work presented in this paper, some of the key observations from the past are repeated here. The initial

K. Kailasanath (✉)
Lorton, VA, USA

A. Gupta et al. (eds.), *Advances in Energy and Combustion*, Green Energy and Technology,
https://doi.org/10.1007/978-981-16-2648-7_1

emphasis on the development of detonation engines was focussed on "intermittent" or "pulsed" detonation engines, called PDE. The basic concept and early work has been reviewed by Eidelman et al. [3]. There was uncertainty even in the estimates of the ideal performance of these engines [4]. Advances in laser diagnostics and numerical simulations enabled better performance estimates as well as development of laboratory-scale devices that produced close to the estimated performance. Research in the 1990s and early 2000s also identified key factors that control the performance and operation of pulse detonation engines. The inherent intermittency of the PDE, as well as the need to repeatedly initiate detonations rapidly and with the use of very little energy, are major obstacles in the development of PDEs. The attention then shifted to continuous detonation engines with a single initiation event to overcome the above obstacles. Rotating detonation engines (RDE) are a form of continuous detonation engine, where a detonation moves continuously in a annular, cylindrical chamber, as long as it is supplied continuously with fresh detonable mixture. Detonation needs to be initiated only once at the beginning of the operation. Hence, this concept overcomes two of the key difficulties in the development of PDEs. However, this does introduce new issues that need to be researched and resolved. One such issue is pressure gain in these engines.

In this paper, key factors that control the pressure gain in these devices are explored in more detail using numerical simulations of a generic rotating detonation engine (RDE). A generic configuration has been chosen so that the results can have wider applicability than choosing a specific engine under current development. The focus is also on identifying key factors that may be useful in further development of this type of engine. The fuel considered in the simulations is hydrogen, because its chemistry is well understood and can be represented in a tractable manner in complex three-dimensional simulations. However, the observations presented are more general and not dependent on the specific fuel.

2 A Rotating Detonation Engine

The generic configuration of a rotating detonation engine used in the simulations is shown in Fig. 1. It consists of an annular cylinder as the combustion chamber with an injector plate at one end (bottom of the figure) and an exhaust nozzle at the other end (top of the figure).

The fresh fuel–air mixture is injected from the bottom and the products of detonation flow out through the nozzle at the top. Once a detonation wave is initiated in the combustion chamber, it keeps moving in a circumferential direction (rotating) as long as the inflow of detonable fuel–air mixture is sustained. Further details of the flow field are not presented here since they have been discussed previously in a number of papers [5, 6].

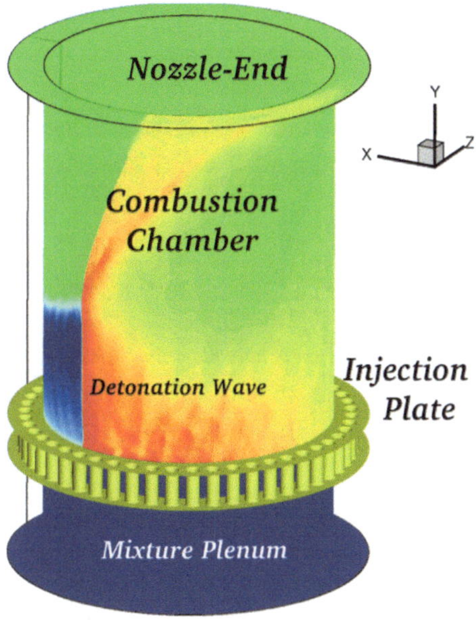

Fig. 1 Simulation of the flow field in a generic rotating detonation engine (RDE)

2.1 Modeling Detonation Engines

The emphasis of this paper is on the impact of the injector geometry on the flow field and dynamics within the engine. Hence, a full three-dimensional simulation that can adequately represent the geometrical details is required. However, the standard unsteady Euler equations can be used since viscous effects have very little influence over the very short time scales involved in detonation propagation. This also allows us to keep the computational cost reasonable and conduct a representative set of simulations to explore various parametric changes.

3 Effect of Area Ratio

Before considering specific types of inflow injectors, let us first consider the generic problem. A key characteristic defining an inflow injector is the open area as a fraction of the total area available. We call this the area ratio for the purpose of our discussion here. A series of numerical simulations was conducted in which this area ratio was varied from 0.1 (small opening) to 0.75 (large opening). The pressure in the upstream plenum chamber was held constant at 10 atmospheres for all the cases. The pressure in the combustor chamber just at the exit of the injector and the pressure at the exit of the combustor after combustion were noted in each of the cases. These results are presented in Table 1, along with those holding the mass flow constant. Of course in all cases, the pressure at the exit plane of the combustor is higher than at the inlet, because

Table 1 Effect of inlet area-ratio on pressure gain

Area ratio	P_{plenum} (atm)	$P_{t,inlet}^{*}$	$P_{t,exit}$ (atm)	% inlet choked
Set 1. Constant plenum pressure				
0.1	10	1.79	3.92	87.8%
0.2	10	2.99	6.66	90.7%
0.3	10	4.25	9.39	94.4%
0.4	10	5.35	11.91	94.6%
0.5	10	6.27	14.08	94.9%
0.75	10	8.69	18.38	95.0%
Set 2. Approximately hold mass flow constant				
0.1	16.47	3.13	6.59	82.4
0.2	10.00	2.99	6.66	90.7
0.3	6.94	3.15	6.72	72.0
0.4	5.53	3.21	6.55	64.9
0.5	4.71	3.21	6.78	60.3
0.75	3.40	3.39	6.21	26.7

*Inlet pressure only integrated over choked area

of the pressure gained during the detonative combustion process. However, when we also take into account the initial plenum pressure, we observe that the pressure at the exit is lower than the plenum pressure for all cases where the area ratio is smaller than 0.4. This indicated that for these cases, the pressure drop across the injector is larger than the pressure gain in the combustor. Thus, a key injector design parameter to achieve overall pressure gain is the inlet area ratio, which needs to be about 0.4 or larger. The second set of data in Table 1, where the mass flow is held constant, also suggests the same conclusion. These results provide us with some guidance on the overall design of the injector. However, further consideration must be given because with the larger opening there is the possibility of back flow into the plenum chamber. In the case of a flight device, this could result in back flow or pressure oscillations in the inlet. Therefore, a better understanding of how to isolate the pre-mixture plenum from the combustion chamber is essential to practically achiever pressure gain in these types of devices. Let us look at some specific type of injectors to understand this problem better.

4 Impact of Slot Injectors

The basic RDE geometry considered has an inner diameter of 80 mm and an outer diameter of 100 mm, giving a wall thickness of 20 mm for the injectors. First a simple slot type of injector (as shown in Fig. 2) was simulated. A total of 50 injectors were

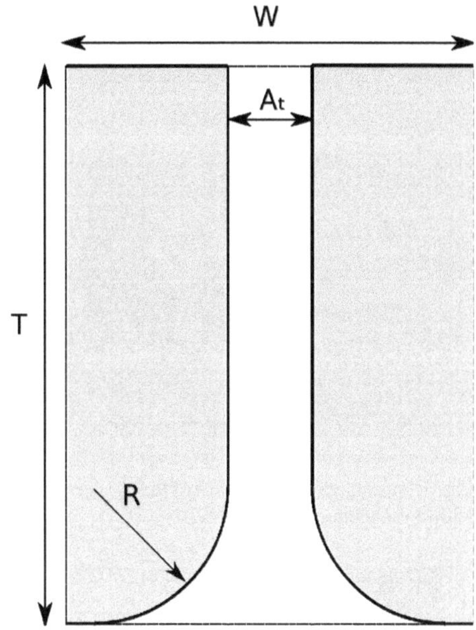

Fig. 2 Schematic of the slot injectors used in the simulations

used with spacing, w of 5.457 mm and a height T of 10 mm. The throat area, A_t, was 1.131 for the area ratio of 0.2 and 2.262 for the area ratio of 0.4.

With these slot injectors, both the temperature and pressure fields within the combustion chamber show the clear presence of relatively cool jets at the injectors and dead zones of higher temperature gas existing between the jets. No back flow into the mixture plenum was observed in the cases simulated. However, pressure waves are seen to propagate into the mixture plenum through the injector opening, as the detonation wave moves past the injectors, as shown in Fig. 3.

5 Impact of Cylindrical Injectors

Next cylindrical injectors were considered. Again 50 injectors were used resulting in the same average spacing as in the previous case. A representation of the installed injectors is shown in a cut-out of the RDE geometry in Fig. 4.

The simulations with cylindrical injectors show the same basic pattern as with the slot injectors. However, the presence of the dead zones between injectors (discussed above) is smaller because of the shape of the injector. As before, there is pressure feedback into the plenum, as seen in Fig. 5. The pressure gradient shows much more structure than with the slot injectors. Multiple shock waves emanate from the injectors into the high-pressure region behind the detonation wave.

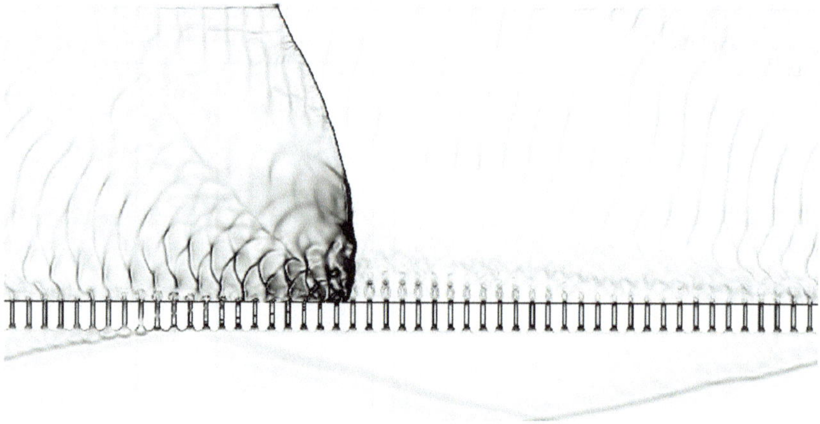

Fig. 3 Instantaneous snapshot of the pressure gradient field showing the presence of pressure waves within the plenum chamber

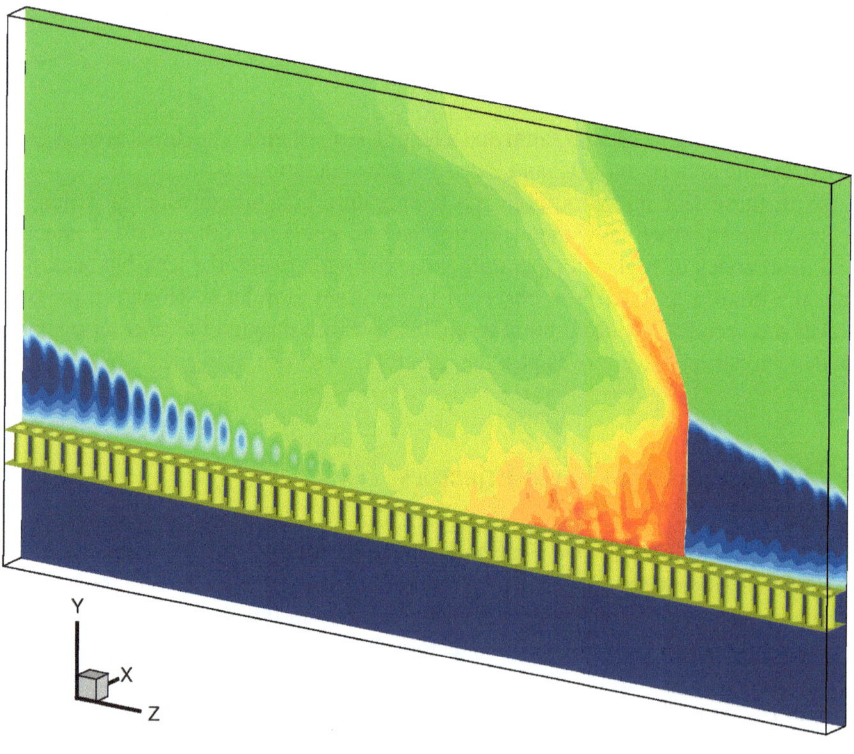

Fig. 4 Cut open view of the RDE geometry showing the cylindrical injectors

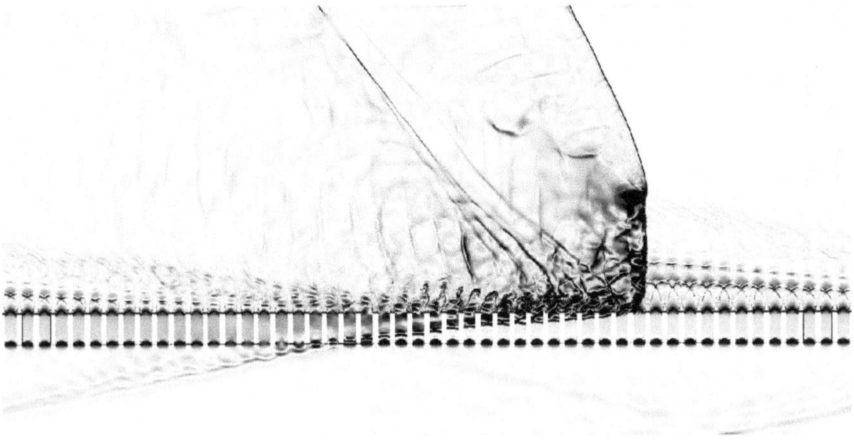

Fig. 5 Instantaneous snapshot of the pressure gradient field for the simulation with cylindrical injectors

6 Impact of Pintel Injectors

The next type of injector considered was of the pintel type, as shown in Fig. 6.

Fig. 6 Schematic of the pintel type of injector

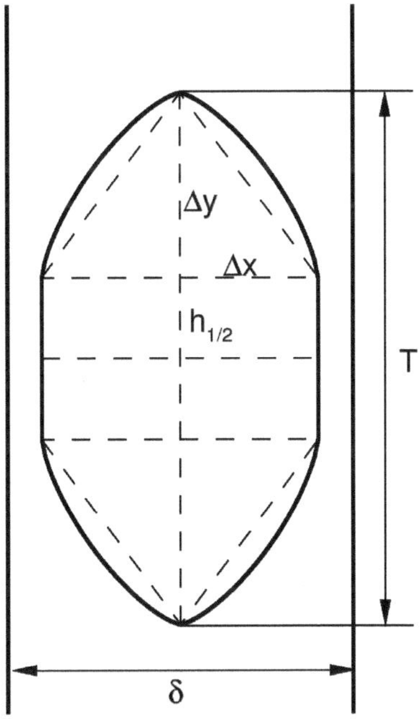

Fig. 7 Instantaneous snapshot of the pressure gradient field for the simulation with pintel injectors

The pintel type of injector resulted in a much smoother temperature and pressure profile within the combustion chamber but it did produce pressure waves propagating into the plenum chamber, as in the previous cases discussed above. As before, the pressure gradient distribution within the plenum chamber is shown in Fig. 7.

7 Performance and Stability

The stability and overall performance with the various types of injectors were compared to simulations where inflow boundary conditions (ideal case, for reference) were used rather than physical injectors of the various types discussed above. There were only small changes in the performance as measured by the specific impulse. With an area ratio of 0.2, the performance for all cases including those with the ideal reference case was within 1%. With the larger area ratio of 0.4, there was a decrease in performance with the physical injectors ranging from 2.4 to 7.7%. The reduction in performance was most noticeable as low (3 atm) and high pressures (15 atm). Similarly, the stability of the waves in the system was also impacted for the extremes in operating pressure. Several different modes of failure including propagation in the reverse direction were observed. For the lower pressure case, detonation wave failure and re-ignition continually happens near the injection plate. Sometimes, the re-ignition event can be strong enough to initiate a detonation wave propagating in the opposite direction to the original direction of propagation.

To visually convey the level of pressure oscillations, the pressure traces in the center of the plenum chamber are compared for the simulations with the different aspect ratios of 0.2 and 0.4 for the slot injector case in Fig. 8.

In general, for any type of injector, the amplitude of pressure oscillations was larger for the larger area ratios.

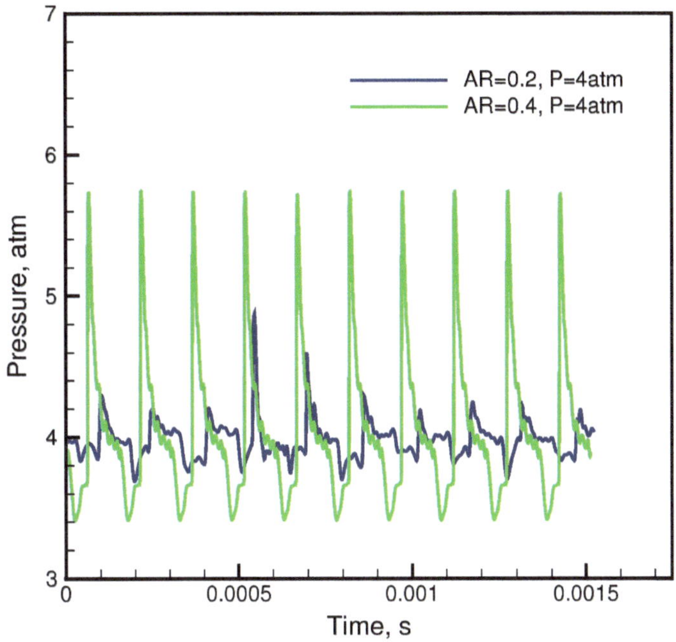

Fig. 8 Comparison of pressure traces in the center of the plenum for the slot injector system with an area ratio (AR) of 0.2 and 0.4. The mean pressure is 4 atm

8 Concluding Remarks

Numerical simulations have been used to explore the impact of various types of injectors on the flow field and performance of rotating detonation engines. It is shown that some choices for the fuel–air injectors could result in a large pressure drop that could negate the pressure gained during the detonative combustion process. The inlet injector area ratio is shown to be a key parameter. Relatively large area ratios (greater than 0.4) will provide better pressure gain. However, with the larger area ratios, the pressure feedback and oscillations in the injector plenum chamber are shown to be larger. In some cases, instability and even failure of the detonation could result. Hence, the choice of the particular type of injector, especially the area ratio, is critical for the successful development of these engines. Further injector choices, including isolators to limit the upstream pressure propagation, are suggested for further research and development.

References

1. Kailasanath K (2000) Review of propulsion applications of detonation waves. AIAA J 38(9):1698–1708
2. Kailasanath K (2020) Recent developments in the research on pressure-gain combustion devices. In: Gupta AK, De A, Aggarwal SK, Kushari A, Runchal A (eds) Innovations in sustainable energy and cleaner environment. Springer, pp 3–21
3. Eidelman S, Grossmann W, Lottati I (1991) Review of propulsion applications and numerical simulations of the pulse detonation engine concept. J Prop Power 7(6):857–865
4. Kailasanath K (2001) A review of PDE research—performance estimates. AIAA paper 2001-0474. AIAA, Reston, VA
5. Schwer DA, Kailasanath K Numerical investigation of rotating detonation engines. AIAA 2010-6880. AIAA, Reston, VA
6. Schwer DA, Kailasanath K Numerical investigation of the physics of rotating-detonation-engines. In: Proceedings of the combustion institute, vol 33. The Combustion Institute, Pittsburgh, PA, pp 2195–2202

Low Emissions Propulsion Engine Characterization Process

Hukam C. Mongia

Abstract Most of the propulsion engines use Low-Emissions rich-quench lean combustors LEC except recently introduced GEnx (2009) and LEAP-X (2016) based on the application of a second-generation lean dome combustion technology TAPS demonstration (2003). These two technologies are targeted for replacing pre-LEC rich domes or the 1st generation lean-dome technology (known popularly as Dual-Annular Combustors DAC) combustors. Sixteen combustors have been selected from three engine manufacturers (CFM International, GE Aircraft Engines, and Rolls-Royce Plc.) involving these four technologies covering a broad range of ICAO takeoff pressure ratios (21–49) and thrust rating (77–514 kN) with engine emissions certification testing conducted during 1985 and 2017. Even though the ICAO emissions regulatory criteria given by engines' landing takeoff emissions of NO_x, CO and hydrocarbons normalized by takeoff thrust, g/kN (LTO NO_x, CO and HC) should be acceptable, they hardly provide any guidance during combustion design and development process. This paper summarizes a low-emissions technology evaluation criterion that originally started in the middle 1970s under the NASA-sponsored Pollution Reduction Technology Program (PRTP). Application of this criteria to the selected 16 propulsion engine combustion products shows that the LEC takeoff NOxEI correlation (NOxEI $= 0.0563 \times \mathrm{PR}^{1.1796}$ with $R^2 = 0.986$) along with idle COEI and HC correlation (COEI $= 8.231 + 7.196 \times$ HCEI with $R^2 = 0.9949$) and idle HCEI entitlement of zero can be used for comparing all rich-dome combustion technologies and products. The corresponding takeoff NOxEI correlations for the 1st and 2nd generation lean dome combustion products (DAC and TAPS) NOxEI $= 1.110\mathrm{E} - 03 \times \mathrm{PR}^{2.950}$ with $R^2 = 0.9772$ and NOxEI $= 7.480\mathrm{E} - 05 \times \mathrm{PR}^{3.4534}$ with $R^2 = 0.9985$, respectively, should be used for guiding future lean-dome combustion products in addition to matching or exceeding idle COEI versus takeoff NOxEI tradeoff correlation established by GEnx product, namely Idle COEI $= 32.183$ Takeoff NOxEI$^{-0.223}$ with $R^2 = 0.8967$ while simultaneously its idle COEI and HCEI is close to that of the LEC correlation COEI $= 8.231 + 7.196 \times$ HCEI.

H. C. Mongia (✉)
CSTI Associates, LLC, Yardley, PA, USA

A. Gupta et al. (eds.), *Advances in Energy and Combustion*, Green Energy and Technology,
https://doi.org/10.1007/978-981-16-2648-7_2

Keywords Gas turbine combustors · Combustion technology · CFD · Design development process · Low emissions combustors

1 Introduction

All the major aero propulsion engine manufacturers have been intensively involved in reducing pollutant emissions starting in the 1970s leading to the 1st generation of low-emissions combustors in several modern engines and enforcement of the International Civil Aviation Organization ICAO 1st set of emissions regulatory standards known popularly as Committee on Aviation Environmental Protection (CAEP) CAEP 1 https://www.icao.int/environmental-protection/Pages/Caep.aspx where from Fig. 1 has been reproduced. The first CAEE committee and subsequent CAEP committee members have maintained a highly admirable professional environment that helped formulate emissions regulatory levels equitably for all Original Equipment Manufacturers OEM's, small, medium, and large size propulsion engines as shown in Fig. 2.

The original CAEP (or CAEP1) regulatory standards for engine exhaust emissions during landing takeoff are given by the following expressions:

$$\text{SAE Smoke Number} = \text{Minimum}\left[50, \frac{83.6}{F_{00}^{0.274}}\right]$$

Hydrocarbons HC = 19.6 gm CH_4/kN.
Carbon monoxide CO = 118 gm/kN.

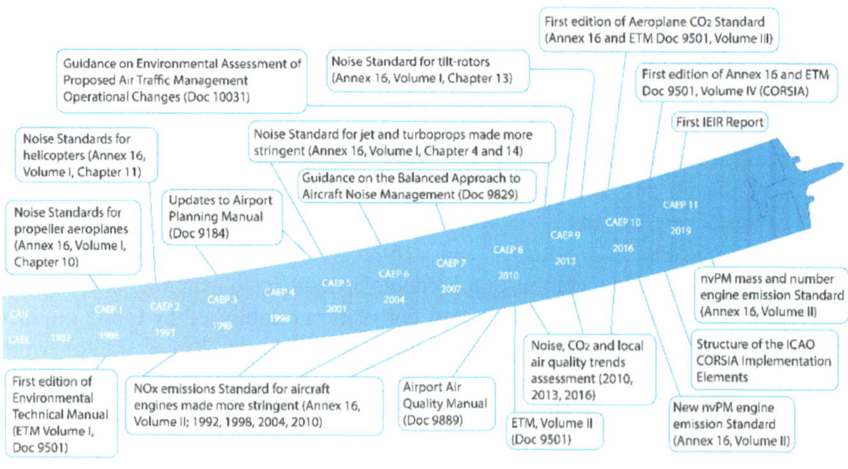

Fig. 1 Major milestones of CAEP since 1986; reproduced from https://www.icao.int/environmental-protection/Pages/Caep.aspx

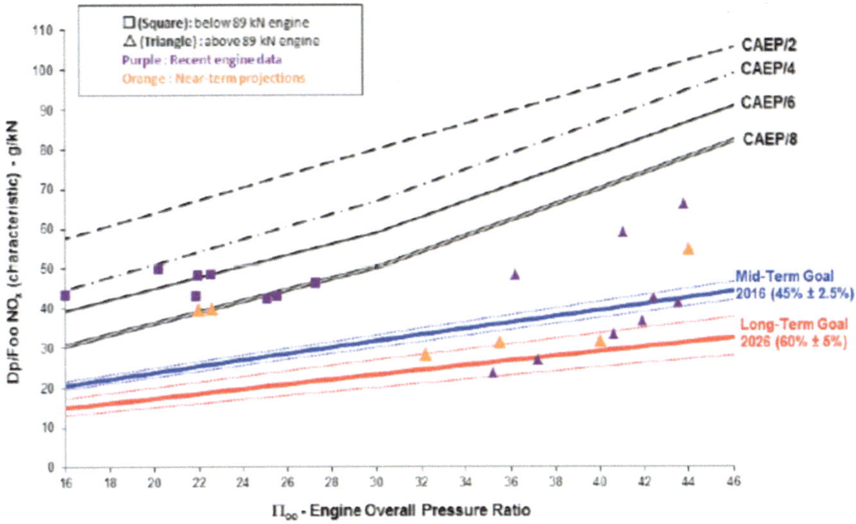

Fig. 2 LTO NOx stringencies from the original CAEP (adopted in 1981 through CAEP2, —CAEP8 (adopted respectively in 1993, 1999, 2005 and 2011) compared to Mid-term and Long-Term goals; reproduced from https://www.icao.int/Meetings/EnvironmentalWorkshops/Documents/2014-Kenya/4-1_LAQ-Technology_notes.pdf

Oxides of nitrogen $NO_x = 40 + 2xPR$ for all engines manufactured before January 1, 2000 with rated thrust $F_{00} > 26.7$ kN. Here, PR is the ICAO-specified engine takeoff pressure ratio.

At 30 PR, the CAEP1 regulatory standard requires that engines with LTO NO_x more than100 gm of NO_2/kN will not be allowed to operate. Therefore, if possible, OEM's would like to have a 30% margin from the standards shown by dotted blue lines in Fig. 3 except for smoke where the preference is generally for 50% margin.

It is educational to pay attention to how low-emissions combustion product evolution took place starting with the engine emissions databank available from the link: https://www.easa.europa.eu/easa-and-you/environment/icao-aircraft-engine-emissions-databank.

We plan to cover engines manufactured by the three OEM (CFM, GE, and RR) listed in this database. For the GE engines, the first engine to incorporate rich-dome Lean-Combustion or Rich-Quench-Lean Combustion LEC/RQL technology was the CF6-80C2 (with emissions test start date of January 13, 1995) followed by quick (but not optimized) transition of LEC technology to GE90 (April 24, 1997). Only minor LEC optimization was undertaken for the GE90-110/115B (November 23, 2003) because of the long-term objective of replacing lean-dome Dual-Annular Combustion system (1st generation) by the 2nd generation lean-dome Twin-Annular partially Premixing Swirlers based concentric flames TAPS [1, 2]. A new centerline CF34-10E was the 4th LEC combustion product (July 24, 2004) followed by drop-in

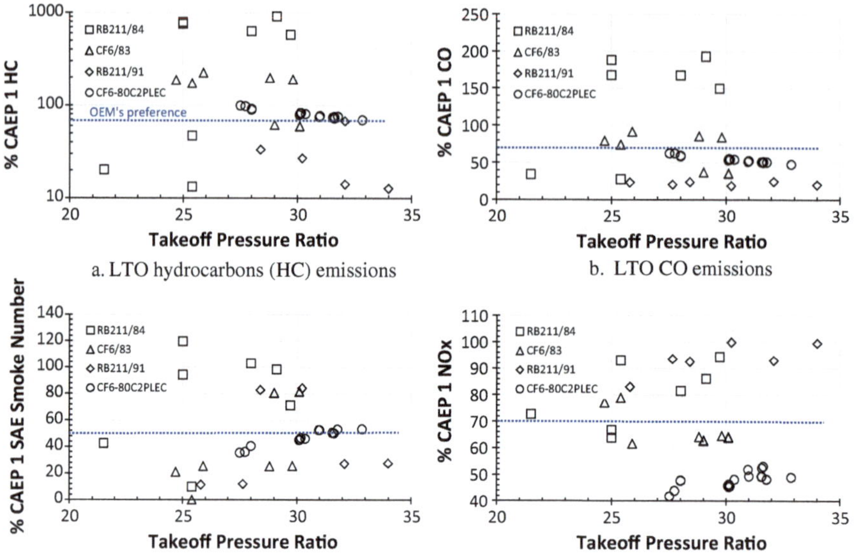

a. LTO hydrocarbons (HC) emissions
b. LTO CO emissions
c. Characteristic smoke emissions
d. LTO NOx emissions

Fig. 3 Remarkable propulsion engine combustion product emissions reduction trends established by the primary leaders of combustor designers community before 1992

LEC combustor for the CFM56-5B and CFM56-7B emissions tested on September 29, 2003.

In summary, the author believes that GE and CFM LEC combustion product development activities were not followed as intensively [4] as for the 1st and 2nd generation lean dome products [1–3], namely DAC and TAPS. The DAC combustion products development during the early 1990s started from an extensive technology know-how and empirical correlations [5–8]. The LEC product development effort relied heavily on the anchored CFD approach [9] since the most reputed semi-analytical correlations [10, 11] didn't produce right trends for NO_x as function of fuel/air ratio FAR [12]. However, the anchored modeling approach needed to be modified for future application in development of lean combustion product [13, 14] which led to formulation and validation of comprehensive models overviewed in [15]. Improved semi-empirical and semi-analytical correlations developed later [16–18], if used, can provide more insight needed for future LEC combustion products.

The RR combustion community is very proud of its RQL combustion products which provided the author motivation to assess emissions characteristics of their rich-domes relative to LEC. It should be pointed out that the author doesn't have first-hand knowledge of the RR RQL combustion products. Therefore, his process of selection and grouping of their engine families was driven mostly by the RR engines' emissions test start date. The legend RB211/84 in Fig. 3 includes all the RB211 engine models with engine start dates falling between March 1979 and July 1984 whereas the legend RB211/91 groups together engine tested between September 1986 and April 1991.

Similarly, the legend CF6-80C2PLEC in Fig. 3 are the engine models tested on May 29, 1983 whereas CF6/83 includes all CF6 engine models tested prior to May 29, 1983. The summary of these engine groupings given in Table 1 highlights the minimum and maximum extent of HC, CO, NO_x and smoke emissions.

It's remarkable for the RR RB211/84 combustion team to step up to reduce HC from its maximum value of 915 to take the challenge of reducing it enough to meet the CAEP1 requirement. Its maximum value was 67.7 as shown in Table 1 for RB211/92. Similarly, good progress was made on meeting CO which came down from 192.9 to 24.4. However, the RR combustion team struggled in bringing down NOx and smoke emissions.

The GE engines group CF6/83 didn't have as much challenging tasks. Here, they had to reduce its maximum HC value of 225 to less than 100, which came down to 99.5 in the CF6-80C2 which achieved or exceeded OEM's target values shown in Fig. 3.

Subsequently, CAEP NO_x regulatory standards have gone through more stringent requirements with introduction of CAEP2, CAEP4, CAEP6, and CAEP8 for engines produced after January 1 of the years 2000, 2004, 2008 and 2014, respectively. The corresponding LTO NOx regulatory values at 30 PR for all engines with takeoff thrust $\geq$89.1 kN are 80, 67, 59, and 50. In other words, relative to CAEP1, the percentage reductions in the regulatory NOx standard are 20%, 33%, 41%, and 50%. Or, the successive NOx reduction steps going from CAEP1 to CAEP8 have been 20%, 16%, 12%, and 15%, respectively.

What started out as a simple expression for CAEP1 NOx standard (namely, 40 + 2PR) has become quite complicated CAEP8 standard as illustrated below. This was driven mostly by one objective: The regulatory standards should be fair to all OEM's whether manufacturing small, medium or large engines in regard to rated static engine takeoff thrust F_{00} and the attendant overall operating takeoff pressure ratio PR. Some engines static takeoff pressure ratio may be medium, high and very high. These groups are currently categorized into the following groups:

$$F_{00} = \begin{matrix} \leq 26.7\,\text{kN} \\ 26.7 - 89\,\text{kN} \\ > 89\,\text{kN} \end{matrix}$$

and

$$PR = \begin{matrix} \leq 30 \\ 30 - 104.7 \\ > 104.7 \end{matrix}$$

The CAEP8 regulatory standards for engines that fall into more than 26.7 kN but not more than 89 kN are quite complex as summarized below:

Table 1 Minimum and maximum range of LTO emissions normalized by regulatory levels of the original CAEP (CAEP1) proposed in 1981 for the selected four engine families

Legends	Hydrocarbons		Carbon monoxide		Oxides of nitrogen		SAE Smoke number	
	Minimum	Maximum	Minimum	Maximum	Minimum	Maximum	Minimum	Maximum
RB211/84	13.3	915.4	27.6	192.9	63.9	94.4	9.8	119.7
RB211/92	3.6	67.7	18.8	24.4	83	99.9	11.4	84.2
CF6/83	59.2	225.3	35.3	91.3	61.6	78.9	20.9	81.6
CF6-80C2PLEC	69.4	99.5	47.5	62.4	41.8	53.1	35.5	53.3

$$\begin{aligned}\text{LTO NO}_x = 40.052 + 1.5681\text{PR} - 0.3615\text{F}_{00} - 0.0018\text{PR} \\ \times \text{F}_{00} \text{ when PR} \leq 3041.9435 + 1.505\text{PR} \\ - 0.5823\text{F}_{00} - 0.005562\text{PR} \\ \times \text{F}_{00} \text{ when } 30 < \text{PR} < 104.7\end{aligned}$$

Fortunately, we will not discuss this class of engines in this chapter. Focus of this chapter is on propulsion engines with takeoff thrust more than 89 kN for which CAEP8 standards are quite simple:

$$\text{LTO NO}_x = \begin{array}{c} 7.88 + 1.408\text{PR when PR} \leq 30 \\ -9.88 + 2\text{PR when } 30 < \text{PR} < 104.7 \\ 32 + 1.6\text{PR when PR} > 104.7 \end{array}$$

As mentioned earlier, CAEP1 NO_x requirement for overall pressure ratio OPR of 30 is 100 g/kN compared to 50 g/kN as per the CAEP8 NOx requirement. It means that the CF6-80C2PreLEC's LTO NOx emission levels $41.8 \leq \text{LTO NO}_x \leq 53.1$ emissions tested in 1985 essentially met CAEP8 requirement applicable for engines manufactured after January 1, 2014. For this reason, in addition to the fact most of this chapter is about TAPS engines, we will present figures when appropriate in term of CAEP8 NO_x requirement, as summarized in Fig. 4. It should be emphasized that applicable NO_x standards for the CF6-80C2 engine families tested in 1985 and 1995 (identified respectively as CF6-80C2PreLEC and CF6-80C2LEC) are CAEP1 and CAEP2, not CAEP8 as shown in Fig. 4. We expressed these in terms of CAEP8 so that it can be compared directly with the emission characteristics of the GEnx and LEAP-X engine families summarized in Fig. 5.

When normalized by CAEP8 regulatory levels, the CF6-80C2PreLEC gives LTO NOx ranging between 91 and 102% which was reduced only slightly to fall between 93 and 97 for the LEC product. However, the resulting low-emission combustion LEC product reduced significantly the unburned hydrocarbons and CO emissions. We will have more discussion on this topic in Sect. 3 which provides the list of 16 engines selected and their emissions normalized by CAEP8 standards.

The author, very proud of his association with the TAPS team that developed the technology, conducted an extensive CFM TAPS engine demonstration program [1, 2] leading to prestigious team recognition award and the 1st GE Aviation's Edison award in 2003, followed by highly successful combustion product development as part of the GEnx combustion, was very disappointed to see the results summarized in Fig. 5a. How can we have 70% increase in LTO NOx of the two models of LEAP-X supposedly having been scaled from GEnx?

The author asked one of his former combustion colleagues, "Is there a typographical error in the ICAO engine emissions databank in regard to LEAP-X with the attendant increase in LTO NOx?" His response - "I wish it was"—is what motivated the author to pull together material for this chapter.

Since the author started paying little attention to the development of rich-dome LEC combustion products after the GE90-110/115B tested in 2003, he out of curiosity

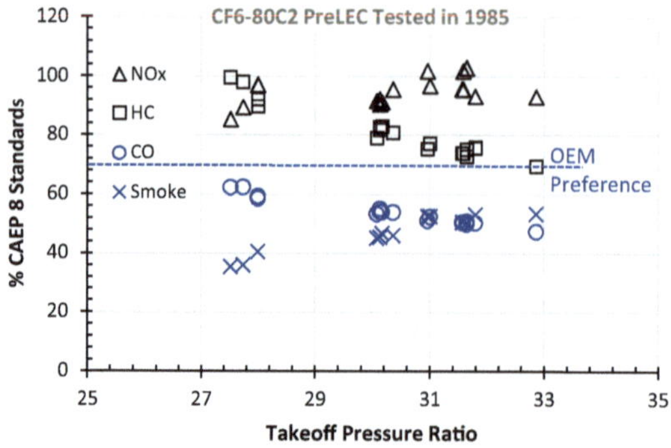

a. CF6-80C2 engine family tested in 1985 identified as CF6-80C2PreLEC

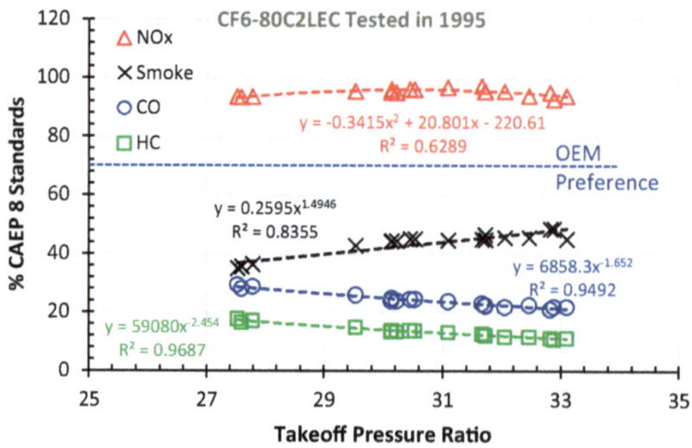

b. CF6-80C2 engine family tested in 1995 identified as CF6-80C2LEC showing that its maximum LTO NOx value of 97.3 doesn't exceed CAEP 8 regulatory level of 100. However, when expressed in terms of its relevant regulatory level, namely CAEP 2, the NOx emission range falls between 57-63% CAEP 2, falling below the desirable 70% level. Smoke emissions given by $0.2595 \times PR^{1.495}$ has more than 50% margin, CO (given by $6858/PR^{1.652}$) more than 70% margin and HC given by $59080/PR^{2.454}$has more than 80% margin from the CAEP 8 standards.

Fig. 4 Back to back comparison of the LTO emissions characteristics of the CF6-80C2 engine families tested in 1985 and 1995 identified as CF6-80C2PreLEC and CF6-80C2LEC, respectively

wanted to compare its LTO NO_x with the RR most recent Trent Extra Wide Body emission tested in 2012 leading him to ask a question. Is RR Trent XWB's 20% lower LTO NO_x compared to GE90-110/115B indicative of RR's continuing improvement in its RQL technology while GE's focus was on developing the 2nd generation lean-dome TAPS technology and product?

Is the LEAP-X setback indicative of innate limitations of lean-domes or simply an anomaly caused by several reasons? How will it impact future TAPS-based lean combustion products including GE9X supposedly a TAPS combustion system? These questions and many more provided the motivation for the author to undertake a

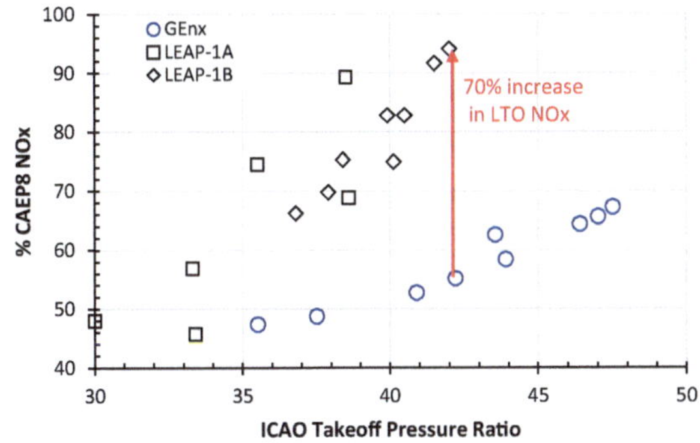

a. LEAP-X (LEAP-1A and Leap-1B) supposedly scaled from GEnx gave 70% higher LTO NOx; why?

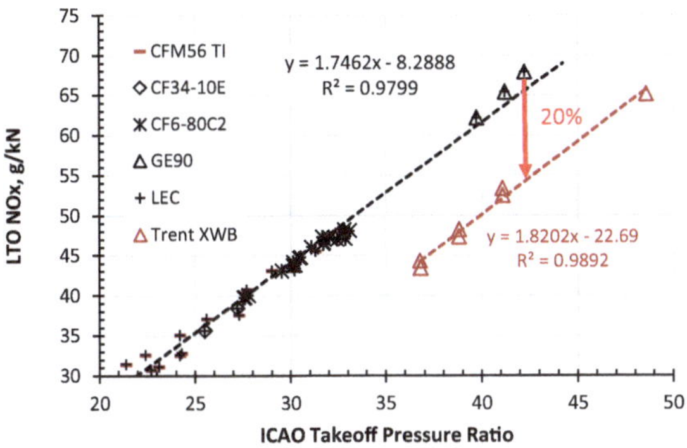

b. Has GEAE's rich-dome LEC combustion products fallen behind RR rich-dome RQL, here Trent Extra Wide Body Trent XWB tested in April 1, 2012? GE90-110/115B (legend GE90) was tested on November 23, 2003. LTO NOx of LEC products represented well by $LEC\ LTO\ NOx = 1.7462 \times PR - 8.2828\ with\ R^2 = 0.9799$ compared to $LEC\ LTO\ NOx = 1.8202 \times PR - 22.69\ with\ R^2 = 0.9892$.

Fig. 5 Back to back comparison between the LTO NOx emissions characteristics of GEnx versus LEAP-X and GE rich-dome LEC versus Trent Extra Wide Body engine models

systematic assessment of the gaseous emissions characteristics of modern propulsion engines.

The low-emissions combustion technology process characterization originally developed as part of [19–21] has worked very well for all the programs the author has been involved since the middle 1970s, as summarized in [25]. Section 2 summarizes the application of this approach for evaluating the CF6-80C PreLEC and LEC combustion products.

A list of 16 modern propulsion engines (Table 3) and their LTO characteristic values expressed in terms of CAEP8 requirement are summarized in Sect. 3 followed by an extensive discussion on their LTO NOx characteristics in Sect. 4.

Application of the high-power NOx methodology of Sect. 2 is demonstrated to all rich-dome combustors of Table 3 in Sect. 5 clearly demonstrates that recent LTO NO_x reductions can be attributed to these engines' reduction in their LTO fuel burn. Their takeoff and climbout NOxEI versus pressure ratio characteristics have not changed since the middle 1990s.

High- and low-power emissions tradeoffs are summarized in Sect. 6 for all 16 rich- and lean-dome combustors followed by deep dive into lean domes in Sect. 7. If TAPS combustion system design is optimized properly, its mission NO_x emissions potentials are significantly superior than rich-domes' as summarized in Sect. 9 preceded by Sect. 8. Here a Data Reduction and Analysis DRA is explored for rich-dome combustor emissions over its sea-level emissions from idle to maximum power followed by its extension over the entire flight envelop. Finally, a brief summary is provided in Sect. 10.

2 Characterization of CF6-80C2 Combustors' Emissions

The ICAO landing takeoff gaseous emissions (pollutants, P_i) is calculated from:

$$\begin{aligned}\text{LTO P}_i = \frac{\text{Charcateristic factor}}{F_{00}}\Big[0.7 \times \dot{W}f_{\text{Takeoff}} \times \text{Takeoff}P_i \\ + 2.2 \times \dot{W}f_{\text{Climbout}} \times \text{Climbout}P_i + 4 \times \dot{W}f_{\text{Approach}} \\ \times \text{Approach}P_i + 26 \times \dot{W}f_{\text{Idle}} \times \text{Idle}P_i\Big]\end{aligned}$$

Here, P_i denotes gaseous emissions hydrocarbons gm CH_4/kg fuel), CO gm/kg fuel and oxides of nitrogen NOx gm NO_2/kg fuel, respectively. $\dot{W}f_i$ denotes the fuel flow rates (kg/min) during the takeoff, climbout, approach, and taxi idle operational modes, respectively, with the corresponding times of 0.7, 2.2, 4, and 26 min. F_{00} (kN) is engine's static takeoff thrust at the standard atmospheric conditions.

Clearly, LTO emissions are determined from engine performance (here, fuel flow rates) and emissions indices of HC, CO, and NO_x, represented in this chapter as HCEI, COEI, and NOxEI, respectively.

Characteristic factors, as explained on p. 106 of [22] take into consideration effect of sample size, namely, the number of engines tested and the repeatability of emissions data of HC, CO, NO_x and smoke number. From [22], we have created Table 2 as an illustration to show effect of number of engines tested, here 1, 2, 3 and in the limiting case of 10 engines.

Most of the ICAO engine emissions data comprise of single engine with three test runs in order to calculate average LTO values of HC, CO, NO_x and Smoke number. However, its characteristic LTO values are expected not to exceed the ICAO regulatory levels. This means that average LTO values have to be multiplied, respectively, by 1.2274, 1.5401, 1.1592, and 1.2872. Sometimes, OEM's might test three engines; and for this case, the multiplication factors reduce to 1.0815, 1.1666, 1.0593, and

Table 2 ICAO recommended characteristic factors for HC, CO, NOx, and smoke number when number of engines tested are 1, 2, 3, and 10

Engines	Characteristic factors for			
Tested	HC	CO	NO_x	Smoke #
1	1.2274	1.5401	1.1592	1.2872
2	1.1393	1.3012	1.0996	1.1727
3	1.0815	1.1666	1.0593	1.1000
10	1.0431	1.0848	1.0316	1.0524

Table 3 List of 16 propulsion engines manufactured by the three engine manufacturers covering a broad range of ICAO takeoff pressure ratios (21–49) and thrust rating (77–514 k N) with engine emissions certification testing conducted during 1985 and 2017

Number	Legend	Takeoff pressure ratio		Takeoff thrust (kN)		Engine test dates	
		Minimum	Maximum	Minimum	Maximum	From	To
Engines by CFM International and GE Aircraft Engines							
1	LEAP-1A	30.0	38.6	106.8	143.1	11/21/2016	12/2/2016
2	LEAP-1B	36.8	42.0	111.3	130.4	3/8/2016	3/9/2016
3	CFM56-5B TI	22.7	32.6	96.1	142.3	9/29/2005	3/23/2006
4	CFM56-7B TI	21.4	29.0	86.7	121.4	9/29/2005	3/23/2006
5	CFM56-7B DAC	22.8	28.8	91.6	121.4	8/30/1997	9/12/1997
6	CFM56-5B DAC	24.4	32.8	102.2	142.4	8/1/1996	9/1/1996
7	GEnx	35.5	47.5	255.3	349.2	6/19/2012	6/21/2012
8	CF34-10E	25.5	27.2	77.4	83.7	6/5/2009	6/6/2009
9	GE90	39.7	42.2	492.6	513.9	11/23/2003	11/23/2003
10	GE90-DAC1	35.2	40.4	360.6	426.7	2/24/1995	7/28/1995
11	CF6-80C2	27.5	33.1	231.1	287.0	1/13/1995	1/17/1995
12	CF6-80C2PLEC	27.5	32.9	231.1	270.0	5/29/1985	6/3/1985
Engines by Rolls-Royce Plc.							
13	Trent XWB	36.8	48.6	334.0	436.7	4/1/2012	4/20/2016
14	T1000	36.2	49.4	268.0	363.9	9/14/2009	11/30/2017
15	TPh5	34.5	45.4	251.9	345.9	3/16/1995	11/30/2017
16	T1994	34.0	41.5	300.3	413.1	9/1/1994	
	Range	21.4	49.4	77.4	513.9	5/29/1985	11/30/2017

1.1. Rarely, the OEM's run10 engines for collecting emissions data; but if required, then these multiplication factors reduce to 1.0431, 1.0848, 1.0316, and 1.0524.

Figure 6 gives a schematic of the CF6-80C2 combustion system cross-section adopted from [4] in order to identify schematically the major changes between the PreLEC and LEC combustors, here, specifically, swirl cup (SC), splash plate, dome,

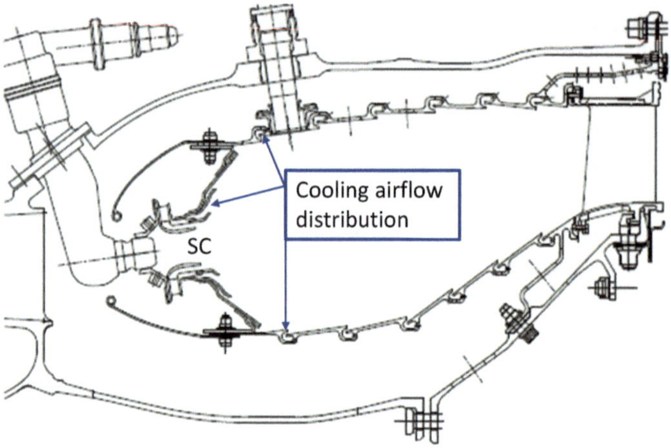

Fig. 6 Schematic layout of the CF6-80C combustion system showing modifications in swirl cup design (for reducing smoke) and cooling airflow distribution in order to optimize wall temperature levels and gradients of the swirl cup splash plate, dome, and primary panels. Modified from [4]

and primary panels' cooling air modifications. These modifications were not intended for changing the spray characteristics or the primary zone flow dynamics. Consequently, LEC takeoff NOxEI was expected to change only slightly from that of PreLEC. The quality of the LEC NOx emission data is considerably better than the data scatter band from the PreLEC as summarized in Fig. 7. Takeoff NOxEI data can be analyzed by several empirical, semi-empirical, or semianalytical correlations, see [10–12, 16–18]. However, here we plot takeoff NOxEI as a function of the ICAO

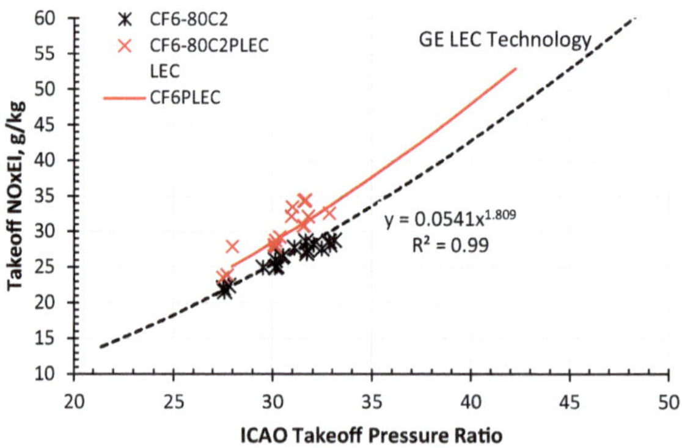

Fig. 7 Takeoff NOxEI versus pressure ratio of the CF6-80C2 with PreLEC and LEC combustion systems compared to GE LEC technology trendline expressed as LEC Takeoff NOxEI = $0.0541 \times PR^{1.809}$ with $R^2 = 0.99$ shown by dotted black curve. Red curve is given by 1.12xLEC Takeoff NOxEI which passes through data scatter of the CF6-80C2 PreLEC

takeoff pressure ratio in Fig. 7 because by doing so does not require any proprietary details about the combustion system.

Details on the development of the GE LEC NOxEI technology will be provided later. Since there is wider data scatter for the PreLEC, we have arbitrarily plotted 1.12 times LEC Takeoff NOxEI trendline for now in order to illustrate that one may be inclined to suggest that PreLEC takeoff NOxEI is only 12% higher than that of the LEC technology. Therefore, at least for this engine model, the emphasis was not on reducing NOx unlike for the RR combustors moving forward after 1991, refer to Table 1.

It is well known that design modifications for reducing smoke and high-power NO_x may be accompanied with challenges for low-power emissions, operability that includes ground and altitude ignition, and lean blowout. Other challenges may include combustor exit temperature quality and combustion system durability. Here, our focus will be on idle CO and unburned hydrocarbon emissions which in addition to generally accepted tradeoff with reducing high-power NO_x can be impacted by atomization characteristics, and entrainment of the cooler film cooling air, and attendant quenching of HC and CO. Alternatively, lack of cooling air for protecting the combustor dome structure can lead to higher liner or dome wall temperature levels and gradients and resulting impact on durability and shop visits. Based on lessons learned from [19–21], we have continued to use since the middle 1970s illustrations similar to Figs. 8 and 9.

If the target takeoff NOxEI goal were 25, the CF6-80C2 LEC combustion product reduced simultaneously NO_x, idle COEI and HCEI emissions by 10%, 58%, and 85% compared to the PreLEC, and hence one can claim success in demonstrating lower emissions technology combustor without increasing smoke emissions as summarized previously in Fig. 4.

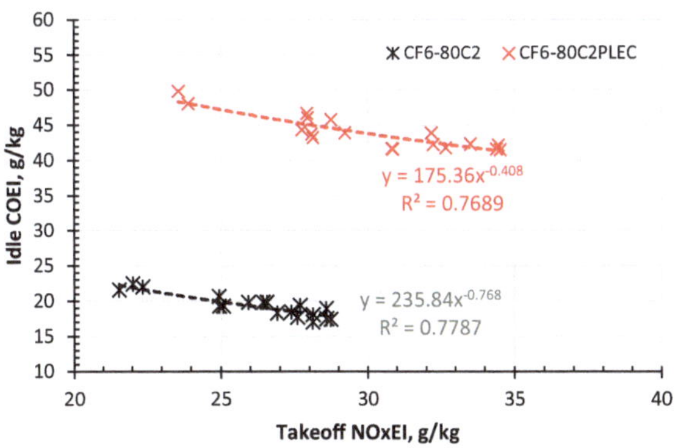

Fig. 8 Idle COEI plotted as function of takeoff NOxEI for the CF6-80C2 PreLEC and LEC combustors are represented well by PreLEC Idle COEI $= \frac{175}{\text{Takeoff NOxEI}^{0.408}}$ with $R^2 = 0.7689$ and LEC Idle COEI $= \frac{236}{\text{Takeoff NOxEI}^{0.768}}$ with $R^2 = 0.7787$ showing that compared to PreLEC, LEC reduces NO_x by approximately 10% while simultaneously at takeoff NOxEI of 25 it reduces idle COEI by 58%

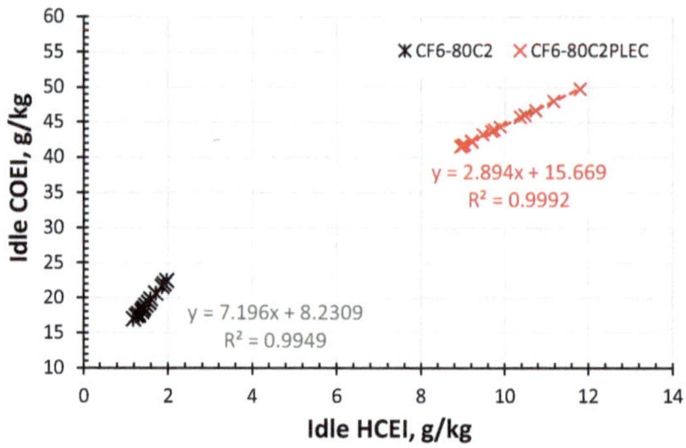

a. Idle COEI plotted as function of idle HCEI giving good quality correlations $PreLEC\ Idle\ COEI = 2.89 \times Idle\ HCEI + 15.67\ with\ R^2 = 0.9992$ and $LEC\ Idle\ COEI = 7.199 \times Idle\ HCEI + 8.231\ with\ R^2 = 0.9949$.

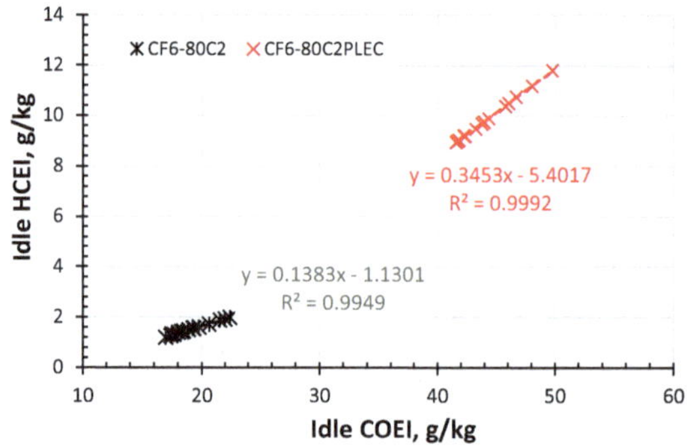

b. Idle HCEI plotted as function of idle COEI giving good quality correlations $PreLEC\ Idle\ HCEI = 0.345 \times Idle\ COEI - 5.402\ with\ R^2 = 0.9992$ and $LEC\ Idle\ HCEI = 0.138 \times Idle\ COEI - 1.13\ with\ R^2 = 0.9949$.

Fig. 9 Idle COEI plotted as function of idle HCEI of the CF6-80C2 PreLEC and LEC combustor (Part a) is preferred over the bottom where idle HCEI has been plotted as function of idle COEI wherefrom we can calculate that for 25 takeoff NOxEI, the LEC combustor reduced NO_x, idle COEI and HCEI by 10%, 58%, and 85%

From Fig. 4b, it is remarkable to see that LEC combustion system emission certified in January 1995 can meet the most stringent CAEP8 NOx standard in a propulsion engine designed in the late 1970's through early 1980s in spite of its higher LTO fuel burn. It also has 50%, 75%, and 85% margin from the regulatory requirements of smoke, CO, and HC emissions. This raises one potential question, (Are the CAEP8 standards not stringent enough?) which is partially answered by Fig. 4c.

A word of caution about the right interpretation of Fig. 9 which comprises of several aspects of the combustion processes taking place simultaneously within a typical propulsion combustor. This includes air atomization characteristics, aero tradeoff between NO_x formation and CO oxidation, quenching of partially burned hydrocarbons and CO by cooling air and or escape of part of spray's excessively large size droplets.

The use of Figs. 7, 8 and 9 combined with the methodology [23, 24] and applicable CFD tools [9, 15] has led successfully to evolving generation of low emissions gas turbine combustors as summarized in [25].

3 LTO Emissions of Selected Propulsion Engines

Focus of this section is to compare characteristic LTO emissions data normalized by the CAEP8 regulatory levels. However, in Sect. 4 we plan to use average LTO emissions data followed by average indices of emissions in Sect. 5. The reason for using CAEP8 NOx standards is simple: it is applicable "for engines of a type or model of which the date of manufacture of the first individual production model was after 1st January 2014." Therefore, this section is intended for young combustion technologist to notice that unless there is further reduction in climbout and takeoff NO_x of rich-dome combustion technology and or LTO fuel burn, it will be difficult to have 30% margin from the regulatory NO_x standard.

Twelve modern propulsion engines manufactured by the CFM International and GE Aircraft Engines were selected in order to compare swirl cup [3] based Low-Emissions Combustors LEC (engines numbered 3, 4, 8, 9, and 11 listed in Table 3), pre-LEC (#12), lean-dome Dual-Annular Combustors DAC (5, 6, and 10), and TAPS #1, 2, and 7. Details on these four combustion technologies are given in [1–4].

All the selected four engine families from RR (#13–16) belong to Rich-quench-Lean RQL category which is essentially similar to LEC technology practiced by GE and the CFM.

As summarized in Table 3, the selected 16 engines cover a very broad area in regard to takeoff pressure ratio (21.4–49.4) and rated takeoff thrust that ranges between 77 and 514 kN. The emissions certification dates of these engines cover a long period of 32 years, 5/29/1985 through 11/30/2017.

Figures 10, 11, 12 and 13 summarize landing takeoff emissions of NO_x, unburned hydrocarbons HC, CO and characteristics SAE smoke numbers of the selected 11 GE and CFM engines showing performance of the three combustion technologies (low-emissions rich-domes LEC, dual annular combustors DAC and lean-dome twin annular partially mixing swirlers TAPS), expressed as % of CAEP8 regulatory standards. Essentially all three technologies meet these requirements, some with considerable margins whereas others approaching or slightly exceeding the limits. But these figures don't shed much light in regard to underlying reasons for explaining the observed wide scatter band. For that we will have to go to Sects. 4–6.

If OEM's had a choice for their future combustion products, they would like to have 30% margin from the regulatory standards for NO_x, 50% for HC, CO, and smoke shown as dotted blue lines in these figures. However, when the products turn out higher than these goals, some of them are willing to contribute toward improving technology. For example, some of the CFM DAC products have margins

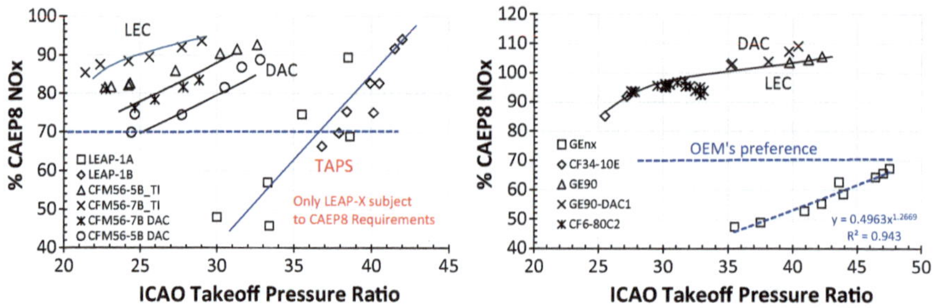

Fig. 10 Characteristic landing takeoff LTO NOx, expressed as % CAEP8 NOx, for the CFM (L) and GEAE (R) engines

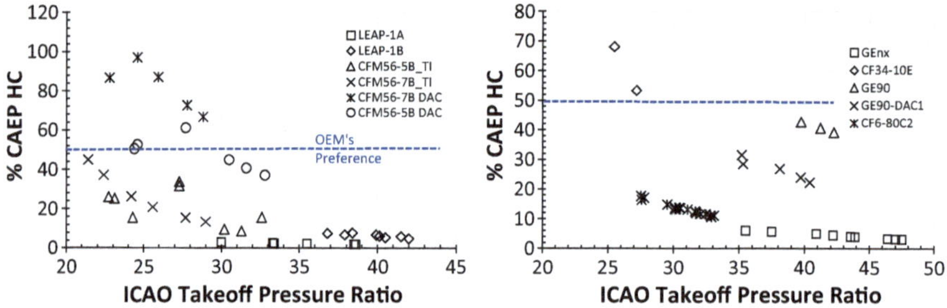

Fig. 11 Characteristic LTO unburned hydrocarbons HC, expressed as % CAEP HC, for the CFM (L) and GEAE (R) engines

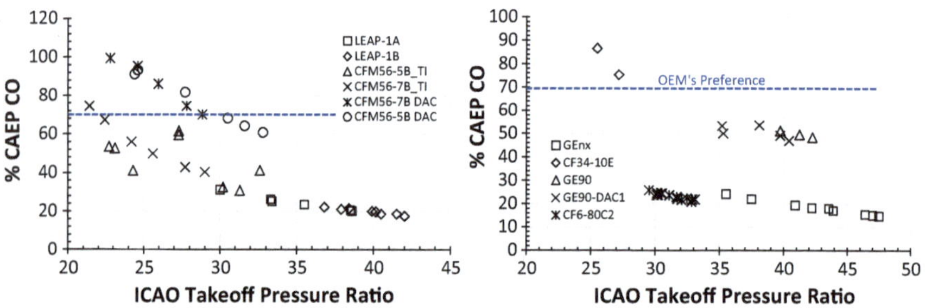

Fig. 12 Characteristic LTO CO, expressed as % CAEP CO, for the CFM (L) and GEAE (R) engines

of 10%, 3%, and 0.4% from NO_x, HC and CO standards, respectively. On the other hand, some of the GE90 DAC models have margins of −9% and 33% for NO_x and smoke. Environmentally conscious passengers don't like to fly in airplanes with visible exhaust plume. During Boeing 777 with the GE90 DAC I flyby over GEAE main building, the author received a phone call from his VP of Engineering asking explanation for his visual observation of exhaust smoke. For the time being, he was satisfied by my explanation that it must be flying outside the LTO cycle.

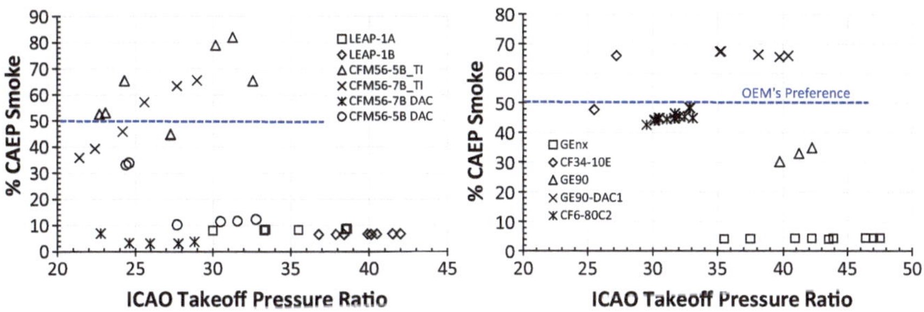

Fig. 13 Characteristic SAE Smoke number, expressed as % CAEP Smoke, for the CFM (L) and GEAE (R) engines

Fearing that 33% margin from smoke standard may not be enough, we increased our goal to more than 50% which we exceeded in the follow-on rich-dome GE90 product with 65% margin, as shown in Fig. 13. But truly lean-dome products are expected to exhibit much higher margins from smoke standards and one may get decent margin from optimized rich domes also. For example, the lean-dome TAPS engines GEnx and LEAP-X have more than 95% margin from smoke standards. The CFM DAC's smoke margin is more than 88% except at 25 overall pressure ratio OPR it reduces to 66%. The rich dome CF6-80C2 has more than 50% margin. But rich domes can have more variability in their smoke emissions as shown by the CFM56 Tech Insertion engines.

Some OEM's are more willing to pursue lean-dome technology and product development activities including the author who prefers concentric lean-flames over radially [5–8] or axially [19–21] staged concepts realizing well that its more complex flow characteristics and unravelling surprises to be discussed later in Figs. 15, 16 and 43. In spite of less than stellar outcome of the DACs, the CFM and GEAE contributed more than $100 million toward development of the TAPS combustion technology, CFM TAPS engine demonstration [1, 2], the GEnx combustion system design, fabrication, limited full annular combustor rig test followed by first engine to test FETT in March 2006 culminating with toll gate TG 6.

In summary, the CFM family of engines with TAPS II technology have only 6% margin for NO_x compared to 32% with GE's GEnx. The author is glad that the LEAP-X combustion used nomenclature TAPS II and not TAPS 2 which according to [1] was intended to get 50% reduction in NOxEI from the 1st generation TAPS technology (TAPS 1) used in GEnx. We will have more discussion on LEAP-X, GEnx, TAPS1, TAPS2, and TAPS3 later.

4 Landing Takeoff NOx Characteristics

Figure 14 presents average landing takeoff NOx (not characteristic values) of the selected eleven GEAE/CFM engine models (Table 3) as function of takeoff pressure ratio along with several lines identified as % of CAEP1 NOx regulatory standard, namely LTO $NO_x = 40 + 2 \times PR$. Here, 50% CAEP1 implies that its value at 30PR is 50 g/kN; and likewise 35 and 20 for 35% and 20%CAEP1 legends, respectively. It should be noted that the slope of these lines is 2.0. It should be recalled that for all practical purposes, 50% CAEP1 is equivalent to CAEP8 standard. Even though the CFM56-5B and CFM56-7B Tech Insertion combustion systems are identical, their LTO values are different; the former has slightly lower LTO NO_x values. We will discuss this apparent anamoly later in Sect. 5. Due to some reasons, there is more data scatter for the CFM56-5B_TI than for the CFM56-7B_TI which lies slightly below 50% CAEP1 line. We observed similar behavior in HC (see Fig. 11), CO (Fig. 12) and smoke emissions as shown in Fig. 13.

In spite of some data scatter of the dual-annular combustion systems for the engine models CFM56-5BDAC and CFM56-7BDAC, it shows slightly higher values for the latter compared to the former, consistent with the rich-dome SACs. There is some tendency for losing advantage of DAC over SAC; but it becomes clearly evident when we combine DAC data from both the CFM56 and GE90 engine families. The CFM engine models lie between 35 and 50% of CAEP1, DAC's giving slightly lower values. But the lean dome dual annular combustion system of the nominal 40 pressure ratio GE90 increases to 50%CAEP1 level a loss of 15 points which is attributed to its lower share of combustion air. As per the conventional wisdom, its primary zone equivalence ratio is closer to 0.9 instead of being closer to the desirable range of less than 0.7 or 0.75.

In summary, DAC technology didn't live up to its expectations inferred somewhat wrongly from its technology development and two engine demonstrator programs summarized in [5–8]. These programs didn't even try to improve fuel/air mixedness quality unlike the small engine pollution reduction technology program PRTP described in [19–21] where more than 70% NO_x reduction was achieved after first modification. DAC technology programs didn't spend enough resources in addressing their low-power challenges unlike PRTP where 75% of the effort was spent in addressing low-power challenges. Moreover, the DAC activities didn't put enough emphasis on addressing significant variation in the combustor exit temperature radial profiles caused by radial staging of the fuel nozzles. Poor fuel/air mixedness, radial staging of the fuel nozzles and stubborn refusal to adopt twin-wall liner cooling technology from [8] were the main reasons for the failure of DAC technology in its transitioning to market competitive low emissions propulsion engine combustion products.

In spite of the great freedom the author enjoyed during his 15 years at GE Aviation, he was unable to get the senior management to buy into developing more cooling air efficient metalic twin-wall technology which would have made more air available for combustion with attendant increased potential for NO_x reduction for GE90 DAC I.

Therefore, the TAPS1 technology development activities proceeded despite the adverse consequences of using state-of-the-art film and multihole cooling techniques used routinely in GE propulsion engine combustion products unlike the competitors with supposedly more efficient twin-wall cooling technologies, Floatall and Tiled. For example, P&W reported first use of Floatwall in its low-emissions engine model PW4168 with emissions test start date of February 14, 1993, two years before the TAPS technology initiative was kicked off. Ten years later, RR reported the use of Tiled constructed combustion system in its engine model Trent900 with engine emissions start date of October 17, 2003.

But lack of twin-wall's use in LEAP-X combustor doesn't explain the failure of the LEAP-X TAPS combustion products summarized in Fig. 14. It starts with 20%CAEP1 at 30 overall pressure ratio OPR, as expected, but increases somewhat erratically and sometimes consistently to 35%CAEP1 by 42 OPR. One cannot miss similarity with the DAC's where we thought we understood losing 15 points going from 30 to 40 OPR engines. Did the scaling from GEnx to LEAP-X process fail for lack of proper hypotheses formulation [23, 24] or unintended consequences of making wrong selection between the two different internal flowfield characteristics

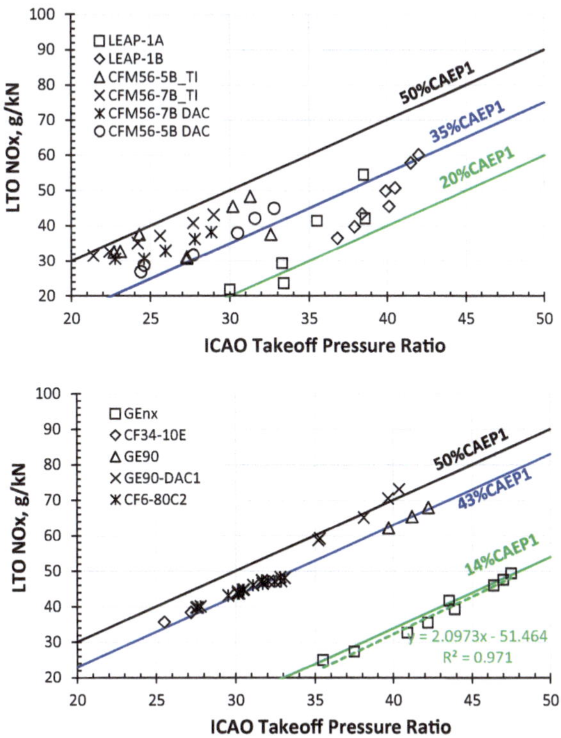

Fig. 14 Average LTO NO_x characteristics for the CFM (T) and GEAE (B) engines

predicted analytically in [27], Figs. 15 and 16? We will come back to this topic later in Sect. 7.

We are inclined to come to very different tentative conclusions from the bottom part of Fig. 14 which summarizes the LTO NOx characteristics of the three technologies of GEAE, namely lean-dome GE90-DAC1, rich-dome LEC products (CF34-10E, CF6-80C2, and GE90 (rated for the ICAO takeoff thrust between 493 and 514 kN) and TAP1 technology-based GEnx for rated takeoff thrust between 255 and 349 kN. It is true that DAC technology product gave LTO NOx falling along 50% CAEP1 compared to the LEC along 43% CAEP1 and TAP1 significantly lower value of 14%CAEP1. But that doesn't mean that combustion technologist can claim that his TAP1 combustion technology produced approximately 1/3rd NO_x emissions compared to the LEC-based rich-dome GE engines for the same reason the Rolls Royce combustion technology engineer can claim that its Trent X wide-body (Trent XWB) has 20% lower NOx emissions compared to the GE90-115B engine model,

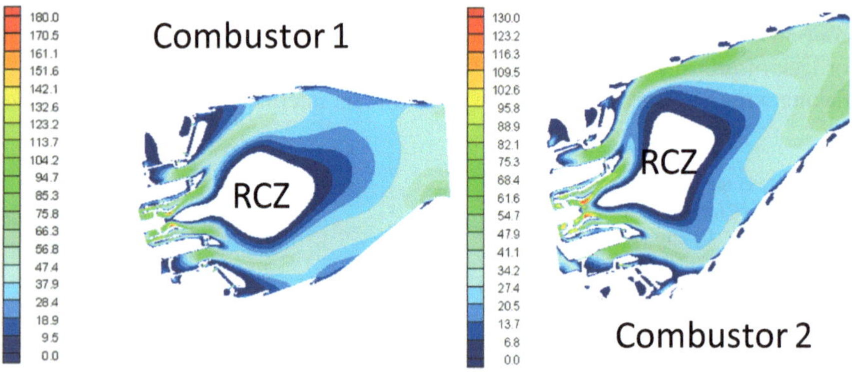

Fig. 15 Reacting flow mean axial velocity contours (m/s) predicted by comprehensive combustion system analytical modeling approach [27]

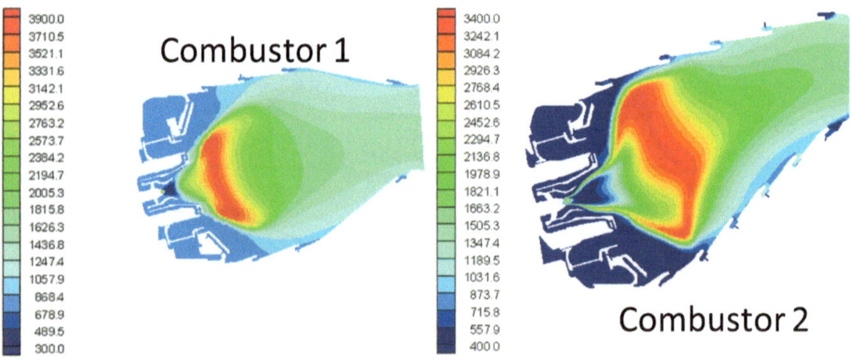

Fig. 16 Mean gas temperature contours (°F) predicted by comprehensive combustion system analytical modeling approach [27]

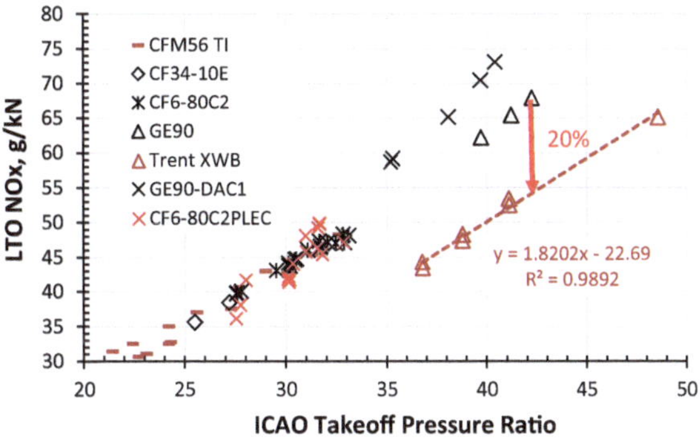

Fig. 17 Average LTO NO_x characteristics of the CFM and GEAE engines' rich domes compared with RR Trent X Wide Body (Trent XWB) showing 20% LTO NO_x reduction compared to GE90-115B engine model

as summarized in Fig. 17. Why? Because the GE90-115B's landing takeoff cycle-specific fuel burn of 3 kg/kN is 6% higher than that of Trent XWB as summarized in Fig. 18 for several engines as function of takeoff pressure ratio and thrust.

There is no simple honest way of filtering out the effect of engine fuel burn out of the LTO NO_x emissions to claim how much contribution combustion technologist has made in reducing emissions. Even more important area of concern for the combustion technologist should be about answering wide scatter band in LTO NO_x of combustors based on supposedly the same technology base, for example, LEC for GE and RQL. And then how much benefit RR got because of its use of more cooling air-efficient double-wall Tiled compared to single wall cooling technology products of GE? These questions will be answered in rest of this chapter starting with the material presented in Sect. 5.

5 Rich-Dome Combustors' NO_x Assessment

Consistent performance in developing combustion technology and products meeting customers' expectations has been of paramount importance to the author since the early 1970s because of what generally happened to Jack Haasis who was his best on-the-job combustion teacher. His superb quality work's productivity was continuously sabotaged by self-proclaimed burned-out combustion experts driven mostly by self-promotional misplaced egos. In order to keep these saboteurs at arm's length, it was very important for the author to reduce data similar to Fig. 17 to some definite conclusions in order to help him for providing guidance in achieving continuous systematic advances in both combustion technology and products. Again, to emphasize what has been mentioned before, the author is not familiar with the Rolls Royce combustion products other than what can be obtained from reliable COMPLETE-STORY publications which hardly exist.

Figure 19b presents a cluster of T1994 that represents LTO NO_x data of the Trent engine models 768, 772, 875, 877, 884, 892, and 895 emission certified on September 1, 1994. This remarkable data set became the benchmark for RQL combustors' NO_x

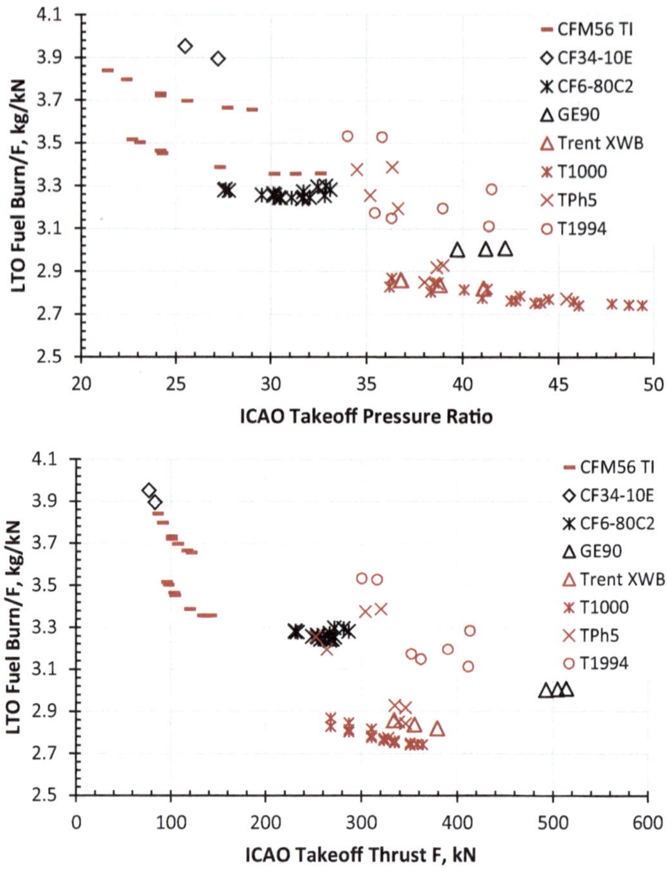

Fig. 18 LTO fuel burn divided by takeoff thrust F versus takeoff pressure ratio (top) and takeoff thrust shown in the bottom

for other OEM's to compete with. As far as the author knows, these combustors didn't use advanced liner cooling 'Tiled'. Tiled were used in the Trent models 7000-72, 7000-72C, 970-84, 972-84, 553-61, 556-61, 970-84, 972-84 and 972E-84 in addition to using 'Phase V' RQL technology. The Trent engine models 768 and 772 also belong to Phase V RQL technology group; but there is no mention of "Tiled' in these two models. All these 11 models have been lumped together under the legend TPh5 in Figs. 19b.

Figures 17 and 19 raises at least four questions:

(1) Is Trent's 20% lower LTO NOx compared to GE90-115B due to its superior RQL combustion technology or Trent's lower LTO fuel burn, Fig. 18?
(2) GEAE/CFM engines cover takeoff thrust range between 77 and 514kN compared to the Trent family of engines with narrower range of 252 to 437kN. The corresponding range of takeoff pressure of the Trent is 34–49 compared to

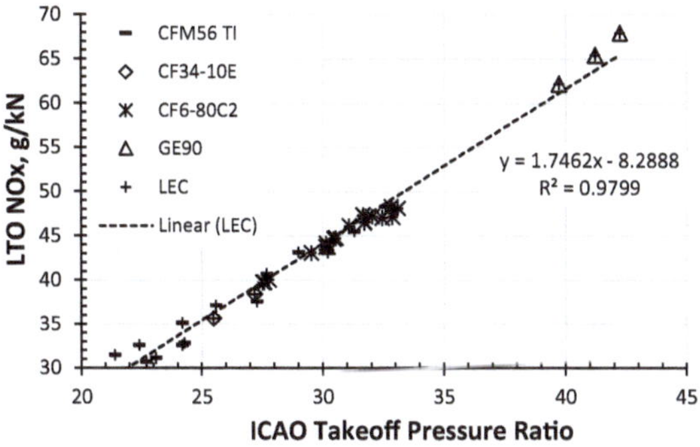

a. Average LTO NOx versus takeoff pressure ratio characteristic described reasonably well by $LEC\ LTO\ NOx = 1.746 \times PR - 8.289\ with\ R^2 = 0.9799$.

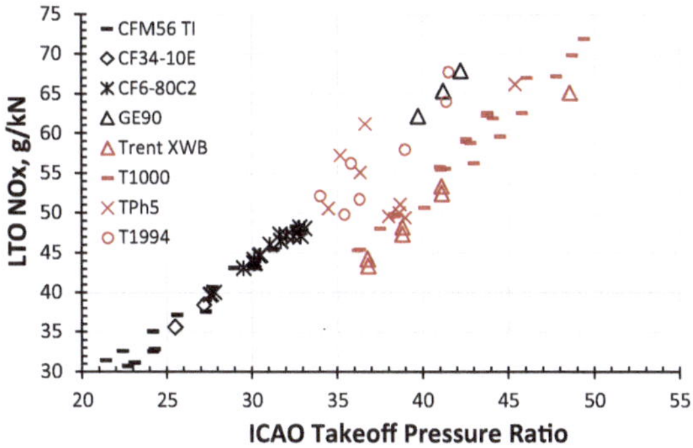

b. Cluster of Trent family of engines that use RR RQL combustion technology compared to the GEAE/CFM LEC engines

Fig. 19 Average LTO NOx versus takeoff pressure ratio characteristic of five GEAE/ CFM LEC combustors fall along a linear trendline (top) while all RR Trent family engines emission certified between 1994 and 2017 divided into four groups form a wide scatter band. Why?

GE/CFM's of 21 to 42. Why more scatter in LTO NOx data of RR compared to GE/CFM, compare Fig. 19b with Fig. 19a?

(3) Is "Phase V" RQL technology the best RR technology or it's simply used as a marketing tool?

(4) How good was RR in transitioning its RQL technology to the pre-LEC combustor of Allison AE3007 combustion product?

The 4th question can be answered at least partially by the author since he was involved in converting a high smoke, industrial engine 570 combustion system with aerodynamic diffuser, dual-orifice pressure atomizer, and double-wall convection-film cooled liner to a dump diffuser, piloted airblast nozzle with converging passage cup to reduce smoke, and effusion cooled liner. No attempt was made to reduce NO_x, and, therefore, the baseline combustor (identified Type 1 in Fig. 20) falls into the pre-LEC category; this engine emissions test start date was March 31, 1994. Since

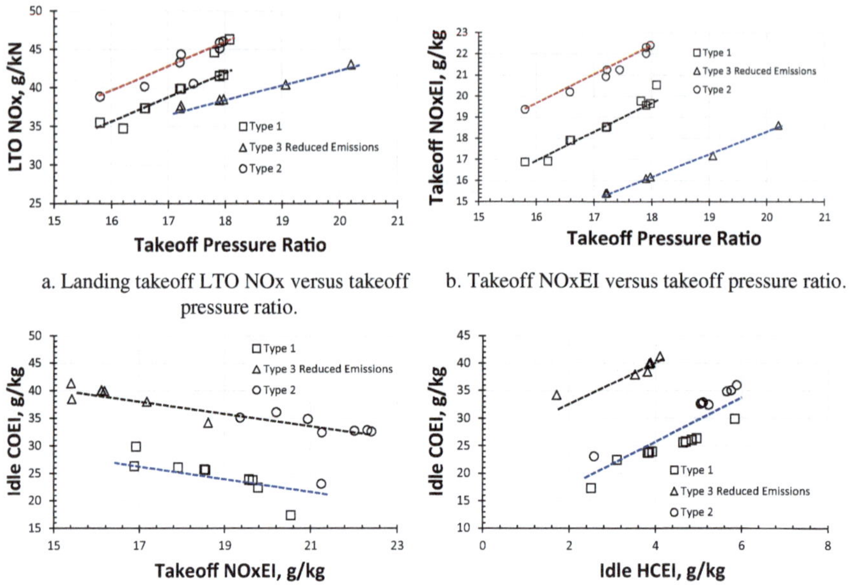

a. Landing takeoff LTO NOx versus takeoff pressure ratio.

b. Takeoff NOxEI versus takeoff pressure ratio.

c. Idle COEI versus takeoff NOxEI.

d. Idle COEI versus idle HCEI.

Fig. 20 On checking RR design process consistency in transitioning RQL technology into pre-RQL Type 1 AE3007 (engine emissions test start date March 31, 1994) through Type 3 Reduced Emissions (June 6, 2001) and Type 2 engine emissions tested on August 28, 2002

the author left Allison during January, 1994, he doesn't know the details about Type 3 Reduced Emissions (with emission test start date June 6, 2001) and Type 2 engine emissions tested on August 28, 2002. As summarized in Fig. 20a, Type 3 Reduced Emissions configuration did reduce LTO NO_x at 17.9PR from 41.53 to 38.49 g/kN, mere 7%. But the baseline combustor has some LTO data repeatability issues varying between 41.53 and 45, approximately 8%. Type 3 engine model's maximum takeoff pressure ratio is 20.2 compared to the baseline's being 17.9. The author can't tell if Type 2 combustor tested in 2002 is a predecessor of Type 3 tested in 2001. But, if we do, then the intended reduction in LTO NO_x happened first by increasing through Type 2 mod followed successfully through Type 3 configuration giving only 8% reduction in LTO NO_x.

As explained next, the preferred output of reduced emissions combustor program should be simultaneous reduction of takeoff NO_x, idle CO and HC or as a minimum reduction in takeoff NO_x without adversely impacting idle CO and HC emissions. Here, for the AE3007 emissions reduction effort, 8% reduction in LTO NO_x was followed by increased level of idle CO and HC emissions as summarized in Fig. 20; a clear contrast to Figs. 7, 8 and 9. The author would be hesitant to categorize this piece of emission reduction effort as spectacular.

A casual look at the NO_x emissions data at the four operating points of the landing takeoff cycle shows that the takeoff NOxEI is generally higher than that of the

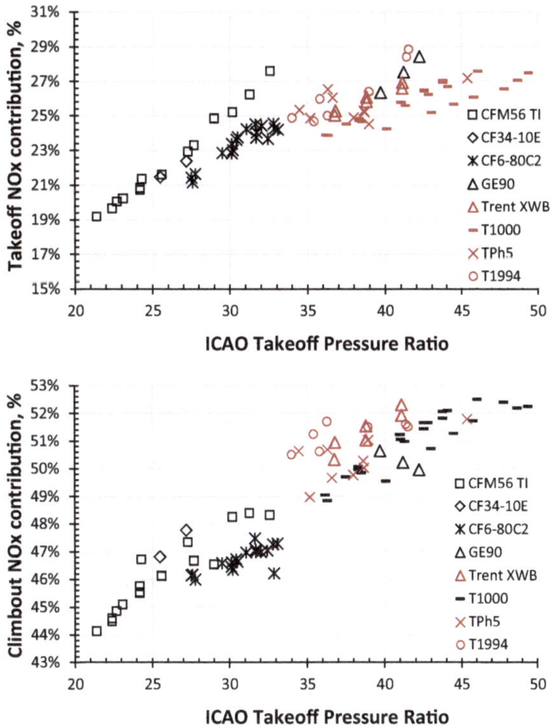

Fig. 21 Contributions of NO_x emissions during takeoff (top) and climbout (bottom) expressed as % of LTO NO_x

climbout phase of the airplane. But the ICAO LTO cycle assumes that the takeoff phase lasts for 0.7 min compared to 2.2 min for the climbout. It is therefore important to know how much contributions to the LTO NO_x emissions come during the takeoff and climbout operations. Figure 21 answers this question for all the engines listed in Table 3 wherefrom we see that the takeoff NO_x emission contribution ranges between approximately 19% and 29%. However, the climbout phase contributes higher percentage of NO_x emissions ranging between approximately 44% and 53%. The remaining 17% to 27% contribution of NO_x comes from the combined taxi-idle and approach operation.

It is very hard to make any conclusions from the RR LTO NOx data scatter of Fig. 19b. Therefore, we will first look at the GEAE/CFM engine data-driven mostly by the fact that the author is very familiar with these engines. From Fig. 19a, we see that the LTO NO_x data of the five LEC engine models (CF34-10E, CFM56-5B and CFM56-7B Tech Insertion (identified as CFM56 TI), CF6-80C2, and GE90-110/115B identified as GE90 can be represented well by a linear equation fit, namely LEC LTO $NO_x = 1.746 \times PR - 8.289$ with $R^2 = 0.9799$.

As discussed earlier in regard to Fig. 14, the slope of the NOx regulatory standard CAEP1 and subsequent more stringent requirements for the medium and large high-pressure ratio engines slope is 2.0. Therefore, if the engine combustors could have slope of ≤ 2.0, the NO_x margin from the regulatory standards will not reduce with increasing pressure ratios. Here, the LEC combustors have slope of 1.75.

Equally good is to have best-fit power law is obtained for the takeoff NOxEI of these five LEC engine models, as summarized in Fig. 22a given by LEC Takeoff NOxEI $= 0.0544 \times PR^{1.806}$ with $R^2 = 0.9788$.

As discussed earlier, the climbout operation contributes more NO_x emissions than the takeoff phase. Therefore, it will be desirable to check whether the NOxEI correlation during climbout operation can be derived similar to that of the takeoff NOxEI. Unfortunately, the engines' climbout overall pressure ratio values unlike takeoff's are not available publicly. Additionally, the ratio of the climbout to takeoff pressure ratios varies from engine to engine and varies as a function pressure ratio.

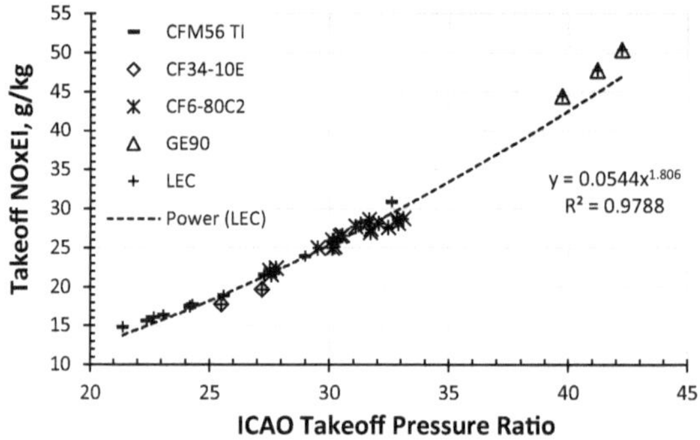

a. Takeoff NOxEI versus takeoff pressure ratio described well by $LEC\ Takeoff\ NOxEI = 0.0544 \times PR^{1.806}\ with\ R^2 = 0.9788$.

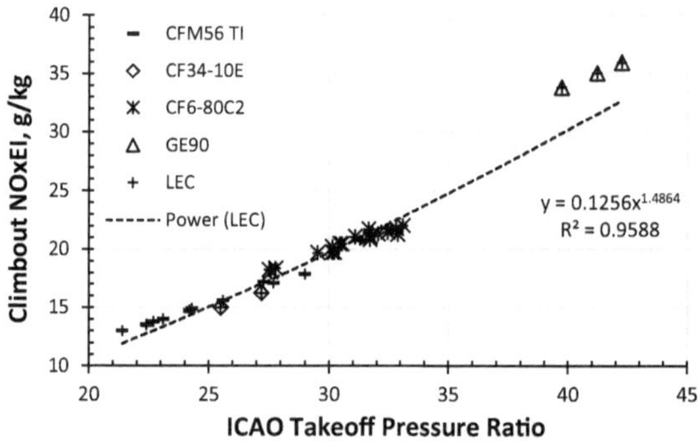

b. Climbout NOxEI versus takeoff pressure ratio described well by $LEC\ Climbout\ NOxEI = 0.1256 \times PR^{1.4864} with\ R^2 = 0.9588$. X-axis value should have been climbout pressure ratio (which is not available publicly), not takeoff pressure ratio explaining qualitatively inferior performance of the best-fit power law.

Fig. 22 Average LTO NOx (Part a), takeoff NOxEI (Part b), and climbout NOxEI (Part c) versus takeoff pressure ratio characteristic of the five GEAE and CFM LEC combustors

Consequently, one cannot get the desired correlation for climbout unless we assume that the ratio Takeoff PR/Climbout PR is essentially constant for these five LEC engines. In that case, we can assume that climbout NOxEI is a function of takeoff pressure ratio. As summarized in Fig. 22b, we see that the climbout NOxEI is also reasonably good, namely, LEC Climbout NOxEI $= 0.1256 \times \mathrm{PR}^{1.4864}$ with $R^2 = 0.9588$ (Fig. 23).

One obvious weakness in the LEC correlations for NOxEI is that these are heavily weighted in favor of nominal 20–30 pressure ratio engines. Since Trent XWB's takeoff and climbout NOxEI are very close to that of the GE90 for its range of takeoff pressure ratio. We, therefore, took the liberty of adding Trent XWB's NOxEI data to the LEC database. The resulting LEC correlations are weighted essentially well for application in engines with PR ranging between 20 and 50. The following NOxEI correlations will be used for the rest of this chapter:

$$\text{LEC Takeoff NOxEI} = 0.0563 \times \mathrm{PR}^{1.796} \text{ with } R^2 = 0.9876$$

$$\text{LEC Climbout NOxEI} = 0.0977 \times \mathrm{PR}^{1.563} \text{ with } R^2 = 0.9741$$

It is encouraging to see that all the Trent engine models' rich-dome combustors which has wide LTO NOx data (shown previously in Fig. 19b) when expressed in takeoff NOxEI, as summarized in top part of Fig. 25 scatter very nicely around the LEC takeoff technology curve (given LEC $\pm$ 2) except for the two engine models shown within a rectangle. A similar behavior is observed for the Trent's climbout NOxEI except all but one model of Trent XWB falls on uniformly biased slightly higher values than the LEC technology given approximately by LEC + 4. Here, again the same two engine models stand out as outliers. However, the Trent engine models show more scatter in their takeoff NOxEI data than the GE/CFM LEC engine models. But one can't help in concluding that in regard to both the takeoff and climbout NOxEI both OEM's rich-domes' NO_x emissions characteristics are essentially similar. It appears that Trent engines' takeoff NOxEI technology has not improved since 1994 in spite of changing single wall film cooling configuration to double-wall tiled construction (Fig. 24).

In summary, even though the landing takeoff emission characteristics are the right metrics for evaluating impact of propulsion engines on the local air quality, high-power NO_x emissions indices provide better guidance in developing low-emissions combustion technology and products. The rich-dome low NO_x products known popularly as LEC or RQL for all the modern engines produced by the three OEM's produce similar levels of takeoff and climbout NOxEI given by the following two correlations which are applicable over a wide range of takeoff thrust and overall pressure ratios, namely, $21 \le$ Takeoff Pressure Ratio ≤ 49, 77kN $\le$ Takeoff Thrust $\le$ 514 kN.

$$\text{LEC Takeoff NOxEI} = 0.0563 \times \mathrm{PR}^{1.796} \text{ with } R^2 = 0.9876$$

$$\text{LEC Climbout NOxEI} = 0.0977 \times \mathrm{PR}^{1.563} \text{ with } R^2 = 0.9741$$

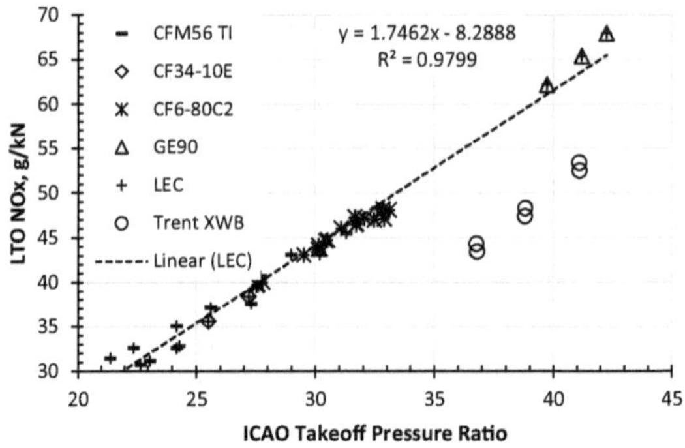

a. LTO NOx showing clearly that LTO NOx of Trent X Wide Body is considerably smaller than the trendline for the GE LEC combustion products.

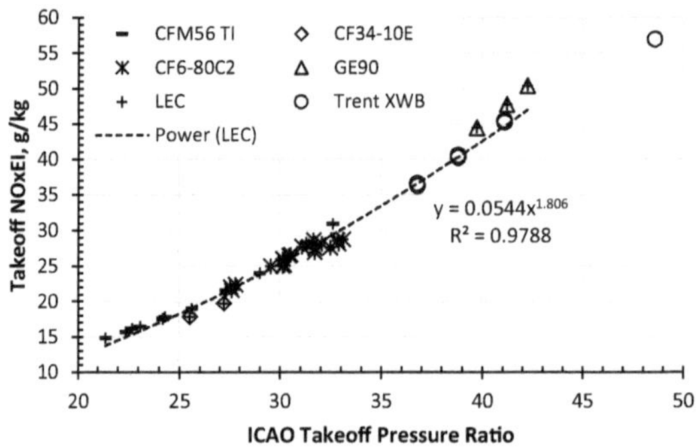

b. Takeoff NOxEI data for Trent X Wide Body Trent XWB is only slightly lower than that of GE90-110/115B.

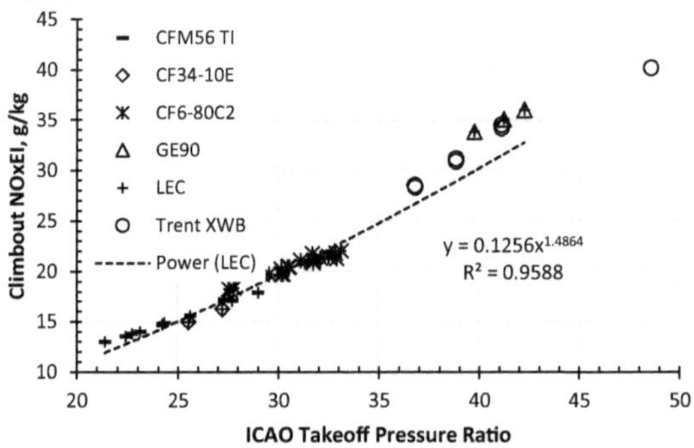

c. Climbout NOxEI data for Trent X Wide Body, Trent XWB is literally sitting on the top of GE90-110/115B leading to updated LEC correlations summarized in Fig. 24.

Fig. 23 Average LTO NOx (Part a), takeoff NOxEI (Part b) and climbout NOxEI (Part c) versus takeoff pressure ratio characteristic of the five GEAE and CFM LEC combustors plus Trent X Wide Body, Trent XWB

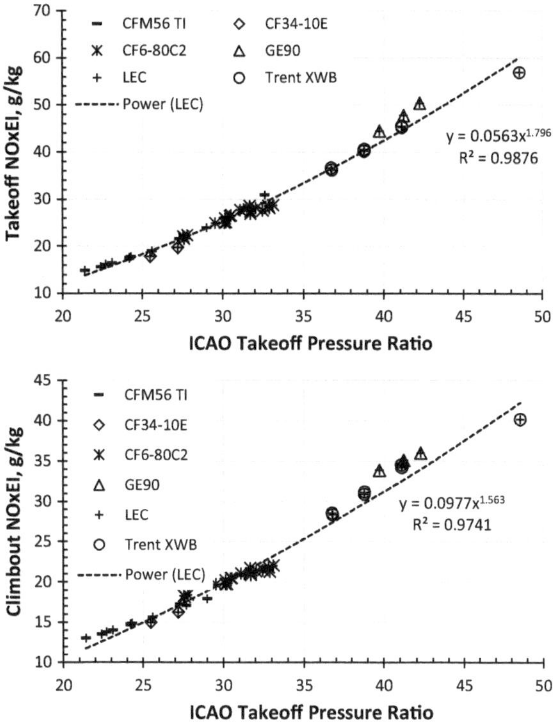

Fig. 24 Takeoff NOxEI (top) and climbout NOxEI (bottom) versus takeoff pressure ratio characteristic of the five GEAE and CFM LEC combustors plus Trent X Wide Body, Trent XWB

6 High- and Low-Power Tradeoffs for Rich and Lean Domes

When the author started working on an axially-staged partially premixing combustion system in November 1974 [19], he went through three major design iterations before settling on the one, followed by its fabrication and testing. The baseline design achieved program goals for idle unburned hydrocarbons and simulated takeoff NO_x. The author's low emissions combustion technology development and test experience gained from the remaining five modifications described in [19] in addition to 5 modifications in [20] and another set of 3 refinements in [21] gave him first-hand appreciation on how hard it is to have a low-emission combustion technology system that meets simultaneously all the design requirements. That is when he learned that getting low NO_x at takeoff is easy but not simultaneously low idle COEI for preferable limit of 30 COEI but not exceeding 45 COEI, as shown in Fig. 26 for the selected eleven GE/CFM engine combustion products we have been discussing. Perhaps less challenging is getting low idle HCEI limits of approximately 3 and 5, as summarized in Fig. 27.

It's very unusual to discover a low-NO_x lean dome combustion product that is comparable with rich-dome combustion products in regard to idle CO-takeoff NO_x

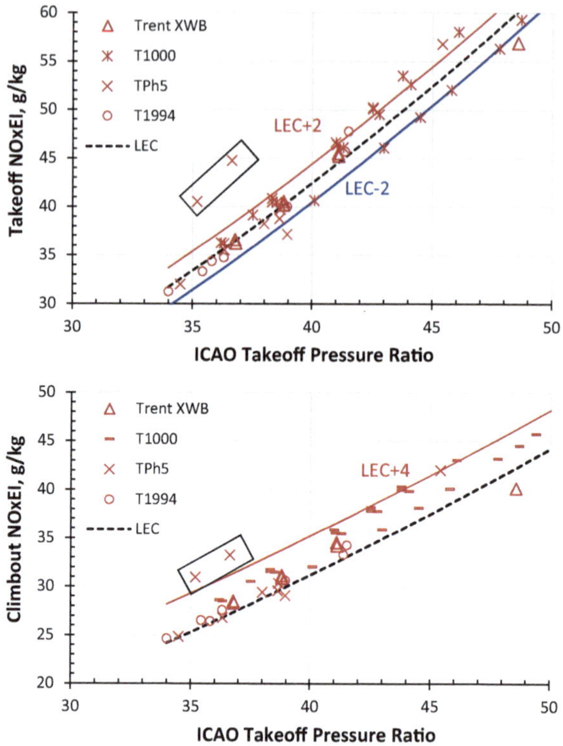

Fig. 25 Takeoff NOxEI (top) and climbout NOxEI (bottom) versus takeoff pressure ratio characteristic of the Rolls Royce RQL combustion products compared with the updated GE Low Emissions Technology trendline

and idle CO and HC characteristics. But here, we have GEnx, the first lean-dome to exceed or meet rich-dome's idle CO and HC emissions. Our benchmark here is the CF6-80C2LEC, and GEnx is even better. Even though CO and HC emissions of LEAP-X are equally low, we don't group all of their engine models with GEnx and CF6-80C2 for the reasons obvious from the data comparison summarized in Figs. 26 and 27. These figures show that three of the LEAP-1A and five of LEAP-1B engine models fall outside of the two boxes, namely Box 1 and Box 2; the former is preferred over the latter. Hypothesis proposed for conducting fundamental investigations in order to improve LEAP-X emissions characteristic will be discussed in the next section.

In summary, GEnx combustor makes the benchmark for the lowest idle CO-HC and idle CO-takeoff NO_x given by the following correlations:

$$\text{Idle COEI} = 32.183/\text{Takeoff NOxEI}^{0.223}$$

$$\text{Idle COEI} = 10 + 7 \times \text{Idle HCEI}$$

The last expression is slightly modified later in Sect. 7.

7 Comparative Evaluation of Lean Domes' Products

In this section, we start with Fig. 28 that provide emissions characteristics of the two combustors, namely the LEC trendline for takeoff NOxEI, and GEnx for showing tradeoff between high-power NO_x and idle CO and HC emissions with the corresponding expressions being:

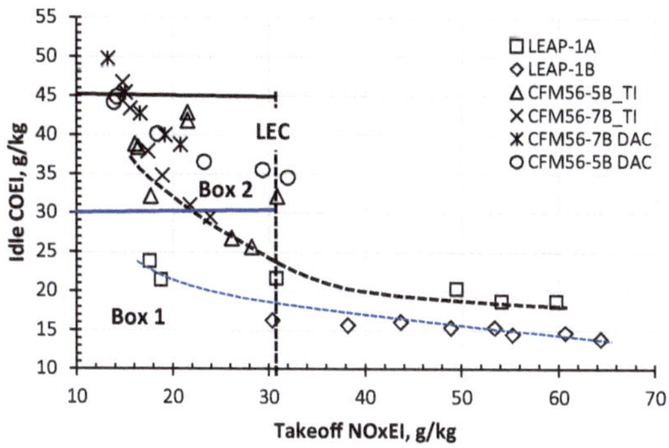

a. CFM combustion products showing right NOxEI limit of 30.9 set by the CFM56-5B Tech Insertion CFM565B_TI. Three engine models of the LEAP-1A and only one engine model of the LEAP-1B fall under Box 1. Two engine models of CFM56-5B_TI and one model of CFM56-7B_TI fall under Box 1; the rest of the models fall under Box 2 where all but three engine models of the CFM56 DAC belong too. The remaining three models of the CFM56 DAC fall outside of Box 2.

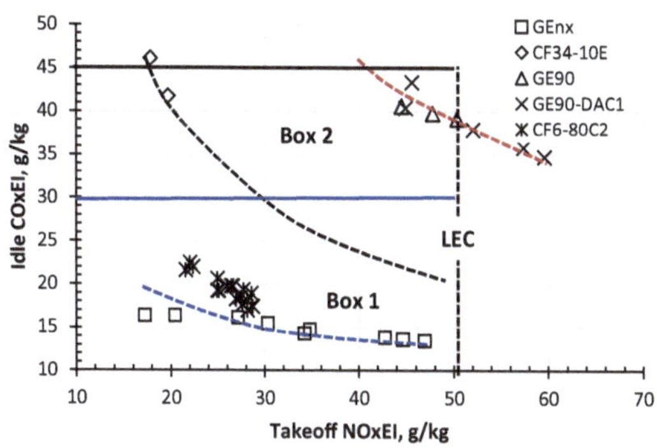

b. GEAE combustion products. Right hand limit value of NOxEI set by the maximum takeoff NOxEI value of 39.1 from the GE90-115B. The three models of GE90 DAC I fall outside of Box 2. Box 2 contains all three engine models of GE90-110/115B, two models of GE90 DAC I and essentially all models of CF34-10E. All engine models of the CF6-80C2 and GEnx fall within Box 1.

Fig. 26 Tradeoff characteristics for idle COEI and takeoff NOxEI for the CFM (Part a) and GEAE (Part b) engines followed by hypotheses-based LEAP-X emissions reduction in Part c; see [24, 25]

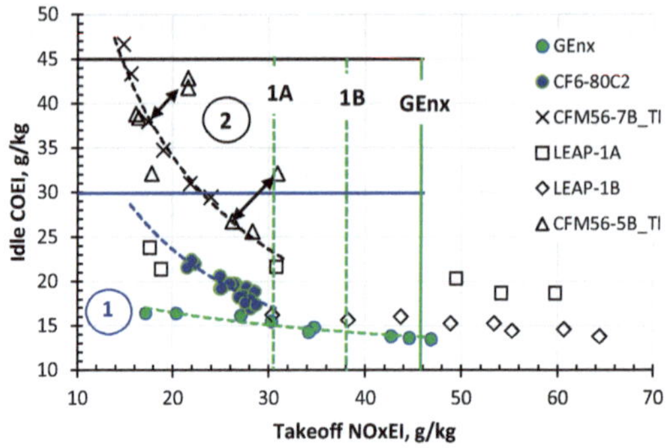

c. High-end NOxEI limits established for GEnx's 46 NOxEI at 47.5PR (for 20% lower NOxEI than LEC), 31 NOxEI at 31 PR (when LEAP-1A has same value of NOxEI as the LEC), and 38 NOxEI at 38PR when NOxEI of LEAP-1B is comparable with that of LEC. This means that only three engine models each of the LEAP-1A and LEAP-1B fall withing the box 1 while others don't exceed takeoff NOxEI of the LEC. On the other hand, all but one of the GEnx engine models fall within box 1.

Fig. 26 (continued)

Fig. 27 Spray quality and quenching effects expressed through idle COEI and HCEI for the CFM (L) and GEAE (R) engines showing that several engines fall in the neighborhood of the lowest idle HCEI-COEI curve given by Idle COEI = $10 + 7 \times$ Idle HCEI. Box 1 is preferred over Box 2

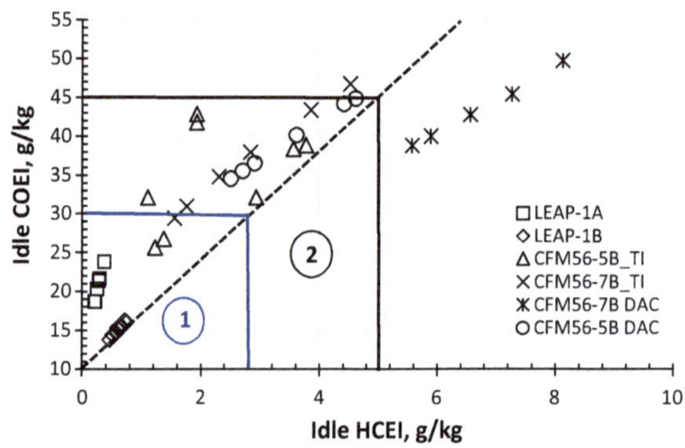

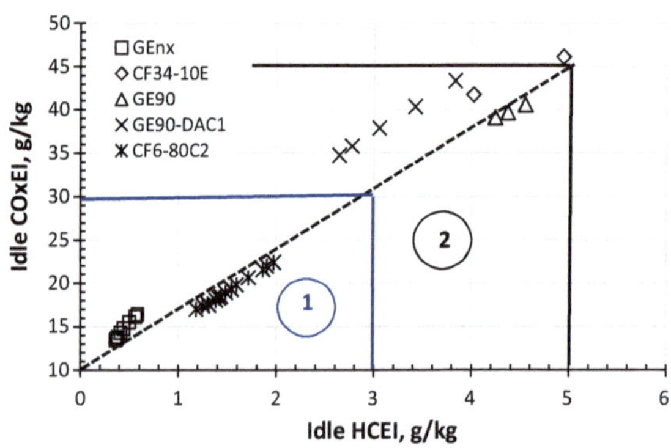

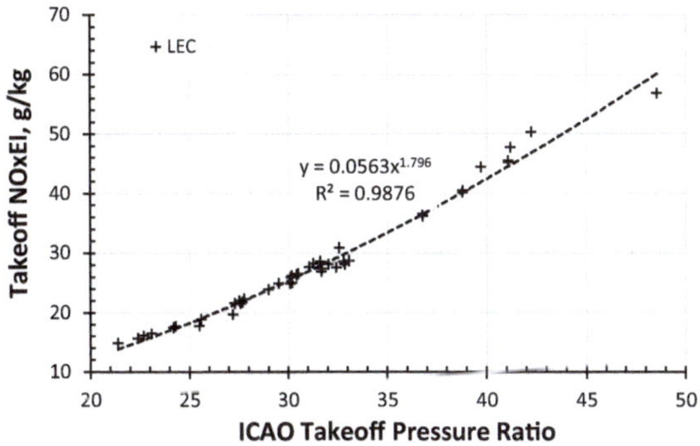

a. Takeoff NOxEI versus takeoff pressure ratio of LEC combustion products given by $LEC\ Takeoff\ NOxEI = 0.0563 \times PR^{1.796}\ with\ R^2 = 0.9876.$

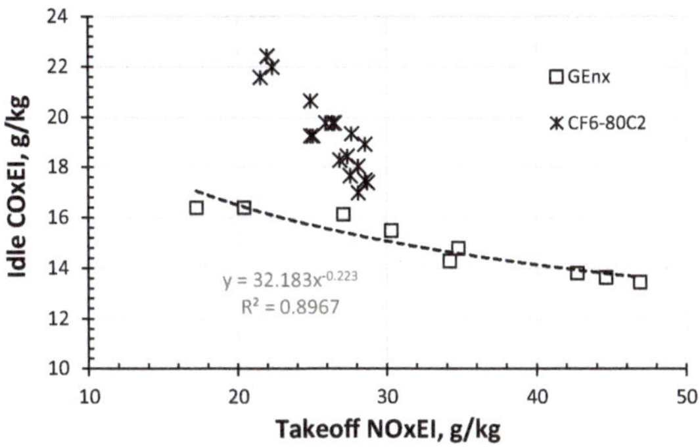

b. Lowest idle COEI versus takeoff NOxEI characteristics of the lowest GEAE combustion products are the rich-dome CF6-80C2 and lean-dome TAPS GEnx; the latter given by $GEnx\ Idle\ COEI = \frac{32.183}{Takeoff\ NOxEI^{0.223}}\ with\ R^2 = 0.897.$

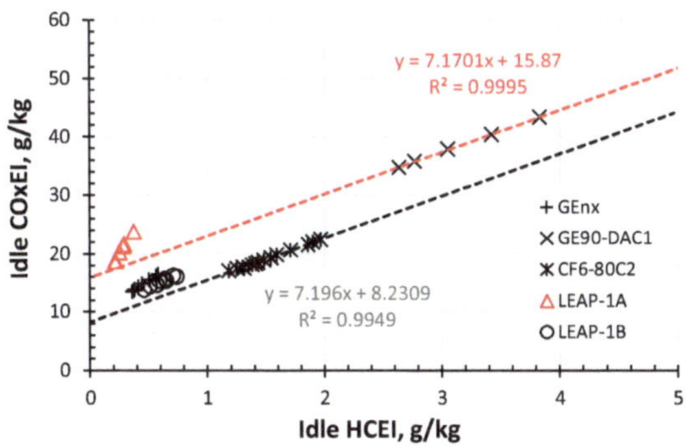

c. Lowest idle COEI versus idle HCEI group comprises of the rich-dome CF6-80C2, lean-domes TAPS GEnx and LEAP-1B given by $CF6 - 80C2\ Idle\ COEI = 8.2 + 7.2 \times Idle\ HCEI\ with\ R^2 = 0.9949.$

Fig. 28 A group of low-emissions (benchmark) combustion products selected for comparison with lean-dome DAC and TAPS combustion products

$$\text{LEC Takeoff NOxEI} = 0.0563 \times \text{PR}^{1.796} \text{ with } R^2 = 0.9876$$

$$\text{GEnx Idle COEI} = \frac{32.183}{\text{Takeoff NOxEI}^{0.223}} \text{ with } R^2 = 0.897$$

$$\text{CF6} - \text{80C2 Idle COEI} = 8.2 + 7.2 \times \text{Idle HCEI with } R^2 = 0.9949$$

One can't miss similarity with GE90 DAC I Idle COEI $= 15.9 + 7.2 \times$ Idle HCEI with $R^2 = 0.9995$. This implies empirically that spray quality of the swirl cups used in both of these engines are essentially similar. However, seeing almost twice the value of intercept, 15.9 versus 8.7 implies empirically that there is more CO quenching in the GE90 DAC I in spite of:

(1) Both engine dome splash plates are convection cooled, see Figs. 6 and 29.
(2) Idle operating conditions T3 and P3 of the GE90 are higher than those of the CF6-80C2.
(3) The centerbody is undercooled.

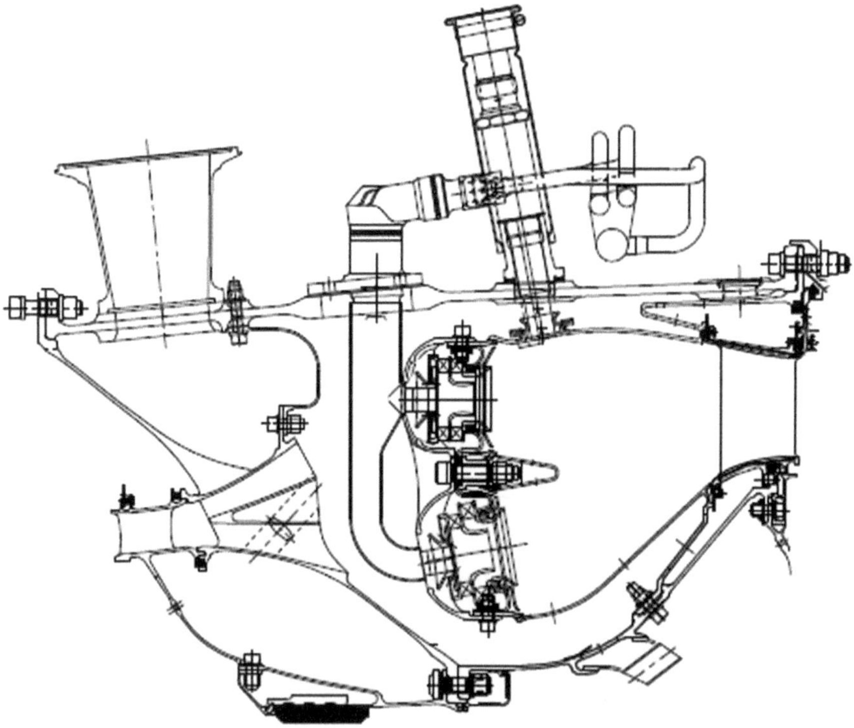

Fig. 29 A schematic representation of the GE90 DAC I combustion system; reproduced from [4]

These three factors lead empirically to suggest that the higher level of CO quenching in GE90 DAC I might have been caused by the mixing of the main combustion air of the inner annular passage and the outer annulus pilot air downstream of the centerbody.

The main reason for selecting the LEC as the benchmark is that its correlations for both the takeoff and climbout NOxEI give good agreement with all rich-dome combustion products of the three major OEM's, namely GE, CFM, and RR; refer to Figs. 24 and 25.

The main reason for selecting GEnx as benchmark for tradeoff between idle COEI and takeoff NOxEI, and idle CO-HC is that it has demonstrated lower values than the best of GE's low emissions combustion products in this regard, namely CF6-80C2, refer to Figs. 26 and 27. And finally, GEnx is currently the lowest LTO NO_x and takeoff NOxEI combustion product.

We start with the DAC combustion products emissions performance keeping in mind that these two concurrently running engine programs, one for the CFM56, a nominal 30 pressure ratio engine, and another one for a new centerline 40 pressure ratio engine were based on the outstanding technology programs conducted by the best of combustion technologists of their era [5–8]. Both of these engine programs were initiated before January 1994 when the author joined GE Aviation to help guide a "world-class combustion section." The first CFM56-5B DAC's engine emissions certification test started on June 15, 1994, followed by the CF6-80C2 LEC engine start date of January 13, 1995, the GE90-DAC I on February 24, 1995, the CFM56-5B DACII on August 1, 1996. The author who regarded only one out of the four-engine programs success was furious. Subsequently, he insisted for more freedom for conducting aerothermal design activities which he was given for the CFM56-7B DAC and the resulting engine emissions certification start date of August 30, 1997. Back to back comparison between the two (CFM56-5B and CFM56-7B) DAC engines' CO, NOx, and HC emissions summarized in Figs. 26 and 27 clearly shows that the author's direct involvement didn't have any impact on the outcome.

It should be pointed out that the turbomachinery technology activities of [8] led to the layout and successful execution of the GE90 turbomachinery product development while simultaneously the GE90 DACI combustion system turned out to be a challenging tradeoff as discussed in the following paragraphs.

We will start with the CFM DAC first because it didn't have to extend beyond the engine operating pressure ratios of the two engine demonstrators reported in [7, 8]. Obviously, comparison of any lean-dome combustion products needs to be made with a benchmark rich-dome combustion system's emissions which in our case is summarized in Fig. 28a for the LEC takeoff NOxEI is given by LEC Takeoff NOxEI $= 0.0563 \times \text{PR}^{1.796}$ with $R^2 = 0.9876$, idle COEI and HCEI given by CF6 - 80C2 Idle COEI $= 8.2 = 7.2 \times$ Idle HCEI with $R^2 = 0.9949$ in Fig. 28c.

Logically speaking, one should select the CF6-80C2 LEC combustor' data on idle COEI versus takeoff NOxEI summarized in Fig. 28b. But we have chosen the GEnx engine instead, the engine that was tested in 2009; because, if someone's rich-dome can't compete with lean-dome's idle COEI versus takeoff NOxEI, his grandmother may not like it. Therefore, GEnx' expression GEnx Idle COEI $= \frac{32.183}{\text{Takeoff NOxEI}^{0.223}}$ with $R^2 = 0.897$ will be used as benchmark.

The three CFM DAC engine products gave only slight reduction in takeoff NOxEI emissions compared to LEC, as summarized in Fig. 30a. The empirical correlation combined for the both 30 and 40 overall pressure ratio OPR engines (CFM56 and GE90) shows that DAC technology gave 8% higher NOx than the LEC technology at 32 OPR. Even with the empirical correlation based the three CFM DAC engines data, it gave us 4% higher NOx value than the LEC technology line. Compared to the CF6-80C2 LEC data, DAC gave us 6% higher NO_x. We have 38% higher NOx with short GE90 DAC1 combustion system compared to LEC technology in addition to inferior performance in regard to idle COEI versus takeoff NOxEI summarized previously in Fig. 27 and 28. Back to back comparison with the lower-cost the Trent engine emissions certified in September 1, 1994, the GE90 DACI gave 35% higher NO_x.

Poor emissions performance of the lean DAC products in spite of the best effort put by the teams regarded very capable by the author who also thought very high of the teams that developed the technology [5–8] should have normally dampened the zeal for pursuing lean-dome technology development for propulsion engines. The driving force for pursuing lean-dome technology was the author's commitment to his customer for developing dual annular TAPS combustion system for achieving 50% reduction from LEC [1, 2]. A similar commitment was made to another customer for transitioning the TAPS technology into the single annular TAPS GEnx combustion product whose takeoff NOxEI is presented in Figs. 30b and 31 that include TAPS2 and TAPS3 technologies. Results are summarized in Tables 4, 5, 6 and 7.

Since the takeoff NOxEI results for the LEAP-1A and LEAP-1B engines are significantly different from those of the GEnx (compare Fig. 30b, c), it's perhaps important to summarize the key aspects of the GEnx design process used by the author who functioned as the team leader from conceptual design through first engine to test (FETT) of GEnx completed by March 2006.

It should be mentioned that all of the GE and CFM LEC propulsion engines can trace to its origin to the CF6-80C2 engine tested on January 13, 1995; similarly, for the RR RQL engines to its origin the Trent engine family tested on September 1, 1994. In summary, all of these LEC/RQL engines emissions certified between 1994 and 2017 are represented well by LEC Takeoff NOxEI $= 0.0563 \times \text{PR}^{1.796}$ with $R^2 = 0.9876$ implying that LEC/RQL should be considered a matured technology.

But that is not the case with the lean-dome propulsion engine products. The development of the lean-dome DAC combustion products stopped after GE90-DACI tested on February 24, 1995 and the CFM56-7B on August 30, 1997 because of its less than stellar performance in regard to low-power HC and CO emissions in addition to takeoff NOx emissions as summarized in Fig. 30a. It gave slightly lower NO_x in nominal 30 OPR (overall pressure ratio) engines than LEC, but significantly higher in 40 OPR engines. When DAC's correlation is extended to 49 OPR, it is projected to have 76% higher takeoff NO_x than its counterpart LEC engines.

TAPS technology initiated in 1995 was expected to address several deficiencies in DAC's products including durability, shifting combustor exit radial profile

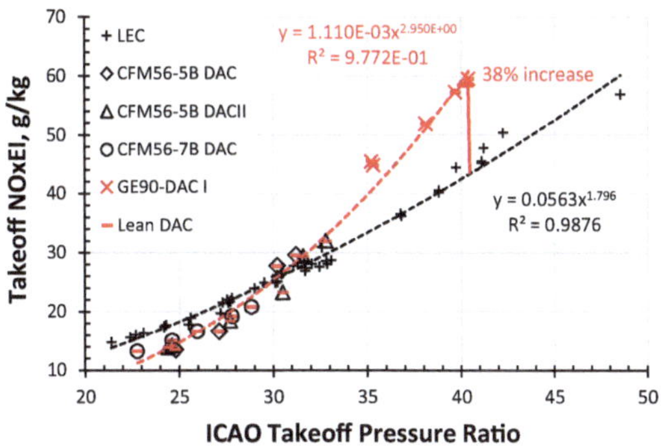

a. Disappointing NOx performance of the 1st generation lean-dome DAC for 40 OPR engine GE90-DAC I giving 38% increase in takeoff NOxEI emissions compared to LEC.

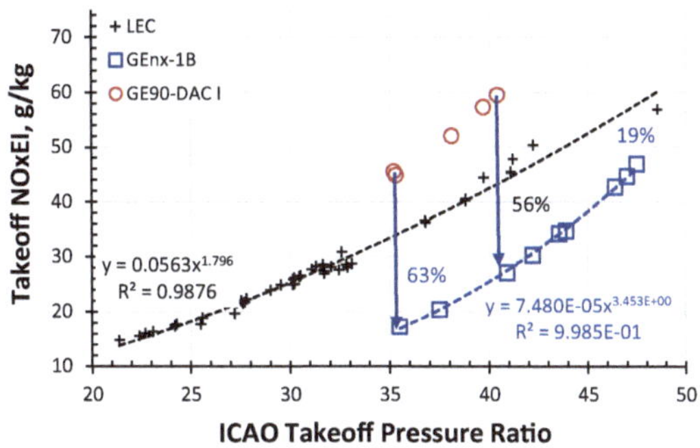

b. TAPS based GEnx combustion system gives 56-63% reduction from the 1st generation lean-dome GE90-DAC I, but only 19% lower NOx than LEC at 47.5 OPR. Refer to Fig. 31 for TAPS2 and TAPS3 technology potentials.

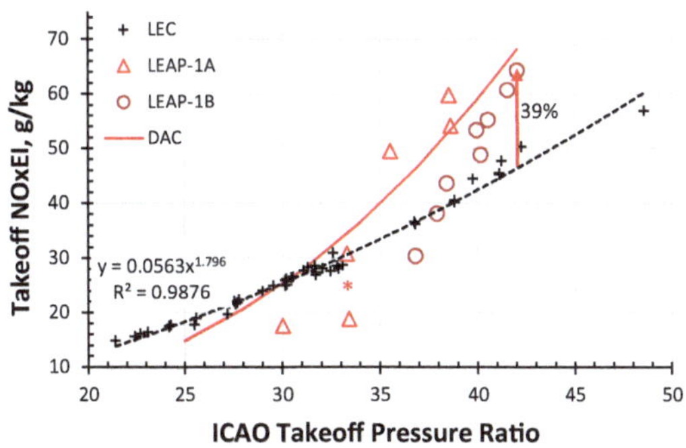

c. What went wrong with TAPS II combustion products of LEAP-1A and LEAP-1B? We are back to NOx performance of the 1st generation DAC technology getting 39% higher NOx than the LEC technology given by $LEC\ Takeoff\ NOxEI = 0.0563 \times PR^{1.796}\ with\ R^2 = 0.9876$.

Fig. 30 Takeoff NOxEI of DAC's based engines (CFM56-5B DAC I, CFM56-5B DAC II, CFM56-7B DAC and GE90-DAC I), LEC engines' best-fit curve, TAPS based engines GEnx, LEAP-1A, and LEAP-1B

Fig. 31 Takeoff NOxEI characteristics of five technologies, DAC based on four engines data covering $24.6 \leq PR \leq 40.4$, six LEC engines data covering $21.4 \leq PR \leq 48.6$ and GEnx engine data covering $35.5 \leq PR \leq 47.5$

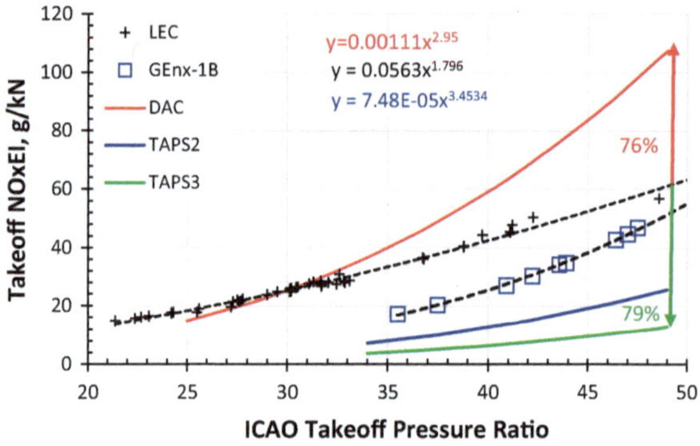

Table 4 GEnx technology exceeded its goals for reducing takeoff NOxEI from GE90-92B DAC I at 40.4 overall pressure ratio

Technology or Product	Goal (%)	Achieved or expected (%)
Rich-dome LEC technology	25	27
GEnx Technology	50	56

Table 5 Reduction from LEC technology and the corresponding overall takeoff pressure ratio

Product or TAPS technology	Demonstrated or expected (%)	Takeoff pressure ratio
GEnx-1B64/P2	39	40.9
TAPS2 Goal	70	40.9
TAPS3 Goal	85	40.9

Table 6 Reduction from LEC technology and the corresponding overall takeoff pressure ratio

Product or TAPS technology	Demonstrated or expected (%)	Takeoff pressure ratio
GEnx-1B76/P2	19	47.5
TAPS2 Goal	59	47.5
TAPS3 Goal	80	47.5

Table 7 Predicted takeoff NOxEI at 49 takeoff pressure ratio and % reduction from LEC technology

Technology	Takeoff NOxEI	% Reduction from LEC
LEC	61.11	0
DAC	107.50	−76%
GEnx	51.38	16%
TAPS2	25.69	58%
TAPS3	12.85	79%

LEC technology curve based on engines data covering $21.4 \leq PR \leq 48.6$ and GEnx engine data covering $35.5 \leq PR \leq 47.5$

from idle to maximum power and attendant adverse impact on fuel burn, higher idle CO/HC emissions, and higher high-power NO_x for the next-generation 40 OPR engines. Summary of TAPS technology development of the two full-scale combustors, Fig. 32, for potential applications in the next-generation medium size and large propulsion engines is provided in [1, 2] including two next-generation TAPS technologies colloquially identified as TAPS2 and TAPS3 implying $\text{TAPS2}_{\text{Takeoff NO}_x} = \text{TAPS1}_{\text{Takeoff NOxEI}}/2$ and $\text{TAPS3}_{\text{Takeoff NO}_x} = \frac{\text{TAPS2}_{\text{Takeoff NOxEI}}}{2}$. The two TAPS2 and TAPS3 curves shown in Fig. 31 are simply takeoff NOxEI expression of GEnx (namely, GEnx Takeoff NOxEI $= 7.480E - 05 \times \text{PR}^{3.453}$) divided by 2 and 4, respectively.

Two of the curves (namely, SAC TAPS engine data, and DAC TAPS full annular rig FAR) shown in Fig. 33 are for the TAPS configurations shown in Fig. 32. Discussion on Fig. 33 will follow after discussing Fig. 34 through xx.

When we started with conceptual design of GEnx, the author had data on:

- The Tech56/CFM TAPS combustion system that successfully completed engine demonstration including emissions, performance, operability, and endurance.
- The DAC TAPS combustion system from its full annular rig test rig at T3, corrected flow rates, fuel/air ratio, and P3 for all the GE90 engine operating conditions except P3 at takeoff was slightly lower. This data set included emissions, performance, operability, and liner wall temperature levels and gradients.

During the conceptual design of GEnx that started in 2004, the team had at its disposal stark reminder that the geometric details of the TAPS' pilot mattered for nominal 30 OPR engines, see Fig. 34. Hypotheses for the CFM TAPS's higher NO_x emissions compared to DAC TAPS (see Fig. 33) while simultaneously lower idle CO and HC of the latter, see [1], were:

1. The CFM TAPS flame was anchored off the pilot's lip recirculation zone LRZ that was discovered later through CFD simulation in 2007, see Figs. 15 and 16, and experimentally in 2008, see Fig. 35.

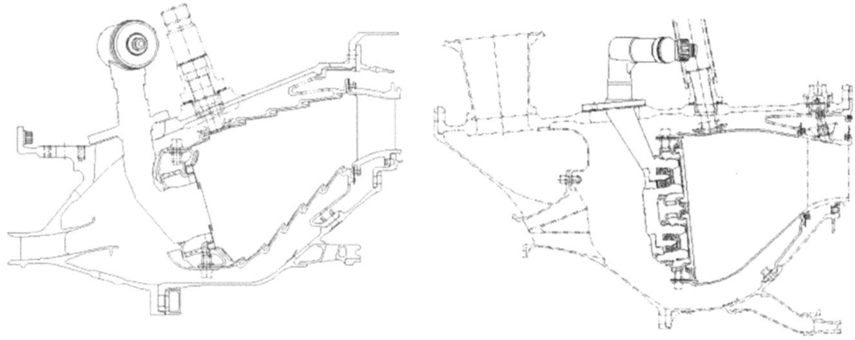

Fig. 32 Schematic layout of TAPS combustion systems; (L) Tech56/CFM TAPS combustion system, (R) GE90 DAC TAPS combustion system. Reproduced from [1]

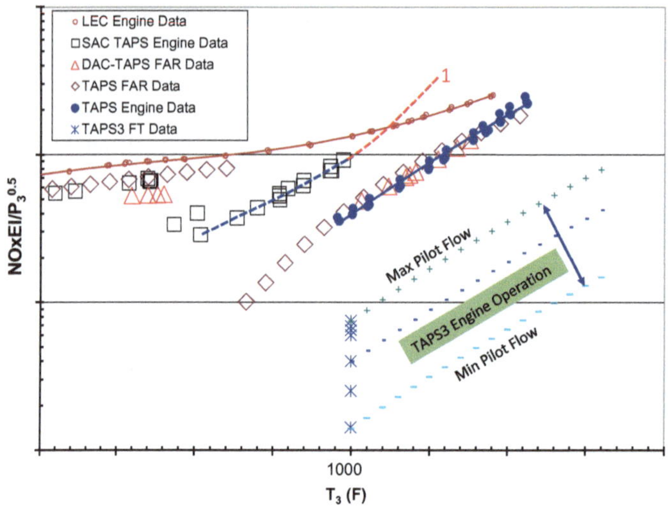

Fig. 33 Typical comparison between a rich-dome low-emission sea-level operation compared with three TAPS1 configurations expressed in terms of $NOxEI/\sqrt{P_3}$ versus combustor inlet temperature T3 compared to TAPS1 and TAPS3 results. There are two TAPS engines data identified as SAC TAPS and TAPS, the latter has data from a full annular rig (FAR) that gives good agreement with engine data. Refer to discussion on Fig. 32 TAPS3 takeoff NOxEI curve regarding projected operational range of TAPS3 based engine which allows variable pilot fuel flow split capability

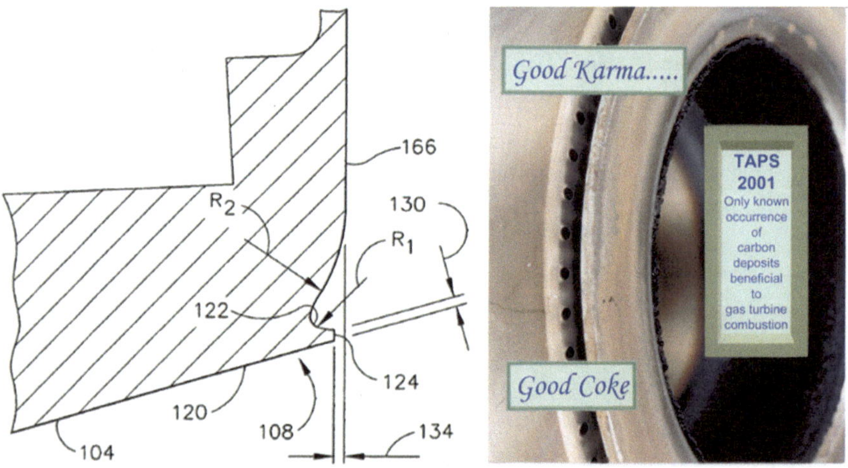

Fig. 34 Pilot exit outer-wall details US Patent 6,865,889, May 19, 2005 (left), "Good Karma, Good Coke" appropriately phrased by Al Mancini, improved idle efficiency from mid-70% to 97%+ on the engine test stand. Reproduced from [2]

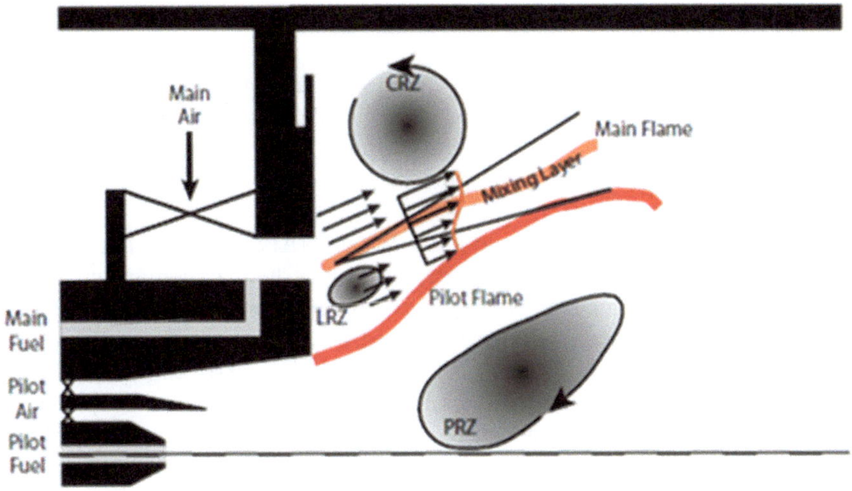

Fig. 35 Schematic representation of TAPS mixer and flow characteristics; here RZ means recirculation zones caused by pilot P, pilot lip L, and sudden expansion corner C formed between the TAPS mixer and combustor dome and liner wall, identified, respectively, as PRZ, LRZ, and CRZ. Reproduced from [17]

2. Mainly because of the smaller size of the DAC TAPS mixer compared to the CFM TAPS, its flame might be a lifted flame with attendant reduction in NO_x emissions. It was verified numerically in 2007, see Figs. 15 and 16.
3. Lower idle CO and HC emissions of DAC TAPS compared to the CFM TAPS could be attributed to the former's higher idle T_3 and P_3 conditions.
4. But wouldn't GEnx TAPS mixer that is bigger than the CFM TAPS lead to its flame anchoring off LRZ? Options:

   (a) Go without the pilot exit lip features of the CFM TAPS, Fig. 34.

   (b) Put more emphasis on TAPS2 mixers.

Therefore, the first pizza-pie full-scale annular combustor rig testing for GEnx included three TAPS2 concepts and only one TAPS1 concept intended for GEnx. Thanks to good secular Karma deeds of the GEnx TAPS team, all three TAPS2 concepts literally fell along the intended TAPS2 characteristics as shown in Fig. 31. TAPS2 configurations had some durability challenges which didn't need to be addressed because the TAPS1 NOx characteristics fell on the top of DAC TAPS. How did that happen? The author has no idea.

TAPS3 conceptual concept screening, selection of a particular design, single-cup test rig and combustor design, fabrication, and instrumentation completed by the GEnx team was followed by testing at the NASA Glenn Research Center covered by Raghavan Pandalai. The parametric testing at simulated cruise condition, P3 = 400 psi, T3 = 1000 °F and FAR = 0.0273 included variations in the pilot fuel flow split for the reasons explained in Sect. 9. According to Raghavan Pandalai, the author told him, "If you don't get test results in line with my TAPS3 expectations, you better hide yourself in a cave."

As hoped for and summarized in Fig. 3 of [15], the pilot fuel flow split had huge impact on NOxEI emissions that ranged between 0.2 g/kg and 7.5 g/kg. It should be obvious that TAPS3 NOxEI less than 1.0 at 400 psi, 1000 °F T3 and 0.027 FAR makes it highly competitive with dry low NO_x mixers used in the LM2500 and LM6000 aeroderivative engine, see [32].

It is, therefore, very important to know for any TAPS mixer configuration the actual fraction of the pilot fuel flow rate and that of the main participating in its so-called flame stabilizing region(s) and their residence time(s), see Fig. 35. TAPS' emissions and performance are impacted strongly by its degree of mixedness, fraction of pilot and the main injectors' fuel flow rates, characteristics of the flowfield including the recirculation zones (PRZ, LRZ, and CRZ) and the mixing layers. CFD simulations are not accurate enough to sort through the complexity of the TAPS flowfield characteristics. Al Mancini, the primary contributor on the TAPS injector technology development, appropriately cast the right-hand side of Fig. 34 as "good Karma and good Coke".

A second set of data for verification of TAPS3 curve is from the sector rig testing results reported in [31]. This sector test program included testing of LTO cycle points of a next-generation engine cycle. The rig test results of LTO NOx value of 20.7 g/kN and 18.9% CAEP6 standard were used to calculate its ICAO takeoff pressure ratio of 55. It's interesting to notice that our TAPS3 expression recklessly extrapolated from 46.9 to 55 OPR gives takeoff NOxEI of 19.1 g/kg compared to the data, 17.6 g/kg, reported in [31].

In summary, for any future TAPS-based combustion products when properly designed, its takeoff NOxEI should fall between data of the GEnx, TAPS2, and TAPS3 curves shown in Fig. 31. The experience gained from the GEnx and LEAP-X TAPS combustion products indicates that idle COEI and HCEI should fall between the data bands presented in Fig. 28c. In cases of high takeoff NOx values (Fig. 30c), the LEAP-X team should:

- Investigate fundamental flow characteristics of LEAP-X TAPS mixer, Fig. 35. But the team should pay attention to the lessons learned by the author as summarized in [26].
- Extend the DRA methodology for rich domes summarized in Sect. 8 to TAPS combustions systems including CFM TAPS, DAC TAPS, GEnx, sector test results reported in [31], LEAP-1A and LEAP-1B.

Hopefully, this process will get the LEAP-X back to the right path for producing low emissions TAPS design.

8 Engine Emissions Data Reduction and Analysis DRA

The author has relied on qualitative guidance from CFD simulations but subject to two verification criteria, namely hypotheses' expectations and empirical data correlations; the latter is discussed first because it has direct relevance to the curve identified as 1 in Fig. 33.

Combustion engineers use empirical correlations, like $\mathrm{NOxEI}/\sqrt{P_3}$ versus T3 from a typical LEC engine operating at sea-level from idle to maximum power shown in Fig. 33. Moreover, instead of using best-fit expressions commonly used in research papers (namely, power law, exponential, logarithmic, or linear functions), industry uses 4th or 6th order polynomials in order to get the highest possible values of R^2. For example, the LEC engine data in Fig. 33 when fitted with a 4th-order polynomial gave $R^2 = 0.9991$. On the other hand, the power-law used for the TAPS engine data gave $\mathrm{NOxEI}/\sqrt{P_3} \propto T_3^{8.497}$ with $R^2 = 0.9887$. The latter approach may not provide good σ value; but it surely sends a strong warning against extrapolating empirical correlations. That was the root cause for failure of seeing through the fog for estimating a priori NO_x emissions of the GE90-DAC I.

The author had the same fear for relying on the Tech56/CFM TAPS NOxEI extrapolated curve 1 in Fig. 33 for scaling up to the higher operating pressure ratio GEnx. On the other hand, DAC TAPS NOxEI data (with 60 TAPS mixers) was significantly lower than the Tech56/CFM TAPS with only 20 mixers. But we wanted to use only 22 mixers in GEnx in order to keep its acquisition cost-competitive with counterpart rich-domes.

Since most of the semi-empirical and semi-analytical correlation works by the research community are mainly focused on rich-dome emissions and operability, we will focus on this topic while fully recognizing that it may not have much relevance to TAPS flames.

Prof. Lefebvre has done outstanding service for the gas turbine engine combustion community by offering several semi-analytical correlations for NO_x, CO, HC, LBO and ignition [10] followed by further refinements by Rizk and Mongia [11]. We will focus on NOx correlation works first.

According to Lefebvre [10], NO_x emission index from propulsion engine combustors can be correlated by:

$$\mathrm{NOxEI} = 9E - 8P_3^{0.25} V_{\mathrm{NO_x}} \exp^{0.01 T_{st}} / (W a_{36} T_{PZ}) \tag{1}$$

Here T_{PZ} is calculated from the primary zone fuel/air ratio, $\mathrm{FAR_{PZ}}$, which is defined as $\mathrm{FAR_{36}}/f$. f is the fraction of combustor air participating in the primary zone. This expression is very simple (which requires best fitting the engine data with two empirical constants, V_{NOx} and f) and elegant for qualitative assessing different combustors in regard to NO_x emissions.

It, however, does not have the accuracy level desired by the designers, as pointed out by Danis, Pritchard, and Mongia [10]. Moreover, it can predict trends very

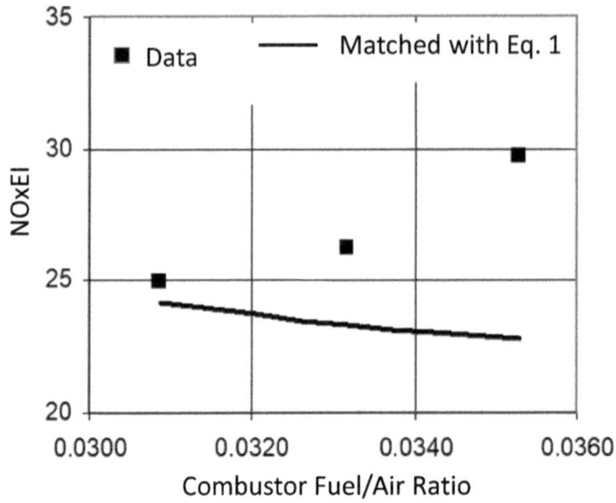

Fig. 36 Typical comparison between test rig NOxEI data at a given P3, T3, and Wa3 conditions with parametric variation in combustor's overall fuel/air ratio and correlation given by Eq. 1. Source [12]

different from data, as shown in Fig. 36 which is very relevant for developing rich-dome combustion technology and products.

Rizk and Mongia [11] recast the Lefebvre's expression into the following equation:

$$\text{NOxEI} = 0.459E - 8P_3^{0.25}\tau_{\text{NO}_x}\exp^{0.01T_{st}} \tag{2}$$

, which implies that with a NO_x characteristic time τ_{NO_x} and the primary zone fraction air f one should be able to fit the NOxEI data.

Rizk and Mongia [11] gave an additionally improved correlation, namely

$$\text{NOxEI} = \frac{0.15E16\left(\tau_{\text{NO}_x} - 0.5\tau_{ev}\right)^{0.5}\exp^{-\frac{71000}{T_{st}}}}{P_3^{0.05}(P/P_3)^{0.5}} \tag{3}$$

The author likes this correlation because his data from the early 1980s could be interpreted only when he assumed the pressure exponent equal to zero.

The author also felt that the root cause for poor agreement with engine data could be the overly simplified assumption of a constant value of the NO_x characteristic time τ_{NO_x}.

This was addressed in [16] by modeling five production engines CFM56-2/3, CFM56-5/7, CF6-80C2 PreLEC and LEC, and GE90 DAC2 by using a 3-reactor model shown in Fig. 37 which allows independent variation of three characteristic times τ_1, τ_2 and τ_3. Embedded in these characteristic times are the spray evaporation times which takes into consideration the spray characteristics of the selected product combustors and their engine operating conditions from idle to maximum power.

As per the standard practice unique to each OEM, the airflow distribution around the combustion system is known. There can be several ways of estimating the axial variation of the equivalence ratios ϕ_1, ϕ_2 and ϕ_3 as function of power setting. Perhaps,

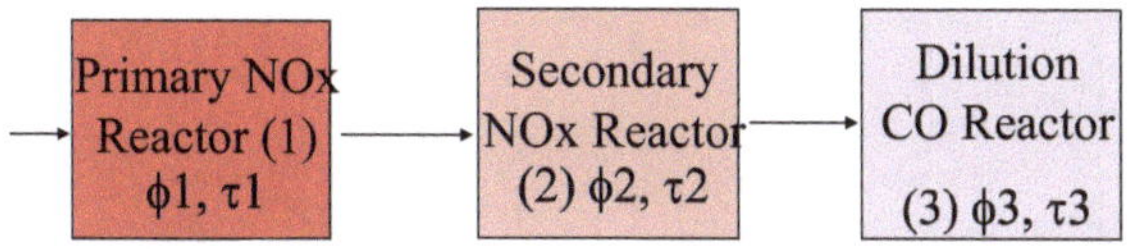

Fig. 37 Three perfectly stirred reactor model used for conducting Data Reduction and Analysis DRA of engine sea-level emissions data from idle to maximum power. Here, we assumed $\tau_1 = \tau_2$. The equivalence ratios $\phi 1$, $\phi 2$, and $\phi 3$ of the selected five production engines were estimated from the combustor airflow distributions and a finite set of axial stations

one-dimensional heat transfer analysis can be used as guide. But the process must be consistent across all the combustors being analyzed. Similarly, a standard practice must be used regarding the location and number of axial stations selected; Fig. 38 can be used as an illustration. It should be made clear that effort described in [16] was called Data Reduction and Analysis, DRA with the main objective being can any model patterned after Equations 1 or 2 get good agreement with HC, CO, and NO_x emissions data of rich-dome propulsion engines from minimum to maximum power?

Readers should refer to [16] for details; here, we summarize the key findings. Figure 39 shows that the calculated τ_1 and τ_2 from idle to maximum power will have to change in order to match the data as summarized in Fig. 39 for NOxEI and COEI.

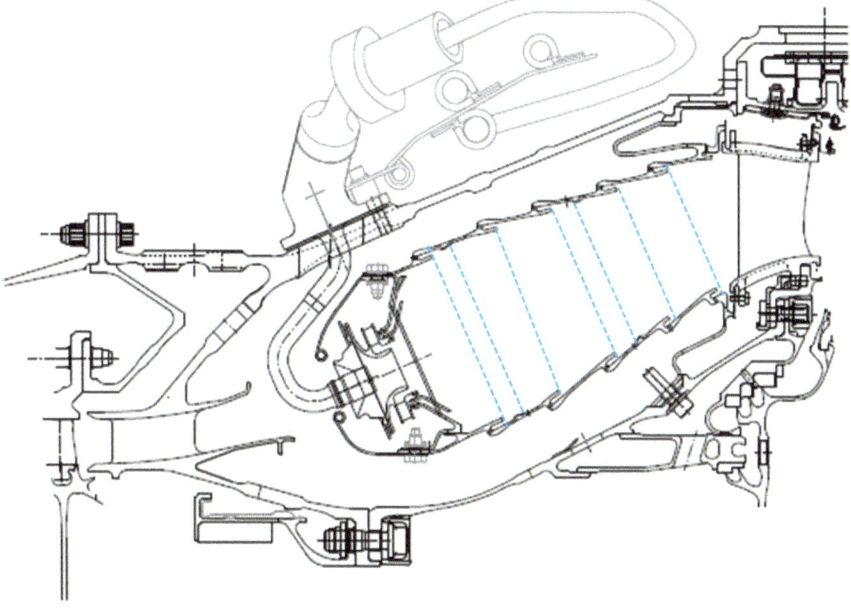

Fig. 38 Typical cumulative axial distribution of combustion zones with power setting

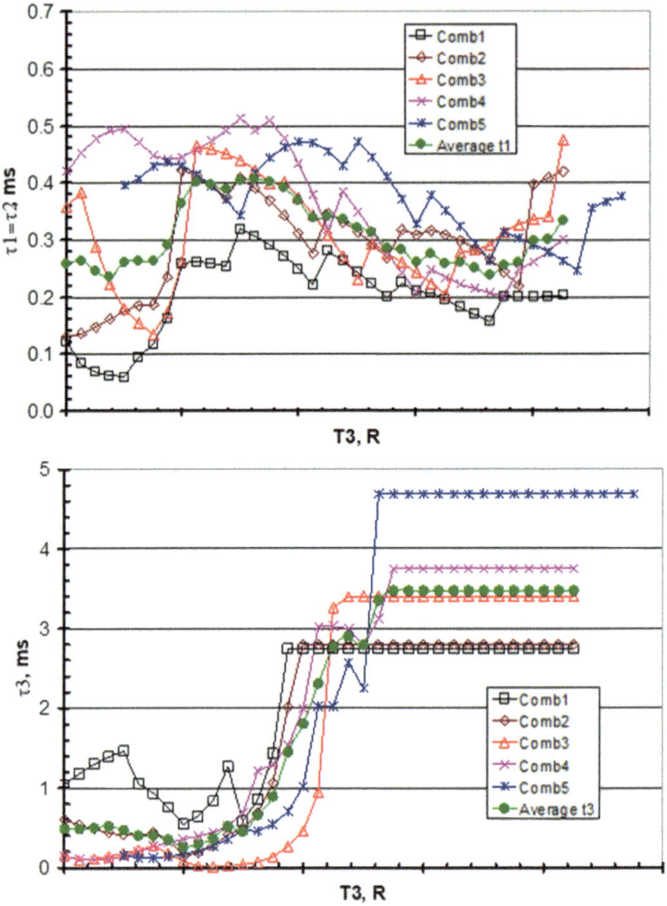

Fig. 39 In order to match engine emissions data for specified variation in ϕ_1, ϕ_2 and ϕ_3, τ_1 and τ_3 vary from idle to maximum power for the five selected production engines

When we went with the average but distinct values of τ_1 and τ_2 for each of the five combustors, the agreement with the data was very poor.

In summary, semi-analytical emissions models with one constant value of τ_{NO_x}, V_{NO_x} and or fraction of combustion air participating in the primary is expected to give only qualitative agreement with engine emission data from idle to maximum power during its sea-level operation. Therefore, state of the art semi-analytical models are not good enough to reduce sea-level engine emissions data to calculate LTO emissions characteristics. That is the main reason why OEM's continue to use an approach similar to Fig. 33 for emissions certification. But, this approach should not be used for calculating mission NO_x emission or for extrapolating to predict emissions from next-generation higher pressure ratio engines.

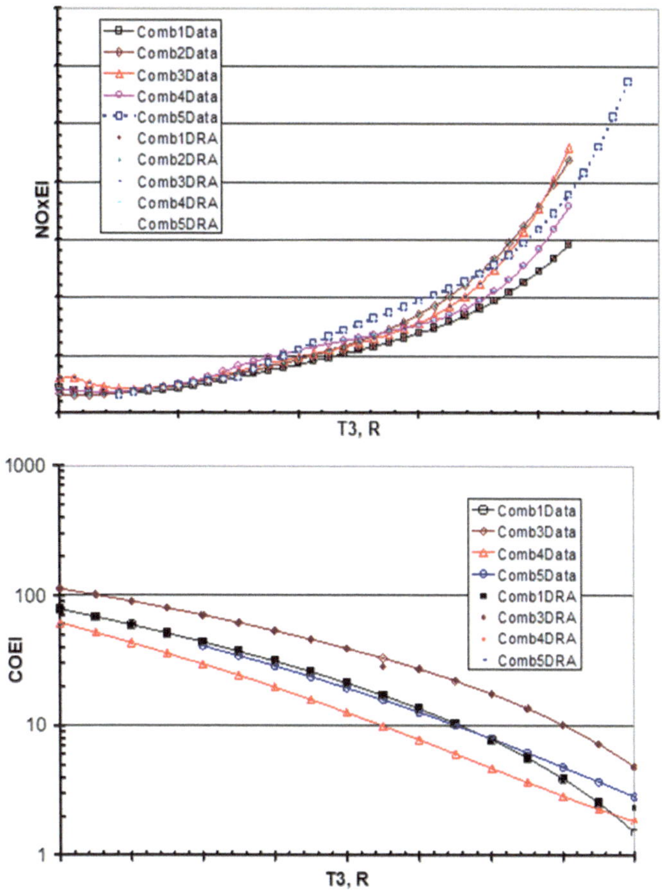

Fig. 40 The data reduction and analysis DRA proposed in [29] matches NO_x and CO emissions data of five production engines

If one were to compare rich-dome with lean-dome combustion product emissions, the criteria used by the author since the middle 1970 [1–4] is not enough; namely takeoff NO_x versus P3, idle CO versus takeoff NO_x, and idle CO versus idle HC. Therefore, an attempt was made to check the validity of the following assumptions for the seven rich-dome combustors out of the twelve overviewed in [15].

a. For the entire engine operation from idle to maximum thrust at sea-level, NOxEI is a function of stoichiometric flame temperature, COEI and HCEI are functions of NOxEI.
b. Assumption (a) can be extended to the propulsion engines' entire flight envelope from sea-level to maximum altitude covering their minimum to maximum thrust capabilities. Altitude emissions data reported by Diehl and Biaglow [33] was used for this exercise.

Readers should refer to [17, 18] for details; here, we have reproduced summary results as presented in Figs. 41 and 42. Agreement between correlations and data was good for all the seven rich-dome combustion products; but the quality of engine emissions was considered to be better for Combustors 1, 5, 6, and 7 which are covered in Fig. 41.

NOxEI correlation versus data, along x-axis, fall between error band of ±4% and the resulting standard deviation σ of 2% which is considered comparable with the

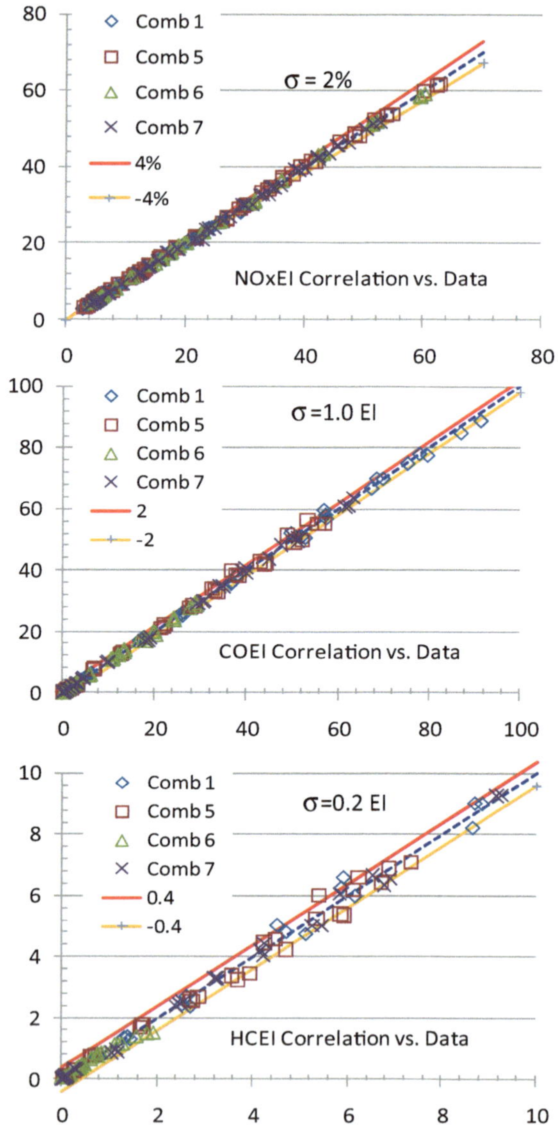

Fig. 41 3rd generation of semiempirical correlations for the entire engine operating points from idle to maximum thrust. Reproduced from [17]

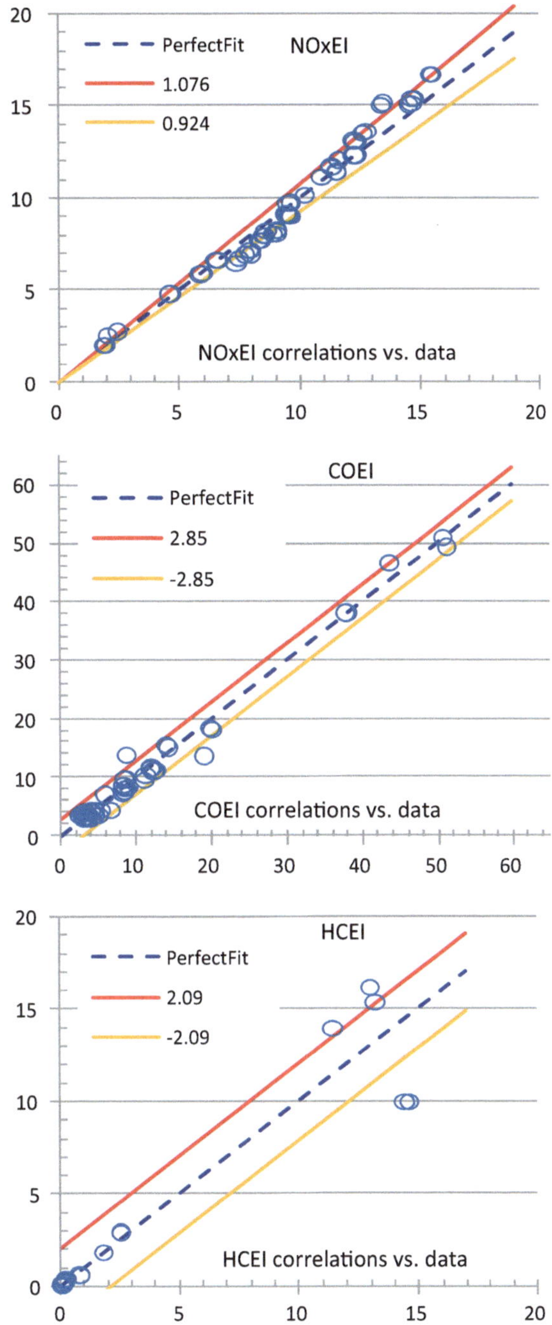

Fig. 42 3rd generation of semiempirical correlations for the TFE731-2 engine operating points from minimum to maximum thrust range at simulated altitude from 640 m to 13,200 m. The root-mean-square errors for NO_x, CO and HC are 3.8%, 1.43EI and 1.04EI, respectively. Reproduced from [18]

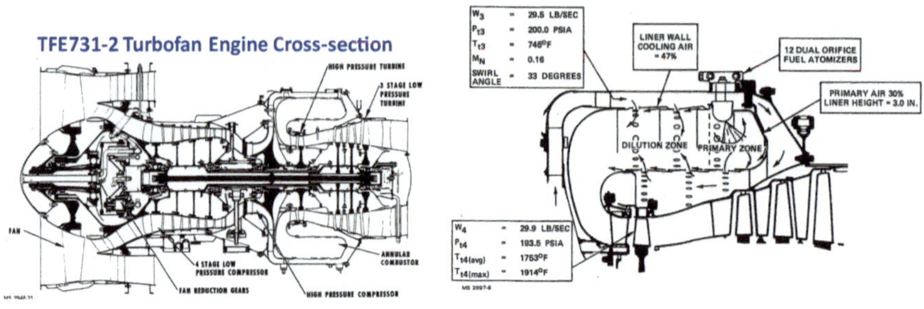

Fig. 43 The TFE731-2 engine cross-section and its reverse-flow annular combustion system. Reproduced from [19]

experimental σ values of these four combustors. When it comes to assessing the quality of correlation for CO and HC, the author prefers to express the error bands in terms of $\pm\delta$COEI and $\pm\delta$HCEI along with the corresponding σ values, as illustrated in the middle and bottom parts of Fig. 41.

In summary, the approach recommended in [17] can reproduce data very well for all the seven combustors. For the four-engine combustors with good quality emissions data, the approach achieved standard deviation σ values of 2%, 1.0 EI and 0.2 EI for NOx, CO, and HC, respectively, for the entire engine operating points from idle to maximum thrust.

It should be emphasized that the emissions data quality for all engine emissions certification programs is expected to be much superior than generic altitude engine test programs similar to the one reported in [33]. Moreover, this engine, small turbofan TFE731-2, Fig. 43, designed in the late 1960s and production start in the early 1970s was going through its growing pains. It is small engine, 15.6 kN rated takeoff thrust with a reverse-flow annular combustor fueled by radially inserted fuel nozzle. The root-mean-square errors for NO_x, CO and HC are 3.8%, 1.43EI, and 1.04EI, respectively, through somewhat higher than for the four medium and large size engines covered in Fig. 41 (namely, σ values of 2%, 1.0 EI, and 0.2 EI) should be regarded quite good considering that it covers engine operation from sea-level to 13,200 m.

In summary, the exploratory investigations summarized in [17, 18] offers an approach based on fundamentals that can be defended significantly better than the currently practiced OEM's approach. Once this approach is refined to meet OEM's expectations, it can be used to extrapolated for predicting emissions from the next-generation rich-dome combustion products. Obviously, a lot needs to be done for extending this approach to lean-dome combustion product, present, and next-generation.

9 TAPS Technology NO_x Emissions Potentials

Lean-dome combustion products for propulsion engines (CFM56-5B, CFM56-7B, GE90, GEnx, and LEAP-X) with varying degree of partial premixing of fuel/air within the mixers comprised of two or more co- or counter-rotating swirlers and nearly perfectly premixed aero-derivative industrial engines (LM2500 and LM6000) starting in the late 1990s have performed safely while simultaneously meeting combustion systems' design requirements including cost of ownership, performance, emissions and operability throughout the operating range of the engines in use without encountering flashback and/or autoignition within the mixers, a generally mentioned area of concern in fundamental research publications. Practical, affordable, and durable solutions for mitigating combustion dynamics when or if encountered during the last phases of product development have also been performing reliably [28–30].

Picking back from Fig. 30a where we see setback (for the GE90-DAC I) and less than stellar reduction in NO_x for the CFM56-5/7B with the DAC technology, and in Fig. 30c for LEAP-X where a similar setback is encountered with the TAPS technology. While simultaneously we see from Fig. 30b superb emissions reduction with GEnx TAPS combustion system and 31 the TAPS technology luring potentials for reducing takeoff NOxEI up to 80% compared to the rich-dome LEC technology. Is the LEAP-X setback due to the team's inability to "see through the fog", micromanagement by burned-out combustion specialists or lack of tools?

Here, we pick up the most boring part, a commonly accepted OEM's process for LTO NO_x calculation from sea-level engine emissions data represented in Fig. 33 Here, engine NOxEI data has been corrected to standard humidity via an agreed to empirical correlation with some nagging questions still unresolved. The assumption $\text{NOxEI}/P_3^{n=0.5} \propto T_3$ over the entire engine operating line is not valid. Even for the takeoff or climbout P_3 and T_3 conditions, it is not unusual to see $0.3 \leq n \leq 0.6$ for different rich-dome combustors. Applicability of this approach for estimating mission NO_x emissions which is an order of magnitude more than LTO NO_x emissions are fraught with several questions.

The alternate approach [16–18] is consistent with fundamentals of NO_x formation [34, 35] allowing linear interpolation between engine data points for calculating NO_x emissions at each of the four LTO operating conditions of a reference engine running in standard ambient conditions with reference fuel, namely, stoichiometric flame temperatures at takeoff, climbout, approach, and idle. No need to know NO_x pressure exponent n which can vary with different rich- and lean-domes. No humidity correction is needed. Effect of fuel properties that can vary even for the same test facility [36] on NO_x emissions can be corrected for the reference fuel agreed to by all OEM's. The sea-level engine emissions data from minimum to maximum power can be used for calculating NO_x emissions for the airplane's entire flight envelope.

Tremendous amount of insight is provided when engine NOx emissions data from the seven rich-dome product combustors are plotted as function of $1000/T_{st}$, Fig. 44 from [7].

Fig. 44 NOxEI versus $1000/T_{st}$ characteristics of the seven rich-dome combustors. Source [15, 17]. Here, T_{st} is the stoichiometric flame temperature which is a function of P_3, T_3 and thermodynamic properties of the fuel

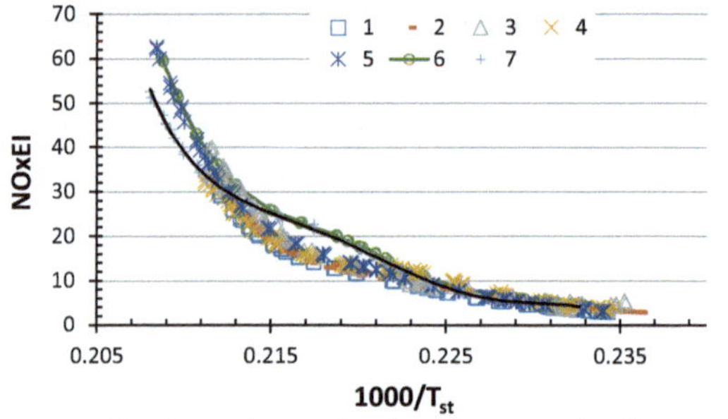

a. For the entire sea-level engine operating lines.

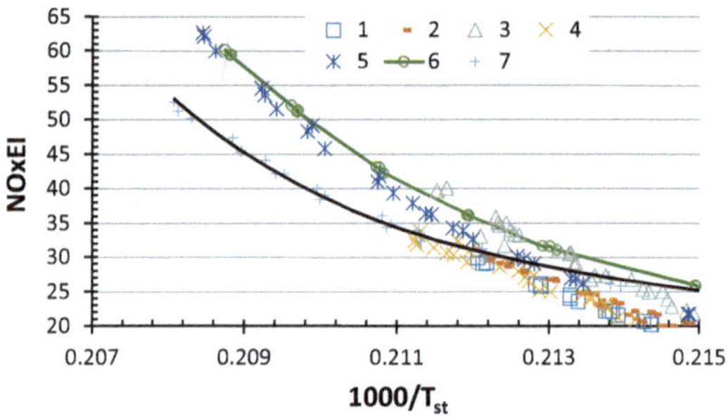

b. Zoomed-in high-power operation.

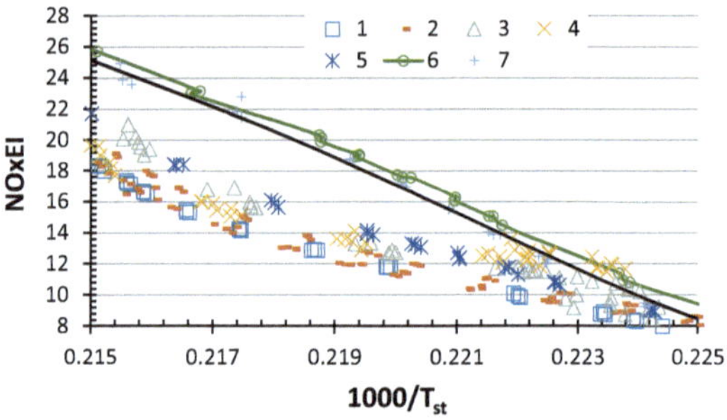

c. Zoomed-in mid-power operation.

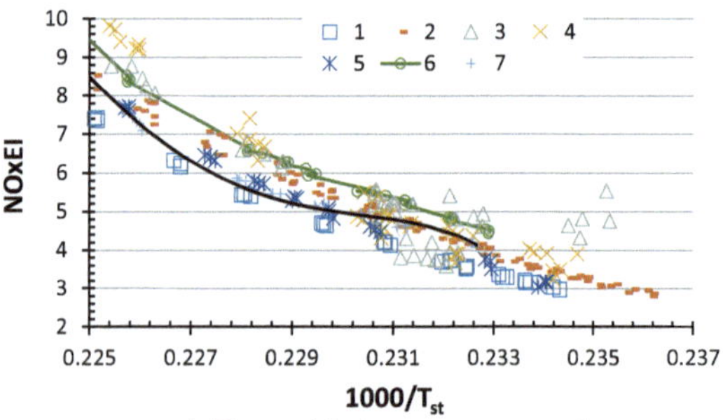

d. Zoomed-in low-power operation.

Figures similar to Fig. 44 for GE90-DAC I, GEnx, LEAP-X, TAPS2, and TAPS3 with the pilot fuel flow split as a variable will provide the insight needed to sort through the root causes leading eventually back to continuing NO_x reduction by using cleverly TAPS technology.

The author felt that by controlling pilot fuel flow split with airblast fueled pilot, Fig. 45a is better than dual-orifice pressure atomizers shown in Fig. 45b especially for the next-generation TAPS combustion products. But if the details of air atomizing devices lead to adversely impacting the intents of the low NOx performing flow characteristics, Fig. 35, the author wouldn't be surprised by their consequences.

TAPS figure similar to Fig. 44 for all the TAPS configurations (CFM TAPS, DAC TAPS, LEAP-1A, LEAP-1B, 55 pressure ratio engine TAPS technology sector [31], flame tube TAPS3 [15], GEnx and its three FAR1 TAPS2 configurations) will provide the insight required to "see through the fog" created by the results of LEAP-1A and LEAP-1B. The TAPS figure can also be used to check the relevance of fundamental investigations [26] and resulting phenomenological description of the critical processes identified in Fig. 35.

In summary, the author feels that TAPS potentials as expressed in Figs. 31 and 33 and summarized in Table 7 are achievable, and regards the LEAP-X setback simply a stark reminder for the team that lean-dome product development is a continuing evolution process.

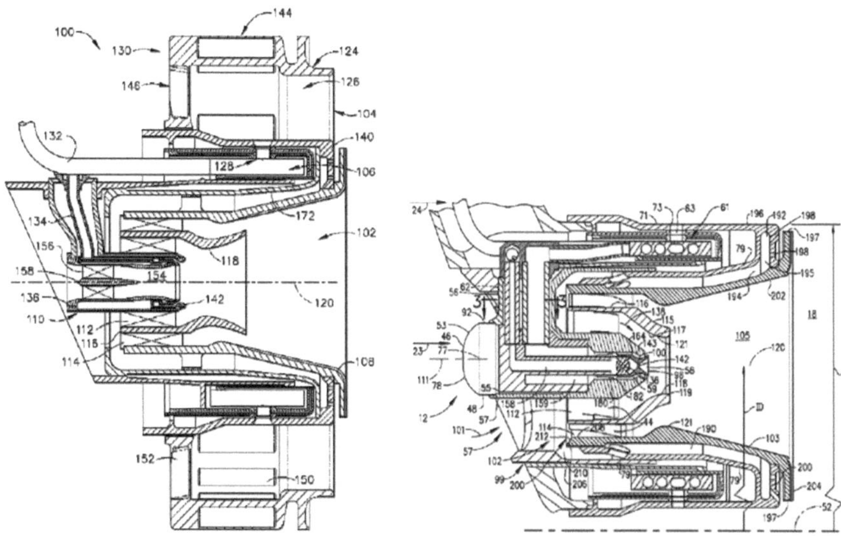

(a) Airblast fueled TAPS pilot (b) Dual-orifice fueled TAPS pilot

Fig. 45 Airblast or dual-orifice fueled TAPS pilot (reproduced respectively from US patents 8,171,735 and 8,387,391) with details adversely influencing TAPS flowfield characteristics (see Fig. 35 can lead to setbacks)

10 Summary

All original equipment manufacturers OEMs of the propulsion engines have been working hard since the early 1970s for reducing undesirable exhaust gaseous emissions, oxides of nitrogen NO_x, unburned hydrocarbons HC and carbon monoxide and smoke. The regulatory agency in conjunction with OEMs have established good metrics, the landing takeoff cycle LTO emissions for quantifying impact of propulsion engines' pollutants on the local air quality. OEM's made rapid progress in reducing HC, CO, and smoke emissions by the middle 1980s and have successfully applied their rich-dome low-emission combustion LEC technology including NO_x developed during the middle 1990s across their entire product line while simultaneously improving safety, cost of ownership, and operability range.

While LTO emissions are the right engine-level metrics, the tradeoffs between low NO_x at high-power versus low-power CO and HC emissions can be quantitatively assessed by comparing characteristics of takeoff NOxEI versus engine overall pressure ratio, idle COEI versus takeoff NOxEI and idle COEI versus idle COEI. 16 engines of the three OEM's, CFM, GE, and RR investigated using publicly available ICAO engine emissions data have shown that rich-dome low NO_x products known popularly as LEC or RQL for all the modern engines produced by the three OEM's produce similar levels of takeoff and climbout NOxEI given by

$$\text{LEC Takeoff NOxEI, LEC TO} = 0.0563 \times \text{PR}^{1.796} \text{ with } R^2 = 0.9876 \quad (4)$$

$$\text{LEC Climbout NOxEI, LEC CO} = 0.0977 \times \text{PR}^{1.563} \text{ with } R^2 = 0.9741 \quad (5)$$

The RR Trent engines takeoff and climbout NOxEI data fall within the following data bounds:ss

$$\text{LEC TO} - 2 \leq \text{Trent engine Takeoff NOxEI} \leq \text{LEC TO} + 2$$

$$\text{LEC CO} \leq \text{Trent engine Climbout NOxEI} \leq \text{LEC CO} + 4$$

These NO_x correlations are applicable over a wide range of takeoff thrust and overall pressure ratios, namely, $21 \leq$ Takeoff Pressure Ratio ≤ 49, 77 kN $\leq$ Takeoff Thrust $\leq$ 514 kN.

The following correlations for idle COEI and HCEI should be used as entitlements for assessing both rich-dome and TAPS emissions capability:

$$\text{Idle COEI} = \frac{32.183}{\text{Takeoff NOxEI}^{0.223}} \text{ with } R^2 = 0.8967 \quad (6)$$

$$\text{Idle COEI} = 7.199 \times \text{Idle HCEI} + 8.231 \text{ with } R^2 = 0.9949 \quad (7)$$

The use of these correlations combined with the methodology [23, 24] and applicable CFD tools [9, 15] has led successfully to evolving generation of low emissions rich-dome combustors with insignificant reduction from the takeoff NOxEI expression (4) since the early 2000s. Therefore, any claim in LTO reduction in recently certified propulsion engine should be attributable to the reduction in LTO fuel burn and not advances in rich-dome combustion technology. Spray quality technology should be considered matured in addition to well-practiced rich dome combustors scale-up. Then, COEI values higher than Eq. 5 may be due to overcooled and/or unoptimized dome and/or primary zone combustion structure.

TAPS combustion system scale-up should likewise be considered straightforward for teams that make proper use of the extensive database collected over the last 20 years complimented by comprehensive combustion system analysis and relevant fundamental investigation guided by Data Reduction and Analysis DRA approach [17, 18], a significant advance over the current "black box" OEM's preferred approach. Moreover, this approach is applicable for estimating mission NO_x emission that is generally an order of magnitude higher than LTO NO_x emission.

Passively controllable air atomizing fuel insertion system for the TAPS' pilot is preferred over hybrid or dual-orifice for obvious reasons. Variable pilot fuel flow split (% of total fuel flow rate) with pilot swirlers designed in conformance with TAPS low-emissions flow characteristic should be able to address operability challenges while simultaneously meeting takeoff NOxEI given by Eq. 7 within ±2 EI meeting or exceeding idle COEI vs. takeoff tradeoff given by Eq. 5 and idle emissions by Eq. 6.

$$\text{GEnx Takeoff NOxEI} = 7.480E - 05 \times \text{PR}^{3.453} \text{ with } R^2 = 0.9985 \quad (8)$$

In order to ensure that TAPS engine takeoff NOxEI is at least 25% lower than its counterpart LEC combustion system, TAPS1 based GEnx technology should be limited to takeoff pressure not exceeding 45.

The TAPS2 based combustion products, Eq. 9, should be used for $45 \leq \text{PR} \leq 55$ with resultant reduction of 50% from LEC at 55 PR.

When takeoff pressures are higher than 55, TAPS3 based combustion product, Eq. 10 development should be pursued in order to guarantee at least 50% reduction from LEC.

$$\text{TAPS2 Takeoff NOxEI} = \text{GEnx Takeoff NOxEI}/2 \quad 45 \leq \text{PR} \leq 55 \quad (9)$$

$$\text{TAPS3 Takeoff NOxEI} = \text{GEnx Takeoff NOxEI}/4 \quad \text{PR} > 55 \quad (10)$$

References

1. Mongia HC (2003) TAPS—a fourth generation propulsion combustor technology for low emissions. AIAA2003–2657
2. Mongia HC (2011) Engineering aspects of complex gas turbine combustion mixers Part V: 40 OPR. AIAA2011–5527
3. Mongia HC (2011) Engineering aspects of complex gas turbine combustion mixers Part IV: Swirl cup. AIAA2011–5526
4. Mongia HC (1997) Recent progress in low-emissions gas turbine combustors. ISABE
5. Bahr DW, Gleason CC (1975) Experimental clean combustor program Phase I—final report. NASA CR-134737
6. Gleason CC, Roger DW, Bahr DW (1976) Experimental clean combustor program Phase II—final report, NASA CR-134971, August 1976
7. Gleason CC, Bahr DW (1979) Experimental clean combustor program Phase III—final report. NASA CR-135384, June 1979
8. Stearns EW et al (1982) Energy efficient engine core design and performance report. NASA CR-168069, December 1982
9. Danis AM, Burrus DL, Mongia HC (1996) Anchored CCD for gas turbine combustor design and data correlation. ASME1996-GT-143
10. Lefebvre AH (1984) Fuel effects on gas turbine combustion liner temperature, pattern factor and pollutant emissions. AIAA J Aircraft, v 21, no 11, pp 887–898
11. Rizk NK, Mongia HC (1993) Semianalytical correlation for NO_x, CO, and UHC emissions. J Eng Gas Turbines Power 115:612–619
12. Danis AM, Pritchard BP, Mongia HC (1996). Empirical and semi-empirical correlation of emissions data from modern turbopropulsion gas turbine engines. ASME96-GT-86
13. Hura HS, Joshi ND, Mongia HC (1998) Dry low emissions premixer CCD modeling and validation. ASME1998-GT-444
14. Hura HS, Mongia HC (1998) Prediction of NO emissions from a lean dome gas turbine combustor. AIAA1998-3375
15. Mongia HC (2008) Recent progress in comprehensive modeling of gas turbine combustion. AIAA2008-1445
16. Mongia HC, Vermeersch M, Held TJ (2001) A DRA modeling approach for correlating gaseous emissions of turbo-propulsion engine combustors. AIAA2001–3419
17. Mongia HC (2010) On initiating 3rd generation of correlations for gaseous emissions of aero-propulsion engines. AIAA2010-1529
18. Mongia HC (2010) Correlations for gaseous emissions of aero-propulsion engines from sea-level to cruise operation. AIAA2010-1530
19. Bruce TW, Davis FG, Kuhn TE, Mongia HC (1977) Pollution reduction technology program small jet aircraft engines Phase 1. Final report. NASA CR-135214
20. Bruce TW, Davis FG, Kuhn TE, Mongia HC (1978). Pollution reduction technology program small jet aircraft engines Phase 2. Final Report. NASA CR-159415
21. Bruce TW, Davis FG, Kuhn TE, Mongia HC (1981) Pollution reduction technology program small jet aircraft engines Phase 3. Final Report. NASA CR-165386
22. Annex 16 to the Convention on International Civil Aviation Environmental Protection Volume II Aircraft Engine Emissions, International Civil Aviation Organization
23. Mongia HC (2016) An art of developing aviation engine combustion technologies and products. Energy, combustion and propulsion new perspectives, pp 239–268, ISBN: 978-19-1039-029-0. Eds. Agarwal etc. Athena Academic Ltd
24. Mongia HC, Ajmani K, Sung CJ (2020) Hypotheses driven combustion technology and design development approach pursued since early 1970's. in 50 Years of CFD in engineering sciences: a commemorative volume in memory of D. Brian Spalding, pp 439–484, Ed. A. Runchal, ISBN 978-981-15-2669-5. Springer Nature Singapore Pte Ltd

25. Mongia HC, Ajmani, K, Sung CJ (2020) Fundamental combustion research challenged to meet designers' expectations. In: Chapter 11 in a Springer book, Ashoke De et al (eds) Sustainable development for energy, power, and propulsion, expected to be published in 2020
26. Mongia, HC (2020) Synopsis of propulsion engines combustion technology/product development and analysis substantiated during last 47 years. In: Gupta et al (eds) Advances in IC engines and combustion technology, Select Proceedings of NCICEC 2019. Springer Nature Singapore Pte Ltd
27. Mongia H, Krishnaswami S, Sreedhar PSVS (2007) Comprehensive gas turbine combustion modeling methodology. Fluent's Int Aerospace CFD Conference, June 18, 2007, Paris
28. Mongia HC, Held TJ, Hsiao GC, Pandalai RP (2003) Challenges and progress in controlling dynamics in gas turbine combustors. J Propulsion Power 19: 822–829
29. Lieuwen TC, Yang V (2005) Progress in astronautics and aeronautics, vol 210
30. Anderson WE, Lucht RP, Mongia H (2015) Integrated physics-based modeling and experiments for improved prediction of combustion dynamics in low-emission systems. NASA/CR-2015-NNX11AI62A
31. Lee CM et al (2013) NASA project develops next generation low-emissions combustor technologies. AIAA2013-0540
32. Joshi ND, Mongia HC, Leonard G, Steggmaier JW, Vickers EC (1998) Dry low emissions combustor development. ASME 98-GT-310
33. Diehl LA, Biaglow JA (1974) Measurement of gaseous emissions from a turbofan engine at simulated altitude conditions. NASA TM X-3046
34. Turns SR (2000) An introduction to combustion. The McGraw-Hill Companies, Inc
35. Law CK (2006) Combustion physics. Cambridge University Press, Cambridge
36. World Fuel Sampling Program, CRC Report No. 647, June 2006

Aerodynamic and Aero-Acoustic Performance of an Adjustable Pitch Axial Flow Fan

Rajat Arora, Ramraj H. Sundararaj, T. Chandrasekar, Sharad Saxena, and Abhijit Kushari

Abstract To increase the operational regime of an axial flow fan has been of immense focus lately. Variable-pitch rotor blades provide a crucial means of adapting to various inlet flow conditions. The operating envelope of an axial flow fan is characterized by flow instabilities. Experiments were conducted on a six-bladed scaled down version of an inducer-type industrial fan. The performance was evaluated at three different blade pitch configurations of 8, 16, and 22° with a speed variation of 460, 560, 660, 760, 860, and 960 RPM. Static pressure measurement was carried out at the inlet of fan and exit of fan at sampling rate of thirty times the rotor order of maximum RPM. Sound pressure was measured at the inlet of test facility. Experiments conducted revealed that with increase in the pitch angle, flow rate, and static pressure rise increases but the trend remains nonlinear, exception being the low flow rates operating regime. Total efficiency increased and then decreased with increase in pitch angle. Noise levels were highest at high pitch angles and low flow rates indicating the existence of instabilities in the system. The presence of high amplitude pressure fluctuations was confirmed by power spectral density plots of static pressure and microphones data. The dominant frequency of 66 Hz appeared at RPM of 860 at near to fully closed operation of exit throttle. Other frequencies 22 Hz and 45 Hz were found to be dominant at lower speeds.

Keywords Axial flow fan · Aero-acoustics · Variable-pitch fan · Aerodynamic performance

Nomenclature

•

B Number of blades

R. Arora · R. H. Sundararaj · T. Chandrasekar · S. Saxena · A. Kushari (✉)
Indian Institute of Technology Kanpur, Department of Aerospace Engineering, Kanpur, India
e-mail: akushari@iitk.ac.in

A. Gupta et al. (eds.), *Advances in Energy and Combustion*, Green Energy and Technology,
https://doi.org/10.1007/978-981-16-2648-7_3

BPF	Blade passage frequency
C_a	Axial flow velocity (m/s)
C_d	Discharge coefficient of bell mouth 0.98 assumed
D	Diameter of the casing
FSP	Fan static pressure
N	Rotor speed in revolutions per minute
N_d	Non dimensional speed
N_{ref}	Reference or designed 960 RPM
OASPL	Overall sound pressure level in decibel (dB)
P	Static pressure in Pascal
ΔP	Static pressure rise in Pascal (SPR)
$\dot{P}$	Power in Watts
RO	Rotor order N/60
RPM	Revolutions per minute
U	Blade velocity at mean (m/s)
V	Voltage in Volts
i	Current in Ampere
α	Blade pitch angle in degree
η_m	Mechanical efficiency of motor
η_t	Total to total efficiency
$\forall$	Volumetric flow rate in m^3/s
Ø	Flow coefficient at mean
Ψ	Work done coefficient at mean

Subscript

1	Inlet of the fan
2	Outlet of the fan
ref	Reference Pressure for air (20 μ Pa)
rms	Root mean square

1 Introduction

Axial flow fan in an industry application is desired to operate efficiently at different operating regimes, e.g., full load and part load operation. The operation is intrinsically limited by aerodynamic instabilities incurred during distortions in the inlet flow, mechanical deterioration, and transient operations [1]. The instabilities can be characterized as rotating stall or surge [2], where surge is associated with the global oscillations [3], and rotating stall is the local flow adaptation technique leading to non-uniformity in the flow around the annulus. A disturbance hence generated can be further amplified [4, 5], which will be manifested by overall drop in overall performance, increased noise and vibration, lower efficiencies, and catastrophic mechanical failures.

In contrast to a fixed-pitch fan, a variable-pitch axial flow fan has several benefits such as high operating efficiency, a large flow rate, and a larger operating range [6]. Variable-pitch axial flow fans find applications in the thermal power stations where fresh air is needed (forced draft) or process involving high temperatures such as in boilers (induced draft). These types of fans are proven to have a negligible loss in power at the output, and hence, they are also used in pulverized air fans and flue gas desulphurization (FGD) applications. Even though the variable-pitch axial fan provides and an efficient alternative, the increase in pitch angle leads to reduction in stall-free operating margin [1]. The limiting factor that imposes an upper limit on the pitch angle is not reported systematically.

The work presented here is directed at providing a depiction in the variation of aerodynamic and acoustic performance when pitch angle is increased beyond the designed values. The results of detailed experimentation at two measurement planes are provided. The diagnostic for detecting the instabilities is discussed on the basis of unsteady pressure measurement and non-intrusive sound measurements technique of sound signals [7, 8]. The details pertaining to experimental facility, instrumentation implemented, methodology adapted and experimental results are presented in subsequent sections.

2 Experimental Setup and Instrumentation

The experimental setup shown in Fig. 1 is a six-bladed axial flow fan. The setup is equipped with a bell mouth to establish clean flow at inlet of the fan without an appreciable loss in total pressure. The flow enters through the bell mouth and exits toward the throttle cone side. The pitch angle is altered manually; pre-calibrated slots for different pitch angle are provisioned in the hub of the rotor. The diameter of the fan casing is 1 m, and the length of the duct is 6D. Total length if the test including bell mouth is 7D. Tip diameter and hub diameter for the current configuration are

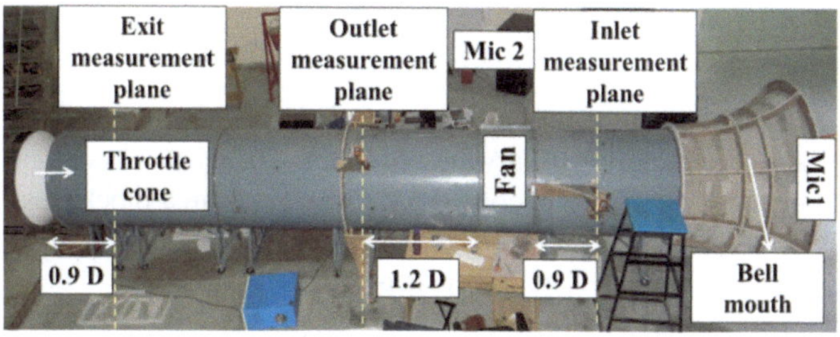

Fig. 1 Photograph of the experimental test facility

0.98D and 0.12D, respectively. The designed speed and designed pitch angle for the current fan are $N_{ref} = 960$ RPM and $\alpha = 8$ degree, respectively. The setup has a provision for flow rate variation via a throttle cone of length 0.5D. The throttle cone is mounted on a linear traverse, operated with a stepper motor that facilitates the accurate control of valve location from fully open to fully closed condition. The facility is equipped with instrumentation allowing the measurement of quantities to derive the aerodynamic and acoustic performance of the fan.

The inlet measurement plane is positioned at 0.9D ahead of the fan, whereas the outlet plane measurement is located at 1.2D aft of the fan. Total pressure at the inlet is measured by Pitot static probe aligned with the axis of the setup. Total pressure at the exit is measured by 4 Kiel probes also aligned with the axis of the test facility. Static pressure measurement at the inlet and exit is measured by six equally spaced pressure ports along the circumference of the casing. The circumferentially averaged value of the pressure is used for estimation of flow properties. All the pressure ports are connected to electronically scanned pressure (ESP) scanner with a range of 1000 Pa through flexible tubes. These scanners consist of an array of 32 silicon piezo-resistive pressure sensors, one for each pressure port, whose analog output is electronically multiplexed at the rate of 16,000 Hz through a built-in instrumentation amplifier. The data is acquired at a rate of 500 samples per second for duration of 10 s. The Nyquist criterion for the above sampling rate is 250 Hz which is 15 times the rotor order at designed speed (16 Hz) of the fan.

Sound measurements were ascertained with the assistance of a PCB free field 130A24 microphones having a sensitivity of 10 mV/Pa. Microphone was placed at a distance of 1D ahead of the bell mouth aligned with the axis of the tunnel. The voltage was acquired through NI-6211 DAQ at a sampling rate of 50,000 samples per second for a duration of 4 s. The fan is operated with a calibrated DC shunt motor with maximum capacity of 3000 W. The RPM of the fan is regulated using a variable frequency drive (VFD). The input voltage to motor is measured using multimeter, and the input current is measured using clamp meter. All the measurements are interfaced with LabVIEW application software. The errors in performance variables of fan map are estimated to be less 0.1% in static pressure and flow rates and within $\pm$ 1 dB in overall sound pressure level.

3 Methodology

The fan was operated and maintained constant at a selected speed, and the throttle cone was inserted in the casing to vary the exit area and hence the flow rate. The values static pressure and sound were recorded for several exit area blockages from fully open to fully close condition. The experiments were repeated for various rotational speeds as listed in Table 1. The experiments were conducted for designed pitch angle $\alpha = 8°$ and two other settings of $\alpha = 16°$ and $\alpha = 22°$. Table 1 summarizes the operating speed and characteristic frequencies of the system at these conditions. It is to be noted that pitch angle was not varied continuously during the experiments. The

Table 1 Test conditions and characteristic frequencies

Non-dimensional speed N_d	Rotor order RO (N/60 in Hz)	BPF (N × B/60 in Hz)
0.48	7.6	46
0.58	9.3	56
0.68	11	66
0.79	12.6	76
0.89	14.3	86
1	16	96

two set of experiments were repeated for a given pitch setting, followed by change in pitch setting, and the procedure was repeated for the next case.

Plots of static pressure rise versus flow rate, total efficiency versus flow rate, work done coefficient versus flow coefficient, and overall sound pressure level constitutes fan aerodynamic and acoustic fan characteristics. These plots were generated using the standard formulas listed in Eqs. 1–10. Fan static pressure is the capability of the fan to push air against a resistance (defined as fan total pressure rise—velocity pressure), which in current set of experiments is represented by P_2 since the inlet stagnation pressure was always at ambient condition. Frequency domain analysis of times series data obtained from pressure scanner and microphone was separately performed using Welch's power spectral density estimate with a frequency resolution close to 1 Hz. This analysis would aid in identifying the frequency bands of instabilities if present.

$$\forall = C_d A \sqrt{\frac{2(P_0 - P_1)}{\rho}} \tag{1}$$

$$\Delta P = P_2 - P_1 \tag{2}$$

$$\dot{P}_{\text{in}} = V \times i \times \eta_m \tag{3}$$

$$\dot{P}_{\text{out}} = \forall (P_{O2} - P_{O1}) \tag{4}$$

$$\eta_t = \frac{\dot{P}_{\text{out}}}{\dot{P}_{\text{in}}} \tag{5}$$

$$C_a = \sqrt{\frac{2(P_{O1} - P_1)}{\rho}} \tag{6}$$

$$U = \frac{\pi \text{ND}_{\text{mean}}}{60} \tag{7}$$

$$\emptyset = \frac{C_a}{U} \tag{8}$$

$$\Psi = \frac{(P_{O2} - P_{O1})}{\rho U^2} \tag{9}$$

$$\text{OASPL} = 10 \log_{10} \frac{\sum P_{\text{rms}}^2}{P_{\text{ref}}^2} \tag{10}$$

4 Experimental Results

In the current section, aerodynamic and acoustic results for designed blade pitch angle of 8° are discussed, and effect of change in pitch angle on performance parameter is further discussed for non-dimensional speed of $N_d = 0.89$. Figure 1 shows the variation of static pressure rise at different off design conditions. For a given rotor speed, SPR increases from fully open to fully closed throttle valve position. With increase in rotational speed the flow rate as well as SPR increases. The trend obtained from experimental result is consistent with a typical compressor or fan map available in the literature [9, 10]. Figure 2 shows total efficiency increases then decreases with reduction in flow rates with slight variation in the trend at very low speed of $N_d = 0.48$. This nature of curve is also in agreement with the characteristic curves obtained in a fan experiments. It is to be realized at this juncture that maximum efficiency points do not correspond to maximum static pressure rise plot. Operating requirements at higher SPR values necessitate a bargain between static pressure rise and efficiency. The fan operates more efficiently at values close to the designed speed $N_d = 1$ (Figs. 3, 4 and 5).

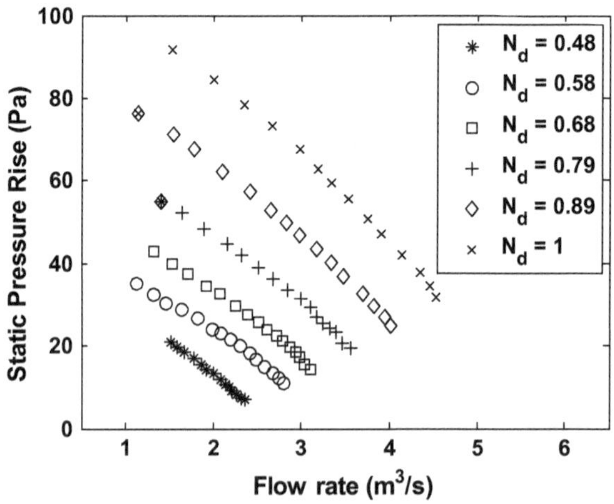

Fig. 2 Static pressure rise versus flow rate at $\alpha = 8°$

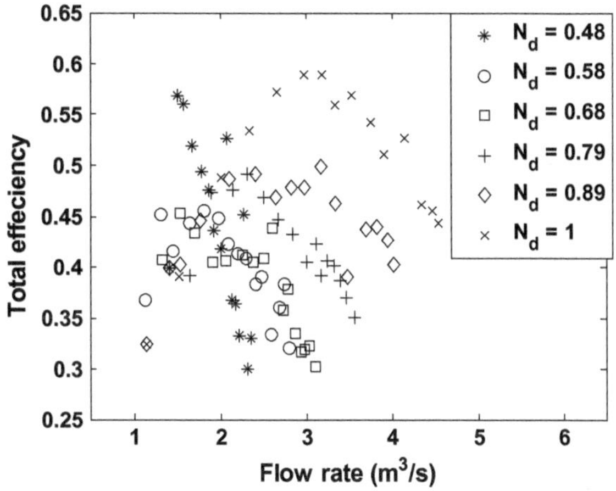

Fig. 3 Total efficiency versus flow rate at $\alpha = 8^{\circ}$

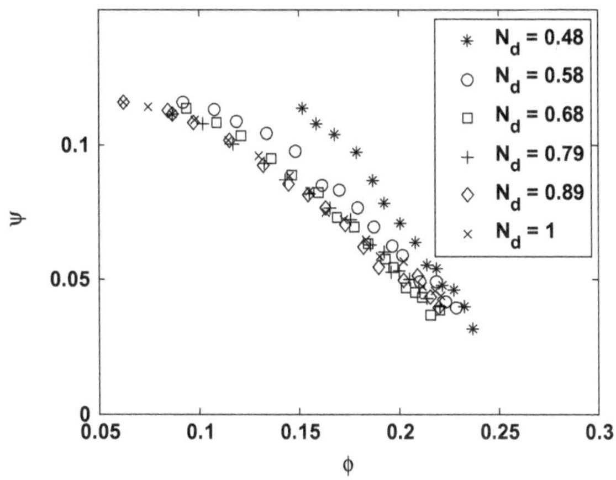

Fig. 4 Workdone coefficient versus flow coefficient at $\alpha = 8^{\circ}$

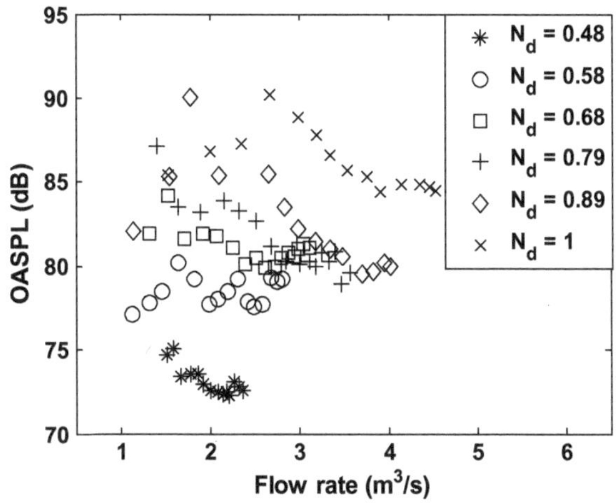

Fig. 5 Overall sound pressure level versus flow rate at $\alpha = 8^{\circ}$

To investigate the effect of pitch angle variation on performance parameters, comparative plots are shown from Figs. 6, 7, 8, and 9 at non-dimensional speed of 0.89. It can be seen from Fig. 6 that for a given value of SPR, flow rate increases from $\alpha = 8°$ to $\alpha = 22^{\circ}$. This increase in flow rate is attributed to decrease in the inlet flow blockage at higher values of blade pitch angle; however, the trend in not a linear function of α. Increase in the values of static pressure rise with increasing α is contributed partially by increase in the FSP and decrease in the value of P_1 at higher α. The SPR values at $\alpha = 22°$ tend to rise gradually as compared to $\alpha = 16°$ finally collapsing at flow rate of 4 m^3/s. The trend of Fig. 8 was found similar to the trend obtained in Fig. 6. At this stage, flow separation seems to be a factor contributing to the above nature of the flow physics. Along the same lines, lower value of efficiency and higher values of noise are observed in Figs. 7 and 9.

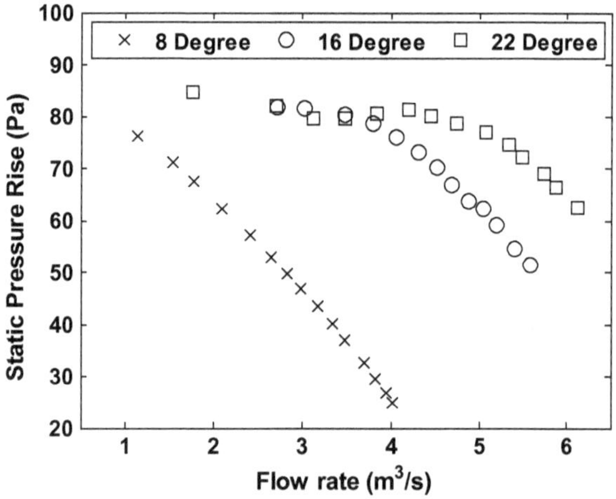

Fig. 6 Plot of static pressure rise versus flow rate at various pitch angles for N_d = 0.89

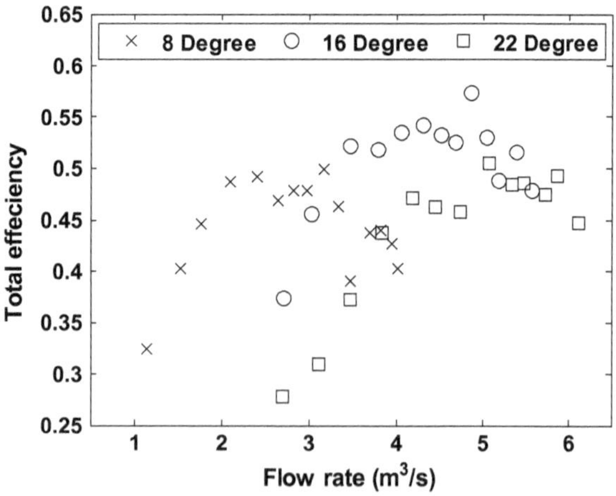

Fig. 7 Plot of total efficiency versus flow rate at various pitch angles for N_d = 0.89

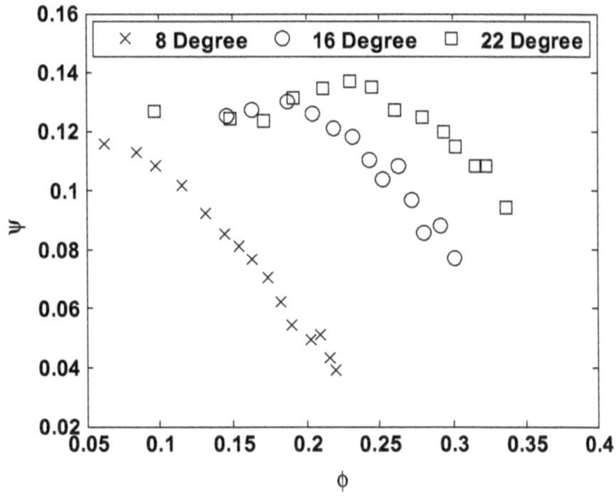

Fig. 8 Plot of work done coefficient versus flow coefficient at various pitch angles for $N_d = 0.89$

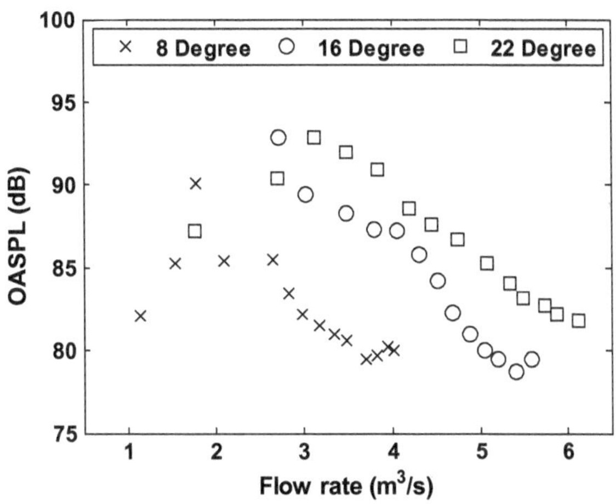

Fig. 9 Plot of overall sound pressure level versus flow rate at various pitch angles for $N_d = 0.89$

To further investigate the cause of reduced efficiency and higher noise level, plots of pressure amplitude variation with frequencies are plotted for $\alpha = 16°$ and $\alpha = 22°$ at N_d 0.89. The graphs are plotted up to frequencies of 150 Hz for better distinction. Legend in Figs. 10, 11, 12, 13, 14, and 15 refers to the values of flow coefficient ϕ. Higher values of ϕ correspond to fully open, whereas lower values correspond to almost fully closed condition of exit valve. The subsequent graphs can be read in conjunction with Figs. 6, 7, 8, and 9 for better interpretation. Figure 10 shows the fluctuations in static pressure P_1 at the inlet plane measurement at $\alpha = 16^{\circ}$. The sharp jump in fluctuations at frequency of 67 Hz was observed at low flow rates corresponding to $\phi = 0.15$. The other small peaks existed at the values of operating range are around 22 and 45 Hz. This point corresponds to very high noise level, lower efficiency and low work done capacity as can be seen as observed in Figs. 7, 8, and 9. Similar frequency was observed in static pressure fluctuations at the exit plane P_2

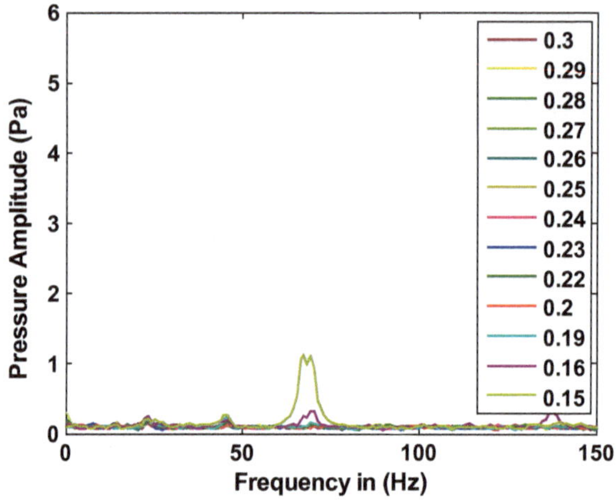

Fig. 10 Plot of pressure fluctuations in P_1 at pitch angle of 16° and $N_d = 0.89$

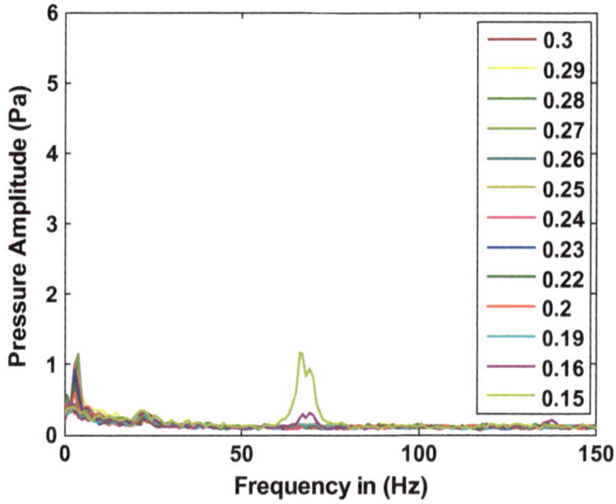

Fig. 11 Plot of pressure fluctuations in P_2 at pitch angle of 16° and $N_d = 0.89$

with an additional frequency of 3.9 Hz. The fact that this frequency exits for all the exit blockages, it manifests the presence of rotating stall for full operating regime. The rotor order for the current rotational speed is 14.3 Hz as listed in Table 1.

Figures 12 and 13 show the fluctuations at $\alpha = 22°$ and $N_d = 0.89$. It depicts very high fluctuations in the static pressure P_1 at 8.8 Hz, corresponding to $\phi = 0.1$. Other frequencies peaks were in the range of 64–66 Hz for values of $\phi < 0.19$. The frequencies (22 and 45 Hz) observed in Figs. 10 and 12 are similar but the magnitude of fluctuations is slightly on the higher side in the case of $\alpha = 22^{\circ}$. The trend of higher amplitude of fluctuations and similar frequencies remains similar for Figs. 11 and 13. The authors conducted similar analysis for other non-dimensional speed, and it was found that for lower speeds the spectrum consisted of similar frequencies 22 and 45 Hz. The frequencies appearing in the spectrum seem to be following a pattern,

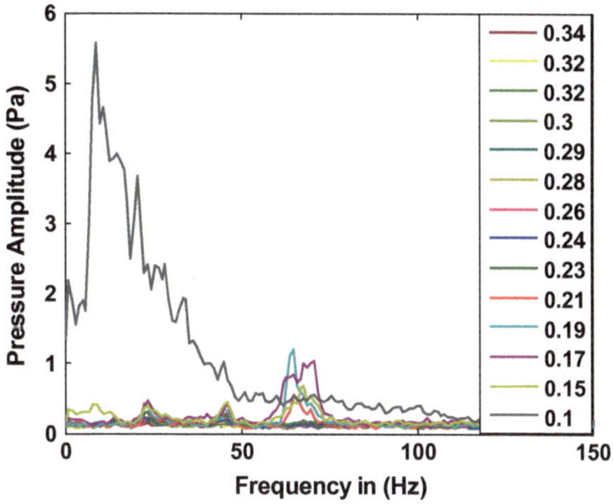

Fig. 12 Plot of pressure fluctuations in P_1 at pitch angle of 22° and $N_d = 0.89$

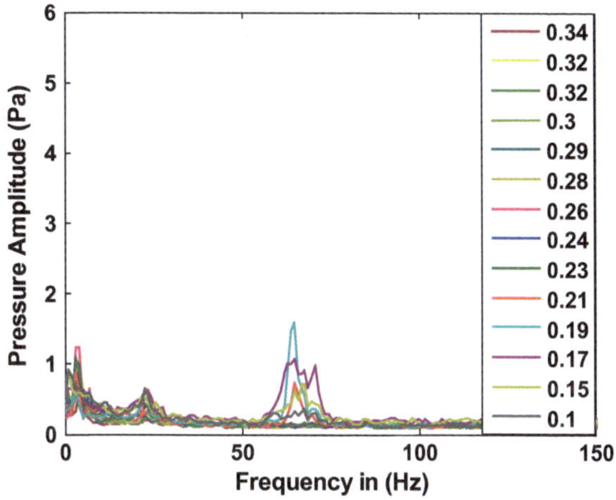

Fig. 13 Plot of pressure fluctuations in P_2 at pitch angle of 22° and $N_d = 0.89$

where 22 Hz appears to be fundamental and 45 and 66 Hz as first and second overtone. The reason for a particular mode getting excited is still unclear at the present stage.

The authors conducted similar analysis to generate power spectral density plots on the data obtained from microphone placed in front of the test facility for the above-mentioned conditions of $\alpha = 16°$ and $\alpha = 22°$ at $N_d = 0.89$. It can be seen that similar frequencies are manifested in sound measurement. The frequencies 22, 45, and 66 Hz getting excited are confirmed by Figs. 14 and 15. The authors conducted a modal analysis to obtain the natural mode shapes and frequencies of a structure. The natural frequency of the bending mode was around 42 Hz, and the second bending mode was 181 Hz eliminating the possibility of structural vibration exciting the frequencies shown by power spectrum.

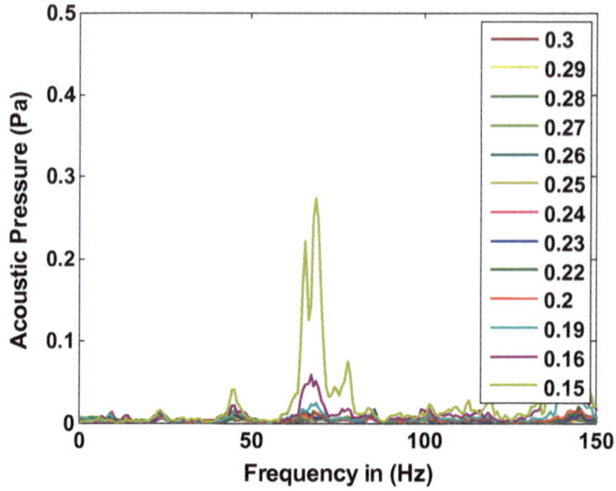

Fig. 14 Plot of acoustic pressure fluctuations in microphone at pitch angle of 16° and $N_d = 0.89$

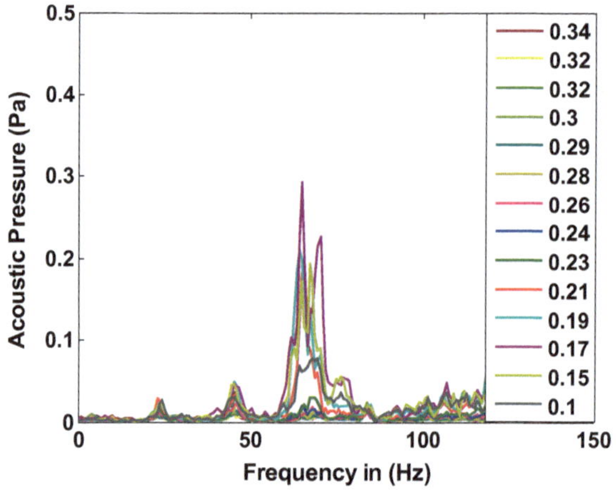

Fig. 15 Plot of acoustic pressure fluctuations in microphone at pitch angle of 22° and $N_d = 0.89$

The possibility of air column contained by the duct vibrating at a forced frequency by the rotor is discussed by Wallis [11]. If the applied frequency is found to be similar to that of natural frequency of the air column as in case of organ pipes, resonance occur leading to intensification in noise levels. Further, investigation at the current stage is required to make any claims regarding the above-mentioned possibility. The main result of the study conducted is the onset of instabilities at high pitch angles fan operating at high speed and reduced flow rates. The detailed study is intended to push the operating envelope further with safe operation.

5 Conclusions

The goal of this investigation was to identify the change in performance map at pitch angles values higher than the designed values. Initial phase of analysis suggests that the fan operation at high pitch angles and lower flow rates is susceptible to flow instabilities leading to inefficient operation. Overall noise levels were found around to be the higher side. The frequencies appearing seemed to be dependent on the exit blockage. Further, improvements in measurement locations with additional measurement technique like vibration measurement using accelerometer and turbulence measurement using hotwire are required at the current stage to better understand the cause associated with instabilities.

Acknowledgements The work was carried out using the axial flow fan facility developed and located within National Wind Tunnel Facility (NWTF) premises at IIT Kanpur. The authors gratefully acknowledge the support of Head NWTF for facilitating the current research.

References

1. Bianchi S, Corsini A, Sheard AG (2013) Experiments on the use of signal visualization technique for in-service stall detection in industrial fans. Adv Acoust Vibr 2013:10. Article ID 610407
2. Lawless PB, Fleeter S (1993) Rotating stall acoustic signature in a low speed centrifugal compressor: Part 1-vaneless diffuser. ASME Paper no 93-GT-297
3. Greitzer EM (1976) Surge and rotating stall in axial flow compressors—Part I: theoretical compression system model. J Eng Power 98(2):190–198
4. Velarde-Suarez S, Ballestesteros-Tajadura R, Santolaria-Morros C, Gonzalez-Perez J (2001) Unsteady flow pattern characteristics downstream of a forward-curved blades centrifugal fan. J Fluid Eng 123/265
5. Cumpsty NA (1989) Part-circumference casing treatment and the effect on compressor stall. ASME Paper No 89-GT-312
6. Ye X, Fan F, Zhang R, Li C (2019) Prediction of performance of a variable-pitch axial fan with forward-skewed blades. Energies 12(12):2353
7. Bently D, Goldman P, Yuan J (2001) Rotor dynamics of centrifugal compressor in rotating stall. Orbit 2Q01, pp 40–50
8. Bently D, Goldman P (2000) Vibrational diagnostics of rotating stall in centrifugal compressors. Orbit 1Q01, pp 32–40
9. Dixon S, Hall C (2014) Fluid mechanics and thermodynamics of turbomachinery. Butterworth-Heinemann
10. Cohen H, Rogers GFC, Saravanamuttoo HIH (1996) Gas turbine theory. Longman group Limited
11. Wallis RA (1961) Axial flow fans design and practices. Academic Press, New York, London

Proposed Thrust Profile Design of Pulse Detonation Engine (PDE) for Aerospace Applications

Subhash Chander and T. K. Jindal

Abstract The paper deals with challenging assignment of development and optimization of pulse detonation engine (PDE) to test it in aerospace system configurations. Our researchers have successfully developed and deployed static test bed for PDE firing earlier. This has been trial proven in dynamic PDE firings in last one decade. In the present case of design of propulsion system for aerospace vehicle, which includes trial setup for prolonged testing in autonomous mode, pneumatic air/oxygen supply (compressed), fuel supply, ignition system, command generation system (wired/remote), recorders/telemetry, safety systems including thrust stand hardware need to be system engineered meticulously. Extensive testing of subsystem/subassembly levels need to incorporated before to get confidence to function in dynamic environment. This included 150 bar compressed air delivery system, which needs to operate in highly pulsating environment at test bed. A small recorder system is also installed to record four channels of data acquisition system (DAQ) to validate the performance of PDE; full function trial run in autonomous configuration resulted in successful test runs, repeatedly. It generated a lot of data to predict system performance to make it function better. All these exercises are executed simultaneously and efforts are still on in this regard. Careful data analysis has resulted in validation of predicted mission parameters in actual trials at various test conditions. The same data is utilized to inject in experimental aerospace vehicle (EAV) six DOF model and tailor it to meet these applications. The performance of the vehicle in terms of speeds, range, intercept dynamics, weight advantage and compact form factor achievement, etc. need to be explored in this paper.

Keywords Autonomous PDE rig · Detonation · Ignition system · Thrust generation · Measurements

S. Chander (✉)
DRDO, Terminal Ballistics Research Lab, Chandigarh, India
e-mail: subashchander@tbrl.drdo.in

T. K. Jindal
Punjab Engineering College (Deemed To Be University), Sector 12, Chandigarh, India
e-mail: tkjindal@pec.ac.in

A. Gupta et al. (eds.), *Advances in Energy and Combustion*, Green Energy and Technology,
https://doi.org/10.1007/978-981-16-2648-7_4

1 Introduction

Pulse detonation engine is a modern propulsion system that utilizes repetitive detonations to produce thrust. PDE offers the potential to provide increased performance, while simultaneously reducing the engine weight, miniaturization, cost and complexity relative to conventional propulsion systems currently in service. Due to its obvious advantages, attention has been paid to the scientific and technical issues concerning pulse detonation engine worldwide. Despite extensive research in PDE over the past several decades, it is not yet ready to be used for practical propulsion applications [18–20]. Many reputed institutions like NASA, DAPRA, DRDO, GE, P&W, EADS and many others are seriously engaged in active research in this area and are in a mature stage of development of such technologies and systems. Many academic organizations like University of Taxas, Arlington, USA, Nogaya University, Japan, N.N. Semenov Institute of Chemical Physics, Russia, NUS, Singapore and Punjab Engg. College, Chandigarh, etc. are engaged in active research in this area.

One of the key barriers to the realization of an operational PDE lies in the difficulty to initiate the detonation wave in the engine, in a reliable and controllable manner. Detonations are difficult to achieve within a practical length in hydrocarbon/air mixtures, which are less sensitive to detonation initiation taking atomization, vaporization and detonation into account. There are challenges in reducing noise and dampening the vibrations due pulsating power coupled by handling thermal management for prolonged runs.

2 The Setup

The dedicated test setup for experimental trials is used to investigate, address the issues and observe the performance improvements due these modifications as shown in Fig. 1 [20, 23, 31].

For this purpose, detonation process details are being explored extensively. Detonation can be initiated either directly or by deflagration followed by deflagration to detonation transition (DDT). Direct detonation initiation energy and power requirements are function of cell size or cell width (λ) of a detonable mixture, which is the measure of the combustion reaction rates of the fuel mixture. Minimum energies are of the order of 250 kJ for several hydrocarbon/air mixtures, which are prohibitive for practical pulse detonation engines [31]. It is desirable to use low initiation energy system to produce fully developed detonations with in an acceptable distance [21]. An important diagnostic tool for such exercise is sensor networking, which is explained below.

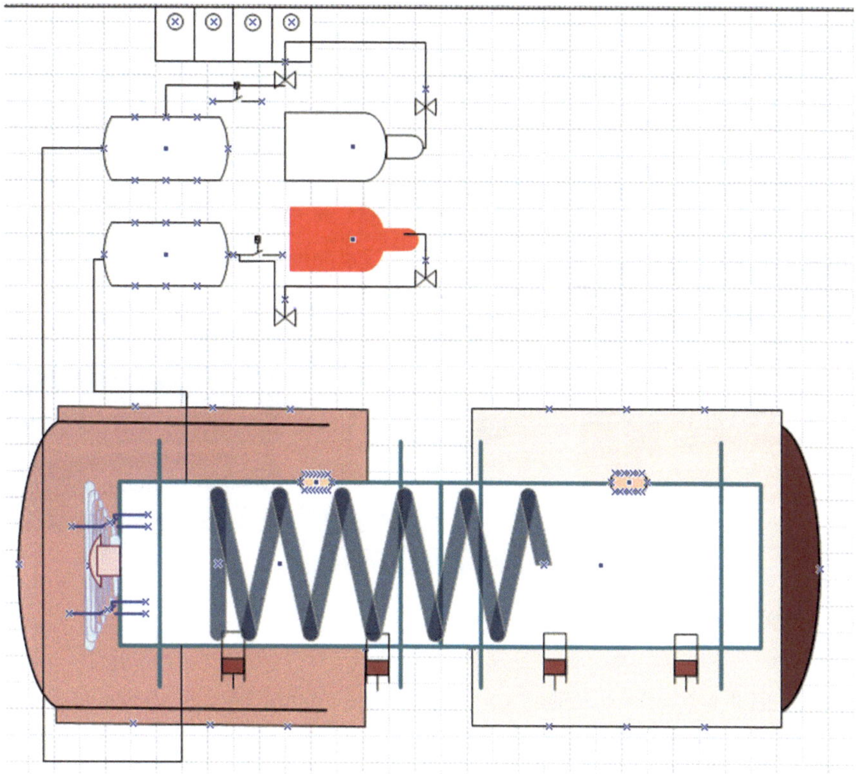

Fig. 1 Schematic of demonstrator test configuration

3 Sensor Networking

A sensor network is designed and powered using various rating of SMPS to meet the diagnostics requirements of the setup. Main sensors used for this purpose are [22, 37].

1. MEAS Pressure sensors (4/10 bar) for line pressure measurements
2. Analogue gauges (20/100 bar) for indications and manual recording
3. PCB/Zuma pressure gauge (10/20 bar)
4. Max load cells (10 kN) for force measurement
5. K/J type thermocouple network for hot mixture and tube surface temperature measurements.

Some other devices are also used:

1. Solenoid valves for gas filling to settling chambers/PDE tube
2. Flow controllers for auto filling of gases to settling chambers.

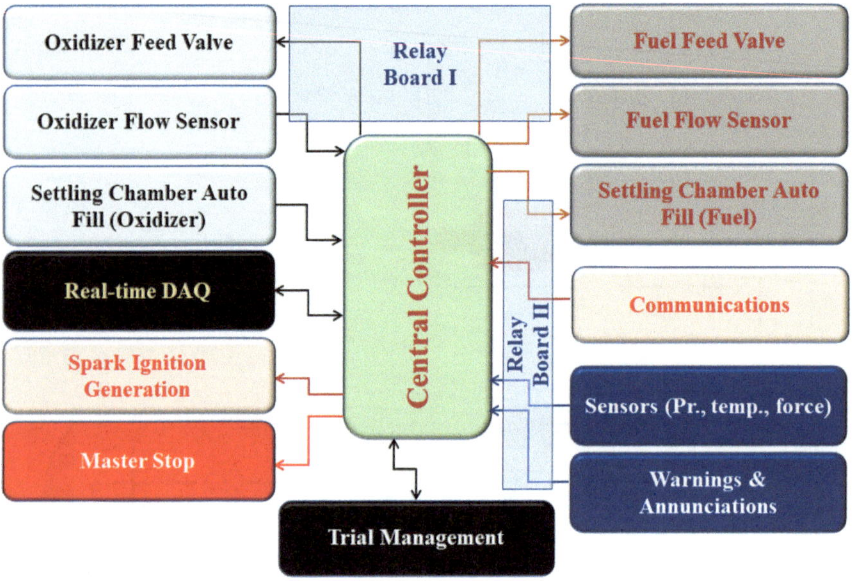

Fig. 2 PDE structured control plan

A special network is created to establish the connectivity is shown in Fig. 2 [31]. Organizing the sensors and control components has helped team to achieve a structured architecture of the plan. It has created a modular approach to highly reliable electronics and instrumentation setup to ease the rapid recording/processing of data in an organized pattern. These modifications have also helped team to easy detection of problems and resolving the major issues in a simplified diagnostics approaches.

4 The Engine

The PDE developed is based on C_2H_2 as fuel and O_2 as oxidizer. A typical data set obtained is plotted (Fig. 3) and is used as reference thrust input for our problem. Same data is assumed for all four tubes. The thrust of total system is based on this content of PDE-based sustainer. This study will be able to make researchers understand the acceleration growth, velocity and down range extension trends. This preliminary work will become a vital study for identifying the suitability of PDE engine as sustainer as compared to conventional system. Further improvements can be reinforced after seeing the performance enhancements observed and its verification in detailed complex 6° of freedom model. Its effect on other systems and handling techniques is to be explored further [22, 45, 30].

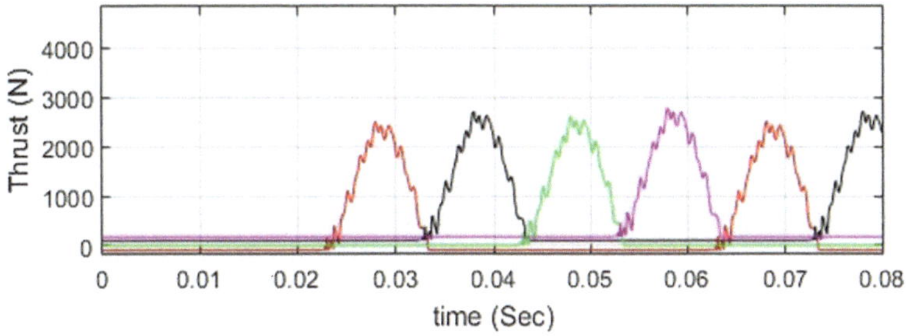

Fig. 3 Multiple tube PDE thrust profile

5 The Ignition System Hardware

The ignition control system (ICS) used for this purpose is shown in Fig. 4 [23]. The objective of this research is to study the thrust generation algorithms to obtain solutions of pulsating thrust produced by PDE.

The cyclic PDE engine system is centrally controlled by a custom built software program developed and installed on a computer controller-based data acquisitions and

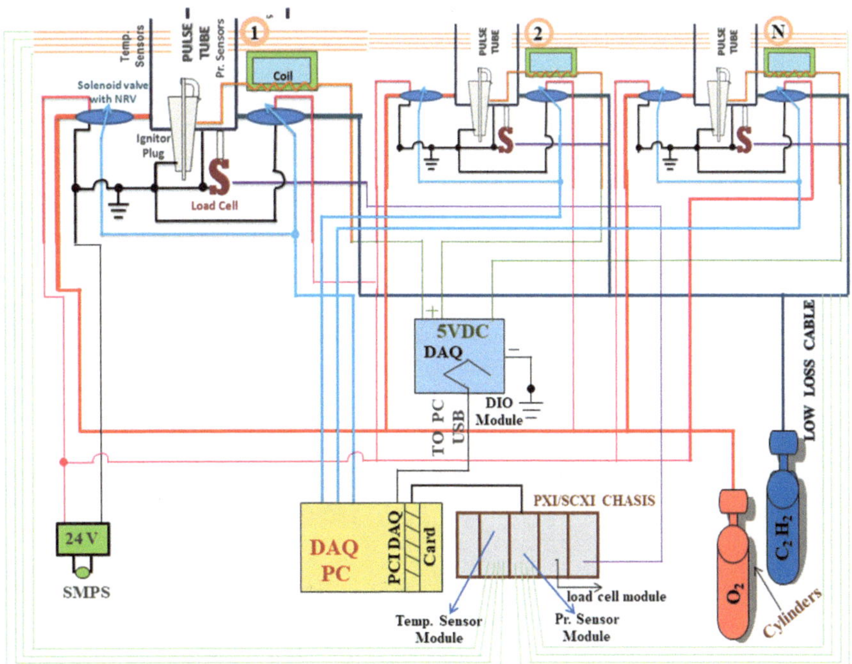

Fig. 4 Multitube multicycle PDE AFCS—electrical connectivity line diagram multicycle PDE control

control hardware installed in robust environment PDE test bed with structured terminations. The controller computer is linked to a data processing, analysis, modeling and simulation systems in the rig through a network connection to the especially built PDE domain.

The software program controlled engine operation by managing engine supply acetylene gas operation via solenoid valves using opening time programmed located in the rig. A D-sub μ-connector-based pulse generator also located in the control room controlled timing of trigger signals sent to both the mixture igniters via electrical relays (12 VDC) of the system. 24 VDC and 220 VAC master power relay modules are also installed in control room. Further, an emergency shut-off button coupled with emergency stop lever too. These electrical relay switches controlled power within the rig for engine control and instrumentation. These instrumentations include load cells, temperature and pressure transducers connected through termination boxes. The flow of gases inside pulse detonation engine is controlled by solenoid-based flow control valves with non-return valve for safety purposes coupled by flash arrestors. The emergency shut-off button would disable rig by closing all supply gas ball valves, it will also interrupt fuel injection coupled with stoppage of ignition trigger signals.

The typical PDE cycle with details is shown in Fig. 5.

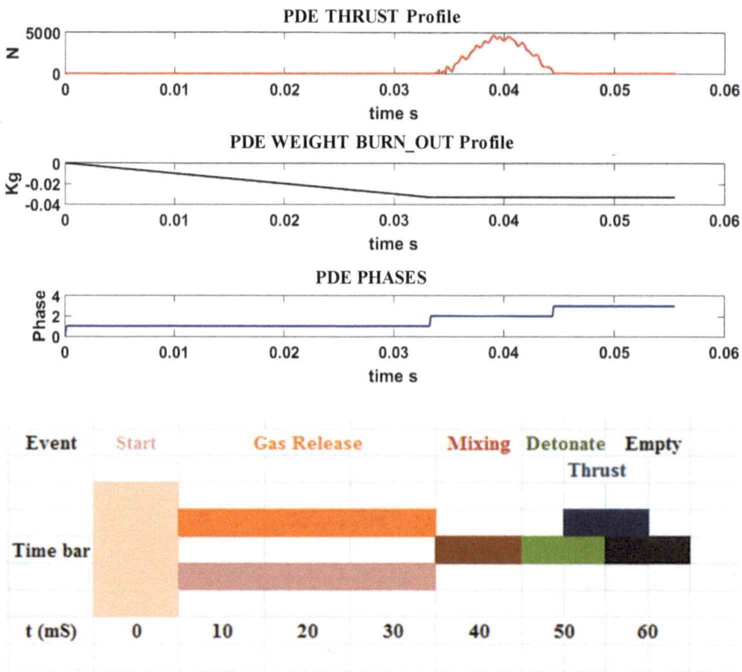

Fig. 5 EAV embedded PDE phases in each cycle

6 Automated Multicycle ICS Software

Software is especially designed to give the commands for firing of PDE and record the outputs in correct sequences and timings [31]. The software is written in NI Labview working in robust test environment with NiDAQ and DOP libraries. The detailed analysis and plotting is carried out in MATLAB.

The software is divided into three parts as below:

- Firing control software
- DAQ analog input/output software module
- MATLAB-based analysis utility.

These are explained as below:

6.1 *Firing Control Software*

The flow chart of software is shown in Fig. 6 and software screen of National Instruments Labview Firing Control Software is shown in Fig. 7, hich are self expantory. The software allow user to select the gas release time in GUI Environment. Then Gas release button knob is pressed.

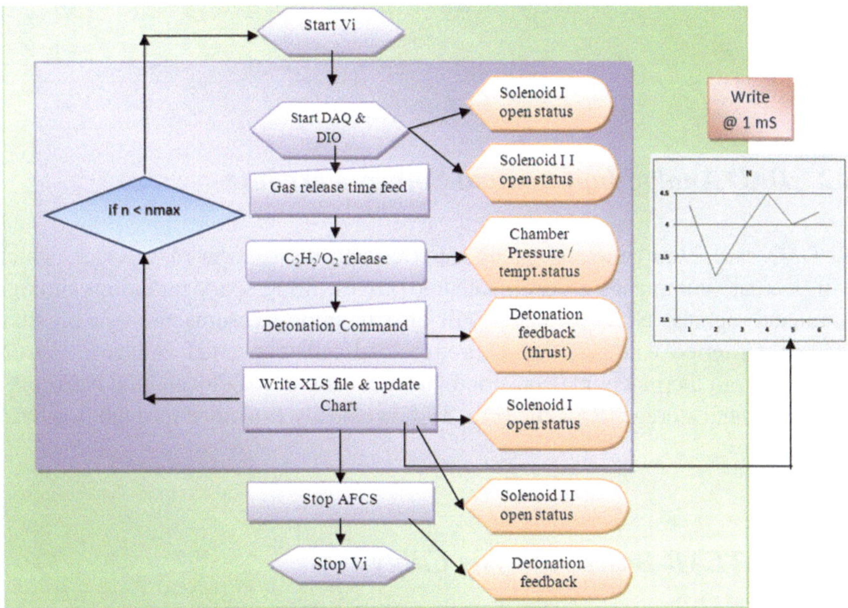

Fig. 6 Multicycle ICS software flowchart

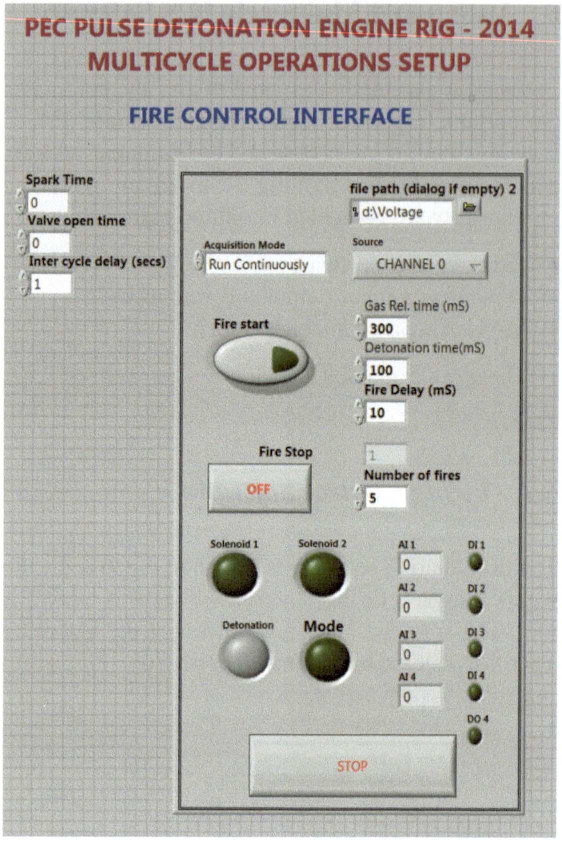

Fig. 7 Firing software GUI

6.2 DAQ Analog Input/output Software Module

NI DAQ cards will generate analog input commands in form of voltages, which will generate commands to open solenoid valves, firing spark initiations through command signals. NI DAQ cards will also monitor pressures through pressure sensors converted voltages, thrust through load cells converted voltages, temperatures through thermocouple converted voltages sensed by DAQ and fed back to VI for display and storage to.xls file (Fig. 8). It is visually explained through flowchart [31].

6.3 MATLAB-Based Analysis Utility

This explains the development of MATLAB-based software AFCS for PDE analyzing detonation, signal conditioning, display and .XLS file save for further analysis.

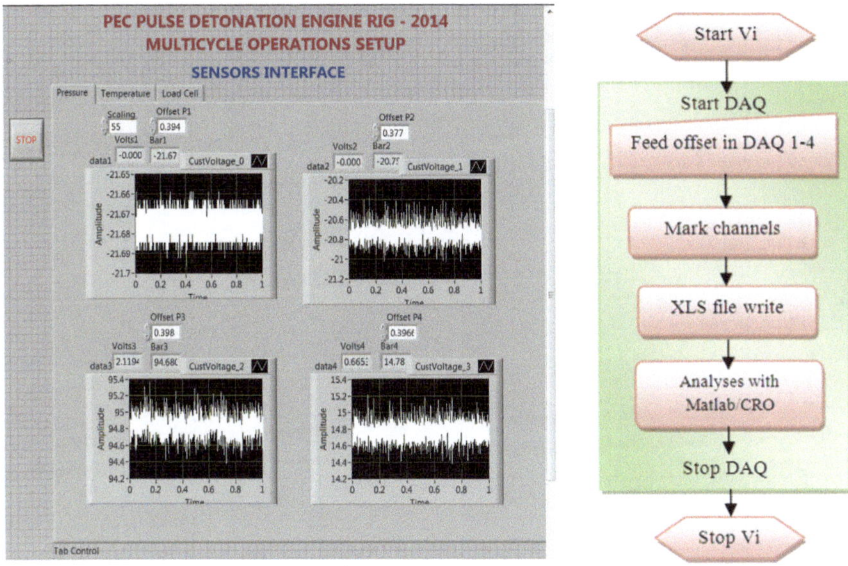

a. Sensor online display b. Flow chart

Fig. 8 Sensor output online plots generation

MATLAB program was written to analyze, filter, plot, compare and conclude the test achievements for further collaborations (Fig. 9). Since it is involving highly flammable fuels at closed vicinity of PDE system so appropriate check lists were developed for further enhancing g safeties.

7 Conversion of Single Shot ICS to Multitube and Multicycle ICS

IFCS for multicycle multitube operations is a mission critical hardware. The derived scheme for multitube multicycle operations [23].

The design is similar to single shot, but large no. of components makes it impossible to ensure high reliability and safety. The system has been trial proven after repetitive testings.

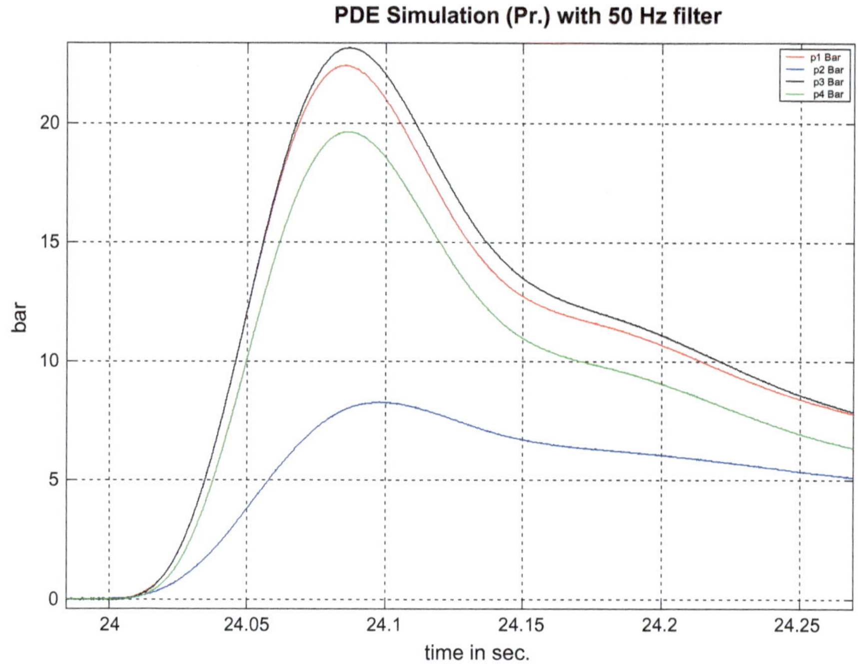

Fig. 9 Depicting P_1, P_2, P_3 and P_4 cross plot of one typical run in MATLAB

8 EAV Configurations with Pulse Detonation Engine Option

The area of research was identified as use of PDE as secondary propulsion system (sustainer) for tactical EAV application. The technical concepts of design, operations and integration are as follows [31].

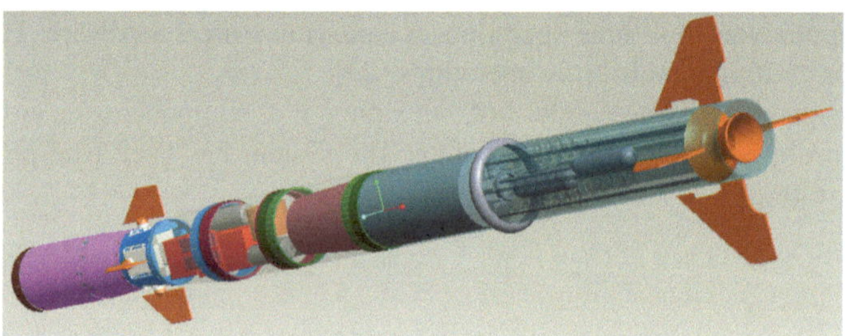

Fig. 10 Proposed tactical missile with PDE sustainer

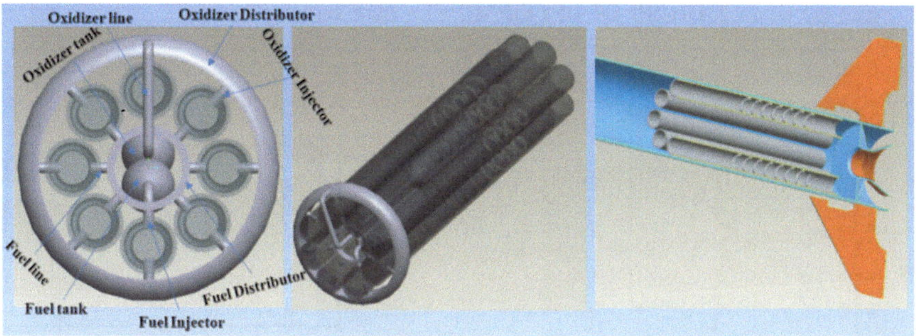

Fig. 11 Proposed EAV configuration with PDE sustainer

A typical tactical EAV (Fig. 10) is being utilized to experiment the trajectory variations due PDE incorporations. The basic existing booster sustainer configuration is studied and their forces, accelerations, velocities and down range are being analyzed.

The experimental aerospace vehicle (EAV) launch weight is <130 kg. The booster is having <4 s run and sustainer has got about <7 s run. It is launched from ground against incoming missiles, UAVs and aircraft etc. It is a supersonic 12 km+ range weapon. It is a quick reaction seeker/INS guided missile. It has got classical warhead. It can be also fired in many other modes too. Finalization of PDE replacement of rocket sustainer, no. of tubes and frequency of firing including its associated modifications/corrections required to be carried out in various subsystems (Fig. 11):

- Since the thrust is sufficient for the current mission, so the feasibility part is concluded
- Since the system has to fit in EAV, so, sizing and shaping has been designed analytically and through modeling and simulation
- The need of design of fuel and oxidizer tanks also was considered and hence implemented
- The piping for transport of fuel and oxidizer from tank to PDE injector was also forecasted and realized
- Mounting mechanism and gas path design was also carried out.

9 Thrust Analysis of EAV Propulsion

The thrust of EAV with existing configuration and PDE modified configuration is examined in detail. The thrust profile of both the configurations is cross plotted for comparison (Fig. 12).

It is clear that PDE is not able to take parallel performance to rocket sustainer in the present form, so it is planned to apply some changes to suit the requirements, which are covered below.

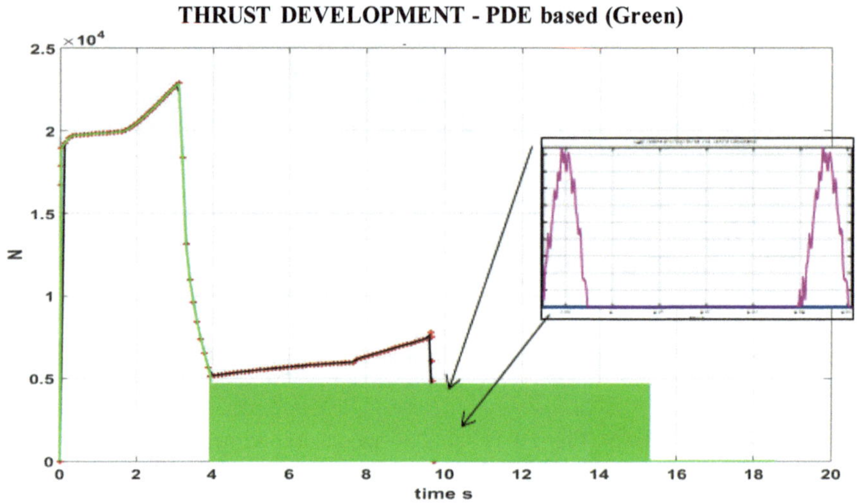

Fig. 12 Pulsating thrust handling plan

10 Staggered Firing Techniques

It was tried to fire PDE in staggered mode to reduce the dead thrust band, which did improve the performance; however, it was still not meeting the performance criteria in totality (Fig. 13). Now, the frequency increase is injected gradually, at about 27 Hz per tube in staggered mode did give promising results (Figs. 14, 15); however, there is a 10% pulsating effect, but no dead thrust band. Additionally, it had super-imposition of thrust cycles, resulting in higher thrust than before. This is simply exceptional achievement in performance of the EAV as it did improve the range, intercept speeds and kill mechanisms, which is a game changer for PDE in future [32].

- Modified EAV is being utilized to experiment the trajectory variations due PDE incorporations in MATLAB simulation as follows:
- Basic existing booster sustainer configuration

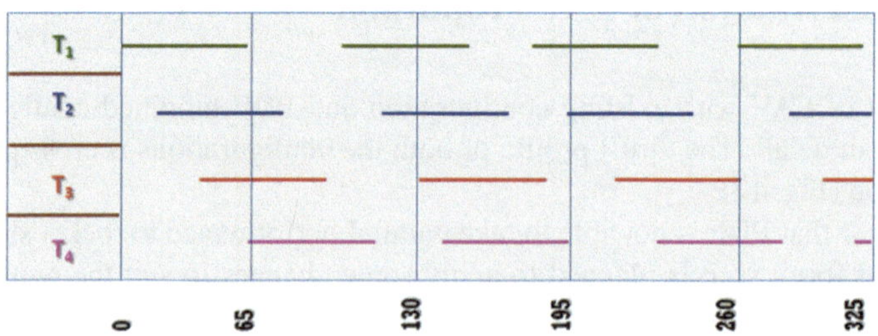

Fig. 13 Staggered firing sequence timing events in each PDE cycle

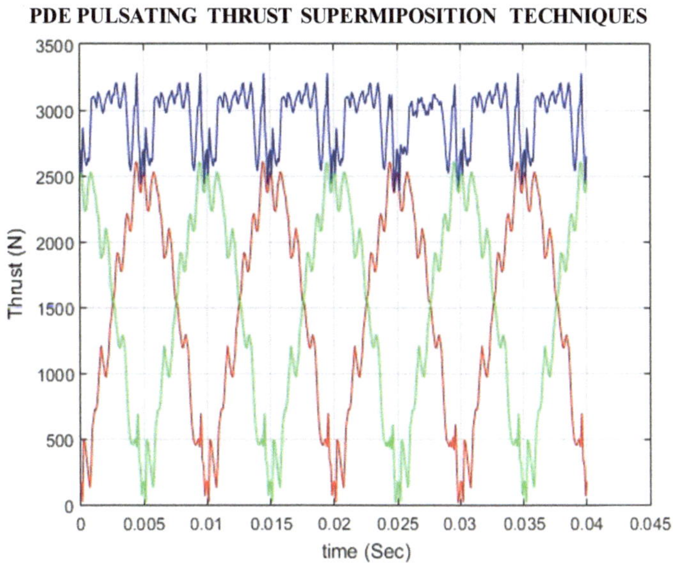

Fig. 14 Superimposed staggered firing sequence timing events in each PDE cycle

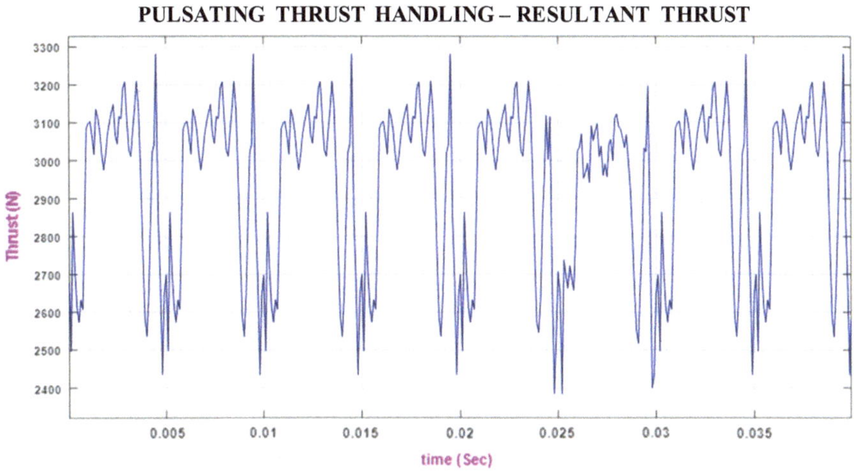

Fig. 15 EAV configuration with PDE-based sustainer augmentation

- PDE sustainer firing configuration—forces, accelerations, velocities and down range.

This research promise better latex, intercept velocities and down range etc. Multi-tube and multicycle operations hold key to the solution. This novel approach with change in fire control system will help to optimize the performance and handling characteristics too.

The effects of PDE-based EAV system performance is deteriorated after PDE introduction due pulsating power. I explored techniques to improve the performance through configurations and physical parameter variations and reduction of pulsating power effects to open up new option available for propulsion, which is having **better power/weight ratio, form factor and easier to maintain/operate** in future.

Consequent to our work on PDE in single shot/multicycle operations in ground demonstration and experimentation rigs, we have optimized the performance of PDE. Our team too explored the use of PDE as sustainer for single shot EAV. Existing AV system is being adapted for this purpose and PDE developed data is used for this highly ambitious iteration. The removal of rocket-based sustainer propulsion and insertion of PDE alternative is being experimented in theoretical work and 3D modeling. The PDE-related hardware is also being inserted inside the custom design of EAV. The space constraints, routing of components, CG/MI distribution and resulting re-orientation of various components are being finalized. This work is being limited by various other aspects due pulsating flows options, longer duration run challenges, etc. need to be tried. General system engineering (SE) aspects and its effect on the system performance also force designers to study into change of key parameters very meticulously. The current paper deals with these aspects and lead to finalization of weapon system propulsion by using advanced features of modeling and simulation tools coupled with state-of-the-art integration/assembly techniques with analytical backing. The improvements carried out thereon, after meticulous selection of vital parameters in the engine, ground rig and instrumentation setup, etc. It also yielded in robust optimal design coupled by better consistency of dynamic performance of PDE powered AV and optimization of main subsystems of PDE. Further, team will also reduce the development cycle and trial complexity also, in addition to generate data to look further diversified use.

The PDE related hardware is also being inserted inside the custom design of EAV. The space constraints, routing of key piping, wiring of ignition system, fuel/oxidizer tanks, reorienting exhaust of exit nozzle, CG/MI distribution and resulting re-orientation of various components are being finalized. This work is being limited by various other aspects including channelization of hot gases, vibration/shock mounting due pulsating flows options, longer duration run challenges etc. need to be tried. General system engineering (SE) aspects and its effect on the system performance also force designers to study into change of key parameters very meticulously. The current paper deals with these aspects and lead to finalization of weapon system propulsion by using advanced features of modeling and simulation tools coupled with state-of-the-art integration/assembly techniques with analytical backing. Researchers are actively engaged in PDE research to optimize the thrust and suppress sound. Handling power inconsistency, thermal management DDT are the other areas of focused research due issues. The improvements carried out thereon, after meticulous selection of vital parameters in the engine, ground rig and instrumentation setup, etc.

By using this setup, trials have been carried out in detail and results are examined. This propulsion data is inserted in 6° of freedom model of aerospace vehicle to assess the performance of EAV [20, 21].

With the above modifications in EAV to incorporate PDE, overall system performance has improved, as key problem areas have been addressed by improving fin deflection limits δ from 15 to 18°. Effectivity of control is also improved by tweaking deflection rates δ' from 300 to 350°/s. Robustness and response of system can be improved through deflection accelerations δ" from 3.75×10^6 to 4.0×10^{6}°/s. Finally, after computing other parameters for EAV, 6° of freedom model is run to estimate the performance evaluation of the EAV and suggest corrective actions to this effect [30].

11 Conclusion and Scope for Future Work

- PDE-based secondary propulsion successfully designed for EAV mission. It is proved to be superior to current sustainer in term of launch weight, intercept zone enhancement, better interception, controllability and agility.
- It is using simpler PDE technology, which has paved a new concept of pulsating power usage in aerospace propulsion.
- **EAV configuration with PDE-based sustainer augmentation**—Modified EAV is being utilized to experiment the trajectory variations due PDE incorporations in MATLAB simulation as follows:
- Basic existing booster sustainer configuration.
- PDE sustainer firing configuration—forces, accelerations, velocities and down range.
- This research promises better latex, intercept velocities and down range, etc. Multitube and multicycle operations hold key to the solution. This novel approach with change in fire control system will help to optimize the performance and handling characteristics too.
- The effects of PDE-based EAV system performance is deteriorated after PDE introduction due pulsating power. I explored techniques to improve the performance through configurations and physical parameter variations and reduction of pulsating power effects to open up new option available for propulsion, which is having **better power/weight ratio, form factor and easier to maintain/operate** in future.
- Cost of weapon system got reduced extensively due classical hardware, especially for single shot devices.
- Fire control system safeties are now embedded in design due remote operations of stand-offs.
- Due multiuse remote operations, launch operations become affordable.
- Operations of rocket mode PDE do not get affected by height or other environmental changes.

12 Future Works Includes

- Creating working SAM model to prove concepts
- Sensor networking
- Trying similar concepts on UAV and satellite launch vehicle
- Trying liquid-based/air breathing PDE/RDE
- Thermal management for longer runs
- Optimizing PDE power for better applications
- Optimizing ignition energies and fuel/oxidizer flows.

Acknowledgements Authors are thankful to the Director, PEC University of Technology and the Director, TBRL, Chandigarh, for their support and guidance.

References

1. Zhang F-Y, Fujiwara T, Miyasaka T, Nakayama E-I, Hattori T, Azuma N, Yoshida S, Tamugi A (2003) Experimental study of key issues on pulse detonation engine development. Trans Japan Soc Aero Space Sci 45(150):243–248. https://doi.org/10.2322/tjsass.45.243
2. Schauer F, Stutrud J, Bradley R (2001) Detonation initiation studies and performance results for pulsed detonation engine applications. In: 39th AIAA aerospace sciences meeting and exhibit. AIAA, pp 2001–1129
3. Chander S, Jindal TK (2012) Integration challenges in design and development of pulse detonation test rig. Int J Adv Res Electr Electron Instrum Eng 1(4), ISSN 2278-8875. https://doi.org/10.15662/ijareeie
4. Lu FK (2016) Progress and challenges in the development of detonation engines for propulsion and power production. Appl Mech Mater 819:3–10. ISSN 1662-7482
5. Chander S, Jindal TK (2014) Performance enhancement of surface to air missile by application of pulse detonation engine based secondary propulsion system. Int J Mech Engg (IJME). Recent Science Publications
6. Chander S, Jindal TK (2013) Design of automated fire control system for C_2H_2/O_2 pulse detonation rig. Int J Adv Res Electr Electron Instrum Eng 2(2). ISSN 2278-8875. https://doi.org/10.15662/ijareeie
7. Jindal TK (2012) Pulse detonation engine—a next gen propulsion. Int J Modern Eng Res (IJMER) 2(6). ISSN 2249-6645
8. Jindal TK (2012) Single cycle pulse detonation engine testing. Int J Mech Eng 40(10). ISSN 2051-3232
9. McManus K, Furlong E, Leyva I, Sanderson S (2001) MEMS-based pulse detonation engine for small scale propulsion applications. AIAA paper, pp 2001–3469
10. Valli DM, Jindal TK (2013) Thrust measurement of single tube valveless pulse detonation engine. Int J Sci Eng Res (IJSER) 4(3). ISSN 2229-5518
11. Soni SK, Singh A, Sandhu M, Goel A, Sharma RK (2013) Numerical simulation to investigate the effect of obstacle on detonation wave propagation in a pulse detonation engine combustor. Int J Emerg Technol Adv Eng 3(3): 458–464. ICERTSD-2013. ISSN 2250-2459
12. Ostertagová E, Ostertag O (2013) Methodology and application of oneway ANOVA. Am J Mech Eng I(7): 256—261. ISSN (Print) 2328-4102, ISSN (Online) 2328-4110, 10.12691
13. Chander S, Jindal TK (2014) A study of possible technical aspects needed to be addressed in respect of system engineering of surface to air missile by application of pulse detonation engine. Int J Mech Eng (IJME) 1(1): 40–43. ISSN 1694-2302

14. Chander S, Jindal TK (2018) Performance analysis of surface to air missile by application of pulse detonation engine based secondary propulsion system. Ph.D. thesis submitted to Punjab Engineering College (Deemed to be University), Chandigarh
15. Robbins TJ, Sinibaldi JO (2006) Fuel injection strategy for a next generation pulse detonation engine. Naval Postgraduate School, Monterey, California
16. Rodriguez J (205) Investigation of transient plasma ignition for a pulse detonation engine. Master's thesis, Naval Postgraduate School, Monterey, California
17. Kailasanath K (2000) Review of propulsion applications of detonation waves. AIAA J 38(9)

Proceedings/Transactions Articles

18. Smirnov N, Nikitin VF, Biochinto AP, TyurniKov MV, ShevtSova VM, VBaskaKov V (1999) In: Proceedings of the 12th ONR propulsion meeting
19. Roy GD, Frolov SM, Borisov AA, Netzer DW (2014) Pulse detonation propulsion: challenges, current status, and future perspective. Prog Energ Combust Sci 30. www.sciencedirect.com
20. Chander S, Valli DM, Jindal TK (2013) Multiple option levels based full function PEC aero simulation (PAeroSim) development for pulse detonation engines and systems. In: Symposium on applied aerodynamics and design of aerospace vehicle (SAROD 2013), Hyderabad, India
21. Schauer F, Stutrud J (1999) AFRL's in house research pulse detonation engine. In: 11th annual symposium on propulsion
22. Falempin F, Bouchaud D, Forrat B, Desbordes D, Daniau E (2001) Pulsed detonation engine possible application to low cost tactical missile and to space launcher. In: AIAA-2001-3815, 37th AIAA/ASME/SAE/ASEE joint propulsion conference, Salt Lake City, Utah
23. Cloutier JR, Evers JH, Feeley JJ (1988)Assessment of air-to-air missile guidance and control technology. In: American control conference, Atlanta, Georgia
24. Panicker PK, Lu FK , Wilson DR (2009) Practical methods for reducing the deflagration-to-detonation transition length for pulse detonation engines. In: Proceedings of the 9th international symposium on experimental and computational aerothermodynamics of internal flows (ISAIF9), Gyeongju, Korea
25. Falempin F, Bouchaud D, Daniau E (2000) Pulsed detonation engine towards tactical missile application. In: AIAA-2002-3473, 36th AIAA/ASME/SAE/ASEE joint propulsion conference, Huntsville, Alabama

Company or NASA Report and Patents

26. Chao TW, Winterberger E, Shepherd JE (2001) On the design of pulse detonation engine. Galcit Report FM00–7, Graduate Aeronautics Laboratories, California Institute of Technology
27. Jindal TK, Chauhan YS (2012) Development of pulse detonation rig. A Report on sponsored research by TBRL
28. Chander S, Singh S, Ray SK et al (2008) System engineering for Antiship missile—a report. RCI, Hyderabad
29. Lam M, Tillie D, Leaver T, McFadden B (2004) Pulse detonation engine technology: an overview. Submitted to: Michael Schoen, Applied Science 201, The University of British Columbia
30. Chander S (2007) Pointmass study of guided bomb—a report. RCI, Hyderabad
31. Chander S, Chaudhuri SK et al (2008) Missile launched PGMs—a report. RCI, Hyderabad
32. Chander S, Jindal TK (2019) Pulse detonation engine—pulsating power handling schemes. A petent filed with Department of Patents, Govt. Of India, 2019 (Patent Pending)

AIAA and Other Book Series

33. Travis J, Kring J (2009) LabVIEW for everyone. Pearson Education
34. Thomson WT, Dahleh MD (1998)Theory of Vibration with applications, 5th ed. Prentice Hall, Inc
35. Fleeman E (2006) Tactical missile design. AIAA Edu. ISBN-10 1563477823
36. Operations and maintenance manual, six component, 500 pound pulse detonation measurement system. Pacific Press Co., N00244-02-C-0060, Manual No. 110028, September 2002
37. Chin SS (1961)Missile configuration design. McGraw-Hill Book Company, Inc., New York. (OCoLC) 602683910
38. Garnell P (1980) Guided weapon control systems. Brassley's Defence Publishers, London
39. SK Ray (2007) Missile control system. DESIDOC, DRDO, Delhi
40. Ramamurti K (2010) Explosions and explosion safety. Tata Mcgraw Hills Publishing Co. Ltd. ISBN 9780070704473
41. Kothari CR, Garg G (2014) Research methodology: methods and techniques. New Age Publishers. ISBN-13 978–8122436235
42. Montgomery DC, Runger GC (2003) Applied statistics and probability for engineers. Wiley
43. Royal Military College of Science Guided weapons handbook, 14th ed., pp 1–1
44. JF Rouse Guided weapons. Royal Military College of Science, Shrinavenham, UK, Brassey's, London
45. Mullins SS, Penner BP (1959) Explosions, detonations, flammability and ignition. Pergamon Press

Fuels

The Formation of PAH Compounds from the Combustion of Biofuels

A. Williams, A. R. Lea-Langton, and K. D. Bartle

Abstract Polycyclic aromatic hydrocarbons (PAH) are byproducts resulting from the incomplete combustion of hydrocarbon-based fuels such as fuel oils, coal and biofuels. PAH are implicated in the formation of soot, another undesirable byproduct formed during combustion. Such compounds are harmful to human health, and soot plays an important role in climate change. Black carbon is known to absorb visible solar radiation, and the associated organic carbon fraction, including PAH, that is co-emitted can absorb radiation at shorter wavelengths leading to climate forcing. As a result of climate change considerations and the need to reduce greenhouse gas emissions, fossil fuels are slowly being replaced by lower carbon renewable biofuels. However, these biofuels are often carbon based, and so, their incomplete combustion produces a similar range of pollutant species to fossil fuels. These substitute fuels can form PAH compounds, although often to a lesser degree due to differences in the fuel chemistry. The EPA 16 list is widely used to determine the extent of pollution due to PAH but does not include other types of polycyclic aromatic compounds such as oxygenated species (O-PAC) or nitrogenated species (N-PAC) so does not reflect many of the priority pollutants.

Keywords Biomass · PAH · O-PAH · Renewable fuels

1 Introduction

There has been considerable interest in the emission of PAH species from the combustion of fossil fuels such as coal, diesel and heavy fuel oils because many species have been shown to be harmful to human health [1–6]. Soot is an associated co-product with adverse health and climate impacts. Because of climate change concerns

A. Williams (✉) · K. D. Bartle
School of Chemical and Process Engineering, University of Leeds, Leeds LS2 9JT, UK
e-mail: A.Williams@leeds.ac.uk

A. R. Lea-Langton
Department of Mechanical Engineering, Aerospace and Civil Engineering, School of Engineering, University of Manchester, Manchester M13 9PL, UK

A. Gupta et al. (eds.), *Advances in Energy and Combustion*, Green Energy and Technology,
https://doi.org/10.1007/978-981-16-2648-7_5

fossil fuels are being replaced by biofuels in some applications, which includes mainly solid, liquid and gaseous forms. The solid biofuels involve wood, agricultural residues, charcoal and biochar; the liquids are mainly ethanol and biodiesel; and the gaseous fuels are biogas or products from gasification of biomass. Biomass accounts for about 10% of the global energy supply and 80% of the renewable energy [7]. Biomass fuels have the advantage that they share many of the properties of fossil fuels and can easily substitute for them for heating and electricity generation, road vehicles and as aircraft fuels.

The use of solid renewable fuels has increased markedly in recent years [7, 8]. Solid fuels, mainly wood logs, pellets from waste wood or agricultural residues have been widely used for many years, but there has been a considerable increase because of climate change concerns; however, there are consequential pollution issues particularly with regard to smoke pollution [4, 5, 8]. There has also been an increase in use of liquid biofuels where the fast pyrolysis of biomass feedstock is potentially an economic method of producing renewable liquid fuels.

Road vehicles have proved to be a suitable case for the substitution of fossil fuels by the biofuel equivalent either in total or by blending. This is the case for petrol or diesel, and the substitutes may be bioethanol/biobutanol or biodiesels derived from transesterification of bio-oils, e.g. from rape seed oil [9, 10].

Aviation bio-jet fuels present different challenges although the same types of fuels may be used [11, 12]—Alcohols are too volatile to be used alone but could be used in blends with Jet A1. This is also the case with bio-oils which can be blended with petroleum fuels although currently the fuel specifications are very strict.

Much depends on the fuel refining operations which can result in a range of fuels types [11, 13] including heavy fuels which can be burned as fuel sprays in furnaces or marine engines as substitutes for heavy fuel oils [3, 14, 15]. Another source of these heavy oils could be via the processing of algae [16].

Oxygenated PAH (O-PAH) species can be formed from the combustion of solid and liquid biofuels [e.g. 17]. There is considerable interest in this because of the importance in relation to the health aspects [10–16] and due to the role that they play in climate chemistry [4, 18, 19]. It is known that black carbon absorbs visible solar radiation; however, the organic fraction has also been found to absorb at the shorter visible wavelengths. Some of these species are thought to arise by secondary reaction in the atmosphere, which comes about because of the formation of oxygenated PAH species that are able to react to give larger molecular structures.

In all these applications, the bio-replacements, although renewable and sustainable, can be carbon containing, and consequently, there is a route available leading to the formation of PAH species. The key to the understanding of this is a knowledge of the mechanism of formation of PAH from the hydrocarbon component, although some petroleum-based fuels already contain a small amount of PAH species. Thus, fuels from bio-oils are more likely to produce PAH than an alcohol such as ethanol; butanol containing a higher carbon content than ethanol would produce slightly more PAH.

2 The Formation of Polyaromatic Hydrocarbons (PAH) from Hydrocarbons

The pyrolysis and oxidation reactions that take place during the combustion of fossil fuels result in the generation of a number of species which can react and form into benzene. This then reacts to produce small PAH molecules which grow to produce larger PAH molecules and subsequently into soot particles. This only occurs if there are sufficient building block materials available; that is, if the fuel–air ratio is rich enough for soot to be formed. The routes to PAH formation and to soot are thus intertwined; PAH in combustion products are essentially byproducts of the soot forming process and consequently also linked in steady-state combustion equipment to CO emissions.

Historically, a number of mechanisms have been postulated based on acetylene (ethyne) as the main precursor. The most widely used chemical model is the Hydrogen Abstraction C_2H_2 Addition (HACA) set of reactions [20], and together with the intermediate CPDyl (cyclopentadienyl) gives the soot precursors, some of which have five-membered rings and the PAH species. It should be noted that the PAH species act as the precursor to soot formation and may be produced and emitted even below the critical limit for soot formation.

In the HACA mechanism [20], the initial pyrolysis product of fossil fuels is acetylene which polymerizes to give benzene, which go on to form larger PAH species [20–23]. The structures of the PAH present in the combustion products are determined by the thermodynamic stabilities of the molecules produced.

The HACA mechanism is set out below in a simplified form:

$$\begin{aligned} &\mathrm{Ai} + \mathrm{H} = \mathrm{Ai}^{-} + \mathrm{H}_2 \\ &\mathrm{Ai}^{-} + \mathrm{C}_2\mathrm{H}_2 = \mathrm{AiC}_2\mathrm{H}_2 \\ &\mathrm{AiC}_2\mathrm{H}_2 + \mathrm{C}_2\mathrm{H}_2 = \mathrm{Ai}_{+1} + \mathrm{H}\cdot \end{aligned}$$

where Ai is an aromatic molecule with i-peri-condensed rings and Ai^{-} is its radical. So, overall:

$$\mathrm{Ai}^{-} + \mathrm{C}_2\mathrm{H}_2 = \text{products} = \text{PAH growth}$$

where A_1 would be benzene and H are the hydrogen atoms produced during the combustion and pyrolysis processes. A_2 would be naphthalene and so on.

A fuller mechanism leading to the formation of PAH species is shown in Fig. 2 including the steps up to the formation of soot. Five-membered rings in the precursors can then lead to pentagonal curved network regions and hence to bowl-shaped structures (so-called protofullerenes) and curved carbon atom sheets consistent with the observed morphology of the soot of the 'protographene' PAH such as anthanthrene which act as 'dead-end' PAH and terminate HACA reaction sequences.

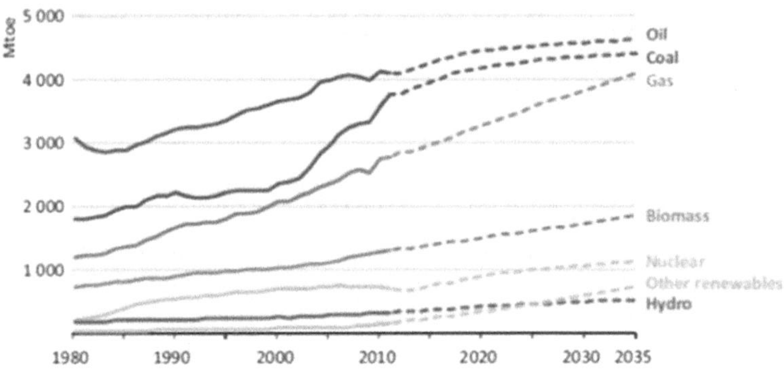

Fig. 1 IEA estimates of world energy use until 2035 (based of ref 7)

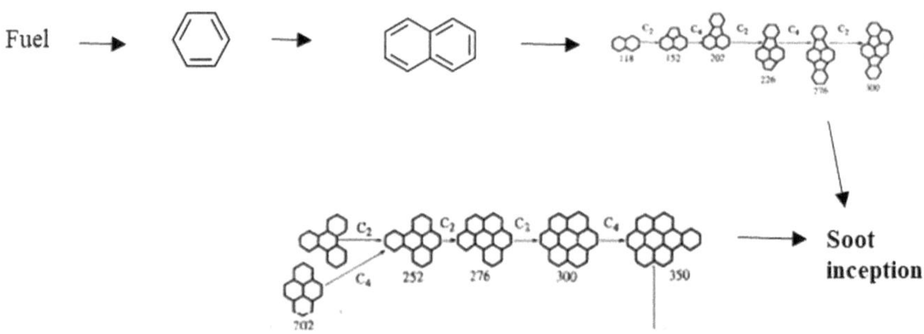

Fig. 2 Schematic of the formation of PAH and the pathway to the point of soot from [20, 22, 23]

There are a number of PAH forming/soot models which include the original HACA model [20], but have additional growth chemistry or invoke the role of aromatic molecules present in the fuel such as naphthalene and indene which react further with CPD or CPDyl to form larger PAH. These may react with five-membered ring PAH yielding open structures or undergo ring closure to yield large PAH. There is also a more recent model [24] which involves a different growth mechanism incorporating resonance-stabilized hydrocarbon radical chain reactions (CHRCH mechanism), and these routes result in the formation of clusters. Some evidence for this route also comes from ATOF-MS studies [25]. Another recent model for soot formation includes the formation of reactive clusters and may involve ions [26].

The more general model used for the formation of soot involves stacked PAH layers as precursors which bind together through pi-pi interactions between graphene sheets to give crystallites in flame gases of size approximating to that of circumcoronene (C_{54} H_{18}). This approach is briefly outlined Fig. 3 and may involve the icospiral mechanism described in [23] to build up the soot structure.

Most of the early understanding of the mechanism of PAH formation comes from studies associated with diesel fuels, coal or fuel oil combustion and industrial operations involving the pyrolysis of hydrocarbons. Investigations of soot formation

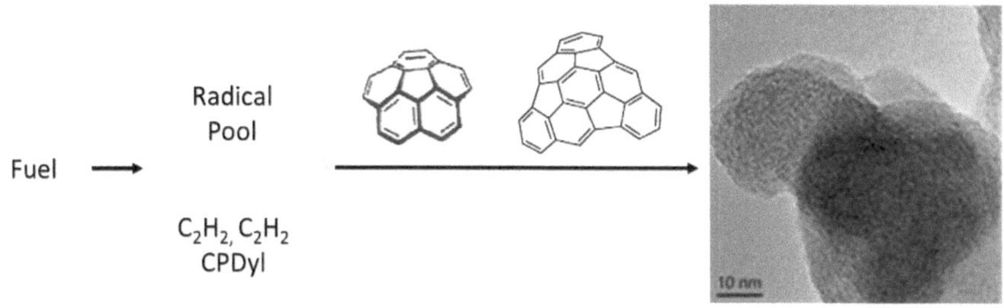

Fig. 3 Schematic of soot formation through icospiral mechanism, based on [23]

during the combustion of a wide variety of fuels, gaseous and liquid hydrocarbons, coal and biomass have allowed us to classify the properties of a number of soot forms. These range from the nature of the structure of black carbon (BC) and the associated organic material (OC) which arises from the partial combustion of diesel fuels and wood.

Although hundreds of PAH exist in the environment, only 16 PAH, i.e. naphthalene, acenaphthylene, acenaphthene, fluorene, phenanthrene, anthracene, fluoranthene, pyrene, benzo[*a*]anthracene, chrysene, benzo[*b*]fluoranthene, benzo[*k*]fluoranthene, benzo[*a*]pyrene, dibenzo[*a,h*]anthracene, benzo[*ghi*]perylene, and indeno[1,2,3-cd]pyrene, have been selected by the US Environmental Protection Agency (US EPA) as priority pollutants, which are monitored routinely for regulatory purposes. The toxicity of these varies from species to species.

There is still uncertainty about the process of early growth relating to PAH formation although most research has been concerned with the process of soot inception [23, 24]. Soot inception include the dimerization of PAH species [20], clustering of resonance-stabilized radicals [24], or ion-induced nucleation of curved PAH species [26]. This latter route would involve other flame ions such as K+ which would mainly be present during the combustion of solid biomass materials such as wood.

The well-developed sequence of PAH then leads to planar carbon sheets through the addition of C_2 and C_4 to the carbon network in the HACA mechanism. Recently, however, Johansson et al. [24] have pointed out that the HACA mechanism incorporates hydrogen abstraction with high energy barriers, whereas a scheme in which highly reactive resonance stabilized radicals (RSR) are clustered with PAH molecules results in the regeneration and growth of the RSR and the effective dimerization of the PAH and extension of the protographene structures.

The early stages of the clustering of hydrocarbons by radical chain reactions (CHRCR) mechanism are shown in Fig. 4. The first stage involves molecular growth of the pool of RSR that acts as seeds for initiating hydrocarbon clustering. Electronic structure calculations predict a sequence of RSR which are readily generated through radical chain reactions. Each reaction adds acetylene (C_2H_2) or vinyl (C_2H_3) to an RSR and generates a new RSR. Thus, these radicals are not consumed and grow to larger species. For instance, propargyl (C_3H_3) is an RSR that can generate cyclopentadienyl by reaction with acetylene.

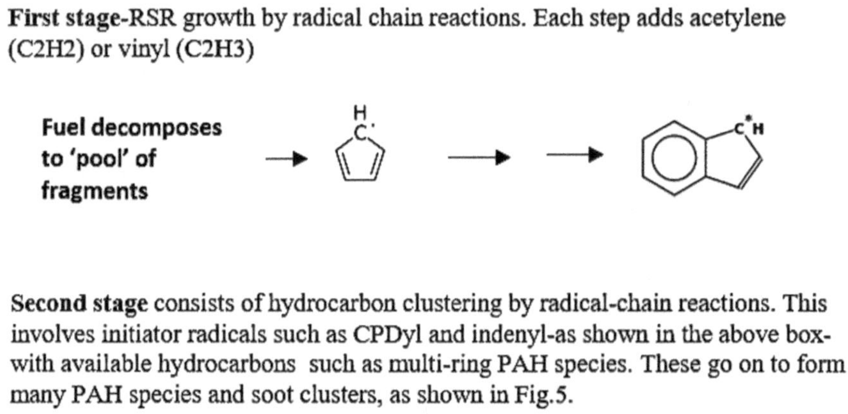

Fig. 4 CHRCH (clustering of hydrocarbons by radical chain reactions) Mechanism (from [24])

The second stage comprises the clustering of hydrocarbon caused by radical chain reactions. This mechanism provides a satisfactory way of describing the formation of PAH species, but the clustering step has also to explain how the well-defined onion structures occur which are seen by electron microscopy (see Fig. 3).

The formation of larger PAH species from the small precursors via the different routes can have different outcomes as summarized in Fig. 5.

In addition, the PAH species may be methylated depending on the combustion condition and the temperature of the cooling process of the exhaust gases, and this process is significantly different between vehicles and stoves [27–30]. The nature of different types of PAH, whether methylated or demethylated species from specific sources-marine, heavy diesel and gasoline car engines have been identified [28] and are given in Table 1.

Several studies have been made of emission factors for a number of individual PAH species for different fuel types and combustion applications [31–36]. For practical purposes, the total PAH emission are given, and this includes both particle bound and

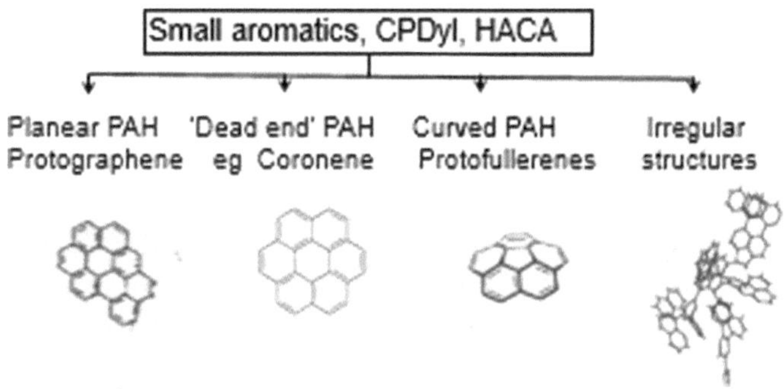

Fig. 5 Different types of PAH species that can occur

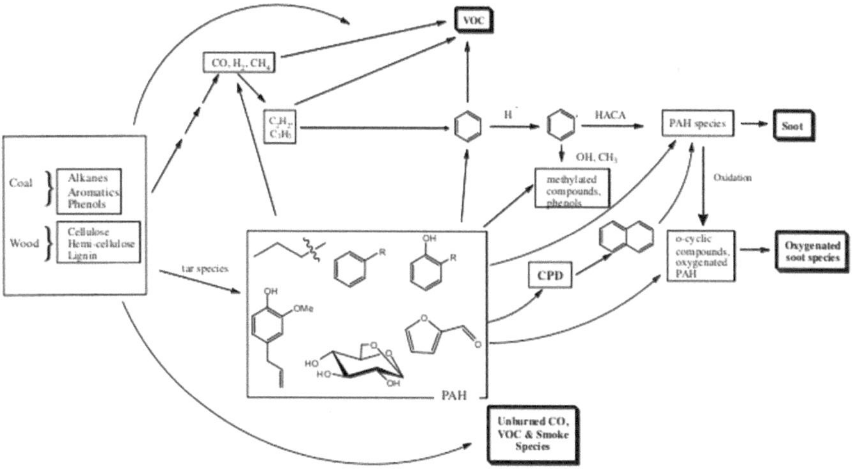

Fig. 6 Mechanism of biomass pyrolysis leading to PAH and soot formation [39, 40]

Fig. 7 Lignin units, from left to right: p-hydroxyphenol (H), guaiacyl (G) and syringyl (S)

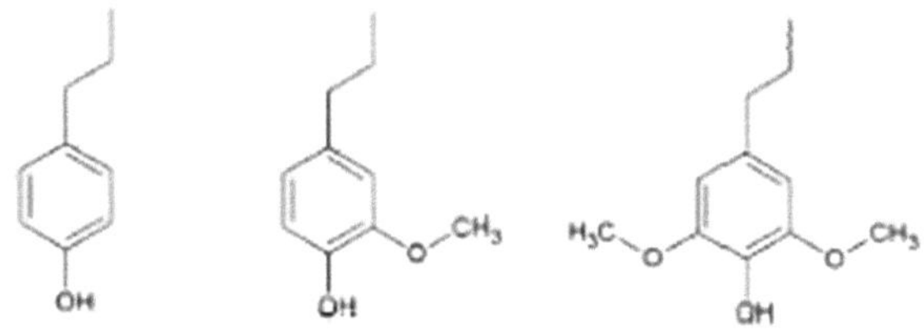

Table 1 Types of PAH compound emitted by different types of engines together with emissions from wood stoves, based on [28]

Diesel cars and wood stoves	Marine engines, heavy-duty diesel vehicles	Gasoline vehicles
2–5 rings not methylated naphthathalenes, phenanthrene, pyrene e.g. pyrene 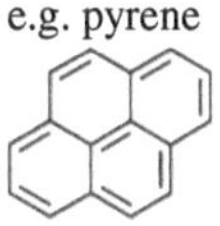	2–3 rings, substituted with methyl or ethyl groups, e.g. 7-ethyl-1-methyl phenanthrene 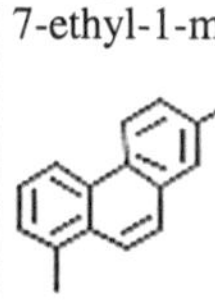	1 or 2 ring methylated species: toluene, xylene, methyl naphthalene e.g. toluene 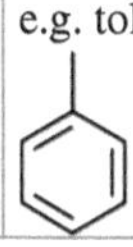

gaseous PAH; the amount of particle bound PAH varies from system to system but is usually about 10% mass. This includes the volatile naphthalene and acenaphthalene, and the results are greater than some other methods.

A selection of emission factors is shown in Table 2. It should be noted that emissions from engines and boilers are not measured using the same sampling technique and it is not appropriate to directly compare the emission factors.

Table 2 Selection of total PAH emission factors from different combustion systems

System	PAH emission factor, μg/MJ	References
Coal, domestic burning	40	[31]
Oil boiler, No. 2 fuel oil	1.4	[32]
Oil boiler soy biodiesel	1.1	[32]
Oil boiler, animal biodiesel	0.6	[32]
HFO boiler, older design	60	[31]
Diesel truck, heavy duty, DI diesel fuel	26	[33]
Same engine, with rapeseed fuel	17	[33]
Wood stove, pellet stove	2	[36]
Wood stove, modern, logwood	36	[36]
Wood stove, older design logwood	1000	[35]

From the data given here, it is seen that for the same type of equipment and on the same thermal basis that the biofuels emit less that the fossil fuel.

3 The Formation of PAH from Biomass

PAH species can be formed during the combustion of all types of biofuel. From the early development of PAH species to the inception of soot, there are also opportunities to form O-PAH species, which are discussed in the next section. Studies have been concerned with the interpretation of the identity and sources of PAH in emissions from burning biomass together with evidence from other sources [5, 17, 28, 35–38]. Particular attention is directed to those species which play an important role a determining the toxicity of combustion products.

The pyrolysis of biofuels during combustion leads to a number of PAH species resulting from the carbon content of the biofuel. Thus, biogas which will be very low carbon containing will produce very low amounts of PAH, whilst solid biofuels such as wood produce greater amounts. They also produce oxygenated PAH as a result of being highly oxygen-containing compounds. The two main classes of O-compounds: firstly, cellulose products form small PAH via HACA reactions and together with CPDyl, etc., give the soot precursors, some of which have five-membered rings [17,

35, 39, 40]. The initial pyrolysis products of the lignin are phenols, e.g. 4-methyl-2-methoxyphenol from eugenol which goes on to form precursors of larger PAH species [20, 22]. The five-membered rings in these PAH compounds formed by both routes can lose an H by H-abstraction and react with an oxygen atom forming compounds such as 1-indanone, 9-fluorenone and peri-naphthalenone. Five-membered rings in the precursors can then lead to pentagonal-curved network regions and hence to bowl-shaped structures (so-called protofullerenes) and curved carbon atom sheets consistent with the observed morphology of the soot, as previously described. Some of these O-PAH are the analogues of the 'protographene' PAH such as anthanthrene which act as 'dead-end' PAH and terminate HACA reaction sequences.

The smaller O-PAH also plays a role in the reactions leading to the formation of the larger molecules. It is suggested that the formation of soot involves stacked PAH layers as precursors which bind together through pi-pi interactions between graphene sheets to give crystallites in flame gases of size approximating to that of circumcoronene ($C_{54}H_{18}$). As well as growing in size they can, under certain circumstances, they can be oxidized by reaction with O or O_2 to give ketones. From the early development of PAH species to the inception of soot, there are many opportunities to form larger O-PAH and some of these become trapped in the soot matrix leading to a disordered soot structure, which in turn are more reactive to oxidation. This may be advantageous in the case of regenerative diesel soot traps.

The early stages involve the liberation of small aromatics, especially indene and naphthalene, formed from cyclopentadienyl radical which is the product of elimination of CO from phenols [41–43]. This sequence is particularly relevant to biomass combustion in the light of the contribution of lignin. The reactive hydrocarbon acenaphthylene has also been implicated as a product from further reactions here.

Lignin → coumaryl, coniferyl, sinapyl alcohols → phenols → CpDyl

This sequence produces aromatic ketones in the ratio of O-PAH to PAH given by the thermodynamic stabilities. Such reaction in the precursors can lead to pentagonal regions and hence to bowl-shaped structures, protofullerenes and curved C atom sheets consistent with the observed morphology of the soot. Some of these O-PAH are structural analogues of protographenes such as coronene and anthanthrene which act as dead ends and terminate HACA reaction sequences.

However, extension to smaller PAH along with CPDyl accounted for the soot precursors, some of which had five-membered rings and could lose H by abstraction through reaction with O or O_2.

The mechanism is based on two main studies biomass studies [39, 40], and a more detailed biomass pyrolysis model is now available [42]. Routes leading to the reactions of the lignin decomposition model have also been studied [42, 43].

4 OXY-PAH (Designated Here as O-PAH)

O-PAH species can be formed during the combustion of all types of biofuel as well as fossil fuels. From the early development of PAH species to the inception of soot, there are also opportunities to form O-PAH species which are discussed in the next section. Studies have been concerned with the interpretation of the identity and sources of PAH in emissions from burning biomass together with evidence from other sources [5, 17, 28, 35, 39, 40] Particular attention is directed to those species which play a determining the toxicity of combustion products [44–47]. The ratio of the O-PAH to the PAH present in the combustion products is determined by the thermodynamic stabilities of these molecules; these can act as precursors of the oxygenated involatile atmospheric PAH species observed. Here, we particularly consider the interpretation of the identity and sources of oxygen-containing PAH in emissions from burning biomass, which is a major source, together with evidence from other sources.

The experimental data used to draw conclusions about the formation of O-PAH are taken from the following sources:

1. Samples of soot were taken from a wood burning stove. Soot formation was determined firstly by collecting and filtering all particulate pine wood combustion products [39, 40].
2. Fitzpatrick et al. [17] studied the emission of oxygenated species from the combustion of pine wood and its relation to soot formation using a firebox. Samples were analysed by GC–MS.
3. Liquid biomass model compounds were burned in a wick burner [41] using a 2 mm wick diameter, a height of 7 mm and surrounding uniform flow.
4. Shen et al. [37, 38] measured the emission factors for 9-fluorenone, anthracene-9,10-dione, benzanthrone and benzo[*a*]anthracene-7,10 dione from wood stoves in rural China. They also measured the emission of N-PAH.
5. Measurements have also been made of O-PAH in the atmosphere, and some of this work is used here [48, 49]. The data are complicated because samples come from many sources, and some have undergone oxidation in the atmosphere. Typically, samples of urban air particulates were collected on quartz filters and analysed by a variety of methods: Py-GC-MS, Py-GC-AED for selective detection of oxygen compounds; or samples were separated by extraction with supercritical carbon dioxide, and multi-dimensional chromatography and detection by high-resolution MS to give molecular formulae of individual compounds.

4.1 O-PAH Sources

Broadly, classified experimental data from analysis of biomass emissions revealed the information shown in Tables 3 and 4. There are interesting differences in the combustion of wood in a stove, which are shown in Table 3, between the emissions from the lump wood and the smaller particles of wood in briquettes due to differences

Table 3 Relative masses of typical emissions from a pine wood-fired unit [35]. Pine pyrolysis products are given in the RH column

Emission mg/MJ	Stove lump wood	Stove sawdust briquette	Pine pyrolysis products
PAH	14.0	12.0	3200
Alkyl PAH	7.6	4.7	1900
Phenols	13.0	2.5	29,100
O-PAC	1.7	1.2	

Table 4 Emission factors for O-PAH in flue gas from a wood stove, (μg/MJ) [35]

O-PAH species	B.Pt °C	Lump wood
1-naphthaldehyde 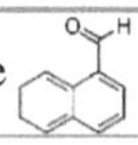	160	310
2-Me-1,4-naphthoquinone 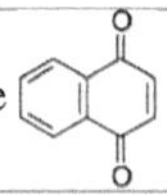	212	100
1-indanone 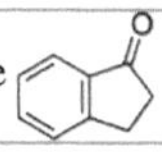	243	550
Perinaphthenone (phenalen-1-one) 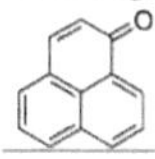	273	160
9-fluorenone 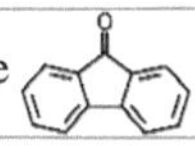	342	350
Anthraquinone (anthracene-9,10-dione) 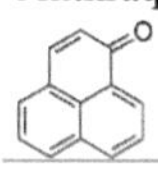	380	120
Pyrene-1-carbaldehyde 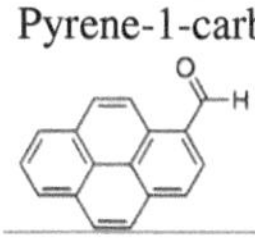		160

in the way in which the biomass decomposition products are released and burn from the reacting solid source.

Wood that burns in an open crib or in wild fires do so with a considerable amount of excess air at least at the flame extremity because they are not enclosed. The temperatures in a wood burning stove are much less. In terms of mass, PAH makes the most important contribution to emissions, but there are also significant amounts of phenols, phenoxy compounds and alkyl PAH.

In Table 3, the 'phenols' were found to consist of 33% phenol, 39% cresols (o, m and p) and 10% dimethyl phenols. The amounts of the emitted PAH and alkyl PAH combustion products follow that of the pyrolysis products, but at lower concentrations. This suggests either that they result from 'slippage' or that the combustion products are in partial equilibrium. The term O-PAC includes the polyaromatic

Table 5 O-PAH in urban air particulates (ng/m^3) and EFs for wood stove particulates (mg/kg) [38]

Species	Urban air particulates [48]	Afghanistan [46]	EFs [13] mg/kg
9-Fluorenone	0.05	0.27	0.84
Phenanthrene-9-carboxaldehyde	0.06		
10-ethylanthracene-9-carboxaldehyde	0.05		
Benz[*a*]anthracene-7,12-dione	0.19	0.07	
Anthraquinone	0.21	0.46	0.61
2-methyl anthraquinone	0.20	0.36	
Benzanthrone	1.01	6.3	0.67
Naphthacenequinone	0.14		

compounds as well as the O-PAH. Table 4 gives emission factors for O-PAH in flue gas from the same wood stove as in Fig. 3, (μg/MJ) [38].

The O-PAH found in the flue gases from this unit forms about 5% of the total PAH. Table 5 gives data obtained for O-PAH on particulates in urban air in a UK city [48, 49] and in rural Afghanistan [46] are shown in Table 5, together with emission factors (EF) from a wood burning stove in China (Shen et al. [38]). Some of the pollutants are clearly from wood burning, and others (aldehydes) are from diesel fuelled vehicles.

It is interesting to note that fast pyrolysis and the analysis of bio-oils produced by Stas et al. [50] give O-PAH compounds, but they are different to those emitted in the flue gases. The pollutants found in the environment (Table 5) are also different to those found in bio-oil. There are differences also between the total O-PAH emitted from a wood stove and those present on air particulates, presumably due to the volatility of the lower O-PAH species. The O-PAH compounds observed in the combustion products of biomass, especially those on the particulates cover only a small fraction of the range produced-small volatile molecules are not collected and the larger species exist as organic aerosols. The O-PAH species formed by combustion can be classified into four broad groups (but with considerable overlap) which are set out below.

Group 1. Phenolic Products

Lignin is a major polymeric component of biomass which decomposes in a complex manner as a result of differences in the thermal stability of its oxygen functional groups. Early decomposition products are guaiacol and syringol derivatives followed by highly substituted phenols which lose alkyl substituents [51].

A number of phenols, 30 monohydric, 14 dihydric and 44 methoxy and dimethoxy phenols have been identified in bio-oil [50, 51] and can be considered as the primary products input to the biomass combustion process. Lignin decomposes in the temperature range 240–500 °C, and the process is complex [51–53]. These early products are shown in Fig. 8 and can react further giving small furans and PAH; some of these have five-membered rings. The products will depend on the pyrolysis time [52, 53].

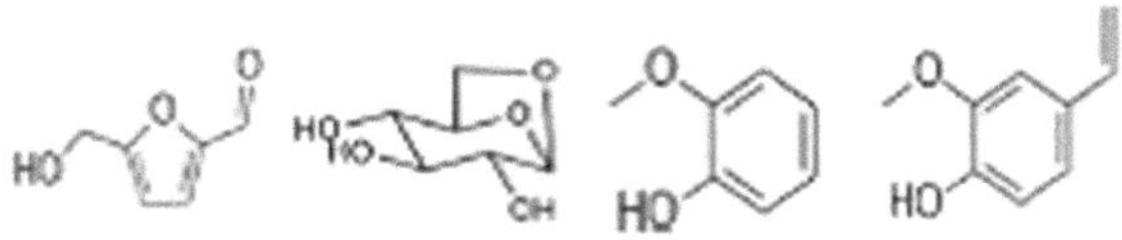

Fig. 8 Pyrolysis of pine as a function of pyrolysis time at 500 °C (increasing from left to right): hydroxyfurfural; levoglucosan; methoxyphenol; 4-vinyl-2-methoxyphenol

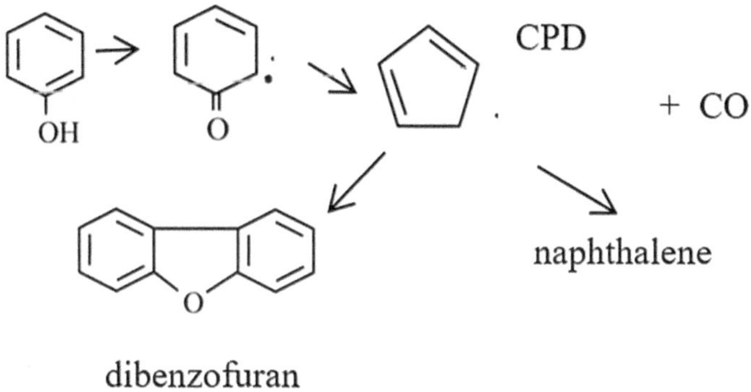

Fig. 9 Outline of reaction paths from phenol via CPD to naphthalene or to dibenzofuran

Fig. 10 Methylbenzo[*b*]naphtho[1,2-d]furan,

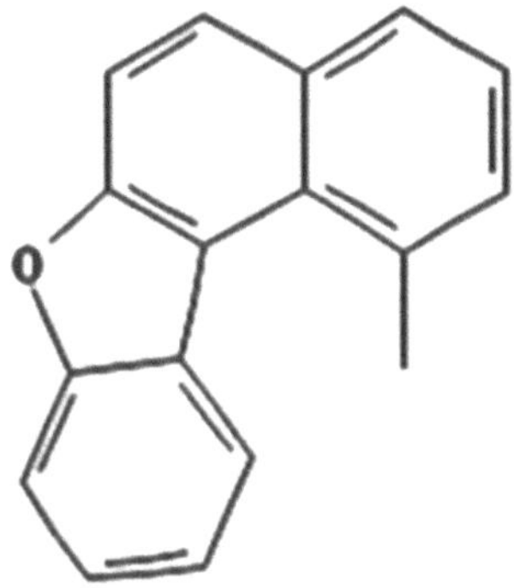

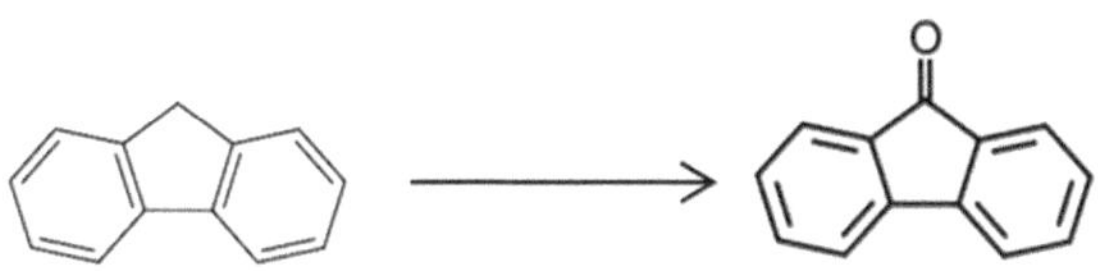

fluorene [ΔHf(g)= 177 kJ/mol] 9-fluorenone [ΔHf (g) 52 kJ/mol]

Fig. 11 Oxidation of a 5-ring PAH species to a ketone

Detailed analyses have been made of the fission of the bridges in lignin leading both to small molecules and to heavier condensates the tars [52]. Many of these products are found in the emission products from the combustion of biomass, for example [5, 17, 39].

Group 2. Compounds derived from phenols

The phenols produced from the initial decomposition step of lignin pyrolysis can themselves react further with other phenolic products or can react further producing small furans and PAH compounds; some of these have five-membered rings [54]. The most important biomass products are then a series of PAH some of which become oxidized forming compounds such as 1-indanone, 9-fluorenone, benzofluorene-ones and larger O-PAH formed by the decomposition of lignin. In addition, there are a range of cyclic compounds, but these are treated as a special case and mainly discussed in Group 3.

An example of a product from lignin pyrolysis is eugenol [41, 43]. This in turn undergoes thermal decomposition giving a range of oxygenates, some with a single ring but also some are O-PAH compounds [43] which in turn decompose reaching a maximum at about 700 °C. The main phenolic product of thermal decomposition is 2-methoxy-4-methylphenol [41, 43]. Progressive degradation of the side chain of eugenol produces simpler phenols which are sources of substituted CPDyl compounds (C_5H_5). The most important biomass products are then a series of PAH some of which become oxidized forming compounds such as 1-indanone, 9-fluorenone (see Tables 2 and 3), benzofluorenones and larger O-PAH formed by the decomposition of lignin.

A number of studies have been made of the rapid pyrolysis of a variety of the phenols produced by the pyrolysis of biomass [54–56]. It has been found that phenol produces, amongst other PAH species, dibenzofuran, m-cresol produces 1,2-dimethoxydibenzofuran, guaiacol produces benzofuran; pyrocatechol produces 1-dihydroindanone; and syringol produces related polymethoxyphenols and ethyl phenol-5-methylbenzofuran and analogues.

Phenols initially lose H atoms mainly by abstraction processes, and the phenoxy groups so formed rearrange to form cyclic-O compounds (such as furans) by condensation reactions or phenols decompose with elimination of CO to form cyclopentadiene (CPDyl) with subsequent reaction [54]. The is cyclohexadiene from which are derived CPDyl structures; these in turn isomerise to naphthalenes and indenes which are the source of a number of O-PAH compounds such as dibenzofuran, (see Tables 4 and 5), alkylated dibenzofurans and also benzonaphthofurans. But this route also leads to other larger PAH species and the formation of soot precursors [20, 22, 23].

Larger cyclic-O compounds are formed by condensation reactions of phenols, e.g. methylbenzo[*b*]naphtho[1,2-d]furan, which can be formed by the interaction of phenolic products [55].

Kojima et al. suggested [55] that the ortho position of phenol is important in the formation of O-PAH and that such phenols are abundant in biomass emissions.

Group 3. O-PAH

As discussed in the earlier section, a group of oxygen-containing PAH represented by benzofuran, dibenzofuran and benzonaphthofuran and their substituted derivatives can be formed by reactions of phenoxy in which CO is not eliminated. Thus, the model substance for lignin, guaiacol [53, 56] yields benzofuran. These can also arise from the formation of PAH compounds formed by HACA [20, 24]; (ii) through cyclopentadiene (CPD) generated by decomposition of phenoxy radical; but with further reaction with naphthalene and indene to form larger PAH which may be further condensed with CPD, cyclodehydrogenation and ring closure to yield large PAH. After early growth, these give flat sheets of PAH. The routes can also lead to curved PAH with cyclopenta groups on the periphery.

Group 4. PAH aldehydes and ketones

It is clear the aldehydes are formed during the combustion of wood (Table 2) but is also abundant in the atmosphere (Table 5) because of the contribution of emissions from vehicles, particularly diesel vehicles. Diesel vehicles fuelled by conventional or biodiesel fuels emit many lower aldehydes as well. In the case of diesel emissions, these are derived either from unburned fuel (alkyl derivatives) or PAH originating as combustion products in the aromatic fraction of diesel fuels which may contain two to three PAH with short alkyl side chains [33]. Not all such alkyl compounds are burned, see for example Table 3 which lists some of the alkyl anthracene/phenanthrene, alkyl fluoranthene, pyrene carboxyaldehydes and benzofluorenones of urban air particulates collected during a pollution episode. The routes leading to their formation are complex because they can involve atmospheric oxidation.

The ketones are formed by the oxidation of PAH which contain 5-membered rings, e.g. PAH ketones such as 9-fluorenone, naphthalenone and cyclopentaphenanthrenone, or alkyl PAH. Taken with the presence of ketone derivatives of 5-membered ring PAHs, such as the parent [Shen et al. 37, 38] and numerous 9-fluorenone derivatives found in urban air, examples such as those in Table 5, provide abundant evidence for the participation of oxidized hydrocarbons in biomass combustion emission chemistry, including soot formation. Table 5 compares concentrations of some representative O-PAH levels of the order of 0.01 ngm^{-3} are accessible to quantitation with these methods. A typical route leading to formation is by oxidation.

The ratio of the O-PAH to the PAH present in the combustion products is determined by the thermodynamic stabilities of these compounds; these can act as precursors of the oxygenated involatile PAH species observed in the atmosphere. Five-membered rings in the precursors can then lead to pentagonal-curved network regions and hence to bowl-shaped structures (so-called protofullerenes) and curved carbon atom sheets consistent with the observed morphology of the soot. Some of these O-PAH are the analogues of the 'protographene' PAH such as benzo[*ghi*]perylene which act as 'dead-end' PAH and terminate HACA reaction sequences.

5 Toxicity of PAH and O-PAH Compounds

The toxicity of the combustion products from biomass burning is associated with the gas phase contribution and from the fine particles (<PM2.5) and the PAH compounds associated with it. The toxicity of the combustion products can be determined by the B[a]P-eq-based PAH concentration (see Polycyclic Aromatic Hydrocarbons Factsheet https://ec.europa.eu/jrc/sites/jrcsh/files/Factsheet%20PAH_0.pdf).

The presence of ketone and quinone derivatives of PAH may pose significant health hazards in addition to the carcinogenesis properties of some PAH [45, 57]. Indeed, O-PAH may be the key compounds in the toxicity of particulate matter; it has been suggested by Walgraeve et al. [45] and others that PAH quinones are more toxic than the corresponding PAH. Moreover, measurements on NIST Standard Reference Samples (SRMs) revealed [44] that levels of O-PAH similar to or greater than those of PAH, with emission factors for residential stoves 1–2 orders of magnitude greater than for open burning, Shen et al. [37, 38].

There are seasonal effects which are indicated by variations with time of the ratio O-PAC/parent PAH, and of vapour/particle phase equilibria [27]. The mode of action of quinones is thought to differ from that of PAC in general in terms of mutagenicity and carcinogenicity.

It follows that candidates must be sought for extending the EPA16 list to make it more relevant to the products of biomass combustion. Andersson et al. [6] have proposed that benzofurans and quinones are representative O-PAH and the 10 proposed [6] are widespread in the environment. Other oxygen compounds in emissions are the aldehyde and keto (quinone) derivatives of PAH and O-PAH resulting from the oxidation of alkyl 2–4 ring aromatics as previously discussed.

The most toxic PAH compounds are the nitro-PAH [58], but they are present in only small quantities in the atmosphere. It was shown [58] that in a location in China biomass burning was the main source of cresols, 2,6-dimethyl-4-nitrophenol, 4-nitrocatechol and methyl-nitro-catechol, but other major sources were coal combustion and vehicles.

5.1 Control of PAH and O-PAH Emissions

It is seen from the previous discussion that the routes leading to both smoke formation and to PAH are linked in the combustion processes, and to a lesser extent to O-PAH since this may also be formed post-combustion. Not only are there toxicity effects from the PAH family as discussed above, but these molecules can have an effect on the climate [4, 18, 19, 58]. The total PAH emitted together with O-PAH and the soot can be minimized by better combustion control. Thus, in the case of biomass burning in a pellet stove the amount of PAH formed is 0.1% of the soot formed [5] and it is likely, but there is a lack of evidence that the PAH emission is proportional to the soot emitted. The O-PAH is about 5% of the total PAH formed (see Table 4) but can increase by reactions in the atmosphere. The use of alcohol fuels in vehicles is advantageous because of the lower carbon content and lower soot forming tendency though PAH can be formed from ethanol [59].

In all cases, they can be further reduced by catalytic combustion, for example as described in reference [60].

6 Conclusions

An analysis has been made of the extent of PAH and the oxygenates (O-PAH and phenols) present in biomass combustion emissions in exhaust gases and in urban air by a number of techniques. It is difficult to compare the toxic impact from different sources because measurement techniques are different; products from engines are measured at a specific filter temperature whilst different filter temperatures are used for stationary units such as stoves or furnaces. Nevertheless, it is seen that PAH is linked to the formation of smoke (and CO) and is best controlled by optimizing the overall combustion process, and this would involve any staged combustion and the cooling of the combustion products. Oxygenated PAH products such as O-PAH could be decreased in this way in the exhaust gases, but additional reactions in the atmosphere may influence the atmospheric concentrations.

The choice of fuel can be advantageous depending on the application. In the case of bio-oils, there could be significant differences between raw and processed bio-oils, the final refined products of flash pyrolysis and alcohols, much depending on the carbon and oxygen content and the aromatic content. In the case of the combustion of solid fuels, the principal contributors are the numerous phenols liberated by pyrolysis of lignin and their secondary pyrolysis products, especially if they are only partially combusted.

Acknowledgements We acknowledge helpful discussions with our colleagues.

References

1. IARC Diesel engine exhaust carcinogens. The Lancet Oncology (online). 15 June 2012. https://doi.org/10.1016/S1470-2045(12)70280-2
2. Abdel-Shafy HI, Mansour MSM (2016) A review on polycyclic aromatic hydrocarbons: source, environmental impact, effect on human health and remediation. Egypt J Pet 25:107–123
3. Wu D, Li Q, Ding X, Sun J, Li D, Fu H et al (2018) Primary particulate matter emitted from heavy fuel and diesel oil combustion in a typical container ship: characteristics and toxicity. Environ Sci Technol 52:12943–12951
4. Samburova V, Connolly J, Gyawali M, Yatavelli RLN, Watts AC et al (2016) Polycyclic aromatic hydrocarbons in biomass-burning emissions and their contribution to light absorption and aerosol toxicity. Sci Total Environ 568:391–401

5. Orasche J, Seidel T, Hartmann H, Schnelle-Kreis J, Chow JC, Ruppert H, Zimmermann R (2012) Comparison of emissions from wood combustion. Part 1: emission factors and characteristics from different small-scale residential heating appliances considering particulate matter and polycyclic aromatic hydrocarbon (PAH)-related toxicological potential of particle-bound organic species. Energ Fuels 26:6695–6704
6. Andersson J, Achten TC (2015) A critical look at the 16 EPA PAHs. Polycycl Aromat Comp 35:143–146
7. World Energy Outlook (2013) © OECD/IEA, 2013, fig. 2.5, p 63
8. Williams A, Jones JM, Ma L, Pourkashanian M (2012) Pollutants from the combustion of solid biomass fuels. Prog Energy Combust Sci 38:113–137
9. Knothe G, Razon LF (2017) Biodiesel fuels. Prog Energ Combust Sci 58:36–59
10. Zabed H, Sahu JN, Suely A, Boyce AN, Faruq G (2017) Bioethanol production from renewable sources: current perspectives and technological progress. Renew Sustain Energ Rev 71:475–501
11. Wang M, Dewil R, Maniatis K, Wheeldon J, Tan T, Baeyens J, Fang Y (2019) Biomass-derived aviation fuels: challenges and perspective. Prog Energ Combust Sci 74:31–49
12. International Air Transport Association (2019) IATA Fact Sheet: Sustainable Aviation Fuels, May 2019
13. Sharifzadeh M, Sadeqzadeh M, Guo M, Borhani TN, Murthy Konda NVSN, Garcia MC, Wang L, Jason Hallett J, Shah NP (2019) The multi-scale challenges of biomass fast pyrolysis and bio-oil upgrading: review of the state of art and future research directions. Energ Combust Sci 71:1–80
14. Bartle KD, Fitzpatrick EM, Jones JM, Kubacki ML, Plant R, Pourkashanian M, Ross AB, Williams A (2011) The combustion of droplets of liquid fuels and biomass particles. Fuel 90:1113–1119
15. Tótha P, Ögrena Y, Sepmana A, Vikströma T, Grenc P, Wiinikkaa H (2019) Spray combustion of biomass fast pyrolysis oil: experiments and modeling. Fuel 237:589–591
16. Lea-Langton AR, Biller P, Ross AB (2013) Investigation of HTL derived micro-algal bio-crudes as substitutes for marine shipping fuels. In: The 3rd International conference on algal biomass, biofuels and bioproducts, Toronto, June 2013
17. Fitzpatrick EM, Ross AB, Bates J, Andrews GE, Jones JM, Phylaktou H, Pourkashanian M, Williams A (2007) Emission of oxygenated species from the combustion of Pine wood and its relation to soot formation. Proc Saf Environ Protect 85:430–440
18. Trostl J, Chuang WK, Gordon H, Heinritzi M, Yan C, Molteni U, Ahlm L, Frege C, Bianchi F et al (2016) The role of low-volatility organic compounds in initial particle growth in the atmosphere. Nature 533(7604):527–531
19. Barsanti KC, Kroll JH, Thornton JA (2017) Formation of low-volatility organic compounds in the atmosphere. Recent advancements and insights. J Phys Chem Lett 8:1503–1511
20. Frenklach M, Wang H (1994) In: Bockhorn H (ed) Soot formation in combustion: mechanisms and models. Springer, Berlin, pp 162–192
21. Colket MB, Hall RJ (1994) In: Bockhorn H (ed) Soot formation in combustion: mechanisms and models. Springer, Berlin, pp 442–470
22. Smedley JM, Williams A, Bartle KD (1992) A Mechanism for the formation of soot particles and soot deposits. Combust Flame 91:71–82
23. Wang Y, Chung SH (2019) Soot formation in laminar counterflow flames. Prog Energ Combust Sci 74:152–238
24. Johannsson KO, Head-Gordon MP, Schrader PE, Wilson, KR, Michelsen HA (2018) Resonance-stabilized hydrocarbon radical chain reactions may explain soot inception and growth. Science 361:997–1000
25. Wilson JM, Baeza-Romero MT, Jones JM, Pourkashanian M, Williams A, Lea-Langton AR, Ross AB, Bartle KD (2013) Soot formation from the combustion of biomass pyrolysis products and a hydrocarbon fuel, n-decane: an ATOFMS study. Energ Fuels 27:1668–1678
26. Bowal K, Martin JW, Misquitta AJ, Kraft M (2019) Ion induced soot nucleation using a new potential for curved aromatics. Combust Sci Tech 191:747–765

27. Muñoz M, Haag R, Honegger P, Zeyer K, Mohn J, Comte P, Czerwinski J, Heeb NV (2018) Co-formation and co-release of genotoxic PAHs, alkyl-PAHs and soot nanoparticles from gasoline direct injection vehicles. Atmos Environ 178:242–254
28. Czech H, Stengel B, Adam T, Sklorz M, Streibel T, Zimmermann R (2017) A chemometric investigation of aromatic emission profiles from a marine engine in comparison with residential wood combustion and road traffic: implications for source apportionment inside and outside sulphur emission control areas Atmos. Environ 167:212–222
29. Lea-Langton AR, Ross AB, Bartle KD, Andrews GE, Jones JM, Pourkashanian M, Williams A (2012) Low temperature PAH formation in diesel combustion. JAAP 103:119–125
30. Lea-Langton AR, Andrews GE, Bartle KD, Jones JM, Williams A (2015) Combustion and pyrolysis reactions of alkylated polycyclic aromatic compounds: the decomposition of 13C methylarenes in relation to diesel engine emissions. Fuel 158:719–724
31. Ravindra K, Sokhi R, Van Grieken R (2008) Atmospheric polycyclic aromatic hydrocarbons: source attribution, emission factors and regulation. Atmos Environ 42:2895–2921
32. Miller CA Characterizing Emissions from the Combustion of Biofuels U.S. Environmental Protection Agency Office of Research and Development National Risk Management Research Laboratory Air Pollution Prevention and Control Division Research Triangle Park, NC
33. Lea-Langton AR, Li H, Andrews GE (2008) Comparison of particulate PAH emissions for diesel, biodiesel and cooking oil using a heavy-duty DI diesel engine. SAE Technical Paper Series 2008-01-1811
34. Dandajeh HA, Talibi M, Ladommatos N, Hellier P (2019) Influence of combustion characteristics and fuel composition on exhaust PAHs in a compression ignition engine. Energies 12:2575
35. Ross AB, Jones JM, Chaiklangmuang S, Pourkashanian M, Williams A, Kubica K, Andersson JT, Kerst M, Danilhelka P, Bartle KD (2002) Measurements and prediction of the emission of pollutants from the combustion of coal and biomass in a fixed bed. Fuel 81:571–582
36. Czech H, Miersch T, Orasche J, Abbaszade G, Olli Sippula O, Tissari et al (2018) Chemical composition and speciation of particulate organic matter from modern residential small-scale wood combustion appliances. Sci Total Environ 612:636–648
37. Shen G, Tao S, Wang W, Yang Y, Ding J, Xue M, Min Y et al (2011) Emission of oxygenated polycyclic aromatic hydrocarbons from indoor solid fuel combustion. Environ Sci Technol 45:3459–3465
38. Shen G, Tao S, Wei S, Zhang Y, Wang R et al (2012) Emissions of parent, nitro, and oxygenated polycyclic aromatic hydrocarbons from residential wood combustion in rural China. Environ Sci Technol 46:8123–8130
39. Fitzpatrick EM, Jones JM, Pourkashanian M, Ross AB, Williams A, Bartle KD (2008) Mechanistic aspects of soot formation from the combustion of pine wood. Energ Fuels 22:3771–3718
40. Fitzpatrick EM, Jones JM, Pourkashanian M, Ross AB, Williams A, Bartle KD, Kubica K (2009) The mechanism of the formation of soot and other pollutants during the co-firing of coal and pine wood in a fixed-bed combustor. Fuel 88:2409–2417
41. Atiku FA, Lea-Langton AR, Bartle KD, Jones JM, Williams A, Burns I, Humphries G (2017) Some aspects of the mechanism of formation of smoke from the combustion of wood. Energ Fuels 31:1935–1944
42. Ranzi E, Debiagi PEA, Frassoldati AA (2017) Mathematical modeling of fast biomass pyrolysis and bio-oil formation: kinetic mechanism of biomass pyrolysis. ACS Sustain Chem Eng 5:2867–2881
43. Ledesma EB, Campos C, Cranmer DJ, Foytik BL, Ton MN et al (2013) Vapor-phase cracking of eugenol: distribution of tar products as functions of temperature and residence time. Energ Fuels 27:868–878
44. Layshock JA, Wilson G, Anderson KA (2010) Ketone and quinone-substituted polycyclic aromatic hydrocarbons in mussel tissue, sediment, urban dust, and diesel particulate matrices. Environ Toxicol Chem 29:2450–2460

45. Walgraeve C, Demeestere K, Dewulf J, Zimmermann R, Van Langenhovewingbows H (2010) Oxygenated polycyclic aromatic hydrocarbons in atmospheric particulate matter: molecular characterization and occurrence. Environ 44:1831–1846
46. Wingfors H, Hägglund L, Magnusson R (2011) Characterization of the size-distribution of aerosols and particle-bound content of oxygenated PAHs, PAHs, and n-alkanes in urban environments in Afghanistan. Atmos Environ 45:4360–4369
47. Zhou J, Zotter P, Bruns EA, Stefenelli G, Bhattu D, Brown S et al (2018) Particle-bound reactive oxygen species (PB-ROS) emissions and formation pathways in residential wood smoke under different combustion and aging conditions. Atmos Chem Phys 18:6985–7000
48. Lewis AC, Askey SA, Holden KM, Bartle KD, Pilling MJ (1997) On-line. LC-GC-and LC-GC-MS—multidimensional methods for the analysis of PAC in fuels, combustion emissions and atmospheric samples. J High Res Chromatogr 20:109–114
49. Lewis AC, Kupiszewska D, Bartle KD, Pilling MJ (1995) City centre concentrations of polycyclic aromatic hydrocarbons using supercritical fluid extraction. Atmos Environ 29:1531–1542
50. Staš M, Kubička D, Chudoba J, PospíšI M (2014) Overview of analytical methods used for chemical characterization of pyrolysis bio-oil. Energ Fuels 28:385–402
51. Pelucchi M, Cavallotti C, Cuoci A, Faravelli T, Frassoldati A, Ranzi E (2019) Detailed kinetics of substituted phenolic species in pyrolysis bio-oils. React Chem Eng 4:490–506
52. Zhang T, Li X, Qiao X, Zheng M, Guo L, Song W, Lin W (2016) Initial mechanisms for an overall behavior of lignin pyrolysis through large-scale ReaxFF molecular dynamics simulations. Energ Fuels 30:3140–3150
53. Maduskar S, Facas GC, Papageorgiou C, Williams CL (2018) Five rules for measuring biomass pyrolysis rates: pulse heated analysis of solid reaction kinetics of ignocellulosic biomass. ACS Sustain Chem. Eng. 6:1387–1399
54. Akazawa M, Kojima Y, Kato Y (2015) Formation of polycyclic compounds from phenols by fast pyrolysis. EC Agric 1:67–85
55. Kojima Y, Inazu K, Hisamatsu Y, Okochi H, Baba T, Nagoya T (2010) Comparison of PAHs, nitro-PAHs and oxy-PAHs associated with airborne particulate matter at roadside and urban background sites in downtown Tokyo, Japan. Polycycl Aromat Comp 30:321–333
56. Liu C, Ye L, Yuan W, Zhang Y, Zou J, Yang J et al (2018) Investigation on pyrolysis mechanism of guaiacol as lignin model compound at atmospheric pressure. Fuel 232:632–638
57. Barradas-Gimate A, Murillo-Tovar MA, Díaz-Torres JJ, Hernández-Mena L, Saldarriaga-Noreña H et al (2017) Occurrence and potential sources of quinones associated with PM2.5 in Guadalajara, Mexico. Atmosphere 8:140
58. Li M, Wang X, Lu C, Li R, Zhang J, Dong S, Yang L, Xue L, Chen J, Wang W (2020) Nitrated phenols and the phenolic precursors in the atmosphere in urban Jinan, China. Sci Total Environ 714:13670
59. Viteri F, López A, Millera Á et al (2019) Influence of temperature and gas residence time on the formation of polycyclic aromatic hydrocarbons (PAH) during the pyrolysis of ethanol. Fuel 236:820–828
60. Soufi J, Pitault I, Gelin P, Meille V, Vanoye L et al (2016) Catalytic combustion of polycyclic aromatic hydrocarbons (PAH) over zeolite type catalysts: effect of Si/Al ratio, structure and acidity. In: French conference on catalysis (FCCat), May 2016, Frejus, France

Review of Biomass Energy Resources with Livestock Manure

Osama M. Selim, Juan Espindola, and Ryoichi S. Amano

Abstract This paper presents different thermochemical conversion processes on different biomass species. The results show it is recommended to use the lowest heating rate to allow a quasi-equilibrium state through slow heating, hence avoiding measurement errors. Thermal degradation of the three main components of the chicken manure was obtained. The initial results show that for the slow heating rates, 5 °C/min, the thermal degradation of the cow manure is different compared to that one obtained from chicken manure. The Hemicellulose decomposition took place at 250 and 300 °C for the chicken manure and cow manure, respectively. The Cellulose decomposition was started at 300 °C for chicken manure and 470 °C for cow manure. Rice husk, the peak of the Pyrolysis of rice husk, is less distributed than the Pyrolysis of chicken manure due to the absence of Hemicellulose and less Lignin content. For CO_2 gasification, the chemical reactions, for all different heating rates tested, were endothermic. Consequently, the energy must be supplied in terms of heating to sustain the reaction, while air gasification was exothermic, which means that the reaction can be sustained without external heating where the self-ignition was observed between 450 and 600 °C. In addition, it was observed that carbon dioxide had the most complicated mechanism with four stages. CO-Pyrolysis, results show the 40% RH-60%CH decreasing the activation energy by 12% compared to Chicken manure. In addition, an increase in the mass conversion by more than 3% was achieved. The 40% CM-60% CH shows a positive result in terms of keeping an exothermic reaction over the co-Pyrolysis process.

Keywords Thermochemical · Pyrolysis · Gasification · Co-Pyrolysis · Chicken manure · Cow manure · Rice husk

---

O. M. Selim · J. Espindola · R. S. Amano (✉)
University of Wisconsin Milwaukee, Milwaukee, WI 53211, USA
e-mail: amano@uwm.edu

A. Gupta et al. (eds.), *Advances in Energy and Combustion*, Green Energy and Technology,
https://doi.org/10.1007/978-981-16-2648-7_6

1 Introduction

Energy consumption in the world is predominantly from fossil fuels for transportation and electricity generation [1]. Fossil fuels cause problems such as air pollution, greenhouse gases, and environmental issues from the extraction. Also, high price fluctuation and possible fuel depletion are threats to the fossil-fuel-based economy. Furthermore, society is turning into energy consumption monitoring and low carbon practices [2]. Consequently, interest in alternative and renewable energy sources has been heightened. In 2020, the generation of electricity was 19% from renewable sources, and it is expected to take 36% of the share by 2050 [1]. Biomass raises as one of the alternatives to meet the increasing energy demand because it is abundant, carbon-neutral, and clean. Biomass is defined an organic material that comes from living organisms and can be converted to energy. Biomass can be classified into first-, second-, and third-generation biomass. The first-generation biomass converts carbohydrates and starches into biofuel (e.g., corn). The second generation includes agricultural, animal waste, and wood, while the third generation refers to algae [3]. To convert biomass into usable bioproducts, three main technologies can be used: (1) Mechanical conversion, (2) biochemical conversion and, (3) thermochemical conversion [4, 5]. In mechanical conversion, mechanical forces reduce the size of the biomass. This step usually prepares the fuel for combustion or another thermochemical process. The biochemical process uses microorganisms to degrade biomass. Among the technologies used for biological decomposition are anaerobic digestion and hydrolytic fermentation. Anaerobic digestion produces mainly methane, while fermentation produces ethanol [6, 7]. Thermochemical process uses heat to promote chemical reactions that decompose the organic matter in the biomass, yielding bio-oils and/or Syngas (depending on the temperature and atmosphere) [8]. Pyrolysis uses an inert atmosphere such as (nitrogen or argon) to decompose the biomass in a reactor at elevated temperatures, the products of Pyrolysis are bio-oils, non-condensable gases, and biochar [9], while gasification uses controlled amounts of oxygen to partially oxidize the fuel, producing Syngas and ashes [9].

There are three main technologies for Pyrolysis reactors, batch reactors, moving bed, reactors with movement caused by mechanical forces, and reactors when the movement I caused by fluid flow (fluidized bed) [10]. The gas yield depends on several factors: (1) type of biomass, (2) particle size, (3) heating rate, and (4) temperature and residence time [11]. This study aims to review recent advances in the Pyrolysis of feedstock from the perspective of the kinetics and product yields.

1.1 Pyrolysis Process

Pyrolysis is a thermochemical process that decomposes the livestock in the absence of oxygen. Pyrolysis happens at medium temperatures (300–800 °C) or high temperatures (800–1300 °C) [5]. Pyrolysis products are (1) bio-oil (tar), (2) non-condensable gases (or Syngas) that is a mix of different gases such as H_2, CH_4, CO, C_2H_X, and C_3Hx, and (3) char. The yield of each of these products depends on several factors, such as the kind of feedstock, temperatures, heating rates, and residence time [11, 12]. According to the operating conditions, Pyrolysis can be classified as slow, fast, or flash Pyrolysis [13].

Slow Pyrolysis uses low heating rates (usually less than 10 °C/s) and long residence times (from 5 to 30 min) [14]. Hussein showed that low hearing rate and high residence time promote the completion of chemical reactions [15], giving as a result higher yields of Syngas and biochar. Vieira et al. [16] optimized slow Pyrolysis parameters for a fixed bed reactor. Results showed that the optimal conditions were given for the longest residence time and the slowest heating rate (5 °C/min and 120 min).

Fast Pyrolysis uses heating rates higher than 100 °C/s and residence times from 0.5 to 2 s. Fast Pyrolysis has the advantage of low residence time, reducing the overall cost of the process [14, 17]. Besides, biochar yields are much lower in fast Pyrolysis compared with slow Pyrolysis reducing the solid handling. The main product of fast Pyrolysis is bio-oils.

Flash Pyrolysis uses heating rates greater than 500 °C/s and residence times of less than one second. This technology requires special reactor configuration and very small particle size to ensure temperature uniformity [18].

Abundant research compares the yields of each technology. Yuan [17] compares the biochar yields of fast and slow Pyrolysis of walnut shells, showing that the mechanism of biochar formations is similar in slow and fast Pyrolysis, but the yields of biochar in slow Pyrolysis are much higher than the yields of biochar in fast Pyrolysis.

Al Arni [19] studied the conversion of bio-oil from sugarcane bagasse, finding that slow Pyrolysis produces more Syngas and less bio-oil compared to fast Pyrolysis. Pyrolysis follows three main steps: drying, primary Pyrolysis, and secondary Pyrolysis [11, 20].

Drying

In the drying step, the moisture on the biomass is removed, and moisture in livestock varies between 5 and 20% [21]. Biomass drying is mainly done to reduce the volume of the biomass and increase the efficiency of the thermochemical process. Reed et al. [22] showed that high moisture reduces the value of the fuel. The removal of moisture for biomass is usually done by thermal drying. However, thermal drying can be an expensive process, such as fuel's value, and energy cost from drying must be a trade-off. Researchers have focused on modeling drying processes to optimize drying conditions for biomass [23, 24]. Studies conclude that small particle size and low heating rates are the most effective ways to dry.

Primary Pyrolysis

Primary Pyrolysis happens at temperatures between 250 and 600 °C [11]; the chemical reactions are associated with the thermal cracking of Hemicellulose, Cellulose, and Lignin. Gaur et al. [25] compiled the degradation for different kinds of biomass, including the independent three components (Hemicellulose, Cellulose, and Lignin) using thermal gravimetrical analysis (TGA). Gaur showed that Hemicellulose and Cellulose decompose at a faster rate than Lignin. Also, the peaks of degradation occur at temperatures between 270, 340, and 380 °C, respectively. Pyrolysis of biomass occurs in different sets of competitive reactions (such as char formation, depolymerization, or methanation) [26]. Yang [27] showed that the degradation of Hemicellulose and Lignin is highly exothermic for temperatures below 500 °C; after that temperature, the endothermic reactions are dominant, and the degradation of Hemicellulose and Lignin becomes highly endothermic. The opposite behavior was observed for Cellulose. For temperatures below 400 °C, the degradation of Cellulose was highly endothermic. However, after 400 °C the exothermic reactions dominate the decomposition of Cellulose.

Secondary Pyrolysis

Secondary Pyrolysis occurs at temperatures from 600 °C to 1,400 °C. This stage further decomposes the products (tars) from the primary Pyrolysis, yielding Syngas and char. Several publications [11, 15, 28] showed that hydrogen formation is promoted at higher temperatures, which increases the heating value of the Syngas. In addition, the char formed in the primary Pyrolysis can work as a catalyzer increasing the total degradation at higher temperatures.

1.2 Gasification

Gasification is an efficient thermochemical decomposition of solid carbonaceous material to produce gases of higher heating value compared to the solid precursor in the presence of gasifying agents such as steam, air, or CO_2. The major components of product gas are H_2, CO, CO_2, CH_4, and other higher hydrocarbons. Undesirable by-products containing a mixture of heavy aromatic hydrocarbon residues, referred to as tar, are also formed, which must be further pyrolyzed or reformed to enhance the Syngas yield. The factors that affect the composition of Syngas evolved from gasification are gasifying agent, temperature, and feedstock composition, and other reactor operational parameters affect the composition of Syngas evolved from gasification. Production of homogeneous fuel (Syngas as intermediate) from heterogeneous biomass is a combination of sources including chicken manure, poultry litter, biomass, and solid wastes, and coal makes it an attractive solution over direct combustion of these low-value heterogeneous feedstocks providing feed flexibility along with versatility. Abundant literature is available on the effects of using different catalysts and gasifiers, and favorable temperatures on biomass gasification [29–44]. Equations (1–5) show the major governing reactions (considering C to represent the biomass) that occur during gasification [45].

$$\text{Char Oxidation} : C + O_2 \Rightarrow \text{CO}_2 \tag{1}$$

$$\text{Boudouard Reaction}: CO_2 + C \rightleftharpoons 2CO \quad (2)$$

$$\text{Water} - \text{Gasshift Reaction}: CO + H_2O \rightleftharpoons CO_2 + H_2 \quad (3)$$

$$\text{Water} - \text{Gas Reaction}: +H_2O \rightleftharpoons CO + H_2 \quad (4)$$

$$\text{Thermal Cracking}: C_nH_m \Rightarrow C + C_xH_y + H_2 \quad (5)$$

Chicken droppings, waste beddings, waste food, and feathers from the coops form the major source from poultry litter. This biomass has high nitrogen content due to proteins and amino acids, high phosphorus, and ash content, making poultry litter a low-grade feedstock that is different from other conventional biomasses such as wood. Such differences endeavor the motivation for an investigation into the pyrolytic behavior of chicken litter [46]. One of these differences investigated from thermal gravimetric analysis (TGA) was the start of decomposition at lower temperatures compared to coal, wherein the maximum weight loss rate was observed at 643 K [47]. The TGA results on the Pyrolysis of chicken litter showed three stages of weight loss, unlike woodchip Pyrolysis, which occurs in two stages because of lignocellulose content. Three main peaks of chicken litter decomposition were observed during pyrolysis process which are corresponding to the decomposition of Cellulose and Hemicellulose with a peak at a temperature less than that of wood chips, followed by manure and Lignin decomposition and the last stage of residual char devolatilization. This is unlike TGA of woodchip Pyrolysis, which showed only two stages [46]. Higher manure content is assisted in the faster decomposition of waste litter. An increase in temperature increased Syngas yields and decreased liquid and char yields [48]. CO2 yields increased due to decarboxylation of mineral carbonates in the ash at temperatures above 973 K until a stable value above 1273 K [49]. Highly viscous bio-oils produced from fast Pyrolysis of chicken litter had less acidity and higher heating value compared to bio-oils from hardwood due to the decomposition of protein content in a chicken litter into hydrocarbons [50].

Although the high inorganic content in the chicken litter can provide fertilizing capabilities and catalytic activity in Syngas production, it causes handling difficulties due to the low fusion temperature of inorganics present since the chicken litter ash contains high amounts of Ca, Na, and K. Such high amounts of ash content (>20%) in a chicken litter with phosphorous and potassium oxides make it an efficient fertilizer even after gasification [51]. The gas evolution from inorganic matter in manure, such as decarboxylation of carbonates and the low melting points of its constituents, induces error into the modeling of product gases from high-temperature gasification and Pyrolysis if ash is considered inert as in conventional models [52]. Chicken litter gasification was demonstrated by blending with coal and found no change in the heating value of product gas [53]. The ultimate analysis showed enough oxygen present in a chicken litter to gasify its carbon contents; such a method was called “auto-gasification” [54]. This investigation also found ineffective availability of carbon to react with oxygen due to reactions with other hydrocarbons and water–gas reaction, causing oxidation of CO evolved to form CO_2. Use of catalysts in steam gasification of chicken manure in a fluidized bed at temperatures around 873 K produced Syngas rich in H_2 content and total gas yields with the addition of Ni-Al_2O_3 catalyst [55]. Supercritical gasification of chicken manure was also investigated [56, 57]. The abundant literature on biomass, although helpful in assessing chicken litter, concrete understanding of the differences with biomass, and effects of other gasifying parameters, is still lacking and much desired. Using steam as the gasifying agent is known to generate a higher-quality Syngas with high amounts of hydrogen content [58–60]. The high latent heat required for evaporation makes steam the most expensive gasifying agent, and thus long reaction time is undesirable. On the other hand, air, which is the cheapest agent, generates the lowest quality Syngas and fastest reaction rates. Using a mixture of air (oxygen) and steam as the gasifying agent provides the benefit of improved Syngas quality, while the exothermic reaction of air boosts the reaction rates. Steam can be injected at later stages in gasifiers [61] to reform the Syngas and improve its quality, or it can be mixed with air at the early stages of the gasification [62], to give even improved concentrations of hydrogen. The concentration of oxygen in the gasifying agent plays an important role in the quality of Syngas evolved and the rates of reaction. Thus, the oxygen concentration needs to be optimized to seek for the most feasible reaction conditions [63].

1.3 Co-pyrolysis and Co-gasification

Co-Pyrolysis and co-gasification are viable options to enhance the performance of the thermochemical processes. Depending on the biomass mixed (and the ratio), synergetic effects can promote the Syngas yield, the energy of activation, or degradation rate. Guo et al. [64] demonstrated that the extractives in singular biomass could improve the thermal degradation of the biomass components (Hemicellulose and Cellulose). Mallick [65] showed that the binary conversion of the biomass, depending on the ratio, can be considered as an additive equation.

Research has focused on mixing lignocellulosic biomass with coal [66, 67]. Results show that some biomasses can produce positive synergetic effects increasing the yield of volatiles without affecting the process. He [68] mixed cotton and coal, and results showed that the rate of decomposition increased for the co-Pyrolysis blends compared to the biomasses stand-alone.

Mixtures of lignocellulosic biomasses have been largely studied. Another common path is to mix lignocellulosic biomass with plastics. Burra et al. [69] studied the co-Pyrolysis of pinewood with plastic waste. Results show non-synergetic effects between the two biomasses. The best performance was obtained by mixing pinewood with BPC. Hossain et al. [70] studied the co-Pyrolysis of a solid tire with rice husk to show the possibility of getting liquid products that are comparable with petroleum fuels. However, these liquid products are only produced if the Pyrolysis conditions are correctly selected. Costa et al. [71] investigated mixing the rice husk with plastic waste using different pressures and residence time; an enhancement in the biomass conversion was observed. Also, the best performance was obtained for lower pressure, high temperatures, and short residence times, or low pressure, low temperature, and high residence time.

Co-gasification targets to promote reactions that yield hydrogen to increase the quality of the produced Syngas. Similarly, to Pyrolysis, mixtures of lignocellulosic biomasses have been largely studied. Ng et al. [72] studied air gasification of chicken manure and woody biomass, and an increase on the lower heating value (LHV) of the Syngas was accounted to the properties of chicken manure (softer, smaller size, and loosely packaged). Dayananda [73] experimentally studied the co-gasification of chicken manure and rice husk. Results show that the mixing ratio of 70% chicken manure and 30% rice husk produced the Syngas with the highest energy value. Seçer [74] studied the efficiency of co-gasification of biomass hydrolysates (products from hydrolysis), lignocellulosic biomass, and coal. Results conclude that biomass hydrolysates produce more hydrogen than the other biomasses because hydrolysis breaks down the structure of the biomass in smaller monomers, leading to higher Syngas yields. Co-gasification yields higher CO_2 due to the higher oxygen on biomass. However, this CO_2 is consumed on further reactions such as Boudouard reaction, which is later converted into hydrogen in other consecutive reactions (water–gas shift reaction).

The main objective of this paper is to explain the chemical kinetics of the thermochemical conversion process for different biomass species. Consequently, it helps in explaining and gives a better understanding of some of the complicated thermochemical processes, such as co-Pyrolysis. Due to the mass production of these biomass species (chicken manure, cow manure, and rice husk), the chemical kinetics of different thermochemical processes were studied.

2 Experimental Setup and Procedures

Figure 1 shows the differential thermal gravimetry (DTG) apparatus, which is used simultaneously to perform the thermal gravimetric analysis (TGA) and differential thermal analysis (DTA). The device is consisting of three main parts. The first part is the electric furnace, which supplies the required heat of reaction and can be controlled to generate different heating rates and then maintain the reaction at a constant desired temperature. The second part is the measurement system, which consists of rods that are each fitted with thermocouples; then, both rods are connected to a sensitive balance mechanism. One of the detector rods is used as a reference/control, and it carries an empty sample cell, while the other rod is used to measure the changes of mass and temperature of the active sample cell relative to the control cell. The third part of the device is the autosampler, which is a robotic system capable of loading and unloading the samples on the detectors automatically. The weight difference between the two detectors, obtained by the delicate balance, indicates the sample

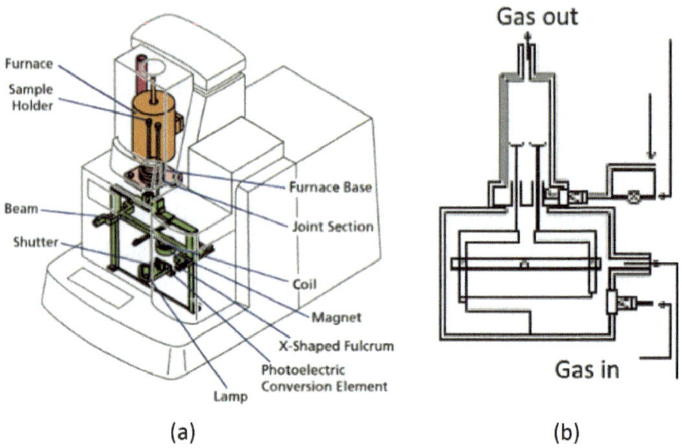

Fig. 1 Shimadzu DTG-60AH; **a** main device components and **b** gas flow through the device

weight. The difference between the voltage readings of the two sensors is measured for the differential thermal analysis. Any dry non-corrosive gas agent can be used with the device. The measurable range of the mass is ± 500 mg, with a resolution of 1 µg and ± 1% accuracy. The device can provide up to 1500 °C with ± 1 °C uncertainty. The thermocouples used here are Pt-10% pt/Rh thermocouples. The measurable range for the differential thermal analysis is 1000 µV with ± 1% µV accuracy and noise level less than 1 µV.(Table 1)

To minimize the effect of gas turbulence on the readings, a regular calibration was performed on the device all over the temperature range and operating gas flow rate. Alumina pans were used for all case studies. Equation () is used to calculate the extent of reacti6on, while its rate is calculated by using Eq. (7). The Arrhenius rate expression was used as the first step of any kinetics model [75]. To calculate

Table 1 Proximate and ultimate analysis of chicken manure and cow manure [76, 77]

	Chicken manure	Cow manure
Proximate analysis (wt.% dry)		
Volatile content	65.56	62
Ash content at 550 °C	21.65	23
Fixed carbon	12.8	15
Ultimate analysis (wt.% dry)		
Carbon	35.59	35.2
Hydrogen	4.57	3.1
Nitrogen	4.98	2.2
Sulfur	1.45	0.7
Oxygen	35.52	33.3
HHV (in MJ/kg)	13.15	16

Table 2 Kinetic parameters for nitrogen Pyrolysis with different heating rates [80]

β (°C /min)	n	Log(A/β)	Ea (kJ/mole)
5 (250–360 °C)	5	14.9	84
5 (>360 °C)	5	10.9	63.1
10 (250–360 °C)	5	15	88.1
10 (>360 °C)	5	11	65.1
15 (250–360 °C)	5	15.3	89.6
15 (>360 °C)	5	11.2	69.0
20 (250–360 °C)	5	15.6	93.2
20 (>360 °C)	5	11.4	72.3
25 (250–360 °C)	5	16	95.8
25 (>360 °C)	5	11.7	76.4
30 (250–360 °C)	5	16.5	96.5
30 (>360 °C)	5	11.8	80.9
35 (250–360 °C)	5	16.8	98.3
35 (>360 °C)	5	11.9	84.5
40 (250–360 °C)	5	17	87.8
40 (>360 °C)	5	11.9	99

the activation energy, the following procedure was followed; Eqs. (9) and (10) were combined to get Eq. (11); then by taking the natural log, Eqs. (9) through (12) are used to calculate the activation energy where Eqs. (9) and (10) are combined to get Eq. (11). Equation (12) is obtained by taking the natural log of Eq. (11), where the activation energy E_a and the exponent constant A can be obtained as shown in Tables 2, 3, 4, 5, and 6.

$$\alpha = \frac{w_0 - W_t}{w_0 - w_f} = \frac{v_t}{v_f} \tag{6}$$

Table 3 Kinetic parameters for air gasification with different heating rates [80]

β (°C/min)	n	Log(A/β)	Ea (kJ/mole)
5	3	11.2	70.3
10	3	10.1	66.7
15	3	10.1	64.5
20	3	10.9	70.8
25	3	10.3	67.6
30	3	10.9	70.3
35	3	10.3	67.9
40	3	10.2	69.0

Table 4 Kinetic parameters for carbon dioxide gasification with different heating rates [80]

β (°C /min)	n	Log(A/β)	Ea (kJ/mole)
5 (250–360 °C)	5	9.8	63.2
5 (360–630 °C)	5	4.3	39
5 (>630 °C)	5	72.5	575.5
10 (250–360 °C)	5	8.9	60.2
10 (360–630 °C)	5	5.0	38.7
10 (>630 °C)	5	68.1	523
15 (250–360 °C)	5	9.1	61.5
15 (360–630 °C)	5	3.5	32.9
15 (>630 °C)	5	68	556
20 (250–360 °C)	5	8.9	61.3
20 (360–630 °C)	5	3.8	34.8
20 (>630 °C)	5	67.3	551
25 (250–360 °C)	5	8.0	57.1
25 (360–630 °C)	5	3.0	305
25 (>630 °C)	5	66.5	549
30 (250–360 °C)	5	8.7	60.3
30 (360–630 °C)	5	3.0	33.3
30 (>630 °C)	5	68.1	561.5
35 (250–360 °C)	5	8.0	56.4
35 (360–630 °C)	5	2.9	32.6
35 (>630 °C)	5	63.6	544.7
40 (250–360 °C)	5	7.5	56.4
40 (360–630 °C)	5	3.7	37.6
40 (>630 °C)	5	62	519.7

Table 5 Biochar percentage for different heating rates and mix ratios [70]

%CM	% Biochar		
	5 °C/min	10 °C/min	15 °C/min
100	34	39.3	36
90	39	35	44
80	37	37	38
70	30	33	35
60	26	36	37
50	33	32	32

$$\frac{d\alpha}{dt} = \frac{(\alpha_t - \alpha_{t-1})}{\Delta t} \tag{7}$$

$$k(T) = A\exp\left(\frac{-E_a}{RT}\right) \tag{8}$$

$$\frac{\mathrm{d}\alpha}{\mathrm{d}T} = \frac{k(T)}{\beta} f(\alpha) = \frac{A}{\beta}\exp\left(\frac{-E_a}{RT}\right) f(\alpha) \tag{9}$$

$$\frac{\mathrm{d}\alpha}{\mathrm{d}T} = k(T)(1-\alpha)^n \tag{10}$$

$$\frac{\mathrm{d}\alpha}{\mathrm{d}T} = \frac{k(T)}{\beta} f(\alpha) = \frac{A}{\beta}\exp\left(\frac{-E_a}{RT}\right)(1-\alpha)^n \tag{11}$$

$$\ln\left(\frac{\frac{\mathrm{d}\alpha}{\mathrm{d}T}}{(1-\alpha)^n}\right) = \ln\left(\frac{A}{\beta}\right) - \left(\frac{E_a}{\mathrm{RT}}\right) \tag{12}$$

3 Operation Conditions

Chicken manure used in this study was a dried chicken manure fertilizer of irregular form, with sizes ranging from 1 to 3 mm. The apparent density of particles measured from fluid immersion was 1.5 $g.ml^{-1}$. The ultimate and proximate analysis, along with the higher heating value (HHV) [76] of the sample, is given in Table 1.

4 Results and Discussion

Three thermochemical processes were investigated in this paper: Pyrolysis, gasification, and co-Pyrolysis.

4.1 Pyrolysis of Chicken Manure

The thermal gravimetric and differential thermal analyses were conducted for the chicken manure using nitrogen as gas agent with eight different heating rates from room temperature up to 1000 °C. The three main components in the chicken manure are Hemicelluloses, Cellulose, and Lignin. The thermal degradation of Hemicelluloses is known to peak at 240 °C and Cellulose at 380 °C, while Lignin has more of a steady degradation with a small peak at high temperature 600–800 °C [78].

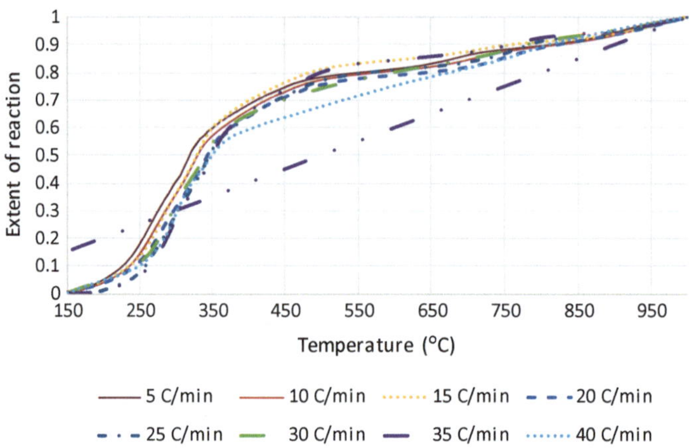

Fig. 2 Extent of the reaction of chicken manure at different heating rates [80]

Figure 2 shows the extent of reaction of chicken manure at different heating rates when nitrogen is used. It can be observed that all heating rates had the same trend with the increase of the temperature, and the extent of reaction increases. In addition, when the heating rate increased, the progress of the extent of reaction seems to be delayed to a higher temperature. Furthermore, the figure shows that the faster the heating rate, the more delayed is the progress of each reaction with respect to temperature. The main reason is that increasing the heating rate does not allow enough time for each reaction to be completed before increasing the temperature. Another plausible reason can be the response of the measuring device. This delay can be observed as well when using different gas agents. This kind of behavior was also observed for wood [79]. The Pyrolysis reaction was best represented by dividing the decomposition into three main stages. The first stage begins with the start of thermal cracking up to 250 °C, the second stage with a faster reaction between 250 and 350 °C, and finally, the third stage with a steady reaction from 350 °C to the end of the reaction. These three distinct stages were used to find the kinetics of the reaction. Starting from 350 °C and up to 700 °C, the thermal degradation is a complicated overlap between the three main components (Hemicellulose, Cellulose, and Lignin) decomposition, and the effect of heating rate on the extent of reaction is more significant compared to the other temperature ranges. At the end of each reaction, a residual mass that failed to react because of either insufficient activation energy or the absence of suitable reacting media was left over.

The rate of change of the extent of reaction of chicken manure with the temperature at different heating rates is shown in Fig. 3. Similar to Fig. 2, it can be observed that the trend is the same for all tested heating rates. At 150 °C for the low heating rates, all the moisture content has evaporated from the sample. For the highest two heating rates, 35 and 40 °C, two peaks take place near 150 °C due to the evaporation of any moisture in the sample. The main reason is that the heating rate is faster than the evaporation rate of the moisture content.

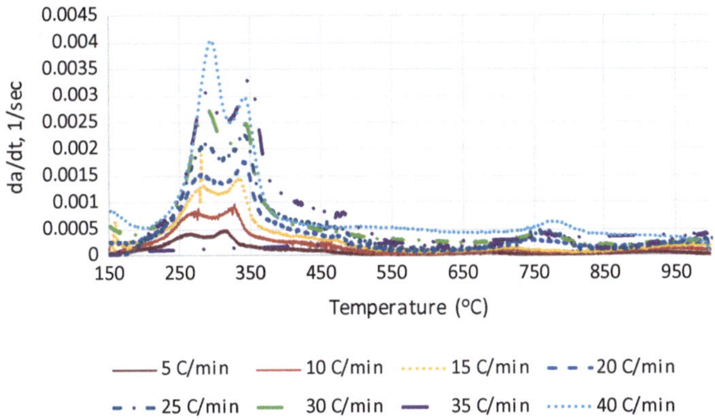

Fig. 3 Rate of Change of the extent of reaction with the temperature at different heating rates [80]

For all heating rates, the first two peaks are within the temperature range of 250–350 °C, and then the third peak started to occur at 670 °C for the lowest heating rates (5 °C /min).

It can be observed that the peak value is increased with the increase in the heating rate, as well as the temperature at which the peak occurs. Starting with 5 °C/min, the first peak took place at 255 °C, and after increasing the heating rate, the temperature at which peak occurs is increased as well to reach 298 °C for the heating rate of 40 °C/min. The same scenario for the second peak is 320, 325, 335, 345, and 343 °C for the heating rates 5, 15, 20, 30, and 40 °C /min, respectively. The three peaks are indications for the three main components: Hemicellulose, Cellulose, and Lignin, respectively. The values obtained in this study are matching the values obtained by Yang et al. for determining the degradation temperature range of these main three components [81].

Kinetics of Reaction: Table 2 shows the activation energy and the Arrhenius constant at different heating rates. For the temperature range from 250 to 360 °C, it can be observed that the average activation energy is 91 kJ/mole and the average value for $\log A = 14.5\ s^{-1}$. For the range greater than 360 °C, the average activation energy is 75.6 kJ/mole while $\log A = 10\ s^{-1}$.

Differential Thermal Analysis (DTA): Figure 4 shows the DTA variation with the temperature at different heating rates. For all the heating rates, it can be observed that the trend is the same for all the heating rates. The reaction is relatively steady except for a large peak downward for the faster heating rate cases by activating the endothermic reaction due to the quick breakdown of Cellulose and Hemicelluloses below 350 °C. With the increase in the heating rate, the magnitude of the peaks increases. The more massive peak is a result of a fast increase in the temperature of the control (empty) pan, while the active pan is enduring high rates of endothermic reactions hindering the sample temperature increase. Consequently, for 30 °C/min and higher, the trend looks different from the other cases.

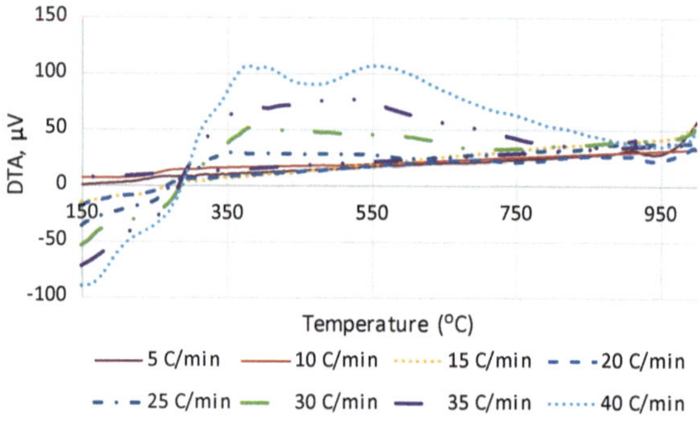

Fig. 4 Change of the DTA with the temperature at different heating rates [80]

4.2 *Pyrolysis of Cow Manure*

Figure 5 shows the extent of the reaction for the cow manure at eight different heating rates. It can be observed that all the heating rates have the same trend and increasing the heating rates responsible for delaying the extent of the reaction to a higher temperature. The Pyrolysis reaction was best presented by dividing the decomposition into four main stages. First stage is up to 255 °C, which represents the thermal cracking. A second faster reaction is from 255 to 315 °C. Then, from 315 to 745 °C represents the thermal degradation of the three main components: Hemicellulose, Cellulose, and Lignin.

The degradation temperature of the three main components can be obtained from Fig. 6. The first peak represents the Hemicellulose degradation. For all the heating rates, the first peak occurs at 321 °C. At 475 °C, the thermal degradation of the Cellulose takes place for the lower heating rates and 500 °C for the 40 °C/min. A plausible reason for the small fluctuation between 350 and 500 °C is the degradation of impurities such as bedding. Relatively, more minor bumps appear at 760 °C for 35

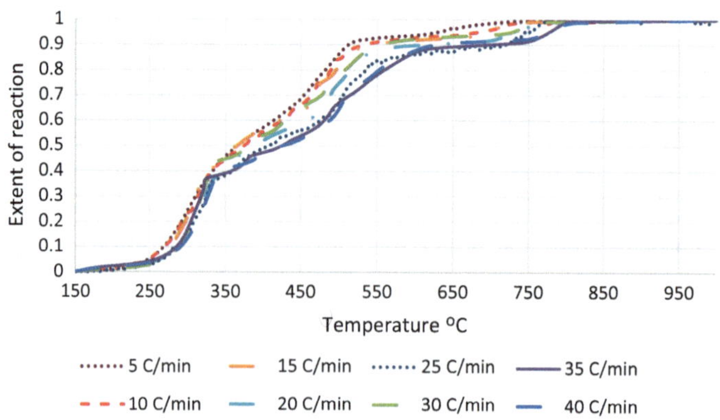

Fig. 5 Extent of reaction variation with temperature for cow manure at different heating rates [80]

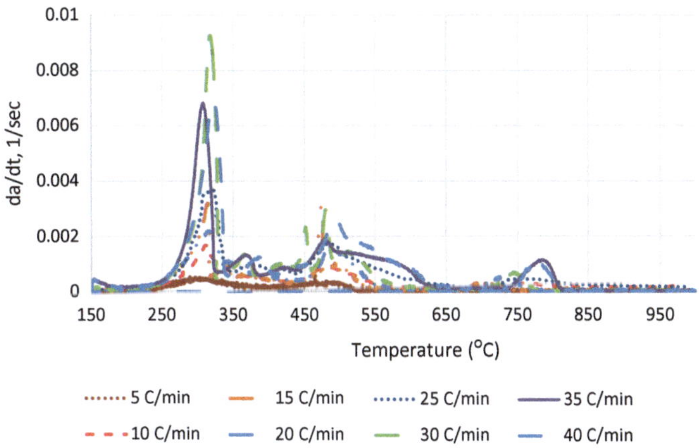

Fig. 6 Rate of change of the extent of reaction with the temperature for cow manure at different heating rates [80]

and 40 °C/min, which is due to the degradation of the low residuals of Lignin after combustion.

Figure 7 shows the DTA variation with the temperature at different heating rates. All heating rates are following the same trend. All the heating rates are exothermic, which means that the reaction is self-sustainable. For the higher heating rates, the sharp decline after the first peak implies that the reaction turns into endothermic. The main reason is that the faster heating rate does not allow enough time to thermal equilibrium besides the heat transfer limitation. Based on this conclusion and for figure clarity, the 5 °C/min is used to compare between the chicken manure and cow manure.

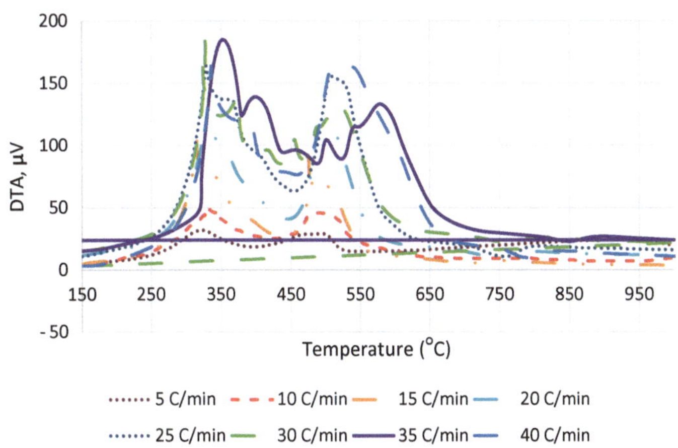

Fig. 7 Change of the DTA with the temperature for cow manure at different heating rates [80]

4.3 *Pyrolysis of Rice Husk*

Rice husk-chemical content consists of 50% of Cellulose, 25–30% Lignin, and 15–20% silica [25]. No peak below 270 °C is expected because the degradation peak of Hemicellulose is at temperatures between 240 and 270 °C [69]. The mass degradation of rice husk can be divided into three stages. The first stage is the same as that for chicken manure—the water is evaporated from 100 to 225 °C. The second stage starts around 250 °C and peaks at a temperature between 323 and 340 °C, thus corresponding to the degradation of the Cellulose. Approximately, 50% of the biomass is degraded in this stage. The peak of the Pyrolysis of rice husk is less distributed than the Pyrolysis of chicken manure due to the absence of Hemicellulose and less Lignin content. The third stage occurs within the 400–650 °C range. In this stage, the Lignin degrades. The percentage of Lignin in the studied rice husk prior to degradation is small because there is no peak at this stage. Also, the Lignin is more thermally stable than the Cellulose. Thus, degradation occurs at a decreased rate. The total conversion for this biomass is 60% wt, signifying that 40% of the biomass remained as biochar (see Fig. 8).

When the heating rate is increased, the peak of the degradation rate and the temperature (at which the peak occurs) increases (Fig. 9). This result is the same as obtained for the chicken manure case in Fig. 3. It can also be observed that the DTG curve has only one dominant peak, which takes place at 330, 340, and 350 °C, for the 5, 10, and 15 °C/min, respectively. A plausible reason for that result is that the rice husk has more Cellulose content than Hemicellulose. This explanation is not applicable in the case of chicken manure, where Hemicellulose and Cellulose percentages are closer to each other. When the heating rate was increased, the TG curves shifted slightly to the higher temperature side in the second stage but merged in the third stage. Other than that, no significant changes in degradation were observed.

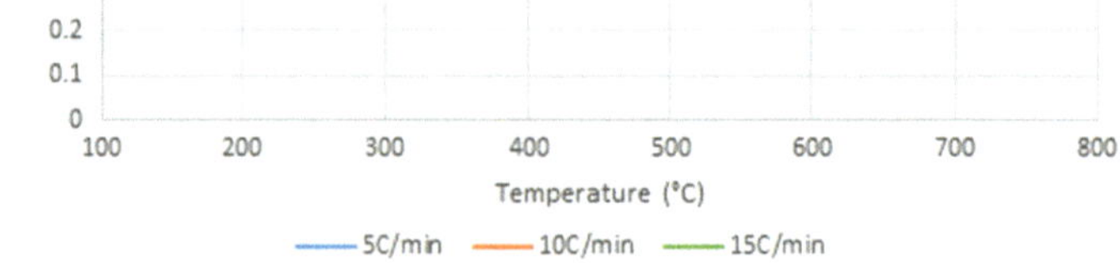

Fig. 8 TG curves for rice husk at different heating rates [70]

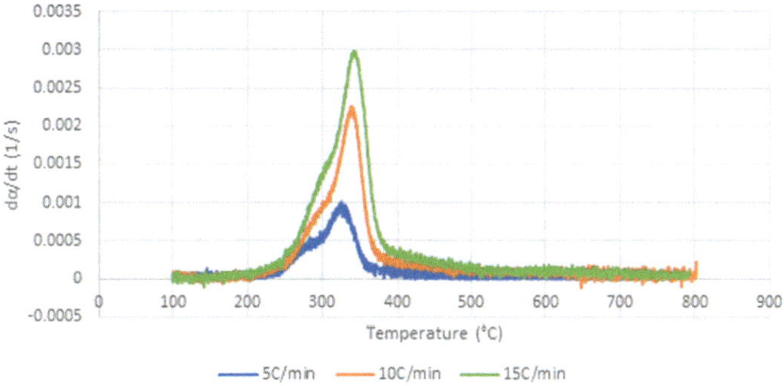

Fig. 9 DTG curves for rice husk at different heating rates [70]

4.4 Gasification of Chicken Manure

Air Gasification

In this case, the air was used as a gas agent. As a result, the chemical reactions were not only limited to the Pyrolysis process but also gasification to take place because of the incomplete combustion of the gases and fixed carbon in the presence of O_2. With the same as nitrogen Pyrolysis, the same parameters were studied at the same heating rates. The same remarkable results are obtained with case A, as seen in Fig. 10, where all different heating rates had the same trend. As the temperature increases, the extent of reaction increases. When the heating rate increased, the progress of the extent of reaction seems to be delayed to a higher temperature. For the gasification part, the reaction can be divided into five main stages. The first stage takes place, starting from the thermal cracking up to 250 °C. A faster second reaction takes place between 250 and 350 °C and then from 350 to 450 °C another fast stage with a mild slope compared

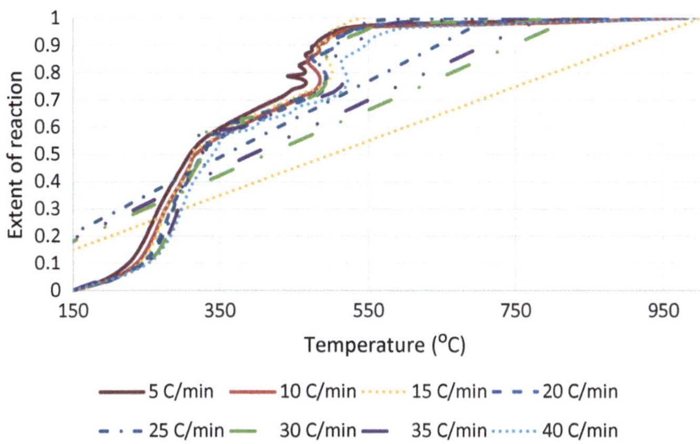

Fig. 10 Extent of the reaction of chicken manure at different heating rates [80]

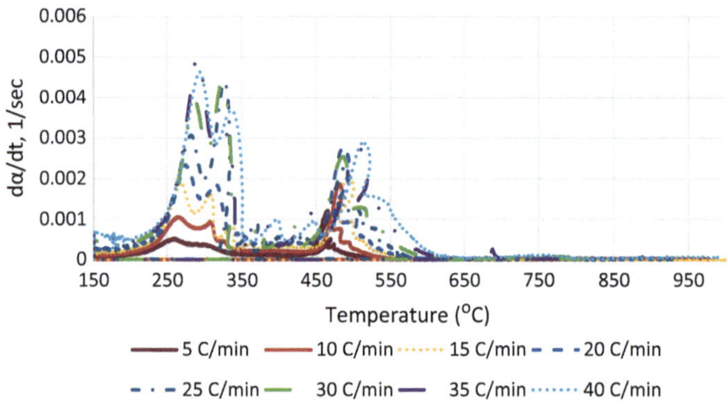

Fig. 11 Rate of change of the extent of reaction with the temperature at different heating rates [80]

to the previous region. The fluctuation in temperature between 450 and 600 °C is because of the self-ignition. Right after 600 °C, the extent of reaction is more than 95%, and the reaction tends to be very slow.

Again, three significant peaks can be shown in Fig. 11: 255, 340, and 500 °C. As expected, the magnitude of the peak is increased as the heating rate increased as well as the temperature at which the peak occurs. As shown in the figure, for 5 °C/min, the first peak occurs at 255 °C. For 10 °C/min, one can see that the first peak is shifted to the right (higher temperature) as well, and the peak value is higher. For 40 °C/min, the peak value is 0.005 s^{-1}, which is the highest peak, and it happened at 293 °C, and so on. The increase in the peak value is resulting from the faster change in temperature. The characteristics of hemicellulose and Cellulose can be seen on the first two peaks. Another indication of self-ignition can be seen from the temperature fluctuation between 350 and 450 °C, which was already observed in case A. A tiny bump appears at 680 °C for 35 and 40 °C/min, which is due to the degradation of the low residuals of Lignin after combustion.

Kinetics of reaction (TGA). The values for the Arrhenius equation constant and the activation energy are shown in Table 3. Even though the reaction appeared more complicated than the N_2 Pyrolysis case, single kinetic reaction constants were calculated for the whole conversion reaction where the average activation energy is 68.4 kJ/mole and the average $\log A = 9.3\ s^{-1}$.

Differential thermal analysis (DTA). In Fig. 12, the change of the DTA with the temperature at different heating rates is shown; the same peaks are formed at the same temperatures obtained from Figs. 10 and 11, and these peaks indicate the thermal degradation of the Hemicellulose, Cellulose, and self-ignition. The fact that these peaks are positive (upward) is due to the oxidation of some of the evolving gases rendering the reaction as exothermic. After 700 °C, no significant reaction was recorded, and the weight of the sample was stable.

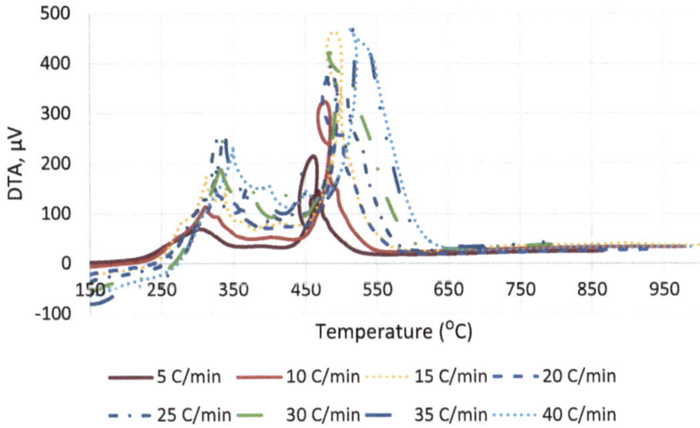

Fig. 12 Change of the DTA with the temperature at different heating rates [80]

Carbon Dioxide Gasification

The extent of reaction. Another gasification case was studied when CO_2 is used. In this case, as shown in Fig.13, the four stages of the gasification reaction are as follows: start of thermal cracking up to 250 °C, the faster reaction between 250 and 360 °C, steady reaction from 360 to 700 °C, and finally, a quick reaction at 700 °C to the end of the reaction. These four distinct stages were used to find the kinetics of the reaction.

In Fig.14, three significant peaks can be observed close to the following temperature: 250, 360, and 700–800 °C. The first two peaks are characteristics for Hemicellulose and Cellulose, while the third peak (700–800 °C) is an overlap between the peak for Lignin and the peak for the Boudouard reaction between fixed carbon and CO_2, respectively. The peak value increases with the heating rate due to the faster increase of the temperature with time.

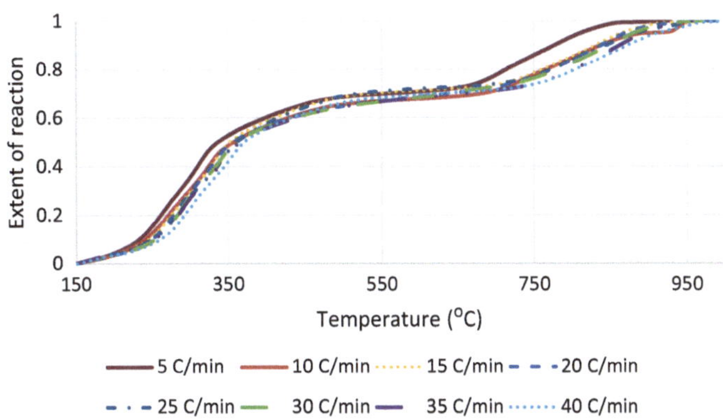

Fig. 13 Extent of the reaction of chicken manure at different heating rates [80]

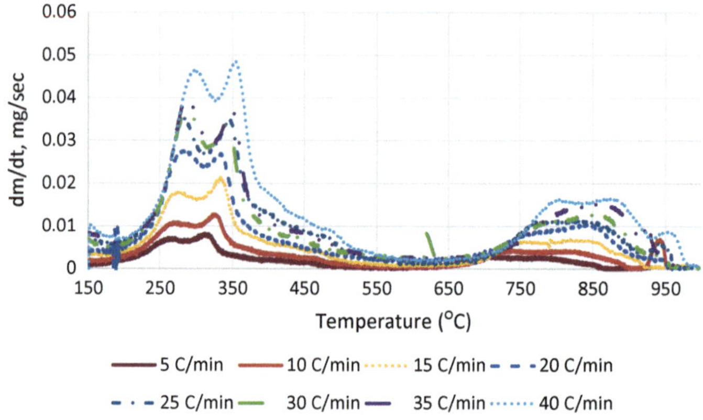

Fig. 14 Rate of change of the extent of reaction with the temperature at different heating rates [80]

Kinetics of reaction (TGA). Table 4 shows the values of the Arrhenius equation constant and the activation energy. For the range 250–360 °C, the average value for Ea = 59.6 kJ/mole and the average value for log $A = 9.8\ s^{-1}$; for the range 360–630 °C, the average value for Ea = 34.9 kJ/mole and the average value for log $A = 4.8\ s^{-1}$; and for > 630 °C, the average value for Ea = 547.6 kJ/mole and the average value for log $A = 68\ s^{-1}$.

As shown in Fig. 15, the reaction is relatively steady from 350 to 700 °C with an exothermic reaction where the DTA values are positive. The endothermic reactions took place below 350 °C due to the fast breakdown of Cellulose and Hemicelluloses, and another endothermic reaction of CO_2 with the fixed carbon in the chicken manure at a temperature greater than 700 °C. After 700 °C, when the heating rate increases, the curve started to drop sharply, which means that the reaction is endothermic. Also, the magnitude of the peak increases. Comparing the different heating rates,

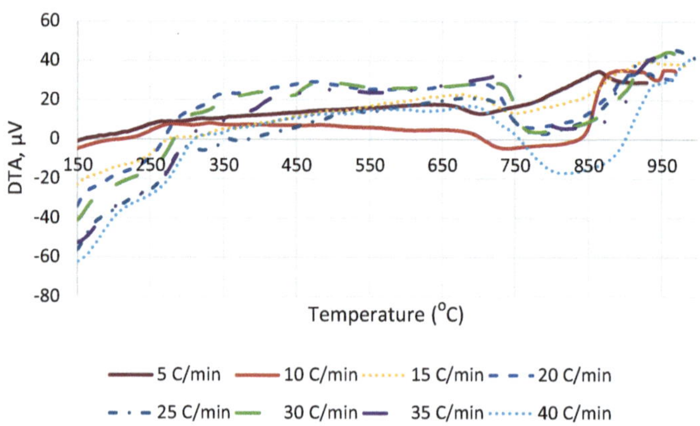

Fig. 15 Change of the DTA with the temperature at different heating rates [80]

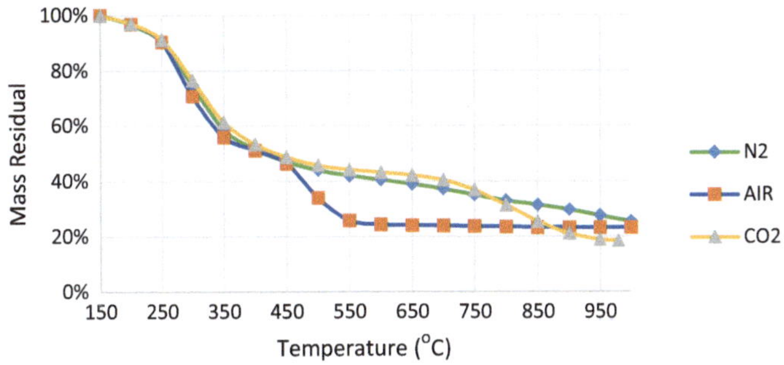

Fig. 16 Mass residual change versus temperature for the three tested gas agents [80]

after 700 °C, one can observe that, for 5 °C/min, the reaction is exothermic, where the curve goes up, while for the 40 °C/min, it sharply goes down. The faster rate of change in temperature does not allow enough time for thermal equilibrium between the reference and sample, and thus the peak value appears larger.

In terms of mass residual, Fig.16 shows the average mass residual for the three gas agents. The average residual mass in the case of N2 Pyrolysis is 25% of the total mass, and it is independent of the heating rate, while for the air gasification and CO_2 gasification, the average residual mass is 23% and 19%, respectively.

4.5 *Co-pyrolysis*

Co-pyrolysis of chicken and cow manure

The two livestock manures have been mixed at four different ratios: 20, 40, 60, and 80% cow manure of the total mixture weight. The same three parameters are used to give a better idea of the chemical kinetics for the mixture of thermal degradation. Starting with the extent of reaction, Fig. 17 shows that the 40 and 60% cow manures give a slower weight consumption compared to 20 and 80% cow manures. Figure 18 shows that the extent of reaction rate is increasing with increase in the cow manure percentage, which was expected as it approaches the cow manure characteristic to stand alone.

Regarding the DTA, from Fig. 19 we see that all the tested samples are exothermic except for the 60% cow manure, whose reaction turns into an endothermic reaction after 620 °C, which means further energy is needed to complete the Pyrolysis process. There are several reasons for the inconsistency in the trend with 60% cow manure in comparison with other cases. The first one could be the bedding of cow manure (some wood straws). Another reason, as you can see, increasing the cow manure percentage makes the reaction to be endothermic, and this is mainly because of the protein and amino acids that need a large amount of energy to be lost. In general, the thermal degradation of the cow manure is hard to predict because of the variation of

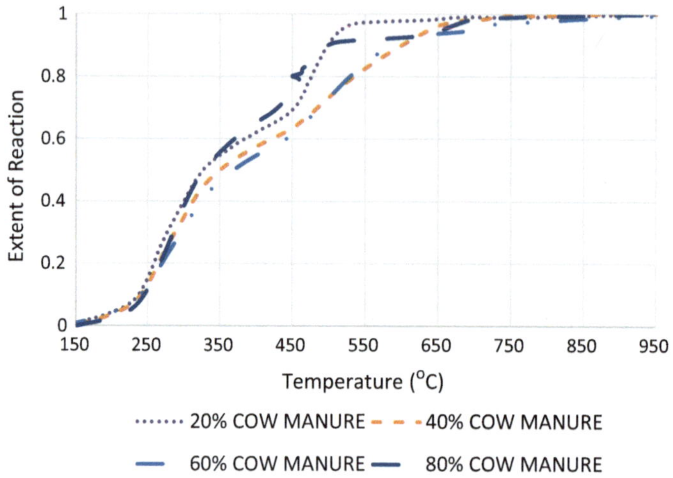

Fig. 17 Extent of reaction variation with the temperature at different cow manure percentages [25]

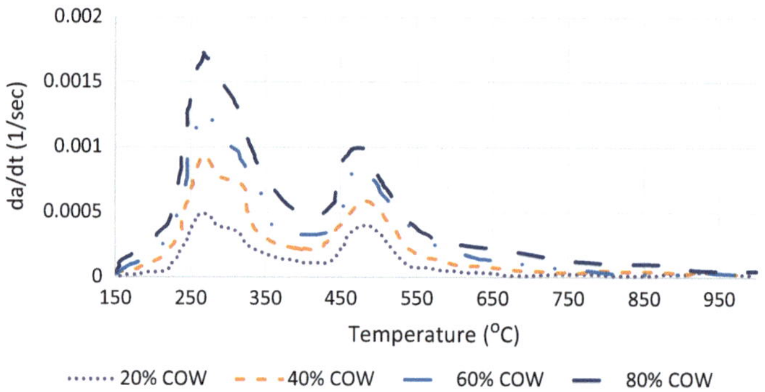

Fig. 18 Rate of change of the extent of reaction with the temperature at different cow manure percentage [25]

its components. All of these reasons could participate in this discrepancy. Overall, it can be observed that the 40% cow manure gives a stable exothermic reaction to the end of the process.

Co-pyrolysis of Chicken Manure and Rice Husk

For the biomass blends, the TG curves for the mixtures are shown in Fig. 20. For clarity, only the heating rate with the most noticeable effect (5 °C/min) is shown. However, it is worth mentioning that all the heating rates had the same effect on biomass degradation. The first stage occurs at temperatures between 100 and 220 °C, which corresponds to the humidity extraction. The humidity content of the blends was 10%wt. The second stage is given at temperatures between 250 and 360 °C. In this

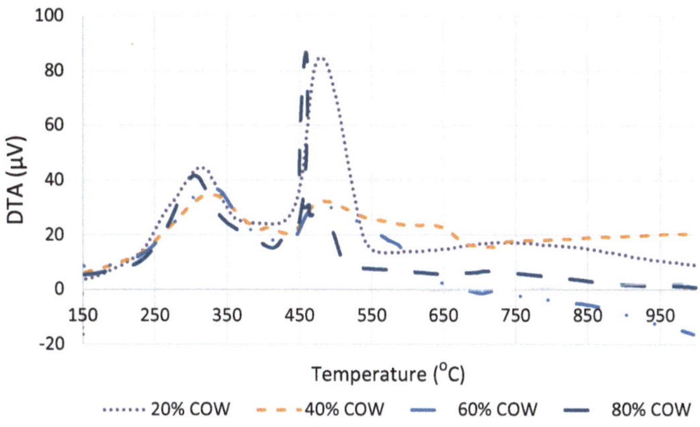

Fig. 19 Change of the DTA with the temperature at different cow manure percentages [25]

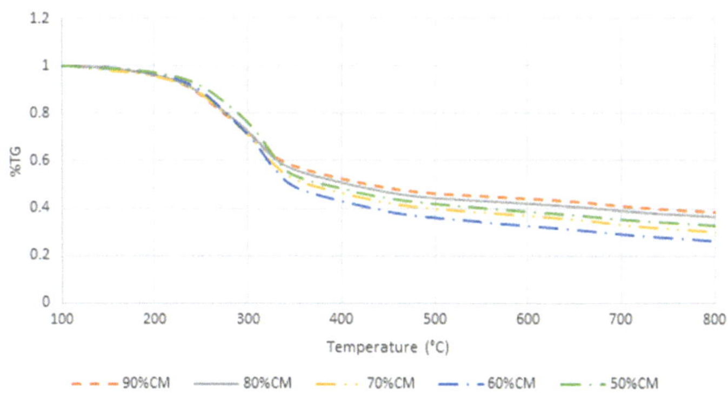

Fig. 20 TG curves for different mixing ratios at 5 °C/min [70]; you can use the color version

stage, the Hemicellulose and Cellulose from both biomasses decompose. This stage has the most massive mass degradation with around 40%wt on average. Therefore, the kinetic parameters are estimated at only this single stage. The third stage occurs at temperatures higher than 360 °C, during which the crystalline Cellulose and Lignin are slowly degraded. This stage decomposes about 20% of the total volatiles.

Figure 21 shows that the degradation curve follows a straight line during the second stage. In order to evaluate the decomposition, the mass degradation and temperature between 250 and 340 °C are approximated to a straight line (Fig. 21). The degradation increased when more rice was added. At 250 °C, the lost mass is almost identical within all blends; an acceleration occurs in the degradation at 340 °C, causing increased degradation from 0.7 to 0.51 (marked in Fig. 21). However, the lines are almost parallel—the decomposition of the different blends takes place at the same temperatures. This result is because both the chicken manure and rice husk have similar volatile matter [25, 70].

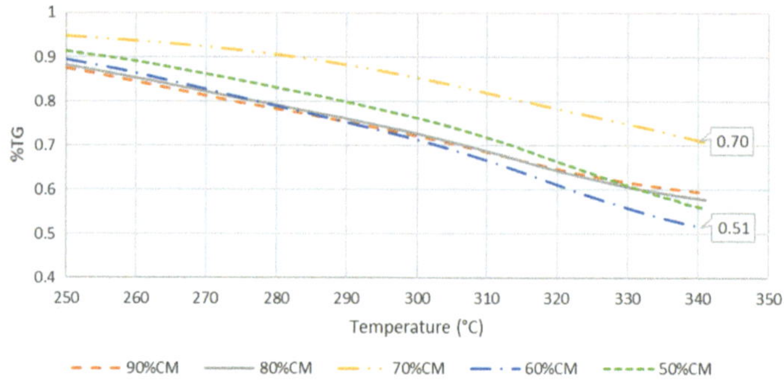

Fig. 21 TG curves between 250 and 340 °C at 5 °C/min [70]

The first and third stages of degradation remained relatively unchanged. For the second stage, the degradation rate increases when the percentage of rice husk increases. The chicken manure's degradation rate is increased because there is additional heating provided by the exothermic reaction that comes from rice husk's char gas formation. However, the degradation peak is slightly delayed, causing a need for higher temperatures to reach the degradation peak (Fig. 22).

Table 5 shows the remaining biomass for different heating rates. The increase in the heating rate reduced the conversion percentage for all the blends except for a few cases where the remaining mass was higher than the mass of singular biomass. The effect of the synergetic effect is also seen in the energy of activation, which is discussed in the next sections.

Differential Thermal Analysis

The differential thermal analysis curves for different mixing ratios are shown.

The curves are consistent with the DTG curves, the maximum degradation rate, and the peaks in the thermal energy, which all occur at the same temperature. Three peaks can be seen. The first peak represents water degradation at a temperature of

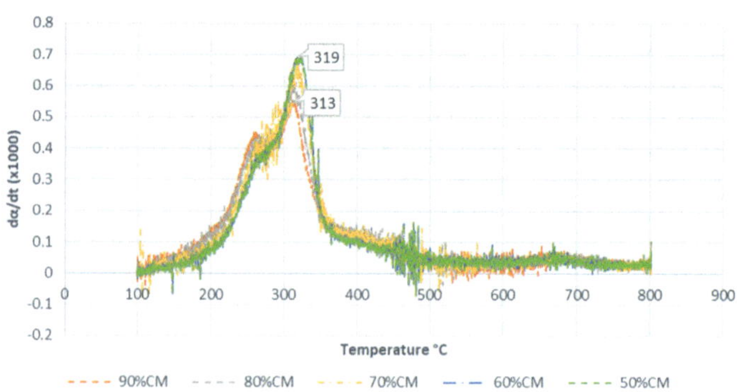

Fig. 22 DTG curves for different blends [70]

120 °C. This stage corresponds to water extraction degradation of Hemicellulose and Cellulose and additionally peaks at a temperature of about 350 °C. The energy extracted remains stable up to 380 °C. Combining the two samples increases the amount of energy released during the reaction than it would be if each biomass sample stood alone because the two samples each provide the same exothermic reaction over the same temperature range. At the early stages of Pyrolysis, chicken manure has the least energy extraction mainly because all the energy supplied by the furnace is being used to remove moisture content and breaking up Hemicellulose. However, for 0% chicken manure (100% rice husk), it can be observed that the existing energy is higher due to lesser moisture content and the absence of Hemicellulose. By adding more rice husk (10 and 20%), the DTA curve is then located between 0 and 100% chicken manure, which emphasizes this conclusion. When more rice husk is added, the exothermic reaction is dominant (Fig. 23).

Starting from 260 °C (which is corresponding to thermal degradation of Hemicellulose in chicken manure) to 480 °C (which is the start of Lignin degradation), it is observed that the amount of energy released from the reaction for all mixture blends is higher than each sample alone. The main reason behind this result is the overlap between two exothermic reactions taking place at the same time. Among all cases, the 60% chicken manure mixture gives the highest exothermic reaction, which makes the reaction self-sustaining to the end of Pyrolysis.

Pyrolysis Kinetics

The kinetic parameters for the reaction between 250 °C and 360 °C were calculated for each blend at different heating rates. The average values for each combination are shown in Table 6. The average energy of activation of chicken manure and rice husk is 108 and 95.5 kJ mol^{-1}, respectively. The pre-exponential factor is 2.05 $\times 10^{10}$min $^{-1}$ and 1.17 $\times 10^{8}$min^{-1} for chicken manure and rice husk, respectively. The order of reaction of rice husk was 2.0, while the order of reaction for chicken manure was

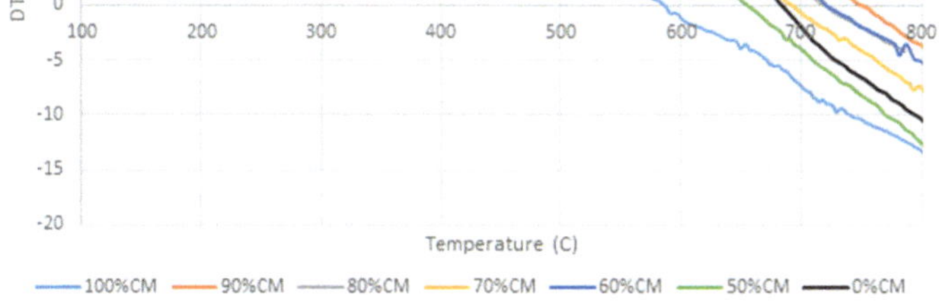

Fig. 23 DTA curves for different blends [70]; you can use the color version

Table 6 Averaged kinetic parameters [70]

% cm	*n*	ln(A/b)	Ea (*Kj*/mol)	A (min-1)
100	5	17.7	108.1	2.05E + 10
90	4.7	21.1	112.7	2.44E + 10
80	4.4	21.3	109.7	5.40E + 09
70	4.1	21.3	106.8	3.78E + 09
60	3.8	13.3	95.0	5.87E + 08
50	3.5	17.3	101.1	6.97E + 08
0	2	16.8	95.5	1.17E + 08

5.0. For the blends, the order of the reaction was found to vary linearly with the blend ratio.

The addition of rice husk increased the energy of activation compared with individual types of biomass for the first two mixing ratios. This increase could be due to the difference in particle sizes of rice husk compared to chicken manure. That size difference negatively affects the Pyrolysis reactions due to the temperature gradients, also leading to lower conversion. However, after adding more than 20% RH, the energy of activation and pre-exponential factor starts decreasing and reaches its minimum energy of activation—95.0 kJ mol^{-1} for the blend (60%CM + 40%RH). This minimum value signifies a decrease of 12.1% of the energy of activation when compared with chicken manure's energy of activation. This decrease means that the synergetic effects overcome the heat and mass transfer effects of the mixture (Table 6). The comparable behavior was observed by Dayananda [74] for the gasification of chicken manure and rice husk under a fluidized bed.

The increase in the heating rate, in turn, increases the energy of activation, which is in agreement with the literature for a different biomass Pyrolysis [35, 36]. Several factors affect the kinetics of the biomass. First, the particle size of the biomass plays a substantial role because the rapid change in temperature prevents the particles from having an even temperature distribution. Also, because less time is given before the temperature increases, the reactions in the Pyrolysis cannot be complete, increasing the energy of activation (Fig. 24). Also, the pyrolytic reactions cannot be perfect because less time is given before the temperature decreases, thus increasing the energy of activation.

5 Conclusions

The thermal gravimetric and differential thermal analyses were conducted for chicken manure using three different gasifying media (nitrogen, air, and carbon dioxide) with eight different heating rates (5, 10, 15, 20, 25, 30, 35, and 40 °C/min.) from room temperature to 1000 °C. The main difference between the nitrogen Pyrolysis and carbon dioxide gasification occurred after the temperature regime (>700 °C). At this

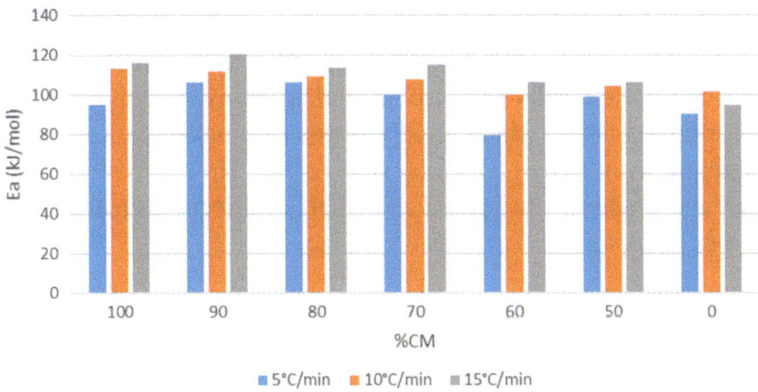

Fig. 24 Energy of activation for different cases [70]; you can use the color version

regime, the reaction turned into the endothermic reaction for the CO2 as it is governed by the Boudouard reaction, where the reaction turned into the endothermic reaction for the CO2. Air gasification is exothermic, and ignition was observed between 450 and 600 °C. Therefore, with air gasification, the reaction has the potentials to be self-sustainable with no external heating. The chicken manure thermal degradation implied the presence of the three components: Hemicellulose, Cellulose, and Lignin. The highest reaction rates were observed at temperatures corresponding to known peak characteristics of the three components. Gasification with air had an extra peak for ignition at 600 °C. The reaction kinetic parameters for the conversion reaction were calculated for the different media. The nitrogen Pyrolysis was divided into two regions at 360 °C with the order of reaction of five for both regions. Kinetic parameters for air gasification were calculated using a third-order single region reaction. CO_2 had the most complicated mechanism of the three cases and was divided into three regions at 360 and 630 °C. The kinetic parameters varied with the heating rate. It is recommended to utilize values generated by the lowest heating rate because the slow heating rate allows a quasi-equilibrium state, thus decreasing the effects of measurement errors due to delay in response or any transient condition error. Four different case studies were examined where the cow manure was added to chicken manure. The 40% cow manure of the total sample gives a positive result where the reaction is steady exothermic reaction all over the Pyrolysis process. Further research is required to obtain the amount of the produced Syngas and cumulative energy yield. Rice husk is compatible with chicken manure, causing an observable positive synergetic effect and increasing the conversion of the biomass. This effect is observed in the energy of activation. The energy of activation decreases when the rice husk ratio increases, thus decreasing the energy needed to begin Pyrolysis. The energy of activation decreases due to the synergetic effect of the rice husk on the blends. Moreover, adding rice husk increases the degradation rate, thus reducing the length of time that the biomass has to be in the reactor, resulting in an overall decrease in the energy consumed during Pyrolysis. The blend with the best performance was 60%CM + 40%RH. This blend decreased the energy of activation by 12.1% with respect to chicken manure. Additionally, a maximum increase in the conversion of 28% was achieved when the heating rate was 5 °C/min. These optimistic results need to apply more gas analysis to give a detailed explanation of the Syngas reactions taking place. A gas analysis of these results needs to be studied to give a detailed explanation of the Syngas reactions taking place. Co-Pyrolysis of chicken manure and rice husk demonstrated improved Pyrolysis of chicken manure. However, more research must be conducted in evaluating other parameters that can affect the co-Pyrolysis between these two types of biomass, such as particle size.

References

1. Outlook AE (2020). Energy information administration. Department of Energy, Washington, DC 20585, pp 1–161
2. Lv Y, Bi J, Yan J (2018) State-of-the-art in low carbon community. Int J Energy Clean Environ 19(3–4)
3. Ryu HW, Kim DH, Jae J, Lam SS, Park ED, Park Y-K (2020) Recent advances in catalytic co-Pyrolysis of biomass and plastic waste for the production of petroleum-like hydrocarbons. Bioresource Technol 123473
4. Ong HC, Chen W-H, Farooq A, Gan YY, Lee KT, Ashokkumar V (2019) Catalytic thermochemical conversion of biomass for biofuel production: a comprehensive review. Renew Sustain Energy Rev 113:109266
5. Pandey A (ed) (2011) Biofuels: alternative feedstocks and conversion processes. Academic Press
6. Chen H, Wang L (2016) Technologies for biochemical conversion of biomass. Academic Press
7. Pandey A, Bhaskar T, Stöcker M, Sukumaran R (eds) Recent advances in thermochemical conversion of biomass. Elsevier
8. Rosendahl L (ed) In: Biomass combustion science, technology and engineering. Elsevier
9. Verma M, Godbout S, Brar SK, Solomatnikova O, Lemay SP, Larouche JP (2012) Biofuels production from biomass by thermochemical conversion technologies. Int J Chem Eng
10. Molino A, Chianese S, Musmarra D (2016) Biomass gasification technology: the state-of-the-art overview. J Energy Chem 25(1):10–25
11. Neves D, Thunman H, Matos A, Tarelho L, Gómez-Barea A (2011) Characterization and prediction of biomass Pyrolysis products. Prog Energy Combust Sci 37(5):611–630
12. Guedes RE, Luna AS, Torres AR (2018) Operating parameters for bio-oil production in biomass Pyrolysis: a review. J. Anal Appl Pyrolysis 129:134–149
13. Pandey A (2008) Handbook of plant-based biofuels. CRC Press
14. Heidari A, Eshagh K, Habibollah Y, Lu HR (2019) Evaluation of fast and slow Pyrolysis methods for bio-oil and activated carbon production from eucalyptus wastes using a life cycle assessment approach. J Cleaner Prod 241:118394
15. Hussein MH (2016) Experimental investigation of chicken manure pyrolysis and gasification. Ph.D. Thesis, University of Wisconsin-Milwaukee
16. Vieira FR, Romero Luna CM, Arce GLAF, Ávila I (2020) Optimization of slow Pyrolysis process parameters using a fixed bed reactor for biochar yield from rice husk. Biomass Bioenergy 132:105412
17. Yuan T, He W, Yin G, Shiai Xu (2020) Comparison of bio-chars formation derived from fast and slow Pyrolysis of walnut shell. Fuel 261:116450
18. Al Arni S (2018) Comparison of slow and fast Pyrolysis for converting biomass into fuel. Renew Energy 124:197–201
19. Goyal HB, Seal D, Saxena RC (2008) Bio-fuels from thermochemical conversion of renewable resources: a review. Renew Sustain Energy Rev 12(2:504–517

20. Tanoue K-I, Hinauchi T, Oo T, Nishimura T, Taniguchi M, Sasauchi K-I (2007) Modeling of heterogeneous chemical reactions caused in Pyrolysis of biomass particles. Adv Powder Technol 18(6):825–840
21. Dry ME (1996) Practical and theoretical aspects of the catalytic Fischer-Tropsch process. Appl Catal A 138(2):319–344
22. Reed TB (2002) Encyclopedia of biomass thermal conversion: the principles and technology of Pyrolysis, gasification and combustion. Biomass Energy Foundation Press
23. Gebreegziabher T, Oyedun AO, Hui CW (2013) Optimum biomass drying for combustion–a modeling approach. Energy 53:67–73
24. Han J, Choi Y, Kim J (2020) Development of the process model and optimal drying conditions of biomass power plants. ACS Omega 5(6):2811–2818
25. Gaur S, Reed TB (1995) An atlas of thermal data for biomass and other fuels. No. NREL/TP-433–7965. National Renewable Energy Lab., Golden, CO (United States)
26. Balat M (2008) Mechanisms of thermochemical biomass conversion processes. Part 1: Reactions of Pyrolysis Energy Sources Part A: Recovery Utilization Environ Effects 30(7):620–635
27. Yang H, Yan R, Chen H, Lee DH, Zheng C (2007) Characteristics of hemicellulose, cellulose and lignin Pyrolysis. Fuel 86(12–13):1781–1788
28. Uddin MN, Daud WMAW, Abbas HF (2014) Effects of pyrolysis parameters on hydrogen formations from biomass: a review. RSC Adv 4(21):10467. https://doi.org/10.1039/c3ra43972k
29. Kumar A, Jones DD, Hanna MA (2009) Thermochemical biomass gasification: a review of the current status of the technology. Energies 2:556–581. https://doi.org/10.3390/en20300556
30. Bahng M-K, Mukarakate C, Robichaud DJ, Nimlos MR (2009) Current technologies for analysis of biomass thermochemical processing: a review. Anal Chim Acta 651:117–138. https://doi.org/10.1016/j.aca.2009.08.016
31. Kalinci Y, Hepbasli A, Dincer I (2009) Biomass-based hydrogen production: a review and analysis. Int J Hydrogen Energy 34:8799–8817. https://doi.org/10.1016/j.ijhydene.2009.08.078
32. Woolcock PJ, Brown RC (2013) A review of cleaning technologies for biomass-derived Syngas. Biomass Bioenergy 52:54–84. https://doi.org/10.1016/j.biombioe.2013.02.036
33. Sutton D, Kelleher B, Ross JRH (2001) Review of literature on catalysts for biomass gasification. Fuel Process Technol 73:155–173. https://doi.org/10.1016/S0378-3820(01)00208-9
34. Bulushev DA, Ross JRH (2011) Catalysis for conversion of biomass to fuels via Pyrolysis and gasification: a review. Catal Today 171:1–13. https://doi.org/10.1016/j.cattod.2011.02.005
35. Tanksale A, Beltramini JN, Lu GM (2010) A review of catalytic hydrogen production processes from biomass. Renew Sustain Energy Rev 14:166–182. https://doi.org/10.1016/j.rser.2009.08.010
36. Saxena RC, Seal D, Kumar S, Goyal HB (2008) Thermo-chemical routes for hydrogen rich gas from biomass: a review. Renew Sustain Energy Rev 12:1909–1927. https://doi.org/10.1016/j.rser.2007.03.005
37. Puig-Arnavat M, Bruno JC, Coronas A (2010) Review and analysis of biomass gasification models. Renew Sustain Energy Rev 14:2841–2851. https://doi.org/10.1016/j.rser.2010.07.030
38. Chhiti Y, Kemiha M (2013) Thermal conversion of biomass pyrolysis and gasification: a review. Int J Eng Silences 2:75–85
39. Pereira EG, da Silva JN, de Oliveira JL, Machado CS (2012) Sustainable energy: a review of gasification technologies. Renew Sustain Energy Rev 16:4753–4762. https://doi.org/10.1016/j.rser.2012.04.023
40. Parthasarathy P, Narayanan KS (2014) Hydrogen production from steam gasification of biomass: influence of process parameters on hydrogen yield—a review. Renew Energy 66:570–579. https://doi.org/10.1016/j.renene.2013.12.025
41. Mohan D, Pittman CU, Steele PH (2006) Pyrolysis of wood/biomass for bio-oil: a critical review. Energy Fuel 20:848–889. https://doi.org/10.1021/ef0502397

42. McKendry P (2002) Energy production from biomass (part 1): overview of biomass. Bioresour Technol 83:37–46. https://doi.org/10.1016/S0960-8524(01)001183
43. Wang L, Weller CL, Jones DD, Hanna MA (2008) Contemporary issues in thermal gasification of biomass and its application to electricity and fuel production. Biomass Bioenergy 32:573–581. https://doi.org/10.1016/j.biombioe.2007.12.007
44. Kamin´ ska-Pietrzak N, Smolin´ ski A (2013) Selected environmental aspects of gasification and co-gasification of various types of waste. J Sustain Min 12:6–13.https://doi.org/10.7424/jsm130402
45. Higman C, van der Burgt M, Higman C, van der Burgt M (2008) Chapter 1—introduction. Gasification 1–9. https://doi.org/10.1016/B978-07506-8528-3.00001
46. Kim S-S, Agblevor FA (2007) Pyrolysis characteristics and kinetics of chicken litter. Waste Manage 27:135–140. https://doi.org/10.1016/j.wasman.2006.01.012
47. Dejong W, Dinola G, Venneker B, Spliethoff H, Wojtowicz M (2007) TG-FTIR Pyrolysis of coal and secondary biomass fuels: determination of Pyrolysis kinetic parameters for main species and NOx precursors. Fuel 86:2367–2376. https://doi.org/10.1016/j.fuel.2007.01.032
48. Kim S-S, Agblevor FA, Lim J (2009) Fast Pyrolysis of chicken litter and turkey litter in a fluidized bed reactor. J Ind Eng Chem 15:247–252. https://doi.org/10.1016/j.jiec.2008.10.004
49. Agblevor FA, Beis S, Kim SS, Tarrant R, Mante NO (2010) Biocrude oils from the fast Pyrolysis of poultry litter and hardwood. Waste Manage 30:298–307. https://doi.org/10.1016/j.wasman.2009.09.042
50. Joseph P, Tretsiakova-McNally S, McKenna S (2012) Characterization of cellulosic wastes and gasification products from chicken farms. Waste Manage 32:701–709. https://doi.org/10.1016/j.wasman.2011.09.024
51. Font-Palma C (2012) Characterisation, kinetics and modelling of gasification of poultry manure and litter: an overview. Energy Convers Manage 53:92–98. https://doi.org/10.1016/j.enconman.2011.08.017
52. Priyadarsan S, Annamalai K, Sweeten JM, Holtzapple MT, Mukhtar S (2005) Co-gasification of blended coal with feedlot and chicken litter biomass. Proc Combust Inst 30:2973–2980. https://doi.org/10.1016/j.proci.2004.08.137
53. Kirubakaran V, Sivaramakrishnan V, Premalatha M, Subramanian P (2007) Kinetics of auto-gasification of poultry litter. Int J Green Energy 4:519–534. https://doi.org/10.1080/15435070701583102
54. Xiao X, Le DD, Li L, Meng X, Cao J, Morishita K et al (2010) Catalytic steam gasification of biomass in fluidized bed at low temperature: conversion from livestock manure compost to hydrogen-rich Syngas. Biomass Bioenergy 34:1505–1512. https://doi.org/10.1016/j.biombioe.2010.05.001
55. Yanagida T, Minowa T, Nakamura A, Matsumura Y, Noda Y (2008) Behavior of inorganic elements in poultry manure during supercritical water gasification. Nihon Enerugi Gakkaishi/J Jpn Inst Energy. https://doi.org/10.3775/jie.87.731
56. Yanagida T, Minowa T, Shimizu Y, Matsumura Y, Noda Y (2009) Recovery of activated carbon catalyst calcium nitrogen and phosphate from effluent following supercritical water gasification of poultry manure. Bioresour Technol 100:4884–4886. https://doi.org/10.1016/j.biortech.2009.05.042
57. Pinto F, André R, Miranda M, Neves D, Varela F, Santos J (2016) Effect of gasification agent on co-gasification of rice production wastes mixtures. Fuel 180:407–416. https://doi.org/10.1016/j.fuel.2016.04.048
58. Mahinpey N, Gomez A (2016) Review of gasification fundamentals and new findings: reactors, feedstock, and kinetic studies. Chem Eng Sci 148:14–31. https://doi.org/10.1016/j.ces.2016.03.037
59. Karatas H, Olgun H, Akgun F (2012) Experimental results of gasification of waste tire with air and CO2 air and steam and steam in a bubbling fluidized bed gasifier. Fuel Process Technol 102:166–174. https://doi.org/10.1016/j.fuproc.2012.04.013
60. Sharma S, Sheth PN (2016) Air–steam biomass gasification: experiments, modeling and simulation. Energy Convers Manage 110:307–318. https://doi.org/10.1016/j.enconman.2015.12.030

61. Broer KM, Woolcock PJ, Johnston PA, Brown RC (2015) Steam/oxygen gasification system for the production of clean syngas from switchgrass. Fuel 140:282–292. https://doi.org/10.1016/j.fuel.2014.09.078
62. Arabloo M, Bahadori A, Ghiasi MM, Lee M, Abbas A, Zendehboudi S (2015) A novel modeling approach to optimize oxygen–steam ratios in coal gasification process. Fuel 153:1–5. https://doi.org/10.1016/j.fuel.2015.02.083
63. Yuan H, Lu T, Zhao D, Wang Y, Kobayashi N (2015) Influence of oxygen concentration and equivalence ratio on MSW oxygen-enriched gasification syngas compositions. In: Progress in clean energy, vol 2. Springer, Cham, pp 165–176
64. Guo X, Wang S, Wang K, Qian LIU, Luo Z (2010) Influence of extractives on mechanism of biomass Pyrolysis. J Fuel Chem Technol 38(1):42–46
65. Mallick D, Poddar MK, Mahanta P, Vijayan, Moholkar S (2018) Discernment of synergism in Pyrolysis of biomass blends using thermogravimetric analysis. Bioresource Technol 261:294–305
66. Collot A-G, Zhuo Y, Dugwell DR, Kandiyoti R (1999) Co-Pyrolysis and co-gasification of coal and biomass in bench-scale fixed-bed and fluidised bed reactors. Fuel 78(6):667–679
67. Fan Y, Li Y, Zhiqiang Wu, Sun Z, Yang B (2018) Kinetic analysis on gaseous products during Co-Pyrolysis of low-rank coal with lignocellulosic biomass model compound: effect of Lignin. Energy Procedia 152:916–921
68. He Q, Qinghua Guo Lu, Ding JW, Guangsuo Yu (2019) Rapid Co-Pyrolysis of lignite and biomass blends: analysis of synergy and gasification reactivity of residue char. J Anal Appl Pyrol 143:104688
69. Burra KG, Gupta AK (2018) Kinetics of synergistic effects in Co-Pyrolysis of biomass with plastic wastes. Appl Energy 220:408–418
70. Hossain MS, Islam MR, Rahman MS, Kader MA, Haniu H (2017) Biofuel from Co-Pyrolysis of solid tire waste and rice husk. In Energy Procedia, vol 110, pp 453–458. Elsevier Ltd. https://doi.org/10.1016/j.egypro.2017.03.168
71. Costa P, Pinto F, Miranda M, André R, Rodrigues M (2014) Study of the experimental conditions of the co-Pyrolysis of rice husk and plastic wastes
72. Ng WC, You S, Ling R, Gin KY-H, Dai Y, Wang C-H (2017) Co-gasification of woody biomass and chicken manure: syngas production biochar reutilization, and cost-benefit analysis. Energy 139:732–742
73. Dayananda BS, Manjunath SH, Girish KB, Sreepathi LK (2013) An experimental approach on gasification of the chicken litter with rice husk. Int J Innov Res Sci Eng Technol 2(7):2837–2842
74. Seçer A, Küçet N, Fakı E, Hasanoğlu A (2018) Comparison of co–gasification efficiencies of coal lignocellulosic biomass and biomass hydrolysate for high yield hydrogen production. Int J Hydrogen Energy 43(46):21269–21278
75. White JE, Catallo WJ, Legendre BJ (2011) Biomass Pyrolysis kinetics: a comparative critical review with relevant. J Anal Appl Pyrolysis
76. Phyllis2 (2015) database for biomass and waste [Online]. Available at https://www.ecn.nl/phyllis2/Biomass/View/3501
77. Ruiz-Gómez N, Quispe V, Ábrego J, Atienza-Martínez M, Murillo MB, Gea G (2017) Co-Pyrolysis of sewage sludge and manure. Waste Manage 59:211. https://doi.org/10.1016/j.wasman.2016.11.013
78. Selim OM, Hussein MS, Amano RS (2020) Effect of heating rate on chemical kinetics of chicken manure with different gas agents. J Energy Resourc Technol 142(10)
79. Yang H (2007) Characteristics of hemicellulose cellulose and lignin Pyrolysis. Fuel 1781–1788
80. Selim OM, Amano RS (2020) Co-pyrolysis of chicken and cow manure. ASME J Energy Resour Technol 143(1):011301. https://doi.org/10.1115/1.4047597
81. Espindola J, Selim OM, Amano RS (2020) Co-pyrolysis of rice husk and chicken manure. ASME J Energy Resour Technol 143(2):022101. https://doi.org/10.1115/1.4047678

Higher Alcohols as Diesel Engine Fuel

Naveen Kumar, Harveer S. Pali, Ankit Sonthalia, and Sidharth

Abstract India is one of the fastest-growing economies in the world and requires a vast amount of energy which is mainly met by fossil fuels. India, however, spends a huge amount of foreign exchange for importing crude petroleum due to limited indigenous production. The diesel consumption in India is nearly four times higher than gasoline. It is widely used in locomotives, city transport buses, heavy trucks, electric generators, agricultural equipment, and underground mining equipment due to its versatility and higher efficiency. Diesel engines have the main environmental disadvantage of producing more soot particles and nitrogen oxides. They must also meet the stringent Euro VI emission standards, which are difficult to attain just by modifying the engine design. In this situation, higher alcohols such as butanol, pentanol, hexanol, and octanol have vast potential. Previous work on the use of lower alcohols as a blend with diesel in compression-ignition engines has suggested a reduction in emissions. However, problems such as phase separation and increased fuel consumption have also been encountered during the use of lower alcohols, e.g., ethanol in diesel engines. Higher alcohols have the potential to be used along with diesel to make a homogenous blend without any phase separation. They also have molecular oxygen present which improves the combustion and reduces emissions. Higher alcohols are sustainable and have the potential to address the issues of climate change and energy security. However, the use of these fuels in India is still in the early stage and large-scale production and adaptation will take its own time.

N. Kumar (✉) · H. S. Pali · A. Sonthalia · Sidharth
Centre for Advanced Studies and Research in Automotive Engineering, Delhi Technological University, Bawana Road, Delhi 10042, India

H. S. Pali
Department of Mechanical Engineering, National Institute of Technology Srinagar, Srinagar, Jammu & Kashmir, India

A. Sonthalia
Department of Automobile Engineering, SRM Institute of Science and Technology, NCR Campus, Uttar Pradesh, Ghaziabad 01204, India

Sidharth
Department of Mechanical and Automation Engineering, Maharaja Agrasen Institute of Technology, Rohini, Delhi 110086, India

A. Gupta et al. (eds.), *Advances in Energy and Combustion*, Green Energy and Technology,
https://doi.org/10.1007/978-981-16-2648-7_7

Keywords Butanol · Pentanol · Hexanol · Octanol · Higher alcohols · Compression-ignition engine

Abbreviations

ABE	Acetone–Butanol–Ethanol
ATF	Aviation turbine fuel
BTE	Brake thermal efficiency
BSFC	Brake specific fuel consumption
CI	Compression ignition
CO	Carbon monoxide emission
CRDI	Common Rail Direct Injection engine
CuO	Copper oxide
EGR	Exhaust gas recirculation
HC	Unburned hydrocarbon emission
HD	Heavy duty diesel engine
HRR	Heat release rate
LD	Light duty diesel engine
NiO	Nickel oxide
NOx	Oxides of nitrogen emission
PM	Particulate matter emission
ZnO	Zinc oxide

1 Introduction

Humans from primitive times have required energy for their survival. This led to a process of searching and developing energy alternatives, like the employment of animals for waste, biomass, and wind energy along with others. It allowed the exploration of the whole world by different civilizations over many millennia.

Fossil fuels are the main provider of energy in the world. They are formed when the remains of large amount of plants and animals buried deep inside the earth and at the bottom of the seas and lakes are decomposed for millions of years. Fossil fuels are chiefly classified into coal, natural gas, and oil.

Oil was formed mainly from old plants and microorganisms that lived in the ocean and seas of salty water. Oil has been a main source of energy in the developed countries, which will be exhausting very soon. The use of fossil fuels generates a large amount of greenhouse gases into the atmosphere, in addition to nitrogen oxides and sulfur oxides as combustion exhaust. The extraction of fossil fuels also causes oil and fuel spills in the open sea harming the environment.

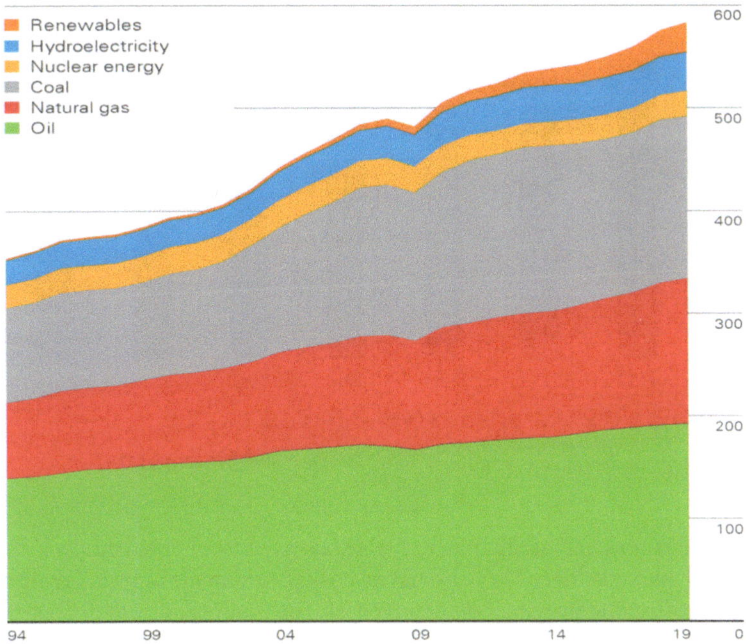

Fig. 1 Global energy consumption [1]

For these reasons, the exploration of alternative fuels has regained interest due to economic, social, and environmental considerations worldwide.

From Fig. 1, it is seen that the world's energy consumption has increased by nearly 1.3% in the year 2019. Fossil fuels are fulfilling most of the energy demand. Oil consumption has increased by 0.9 million barrels per day in the year 2019 when compared to 2018. India accounted for an increase in oil consumption in the year 2019 by 159,000 barrels per day [1].

Indian economy is heavily dependent upon the usage of diesel fuel. It is the livelihood of many people in India. Several applications like tractors, transport buses, trucks, and industries employ diesel engines. Diesel consumption during 2018–19 was 39.46%, whereas petrol consumption was only 13.3% followed by 11.7% for LPG, 10% for petroleum coke, 6.6% for Naphtha, and 3.9% for ATF.

Diesel engines are specifically preferred where the power output and higher efficiency are highly desired. The main applications of diesel engines are the transportation of public and goods, power generation, heavy industries, and the agriculture sector. Diesel despite its better thermal efficiency is not preferred owing to its harmful emissions.

Due to this, alternative sources of energy were found to be capable of replacing or minimizing the consumption of diesel. It is seen from Fig. 2 that the consumption of diesel is many folds to that of petrol in India. Therefore, focus on diesel fuel is necessary for a developing country like India [2].

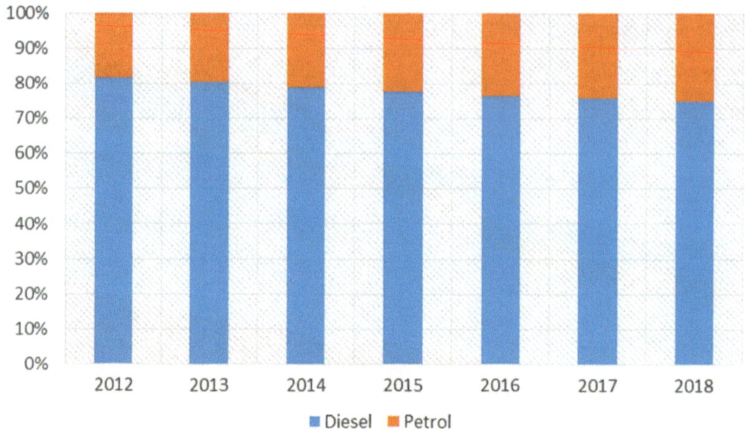

Fig. 2 Diesel versus petrol consumption in India for the years 2012–2018 [2]

It is for this reason that alternative fuels such as alcohol obtained from biomass can be a suitable substitute. They also have a positive economic impact, giving energy independence, and improving competitiveness. Environmentally, they contribute to reducing air pollution, generating biodegradable by-products. Socially, they help the growth and diversification of the rural economy and quality of life.

2 Comparison of Higher and Lower Alcohols

Alcohols are having a general formula as $C_nH_{2n+1}OH$. The most popular ones are methanol and ethanol having numerous applications in the food and industrial sector. Alcohols can also be used in both spark-ignition and compression-ignition engines. However, their suitability is to be checked before usage in the engines. The simplest method for checking the suitability is the comparison of its property with the fuel used in the engine. Several researchers have studied the suitability of lower chain alcohols like methanol, ethanol in a gasoline engine, whereas higher chain alcohols, namely butanol, pentanol, hexanol, and octanol were primarily studied in a diesel engine.

It is observed from Table 1 that the properties of lower alcohols (methanol, ethanol, and propanol) like cetane number and calorific value are quite lower than diesel. Moreover, methanol and ethanol are highly miscible in water [6] making it a non-preferable candidate for the diesel engine.

Several studies carried out in the past with methanol, ethanol, and propanol concluded that at some point in time, there will be phase separation of diesel/biodiesel with these lower alcohols. Therefore, the obvious choice comes out to be higher alcohols as a substitute for diesel. The preferred higher alcohols include butanol, pentanol, hexanol, and octanol.

Table 1 Comparison of several chain alcohols [3–9]

Fuel properties	Methanol	Ethanol	Propanol	Butanol	Pentanol	Hexanol	Octanol
General formula	CH_3OH	C_2H_5OH	C_3H_7OH	C_4H_9OH	$C_5H_{11}OH$	$C_6H_{13}OH$	$C_8H_{17}OH$
Density (at 15 °C)	791.3	789.4	803.7	809.7	814.8	821,8	830
Kinematic Viscosity (at 40 °C)	0.58	1.13	2.96	3.17	3.44	5.32	5.5
Calorific value	20.08	26.83	29.82	32.01	32.16	37.1	38.4
Flash point	12	13	22	35	48	59	81
Cetane number/ index	3.8	8	12	17	20	23	38.4
Oxygen %	49.93	34.73	26.62	21.59	18.15	15.7	12.31

With the increase in the number of carbon atoms in the chain of the higher alcohols, it is observed that the calorific value, flash point, and cetane number of alcohol improves. This makes hexanol and octanol a suitable candidate as a diesel substitute. On the contrary, properties like kinematic viscosity and density also increase with the increase in carbon atoms. It is worthwhile to mention that based on the properties discussed in Table 1, all higher alcohols shall work suitably without any other alteration in a diesel engine.

3 Production of Higher Alcohols

The commercial method for the production of 1-butanol is hydrogenation of n-butyraldehyde. The aldehyde is formed by the Oxo reaction of propylene. In this reaction, a mixture of iso and n-butyraldehyde is obtained which can initially be separated, and the individual aldehydes are then hydrogenated to form butanol. It is also possible that the mixture of butyraldehydes is hydrogenated, and then, the mixture of n- and isobutanol is separated through distillation. The hydrogenation is carried out over a heterogeneous catalyst in the vapor phase. For example, 60 to 40 percent of hydrogen to nitrogen is passed over a mixture of iso and n-butyraldehyde in the presence of CuO–ZnO–NiO catalyst at 25–196 °C and 0.7 MPa, resulting in 98.6% conversion efficiency of the corresponding alcohols [10]. In an old method, ethanol was dehydrogenated to acetaldehyde followed by its condensation to crotonaldehyde and lastly hydrogenation of crotonaldehyde resulting in the formation of 1-butanol. Alternatively, the acetaldehyde can be obtained from ethylene using the Wacker process [11]. In the US, the Zeigler-Natta chain growth reaction is also used for producing small amounts of n-butanol from ethylene [12]. The Acetone–Butanol–Ethanol (ABE) production process is the earliest commercial process for producing butanol [13, 14]. In this process, corn products or molasses are fermented with *Clostridium acetobutylicum* resulting in the formation of acetone, ethanol and butanol which is later separated. Another fermentation route for producing 1-butanol is the use of anaerobic bacteria, namely *Butyribacteriummethylotrophicum* that is based on synthesis gas. As compared to the catalytic processes used for converting syngas to alcohol, this new method is insensitive to sulfur presence. However, only 1 g/L of butanol could be produced from this process which is very low for it to be commercially viable.

A patent for a novel route has been filed in which butadiene is used to produce 1-butanol. In this process, butadiene is first catalytically converted to a mixture of branched and linear butenyl ethers. The branched ethers are converted to linear ethers and then hydrogenated to 1-butanol. Isobutyl alcohol is commercially produced by hydrogenating isobutyraldehyde which in turn is obtained from the propylene hydroformylation. The alcohol is also a co-product of methanol synthesis produced from the Fischer Tropsch process [15]. 2-butanol is produced from Raffinate II type feedstocks, that is, C-4 refinery streams that contain saturated C-4 s and n-butenes after removal of isobutylene and butadiene. In the indirect hydration process, the n-butenes are esterified in sulphuric acid, and intermediate butyl sulfate is hydrogenated. For producing t-butyl alcohol, isobutane is taken as the starting material, and arco propylene oxidation process is used, wherein propylene oxide is the main product, and t-butyl alcohol is the co-product. The process can also be modified to produce only t-butyl alcohol. In this process, the isobutane is fixed oxidized, and the produced t-butyl hydroperoxide can be decomposed under catalytic conditions to give gasoline grade t-butyl alcohol (GTBA) [16–19]. The oxygen released from the decomposition process can be reutilized for the oxidation of isobutane. Another industrial route for producing t-butyl alcohol is acid-catalyzed hydration of isobutylene.

Hydroformylation of butenes is used for the commercial production of pentanol. The syngas reacts with C4 olefins in the presence of rhodium or cobalt-containing catalyst to yield C_5 aldehydes. The C_5 aldehyde is then hydrogenated to yield corresponding alcohol. Figure 3 shows the schematic of the process. From various cracking processes raffinate I is used as a raw material. 1,3 butadiene is separated and Raffinate I contains 44–49 vol.% isobutene, 24–28 vol.% 1-butene, 19 –21 vol.% 2-butene, and 8–11 vol.% C4-hydrocarbons. The isobutene present in the raffinate is reacted with methanol and converted to methyl tert-butyl ether. The remaining olefin mixture is used for producing 1-pentanol and 2-methyl-1-butanol [20]. If the hydroformylation of n-butenes is carried out in the presence of cobalt catalysts, then 70% 1-pentanol and 30% 2-methyl-1-butanol is obtained, whereas with rhodium-triphenyl phosphane catalyst, the produced mixture contains 90% 1-pentanol and 10% 2-methyl-1- butanol [21]. At 130 °C, rhodium phosphite catalyst can convert 2-butene to a mixture of 1-pentanol and 2-methylbutanal in the ratio 1.13:1 [22]. At 110 °C, rhodium tricyclohexyl phosphane catalyst converts 2-butene to a mixture of 1-pentanol and 2-methylbutanal in the ratio 0.2:1 [23]. Using rhodium or cobalt catalyst, isobutene is converted to 3-methyl-1-butanol [24]. With cobalt catalyst, small

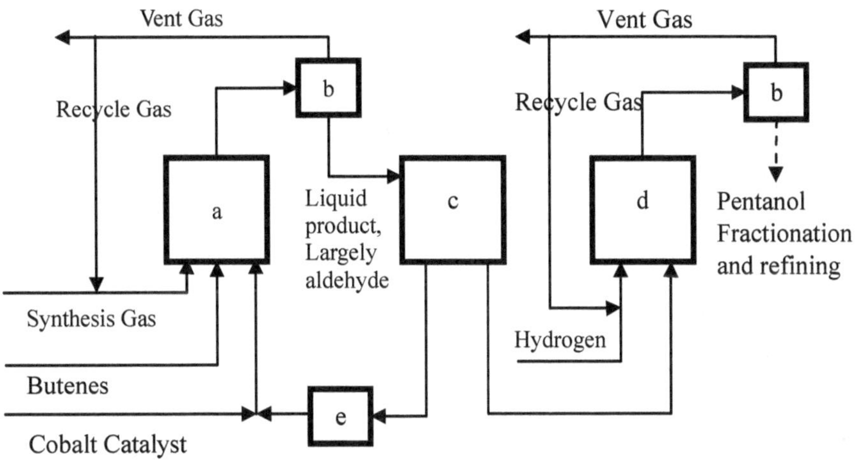

a) Oxo reactor (100 – 200°C, 10 – 45 MPa); b) Separator; c) Catalyst removal(decobalting); d) Hydrogenation unit (50 –250°C, 0.3 – 24 MPa); e) Catalyst recovery

Fig. 3 Production process of pentanol

amount of 2,2-dimethyl-1-propanal is also produced [21]. Apart from pure olefins and raffinate, other olefin mixtures containing four carbon atoms can be used for hydroformylation. At 100 °C, with $Co(CO)_4$ as a catalyst, a mixture of 1-butene, 2-butene, and isobutene in the ratio of 1:1:2 can be converted to a mixture of 1-pentanol, 2-methyl-1-butanol, and 3-methyl-1-butanol in a ratio of 15:5.2:11 using the hydroformylation technique [25]. The catalyst used for the hydrogenation of C_5 aldehydes is CuO-ZnO, Cu-TiO_2, nickel, and zirconia [26–28].

Pentenes can also be hydrated to produce 2-pentanol, 3-pentanol, 2-methyl-2-butanol, and 3-methyl-2-butanol, as shown in Table 2. Apart from the hydration process, modification in the hydroboration reaction can be applied for stereoselective reactions resulting in the formation of pentanol [29]. 2-methyl-2-butene can be

Table 2 Hydration of pentenes

Olefin	Alcohol	Process catalyst	References
1-pentene	2-pentanol	zeolites, aluminosilicates	[37]
2-methyl-2-butene	2-methyl-2-butanol	acid ion exchangers	[38]
2-methyl-1-butene/2-methyl-2-butene	2-methyl-2-butanol	acid ion exchangers	[39]
2-methyl-2-butene	2-methyl-2-butanol	Zeolites	[40]
2-methyl-2-butene	2-methyl-2-butanol	layer silicates	[41]
2-methyl-1-butene/2-methyl-2-butene	2-methyl-2-butanol	acetic acid	[42]

reacted with mono bromoborane–dimethyl sulfide complex resulting in the formation of 3-methyl-2-butanol [30]. Dimesityl borane reacts with 2-pentene at 65 °C to form a mixture of 76% 2-pentanol and 24% 3-pentanol. 2-pentanol, 3-pentanol, and 3-methyl-2-butanol can also be produced by hydrogenating their corresponding ketones [31]. Catalysts for these reactions are nickel, copper, palladium, and specialized catalyst systems such as ruthenium-polysterene, cationic rhodium complexes, and $ClMgAlH_4$ [32]. An acid-stable catalyst such as rhenium (VII) oxide (Re_2O_7) or osmium tetroxide (OsO_4) was developed to produce 2,2-dimethyl-1-propanol from pivalic acid at 90 °C and 100 bar. The yield of the alcohol was 94.3%. Other catalysts used for the process are $HTcO_4$ or $KReO_4$ –$HClO_4$, zirconia [33], and CuO–ZnO–Al_2O_3 [34]. Isobutene can be reacted with formaldehyde (Prins reaction) to form 3-methyl-1-butanol. Depending on conditions, 3-methyl-3-buten-1-ol or 4,4-dimethyl-1,3- dioxane intermediates are formed by reacting isobutene with formaldehyde; then, the hydrogenation of the intermediates in the presence of a catalyst such as $L_2Cu(OAc)_2/Cr_2O_3$–ZnO–BaO results in the formation of 3-methyl-1-butanol. Pentanol can also be produced by fermentation using engineered microorganisms [35] or glucose biosynthesis [36].

1-Hexanol can also be produced by the Ziegler process from ethylene. It is sold under the commercial name of Alfol – 1 or Sasol. It is used as a solvent, an ingredient for the perfume industry, and also a plasticizer. Nitrates of 1-hexanol are also used as a cetane improver. 2-Methyl-1-pentanol is produced by aldol condensation of propionaldehyde followed by hydrogenation of the intermediate 2-methyl-2-pentenal. It is also used as a solvent. 4-Methyl-2-pentanol is produced as a by-product during the synthesis of methyl isobutyl ketone. Dow chemicals sell this alcohol under the name methylamyl alcohol [43]. The paint industry uses this alcohol as a solvent, and it is also used for cleaning semiconductors. Its other uses are as brake fluid, floatation aid, and as a fungicide [44]. 2-Ethyl-1-butanol is prepared by the aldol condensation of 1-butanal and acetaldehyde followed by the hydrogenation of the intermediate. It is a colorless, pleasant-smelling liquid used as a flow improver and solvent for paints and varnishes. It is used as raw material for making corrosion inhibitors and penetrating oils. It is also used as a cleaning agent for the printed circuit boards. 1-Hexanol can also be prepared by engineering microbes. However, Table 3 shows that the yield of alcohol is less with these microbes. 1-Hexanol was synthesized

Table 3 Production of 1-hexanol from microbes

Engineered host organisms	Source	Yield	Reference
Escherichia coli	Glucose	18.5 mg/L	[36]
Escherichia coli	Glucose	47 mg/L	[45]
Escherichia coli	Glucose	302 mg/L	[50]
Clostridium acetobutylicum	Glucose	30 mg/L	[46]
Ralstoniaeutropha	Glucose	280 mg/L	[46]
Clostridium carboxidivorans	Syngas	1 g/L	[49]

from glucose by using a non-natural metabolic pathway in which the chain elongation was carried out by an engineered *E. Coli* [36]. The authors optimized the process and observed the highest yield of 18.5 mg/L. Dekishima et al. [45] modified the 1-butanol pathway that utilizes the *E. Coli* strain *Clostridium* co-enzyme A and used it for synthesizing 1-hexanol from glucose. For producing n-hexanol, Machado et al. replaced the enzyme *Clostridium acetobutylicum* Hbd with Ralstonia*eutropha* homologue PaaH1 [46]. The authors used an anaerobic growth platform that resulted in a ten-fold yield of 1-hexanol. The research group also showcased a 67% increase in 1-hexanol production by applying a selection platform for the PaaH1 evolution. For producing hexanol and butanol from biomass, an indirect method was devised that chemically converted 90% of carbon into carbohydrates [47]. A one-step method for simultaneous production of 1-hexanol and 1-butanol from bioethanol at its supercritical state was developed by Ghaziaskar et al. [48]. The authors achieved 35% ethanol conversion efficiency with the highest selectivity of 62% of 1-butanol and 21% of 1-hexanol. Another method is the gasification of the biomass to syngas (CO + H_2) and passing this gas through the bacteria *Clostridium carboxidivorans* which autotrophically grows using this gas and produces alcohols like butanol and hexanol [49]. Although the yield using this method is low, it is viable from a commercial viewpoint as the medium used is low cost.

The predominant method for producing n-octanol from petrochemical processes is Ziegler alcohol synthesis in which ethylene is oligomerized [43] or the oxo synthesis process [51]. 2-ethylhexanol is also alcohol with eight carbon atoms, and it holds significant use in the manufacture of ester plasticizers for soft polyvinyl chloride. The alcohol is produced in four steps: First, butyraldehyde is first aldolized and then dehydrated. The butyraldehyde is itself produced by hydroformylation of propene. In the reactor carrying out aldolization, butyraldehyde in the presence of aqueous sodium hydroxide reacts to form 2-ethyl-2-hexanal [52]. The reactants are continuously stirred so that secondary reactions taking place due to overheating can be stopped. The primary aldol addition must take place rapidly, and the dehydration of hydroxyaldehyde must be carried out quickly as the aldol is unstable resulting in a decrease in product quality. The reaction takes place mostly between 80–150 °C at a pressure below 0.5 MPa. The ratio of aldehyde to aqueous sodium hydroxide is maintained between 1:10 and 1:20 resulting in 99% conversion rates [53, 54]. The heat produced from aldolization can be used for steam generation. In the second step, in a phase separator, the mixture is separated into a lower aqueous phase containing the aldolized solution and an upper organic phase. Some parts of the aldolized solution can be recycled, and the rest can be removed via a side stream. To the leftover solution, water is added and sodium hydroxide is added to maintain the required concentration. The side stream is then purified by acid treatment, oxidation, and extraction resulting in the partial recycling of valuable products [52]. In the third step, the organic product is hydrogenated on a nickel catalyst in single stage [55]. The hydrogenation can be carried out in multiple stages either in the combination of the gas–liquid phase or liquid–liquid phase [56]. The heat release by breakage of $C = C$ is very high resulting in temperature control problems thereby decreasing yield. In single-step hydrogenation, the product is remixed to dissipate heat, and medium pressure is adequate for conversion [57]. Modern plants use two stages for removing residual carbonyl compounds thus ensuring that high-grade 2-ethylhexanol is obtained [58]. For the hydrogenation, heterogeneous catalysts such as copper [59], nickel [57, 60, 61], or mixed systems [62] are used that give 100% conversion and greater than 99% selectivity. In the final step, the hydrogenated product is distilled. First, light ends are separated that can be used for manufacturing 1-butanol. Then, pure 2-ethylhexanol is collected, and lastly, recyclable intermediate fractions are separated. Octanol can also be produced from biomass. Recently, a multifunctional catalyst was developed by Julius et al. [51] producing C8 linear alcohols and di-octyl ether from a lignocellulosic biomass platform. A yield of up to 93% of linear C8 alcohol products was obtained using this route. Xia et al. [63] achieved a yield of 67% from lignocellulosic biomass using an efficient catalyst system. Even microorganisms have been used for producing octanol from biomass. 1-Butanol pathway was extended by Machado et al. [46] for increasing the carbon chain length of the alcohol and reported a yield of 70 mg/L of 1-octanol. Akhtar et al. [64] used *Escherichia coli* for engineering a biosynthetic pathway for producing 1-octanol.

4 Use of Higher Alcohols as a CI Engine Fuel

Most of lower carbon alcohols are used as fuel for spark-ignition engines due to their high octane number. The cetane number of lower carbon alcohols is very low as compared to petroleum diesel. Therefore, this restricts researchers to use these alcohols for compression-ignition engines. In this scenario, butanol, pentanol, hexanol, and octanol got more attention from researchers for diesel engine fuel. The cetane number and heating value of these alcohols are much higher than methanol and ethanol. Various researches have been published by the number of researchers in which they used higher alcohols as fuel for compression-ignition engines [65–70].

4.1 Application of Butanol in CI Engine

Exhaustive research has been carried out on butanol as a potential CI engine fuel. Armas et al. studied emission of cold and hot start condition of diesel engine fuel blended with butanol with having same oxygen contained in test fuel (~3.3% in mass) and observed that the butanol diesel blends have a positive effect in reducing smoke and PM concentration during hot engine start with similar nitrogen oxide (NOx) emissions. On the cold start, the blends produced combustion instability [71].

Similarly, Ibrahim et al. investigated the performance and emission of a blend, of diesel, biodiesel, and butanol in different ratio and observed that the optimal alternative fuel among all the fuels tested was the D50B50 fuel as its use increased the maximum thermal efficiency of the engine by 6.5% and reduced the specific fuel consumption of the engine brake by 5% less compared to diesel fuel. NO emissions increased significantly with increased engine load and increased slightly with the use of oxygenated fuels, and all fuels tested did not adversely affect engine stability [72]. Karabektas and Hosoz tested isobutanol diesel fuel blends on the single-cylinder diesel engine found that the brake power slightly decreased with the volumetric increase in the isobutanol subsequently BSFC partially increased for similar conditions. Emission results revealed that CO and NOx emissions were decreased, and HC was increased [73]. In another study, Ileri et al. investigated the effect of a diesel-rapeseed oil-n-butanol on a four-stroke, four-cylinder turbocharged direct injection diesel engine. Emissions and engine performance was conducted at full load and variable RPM. Exhaust temperature and brake power got decreased and BSFC increased. Moreover, HC emissions were reduced, and NOx emissions were partially increased. Huang et al. investigated the effect of the pilot on low-temperature combustion in the high-speed diesel engine fueled with n-butanol- diesel blends. The addition of n-butanol to diesel leads pressure rise rate and BSFC. As well as an increase in pilot injection timings retards the HRR [65]. Wang et al. reviewed n-butanol/biodiesel mechanism and predicted exhaust emissions like CO, HC, NOx, soot, and heat release rate under different conditions agreed well with the experimental data. During the ignition, process biodiesel interacts with n-butanol, and biodiesel produced OH radicals in low in-cylinder temperature. These OH radicals are consumed by n-butanol to produce HO_2 radicals. In the successive reaction, it is transformed into H_2O_2, which separates into OH radicals at elevated in-cylinder temperature [74]. In another similar study, Sahin et al. found n-butanol fumigation decrease the NOx [67]. Jamrozik et al. rewired co-combustion of alcohol and found a positive effect on mean effective pressure, thermal efficiency, and stability of how exhaust emissions were significantly reduced [75]. Zheng et al. worked on a dual fuel mode RCCI engine. Investigation showed the difference between two combustion modes at different EGR rates and n-butanol ratios. The optimum EGR rate was 30%. Yilmaz et al. trialed on a four-stroke, single-cylinder indirect injection diesel engine and found lower exhaust gas temperature NOx however a little change in CO and HC. Singh et al. used the n-butanol as an additive eucalyptus biodiesel biodiesel-diesel blends for performance and emission of CI engine [76]. Similarly, other researchers examined the performance of diesel engines at 20% of butanol with petroleum diesel. However, a further increase in the percentage of butanol put some adverse effect on the performance of the engine [66, 68, 77, 78]. This improvement is due to the oxygen content in the butanol which improves the combustion characteristics of blended fuel at higher load. The blend of butanol with diesel fuel also leads to a decrease in exhaust emissions of the diesel engine. However, some emissions like hydrocarbon increase slightly [79].

4.2 Application of Pentanol in CI Engine

Pentanol has good fuel characteristics with good corrosion impediment. As well as it has diesel-like viscosity, calorific value, and stoichiometric properties used in the CI engine. Apart from this, it has good ignition property, better blend stability, and higher oxygen content [68]. Wi et al. explored the pentanol-diesel blends up to 30% in the four cylinders' direct injection diesel engine and found increased ignition delay and similar other combustion characteristics. Moreover, HRR curves of the blends shift closer to the TDC with increasing load [80]. A higher percentage of pentanol-diesel blends is also utilized by few researchers [81, 82]. Kumar et al. tested the CI engine at high load with different settings of RGR and injection timings. The study exhibited combustion characteristics like prolonged ignition delay, lower peak in-cylinder pressure, and more HRR against neat diesel. Some other researchers investigated the effects of pentanol in vegetable oil in the CI engine [83]. Pentanol-diesel blended fuel offers greater atomization and better combustion due to lower viscosity and more oxygen content [84]. Generally, higher emissions of HC and CO was observed by many researchers; moreover, drop in smoke emission was also observed. NOx emission showed mixed trends in the working conditions and specific engine type [85, 86]. Higher pentanol content minimizes particulate mass concentration and total particle number concentration [87]. Radheshyam et al. investigated the effects of pentanol blends and EGR on the CRDI diesel engine and found a higher delay period for all blends. BSFC increases and BTE decreases with an added concentration of pentanol. CO and HC emissions were increased, and NOx emissions were reduced with the addition pentanol blends and EGR [88]. In another study, Yesilyurt et al. scrutinized binary and ternary blend of pentanol of a diesel engine. After the study, pentanol proved its potential as a promising higher alcohol fuel to fulfill the future energy burden of the automobiles [7, 89].

4.3 Application of Hexanol in CI Engine

Hexanol is a high-carbon bio-alcohol, and it has high energy density and cetane number. So it is partially used as diesel engine fuel for the last decade. Duncan et al. explored diesel hexanol blends and suggested potential fuel for CI engine [90]. Sathiyagnanam et al. used the hexanol as a co-solvent to prevent the separation of ethanol from diesel and improved performance and lower emissions [91]. De poures et al. investigated the influence of injection timing and exhaust gas recirculation on hexanol as a CI engine fuel. It has resulted in longer ignition delay, higher peak pressure, and HRR. Authors suggested that 30% hexanol blend was advantageous in terms of smoke density up to 35.9%; however, it had partially increased in NOx emission up to 3% [9]. In another assessment, Ashok et al. examined *calophyllum inophyllum* biodiesel with hexanol and decanol. All ternary blends diesel, biodiesel, and higher alcohol revealed higher BTE and lower BSFC. Moreover CO, HC, and smoke were lower. Furthermore, peaks in-cylinder pressure and HRR were significantly lower than neat diesel [92]. Pandian et al. dopped hexanol into the cashew nut shell oil biodiesel and found a reduction in emissions. Along with significant improvement in brake thermal efficiency and reduction in brake-specific fuel consumption [93]. A similar study was carried out by Raju et al. on the mahua oil biodiesel with hexanol and concluded that BSFC, HC, and smoke were lower to pure diesel, and BTE and NOx were slightly higher [94]. In another study, Suhaimi et al. analyzed performance, emission, and combustion of CI engine using hexanol blend up to 20%. The peak pressure of hexanol blends was more which improved the fuel–air mixing ratio. Moreover, it increased the HRR due to enriched oxygen content [95].

4.4 Application of Octanol in CI Engine

Octanol is higher carbon chain alcohol which is explored extensively in the present days. Now, researchers started using octanol as fuel for diesel engines. The results quoted by the researchers are pretty impressive and attractive as well [81, 96–98]. In the initial stage of octanol production, octanol is used to produce in industries through chemicals and other alcohols and is mostly used in perfumes and cosmetic products [64]. Once the properties of octanol were tested and found impressive as fuel and the researchers start working on it as an additive and as the main fuel, the producer starts working on alternative methods of octanol production. Deep et. al used diesel blended with octanol and observed that increase in the volumetric blending percentage of octanol in mineral diesel, improved the BTE, diminished the HC and CO emissions [98]. Rajesh et al. analyzed the performance of the engine using the blend of diesel and n-octanol up to 30% and compared with diesel under both naturally aspirated and exhaust gas recirculation mode and found significant improvement in the engine performance. Due to prolong ignition delay, it developed higher peaks of in-cylinder pressure and higher heat release rates during the premixed combustion phase. The thermal efficiency is increased while the specific fuel consumption decreased with an increase in the n-octanol fraction. The emissions of smoke, NOx, hydrocarbons, and CO decreased with the addition of n-octanol. NOx and smoke emissions also remained low at all EGR rates. Both BTE and BSFC suffered higher EGR rates. HC and CO emissions increased as EGR rates increased. N-octanol has shown great promise for the replacement of fossil diesel by up to 30% [81]. Zhang et. al conducted several experiments for engine performance and emissions on both a light engine (LD) and a heavy engine (HD). Besides, cold start tests of mixtures were performed on a multi-cylinder LD diesel engine. Blends of four different alcohols were used with diesel, n-butanol, isobutanol, 2-ethyl hexanol, and n-octanol. To keep the cetane number close to diesel, the blend is completely mixed with a cetane index improver, di-tertiary butyl peroxide, or hydrogen treated vegetable oil. The results indicated that with a similar CN and the same engine configuration, the alcohol/diesel blends and diesel fuel showed the same onset of combustion and identical heat release processes. The blends generated slightly faster combustion and higher indicated thermal efficiency than diesel fuel under most conditions tested and in both engines. The diesel blends of n-butanol and 2-ethyl hexanol showed good cold-start performance in the multi-cylinder LD engine. Emissions showed that the total amount of particulate matter (PM) decreased when alcohol/diesel blends were used, and the diameter of the PM appears to decrease as the amount of oxygen in the fuel increases in the LD engine. It was concluded that alcohol/diesel blends produce much less soot than diesel fuel in both types of engines owing to the higher oxygen content in the blends but cause slightly increased NO formation [6]. Bharti et al. analyzed the combustion behavior of n-octanol by reactive force field simulation [97]. Ashok et al. investigated the effect of n-octanol with calophyllum inophyllum biodiesel on compression-ignition engine characteristics. Five different fuel blends of n-octanol from 10 to 50% with calophyllum inophyllum biodiesel were prepared on volume basis. The experiments showed better results compared to diesel [99].

5 Conclusion

Higher alcohols produced from microorganisms are considered third-generation biofuels. They have the capability of addressing energy security and environmental problems. However, huge capital is required presently for producing them but higher benefits in terms of animal life, vegetation, and human health can be reaped from their use. In compression-ignition engines, higher alcohols can be easily utilized which would reduce the reliance on food crops and simultaneously increase the fraction of renewables in the fuel. The properties of alcohol play a major role in engine performance and emission. Due to the low viscosity and density of the alcohols, the spray characteristics were improved. Due to the low cetane number, the ignition delay increases that result in higher pressure rise. The lower calorific value of alcohol tends to increase fuel consumption. As the alcohols contain oxygen, it improves the local equivalence ratio resulting in lower soot emissions. It is seen that 20% blending of butanol, pentanol, and hexanol results in improved engine performance and lower emissions, whereas 30% blend of octanol is well suited for engine applications.

References

1. BP Stats BP Statistical Review of World Energy (2019)
2. Government of India Indian Petroleum and Natural Gas Statistics 2018–19 (2020)
3. Jeevahan J, Lakshmi Sankar S, Karthikeyan P, Sriram V, Britto Joseph G (2020) Comparative investigation of the effects of lower and higher alcohols/bio-diesel blends on engine performance and emissions characteristics of a diesel engine. Int J Ambient Energy 41:652–658
4. Fernández JC, Arnal JM, Gómez J, Dorado MP (2012) A comparison of performance of higher alcohols/diesel fuel blends in a diesel engine. Appl Energy 95:267–275
5. Gautam M, Martin DW (2000) Combustion characteristics of higher-alcohol/gasoline blends. Proc Inst Mech Eng Part A J Power Energy 214:497–511

6. Zhang T, Nilsson LJ, Björkholtz C, Munch K, Denbratt I (2016) Effect of using butanol and octanol isomers on engine performance of steady state and cold start ability in different types of diesel engines. Fuel 184:708–717
7. Babu D, Anand R (2017) Effect of biodiesel-diesel-n-pentanol and biodiesel-diesel-n-hexanol blends on diesel engine emission and combustion characteristics. Energy 133:761–776
8. Atmanli A (2016) Comparative analyses of diesel–waste oil biodiesel and propanol, n-butanol or 1-pentanol blends in a diesel engine. Fuel 176:209–215
9. De Poures MV, Sathiyagnanam AP, Rana D, Rajesh Kumar B, Saravanan S (2017) 1-Hexanol as a sustainable biofuel in DI diesel engines and its effect on combustion and emissions under the influence of injection timing and exhaust gas recirculation (EGR). Appl Therm Eng 113:1505–1513
10. Cropley JB, Burgess LM, Loke RA (1984) Butyraldehyde hydrogenation: a case study in process design. Chemtech (United States) 14:374–380
11. Crabtree RH (1988) Butyl alcohols. In: Organometallic chemistry transition metal first. Wiley Inc, New York, NY, pp 173–176
12. Consulting S (1950) In: Chemical economics handbook, California
13. Donmez S, Ozcelik F, Pamir HH (1990) Optimal conditions for acetone-butanol production from molasses. Doga Turk Tarim Orman Derg 14:71–81
14. Kwon GS, Kim BH, Ong ASH (1989) Studies on the utilization of palm oil wastes as the substrates for butanol fermentation. Elaeis 1:91–102
15. Keim W (1989) Carbon monoxide: feestock for chemicals, present and future. J Organomet Chem 372:15–23
16. Worrell G (1980) Tertiary butyl alcohol production, US4296263A
17. Grane HR, Jr. Jubin JC, Worrell GR (1979) Removing water from tertiary butyl alcohol, US4239926A
18. Grane HR, Jr. Jubin JC, Worrell GR (1979) Preparing oxygen-containing fuel wherein tertiary butyl alcohol is the major product of the process, US4294999A
19. Grane HR, Jr. Jubin JC, Worrell GR (1980) Manufacture of tertiary butyl alcohol, US4296262A
20. Arpe H-J, Hawkins S (2010) Industrial organic chemistry, 5th edn. Wiley-VCH Verlagsgesellschaft, Weinheim
21. Scheidmeir WJ (1972) Cheminform abstract: hydroformylation of butenen and pentenen, synth, products and possibilities of their use. Chem Inform sd. 3:383–387
22. Billig E, Abatjoglou AG, Bryant DR, Murray RE, Maher JM (1986) Transition metal complex catalyzed reactions, US4717775A
23. Young DA (1986) Improved rhodium catalyzed hydroformylation of internal olefins, EP0195656B1
24. Imai T (1982) Preparation of alcohols, US4438287A
25. Barker GE, Forster D (1983) Preparation of alcohols from olefins having from 3 to 7 carbon atoms, EP0094456A1
26. Bonnier J-M, Damon J-P, Masson J (1987) Raney nickel as a selective catalyst for aldehyde reduction in the presence of ketones. Appl Catal 30:181–184
27. Delk FS, Vāvere A (1984) Anomalous metal-support interactions in $CuTiO_2$ catalysts. J Catal 85:380–388
28. Pai C-C (1979) A heterogeneous vapor phase process for the catalytic hydrogenation of aldehydes to alcohols, EP0008767A1
29. Aoyama Y, Tanaka Y, Fujisawa T, Watanabe T, Toi H, Ogoshi H (1987) Catalytic reactions of metalloporphyrins 3 Catalytic modification of hydroboration-oxidation of olefins with rhodium(III) porphyrin as catalyst. J Org Chem 52:2555–2559
30. Brown HC, Ravindran N (1977) Monochloroborane-methyl sulfide, H2BClS(CH3)2, and dichloroborane-methyl sulfide, HBCl2S(CH3)2, as new stable hydroborating agents with high regiospecificity. J Org Chem 42:2533–2534
31. Pelter A, Singaram S, Brown H (1983) The dimesitylboron group in organic chemistry 6 Hydroborations with dimesitylborane. Tetrahedron Lett 24:1433–1436

32. Fujitsu H, Matsumura E, Takeshita K, Mochida I (1981) Reactivity of ketones in homogeneous catalytic hydrogenation with cationic rhodium complexes. J Chem Soc Perkin Trans 1:2650–2655
33. Matsuhita H, Shibagaki M, Takahashi K (1988) Verfahren zur Herstellung von Alkoholen, EP0285786B1
34. Butter SA, Stoll I (1986) Synthese von Neoalkoholen, EP0180210A2
35. Cann AF, Liao JC (2010) Pentanol isomer synthesis in engineered microorganisms. Appl Microbiol Biotechnol 85:893–899
36. Zhang K, Sawaya MR, Eisenberg DS, Liao JC (2008) Expanding metabolism for biosynthesis of nonnatural alcohols. Proc Natl Acad Sci 105:20653 LP–20658
37. Okumura Y, Kamiyama S, Furukawa H, Kaneko K (1984) Process for hydration of olefins, EP0127486A1
38. Malessa R, Schleppinghoff B (1989) Process for the preparation of tert-amylic alcohol, EP0325144A3
39. Giles JH, Stultz JH, Jones SW (1980) Process for hydration of olefins to produce alcohols, US4182920A
40. Maksimov AI, Levitskii II, Minachev KM (1981) Dehydration of 2-methyl-2-butanol on zeolites. Bull Acad Sci USSR Div Chem Sci 30:2275–2278
41. Ballantine JA, Davies M, O'Neil RM, Patel I, Purnell JH, Rayanakorn M, Williams KJ, Thomas JM (1984) Organic reactions catalysed by sheet silicates: ester production by the direct addition of carboxylic acids to alkenes. J Mol Catal 26:57–77
42. Henke AM, Odioso RC, Schmid BK (1962) Hydration of olefins in the presence of a solvent, US3285977A
43. Falbe J, Bahrmann H, Lipps W, Mayer D, Frey GD (2013) Alcohols Aliphatic. Ullmann's Encycl Ind Chem
44. Lewis HL (1973) Control of fungi on cotton plants, US3778509A
45. Dekishima Y, Lan EI, Shen CR, Cho KM, Liao JC (2011) Extending carbon chain length of 1-butanol pathway for 1-Hexanol synthesis from glucose by engineered *Escherichia coli*. J Am Chem Soc 133:11399–11401
46. Machado HB, Dekishima Y, Luo H, Lan EI, Liao JC (2012) A selection platform for carbon chain elongation using the CoA-dependent pathway to produce linear higher alcohols. Metab Eng 14:504–511
47. Verser DW (2012) Method for the indirect production of butanol and hexanol, US8252567B2
48. Ghaziaskar HS, Xu C (2013) (Charles) One-step continuous process for the production of 1-butanol and 1-hexanol by catalytic conversion of bio-ethanol at its sub-/supercritical state. RSC Adv 3:4271–4280
49. Phillips JR, Atiyeh HK, Tanner RS, Torres JR, Saxena J, Wilkins MR, Huhnke RL (2015) Butanol and hexanol production in clostridium carboxidivorans syngas fermentation: medium development and culture techniques. Bioresour Technol 190:114–121
50. Marcheschi RJ, Li H, Zhang K, Noey EL, Kim S, Chaubey A, Houk KN, Liao JC (2012) A synthetic recursive "+1" pathway for carbon chain elongation. ACS Chem Biol 7:689–697
51. Julis J, Leitner W (2012) Synthesis of 1-Octanol and 1,1-Dioctyl ether from biomass-derived platform chemicals. Angew Chemie Int Ed 51:8615–8619
52. Bahrmann H, Hahn H-D, Mayer D (2011) 2-Ethylhexanol. Ullmann's Encycl Ind Chem
53. Lemke H, Duval R (1972) A process for the preparation of 2-ethylhexanol, DE1966388A1
54. Nienburg HJ, Nicolai E, Hagen W (1955) A process for preparing ethyl—isopropylacrolein. DE927626C
55. Matthey J (1977) Low-pressure oxo process yields a better product mix. Chem Eng J 84:110
56. Egerev, Pritsker, Tkachenko, Shapiro, Kustov, Gurevich, Sedova (1984) Ledovskikh The method of obtaining 2-ethylhexanol, SU 1084268
57. Dümbgen G, Neubauer D (1969) Large-scale production of oxo alcohols from propylene at BASF. Chemie Ing Tech 41:974–980
58. Cornils B, Mullen A (1980) 2-EH: what you should know. Hydrocarb Process 59:93
59. Horn G (1978) C.D.F. A process for producing copper traegerkatalysatoren, DE2538253B1

60. Corr H, Haarer E, Hoffmann H (1968) S.W. A process for the preparation of saturated aliphatic alcohols, DE1277232B
61. Kubicka R, Veprek J, Hum M, Macek Lv, Jasansky Z (1981) Method of making the 2-ethylhexanole, CS207933B1
62. Egerev, Pritsker, Tkachenko, Shapiro, Kustov, Gurevich, Sedova (1979) Ledovskikh Method of preparing 2-ethylhexanol, SU692824A1
63. Xia Q, Xia Y, Xi J, Liu X, Wang Y (2015) Energy-efficient production of 1-octanol from biomass-derived furfural-acetone in water. Green Chem 17:4411–4417
64. Akhtar MK, Dandapani H, Thiel K, Jones PR (2015) Microbial production of 1-octanol: a naturally excreted biofuel with diesel-like properties. Metab. Eng. Commun. 2:1–5
65. Huang H, Liu Q, Yang R, Zhu T, Zhao R, Wang Y (2015) Investigation on the effects of pilot injection on low temperature combustion in high-speed diesel engine fueled with n-butanol-diesel blends. Energy Convers Manag 106:748–758
66. Sinha SK, Kumar N (2019) Utilization of blends of biodiesel and higher alcohols in a small capacity diesel engine. SAE Tech Pap 2019-April (2019), pp 1–8
67. Şahin Z, Durgun O, Aksu ON (2015) Experimental investigation of n-butanol/diesel fuel blends and n-butanol fumigation—evaluation of engine performance exhaust emissions heat release and flammability analysis. Energy Convers Manag 103:778–789
68. Vinod Babu VBM, Madhu Murthy MMK, Amba Prasad Rao G (2017) Butanol and pentanol: the promising biofuels for CI engines—a review. Renew Sustain Energy Rev 78:1068–1088
69. Mack JH, Flowers DL, Buchholz BA, Dibble RW (2005) Investigation of HCCI combustion of diethyl ether and ethanol mixtures using carbon 14 tracing and numerical simulations. Proc Combust Inst 30:2693–2700
70. Zhang Q, Yao M, Zheng Z, Liu H, Xu J (2012) Experimental study of n-butanol addition on performance and emissions with diesel low temperature combustion. Energy 47:515–521
71. Armas O, García-Contreras R, Ramos Á (2012) Pollutant emissions from engine starting with ethanol and butanol diesel blends. Fuel Process Technol 100:63–73
72. Ibrahim A (2016) Performance and combustion characteristics of a diesel engine fuelled by butanol-biodiesel-diesel blends. Appl Therm Eng 103:651–659
73. Karabektas M, Hosoz M (2009) Performance and emission characteristics of a diesel engine using isobutanol-diesel fuel blends. Renew Energy 34:1554–1559
74. Wang X, Liu H, Zheng Z, Yao M (2015) Development of a reduced n-butanol/biodiesel mechanism for a dual fuel engine. Fuel 157:87–96
75. Jamrozik A, Tutak W, Pyrc M, Gruca M, Kočiško M (2018) Study on co-combustion of diesel fuel with oxygenated alcohols in a compression ignition dual-fuel engine. Fuel 221:329–345
76. Singh R, Singh S, Kumar M (2020) Impact of n-butanol as an additive with eucalyptus biodiesel-diesel blends on the performance and emission parameters of the diesel engine. Fuel 277:118178
77. Sidharth, Kumar N (2019) Performance and emission studies of ternary fuel blends of diesel, biodiesel and octanol. Energy Sources Part A Recover Util Environ Eff 0:1–20
78. Kumar P, Kumar N (2018) Study of ignition delay period of n-Butanol blends with JOME and diesel under static loading conditions. Energy Sources Part A Recover Util Environ Eff 40:1729–1736
79. Goga G, Chauhan BS, Mahla SK, Cho HM (2019) Performance and emission characteristics of diesel engine fueled with rice bran biodiesel and n-butanol. Energy Rep 5:78–83
80. Wei L, Cheung CS, Huang Z (2014) Effect of n-pentanol addition on the combustion, performance and emission characteristics of a direct-injection diesel engine. Energy 70:172–180
81. Rajesh Kumar B, Saravanan S, Rana D, Anish V, Nagendran A (2016) Effect of a sustainable biofuel - N-octanol - on the combustion, performance and emissions of a di diesel engine under naturally aspirated and exhaust gas recirculation (EGR) modes. Energy Convers Manag 118:275–286
82. Rajesh Kumar B, Saravanan S (2016) Use of higher alcohol biofuels in diesel engines: a review. Renew Sustain Energy Rev 60:84–115
83. Sivalakshmi S, Balusamy T (2011) Performance and emission characteristics of a diesel engine fuelled by neem oil blended with alcohols. Int J Ambient Energy 32:170–178

84. Li L, Jianxin W, Zhi W, Jianhua X (2015) Combustion and emission characteristics of diesel engine fueled with diesel / biodiesel / pentanol fuel blends. Fuel 156:211–218
85. Santhosh K, Kumar GN, Radheshyam, Sanjay PV (2020) Experimental analysis of performance and emission characteristics of CRDI diesel engine fueled with 1-pentanol/diesel blends with EGR technique. Fuel 267:117187
86. Yilmaz N, Atmanli A (2017) Experimental assessment of a diesel engine fueled with diesel-biodiesel-1-pentanol blends. Fuel 191:190–197
87. Huang H, Guo X, Huang R, Lei H, Chen Y, Wang T, Wang S, Pan M (2020) Effect of n-pentanol additive on compression-ignition engine performance and particulate emission laws. Fuel 267:117201
88. Radheshyam, Santhosh K, Kumar GN (2020) Effect of 1-pentanol addition and EGR on the combustion, performance and emission characteristic of a CRDI diesel engine. Renew Energy 145:925–936
89. Yesilyurt MK, Yilbasi Z, Aydin M. (2020) The performance, emissions, and combustion characteristics of an unmodified diesel engine running on the ternary blends of pentanol/safflower oil biodiesel/diesel fuel, Springer International Publishing
90. Aloko D, Adebayo GA, Oke OE (2007) Evaluation of diesel-hexanol blend as diesel fuel. Leonardo Electron J Pract Technol 10:151–156
91. Sathiyagnanam AP, Saravanan CG, Gopalakrishnan M (2010) Hexanol-ethanol diesel blends on DI-diesel engine to study the combustion and emission. In: WCE 2010-World Congress Engineering, vol 2. pp 1602–1606
92. Ashok B, Nanthagopal K, Darla S, Chyuan OH, Ramesh A, Jacob A, Sahil G, Thiyagarajan S, Geo VE (2019) Comparative assessment of hexanol and decanol as oxygenated additives with calophyllum inophyllum biodiesel. Energy 173:494–510
93. Pandian AK, Munuswamy DB, Radhakrishanan S, Devarajan Y, Ramakrishnan RBB, Nagappan B (2018) Emission and performance analysis of a diesel engine burning cashew nut shell oil bio diesel mixed with hexanol. Pet Sci 15:176–184
94. Dhana Raju V, Kirankumar K, PSK Engine performance and emission characteristics of a direct injection diesel engine fuelled with 1-Hexanol as a fuel additive in mahua seed oil biodiesel blends. Int J Therm Environ Eng 13:121–127
95. Syuhaida N, Asri M (2018) Analysis of combustion characteristics. Engine Perform Emissions of Diesel-Organ Germanium Fuel Blends 220:1–17
96. Kumar N (2018) Sidharth some studies on use of ternary blends of diesel, biodiesel and n-octanol. Energy Sources Part A Recover Util Environ Eff 40:1721–1728
97. Bharti A, Banerjee T (2016) Reactive force field simulation studies on the combustion behavior of n-octanol. Fuel Process Technol 152:132–139
98. Deep A, Kumar N (2014) Assessment of the performance and emission characteristics of 1-Octanol / diesel fuel blends in a water cooled compression ignition engine
99. Ashok B, Nanthagopal K, Anand V, Aravind KM, Jeevanantham AK, Balusamy S (2019) Effects of n-octanol as a fuel blend with biodiesel on diesel engine characteristics. Fuel 235:363–373

Photocatalytic Hydrogen from Water Over Semiconductors

Siow Hwa Teo, Aminul Islam, and Yun Hin Taufiq-Yap

Abstract Water splitting to produce hydrogen (H_2) over a semiconductor photocatalyst using solar energy is a promising process for the large-scale production of clean chemical energy in the form of H_2, a clean and renewable energy carrier. Many efforts have been implemented to discover photocatalysts that workable under visible-light irradiation to efficiently utilize solar energy. Generally, a suitable cocatalyst to provide an active redox site, is a recognize manner to modify the water-splitting photocatalysts. Apart from that, the constructed two-step photoexcitation photocatalysts, mimicking the natural photosynthesis system, using two different semiconductor powder, have many merits, including amplified light harvesting, spatially separated reductive and oxidative active sites, and perfectly preserved powerful redox ability, which advantage the photocatalytic performance. Especially, a photocatalytic framework called Z-scheme, which is fundamentally comprises of a H_2-photocatalyst and an O_2-photocatalyst to perform water reduction and oxidation, individually. This chapter describe the fundamental development of water-splitting photocatalytic systems.

Keywords Semiconductor · Water-splitting photocatalyst · Cocatalyst · Z-scheme · H_2 evolution

S. H. Teo (✉) · Y. H. Taufiq-Yap (✉)
Faculty of Science and Natural Resources, Universiti Malaysia Sabah, Jalan UMS, 88400 Kota Kinabalu, Sabah, Malaysia
e-mail: tony@ums.edu.my

Y. H. Taufiq-Yap
e-mail: taufiqyap@ums.edu.my

A. Islam
Department of Petroleum and Mining Engineering, Jashore University of Science and Technology, Jashore 7408, Bangladesh

A. Gupta et al. (eds.), *Advances in Energy and Combustion*, Green Energy and Technology,
https://doi.org/10.1007/978-981-16-2648-7_8

Nomenclature

•

$A_{Geometric}$	Area of reactor
E_c	Conductive band minimum
E_F	Fermi level
EH_2/H_2O	Hydrogen evolution potential
EO_2/H_2O	Oxygen evolution potential
E_v	Valence band maximum
E_v	Planck's constant
I_{light}	Irradiance intensity at λ
I_o	Power density of incident light
j_p	Photocurrent density at the measured potential
J	Photocurrent density at 0 V at a certain wavelength
P_{sun}	Energy free of sunlight
v	Frequency
V	Voltage
v_{ac}	Vacuum level
rH_2	Hydrogen production rate
V	Bias potential
W	Work function
ΔG	Gibbs free energy
λ	Wavelength
λ	Wavelength

Abbreviations

Ag	Silver
Au	Gold
Br	Bromide
Br_2	Bromine
C	Carbon
CB	Conductive band
CH_4	Methane
CNTs	Carbon nanotubes
CO_2	Carbon dioxide
Cu	Copper
Cu_2O	Copper oxide
DMF	Dimethyl fluoride
EQN	Equation
g-C3N4	Graphitic carbon nitride

GO	Graphene oxide
HER	Hydrogen evolution reaction
HI	Hydrogen iodine
H_2	Hydrogen
H_2O	WaterWater
IPCE	Incident photon to current conversion efficiency
MoS_2	Molybdenum disulfide
NHE	Normal hydrogen electrode
Nd_2O_5	Neodymium oxide
NiO	Nickel oxide
NPs	Nanoparticles
OER	Oxygen evolution band
O_2	Oxygen
pH	Potential of hydrogen or power of hydrogen
PC	Photocatalyst
PCE	Photo conversion efficiency
Pt	Platinum
QE	Quantum efficiency
RuO_2	Ruthenium oxide
SHT	Solar to hydrogen
Si	Silicon
$SrTiO_3$	Strontium titanium oxide
THF	Tetrahydrofuran
TiO_2	Titanium oxide
UV	Ultraviolet
VB	Valence band
WO_3	Tungsten oxide
vs.	Versus
Zno	Zinc oxide
2D	Two dimensional

1 Introduction

Solar energy is considered as most promising candidate to substitute the fossil fuels. It also leads us to achieve clean, sustainable and environment friendly fuel and it is predicted that only 0.01% solar energy conversion can be enough to solve the world's energy crisis [1, 2]. The most abundant component across the universe, hydrogen (H_2) is burned to generated only water (H_2O) and hence it is reflected as a perspective clean energy source for the future once it can be made inexpensively and expeditiously [3, 4]. One of the best way to keep and manage the solar energy is photocatalysis which stores energy in the form of chemical fuel [5]. Therefore, solar H_2 production can attain directly from H_2O splitting process using solar energy as driving force to split water molecules to H_2 and oxygen (O_2) on the surface of catalyst. Additionally, the method is also as source for O_2, carbon dioxide (CO_2) and methane (CH_4), which are the products for other fuel production routes [6].

In 1972, photocatalytic water splitting was emerged, which the first report is authorized by Fujishima and Honda [9] using a photoelectrochemical system. Through the report published in Nature, suggesting that the H_2O molecule is oxidized via photoinduced holes on TiO_2 surface with the assistant of small electrical voltage, until near recent twentieth century, it witnessed the nonstop progression of photocatalysis [2, 5, 8, 10] including photocatalytic water splitting on powder photocatalyst particles. An overall photocatalytic water splitting process (based on semiconductor) typically includes three essential steps [10], (1) absorption of light followed by electron–hole separation inside the semiconductor, (2) migration of the charge carries toward the surface, and (3) chemical reactions at the surface of the semiconductor making H_2 and O_2 evolution. Those processes demonstrated both the oxidizing sites and the reduction sites on the same catalyst (Fig. 1a). The photons with appropriate energy are harvested by semiconductor to excite electrons from valence band (VB) to conduction band (CB), leaving holes in the VB. Consequently, both electrons and holes are separated and then move towards catalyst's surface, unless they recombine in the bulk or on the surface. At the end, the electrons and holes on surface will allocate to the adsorbed species to initiate corresponding water redox processes, respectively.

Water splitting is thermodynamically not favourable process and an absorption of light having an energy more than (or equal to) the band gap of the photocatalysts that consist of the semiconductor materials in order to drive the reaction forward [6, 8]. Conceptually, the band gap energy should be at least 1.23 eV for overall water oxidation process [9]. Practically, an activation barrier is created when the electron moving between molecules photocatalysts and water. Therefore, the photon energy requires should be greater than the minimum band gap energy in order to perform photocatalytic water splitting [11]. For the hydrogen evolution reaction (HER) and the oxygen evolution reaction (OER) to occur on the catalyst's active sites, the minimum/ bottom of the CB energy level should be located at a more

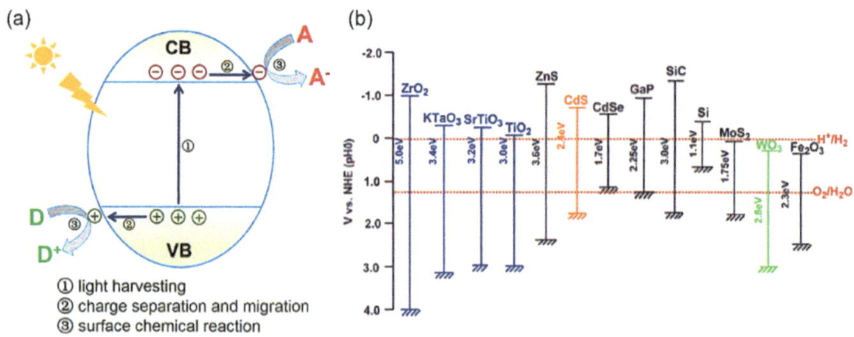

Fig. 1 **a** Schematic illustration for a typical photocatalytic process based on semiconductor [7]; **b** Band structure of semiconductor and redox potential of water splitting [8]

negative than the H_2 evolution potential ($E_{H2/H2O}$, 0 V versus NHE, pH = 0) (Eq. (1)), whereas the maximum/ top of the VB must be more positive than O_2 evolution potential ($E_{2/H2O}$, 1.23 eV vs. NHE, pH = 0) [9] (Eq. (2)).

$$4H^+ + 4e^- \rightarrow 2H_2 \tag{1}$$

$$2H_2O + 4h^+ \rightarrow 4H^+ + O_2 \tag{2}$$

In Fig. 1b, the band positions of various materials are shown. Band engineering is one of the approaches used to design the visible light sensitivity of water splitting materials. Low band gap photocatalytic materials are regularly more sensitive to photon-corrosion as compared to larger band gap materials, but higher band bap are not adept of harvesting visible light [12]. To achieve the target, many researchers are concentrated in band gap modification, including coupled with narrow band gap semiconductors, surface plasmonic effect, dye sensitization in order to improve visible light activity of the semiconductors [13–17]. Meanwhile, heterojunction composite-based photocatalysts can be boosted to obtained apposite band gap alignment for visible light susceptible water splitting [5, 11]. The non-single compound-based photocatalysts are expedient over single material-based photocatalysts as they capable to harvest more visible light than the individual materials. Other than that, heterojunction architecture with appropriate band edge also facilitates an enhanced charge separation [2, 6, 8]. Additionally, the efficient photon absorption, charge carrier transportation, and separation may be attained through physical improvements such as growth of nanostructured photocatalytic materials with various morphologies for high crystallinity and stability, lesser defects, and minimum charge diffusion length [13, 18, 19].

The photocatalytic activity of a substantial is evaluated by solar to hydrogen (STH) energy conversion for overall redox reaction. The STH efficiency is identified as:

$$\text{SHT} = \frac{\text{Output energy}}{\text{Incident solar light energy}} = \frac{r_{H_2} \times \Delta G}{P_{\text{Sun}} \times A_{\text{Geometric}}} \tag{3}$$

where

r_{H_2}	H_2 production rate
ΔG	Gibbs free energy
P_{Sun}	Energy flux of sunlight
$A_{\text{Geometric}}$	Area of reactor

Theoretical r_{H_2} can be counted from the photon's number in the solar spectrum at a different quantum efficiency (QE).

$$\text{QE}(\text{h}v) = \frac{2 \times r_{H_2}}{I_o(hv)} \tag{4}$$

Hence, hypothetical STH at different QE can be predicted by Eqs. (3) and (4), respectively. For sacrificial agent-assisted water splitting, photo-conversion efficiency (PCE) and incident photon to current conversion efficiency (IPCE) are used which are calculated by Eqs. (5) and (6).

$$\mathrm{PCE} = j_\mathrm{p} \times \left(1.23 - \frac{|V|}{I_\mathrm{o}}\right) \tag{5}$$

where

j_p Photocurrent density at the measured potential
V bias potential versus NHE
I_o power density of incident light

$$\mathrm{IPCE} = \frac{1240 \times J}{\lambda \times I_\mathrm{light}} \tag{6}$$

where

J Photocurrent density at 0 V at a certain wavelength (λ)
I_light Irradiance intensity at λ.

The multiple combines different auspicious properties from various materials, which may also improve physiochemical properties and electronic configuration, hence slow recombination of photo-generated charges, sustain to photo-corrosion, and an appropriate band gap can be achieved, which is possible to reach a targeted STH efficiency. Furthermore, the suitable band alignment and high stability of the photocatalytic material in an aqueous mixture are also tremendously important for the water splitting reaction.

2 Different Type of Junction Used in Photocatalysis

Designation of a suitable semiconductor photocatalyst is the key factor to solve the current energy and environmental issues owing to its ability to maximize solar energy to stimulate various photocatalytic reactions. Hence, it is essential for the beginners/researchers in this field to understand the different type of heterojunctions used in photocatalysis, which include Type-II heterojunction and Z-schemes. Next, we will discuss the difference of each heterojunction.

For Type-II heterojunction, this type of junction is composed of two semiconductors with staggered band structure configuration, as shown in Fig. 2a. Under the incident light with sufficient energy, the electron in photocatalyst I (PC I) (with

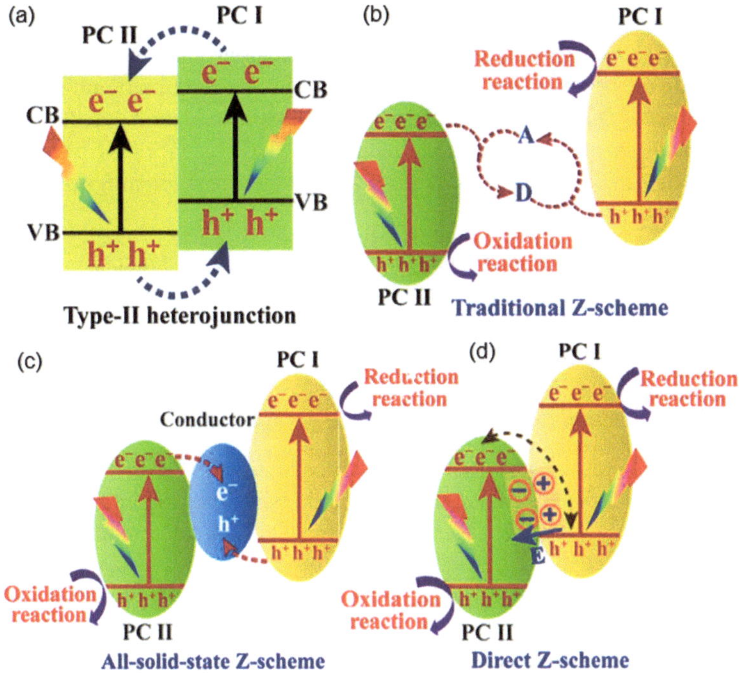

Fig. 2 **a** Schematic illustration of Type-II heterojunction. CB, VB, PC I, and PC II stand for the conduction band, valence band, photocatalyst I and photocatalysts II, respectively, **b** schematic illustration of charge carrier transfer in traditional Z-scheme photocatalysts; A and D stand for electron acceptor and donor, respectively, **c** schematic illustration of charge carrier transfer in all-solid-state Z-scheme photocatalysts, **d** schematic illustration of charge carrier transfer in direct Z-scheme photocatalysts. E means the electric field [20]

higher CB position) will be migrating to photocatalyst II (PC II) (lower CB position). Whilst the holes move in a reverse way, gave rise to an accumulation of holes at PCI for an oxidation reaction and the accumulation of electrons at PC II for a reduction reaction. Although the mentioned effect promotes the spatial separation and suppressed charge recombination, however, this kind of heterojunction may fail to drive a specific photocatalytic reaction [21].

A Z-scheme photocatalyst, in contrast could perform excellent photocatalytic activity owing to (1) simultaneous preservation of strong reduction and oxidation capabilities; (2) spatial separation of oxidative and reductive active sites; (3) high charge separation efficiency; (4) wide spectrum of photocatalysts, and (5) extended light harvesting range. It should be notable that a semiconductor with narrow bandgap should be selected in order to achieve points (4) and (5).

Depending on the type of charge carrier mediator introduced, the Z-scheme photocatalysts are classified into three, one of them is traditional Z-scheme photocatalyst, which is usually composed of two different photocatalysts coupled through a reversible redox ion pair (Fig. 2b), taking examples: Fe^{3+}/Fe^{2+} and IO^{3-}/I^- are used as the charge carriers transfer medium. When the photocatalysts are exposed to light, the electrons generated in CB of PC II are consumed by electron acceptor species while the holes in VB of PC I are consumed by electron donor species. The photogenerated holes in VB of PC II participated in the oxidation reaction, while the electrons in the CB of PCI performed a reductive reaction, hence gave rise to enhance photocatalytic performance. Nevertheless, there are some shortenings of this type of photocatayst, as listed below [22–24]:

i. Redox mediator-induced back reactions are thermodynamically feasible due to the generated electrons and holes with strong redox power are consumable by shuttle redox ion pairs.
ii. Light shielding effect in solution systems.
iii. Slow charge carrier transfer and solution pH sensitivity
iv. Reduced reaction rate due to unstable redox mediators.

While, for all-solid-state Z-scheme photocatalyst, an electron solid conductor (Au, Ag, and Cu NPs) is utilized as a charge transfer bridging in all-solid-state Z-scheme photocatalyst (Fig. 2c). Typically, noble metals are commonly employed as excellent electron mediators, in addition, graphene, CNTs, etc. are also good for photocatalytic performance and stability performance. The all-solid-state Z-scheme photocatalysts are viable to work in both liquid and gas phases, ascribed to the solid conductor prompted for a fast charge carrier transfer. However, the problems of this photocatalyst would be the expensive noble metals and its shielding effect [23, 25–28].

Lastly, a direct-Z-scheme photocatalyst is meant for a direct contact between two semiconductors without the need of charge carrier transfer mediator [29–33]. As shown in Fig. 2d, both the semiconductors are in close contact without any mediator, which resulted in suppressed backward reactions, shielding effects, and resistant to corrosion. In this case, a work function difference between both photocatalysts is important to induce charge redistribution and to form internal electric field, which is significantly affect the charge carrier separation and transfer process [34, 35].

To fully understand the mechanisms, there are two types of photocatalytic systems to be discussed, which are (i) PC I has a higher CB and VB positions and smaller work function (higher Fermi level) than PC II, and (ii) typical p–n junction. For the first case (Fig. 3a), free electrons of PC I can be easily transfer to PC II when PC I and PC II are in contact until their Fermi levels are equilibrated (Fig. 3b). Hence, a built-in electric field and a band edge bending are formed, attributed to the PC I side is positively charged, while negatively charged at the PC II interface. The Z-scheme charge transfer mode is much in favour for this case. The existence of internal electric field favors the recombination between the photogenerated electrons in the CB of PC II and photogenerated holes in the VB of PC I [36] (Fig. 3c). The factors such as internal electric field, the extra potential barrier induced from band bending, and Coulomb repulsion hinder the transferred of photogenerated electrons from the CB of PC I to PC II CB, as well as the transfer of photogenerated holes (Fig. 3d).

Meanwhile, the electrons from the CB of PC I and holes from the VB of PC II are maintained and spatially separated, in addition, they can participate in specific

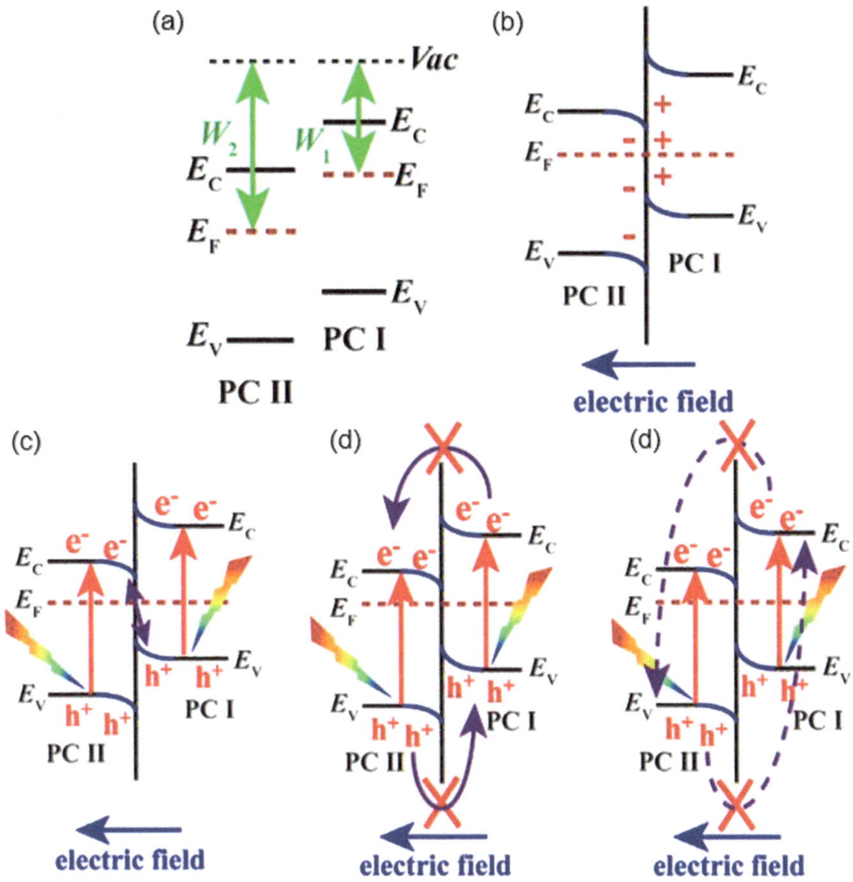

Fig. 3 Schematic illustration of semiconductor–semiconductor junction with staggered band configurations ($W_1 < W_2$) [18]: **a** before contact, **b** in contact, **c** photogenerated charge carrier transfer process in direct Z-scheme mode, **d** photogenerated charge carrier transfer process in Type-II mode, and **e** photogenerated charge carrier recombination. W_1 and W_2 denote the work function of PC I and PC II, respectively. V_{ac}, E_C, E_V, and E_F stand for the vacuum level, conduction band minimum, valence band maximum, and Fermi level, respectively

reductive and oxidative photocatalytic reactions, respectively. It is worth noting that the induced electric field can suppress the recombination between the photogenerated electrons in the CB of PC I and the photogenerated holes in the VB of PC II, as shown in Fig. 3e. In contrast, the second type of the system, p–n junction (Fig. 4a), is significantly different from the first one (Fig. 3a). When p-type semiconductor is contacted with the n-type semiconductor, free electrons of n-type semiconductor can be transferred to p-type semiconductor to form a built-in electric field (Fig. 4b). Under this circumstance, a p–n junction charge carrier transfer mode (Fig. 5c) is formed rather than direct Z-scheme charge carrier transfer mode (Fig. 4d) owing to the existed built-in electric field [37, 38]. Thus, the photogenerated charge carrier

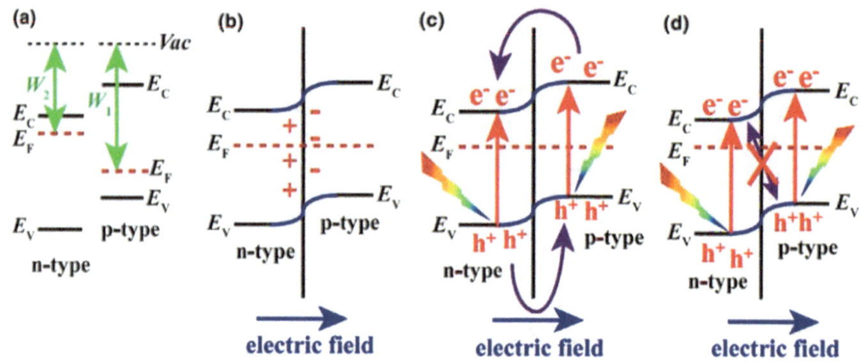

Fig. 4 Schematic illustration of p–n junction [18]: **a** before contact, **b** in contact, **c** transfer of photogenerated charge carriers in p–n junction mode, and **d** not-allowed transfer of photogenerated charge carriers in direct Z-scheme mode

transfer mode in p–n junction is unflavored for direct Z-scheme mode (Fig. 4c and d).

Similar to direct Z-scheme photocatalysts, all-solid-state photocatalysts exhibiting similar VB and CB edge bending. At the metal–semiconductor interface of all-solid-state photocatalysts, the free electrons will be migrated from a semiconductor to a metal when metal has larger work function than the semiconductor, resulted in an upward bent semiconductor band induced by an electric field [23, 25]. Nevertheless, when metal has smaller work function than semiconductor, the free electrons will then be migrated from a metal to a semiconductor, which led to a downward bending of semiconductor band edge. While for the traditional Z-scheme photocatalysts, free electron redistribution would not be occurring (no internal electric field between them) because there isn't direct contact between two different photocatalysts in the presence of an appropriate shuttle redox ion mediator [31]. Hence, the VB and CB edges of each component are not affected when traditional Z-scheme photocatalysts are formed.

3 Why Semiconductor in Photocatalytic Hydrogen Production

Why semiconductor for photocatalysis? Semiconductor-based photocatalysis such as TiO_2 (the most commonly reported) has emerged as an effective photocatalyst for various applications owing to its renewable, clean, and safe technology where photocatalyst requires only solar energy as the prime energy source [39, 40]. The pioneering work of TiO_2 semiconductor photocatalysis was initiated by Fujishima and Honda for photochemical water splitting [9]. TiO_2 based two dimensional (2D) nanosheets are derived from the exfoliation of titanate layer. Although it exhibits

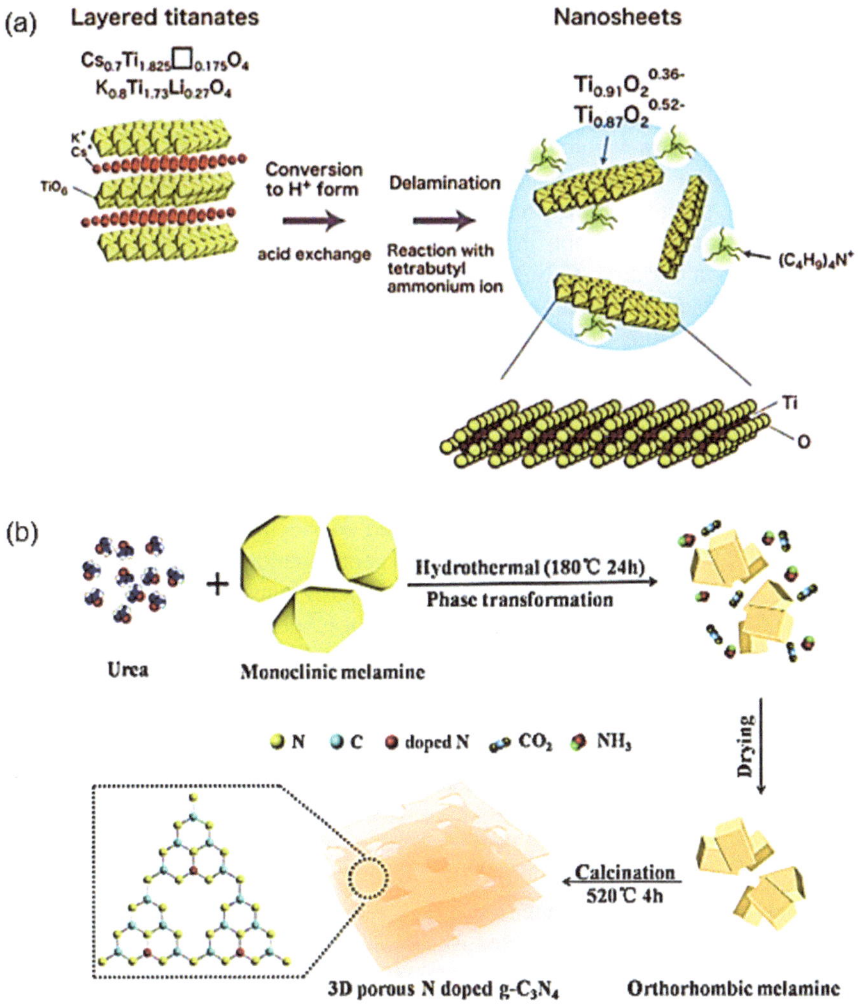

Fig. 5 **a** Schematic illustration of the crystal structure of a typical lepidocrocite-type titanate and its exfoliation into 2D titania nanosheets [43], **b** Schematic illustration for the formation of 3D ultrathin porous N-doped g-C_3N_4 [44]

similar properties as rutile and anatase TiO_2, however, it has larger band gap due to the size quantization. Taking an example, $Ti_{0.91}O_2^{0.36-}$ nanosheets has a band gap of 3.8 eV, as compared to anatase TiO_2 with a band gap of 3.2 eV [41]. Usually layered titanates are prepared through high temperature solid state reaction of a mixture of TiO_2 and alkali metal carbonates, subsequently processed with acidic solution to form protonated intermediate via ion-exchange process (Fig. 5a). In addition, other layered metal oxides nanosheets such as WO_3, titanoniobate, and perovskite oxides are also prepared through high temperature solid state reactions and wet-chemical exfoliation processes, where high temperature synthesis process is unfavouring industrial purpose.

Doping of TiO_2 has been showing effective approach to extend light absorption to visible region, however, low photocatalytic efficiency due to rapid charge recombination could be the main challenge of the existing photocatalysts [42]. Thus, to overcome the recombination issue and to improve charge separation, developing a suitable semiconductor composites or engineering a 2D material such as graphene or graphitic carbon nitride (g-C_3N_4) is one of the best options. The g-C_3N_4 is denoted as the most stable allotrope among the carbon nitride and g-C_3N_4 exists in the form of 2D sheets consisting of tri-s-triazines interconnected with tertiary amines. What is the difference between TiO_2 and g–C_3N_4? Unlike TiO_2 (active in UV region), g-C_3N_4 is a visible light active photocatalyst due to its bandgap of ca. 2.7 eV with a suitable conduction and valence band positions. In addition, g-C_3N_4 is thermally and chemically stable in most solvents such as water, alcohols, DMF, THF, diethyl ether, and toluene.

Recently, ultrathin 2D g-C_3N_4 has attracted consideration attentions due to its large specific surface area, and low carrier recombination properties, which thus demonstrated superior potential for electronics, energy conversion and storage applications. A highly anisotropic 2D g-C_3N_4 nanosheets prepared through a top-down strategy (thermal oxidation etching of bulk g-C_3N_4 in air) has demonstrated a high specific surface area of 306 m^2g^{-1}, accompanied with larger band gap, thus improved electron transport ability along the in-plane direction, increased photoexcited charge carrier lifetime, and excellent photocatalytic activities than bulk g-C_3N_4 [45]. Nevertheless, the 2D g-C_3N_4 nanosheets usually suffer from severe restacking problem, which is highly detrimental to the photocatalytic activity. Thus, owing to its tunable morphologies, g-C_3N_4 is highly favored in producing an enlarged specific surface area, improved charge transfer efficiency and surface catalytic reaction. In accord to the advantages of three dimensional (3D) porous structure, superior photocatalytic performance has been reported for a 3D mesoporous g-C_3N_4. This photocatalyst, in short denoted as UM3 has shown remarkable photocatalytic H_2 evolution performance and stability, as compared to the pristine (Fig. 5b). Hence, the research on developing more potential semiconductors which is cost-effective, efficient in photocatalytic activity and charge transport is still ongoing.

4 Facile Developments of Photocatalyst for Water Splitting

Despite the TiO_2 has demonstrated improved photocatalytic activity towards water-splitting, its efficiency is still far underneath the prerequisite for commercialization due to the inherent constraint of TiO_2. Hence, the researchers have been concentrated to create other potential cocatalysts to improve the productivity of water-splitting reaction. The photocatalysts combined with at least two elements for H_2 evolution through photocatalytic water splitting that are more active than

TiO_2. For instant, Liu et al. [46] reported efficient Au–Cu/TiO_2/MoS_2 composite photocatalyst for H_2 production through water splitting. Firstly, TiO_2 nanosheets were blended, at that point hybridized them with MoS_2 nanosheets by a simple aqueous solution. Subsequently, the Au-Cu nanoparticles were effectively inserted into the TiO_2/MoS_2 composite by a refluxing technique followed by calcination, lastly obtained TiO_2/MoS_2/Au-Cu ternary composite catalyst. The balanced mixture of the ternary composite improves the photocatalytic efficiency and leds the separation of photo-induced electron–hole separation. Zhao et al. [47] synthesized NiO-TiO_{2-x}/C photocatalysts for H_2 production through water splitting. Because of the cooperative energy of carbon nanosheets, NiO, and TiO_2, the optimum (1 wt.%) NiO-TiO_{2-x}/C-T650 nanocomposites was demonstrated efficient photocatalytic hydrogen evolution from water, beating 18-fold in efficiency higher than TiO_2/C. However, NiO/TiO_2 nanoparticles generates agglomerate into enormous bunches, obstructing effective photo-induced electron transfer, which reduces a space of NiO/TiO_2 p-n heterojunctions for the photocatalytic reaction.

Stabilization of Cu_2O onto TiO_2 was reported by Wei et al. [48] by adjustment of crystal facets and defects of TiO_2 (Fig. 6). Both experimental and hypothetical conception demonstrated that 101-faceted TiO_2 could make electron importing channel into the Z-scheme system. The Cu_2O/TiO_2 heterostructures with {101} facet was reported to be 251-fold activity for photo-induced water splitting reaction compared with other facets, along with high stability. However, the impact of defect in Z-scheme system on the stability of Cu_2O should be investigate in detail for H_2 evolution.

Some different semiconductors, for example, Nb_2O_5, NiO–GO, NiO–TiO_2, Pt/RuO_2 and $SrTiO_3$ were demonstrated to show photocatalytic water splitting [49–51]. However, the proficiency of these mass photocatalysts is restricted by active sites of reaction and low change transfer efficiency because of the electron/hole recombination. In this manner, discovering steady and productive photocatalyst materials actually stays challenging. Two dimensional materials which have been an interesting issue in a decade ago, presenting tunable electronic and optical properties that make them promising candidates for water splitting [52]. A few exploratory examinations have demonstrated that 2D material-based photocatalysts display exceptionally

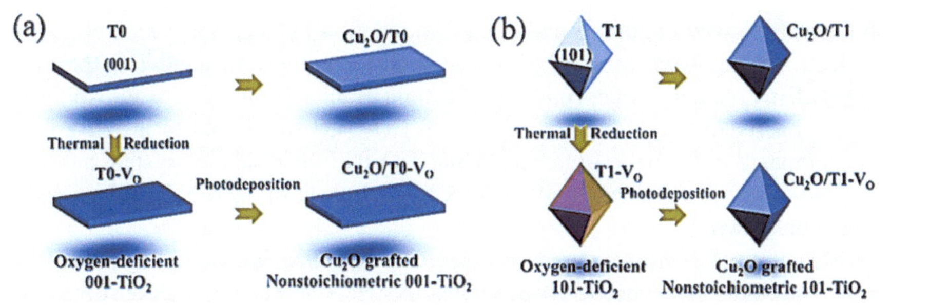

Fig. 6 Schematic illustration of synthetic routes for Cu_2O/TiO_2 heterostructures [48]

improved photocatalytic activities contrasted with the conventional bulk materials [53, 54].

To overcome the photocorrosion issue of composite photocatalyst while keeping up its high effectiveness, a photocatalytic framework called Z-scheme, which imitates the Z-scheme system in the characteristic photosynthesis of green plants, has been created to produce H_2 and O_2 at the same time. The Z-scheme is a double photocatalyst framework that fundamentally comprises of a H_2-photocatalyst and an O_2-photocatalyst to perform water reduction and oxidation, individually. Wen et al. [55] synthesized ZnO coupled with g-C_3N_4 to create direct Z-scheme, photocatalytic systems attributable to their well-matched band positions. The outcomes affirmed that the inserted Au NPs act as an electron mediator to advance vectorial electron move in direct Z-scheme ZnO/g-C_3N_4 heterojunction. Recently, some works reported that the ZnO/TiO_2 heterojunctions demonstrated increased photocatalytic activity due to their matched energy band alignments and high electron mobility of ZnO that efficiently promote the carrier separation and transfer [56]. However, the reported ZnO/TiO_2 heterostructures were usually sealed structures with lower surface areas, such as the core–shell structured ZnO/TiO_2 or branched ZnO/TiO_2 and disordered ZnO rods due to the aggregation of ZnO seeds on the TiO_2 nanorods, which result in the obstruction of light absorption for TiO_2, usually demonstrating unsatisfactory water splitting performance. Besides, ZnO also suffers from the low photocorrosion resistance in most electrolytic media and the morphologies and structures of ZnO seriously affect the photostabilities of the ZnO/TiO_2 heterojunctions.

Niu et al. [57] evaluated the photogenerated electrons/holes recombination properties of C_3N/C_3B catalyst and compared the performance for H_2 generation through water splitting. The 2D C_3N/C_3B p-n heterostructure was found to be promising metal free Z-scheme photocatalyst for water splitting reaction. The strong redox ability of C_3B and C_3N layers are driving force to split water without addition of overpotential or co-catalyst. Bismuth-based oxides have attracted attention in photocatalytic water splitting because of their high dielectric permittivity, reasonable band structure (2.8 eV) and position of valence band [58]. The Z-scheme photocatalyst offer successful charge seperation and move without adding any sacrificial reagent. Nitrogen/fluorine codoping decreases the defect concentration, which reduced the recombination of photoexcited electrons, contrasted with nitrogen-doped photocatalyst. It has been suggested that fluorine doping into TiO_2 improved its photocatalytic water splitting under UV illumination by supressing recombination of electron/hole. Miyoshi et al. [59] compared the activity of nano-TiO_2:N,F and TiO_2:N,F in Z-scheme system under visible light irradiated Z-scheme water-splitting. It was concluded that Z-scheme system of TiO_2:N,F was found to increase fourfold compared to the bulk TiO_2:N,F. While TiO_2:N,F in Z-scheme system has been reported more effective than bulk TiO_2:N,F, the impact of TiO_2:N,F molecule size remain unknown.

Utilization of Nanostructured Si essentially improves the solar H_2 conversion proficiency of an ideal photoelectrochemical system. Oh et al. [60] demonstrated that the water splitting efficiency of nanoporous 'black' Si was more than the polished Si electrode. In similar, n-type nano-Si appeared about 20% improvement in quantum efficiency because of the decreased reflectance and high photocorrosion in concentrated HI. Recently, n-type Si nanowire demonstrated the high water splitting activity, contrasted with the planar Si, in the presence of Br_2/Br solution. Zhang et al. [61] investigated the activity of water splitting by running Si photocatalyst. It has been reported that the photocatalytic hydrogen development execution of silicon catalyst incredibly impacts on a surface functional group of catalyst. This outcome is consistent with the information proclaimed by Rye et al. [62] who endeavored to synthesized Pt/Si photocatalyst for H_2 production through water splitting. Nonetheless, the costly platinum co-catalyst remains challenge for real application.

There is a basic postulation that the H atoms on Si surface are electron-deficient and may fill in as an electron sink to improve charge separation. Similar phenomenon has been discussed by other researchers and reported that oxidation occurs on Si–OH and H_2 produces on Si–H surfaces by reduction process, as shown the Eqs. (7–9):

$$\text{Oxidation}: h^+ + Si + H_2O \rightarrow Si - OH + H^+ \tag{7}$$

$$\text{Reduction}: e^- + Si - H + H^+ \rightarrow Si + H_2 \tag{8}$$

$$\text{Reduction}: e^- + Si - OH + H^+ \rightarrow Si + H_2O \tag{9}$$

The function of photogenerated electron and holes on the photocatalyst is schematically shown in Fig. 7. The function of photogenerated electrons on Si photocatalyst could be elucidated based on the dynamic charge separation [63, 64]. As recently revealed the ultrathin layer of catalyst led the detachment of photogenerated electrons

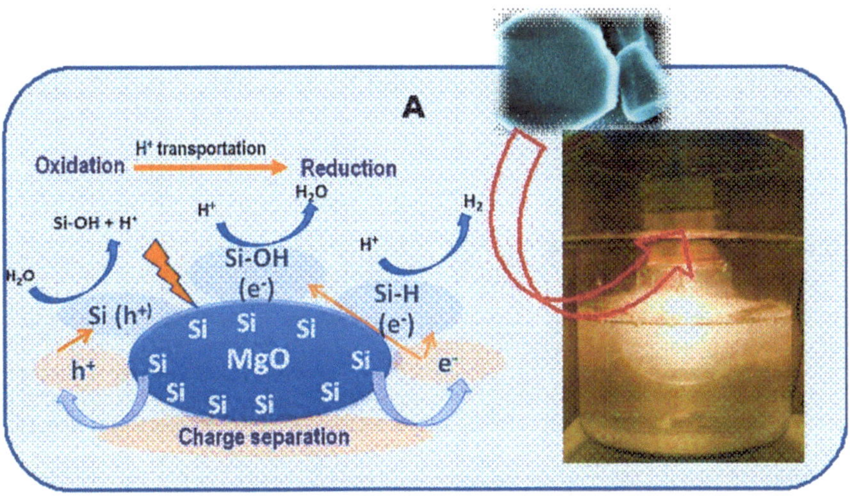

Fig. 7 Water splitting mechanism on Si photocatalyst [63]

because of intrinsic effect of Si [65]. While conceding to this conceivable explanation, there is still need to investigate the molecular interaction and kinetic of carriers for photocatalytic water splitting reaction.

5 Conclusion

In the course of recent years, a few semiconductor materials and photocatalytic frameworks have been produced for the water-splitting irradiation under UV and visible light. It has been seen that photo-induced charge separation, avoidance of water-splitting in reverse reaction, and use of a wide range of light energy are the basic prerequisites for accomplishing high proficiency of water splitting. The high evolution of H_2 energy through water splitting has been evident by the addition of sacrificial reagent that could supress the openings to repress charge recombination. In any case, to accomplish feasible hydrogen evolution, sacrificial reagent must be persistently added during reaction. Besides, photocatalytic reactor design to accomplish separate H_2 and O_2 development, for example, the H-type reactor and the novel Z-scheme, should be required for efficient hydrogen production. Different strategies, for example, potentially doping of metal or metal oxide particles on the photocatalyst have been effectively utilized to improve the performance of photocatalytic reaction. Photocatalytic water splitting is a cross-discipline innovation that requires the contribution of specialists from various fields (i.e., scientific experts, electrical architects, material researchers, and physicists). The advancement of new innovations requires cooperation with a solid hypothetical foundation for a superior comprehension of the hydrogen production catalyst so as to think of a minimal effort for hydrogen evolution.

Acknowledgements The authors acknowledge the financial support of the Impact Research Scheme (SPB, Project Number: SP0005-2020) from UMS Research Grant, Universiti Malaysia Sabah.

References

1. Razek SA, Popeil MR, Wangoh L et al (2020) Designing catalysts for water splitting based on electronic structure considerations. Electron Struct 2:023001. https://doi.org/10.1088/2516-1075/ab7d86
2. Lin L, Hisatomi T, Chen S et al (2020) Visible-light-driven photocatalytic water splitting: recent progress and challenges. Trends Chem 2:813–824. https://doi.org/10.1016/j.trechm.2020.06.006
3. Safari F, Dincer I (2020) A review and comparative evaluation of thermochemical water splitting cycles for hydrogen production. Energy Convers Manag 205:112182. https://doi.org/10.1016/j.enconman.2019.112182

4. Chen Y, Ji S, Sun W et al (2020) Engineering the atomic interface with single platinum atoms for enhanced photocatalytic hydrogen production. Angew Chemie—Int Ed 59:1295–1301. https://doi.org/10.1002/anie.201912439
5. Miseki Y, Sayama K (2019) Photocatalytic water splitting for solar hydrogen production using the carbonate effect and the Z-Scheme reaction. Adv Energy Mater 9:1–15. https://doi.org/10.1002/aenm.201801294
6. Kumaravel V, Mathew S, Bartlett J, Pillai SC (2019) Photocatalytic hydrogen production using metal doped TiO_2: a review of recent advances. Appl Catal B Environ 244:1021–1064. https://doi.org/10.1016/j.apcatb.2018.11.080
7. Zhang N, Gao C, Xiong Y (2019) Defect engineering: a versatile tool for tuning the activation of key molecules in photocatalytic reactions. J Energy Chem 37:43–57. https://doi.org/10.1016/j.jechem.2018.09.010
8. Kudo A, Miseki Y (2009) Heterogeneous photocatalyst materials for water splitting. Chem Soc Rev 38:253–278. https://doi.org/10.1039/b800489g
9. Fujishima A, Honda K (1972) Electrochemical photolysis of water at a semiconductor electrode. Nature 238:37–38. https://doi.org/10.1038/238037a0
10. Wang Y, Vogel A, Sachs M et al (2019) Current understanding and challenges of solar-driven hydrogen generation using polymeric photocatalysts. Nat Energy 4:746–760. https://doi.org/10.1038/s41560-019-0456-5
11. Maeda K (2011) Photocatalytic water splitting using semiconductor particles: history and recent developments. J Photochem Photobiol C Photochem Rev 12:237–268. https://doi.org/10.1016/j.jphotochemrev.2011.07.001
12. Su J, Guo L, Bao N, Grimes CA (2011) Nanostructured WO3/BiVO4 heterojunction films for efficient photoelectrochemical water splitting. Nano Lett 11:1928–1933. https://doi.org/10.1021/nl2000743
13. Kudo A (2013) New materials for photocatalytic and photoelectrochemical water splitting. AIP Conf Proc 1568:7–10. https://doi.org/10.1063/1.4848079
14. Domen K, Kondo JN, Hara M, Takata T (2000) Photo-and mechano-catalytic overall water splitting reactions to form hydrogen and oxygen on heterogeneous catalysts. Bull Chem Soc Jpn 73:1307–1331
15. Domen K, Hara M, Kondo JN et al (2001) New aspects of heterogeneous photocatalysts for water decomposition. Korean J Chem Eng 18:862–866. https://doi.org/10.1007/BF02705609
16. Kudo A (2006) Development of photocatalyst materials for water splitting. Int J Hydrogen Energy 197–202
17. Maeda K, Domen K (2007) New non-oxide photocatalysts designed for overall water splitting under visible light. J Phys Chem C 111:7851–7861. https://doi.org/10.1021/jp070911w
18. Xu Q, Zhang L, Yu J et al (2018) Direct Z-scheme photocatalysts: principles, synthesis, and applications. Mater Today 21:1042–1063. https://doi.org/10.1016/j.mattod.2018.04.008
19. Yuan Y, Lv J, Jiang X et al (2007) Large impact of strontium substitution on photocatalytic water splitting activity of $BaSnO_3$. Appl Phys Lett 91:27–30. https://doi.org/10.1063/1.2778631
20. Low J, Yu J, Jaroniec M et al (2017) Heterojunction photocatalysts. Adv Mater 29:1–20. https://doi.org/10.1002/adma.201601694
21. Bai S, Jiang J, Zhang Q, Xiong Y (2015) Steering charge kinetics in photocatalysis: Intersection of materials syntheses, characterization techniques and theoretical simulations. Chem Soc Rev 44:2893–2939. https://doi.org/10.1039/c5cs00064e
22. Maeda K (2013) Z-scheme water splitting using two different semiconductor photocatalysts. ACS Catal 3:1486–1503. https://doi.org/10.1021/cs4002089
23. Zhou P, Yu J, Jaroniec M (2014) All-solid-state Z-scheme photocatalytic systems. Adv Mater 26:4920–4935. https://doi.org/10.1002/adma.201400288
24. Bala Chandran R, Breen S, Shao Y et al (2018) Evaluating particle-suspension reactor designs for Z-scheme solar water splitting via transport and kinetic modeling. Energy Environ Sci 11:115–135. https://doi.org/10.1039/c7ee01360d

25. Li H, Tu W, Zhou Y, Zou Z (2016) Z-Scheme Photocatalytic systems for promoting photocatalytic performance: recent progress and future challenges. Adv Sci 3. https://doi.org/10.1002/advs.201500389
26. Iwase A, Ng YH, Ishiguro Y, et al (2011) Reduced graphene oxide as a solid-state electron mediator in Z- scheme photocatalytic water splitting under visible light. Am Chem Soc. pp 11054–11057
27. Li H, Yu H, Quan X et al (2016) Uncovering the key role of the fermi level of the electron mediator in a Z-Scheme photocatalyst by detecting the charge transfer process of WO3-metal-gC3N4 (Metal = Cu, Ag, Au). ACS Appl Mater Interfaces 8:2111–2119
28. Li X, Yan X, Lu X et al (2018) Photo-assisted selective catalytic reduction of NO by Z-scheme natural clay based photocatalyst: insight into the effect of graphene coupling. J Catal 357:59–68. https://doi.org/10.1016/j.jcat.2017.10.024
29. Jia Q, Iwase A, Kudo A (2014) BiVO4-Ru/SrTiO3: Rh composite Z-scheme photocatalyst for solar water splitting. Chem Sci 5:1513–1519. https://doi.org/10.1039/c3sc52810c
30. Song H, Yu YT, Norby P (2009) Efficient complete oxidation of acetaldehyde into CO2 over Au/TiO 2 core-shell nano catalyst under UV and visible light irradiation. J Nanosci Nanotechnol 9:5891–5897. https://doi.org/10.1166/jnn.2009.1263
31. Liu X, Chen N, Li Y et al (2016) A general nonaqueous sol-gel route to g-C3N_4-coupling photocatalysts: the case of Z-scheme g-C3N_4/TiO_2 with enhanced photodegradation toward RhB under visible-light. Sci Rep 6:1–16. https://doi.org/10.1038/srep39531
32. Zhou D, Chen Z, Yang Q et al (2016) Facile construction of g-C3N_4 nanosheets/TiO_2 Nanotube Arrays as Z-scheme photocatalyst with enhanced visible-light performance. ChemCatChem 8:3064–3073. https://doi.org/10.1002/cctc.201600828
33. Luo J, Zhou X, Ma L et al (2016) Enhancing visible light photocatalytic activity of direct Z-scheme SnS_2/Ag_3PO_4 heterojunction photocatalysts. Mater Res Bull 81:16–26. https://doi.org/10.1016/j.materresbull.2016.04.028
34. Cui J, Li Y, Liu L et al (2015) Near-infrared plasmonic-enhanced solar energy harvest for highly efficient photocatalytic reactions. Nano Lett 15:6295–6301. https://doi.org/10.1021/acs.nanolett.5b00950
35. Liu J, Cheng B, Yu J (2016) A new understanding of the photocatalytic mechanism of the direct Z-scheme g-C_3N_4/TiO_2 heterostructure. Phys Chem Chem Phys 18:31175–31183. https://doi.org/10.1039/c6cp06147h
36. Zhang L, Li S, Liu B, et al (2014) Highly efficient CdS / WO 3 photocatalysts : Z- scheme photocatalytic mechanism for their enhanced photocatalytic H 2 evolution under visible light
37. Zheng L, Zheng Y, Chen C et al (2009) Network structured SnO_2/ZnO heterojunction nanocatalyst with high photocatalytic activity. Inorg Chem 48:1819–1825. https://doi.org/10.1021/ic802293p
38. Son DY, Im JH, Kim HS, Park NG (2014) 11% efficient perovskite solar cell based on ZnO nanorods: an effective charge collection system. J Phys Chem C 118:16567–16573. https://doi.org/10.1021/jp412407j
39. Bard AJ (1980) Photoelectrochemistry. 207:139–144
40. Hoffmann MR, Martin ST, Choi W, Bahnemann DW (1995) Environmental applications of semiconductor photocatalysis. Chem Rev 95:69–96. https://doi.org/10.1021/cr00033a004
41. Sakai N, Ebina Y, Takada K, Sasaki T (2004) Electronic band structure of titania semiconductor nanosheets revealed by electrochemical and photoelectrochemical studies. J Am Chem Soc 126:5851–5858
42. Cao S, Low J, Yu J, Jaroniec M (2015) Polymeric photocatalysts based on graphitic carbon nitride. Adv Mater 27:2150–2176. https://doi.org/10.1002/adma.201500033
43. Luo B, Liu G, Wang L (2016) Recent advances in 2D materials for photocatalysis. Nanoscale 8:6904–6920. https://doi.org/10.1039/c6nr00546b
44. Tian N, Zhang Y, Li X et al (2017) Precursor-reforming protocol to 3D mesoporous g-C3N4 established by ultrathin self-doped nanosheets for superior hydrogen evolution. Nano Energy 38:72–81. https://doi.org/10.1016/j.nanoen.2017.05.038

45. Niu P, Zhang L, Liu G, Cheng HM (2012) Graphene-like carbon nitride nanosheets for improved photocatalytic activities. Adv Funct Mater 22:4763–4770. https://doi.org/10.1002/adfm.201200922
46. Liu Y, Xu C, Xie Y, et al (2020) Au–Cu nanoalloy/TiO2/MoS2 ternary hybrid with enhanced photocatalytic hydrogen production. J Alloys Compd 820.https://doi.org/10.1016/j.jallcom.2019.153440
47. Zhao X, Xie W, Deng Z et al (2020) Salt templated synthesis of NiO/TiO_2 supported carbon nanosheets for photocatalytic hydrogen production. Colloids Surfaces A Physicochem Eng Asp 587:124365. https://doi.org/10.1016/j.colsurfa.2019.124365
48. Wei T, Zhu YN, An X et al (2019) Defect modulation of Z-Scheme TiO_2/Cu_2O photocatalysts for durable water splitting. ACS Catal 9:8346–8354. https://doi.org/10.1021/acscatal.9b01786
49. Ghouri MI, Ahmed E, Ali A et al (2020) The synergistic role of ternary combined semiconductors (Nb_2O_5–MoS_2-MO) hetrojunctions in photocatalytic water splitting. Mater Chem Phys 255:123320. https://doi.org/10.1016/j.matchemphys.2020.123320
50. Nogueira AE, Silva GTST, Oliveira JA et al (2020) CuO decoration controls Nb_2O_5 photocatalyst selectivity in CO_2 reduction. ACS Appl Energy Mater 3:7629–7636. https://doi.org/10.1021/acsaem.0c01047
51. Fo Y, Wang M, Ma Y et al (2020) Origin of highly efficient photocatalyst NiO/$SrTiO_3$ for overall water splitting: insights from density functional theory calculations. J Solid State Chem 292:121683. https://doi.org/10.1016/j.jssc.2020.121683
52. Faraji M, Yousefi M, Yousefzadeh S et al (2019) Two-dimensional materials in semiconductor photoelectrocatalytic systems for water splitting. Energy Environ Sci 12:59–95. https://doi.org/10.1039/c8ee00886h
53. Fu CF, Sun J, Luo Q et al (2018) Intrinsic electric fields in two-dimensional materials boost the solar-to-hydrogen efficiency for photocatalytic water splitting. Nano Lett 18:6312–6317. https://doi.org/10.1021/acs.nanolett.8b02561
54. Lv X, Wei W, Sun Q et al (2017) Two-dimensional germanium monochalcogenides for photocatalytic water splitting with high carrier mobility. Appl Catal B Environ 217:275–284. https://doi.org/10.1016/j.apcatb.2017.05.087
55. Wen P, Sun Y, Li H et al (2020) A highly active three-dimensional Z-scheme ZnO/Au/g-$C3N_4$ photocathode for efficient photoelectrochemical water splitting. Appl Catal B Environ 263:118180. https://doi.org/10.1016/j.apcatb.2019.118180
56. Zhou T, Wang J, Chen S et al (2020) Bird-nest structured ZnO/TiO_2 as a direct Z-scheme photoanode with enhanced light harvesting and carriers kinetics for highly efficient and stable photoelectrochemical water splitting. Appl Catal B Environ 267:118599. https://doi.org/10.1016/j.apcatb.2020.118599
57. Niu X, Bai X, Zhou Z, Wang J (2020) Rational design and characterization of direct Z-Scheme photocatalyst for overall water splitting from excited state dynamics simulations. ACS Catal 10:1976–1983. https://doi.org/10.1021/acscatal.9b04753
58. Ma Y, Jiang X, Sun R, et al (2020) Z-scheme Bi2O2.33/Bi2S3 heterojunction nanostructures for photocatalytic overall water splitting. Chem Eng J 382:123020. https://doi.org/10.1016/j.cej.2019.123020
59. Miyoshi A, Miyoshi A, Kato K et al (2020) Nano: Versus bulk rutile TiO2:N, F in Z-scheme overall water splitting under visible light. J Mater Chem A 8:11996–12002. https://doi.org/10.1039/d0ta04450d
60. Oh J, Deutsch TG, Yuan HC, Branz HM (2011) Nanoporous black silicon photocathode for H2 production by photoelectrochemical water splitting. Energy Environ Sci 4:1690–1694. https://doi.org/10.1039/c1ee01124c
61. Zhang RQ, Liu XM, Wen Z, Jiang Q (2011) Prediction of silicon nanowires as photocatalysts for water splitting: band structures calculated using density functional theory. J Phys Chem C 115:3425–3428. https://doi.org/10.1021/jp111182c
62. Ryu J, Jang YJ, Choi S et al (2016) All-in-one synthesis of mesoporous silicon nanosheets from natural clay and their applicability to hydrogen evolution. NPG Asia Mater 8:e248–e249. https://doi.org/10.1038/am.2016.35

63. Islam A, Teo SH, Awual MR, Taufiq-Yap YH (2020) Assessment of clean H_2 energy production from water using novel silicon photocatalyst. J Clean Prod 244:1–12. https://doi.org/10.1016/j.jclepro.2019.118805
64. Islam A, Teo SH, Awual MR, Taufiq-Yap YH (2019) Improving the hydrogen production from water over MgO promoted Ni–Si/CNTs photocatalyst. J Clean Prod 238. https://doi.org/10.1016/j.jclepro.2019.117887
65. Islam A, Teo SH, Awual MR, Taufiq-Yap YH (2020) Ultrathin assembles of porous array for enhanced H2 evolution. Sci Rep 10:1–14. https://doi.org/10.1038/s41598-020-59325-4

Flames and Reacting Systems

Evaluation of Hazard Correlations for Hydrogen-Rich Fuels Using Stretched Transient Flames

Michael Peter Meyer and Rune Peter Lindstedt

Abstract The use of sustainably produced hydrogen as part of fuel blends provides a promising route towards reducing the impact of combustion on climate change. The use of such blends also increases the reactivity of the fuel leading to increased fire and explosion risks and can necessitate revised device designs. Applicable models, e.g. for flames in the flamelet regime of combustion, typically rely upon data derived using steady (strained) laminar flames. Parameters include laminar burning velocities and Markstein lengths. However, interactions between turbulence and flame structures are transient in nature. In the current work, physico–chemical properties are extracted from transient spherical laminar flame computations to investigate the impact of strain. It is shown (i) that the response of transient flames to rates of strain is qualitatively similar, but quantitatively different, to strained steady flames and (ii) that the response of flame structures is correlated with the leading edge. The computations have also been used to (iii) identify the impact of the Lewis number on the transient flame response. (iv) Difficulties in relation to the selection of computational flame iso–contours and the non-uniqueness of Markstein lengths are highlighted and it is shown that agreement with experimental data demands that properties are extracted in a consistent manner. Finally, (v) as an application example, the methodology is used to estimate the minimum ignition spark energies and quenching distances for methane/hydrogen/air and propane/hydrogen/air mixtures.

Keywords Hydrogen · Hydrogen - Hydrocarbon blends · Ignition energies · Quenching distances · Hazard classifications · Risk assessment

1 Introduction

The impact of climate change and the need for environmental protection have promoted the development of low carbon footprint and clean energy technologies [1].

M. P. Meyer · R. P. Lindstedt (✉)
Department of Mechanical Engineering, Imperial College, London SW7 2AZ, UK
e-mail: p.lindstedt@imperial.ac.uk

A. Gupta et al. (eds.), *Advances in Energy and Combustion*, Green Energy and Technology,
https://doi.org/10.1007/978-981-16-2648-7_9

State-of-the-art gas turbines operate under fuel-lean conditions and are typically optimized for natural gas [2]. A partial substitution with hydrogen leads to the expansion of the lean operating limit [3] and provides a route towards reduced environmental impact provided the hydrogen is produced using renewable sources. However, the change in fuel composition can lead to fuel flexibility problems for manufacturers [4, 5] and to increased safety concerns [6]. The mixture reactivity is strongly dependent upon the hydrogen content with direct implications for flame propagation speeds, explosion over-pressures, auto-ignition points, and turbulence–chemistry interactions. The latter can lead to differences in flame surface areas [7] and combustion instabilities [1, 5, 8]. Li et al. [6] measured flame speeds and over-pressures generated in an obstructed flame tube for a wide range of binary and ternary H_2, CO, and CH_4 mixtures, with methane showing a stronger inhibiting effect on the mixture reactivity. It was shown that a scaling based on the amount of air required to fully oxidize the mixture correlated the fuel composition impact on explosion over-pressures [6] and turbulent flow fields [9]. Simatos et al. [10] further investigated the impact of hydrogen content for lean ($\phi = 0.80$) binary CH_4/H_2 and CO/H_2 fuel blends on auto-ignition in turbulent shear layers and showed that the stronger inhibiting effect of methane prevailed.

The turbulent burning velocity (S_T) is a key parameter required, for example, to evaluate the strength of explosions [6] and the flashback propensity in gas turbine engines [11]. Investigations seeking to correlate S_T with the turbulence intensity are quite common e.g. [12]. Wang et al. [13] investigated the effect of hydrogen content and explored scaling relations under gas turbine relevant conditions, and found that hydrogen addition alters the correlation due to preferential diffusive-thermal effects [14]. The impact of hydrogen blending on methane was found to be particularly prominent under ultra-lean conditions—of particular relevance to gas turbines operating in premixed or piloted premixed modes under such conditions. The Lewis number effect was also found to be significant for such fuel blends [15] with the impact of pressure also investigated [16, 17]. Multiple turbulent burning velocity definitions (e.g. Bray [18]) have been advanced and Driscoll [19] has shown that different definitions [20, 21] are not equivalent. Yet, the definitions of Bray [18], Lawn and Schefer [21] and a variant proposed by Driscoll [19] can agree within $\pm 20\%$ for premixed twin opposed jet flames [22] with the matter further explored by Hampp et al. [23] for a wide range of hydrogen blends. Models and correlations for S_T inevitably feature the laminar burning velocity with the impact of stretch effects also frequently included.

A large number of computational studies of one–dimensional steady–state flames using detailed chemistry have been performed following the seminal investigations by Dixon–Lewis [24], Miller and Bowman [25] and Warnatz [26], Such studies are typically aimed at understanding pollutant emissions and the thermo–chemical structure of flames. However, practical turbulent flames are inevitably subject to unsteady interactions with the flow field. The transient behavior of curved premixed flame structures is of particular practical relevance due to the coupling with chemistry and transport processes [27]. The non–unique nature of such interactions may be exemplified by steady–state computations of premixed laminar flames in back–to–

burnt geometries (e.g. [28]) wherein the flame is stabilized against burnt gases [29, 30]. Comparatively few detailed computational studies of transient laminar flames have been performed despite the increasing awareness of the importance of time–dependent effects on flame structures. Past studies include the use of asymptotic methods, aimed to clarify the physics (e.g. Buckmaster [31] and Clavin [32]), and numerical studies featuring simplified chemistry e.g. Bradley et al. [33] and Ghoniem et al. [34]. Stahl and Warnatz [35] studied the effects of temporally varying strain rates on premixed one–dimensional flames, while Yoo and Im [36] studied the transient dynamics of edge flames in a non-premixed hydrogen-air counterflow. Malik and Lindstedt [37, 38] studied the transient response of inhomogeneous hydrogen and methane flames to pressure fluctuations and stretch. Ciardiello et al. [39] recently studied the controlling parameters for ignition transients in premixed annular combustors and concluded that the laminar burning velocity is a key parameter.

The advantage of spherically expanding/contracting flames is that the rate of strain is uniformly distributed over the flame surface and the geometry is hence amenable to comprehensive computational studies due to symmetry. A computational method using a mass-based Lagrangian frame of reference is here combined with detailed transport and detailed chemistry featuring 300 reactions and 48 species to study the response of transient flames to rates of strain which is shown (i) to be qualitatively similar, but quantitatively different, to strained steady flames. It is further shown (ii) that the response of flame structures is correlated with the leading edge. The computations have also been used to (iii) identify the impact of the Lewis number on the transient flame response. (iv) Difficulties in relation to the selection of computational flame iso–contours and the non-uniqueness of Markstein lengths are highlighted and it is shown that agreement with experimental data demands that properties are extracted in a consistent manner. Finally, (v) as an application example, the methodology is used to estimate the minimum ignition spark energies and quenching distances for methane/hydrogen/air and propane/hydrogen/air mixtures.

2 Extraction of Stretch Related Parameters

Karlovitz et al. [40] described the influence of a velocity gradient on a flame by introducing the concept of stretch and this well-known concept is illustrated in Fig. 1. Stretch can be either positive or negative. An outwardly propagating spherical flame is an example of the former, whereas a contracting spherical flame experiences a negative stretch. Transient flames experience variable velocity gradients, though stretch–free steady (e.g. one–dimensional flat flames) or stretched stationary cylindrical and spherical flames, maintained by a line or point source, are frequently used to extract data. In order to define stretch, a surface s (e.g. an iso-contour), located within the flame structure moves with a velocity u_L and the surrounding flow field has a velocity v as is shown in Fig. 1. The flame stretch K is then the rate of change of a small element A of that surface normalized by A:

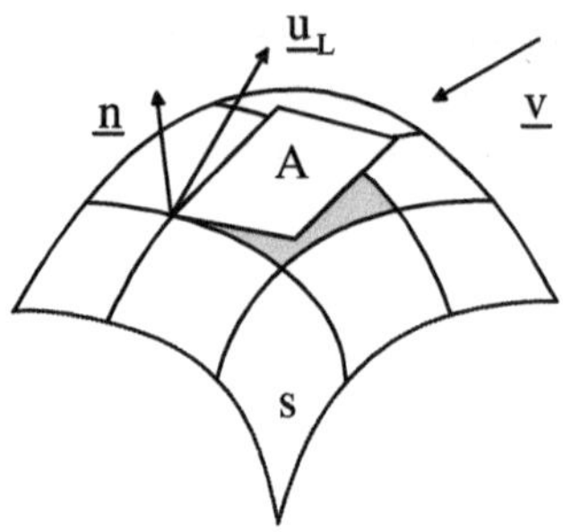

Fig. 1 Sketch of a surface element s showing the velocity u_L of the surface element, velocity v of the fluid and the normal unity vector n

$$K = \frac{1}{A}\frac{\mathrm{d}A}{\mathrm{d}t} \tag{1}$$

Buckmaster [31] and Matalon [41] have shown that K can be expressed as,

$$K = \nabla_t \cdot v_t + (u_L \cdot n)(\nabla \cdot n) \tag{2}$$

where n is the unit normal vector pointing in the direction of the unburnt gas with ∇_t and v_t the tangential components. The first term in Eq. (2) can be written as

$$\nabla_t \cdot v_t = -n \cdot \nabla \times (v \times n). \tag{3}$$

The stretch rate can be recast and expressed in two components, where K_c is due to flame curvature (curvature stretch) [42] and K_s is due to strain (flowfield strain),

$$K_c = (u_L \cdot n)(\nabla \cdot n) \tag{4}$$
$$K_s = -n \cdot v(\nabla \cdot n) + n \cdot n(\nabla v) \tag{5}$$

where u_L is the laminar burning velocity of the flame normal to the surface s. For spherical flames K_c and K_s can be expressed as,

$$K_s = 2\tfrac{u_g}{r} \tag{6}$$
$$K_c = 2n_r\tfrac{u_L}{r} \tag{7}$$

where n_r is the normal unit vector in radial direction and is equal to 1 for outwardly propagating flames and equal to -1 for contracting flames. The gas velocity ahead of the flame is denoted u_g and the radius r is typically defined by the leading edge. The diffusivity of the mixture is usually discussed in terms of the Lewis number (Le).

$$\mathrm{Le} = \frac{\lambda}{\rho c_p \mathscr{D}} \tag{8}$$

For Le = 1 the mixture is diffusion–transport neutral and thus pure stretch has only a small impact. Frankel and Sivashinsky [43] show, via an asymptotic analysis assuming the flame radius to be much larger than the thermal–diffusion thickness, that the upstream properties (e.g. the mass burning rate m_b and flame thickness δ_b) are not affected by stretch when Le $= 1$ with only the downstream burning rate changed from its unstretched value to $m_b(A_u/A_b)$ following the area change caused by thermal expansion and flow divergence. If Le $\neq 1$ the response of a flame towards stretch can change markedly (e.g. Law [44], Sun and Law [45]) as the temperature depends on the relative rates of mass and heat transport. For Le > 1 the mass gain is lower than the heat loss and consequently the burnt gas temperature (T_b) is reduced from the adiabatic value (T_{ad}) so that $T_b < T_{\mathrm{ad}}$ and for Le < 1, conversely, $T_b > T_{\mathrm{ad}}$. The effect of stretch for non–unity Lewis number can be observed, for example, with lean CH_4–air mixtures. Due to the higher diffusivity of the "lighter" CH_4, it diffuses more readily towards the reaction zone and thus changes the local equivalence ratio towards stoichiometric conditions. The burning rate of these mixtures is accordingly enhanced by strain. The opposite would be true for propane. Such effects are explored below under transient conditions.

3 A Lagrangian Calculation Method for Transient Laminar Flames

The current calculation method is based on the simulation of transient laminar flames in a spherically symmetric co–ordinate system. Flames of this type are predominantly influenced by tangential scalar gradients that affect the local balance between chemistry and transport processes. A mass-based (M) Lagrangian frame of reference is adopted and the governing equations are shown below [46, 47].
Continuity (mass conservation)

$$\rho = \frac{1}{R^2}\frac{\partial M}{\partial R} \qquad \text{and} \quad u = \frac{\mathrm{d}R}{\mathrm{d}t} \tag{9}$$

Enthalpy (assuming constant pressure across the flame)

$$\rho\frac{\partial h}{\partial t} = -\rho\frac{\partial}{\partial M}\left\{R^2 \mathscr{J}_h\right\} + \rho\frac{\partial}{\partial M}\left\{\sum_{k=1}^{N_{sp}} R^2 h_k\left[-\mathscr{J}_k + \mathscr{J}_n\right]\right\} \tag{10}$$

Species mass fractions

$$\rho\frac{\partial Y_k}{\partial t} = -\rho\frac{\partial}{\partial M}\left\{R^2 \mathscr{J}_k\right\} + R_k M_k \tag{11}$$

Scalar fluxes

$$\mathscr{J}_k = -\rho^2 D_k R^2 \left(\frac{\partial Y_k}{\partial M} - Y_k \frac{1}{n} \frac{\partial n}{\partial M} \right) - \rho \mathscr{W}_k Y_k - \rho V_c Y_k \tag{12}$$

$$\mathscr{J}_h = -\rho R^2 \frac{\lambda}{C_p} \frac{\partial h}{\partial M} \qquad \text{and} \qquad \mathscr{J}_n = -\rho R^2 \frac{\lambda}{C_p} \frac{\partial Y_k}{\partial M} \tag{13}$$

Thermal diffusion

$$\mathscr{W}_k = -\rho R^2 \frac{D_k \Theta_k}{X_k} \frac{1}{T} \frac{\partial T}{\partial M} \tag{14}$$

Correction velocity (to ensure mass conservation)

$$V_c = -\frac{1}{\sum_{k=1}^{N_{sp}} \rho Y_k} \sum_{k=1}^{N_{sp}} \rho^2 D_k R^2 \left(\frac{\partial Y_k}{\partial M} - Y_k \frac{1}{n} \frac{\partial n}{\partial M} \right) + \rho \mathscr{W}_k Y_k \tag{15}$$

In the above equations, ρ is the density, D_k is the diffusion coefficient for species k, M_k its molecular weight, n the mole number (the inverse of the mean molecular weight), λ the thermal conductivity, C_p the heat capacity at constant pressure, X_k the mole fraction of species k and Θ_k its thermal diffusion ratio. The solution algorithm features second–order temporal and spatial accuracy along with adaptive dynamic grid refinement and a block implicit structure for all solved variables. Flame structures are resolved using a minimum of 100 locally refined computational cells. The transport coefficients for viscosities and binary diffusivities are evaluated as specified by Lindstedt and Maurice [48]. Thermal diffusion (Soret effect) is included using the formula of Chapman and Cowling [49].

The governing equations are discretised using a finite volume approach featuring a staggered grid formulation. The resulting set of equations is solved using a block implicit method. A Newton linearisation technique is utilised for the source terms in the species equations:

$$R_k = \sum_{j=1}^{N_{\mathrm{reac}}} \Xi_{jk} \left[k_j^f \prod_{l=1}^{N_{\mathrm{sp}}} \Phi_l^{\xi_{jk}^f} - k_j^r \prod_{l=1}^{N_{\mathrm{sp}}} \Phi_l^{\xi_{jk}^r} \right] \tag{16}$$

$$R_k^{\nu+1} = R_k^{\nu} + \sum_{l=1}^{N_{\mathrm{sp}}} \frac{\partial R_k}{\partial \Phi_l} \frac{\partial \Phi_l}{\partial Y_l} \left\{ Y_l^{\nu+1} - Y_l^{\nu} \right\} + \frac{\partial R_k}{\partial T} \left\{ T^{\nu+1} - T^{\nu} \right\} \tag{17}$$

Dependent variables, such as temperature and density, are included in the block matrix structure to fully account for dependencies in the transport equations. The forward integration time step (Δt) is controlled by the Courant (Co) number. Second–

order convergence of the iterative process during the solution of the block-matrix structure is maintained for Co ≤ 0.5.

Flames were ignited by defining a radical depleted and isobaric ignition kernel of a radius approximately equal to the flame thickness. The initial temperature and species concentrations in the fully burnt gas were deduced by assuming chemical equilibrium. Initial profiles within the ignition region were set using linear interpolation from reactants to the fully burnt state. The procedure has the benefit of providing weak ignition kernels.

4 Results for Methane–Air and Propane–Air Mixtures

In order to evaluate the accuracy of the current simulation technique, computations of expanding (outwardly propagating) spherical methane–air and propane–air flames were performed. The flame speeds are compared to experimental data obtained by Taylor [50]. The same initial conditions (i.e. P = 101,325 Pa and T_o = 296 K) and equivalence ratio ranges of $0.58 \leq \phi \leq 1.41$ for CH_4–air flames and $0.69 \leq \phi \leq 1.64$ for C_3H_8–air flames were used. Taylor [50] used a schlieren photographic technique to locate the flame front and the evolution of the 305 K isotherm was extracted from the computations (c.f. Bradley et al. [33]).

Figures 2 and 3 show comparisons between simulations and experimental data. The agreement appears satisfactory both qualitatively and quantitatively. Towards the rich and lean flammability limits some discrepancies occur. This may arguably be attributed to a stronger ignition source required in the experimental study.

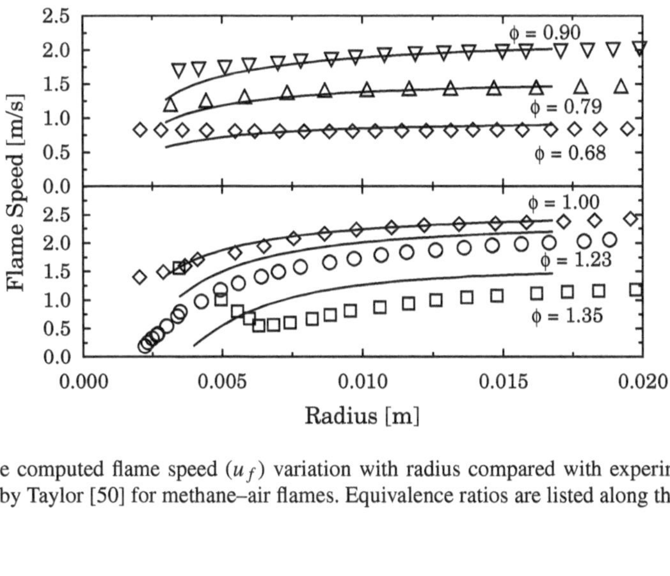

Fig. 2 The computed flame speed (u_f) variation with radius compared with experimental data (symbols) by Taylor [50] for methane–air flames. Equivalence ratios are listed along the curves

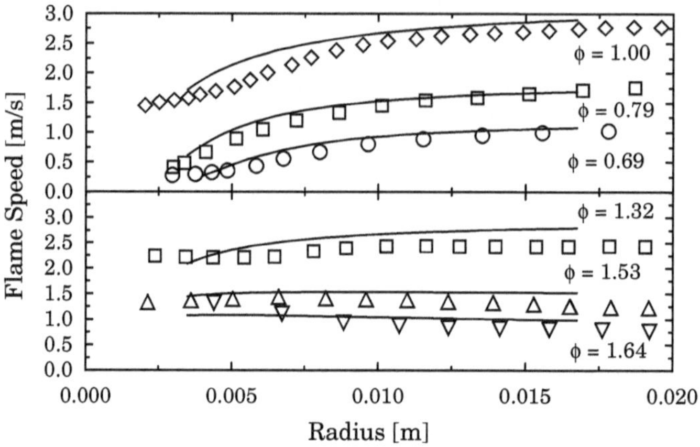

Fig. 3 The computed flame speed (u_f) variation with radius compared with experimental data (symbols) by Taylor [50] for propane–air flames. Equivalence ratios are listed along the curves

5 Impact of the Rate of Strain on Transient Flame Structures

The influence of strain on laminar burning velocities is illustrated in Figs. 4 and 5 for CH_4/H_2 and C_3H_8/H_2 mixtures with air. It can be noted that the effects are proportional to the hydrocarbon fuel content Hence, pure methane and propane flames are more strongly affected by the rate of strain than pure hydrogen flames.

The thermo–chemical structure offers a starting point for the evaluation of the response of flames to transient strain. In particular, the net rate of radical production provides an opportunity for a detailed evaluation. The radical pool considered in the present work features H + O + OH + HO_2 + CH_3 + CH + CHO + $^1CH_2$ + $^3CH_2$ + CH_2O + CH_3 + CH_3O + CH_2OH + C_2O + C_2H + C_2HO + C_2H_3 + C_2H_5 + C_3H + C_3H_2 + C_3H_2O + C_3H_3 + a-C_3H_5 + s-C_3H_5 + t-C_3H_5 + n-C_3H_7 + i-C_3H_7. Figures 6 and 7 show the net rate of radical production versus temperature (rather than physical distance) for lean and rich methane–air flames and for lean and rich propane–air flames, respectively. Distinct points can be identified with T_i the temperature where the net rate of radical production becomes positive (i.e. the ignition temperature). For example, Dixon–Lewis [51] correlated extinction limits with the temperature difference between the maximum flame (T_b) and ignition temperatures ($\Delta T_{\mathrm{bi}} = T_b - T_i$). Other reference planes, e.g. defined by the temperature where the maximum radical production occurs (T_m) or the temperature where the radical consumption rate exceeds production at the hot end of the flame (T_o), can also be defined along with arbitrary iso–contours, e.g. at 400 K, 500 K, 600 K and 1000 K. As shown in Figs. 6 and 7, flame structures respond differently to increasing rates of strain depending on the type of fuel and stoichiometry. The results confirm that the Lewis number of the deficient reactant plays an important role in the response.

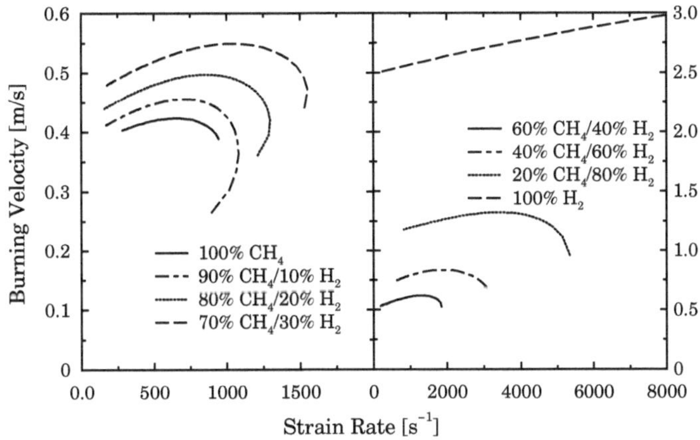

Fig. 4 Laminar burning velocities (u_L) of stoichiometric CH_4/H_2 mixtures based on the propagation of the 305 K isotherm as a function of the rate strain

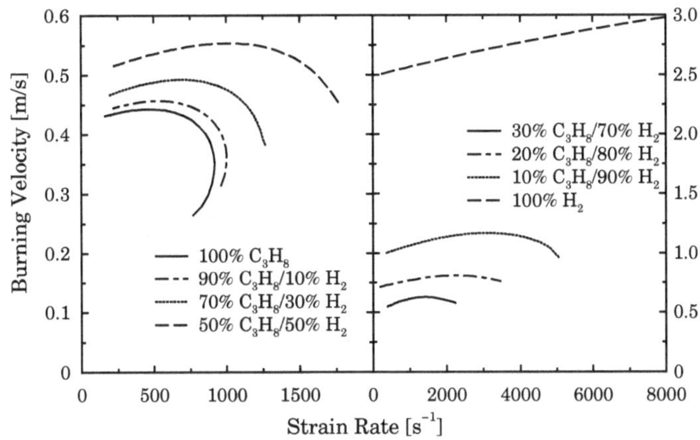

Fig. 5 Laminar burning velocities (u_L) of stoichiometric C_3H_8/H_2 mixtures based on the propagation of the 305 K isotherm as a function of the rate of strain

Due to the higher diffusivity of the comparable "lighter" species, the burning rate increases for lean CH_4–air mixtures and rich propane–air mixtures (where oxygen is the deficient reactant).

Inwardly propagating flames can be used to gain additional information about flame response to curvature. The advantage is that the unburnt gas is not affected by the flame front as it remains quiescent. Therefore, the straining of the *flame front* is only due to curvature effects. However, in the limit of small radii, the flame thickness becomes the same order of magnitude as the radius and the flame can not be assumed to be asymptotically thin. Figures 8 and 9 show the rate of radical pool production for expanding and contracting CH_4–air and C_3H_8–air flames. The radical

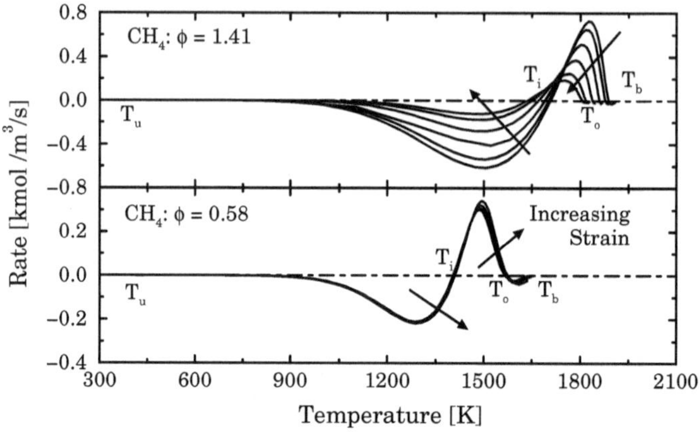

Fig. 6 The variation of net rates of radical production in lean and rich methane–air flames with increasing rates of strain. The radical pool considered here consists of ($H + O + OH + HO_2 + CH_3 + CH + CHO + {}^1CH_2 + {}^3CH_2 + CH_2O + CH_3 + CH_3O + CH_2OH + C_2O + C_2H + C_2HO + C_2H_3 + C_2H_5 + C_3H + C_3H_2 + C_3H_2O + C_3H_3$ + a-C_3H_5 + s-C_3H_5 + t-C_3H_5 + n-C_3H_7 + i-C_3H_7)

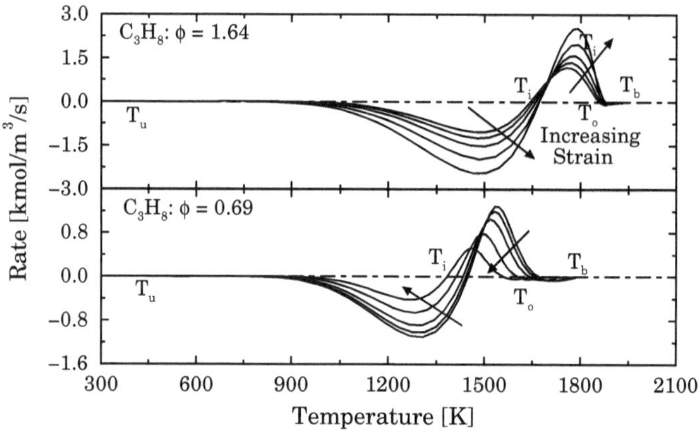

Fig. 7 The variation of net rates of radical production in lean and rich propane–air flames with increasing rate of strain. See Fig. 6 for the radical pool species

consumption in the pre–heat zone (i.e. between T_u and T_i) and radical production in the main reaction zone ($T_i - T_o$) for the expanding flame agrees qualitatively with the influence of strain shown in Figs. 6 and 7. Increasing strain has an opposite effect on the pre–heat zones of expanding and contracting flames. In the main reaction zone, no simple explanation can be given with respect to the response of the rate of radical production due to the narrowing of the reaction zone for contracting flames.

Markstein [52, 53] proposed a relationship, based on small perturbations that permits linearised relations resulting in the following relationship between flame front curvature and the laminar burning velocity,

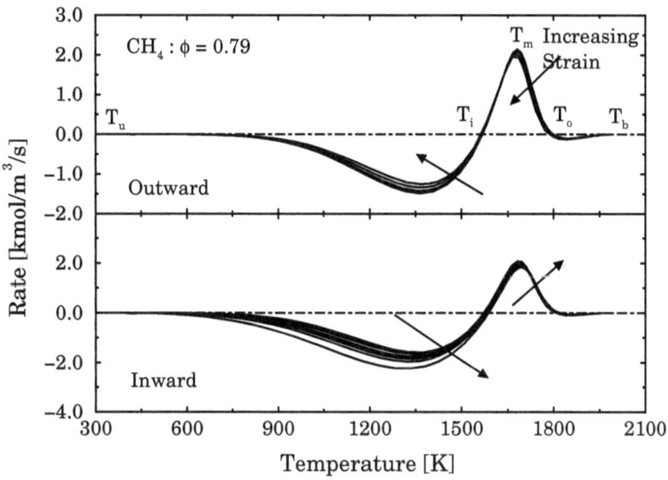

Fig. 8 The variation of net rates of radical production for inwardly and outwardly propagating lean methane–air flames. See Fig. 6 for the radical pool species. The variation with increasing rate of strain for the outwardly and inwardly propagating flames is also indicated

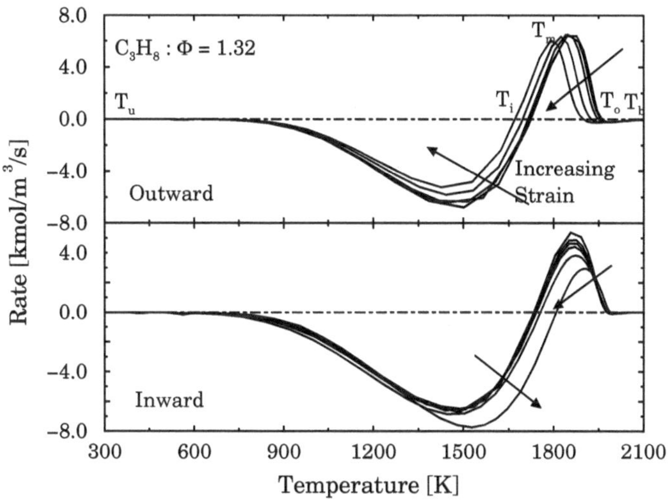

Fig. 9 The variation of net rates of radical production in inwardly and outwardly propagating rich propane–air flames. See Fig. 6 for the radical pool species. The variation with increasing rate of strain for the outwardly and inwardly propagating flames is also indicated

$$u_L = u_L^{\infty}\left[1 + \left(\frac{\mathscr{L}_M}{r}\right)\right] \tag{18}$$

where u_L is the strained burning velocity, u_L^{∞} the unstrained burning velocity and $\mathscr{L}_M$ is the characteristic Markstein length, which is of the order of the flame thickness [53]. The Markstein length implicitly includes the effects of heat conduction, diffusion of the reaction chain carriers and non–unity Lewis numbers. However, in more general terms, $\mathscr{L}_M$ describes not only the response of the burning velocity to the curvature of the flame front, but the influence of stretch on the flame (e.g. Clavin [32], Clavin and Williams [54]). Therefore, the above relationship can be reformulated as

$$u_L^{\infty} - u_L = \mathscr{L}_M\left(\frac{1}{A}\frac{\mathrm{d}A}{\mathrm{d}t}\right) = \mathscr{L}_M K. \tag{19}$$

The Markstein length is a measure of the change in the burning velocity per unit stretch [50]. More than one Markstein length can be defined and these do not vary linearly with strain. Therefore, a "global" Markstein length $\mathscr{L}_M$ can be expressed as a linear combination of the separate lengths due to the rate of strain or stretch due to curvature (Clavin and Joulin [55]),

$$u_L^{\infty} - u_L = \mathscr{L}_s K_s + \mathscr{L}_c K_c \tag{20}$$

where K_s denotes the impact of strain and K_c that of curvature (see Sect. 2).

For an outwardly propagating spherical flame the two Markstein lengths $\mathscr{L}_s$ and $\mathscr{L}_c$ differ quite significantly. According to the definitions of K_s and K_c, the stretch caused the flow field usually yields higher rates of strain than the curvature as the gas velocity is, after the initial phase, much larger than the laminar burning velocity due to the expansion of the burnt gases where $u_f = u_g + u_l$ at the leading edge. As a first–order approximation it is, therefore, possible to ignore the contribution due to curvature in Eq. (20) and to define a single Markstein length as shown in Eq. (21).

$$\mathscr{L}_M = \frac{u_L^{\infty} - u_L}{2u_f(r)/r} \tag{21}$$

Examples of the evolution of the flame speed and the laminar burning velocity for methane/hydrogen/air mixtures are shown in Figs. 10 and 11. Lindstedt et al. [56] used this approach where $u_f(r)$ is the flame speed with the value for the unstretched laminar burning velocity u_L^{∞} obtained by using the asymptotic value extrapolated to zero rate of strain (cf. Vagelopoulos et al. [57]). The evolution of the corresponding stoichiometric flame speeds and laminar burning velocities for propane/hydrogen/air flames are shown in Figs. 12 and 13.

The selection of a reference surface, (e.g., iso-contour) for the determination of the Markstein length is a key issue that has received relatively limited attention. Fristrom [58, 59] used the upstream edge of the luminous zone (i.e. at the beginning of the reaction zone). Bradley et al. [33] selected the 305 K isotherm, while

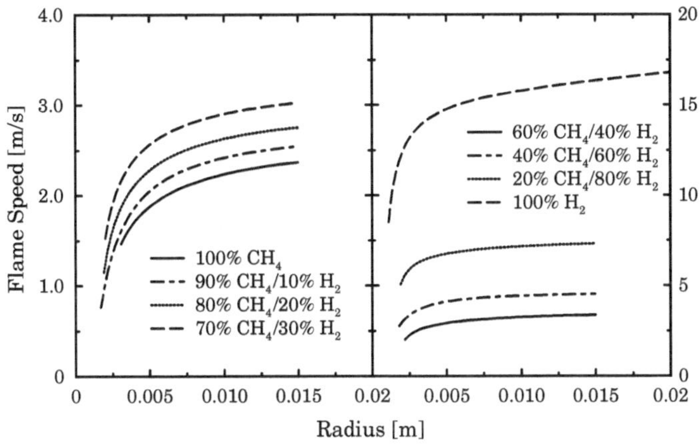

Fig. 10 The flame speed (u_f) as a function of radius for stoichiometric methane/hydrogen/air flames obtained by tracking the 305 K isotherm

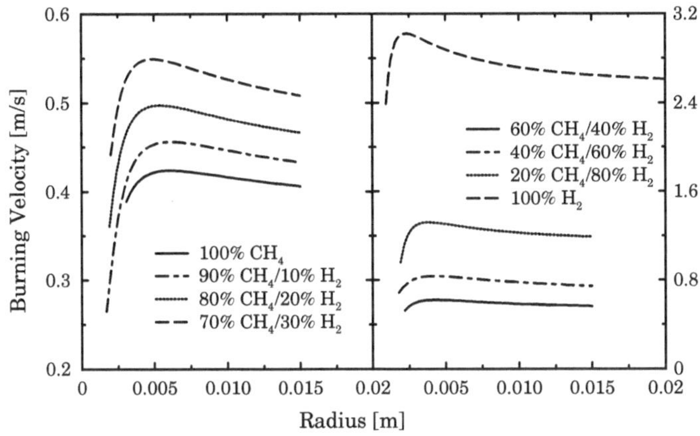

Fig. 11 The laminar burning velocity (u_l) as a function of radius for stoichiometric methane/hydrogen/air flames obtained by tracking the 305 K isotherm

Dowdy et al. [60] used the location where the mass flux in a spherical flame corresponds to that of a steady, one–dimensional planar flame. Figure 14 shows Markstein lengths based on different isotherms as a function of the radius for a stoichiometric methane–air flame. The values for the current first–order approximation depend on the tracked iso-contour to the extent that the qualitative response to strain changes sign. Hence, $\mathscr{L}_M$ is not a unique physico–chemical parameter (Lipatnikov [61]). Tables 1 and 2 show the Markstein lengths for different stoichiometries and isotherms for methane and propane–air flames. The observation of increased burning rates at higher strain rates for lean CH_4–air and rich C_3H_8–air flames are reflected by the Markstein lengths which tend towards negative for these conditions. Qualitatively

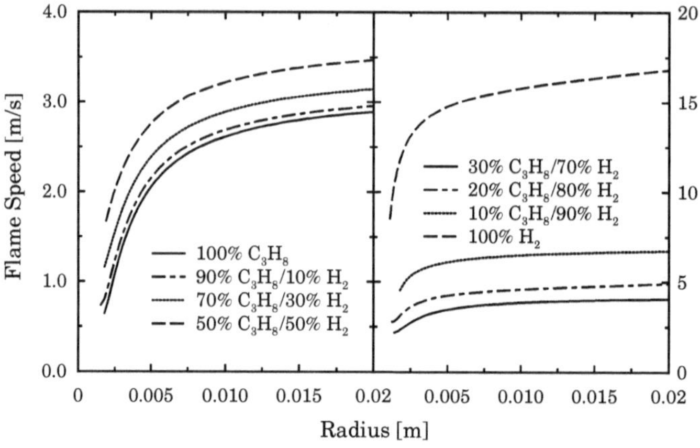

Fig. 12 The flame speed (u_f) as a function of radius for stoichiometric propane/hydrogen/air flames obtained by tracking the 305 K isotherm

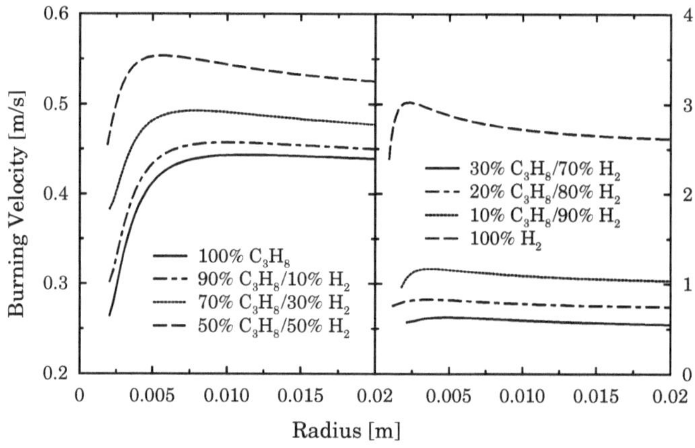

Fig. 13 The laminar burning velocity (u_l) as a function of radius for stoichiometric propane/hydrogen/air flames obtained by tracking the 305 K isotherm

Fig. 14 Markstein lengths ($\mathscr{L}_M$) based on the tracking of different isotherms shown as a function of radius for a stoichiometric methane–air flame at different positions within the flame

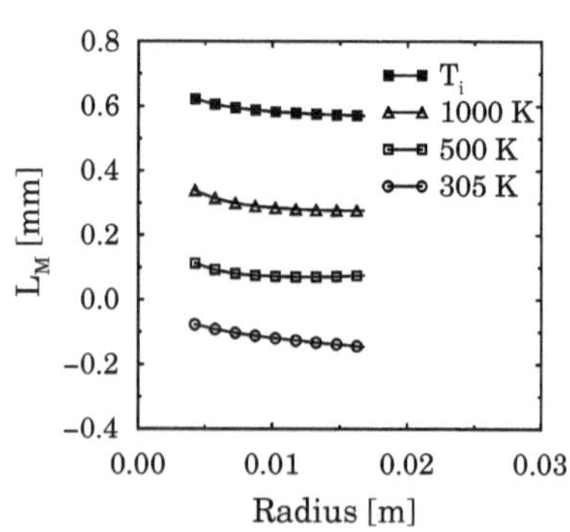

Table 1 Computed Markstein lengths $\mathscr{L}_M$ for ambient ($p = 101{,}325$ Pa and $T = 296$ K) methane–air flames at different equivalence ratios (ϕ) and isotherms

Mixture	ϕ	305 K	500 K	600 K	1000 K	T_i	T_b
CH_4	0.63	−0.46	−0.10	0.04	0.56	0.62	0.66
	0.68	−0.47	−0.04	0.06	0.53	0.62	0.64
	0.79	−0.30	0.01	0.08	0.41	0.53	0.56
	1.00	−0.17	0.05	0.10	0.40	0.55	0.56
	1.23	−0.14	0.05	0.12	0.53	0.85	0.85
	1.41	0.06	0.06	0.17	1.38	2.01	2.23

Table 2 Computed Markstein lengths $\mathscr{L}_M$ for ambient ($p = 101{,}325$ Pa and $T = 296$ K) propane–air flames at different equivalence ratios (ϕ) and isotherms

Mixture	ϕ	305 K	500 K	1000 K	T_i	T_b
C_3H_8	0.69	0.26	0.58	1.08	1.74	1.84
	0.79	0.08	0.34	0.72	1.16	1.21
	1.00	0.01	0.17	0.45	0.74	0.81
	1.21	−0.03	0.11	0.30	0.59	0.71
	1.32	−0.07	0.05	0.21	0.50	0.50
	1.42	−0.14	−0.03	0.09	0.23	0.28
	1.53	−0.21	−0.10	−0.08	0.01	0.01
	1.64	−0.29	−0.20	−0.23	−0.28	−0.32

the trend in $\mathscr{L}_M$ obtained from the different isotherms is similar when proceeding from fuel rich to fuel lean mixtures for each fuel. Furthermore, Markstein lengths based on the thermo–chemical properties of the flame, such as T_i and T_m, are qualitatively and quantitatively similar. However, such properties can not be determined experimentally.

The calculated Markstein lengths for methane–air and propane–air flames are compared in Figs. 15 and 16 with experimental data from Taylor [50], the corrected values from Tseng et al. [62] and the indirect determination by Searby and Quinard [63]. Taylor [50] used a schlieren technique to determine the flame front. As can be seen in Figs. 15 and 16, the trends in the computed and measured Markstein lengths are similar. Weinberg [64] pointed out that the location of the schlieren image in a flame is about 200 K higher than previously thought. Bradley et al. [33] comment that the isotherm obtained by schlieren corresponds better to a 450 K iso-surface. The current computations confirm these observations. Given that flame properties derived either on the basis of the internal structure (e.g. T_i) or integral properties are not accessible experimentally, the leading edge isotherm remains a natural choice. Significant caution is, however, required when comparing experimental and computational data.

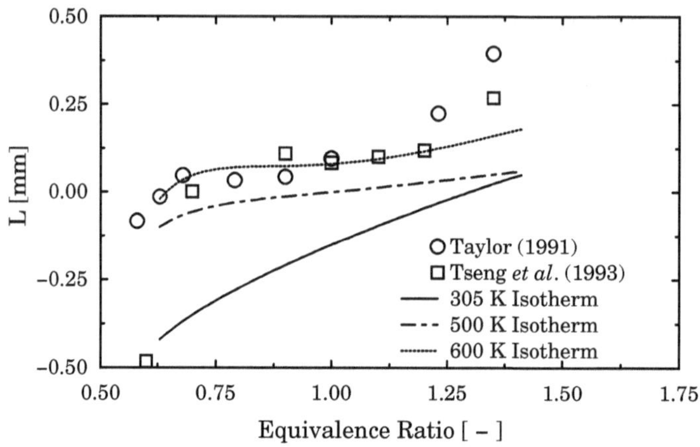

Fig. 15 The computed Markstein lengths $\mathscr{L}_M$ (lines), based on different isotherms, as a function of equivalence ratio in comparison to experimental data (symbols) for methane–air flames

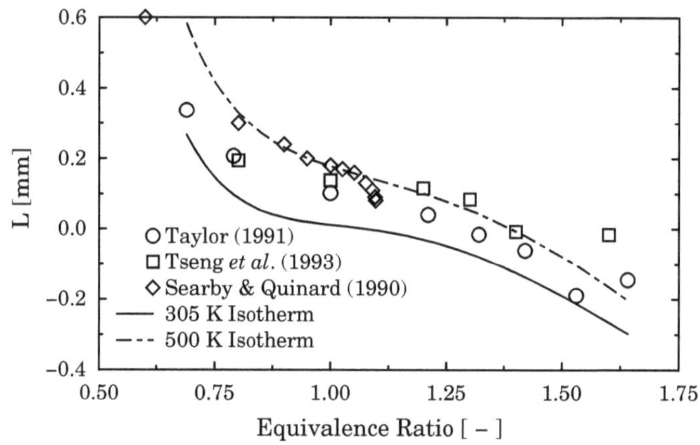

Fig. 16 The computed Markstein lengths $\mathscr{L}_M$ (lines), based on different isotherms, as a function of equivalence ratio in comparison to experimental data (symbols) for propane–air flames

6 Minimum Ignition Energies and Quenching Distances

The response of incipient flame kernels to stretch is critical in the context of minimum ignition energies and quenching distances. As shown above, there is a strong dependency on the reference surface selected when the response is defined in terms of the Markstein length. However, transient computations of the current kind can, irrespective of such concerns, be used directly to estimate values when combined with well-established theory. Transient laminar flames propagating in methane and propane fuel blends with hydrogen were accordingly used to investigate the response to strain and to estimate the minimum ignition energies and quenching distances using the approach of Williams [65] as outlined below. Fuels are categorized into different hazard classes (e.g. [66]) with methane and propane Class IIA fuels with hydrogen a Class IIC fuel.

The Energy Institute applies a hazard criterion that features a 30% hydrogen concentration limit for all types of hydrocarbon/hydrogen mixtures. Mixtures featuring lower hydrogen concentrations are treated as ordinary fuels [67]. Hydrocarbon mixtures with hydrogen from an increasingly important class of process-related fuels and their inherent flammability characteristics comprise a source of concern. The increased explosion hazards associated with syngas-related fuels blends have been explored by Hampp et al. [23] for a wide range of fuel compositions. Li et al. [6, 9] have further shown that methane has a significant mitigating impact on the strength of explosions in CH_4/H_2 fuel blends. However, these studies do not address aspects such as the minimum ignition energy or the quenching distance. The direct determination of the latter through computational means is possible though exceptionally expensive. To enable screening of a large number of mixtures, an alternative approach based on the extraction of relevant properties from transient spherical flame simulations is explored below.

Expressions for the minimum spark energy and quenching distance have been derived by Williams [65] on the basis that the amount of energy provided should be sufficient for a temperature increase that permits the initiation of combustion and that the ignited gas–filament thickness should exceed a critical limit so that flame propagation becomes self–sustaining. Williams [65] showed that the initial spark energy H can be expressed as,

$$H = (A\delta_b)\,\rho_o\,c_p\,(T_\infty - T_u) \tag{22}$$

where A denotes the filament cross sectional area, T_u and T_∞ denote the ambient and maximum flame temperature respectively, and δ_b is the corresponding flame thickness. The latter quantity may be expressed as,

$$\delta_b = \frac{\mu_u}{\rho_u u_L}\frac{1}{Pr} \qquad\qquad Pr = \frac{c_p \mu_u}{\lambda_u} \tag{23}$$

where the flame front values of the properties c_p, μ_u, ρ_u are used. The quenching distance $d_{\min}$ can be assumed to be proportional to the flame thickness δ_b as suggested by Mullins and Penner [69] and Williams [65],

$$d_{\min} = a\delta_b \tag{24}$$

where a is a proportionality constant. Given the above definitions and that the cross sectional area A must be at least equal to the square of the quenching distance (i.e. $A \geq d_{\min}^2$) the minimum initial spark energy H can be expressed as [65],

$$H = a^2\,\delta_b^3\,\rho_u\,c_p\,(T_\infty - T_u) \tag{25}$$

where the proportionality constant a is here assigned the midrange consensus value of 40, see Mullins and Penner [69] and Williams [65]. The maximum flame temperature (T_∞) is extracted from the simulations and the laminar burning velocity (u_L) can be obtained either by tracing an iso–surface (e.g. T = 305 K) or via an integration over the entire flame structure. As strain effects become large for small radii, the tracking of the leading edge iso-contour presents a choice that provides an opportunity for experimental verification. The current simulations permit δ_b to be evaluated via Eq. (23) as a function of the rate of strain by determining the laminar burning velocity.

Minimum ignition energies and quenching distances are strongly dependent on the fuel type, equivalence ratio and ambient conditions (e.g. Calcote et al. [68]), as shown in Fig. 17. The cases considered here are stoichiometric methane/hydrogen mixtures with the following compositions (in volume %) : 0/100, 10/90, 20/80, 30/70, 40/60, 50/50, 60/40, 70/30, 80/20, 90/10, and 100/0. Stoichiometric propane/hydrogen mixtures with the following compositions (in volume %): 0/100, 5/95, 10/90, 20/80, 30/70, 50/50, 70/30, 90/10, and 100/0 where also considered.

The initial conditions are set to standard atmospheric conditions (T = 298 K and P = 101 kPa). Minimum ignition energies and quenching distances for the stoichiometric methane/hydrogen mixtures are shown in Figs. 18 and 19. The agreement between the simulations and experimental data for cases featuring pure methane and hydrogen mixtures is good. Calculated values, including uncertainties, are within 25% of the experimental determinations. It may also be noted that the distribution of minimum spark ignition energies as a function of mixture composition shows a high degree of similarity to the corresponding distribution of quenching distances. This is not surprising, given the formulation of the minimum ignition energy through the quenching distance. The trend is graphically illustrated in Fig. 20.

Figures 18, 19 and 20 permit a comparison of quenching distances and minimum ignition energies obtained for the range of mixtures of CH_4/H_2 with pure ethylene as a typical Class IIB fuel. A similarity can thus be observed between the ignition behavior of pure ethylene and a 20/80% methane/hydrogen mixture.

Computed minimum ignition energies and quenching distances for stoichiometric propane/hydrogen mixtures are shown in Figs. 21 and 22. The computed minimum ignition energy and quenching distance for the pure propane case are in excellent agreement with experimental data.

A comparison with the stoichiometric methane/hydrogen mixtures (Figs. 18 and 19) reveals distinct differences. The quenching distance for propane/hydrogen remains essentially constant over a wide range of propane content ($30 < C_3H_8[\%] < 100$). For mixtures containing less than 30% propane the gradient of the quenching distance and minimum ignition energy curves changes quite abruptly in contrast to the more gradual change for the methane–containing mixtures. The observed differences imply a dependence on the type of hydrocarbon fuel considered.

The corresponding quenching distances and minimum ignition energies, for the range of mixtures considered here, may also be seen in Fig. 23. Comparisons with stoichiometric methane/hydrogen mixtures further highlight the dependency of quenching distances and minimum ignition energies on the specific type of the hydrocarbon component of the mixture.

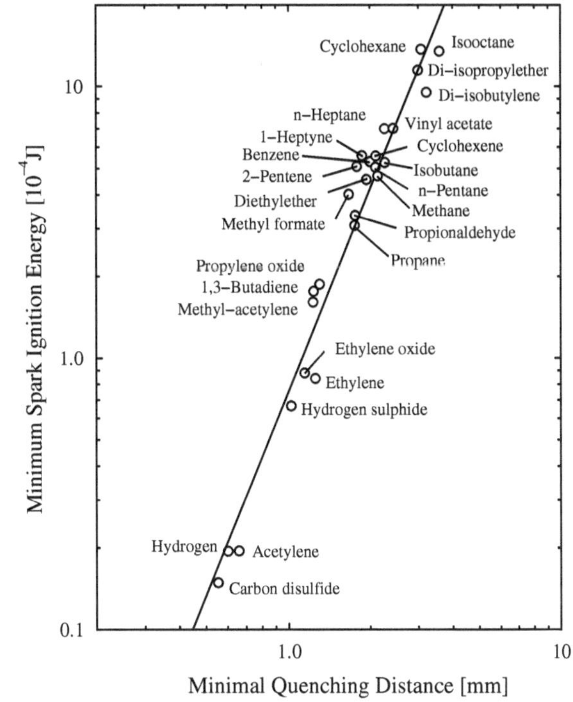

Fig. 17 Correlation of minimum spark ignition energy with quenching distance adopted from Calcote et al. [68]

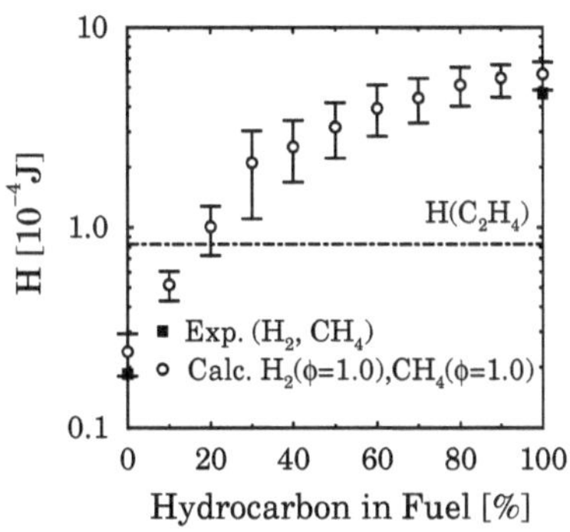

Fig. 18 Computed minimum spark ignition energies as a function for stoichiometric CH_4/H_2 mixtures. The dash-dotted line indicates the value for ethylene and the error bars the variation obtained by extracting data at different radii (see Sect. 7)

7 Evaluation of Uncertainties

The computation of minimum ignition energies and quenching distances through the extraction of relevant flame properties from simulations of transient spherical flames developing from weak ignition kernels requires an assessment of intrinsic uncertainties. Given the significant influence of strain on the evolution of the flame structure at small radii, a source of uncertainty is related to the choice of a specific radius for the extraction of properties of interest. The influence of the chosen radius on computed

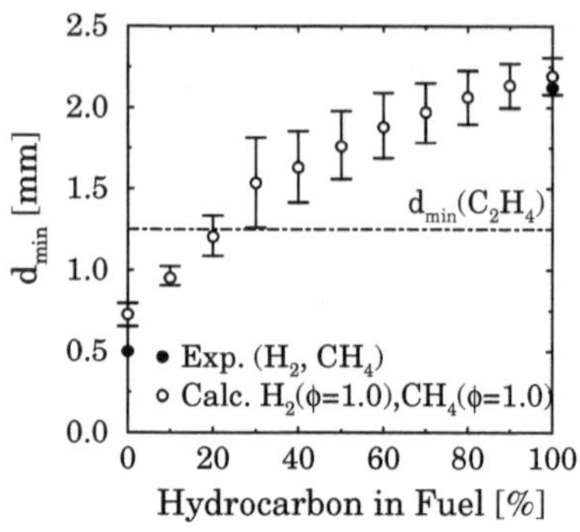

Fig. 19 Computed quenching distances as a function for stoichiometric CH_4/H_2 mixtures. The dash-dotted line indicates the value for ethylene and the error bars the variation obtained by extracting data at different radii (see Sect. 7)

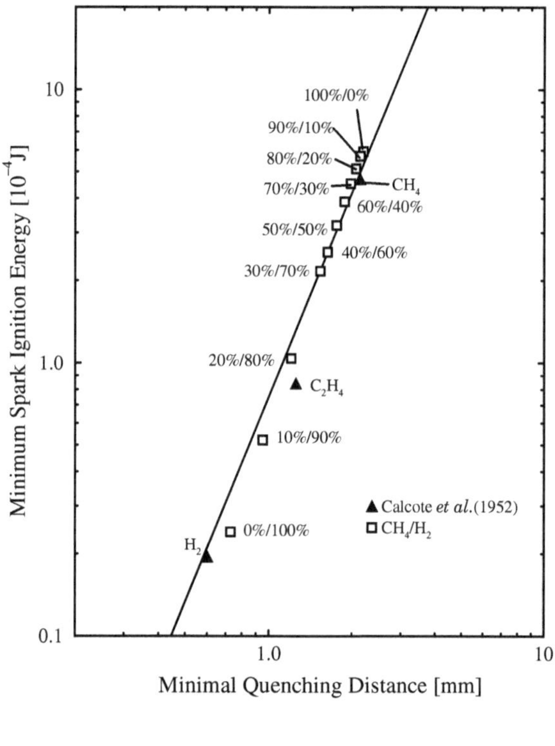

Fig. 20 Correlation of minimum spark ignition energy with quenching distance for CH_4/H_2 mixtures (boxes). Abridged experimental data (triangles) after Calcote et al. [68]

laminar buring velocities for stoichiometric methane/hydrogen mixtures is shown in Fig. 24. Associated uncertainty factors are shown in Table 3 in the form of normalised differences (in %) between values obtained at $r = 5$ mm and $r = 10$ mm and those obtained at $r = 20$ mm. The differences are non–monotonic with respect to the mixture composition and a maximum deviation of about 45% for minimum ignition energies and 18% for quenching distances is deduced for the 30/70% mixture. Similar remarks can be made for the stoichiometric propane/hydrogen mixtures with respect both to the trend and magnitude. Table 4 shows that the maximum observed differences were found for a 20/80% propane/hydrogen mixture and are around 10% for

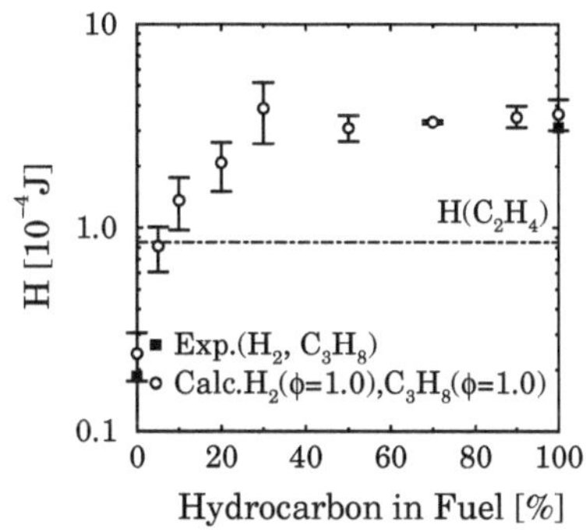

Fig. 21 Computed minimum spark ignition energies as a function for stoichiometric C_3H_8/H_2 mixtures. The dash-dotted line indicates the value for ethylene and the error bars the variation obtained by extracting data at different radii (see Sect. 7)

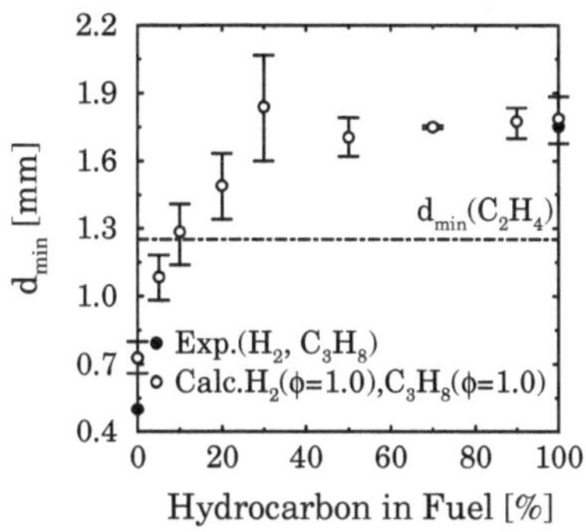

Fig. 22 Computed quenching distances as a function for stoichiometric C_3H_8/H_2 mixtures. The dash-dotted line indicates the value for ethylene and the error bars the variation obtained by extracting data at different radii (see Sect. 7)

the quenching distance and 27% for the minimum ignition energy. This indicates that the current approach is consistent. Furthermore, the maximum observed differences are moderate and arguably within the uncertainties of experimental measurements, which typically feature discrepancies of about a factor of two (e.g. Calcote et al. [68]).

A further issue relates to the use of properties derived using data obtained by tracking the 305 K isotherm. As noted above for Markstein lengths, properties can be obtained at different isotherms or as an integral property of the flame. However, it can be shown that the influence of the chosen radius on computed values based on the integral burning velocity is significantly higher with differences up to 100%. This can be attributed to the fact that the integral burning velocity at small radii is more strongly affected by transient strain effects as the timescale associated with establishing a flame structure is more significant. Accordingly, integral properties deviate more strongly due to the transient behavior at high rates of strain. The observed discrepancies therefore arguably highlight the advantages of a flame front based approach that can also be experimentally verified.

Fig. 23 Correlation of minimum spark ignition energy with quenching distance for C_3H_8/H_2 mixtures (boxes). Abridged experimental data (triangles) after Calcote et al. [68]

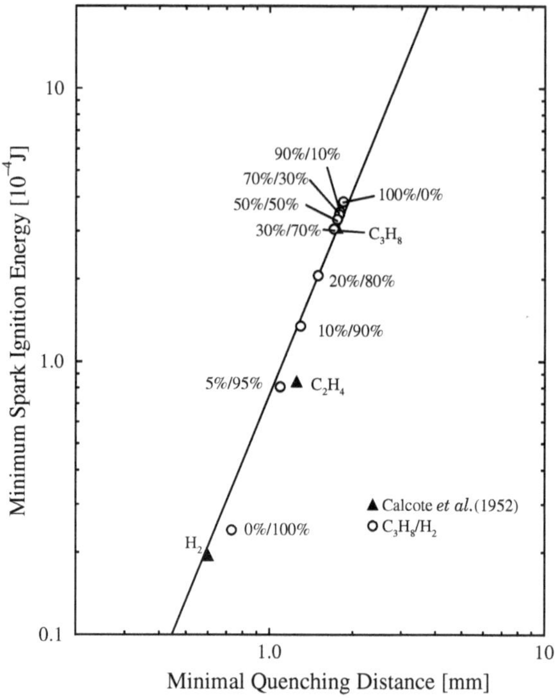

Fig. 24 Extracted flame front burning velocities for stoichiometric CH_4/H_2 mixtures at radii of 5, 10 and 20 mm

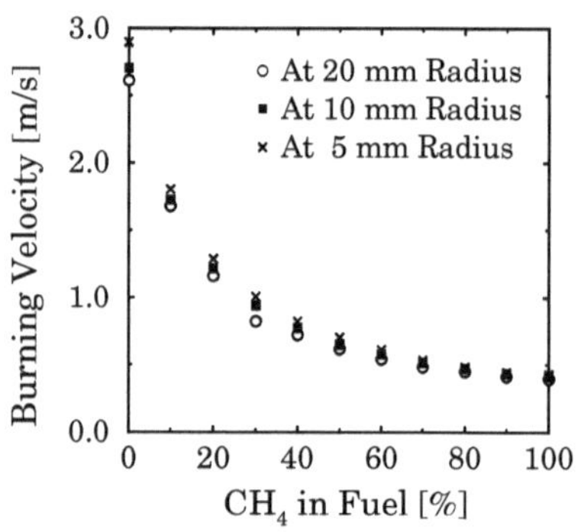

Table 3 Normalised differences (in %) for quenching distance and minimum spark energy values, based on the flame front burning velocity for stoichiometric CH_4/H_2 mixtures, at radii $R = 5\,\text{mm}$ and $R = 10\,\text{mm}$ against values based on $R = 20\,\text{mm}$

Mixture	Radius = 5 mm		Radius = 10 mm	
CH_4/H_2	$\Delta\epsilon(d_{min})$	$\Delta\epsilon(H)$	$\Delta\epsilon(d_{min})$	$\Delta\epsilon(H)$
100–0%	5.21	15.0	3.87	11.3
90–10%	6.40	18.0	4.90	14.0
80–20%	8.15	22.5	5.38	15.3
70–30%	9.40	25.6	5.49	15.6
60–40%	10.9	29.3	5.55	15.8
50–50%	11.7	31.2	5.87	16.6
40–60%	13.3	34.9	7.45	20.7
30–70%	18.0	44.8	12.56	33.2
20–80%	10.3	27.9	4.55	13.1
10–90%	6.48	18.2	3.25	9.44
0–100%	9.66	26.3	3.95	11.4

Table 4 Normalised differences (in %) of quenching distances and minimum spark energies, based on flame front burning velocities for stoichiometric C_3H_8/H_2 mixtures, at radii $R = 5\,\text{mm}$ and $R = 10\,\text{mm}$ against values based on $R = 20\,\text{mm}$

Mixture	Radius = 5 mm		Radius = 10 mm	
C_3H_8 / H_2	$\Delta\epsilon(d_{min})$	$\Delta\epsilon(H)$	$\Delta\epsilon(d_{min})$	$\Delta\epsilon(H)$
100–0%	5.76	18.3	0.86	2.56
90–10%	3.93	12.3	1.73	5.10
70–30%	0.42	1.24	2.85	8.32
50–50%	4.99	14.2	3.41	9.90
30–70%	9.18	23.2	4.88	13.3
20–80%	9.96	27.0	5.18	14.75
10–90%	0.57	28.5	5.11	14.56
5–95%	9.13	25.0	4.06	11.67
0–100%	9.66	3.95	26.3	11.4

8 Conclusions

Computational simulations of spherical transient flames for various fuels and fuel mixtures have been performed using comprehensive detailed chemistry and transport. Changes in the flame speed and burning velocities induced by variations of the hydrogen content and the hydrocarbon fuel type have been extensively investigated. Results featuring flames without hydrogen enrichment are in good agreement with experimental data. Similar comparisons for the whole range and types of mixtures were not possible due to a lack of experimental data.

Given the importance of flow field strain on the development of ignition kernels, the simulation of transient spherical methane and propane flames was also used to examine the uniqueness of Markstein lengths. These lengths, defined on the basis of flame iso–contours (i.e. reference surfaces), were investigated using detailed thermo–chemical flame structures with properties extracted by tracking different iso–contours. The latter included the ignition temperature where the rate of radical production becomes positive when proceeding from the cold front and the temperature at the back end of the flame where the rate of production of the radical pool becomes negative. It was shown that the Lewis number of the deficient reactant at the leading edge of the flame correlates the overall response of the flame towards stretch and that the Markstein length can serve as a qualitative indicator in relation to flame behavior. However, due to the complex behavior of flames such simplified parameters do not permit a quantitative characterization. Moreover, transient flames respond differently to strain than their steady-state counterparts due to the temporal evolution of the flame structure. This is especially the case for ignition events due to the time scale required to establish the radical pool.

As an application example, quenching distances and minimum ignition energies were computed for stoichiometric methane/hydrogen/air and propane/hydrogen/air fuel blends with the current calculation method combined with the theoretical analysis by Williams [65]. The approach provides uncertainty quantification and an efficient methodology with the accuracy expected to be similar to experiments. However, caution is required as in practise issues such as stratification as well as dilution may become crucial.

The advantage of the current approach is that it is applicable under conditions (e.g. high pressure and temperature) where experimental quantification is not feasible. Furthermore, the increasing accuracy of detailed chemical mechanisms for complex fuels creates an opportunity for the accurate theoretical determination of hazard-related properties for emerging fuels and fuel blends. Finally, while ignition energies and quenching distances form an important part of hazard assessments, it must be recognized that the influence of laminar transport properties diminishes in turbulent combusting flows. Hence, the data presented in the current work is complementary to such investigations.

Acknowledgements The authors wish to gratefully acknowledge the financial support of the Energy Institute's Technical Partners for part of the current study along with Members of its Area Classification Working Group that originally technically reviewed that work and the Members that subsequently revalidated it. The technical advice and encouragement of Dr I.A.B Reid and Dr V. Sakthitharan is also greatly appreciated.

References

1. Lieuwen TC, Yang V, Yetter R (eds) (2009) Synthesis gas combustion: fundamentals and applications CRC Press, Boca Raton ISBN 978-1-4200-8535-8
2. Milosavljevic VD, Persson M, Lindstedt RP, Váos EM (2003) Evaluation of the thermoacoustic behaviour of an industrial gas turbine combustion chamber with a second moment method. Cranfield University Press, pp 72–77
3. Korb B, Kawauchi S, Wachtmeister G (2016) Fuel 164:410
4. Milosavljevic VD, Lindstedt RP, Cornwell MD, Gutmark EJ, Váos EM (2006) In Roy G, Yu KH, Whitelaw JH, Witton JJ (2006) Advances in combustion and noise control. Cranfield University Press, pp 149–165
5. Lieuwen TC, McDonell V, Santavicca D, Sattelmayer T (2008) Combust Sci Technol 180(6):1169
6. Li T, Hampp F, Lindstedt RP (2018) Process Safety Environmental Protection 116:663
7. Law CK, Jomaas G, Bechtold JK (2005) Proc Combust Inst 30(1):159
8. Milosavljevic VD, Cornwell MD, Gutmark EJ, Lindstedt RP, Váos EM (2003) Methods and mechanisms for the stabilization of lean partially premixed low-temperature flames. Cranfield University Press, pp 78–83
9. Li T, Hampp F, Lindstedt RP (2019) Comb Flame 203:105
10. Simatos P, Hampp F, Lindstedt RP (2020) in Innovations. In: Gupta A, De A, Aggarwal S, Kushari A, Runchal A (eds) Sustainable energy and cleaner environment. Springer Verlag, Singapore, Green energy and technology, pp 333–356
11. Lin YC, Daniele S, Jansohn P, Boulouchos K (2013) J Eng Gas Turbines Power 135
12. Burke EM, Güthe F, Monaghan RFD (2016) Proceeding ASME Turbo Expo, Seoul, GT2016-57804
13. Wang J, Zhang M, Xie Y, Huang Z, Kudo T, Kobayashi H (2013) Exp Thermal Fluid Sci 50:90
14. Daniele S, Jansohn P, Mantzaras J, Boulouchos K (2011) Proc Comb Inst 33(2):2937
15. Muppala SPR, Nakahara M, Aluri NK, Kido H, Wen JX, Papalexandris MV (2009) Int J Hydrogen Energy 34:9258
16. Venkateswaran P, Marshall AD, Seitzman JM, Lieuwen TC (2013) J Eng Gas Turb Power 136
17. Liu CC, Shy SS, Chiu CW, Peng MW, Chung HJ (2011) Int J Hydrogen Energy 36:8595
18. Bray KNC (1990) Proc Roy Soc A - Math Phy 431(1882):315
19. Driscoll JF (2008) Prog Energy Combust 34(1):91
20. Shepherd IG, Cheng RK (2001) Combust Flame 127(3):2066
21. Lawn CJ, Schefer RW (2006) Combust Flame 146(1):180
22. Goh KHH, Geipel P, Lindstedt RP (2014) Combust Flame 161:2419
23. Hampp F, Goh KHH, Lindstedt RP (2020) Process Safety Environ Protect 143:66
24. Dixon-Lewis G (1990) Proc Combust Inst 23:305
25. Miller JA, Bowman CT (1989) Prog Energy Combust Sci 15:287
26. Warnatz J (1980) Proc Combust Inst 18:369
27. Poinsot T (1996) Proc Combust Inst 26:219
28. Lindstedt RP (1993) Proc Anglo - German Combustion Symposium. The British Section of the Combustion Institute, p 372
29. Hampp F, Lindstedt RP (2017) Combust Flame 182:248
30. Hampp F, Lindstedt RP (2017) Proc Comb Inst 36(2):1911
31. Buckmaster J (1979) Acta Astronautica 6:741
32. Clavin P (1985) Prog Energy Combust Sci 11:1
33. Bradley D, Gaskell P, Gu X (1996) Combust Flame 104:176
34. Ghoniem AF, Soteriou MC, Knio OM, Cetegen B (1992) Proc Combust Inst 24:223
35. Stahl G, Warnatz J (1991) Combust Flame 85:285
36. Yoo CS, Im H (2005) Proc Combust Inst 30:349
37. Malik NA, Lindstedt RP (2010) Combust Sci Technol 182:1171
38. Malik NA, Lindstedt RP (2012) Combust Sci Technol 184:1799

39. Ciardiello R, de Oliveira P, Skiba A, Mastorakos E, Allison P (2020) Combust Flame 221:296
40. Karlovitz B, Dennington D, Knappschafe D, Wells F (1953) Proc Combust Inst 4:613
41. Matalon M (1983) Combust Sci Technol 31:169
42. Candel S, Poinsot T (1990) Combust Sci Technol 70:1
43. Frankel M, Sivashinsky G (1983) Combust Sci Technol 31:131
44. Law CK (1988) Proc Combust Inst 22:1381
45. Sun CJ, Law CK (2000) Combust Flame 121:236
46. Lindstedt RP, Sakthitharan V (1999) Parallel processing and direct simulation of transient premixed laminar flames with detailed chemical kinetics. In: Allen RJ, Guest MF, Simpson AD, Henty AS, Nicole DA (eds) High-performance computing. Kluwer Academic / Plenum Publishers, New York, pp 417–428
47. Meyer MP (2001) The application of detailed and systematically reduced chemistry to transient laminar flames. Ph.D. thesis, Imperial College of Science, Technology and Medicine
48. Lindstedt RP, Maurice LQ (1995) Combust Sci Technol 107:317
49. Chapman S, Cowling TG (1970) The mathematical theory of non-uniform gases. Cambridge University Press, Cambridge
50. Taylor S (1991) Burning velocity and the influence of flame stretch. Ph.D. thesis, University of Leeds
51. Dixon-Lewis G (1996) Proc Roy Soc London, Part A 452:1857
52. Markstein G (1951) J Aero Sci 18:199
53. Markstein G (1964) Non-steady flame propagation. Pergamon Press
54. Clavin P, Williams FA (1982) J Fluid Mech 116:251
55. Clavin P, Joulin G (1989) In: Borghi R, Murthy SNB (eds) Lecture notes in engineering: turbulent reactive flows. Springer-Verlag, Berlin, p 213
56. Lindstedt RP, Meyer MP, Sakthitharan V (1998) EUROTHERM seminar No. 61 on detailed studies of combustion phenomena, p D52
57. Vagelopoulos CM, Egolfopoulos FN, Law CK (1994) Proc Combust Inst 25:1341
58. Fristrom R (1965) Phys Fluids 8:273
59. Fristrom R, Westenberg A (1965) Flame structure. McGraw Hill, New York
60. Dowdy DR, Smith DB, Taylor SC (1990) Proc Combust Inst 23:325
61. Lipatnikov AN (1996) Combust Sci Technol 119:131
62. Tseng L, Ismail M, Faeth G (1993) Combust Flame 95:410
63. Searby G, Quinard J (1990) Combust Flame 82:298
64. Weinberg FJ (1955) Fuel 34:S84
65. Williams FA (1985) Combustion Theory. The Benjamin/Cummings Publ, Company Inc
66. British Standards Institution, Code of Practice for the Selection, Installation and Maintenance of Electrical Apparatus for use in Potentially Explosive Atmospheres. Part 1: Basic Requirements for all Parts of the Code. Technical Report BS 5345: Part 1 (1976)
67. Energy Institute, EI model code for safe practice Part 15: Area classification for installations handling flammable fluids. Tech. Rep. 4th Edition, 61 New Cavendish Street, London W1G 7AR, UK (2015)
68. Calcote H, Gregory C, Barnett C, Gilmer R (1952) Ind Eng Chem 44:2656
69. Mullins BP, Penner SS (1959) Explosions, detonations, flammability and ignition. Pergamon Press, New York

Experimental Investigation of Turbulent Flow/Flame Structure of Double Swirler Burner

Dhanalakshmi Sellan and Saravanan Balusamy

Abstract To develop efficient combustors under various operating conditions, a deeper understanding of the effects of swirl flow mechanisms on gas turbine combustion is essential. The double swirl flow arrangement has an advantage over the single swirl flow by proving better mixing and flame anchoring due to the increased shearing action between the two swirl flows. The flow field of the open-type, double swirler burner at various Reynolds numbers is analyzed using the 2D particle image velocimetry (PIV) technique. The spatial distribution of OH* and CH* chemiluminescence is analyzed for one premixed ($\phi = 0.75$) and three stratified LPG/air flames. Medium swirlers are used for velocity measurements, while two different axial swirlers, defined as low and medium swirlers with geometric swirl numbers 0.5 and 0.8, respectively, are used for chemiluminescence studies. The analysis of non-reacting flow field results shows that the increase in the Reynolds number of outer swirl increases the radial component of the velocity; thus, the swirl flow spreads in the radial direction. As a result, turbulence parameters such as Reynolds stress and turbulence intensity are also dominant in the radial direction. As a consequence of this phenomenon, in the worst case, the spread of the flame front and further local quenching could occur. In comparison, the increase in internal swirl flow increases the axial components of the velocity and the flow field becomes axially elongated. The inner recirculation zone shifts downstream of the flow field, suggesting that the reaction zone could be moved to the downstream under the reacting condition. The acquired OH*/CH* chemiluminescence signals indicate that the premixed LPG flames are elongated while the upstream stratified mixture conditions lead to more compact flames that are appropriate for gas turbine applications. Nonetheless, for medium swirler cases, OH*/CH* chemiluminescence is much more compact in form than low swirl. Intensity variation is independent of the flame strain of propane; thus, the equivalence ratio calculated from the OH*/CH* intensity ratio has helped to understand the interaction between flame and turbulence at various operating conditions.

D. Sellan · S. Balusamy (✉)
Department of Mechanical and Aerospace Engineering, Indian Institute of Technology Hyderabad, Kandi, Sangareddy, Telangana 502285, India
e-mail: saravananb@mae.iith.ac.in

A. Gupta et al. (eds.), *Advances in Energy and Combustion*, Green Energy and Technology,
https://doi.org/10.1007/978-981-16-2648-7_10

Keywords Double swirler burner · Reynolds stress · Turbulent intensity · Stratified flame · Chemiluminescence · OH*/CH* · PIV · LPG/air flames

Nomenclature

ϕ	Equivalence ratio
ϕ_1	Equivalence ratio at an inner annulus
ϕ_2	Equivalence ratio at an outer annulus
Re_i	Reynolds number at an inner annulus
Re_o	Reynolds number at an outer annulus
u_i	Bulk velocity at an inner annulus
u_o	Bulk velocity at an outer annulus
u_c	Co-flow bulk velocity
τ_R	Reynolds stress
TI	Turbulent intensity
I	Chemiluminescence intensity

Abbreviations

LPG	Liquefied petroleum gas
PIV	Particle image velocimetry
CCD	Charge-coupled device
ICCD	Intensified charge-coupled device
SR	Stratification ratio

1 Introduction

In practical combustion devices, stabilization of flame in the lean premixed condition is of critical and complex due to the low heat release rate. The burner featured with a swirler is capable of stabilizing the flame by enhancing the mixing between incoming fresh fuel/air mixture and hot reactants [1, 2]. This is more pronounced in the double swirl flow arrangement [3–5]. The study of turbulent swirl flow under isothermal conditions is important before understanding the case of reaction flow and helps to optimize burner design and validate numerical data [6]. Many combustion devices, such as gasoline and gas turbine engines, operate in a stratified condition to reduce pollution and improve combustion performance [7–10]. Stratified combustion is a special type of partially premixed mode of combustion where the flame propagates through the non-uniform equivalence ratio gradient field [11–14]. Past few decades the stratified combustion research is of interest due to its practical relevance [11, 13, 15–17].

During the combustion process, many reactions take place at the flame front where highly localized heat release occurs. Identifying these reaction zones at the flame is essential for understanding the flame structure and behavior at various operating conditions. The chemiluminescence of the OH* and CH* radicals formed during the chemical reaction provides important information on the structure and nature of the reaction zone of the flame. Therefore, they are used as markers for identifying the reaction zone. Chemiluminescence imaging is one of the reliable methods of non-invasive optical diagnostics techniques used to capture the intermediate free radicals in the reaction zone. Chemically exited OH and CH radicals emit photons at a specific wavelength when returning to their ground level, and this light is captured by intensified charge-coupled device (ICCD). For a perfectly premixed flame, the OH* and CH* chemiluminescence intensity depends linearly on the mass flow rate of the air/fuel mixture and exponentially on the global equivalence ratio in case of globally uniform equivalence ratio distribution [18]. Therefore, the local equivalence ratio of the flow field can be measured from OH*/CH* chemiluminescence ratio [19]. As the chemiluminescence emission gives the reaction zone location and intensity is directly proportional to the species concentration, it can be used as an indicator for lean blow-off [20]. The flame stability was understood from the flame structure, and heat release rate measurements were obtained from chemiluminescence images of premixed and stratified propane flames [21, 22]. Because premixed flames have uniformly distributed species concentration, there are direct methods to measure local equivalence ratio and heat release rate. There are many studies conducted to investigate the properties of premixed flames using chemiluminescence techniques on various fuels [23–27].

In the current study, the effect of Reynolds number on the turbulent structure under non-reacting conditions is being investigated using 2D particle image velocimetry (PIV). Also, the spatial distributions of OH* and CH* are captured by the chemiluminescence technique for turbulent premixed and stratified liquefied petroleum gas (LPG)/air flames. An open-type double swriler burner with bluff body configuration is used in this study. The current study uses LPG due to its availability, combustion performance, and low emission levels [28–32]. Many research groups are focusing on introducing LPG into gas turbine applications [30, 33, 34]. The non-reacting flow PIV data shows the influence of Reynolds number at inner and outer annuli alter the flow structure which helps to predict the flow structure and flame stability. The overall flame structure and stability is understood by OH* and CH* chemiluminescence imaging. The stratified flames are short and compact than the premixed flames and insensitive to stratification ratio (SR).

2 Experimental Methods

This section describes the swirl-stabilized burner unit and diagnostics techniques used in this study.

2.1 Swirl-Stabilized Burner Unit

A swirl-stabilized axisymmetric burner is designed and fabricated to study the premixed and stratified flames. Figure 1a shows the main components of the burner in different colors. The burner mainly consists of three annuli: inner (aqua), outer (orange), and co-flow (green) and two mechanical swirlers, one (red) located in the inner annulus (23 mm diameter) and the other (purple) in the outer annulus (43.5 mm diameter) and a bluff body. The inner annulus provides a premixed fuel–air mixture flow at an equivalence ratio ϕ_1, which is equal or different than the equivalence ratio at an outer annulus ϕ_2 as shown in the cross-sectional view of the burner (Fig. 1b). The mixing of fuel–air is taking place before entering the inner and outer annuli, inside the feed line to the burner.

The premixed fuel–air was made to pass through the bottom of the plenum (gray) to the inner annulus. Similarly, the premixed flow through the outer annulus enters four sideways of the plenum by 90° with reference to the burner axis. The height of the annuli (inner-333 mm, outer-297, co-flow-250) is chosen in such a way that it provides fully developed flow to the burner exit. The entrainment of ambient air is prevented by using a cylindrical co-flow (161 mm diameter) where only filtered air is supplied. The air filtering at the co-flow is achieved by using two 3D printed honeycomb structures

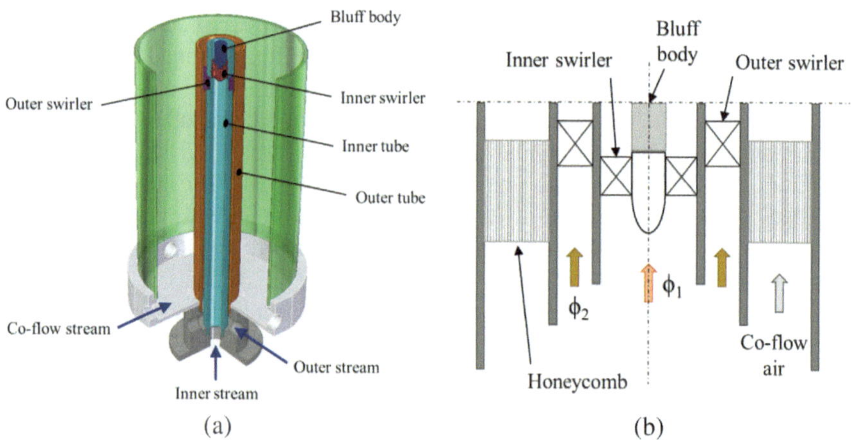

Fig. 1 Swirl-stabilized axisymmetric burner. **a** cut-sectional view **b** cross-sectional view of the burner exit

and two metallic perforated plates. The airflow at the cylindrical co-flow is initially allowed to pass through the two honeycomb structures then through the perforated plates. A collar (metallic silver) type of structure is used to hold the co-flow cylinder where the air is allowed to pass through four sideways, ensuring symmetrical inflow of air.

To achieve different swirling environments, two different swirlers are fabricated, such as the low swirler (L) and the medium swirler (M). The mechanical swirlers are fabricated using a 3D laser metal printing process. Also, this configuration allows us to independently investigate the effect of swirl number (S) and the Reynolds number (Re). Along with the swirlers, flame stabilization is established by bluff body configuration which is shown in Fig. 1b. A bluff body with 6 mm radius placed on inner swirler such a way that it does not affect the flow of inner swirler.

The current burner design concept is similar to the Cambridge/Sandia stratified swirl burner (CSB), and it has been demonstrated that flames can be stabilized over a range of flow rates, stratification ratios, and swirl numbers [35, 36]. The swirl introduced by CSB is basically an aerodynamic swirl where both the swirl number and Reynolds number are dependents on each other. Whereas in this burner both inner and outer annulus flows can be run with the axial flow or on various swirl intensities (low to strong) by using different sets of axial swirlers in both annuli which allows independently to study the effect of swirl and Reynolds number.

2.2 2D-Particle Image Velocimetry (PIV)

PIV is a non-intrusive technique used to measure the instantaneous velocity field of flow. 2D-PIV measures the two components (radial and axial) instantaneous velocity field of flow. Figure 2 shows the schematic diagram of the two-dimensional (2D) PIV setup along with flow configuration. PIV setup consists of a dual-cavity laser (Litron, 532 nm), laser sheet arrangement, and a charged-couple device (CCD) camera equipped with a Nikkor lens and an optical filter (532 $\pm$ 10 nm). The laser system emits two laser pulses with a given delay time between them, and the optical unit expands the beam into a laser sheet, which illuminates the flow field. The micron-level particles are illuminated by a laser sheet, and the laser sheet generated using 532 nm wavelength, 135 mJ Litron double pulsed Nd: YAG laser system. A CCD camera, which is synchronized with the laser system, acquires the image pairs at a given time interval Δt. The captured images are post-processed using 4G Insight software (TSI).

The overall flow setup along with the burner is shown in Fig. 2. Mass flow controllers (MFC-Alicat MCR-D series) are used to regulate the flow rates of the inner and outer annulus air with an accuracy of $\pm 1\%$ of full scale. The airflow through three annuli is seeded with micrometric vegetable oil particles for non-reacting flow PIV measurements using particle seeders. The oil particle generators are fabricated in-house. In these particle generators, Laskin nozzle design is used to efficiently produce oil particles at the micron level. Seeding of these particles is obtained by

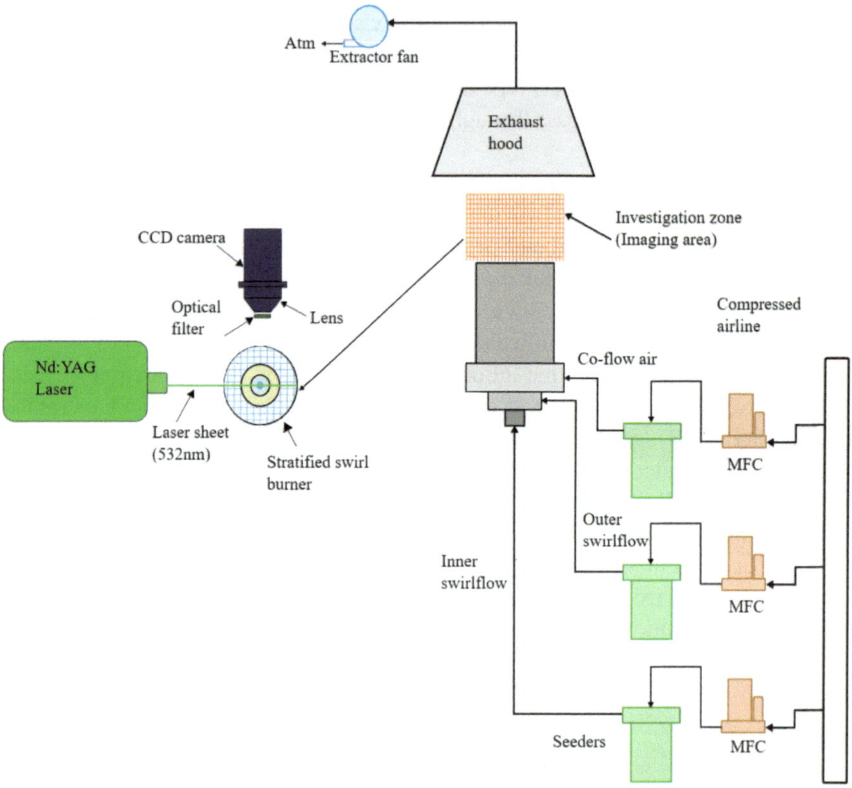

Fig. 2 Schematic of a 2D-PIV experimental setup for non-reacting flow study

using compressed air through a seed particle generator and uniform seed density is achieved by adjusting the needle valve that controls the airflow.

2.3 *Chemiluminescence Imaging*

The experimental setup of measurement of simultaneous OH*/CH* chemiluminescence is shown in Fig. 3. The chemiluminescence imaging is a line-of-sight technique, which captures the signal coming from the chemical excitation. The light emitted from the chemically excited OH* and CH* radicals is recorded by an intensified charge-coupled device (ICCD). Two ICCD (PI-MAX4) cameras equipped with UV lens of focal length 105 mm, providing a 16-bit image of 1024 × 1024 pixels and band-pass filters of 431 nm (CH*) and 308 nm (OH*), are used to capture the simultaneous CH* and OH* chemiluminescence images. Images were captured using the light-field software (Princeton Instruments). Proper calibration was performed, and background corrections were done during the post-processing of images.

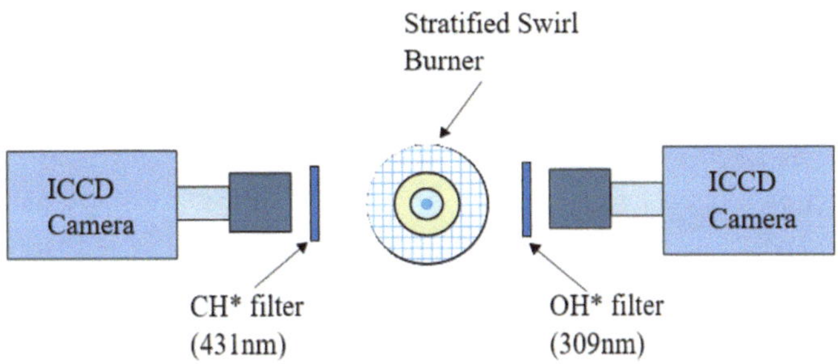

Fig. 3 Schematic of chemiluminescence imaging of OH* and CH* for reacting flow study

2.4 *Operating Conditions*

The 2D PIV for non-reacting flow study was carried out for six experimental cases, which are given in Table 1. The outer Reynolds number (Re_o) and co-flow Reynolds number (Re_c) are maintained constant for three different inner Reynolds numbers (Re_i). Similarly, the inner Reynolds number (Re_i) and co-flow Reynolds number (Re_c) are maintained constant for three different outer Reynolds number (Re_o).

Experiments were performed to investigate the one premixed and three stratified LPG/air flames. To understand the flame structure and stability, four test cases were studied using low and medium swirlers with geometric swirl number of 0.5 and 0.8, respectively. The test matrix is shown in Table 2. The outer swirl bulk velocity is maintained to be twice that of the inner swirl bulk velocity to promote mixing while the co-flow bulk velocity is kept constant at 0.4 m/s and thermal input power is also kept constant. The stratification ratio (SR) is defined as the ratio of inner equivalence ratio (ϕ_1) to the outer equivalence ratio (ϕ_2).

Table 1 Test matrix for PIV study

Case No	Bulk velocity (m/s)			Reynolds number		
	Inner (u_i)	Outer (u_o)	Co-flow (u_c)	Inner (Re_i)	Outer (Re_o)	Co-flow (Re_c)
N1	5	6	0.4	3577	6669	2826
N2	10	6	0.4	7154	6669	2826
N3	15	6	0.4	10,731	6669	2826
N4	7.5	4	0.4	5309	4402	2826
N5	7.5	6	0.4	5309	6581	2826
N6	7.5	8	0.4	5309	8804	2826

Table 2 Test matrix for chemiluminescence study

Case No	Inner swirl flow		Outer swirl flow		Stratification ratio SR $= \frac{\phi_1}{\phi_2}$
	Equivalence ratio (ϕ_1)	Bulk velocity m/s	Equivalence ratio (ϕ_2)	Bulk velocity m/s	
R1	0.75	7.40	0.75	15.27	1.0
R2	1.00	9.17	0.50	18.84	2.0
R3	1.20	9.20	0.40	18.81	3.0
R4	1.40	9.21	0.35	18.89	4.0

3 Results and Discussion

3.1 Velocity Measurements of Non-reacting Turbulent Flow Using 2D-PIV

To understand the isothermal flow structures, an unconfined axial swirl burner with the various Reynolds numbers (Re) (Table 1) of inner and outer flow is used under isothermal condition (298 K). Two components velocity field and turbulent parameters such as Reynolds stress and turbulent intensity are determined using 2D particle image velocimetry (PIV). For constant outer swirl and co-flow Reynolds number, the effect of changing inner swirl Reynolds number is studied. Similarly, for constant inner and co-flow Reynolds numbers, the effect of changing the outer Reynolds number also studied. For each case, 200 images are captured and post-processing was performed using 4G Insight software (TSI). The mean of 200 images is plotted to represent the velocity field and turbulence parameters such as Reynolds stress and turbulent intensity. ‘x’ indicates the axial co-ordinate, and ‘r’ indicates the radial co-ordinates for all the results provided in this section.

Radial velocity contour for six cases is shown in Fig. 4. The top row shows that increasing the inner swirl Reynolds number, the radial component of the velocity field is diminishing because the adverse pressure gradient dominates in axial direction. The radial spread of the flow is prevented by the outer swirl and co-flow when the inner swirl Reynolds number is increased. Therefore, in the case of reacting flow, the flame may elongate axially. The effect of outer swirl flow plays an essential role in the shear action between the inner swirl flow and co-flow conditions, which shows the increase in the radial flow field than the axial flow field. From case N4 to N6, as increasing outer swirl Reynolds number the radial velocity increases, also strong vortex break down was observed which is shown in the bottom row of Fig. 4.

Figure 5 shows the axial velocity vector field for various inner Reynolds number and the center recirculation zone (CRZ) in the velocity field which is shown clearly. It is observed that for the first case the axial component of velocity is radially spread, whereas, in the third case the radial spread is decreased and axial velocity is increased. The CRZ becomes very strong when the inner swirl Reynolds number is increased. The axial velocity component is increased while increasing the inner swirl Reynolds

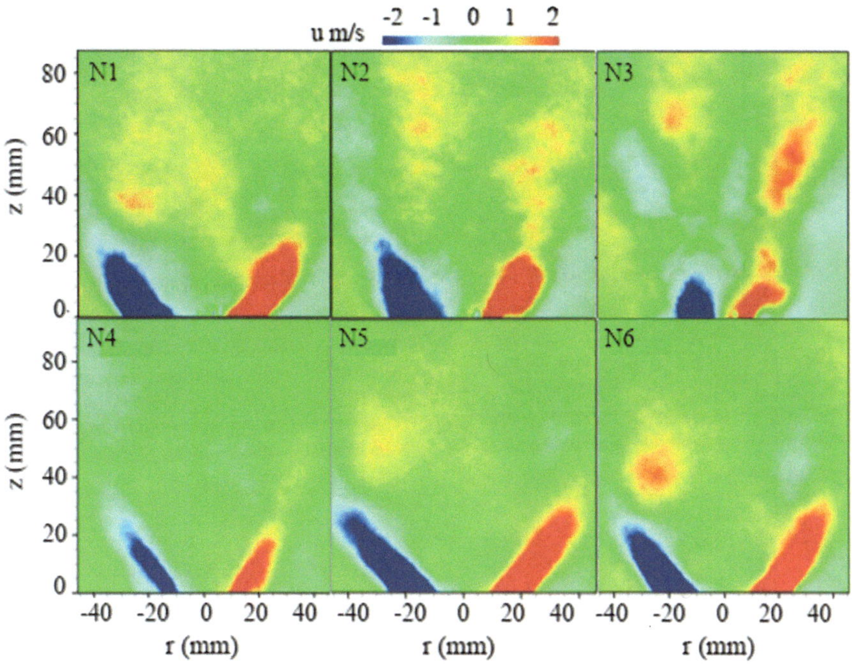

Fig. 4 Mean radial velocity components. Top row: Variation of inner flow Reynolds number. Bottom row: Variation of outer flow Reynolds number

number. As stated before the adverse pressure gradient becomes non-favorable for radial component; therefore, axial components are increasing. In the case of N4, the inner axial Reynolds number dominates the outer flow; therefore, the outer spreading is less, and the magnitude of axial velocity is less due to less overall bulk velocity (inner bulk velocity + outer bulk velocity). But the flow structure becomes wider for the case N5 than N4 and N6. In the case of N6, due to an increase in outer axial bulk velocity, the mean axial velocity starts dominating the outward pull. Therefore in the case of N6, the flow structure has a lesser radial spread compared to case N5 but more than case N4. Due to these flow structures, the flame length will be affected in reacting conditions. When considering the same flow condition (case N5 and case N6) for the reacting case, the flame length may reduce, because there may be an effect of thermal expansion radially.

The streamline plots are very useful in identifying the recirculation zones in the flow field. Figure 6 shows the streamline superimposed with axial velocity components and the inner recirculation zones (IRZ) is clearly identified. From cases N1 to N3, the adverse pressure gradient at the center of the swirl flow generates IRZ and when an increase in inner swirl Reynolds number, the IRZ is moving close to the centerline of the burner exit while shifting toward downstream. When considering the same flow conditions with the reacting case, the reaction zone may shift downstream. Due to an increase in inner Reynolds number, the fuel/air may not mix

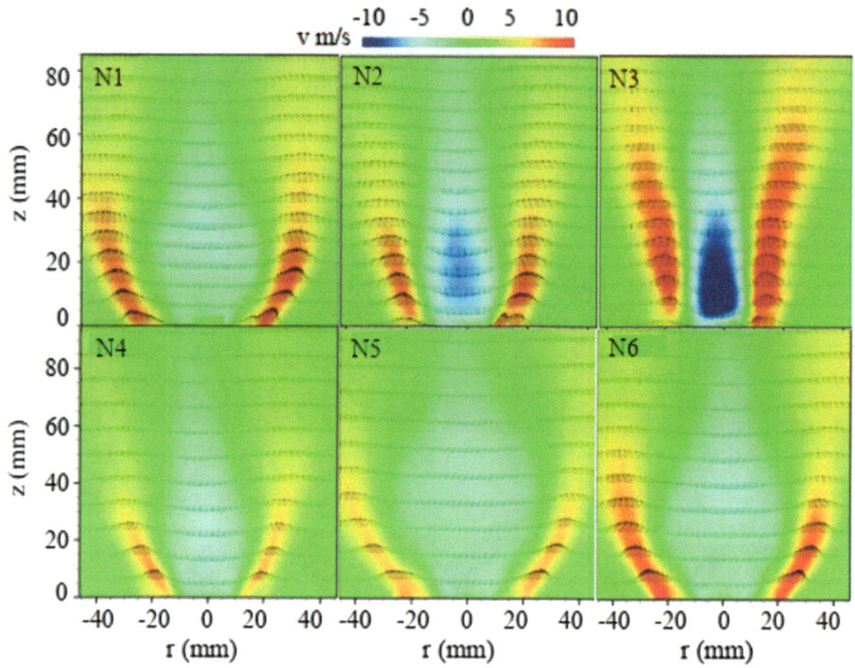

Fig. 5 Mean axial velocity component. Top row: Variation of inner flow Reynolds number. Bottom row: Variation of outer flow Reynolds number

properly at the exit of the burner and flame front may elongate axially. The IRZ becomes strong in the third case compared to the first case. When the inner Reynolds number dominates over the outer (N4), the IRZ is very strong and close to the centre-line of the burner exit. In case N5, when increasing the outer axial bulk velocity, the inner recirculation zones are moved outward and axially upward. On the other hand, when outer swirl Reynolds number dominates on the inner swirl Reynolds number, IRZ becomes weak and moves radially outwards. Also, case N5 shows compact inner recirculation zones, whereas N3 has elongated IRZ.

Figure 7 shows the mean Reynolds stress (τ_R) contour for each case. Reynolds stress is the momentum induced by the turbulent flux that acts as a medium to transfer energy from mean flow to turbulence. From the case, N1 to N3 the increase in Reynolds stress in axial direction indicates that the energy transfer to the turbulence more axially; therefore, the shear between inner and outer swirl flow is more at upstream where combustion is intensive due to velocity gradients. The mean Reynolds stress (τ_R) contour for various outer swirl flow conditions with constant inner swirl flow and co-flow is shown in the bottom row of Fig. 7. An increase in outer axial bulk velocity is contributing to Reynolds stress increment at radially outward as well as axially upward. When the Reynolds stress is high, the energy transfer to the turbulence will also be high, and thus, it promotes the fuel/air mixing for complete combustion.

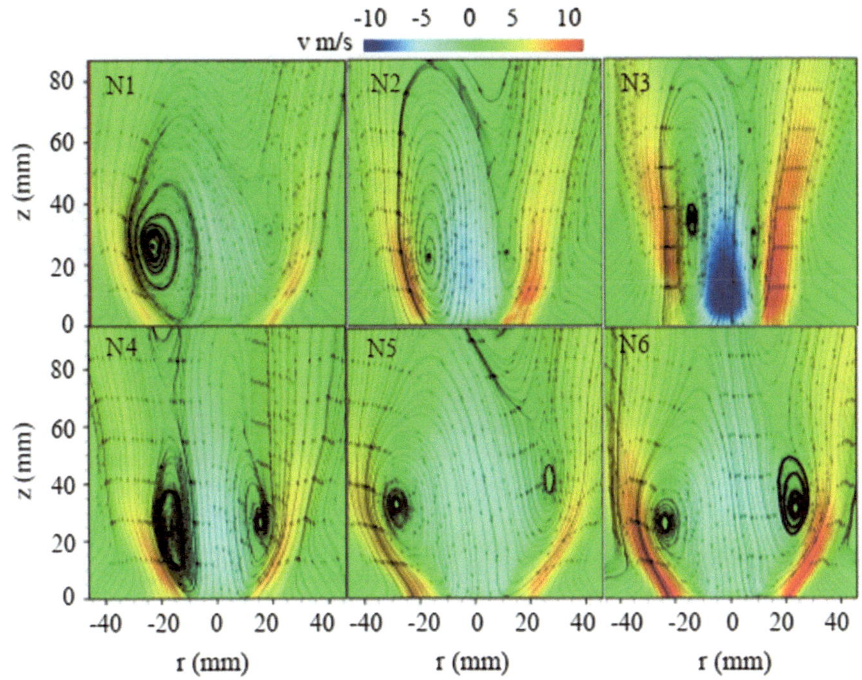

Fig. 6 Streamlines are superimposed with mean axial velocity. Top row: Variation of inner flow Reynolds number. Bottom row: Variation of outer flow Reynolds number

The turbulent intensity is one of the major parameters to characterize the swirl flow under both non-reacting and reacting conditions. Figure 8 shows the total turbulent intensity plot of the flow filed. The top row shows the increase in inner swirl Reynolds number, increases the turbulent intensity axially. The increase in inner swirl Reynolds number increases all the flow parameters axially; therefore, the flame may become elongated axially and the reaction zone is like to happen at downstream of the flow. The turbulent intensity contour for various outer swirl flow conditions with constant inner Reynolds number and co-flow Reynolds number is shown in the bottom row of Fig. 8. With an increase in outer flow, turbulent intensity also increases in both the axial and radial direction. As the Reynolds stress transfers the energy to the turbulence, the increase in Reynolds stress leads to an increase in the turbulence level. However, increasing the swirl Reynolds number at the flow field is not always favorable to enhance flame stabilization like an increase in turbulence level at lean fuel conditions may lead to blow-off. Therefore, it is essential to study the flow characteristics under various operating conditions such as different swirl number, mass flow rate, and equivalence ratio in order to identify the optimized operating conditions.

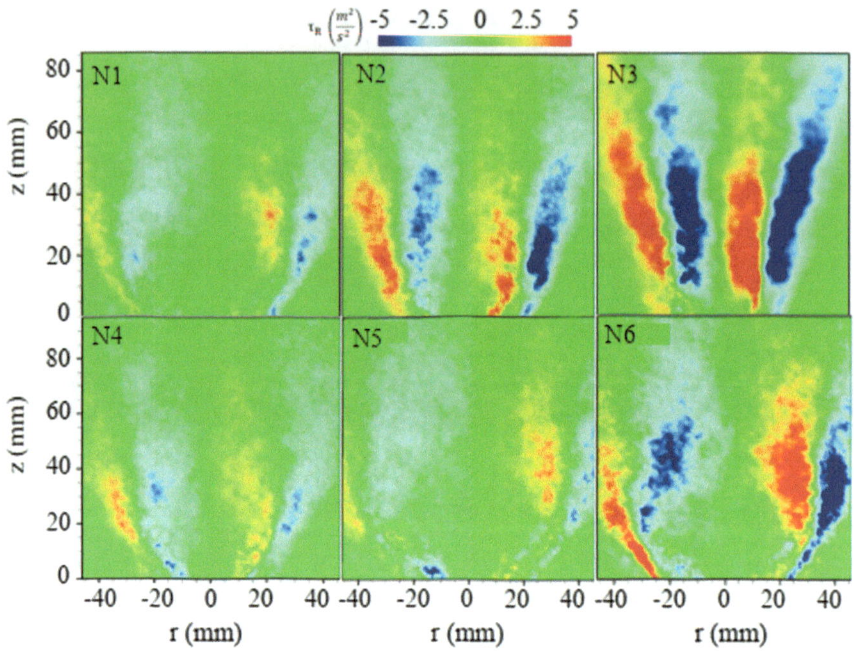

Fig. 7 Mean Reynolds stress for different operating conditions. Top row: Variation of inner flow Reynolds number. Bottom row: Variation of outer flow Reynolds number

3.2 *OH*/CH* Chemiluminescence Study of LPG Flames*

Flame structures and behavior can be investigated by observing intermediate radicals in the reaction zone, and there is a higher local heat release rate in the reaction zone that occurs at the flame front. Intermediate radicals such as OH* and CH* are good indicators for identifying the flame front. Chemiluminescence imaging is one of the reliable methods of non-intrusive optical diagnostics used to study the concentration of intermediate free radicals in the reaction zone. In this work, the spatial distribution of OH*/CH* using chemiluminescence is studied for the premixed and stratified liquefied petroleum gas (LPG) flames using an open-type double swirl-stabilized burner. Two different axial swirler, named as low swirler and medium swirler with geometric swirl numbers of 0.5 and 0.8, respectively, are used in this study. The spatial distribution of OH* and CH* chemiluminescence images are captured simultaneously for one premixed and three stratification conditions, which is shown in Table 2.

To understand the flame–turbulence interaction at premixed and stratification conditions, instantaneous CH* chemiluminescence images are presented in Fig. 9. It is observed that in premixed condition, the reacting wake is spread axially upward. As we approach the lean stratification condition, the reacting wake structure is more compact and the flame front is closer to the burner exit. The premixed LPG flames

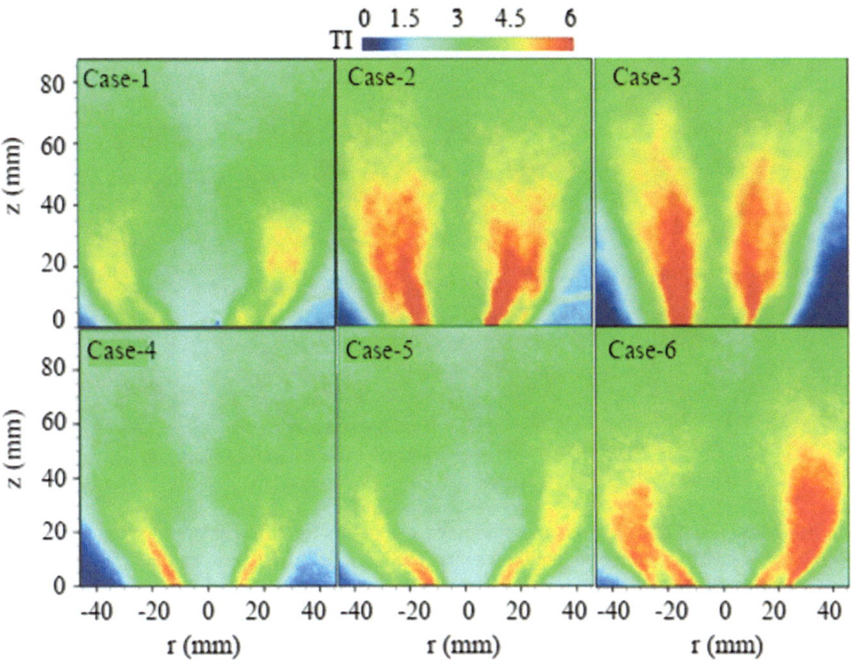

Fig. 8 Mean turbulent intensity plots for various cases. Top row: Variation of inner flow Reynolds number. Bottom row: Variation of outer flow Reynolds number

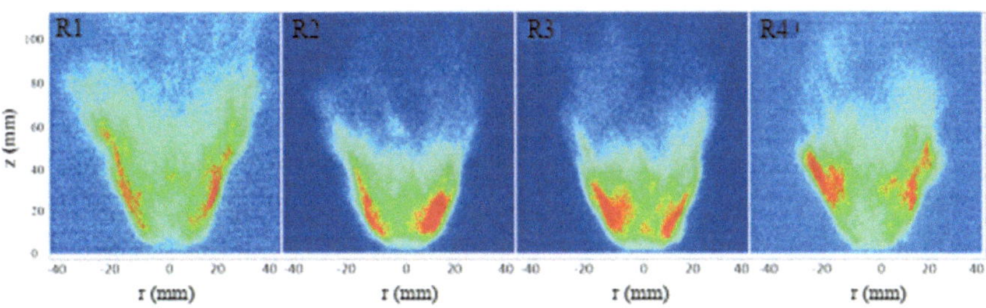

Fig. 9 Instantaneous CH* chemiluminescence images for various stratification ratios at low swirl flow condition (Swirl number 0.5)

are elongated, whereas the stratified conditions flame becomes more compact at the low swirling condition, as shown in Fig. 10. Indeed, the outer equivalence ratio in stratified conditions is below the lower flammable limits, but the combustion is still happening due to the back support from the rich inner flow region. For the medium swirl configuration, OH* and CH* contour plots show much more compact flame shapes than low swirl configuration as we approach the lean stratified condition. The increased swirl strength promotes the mixing of burned gases with incoming LPG/air, leading to faster combustion thus compact in shape. As we look closer, the

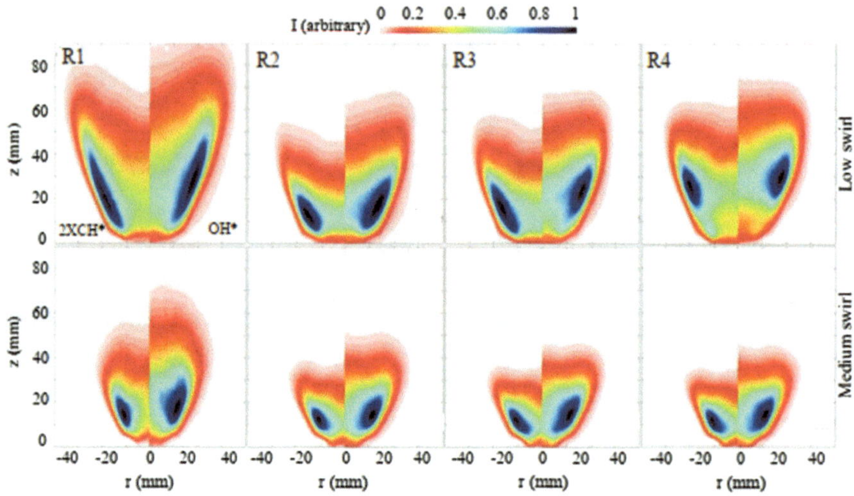

Fig. 10 Mean OH * and CH * chemiluminescence contours for different stratification conditions. Top row: low swirl. Bottom row: medium swirl. The left side is CH* plot, and the right side is the corresponding OH*

spatial distribution of OH* and CH* is very similar between all three stratified cases due to the enhanced mixing, which makes the flame becomes more robust to the local stratifications while the reaction zone stays closer to the exit of the burner.

Previous researchers have shown that intensity variation is independent of flame strain for propane flame [37], and similar assumptions can be made for the LPG flame as well. The flame equivalence ratio is calculated based on Eq. (1).

$$\phi = 0.6 - \ln\left(\frac{\left(\frac{OH*}{CH}\right) - 0.527}{1.135}\right) * 0.171 \quad (1)$$

where OH*/CH* ratio is calculated from the instantaneous images captured simultaneously using two ICCD cameras. The mean contour plots of the equivalence ratio for different stratification conditions are shown in Fig. 11.

From the equivalence ratio contour plots shown in Fig. 11, we can see the combined effect of mixture stratification and swirl strength on flame shapes. As expected, the premixed case (R1) for both swirl configurations shows the more spatial distribution of combustion zones as compared to that of stratified cases (R2-R4). Among the stratified cases, the primary reaction zone is more compact and closer to the exit of the burner for the medium swirl condition because of faster mixing of streams, which balances the slow down of chemical kinetics with the increase of mixture stratification. The corresponding probability distribution of the equivalence ratio for various test conditions is shown in Fig. 12. The premixed cases show the lower distribution as expected, whereas the low swirl is wider than the medium swirl. A similar trend is observed between the stratified cases between low and medium

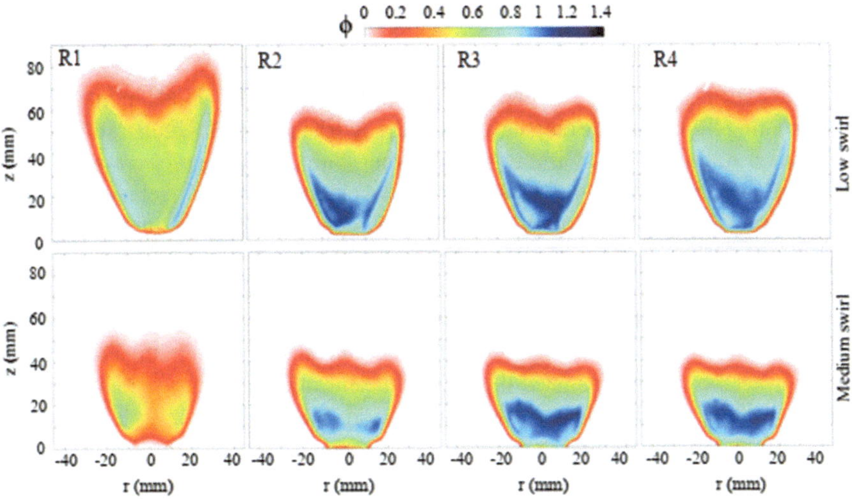

Fig. 11 Equivalence ratio contour plot based on the ratio of OH*/CH* for different stratification conditions. Top row: low swirl. Bottom row: medium swirl

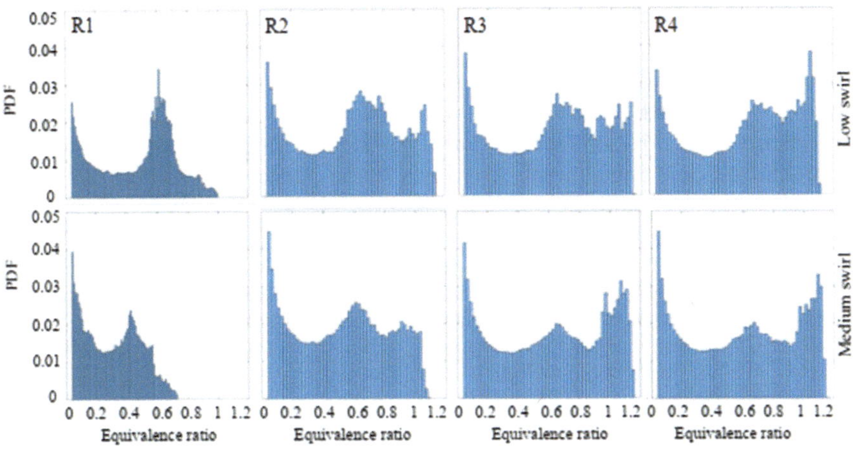

Fig. 12 Probability density function of the equivalence ratio is calculated for different stratification ratios. Top row: Low swirl. Bottom row: Medium swirl

swirl conditions. Indeed, the PDF pattern in almost the same for all three medium swirls stratified cases as compared to the low swirl cases. This experimental investigation on turbulent stratified flames using the chemiluminescence technique shows the potential of using the equivalence ratio derived from the OH*/CH* ratio could help us to understand the flame–turbulence interaction for LPG flames under various operating conditions.

4 Conclusion

Experiments are performed to understand the effect of the Reynolds number on the non-reacting flow structure and the effect of the swirl on the premixed/stratified flame structures. An open-type swirl-stabilized burner with double swirler arrangement is designed, and necessary flow systems are developed. 2D PIV technique is used to measure the flow field, and the chemiluminescence imaging technique is used to capture the flame structures. A series of experiments are performed for premixed and stratified flames. The analysis of non-reacting flow fields showed that the operating parameters play a critical role in the flow structure, such as varying the inner and outer flow rate leads to different recirculation zone size and strength. The increase in the inner flow rate elongates the IRZ while the increase in the outer flow rate widens the IRZ for the MM swirler combination. Similar behavior is observed for other swirler combinations.

The spatial distribution of OH* and CH* is observed from simultaneously acquired OH* and CH* chemiluminescence images under low swirler and medium swirler combinations. Premixed and stratified turbulent flame structures are observed from the instantaneous CH* images. From the intensity of the OH* images, the flame front is located. In the stratified case, the flame front appears compact and closer to the exit of the burner than to the premixed case. Enhanced swirl effect facilitates the mixing of LPG and air; thus, the structure of turbulent LPG/air flames under premixed and stratified conditions is more compact for medium swirler combinations than low swirler combinations, and the lean blow-off limit is delayed. The equivalence ratio plots derived from the OH*/CH* ratio help to understand the effect of stratification and turbulence mixing on the flame characteristics of the LPG under various operating conditions.

Acknowledgements This research is supported by the Science and Engineering Research Board (SERB) of India through grant No. ECR/2015/000343. The optical diagnostic setup is supported by the Fund for Improvement of S&T Infrastructure (FIST) of India through grant No.: SR/FST/ETI-397/2015/(C). The author is grateful to the Ministry of Human Resource Development (MHRD) of India for providing the Ph.D. fellowship.

References

1. Lilley DG (1977) Swirl flows in combustion: a review. AIAA J 15(8):1063–1078. https://doi.org/10.2514/3.60756
2. Ahmed SA, Nejad AS (1992) Swirl effects on confined flows in axisymmetric geometries. J Propuls Power 8(2):339–345. https://doi.org/10.2514/3.23483
3. Chong CT, Hochgreb S (2015) Measurements of non-reacting and reacting flow fields of a liquid swirl flame burner. Chinese J Mech Eng 28(2):394–401. https://doi.org/10.3901/CJME.2015.0109.011
4. Elbaz AM, Moneib HA, Shebil KM, Roberts WL (2019) Low NOX–LPG staged combustion double swirl flames. Renew Energy 138(X):303–315. https://doi.org/10.1016/j.renene.2019.01.070

5. Sellan D, Murugan R, Balusamy S (2019) Experimental and numerical analysis of turbulent swirl flow structure in double swirler burner. In: ASME 2019 gas turbine india conference, GTINDIA 2019, vol 2. https://doi.org/10.1115/GTINDIA2019-2739
6. Vanierschot M, Van den Bulck E (2008) Influence of swirl on the initial merging zone of a turbulent annular jet. Phys Fluids 20(10). https://doi.org/10.1063/1.2992191
7. Zhao F, Lai MC, Harrington DL (1999) Automotive spark-ignited direct-injection gasoline engines. Prog Energy Combust Sci 25(5):437–562
8. Candel S (2002) Combustion dynamics and control: progress and challenges. Proc Combust Inst 29(1):1–28
9. Huang Y, Yang V (2009) Dynamics and stability of lean-premixed swirl-stabilized combustion. Prog Energy Combust Sci 35(4):293–364
10. Karagiannaki C et al (2009) Performance evaluation of a model swirl burner under premixed or stratified inlet mixture conditions. Combust Flame 28(1):1–8
11. Balusamy S, Cessou A, Lecordier B (2009) Direct experimental determination of laminar burning velocity using a new PIV approach pp 8–11
12. Balusamy S, CessouA,Lecordier B (2009) Measurement of laminar burning velocity–A new PIV approach. 4th Eur Combust Meet Vienna 3:1–6
13. Balusamy S, Cessou A, Lecordier B (2014) Laminar propagation of lean premixed flames ignited in stratified mixture. Combust Flame 161(2):427–437
14. Masri AR (2015) Partial premixing and stratification in turbulent flames. Proc Combust Inst 35(2):1115–1136
15. Lauvergne R, Egolfopoulos FN (2000) Unsteady response of C3H8/Air laminar premixed flames submitted to mixture composition oscillations. Proc Combust Inst 28(2):1841–1850
16. Pasquier N, Lecordier B, Trinité M, Cessou A (2007) An experimental investigation of flame propagation through a turbulent stratified mixture. Proc Combust Inst 31(1):1567–1574
17. Balusamy S, Cessou A, Lecordier B (2011) Direct measurement of local instantaneous laminar burning velocity by a new PIV algorithm. Exp Fluids 50(4):1109–1121
18. Guyot D, Guethe F, Schuermans B, Lacarelle A, Paschereit CO (2010) CH*/OH* chemiluminescence response of an atmospheric premixed flame under varying operating conditions. Proc ASME Turbo Expo 2(PARTS A AND B):933–944
19. Kojima J, Ikeda Y, Nakajima T (2000) Spatially resolved measurement of OH*, CH*, and C2* chemiluminescence in the reaction zone of laminar methane/air premixed flames. Combust Inst 28(2):1757–1764
20. Muruganandam TM, Kim B, Olsen R,Patel M, Romig B, Seitzman JM (2003) Chemiluminescence based sensors for turbine engines. In: 39th AIAA/ASME/SAE/ASEE Jt Propuls Conf Exhib July 2003
21. Aleiferis PG, Hardalupas Y, Taylor AMKP, Ishii K, Urata Y (2004) Flame chemiluminescence studies of cyclic combustion variations and air-to-fuel ratio of the reacting mixture in a lean-burn stratified-charge spark-ignition engine. Combust Flame 136(1–2):72–90
22. Karagiannaki C, Paterakis G, Souflas K, Dogkas E, Koutmos P (2015) Performance evaluation of a model swirl burner under premixed or stratified inlet mixture conditions. J Energy Eng 141(2):1–9
23. Taamallah S, Shanbhogue SJ, Ghoniem AF (2016) Turbulent flame stabilization modes in premixed swirl combustion: Physical mechanism and Karlovitz number-based criterion. Combust Flame 166:19–33
24. Jourdaine P, Mirat C, Caudal J, Lo A, Schuller T (2017) A comparison between the stabilization of premixed swirling CO_2-diluted methane oxy-flames and methane/air flames. Fuel 201:156–164
25. Nori VN, Scitzman JM (2009) CH* chemiluminescence modeling for combustion diagnostics. Proc Combust Inst 32(1):895–903
26. Foley C et al (2009) Performance evaluation of a model swirl burner under premixed or stratified inlet mixture conditions. Combust Flame 28(1):1–8
27. Chaudhari RR, Sahu RP, Ghosh S, Mukhopadhyay A, Sen S (2013) Flame color as a lean blowout predictor. Int J Spray Combust Dyn 5(1):49–65

28. Miao J, Leung CW, Cheung CS, Huang Z, Jin W (2016) Science direct effect of H_2 addition on OH distribution of LPG/air circumferential inverse diffusion flame. Int J Hydrogen Energy 41(22):9653–9663
29. Mishra DP (2004) Emission studies of impinging premixed flames, vol 83, pp 1743–1748
30. Shaddix CR, Molina A (2009) Particle imaging of ignition and devolatilization of pulverized coal during oxy-fuel combustion. In: 32nd international symposium on combustion, vol 32 II, pp 2091–2098
31. Makmool U, Jugjai S, Tia S (2013) Structures and performances of laminar impinging multiple premixed LPG-air flames. Fuel 112:254–262
32. Choi GH, Chung YJ, Bin Han S, Performance and emissions characteristics of a hydrogen enriched LPG internal combustion engine at 1400 rpm. Int J Hydrogen Energy, 30(1):77–82
33. Enagi II, Al- KA, Zainal ZA (2017) Combustion chamber design and performance for micro gas turbine application. Fuel Process Technol 166:258–268
34. Elbaz AM, Roberts WL (2019) Stability and structure of inverse swirl diffusion flames with weak to strong swirl. Exp Therm Fluid Sci 112(November 2019):109989
35. Zhou R, Balusamy S, Sweeney MS, Barlow RS, Hochgreb S, Flow field measurements of a series of turbulent premixed and stratified methane/air flames. Combust Flame 160(10):2017–2028
36. Sweeney MS, Hochgreb S, Dunn MJ, Barlow RS (2012) The structure of turbulent stratified and premixed methane/air flames I: Non-swirling flows. Combust Flame 159(9):2896–2911
37. Orain M, Hardalupas Y (2010) Influence du combustible sur la mesure de richesse par chimiluminescence dans les flammes prémélangées. Comptes Rendus Mec 338(5):241–254

Flame Root Dynamics and Their Role in the Stabilisation of Lifted Flames

James C. Massey, Zhi X. Chen, and Nedunchezhian Swaminathan

Abstract This chapter aims to provide a broad overview on the importance of lifted flames in turbulent flows with and without swirl in practical applications for energy production and propulsion. The stabilisation of lifted flames is governed by numerous physical processes that control the behaviour of the flame root or the leading edge. The flame lift-off height, which is the physical distance of the flame root above the burner, is a quantity of practical importance and obtaining accurate predictions using computational fluid dynamics (CFD) is challenging. The large eddy simulation (LES) paradigm has proven to be successful in accurately capturing the flame lift-off height in turbulent reacting flows in simple and complex geometries of practical relevance. The objective of this chapter is to present an overview of the stabilisation mechanisms that have been observed in simulations with relevance to modern applications. An overview of the LES framework for turbulent reacting flows and different sub-grid combustion models are briefly discussed with a focus on the unstrained flamelet combustion model that is used in the case studies presented here. The first part of the simulation results focuses on the canonical jet flame configuration, where it is seen that the lift-off height is sensitive to the jet velocity and the fuel used. The second part of the results focuses on a more complex configuration, which is a gas turbine model combustor with two radial swirlers. The flame root in swirling flows is typically more robust, but failed ignition and local extinction cause the flame root to oscillate and leads to flame lift-off. The amplitude of this oscillation can lead to the occurrence of thermo-acoustic oscillations and flame blow-off under appropriate conditions. The discussion is presented on a physical basis and the observations are compared with measurements. The chapter concludes by summarising the role and importance of modelling flame stabilisation.

Keywords Computational fluid dynamics · Gas turbiness · Lifted flames · Partially premixed flames · Swirling flows · Turbulent reacting flows

---

J. C. Massey (✉) · Z. X. Chen · N. Swaminathan
Department of Engineering, University of Cambridge, Trumpington Street, Cambridge CB2 1PZ, UK
e-mail: jcm97@cam.ac.uk

A. Gupta et al. (eds.), *Advances in Energy and Combustion*, Green Energy and Technology,
https://doi.org/10.1007/978-981-16-2648-7_11

1 Introduction

Chemical reactions in combustion processes are accompanied by a flame, but the type of flame that exists is dependent on the prescribed boundary conditions. Premixed combustion is present when the fuel and oxidiser are mixed prior to ignition, whereas non-premixed combustion is present when the fuel and oxidiser enter through separate inlets, and mixing and burning occur simultaneously [1]. Both combustion processes can exist in laminar and turbulent flows. In practical combustion systems, such as industrial gas turbines and aero engines, turbulent flows are expected and this causes a degree of unavoidable mixing between the fuel and oxidiser before ignition. Therefore, compositionally inhomogeneous mixtures are present and this leads to partially premixed combustion [2]. Flames in large industrial boilers and gas turbines are stabilised using a non-premixed jet configuration and swirling flows are often used, since swirling flows can extend the range of flame stability in the combustion system [3]. Due to the presence of partially premixed combustion and ubiquitous high Reynolds number turbulent flows in practical systems, understanding the underlying mechanisms that lead to flame stabilisation is important.

The jet flame is a simple configuration where partially premixed combustion is present. Fuel enters through a nozzle into quiescent air at atmospheric conditions and hence, compositionally inhomogeneous mixtures that are within and beyond the flammability limits are present. For a low fuel jet exit velocity, the flame is able to stabilise by remaining attached to the nozzle rim. When the jet velocity is increased, the flame is locally quenched due to turbulence induced strain and the flame stabilises at a downstream location where reactants mix with hot combustion products [2]. These are referred to as lifted flames and these are typically preferred to protect the burner nozzle, such as in commercial boilers [4]. Combustion of stratified mixtures in gas turbine engines and diesel engines has elements of lifted flame attributes and hence, laboratory-scale lifted flames remain of high interest to the community [5].

A typical lifted flame configuration is shown in Fig. 1, and the mean flame base and stoichiometric mixture fraction contour are shown in red and grey respectively. The flame base is the stabilisation location of the flame and its position downstream of the fuel nozzle exit is referred to as the flame lift-off height H. The mean flame base is usually marked with a specified temperature value in numerical simulations and OH-PLIF measurements in experimental studies. The stabilisation mechanisms of lifted flames have been hypothesised in detail in the study by Wohl et al. [6] and the review by Pitts [7] addressed the main flame stabilisation mechanisms that have been identified, followed by more recent reviews by Lyons [5] and Lawn [8]. In the study by Vanquickenborne and van Tiggelen [9], it was suggested that the local gas mixture at the flame base is premixed and hence, the flame stabilisation is due to the balance between the local flow velocity and the local burning velocity of the mixture upstream of the flame base along the stoichiometric mixture fraction contour. The burning velocity is known as the turbulent flame speed s_T, which is a function of unstrained laminar flame speed s_L^0 and the local turbulence statistical characteristics, such as the integral length scale and velocity fluctuations. The study by Kalghatgi

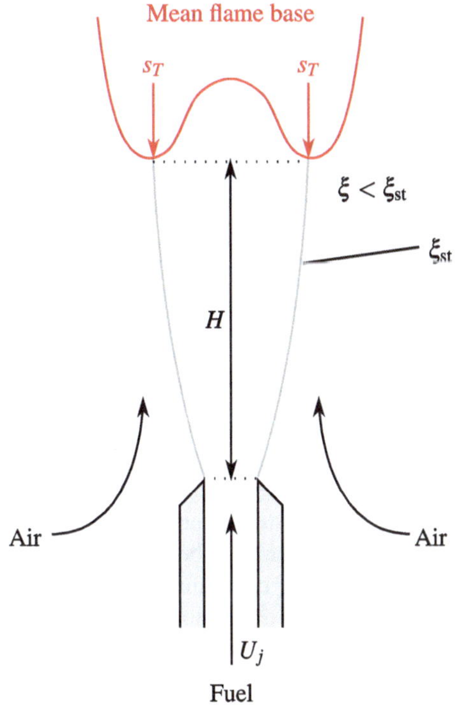

Fig. 1 Schematic of a lifted flame, which is adapted from the schematic shown by Lawn [8]

[10] investigated the turbulent flame speed by using different fuels and jet velocities. An equation for s_T was deduced using simple theoretical arguments and experimental data, and it was suggested that the ratio s_T/s_L^0 is proportional to the square root of the local turbulent Reynolds number Re_T. It was also shown that the lift-off height was linearly dependent on the fuel nozzle jet velocity U_j.

Peters and Williams [11] argued that the major flame stabilisation is related to flamelet quenching at the flame base region due to high scalar dissipation rates and flame stretching. However, the lack of experimental data in the flame base region could not ascertain this theory and the study by Pitts [7] highlights that this theory does not consider the partial premixing of the fuel and oxidiser upstream of the flame. The study by Müller et al. [12] introduced a model for partially premixed combustion that includes both premixed flame propagation and flame quenching by using premixed and non-premixed flamelets respectively. It has since been proposed that high levels of partial premixing lead to the presence of triple flames (or sometimes referred to as edge flames), where the flame base consists of a lean and rich premixed flame and a trailing non-premixed flame between the two premixed flame branches, which is due to excess oxidiser and fuel from the lean and rich branches respectively. These flame structures have been studied and reviewed in significant detail in earlier studies [2, 12–15], but the flame stabilisation mechanisms have been widely hypothesised.

Another explanation for the flame stabilisation mechanism in lifted flames is due to mixing of fuel with hot combustion products in the large-scale eddies formed in the jet flow. The review by Pitts [7] observed using species concentration measurements that the fuel–air mixture composition at the flame base is highly intermittent. For short time periods, mixtures within the flammability limit are present that allows the flame base to propagate upstream, otherwise only fuel or air is present, which prevents the flame base from propagating upstream. Therefore, large-scale neighbouring eddies need to supply hot combustion products to the edges of the jet, along with fresh air and fuel–air mixtures within the flammability range. These large-scale structures are important, since there is no large recirculation zone present that can supply hot combustion products to the edges of the jet [16]. Correlations have also been developed and proposed using experimental observations to determine the blow-off velocity for lifted flames [16, 17]. Lifted flames are highly prone to blow-off, which is treated as the complete extinction of the flame and must be avoided to ensure efficient and smoothly operating combustion systems.

Due to the inherent instability of lifted flames, the risks of blow-off and their influences on system operation, swirling flows are used to improve flame stability because the flame root becomes more robust and improves system stability over a broad range of operating conditions [18]. A schematic of a non-premixed swirl flame configuration is shown in Fig. 2. An inner shear layer (ISL) containing high vorticity is formed from the trailing edge of the burner nozzle. The vortex structures are convected further downstream with the incoming jet stream until vortex breakdown occurs, which leads to the formation of an inner recirculation zone (IRZ). The incoming fuel is entrained into the IRZ and a flame will form along the ISL once the fuel–air mixture is ignited. The IRZ acts as a continuous ignition source since hot products are entrained into the recirculation zone and this continuous supply of heat and radical species makes the flame root more robust [19]. In addition, a hydrodynamic instability that is referred to as a precessing vortex core (PVC) can be present in flows with larger swirl numbers ($S > 0.6$–0.7) [3, 20]. The instantaneous structure of the PVC is shown in Fig. 8 for a gas turbine model combustor simulation, along with the IRZ and the flame. The PVC influences the stability of the flame and can lead to significant changes in the combustion chamber, since the PVC can enhance fuel–air mixing in the flame root region or lead to the presence of thermo-acoustic instabilities and flame lift-off, as shown in a gas turbine model combustor experiment [21, 22].

Swirling flow configurations are beneficial, since they allow quicker mixing of fuel and air, which is conducive for partially premixed combustion and this leads to compact combustion systems [3]. This helps achieving fuel-lean combustion in order to meet the ever-increasing norms on emissions of carbon dioxide (CO_2) and nitrogen oxides (NO_x). Lean combustion is also beneficial as shorter flames are present and the size of combustion chambers can be reduced. However, operating under lean conditions can make combustion systems prone to combustion instabilities, which arise due to close coupling among acoustics, combustion and swirling flow dynamics [23–25]. These instabilities can become detrimental to lean systems through failed ignition, flashback, autoignition and flame blow-off. Moreover, it is near impossible

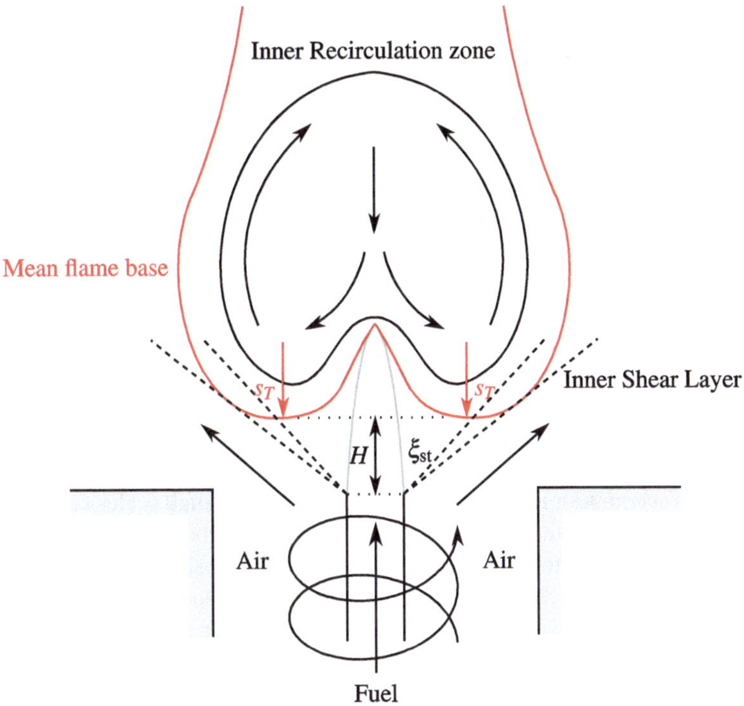

Fig. 2 Schematic of a swirl-stabilised flame, which is adapted from the schematic shown by Feikema et al. [18]

to achieve perfect premixing because the mixing time scales are usually larger than the residence time scales for the reactants and hence, combustion is typically in the partially premixed regime [26]. Therefore, understanding the stabilisation mechanisms for partially premixed flames is imperative for the design of such combustion systems.

The design of combustion systems is increasingly reliant on computational fluid dynamics (CFD), which is now on par with experimental tools for research and development. There are three commonly used CFD approaches, namely direct numerical simulation (DNS), large eddy simulation (LES) and Reynolds-averaged Navier–Stokes (RANS) simulation, in the order of descending computational cost. With the state-of-the-art computing technology, the use of DNS for practical combustion systems is not yet possible. This approach is typically used in simplified configurations (often referred to as "flame-in-a-box") for combustion model development and validation. On the other hand, steady RANS and unsteady RANS (URANS) techniques have been developed as computationally inexpensive alternatives for complex geometries in practical devices. However, these approaches consider ensemble- or time-averaged conservation equations for mass, momentum, energy and scalars. In addition, they resolve only one length scale and one time scale, which are typically (integral scales) larger than the scales of interest to investigate flame behaviour. This means that RANS simulations cannot capture highly unsteady phenomena, such as the small-scale flame stabilisation mechanisms. Hence, there has been an emergence of LES approaches for modelling turbulent combustion over the last two decades, since such approaches can capture the transient phenomena exhibited by turbulent flows and flames.

The device-dependent energy containing large scales and its cascade to smaller scales, down to a cut-off scale Δ, are resolved and captured in LES by using filtered conservation equations. The contributions of device-independent or universal small scales, typically known as the sub-grid scale (SGS) range, are modelled using closure models for the unresolved residual stresses and scalar fluxes [27]. Therefore, the resulting computational cost is largely reduced compared to that for DNS. The LES approach is particularly suitable for physical problems where the transient behaviours are predominantly driven by the resolvable large scales that often vary from one case to another (e.g. geometry, operating conditions, etc.), whereas SGS models accounting for the effect of non-resolvable small scales can be derived prior to the CFD simulations. The turbulent combustion processes occurring in many practical devices fall into this category, especially for those exhibiting highly transient characteristics, such as the lifted flame dynamics and stabilisation, which is of interest here. This makes LES the ideal candidate for the present study. Since combustion is mostly at the SGS level, where the typical chemical length scales ($< \mathcal{O}(10^{-4})$ m) are by orders of magnitude smaller than the device-dependent turbulent flow scales ($> \mathcal{O}(10^{-2})$ m), accurate and robust modelling is required to capture the SGS turbulence–combustion interactions.

The last two decades has seen numerous SGS models developed for turbulent combustion and a detailed review of these models is beyond the scope of this work (see previous studies for extensive reviews [28–30]). A presumed joint probability density function (PDF) approach belonging to the statistical category is used to model the SGS turbulent combustion here. This approach has been tested and validated for various laboratory and practical configurations with stable and unstable conditions using RANS and LES methodologies (see the review by Chen et al. [31]). The aim of this chapter is to present an overview of simulations using this modelling approach for turbulent partially premixed combustion with a focus on flame stabilisation mechanism in various burner configurations.

This chapter is organised as follows. The combustion modelling framework is described in Sect. 2 along with a brief summary of the other models that have been used for the particular target cases presented in later sections. The canonical non-swirling jet flame is first presented in Sect. 3, which focuses on the sensitivity of the lift-off height to the fuel jet velocity for different fuels. A more complex configuration featuring a gas turbine model combustor with swirling flows is investigated in Sect. 4 with detailed analyses on flame root dynamics and its stabilisation mechanisms under various conditions. Concluding remarks are presented in Sect. 5.

2 Turbulent Reacting Flow Modelling

The modelling methodology is introduced in two parts in this section. The first part presents the general LES governing equations for turbulent reacting flow simulations and the SGS modelling for unclosed terms in those equations is summarised. In the second part, the chosen combustion modelling approach involving additional scalar transport equations and closure models is described in appropriate detail, in order to facilitate a better interpretation of the simulation results presented in Sects. 3 and 4.

2.1 Large Eddy Simulation

Simulations with the LES methodology are becoming increasingly popular, due to the continuous advancement of computational resources. However, the computational cost depends on the amount of scales that are inherently resolved in the LES, i.e. the more scales that are resolved, the higher the cost. This is related to the chosen filter size, which is proportional to the local cell volume and is estimated as $\Delta \approx \mathcal{V}^{1/3}$. If Δ is larger, i.e. fewer scales are resolved by the numerical grid, then the SGS model needs to account for the effects of more unresolved scales. As an overall guidance, Pope [27] suggested that 80% of the turbulent kinetic energy should be resolved for an acceptable LES and this is commonly practised.

The transport equations required for a turbulent flow field include the conversation of mass, momentum and energy. When applying the LES filtering to these equations, several unclosed terms appear and these require SGS modelling. To reduce the number of unclosed terms in variable density flows, a density-weighting procedure is used, which is known as Favre filtering. For a physical variable φ, its density-weighted form is written as $\widetilde{\varphi} = \overline{\rho\varphi}/\overline{\rho}$ [32]. Applying this procedure to the filtered transport equations for mass, momentum and the thermochemical enthalpy (sensible plus formation) leads to

$$\frac{\partial \overline{\rho}}{\partial t} + \nabla \cdot \left(\overline{\rho}\widetilde{\boldsymbol{U}}\right) = 0\,, \tag{1}$$

$$\overline{\rho}\frac{\mathrm{D}\widetilde{\boldsymbol{U}}}{\mathrm{D}t} = -\nabla\overline{p} + \nabla \cdot \overline{\boldsymbol{\tau}} - \nabla \cdot \overline{\boldsymbol{\tau}}^{R}\,, \tag{2}$$

$$\overline{\rho}\frac{\mathrm{D}\widetilde{h}}{\mathrm{D}t} = \frac{\mathrm{D}\overline{p}}{\mathrm{D}t} + \nabla \cdot \left(\overline{\rho\alpha\nabla h}\right) - \nabla \cdot \overline{\boldsymbol{\tau}}_{h}^{R}\,, \tag{3}$$

where $\mathrm{D}/\mathrm{D}t = \partial/\partial t + \widetilde{\boldsymbol{U}} \cdot \nabla$ is the material derivative and $\overline{\boldsymbol{\tau}}$ is the molecular viscous stress tensor. The filtered density is obtained using the equation of state $\overline{\rho} = \overline{p}\widetilde{\mathcal{M}}/\Re^0\widetilde{T}$ with $\widetilde{\mathcal{M}}$ and $\Re^0$ being the mixture molecular weight and the universal gas constant respectively. The molecular transport terms, which are the second terms on the right-hand side of Eqs. (2) and (3), are expected to be small for high Reynolds number flows, but these terms cannot be neglected in the LES framework, as they are relevant at the SGS level and for numerical stability in the simulation. The pressure-heating source term $\mathrm{D}\overline{p}/\mathrm{D}t$ in Eq. (3) is included if compressibility effects are expected to be significant in certain cases, e.g. thermo-acoustic instabilities, and if there is bulk compression, e.g. internal combustion engines.

The residual stress tensor $\overline{\boldsymbol{\tau}}^R \equiv (\overline{\rho}\widetilde{\boldsymbol{U}\boldsymbol{U}} - \overline{\rho}\widetilde{\boldsymbol{U}}\widetilde{\boldsymbol{U}})$ in Eq. (2) is decomposed into its isotropic and anisotropic parts respectively as $\overline{\boldsymbol{\tau}}^R = \frac{2}{3}\,\overline{\rho}\widetilde{k}_{\mathrm{sgs}}\,\mathbf{I} + \overline{\boldsymbol{\tau}}^r$, where the SGS kinetic energy $\widetilde{k}_{\mathrm{sgs}}$ is equal to half of the isotropic part of the residual stress tensor and is typically absorbed into the filtered pressure term [27]. The anisotropic residual stress tensor is typically modelled using the Boussinesq eddy viscosity hypothesis [33, 34]:

$$\overline{\boldsymbol{\tau}}^r = -2\,\overline{\rho}\nu_T\left[\widetilde{\boldsymbol{S}} - \frac{1}{3}\,(\nabla\cdot\widetilde{\boldsymbol{U}})\,\mathbf{I}\right], \tag{4}$$

where $\widetilde{\boldsymbol{S}} = \frac{1}{2}\,[\nabla\widetilde{\boldsymbol{U}} + (\nabla\widetilde{\boldsymbol{U}})^{\mathsf{T}}]$ is the symmetric strain rate tensor and ν_T is the SGS eddy viscosity [27].

The residual scalar flux terms are commonly modelled using a gradient transport hypothesis, such as the last term in Eq. (3). This is analogous to Fick's law of diffusion, where the transport of the scalar flux is in the direction of the scalar gradient $-\nabla\varphi$ and is written as $\overline{\boldsymbol{\tau}}^R_\varphi = \overline{\rho}\widetilde{\boldsymbol{U}\varphi} - \overline{\rho}\widetilde{\boldsymbol{U}}\widetilde{\varphi} = -\overline{\rho}(\nu_T/\sigma_\varphi)\nabla\widetilde{\varphi}$, where the dimensionless number σ_φ is a parameter that is physically consistent with the transported variable; for Eq. (3), this dimensionless number is the turbulent Prandtl number $\sigma_\varphi \equiv \mathrm{Pr}_T$, which is taken to be 0.7 in common practice.

The eddy viscosity ν_T requires closure and its model should be carefully chosen. Numerous models for the eddy viscosity have been developed, which include algebraic (zero equation) models or solving additional transport equations to obtain SGS turbulence information. The classical Smagorinsky model [35] is a simple algebraic formulation that is still widely used, and it has an empirical model constant C_s, which can take different values for different cases. Proposed modifications to dynamically compute C_s using resolved quantities have been developed [36–38] to overcome this issue. However, this approach (or any dynamic procedures) should be used with care, as it is highly susceptible to numerical errors leading to unphysical values of C_s, e.g. negative effective (molecular plus turbulent) viscosity. Other popular algebraic models include the eddy viscosity model proposed by Vreman [39] and the wall-adapting local eddy (WALE) model [40], and both approaches give improved predictions of ν_T naturally approaching zero in near-wall regions. The σ-model proposed by Nicoud et al. [41] uses a single-value decomposition on the velocity gradient tensor as a basis for determining ν_T, which also allows for ν_T to scale with the wall distance cubed. The k-equation model [42–44] involves directly obtaining the SGS turbulent kinetic energy $\widetilde{k}_{\mathrm{sgs}}$ by using its transport equation. This approach is commonly referred to as a one equation model, since it is assumed that the length scale for $\widetilde{k}_{\mathrm{sgs}}$ is Δ and therefore, no transport equation is required for its dissipation. For the problems investigated here, the non-dynamic Smagorinsky model is chosen, due to its adequacy and simplicity and also that the region of interest is sufficiently far from the walls.

The framework presented thus far is sufficient for variable density non-reacting turbulent flows, while additional transport equations and closure models are required to capture combustion phenomena, which are described next.

2.2 Combustion Modelling

For the LES of turbulent combustion, the typical grid size is usually larger the laminar flame thickness δ_L^0 and thus, combustion occurs entirely at the SGS level. Developing SGS models to accurately capture the small-scale turbulence–combustion interactions has been the focal point of turbulent combustion modelling research. The state-of-the-art SGS combustion models can be classified into three categories: phenomenological, geometric and statistical models [28]. Phenomenological models evaluate the full thermochemistry on the fly and compute the reaction rate using a chemical mechanism. Such models include the artificially thickened flame model [45, 46] and the linear eddy model [47, 48]. These models often require a fine grid resolution, which lead to high computational costs. Geometric models are inspired and derived from the morphology of turbulent flames, such as the flame surface density [49–51] and the G-equation (or level-set) [32, 52] models. Statistical models typically involve a multi-variant probability density function (PDF), where ideally its shape can be obtained from solving the PDF transport equation [53–56]. This approach is sophisticated as the reaction rate source term naturally appears in a closed form, but requires very high computational costs and a model for scalar mixing physics at the small scales, which are known as micro-mixing models. Other statistical models have been proposed to reduce the computational burden by assigning the shape of the PDF and these models are referred to as presumed PDF approaches. These include the conditional moment closure [57–59] and multiple mapping conditioning [60, 61] models, which are still computationally demanding due to the on the fly chemistry evaluation for the conditional reaction rates. The presumed PDF approach further reduces the cost by assuming that the thermochemical states of laminar flames do not change in a turbulent environment, i.e. the turbulent flame is seen as an ensemble of fluctuating laminar flames. As such, the chemistry can be tabulated prior to the CFD simulation as a function of a number of control variables [62, 63]. This approach is attractive for practical applications because of its high computational efficiency and yet adequate accuracy. Derived from the early attempts by Bradley et al. [64] using a *mixedness-reactedness* concept, many model variants have been developed including the BML [65, 66], FGM [67], FPI [68] and FPV [69] models.

In this work, a presumed joint PDF approach with unstrained premixed flamelets is used for SGS combustion modelling. This approach revisits flamelet based models and the development of these models has demonstrated that the variance of a reactive scalar should be transported, and its dissipation rate should include the influences from turbulent–combustion interactions. This modelling framework is hence referred to as a flamelets revised for physical consistencies (FlaRe) approach. The LES–FlaRe modelling framework described here has been tested and validated for laboratory-scale premixed [70–72] and partially premixed flames [73, 74], and also for gas turbine combustors [75–80]. In this approach, a mixture fraction ξ is used to describe the turbulent mixing process and a reaction progress variable c is used to describe the combustion progress. The mixture fraction is calculated using the definition proposed by Bilger [81]. The normalised progress variable is defined as $c = \psi/\psi^{\mathrm{eq}}$, where $\psi = Y_{\mathrm{CO}} + Y_{\mathrm{CO_2}}$ and the superscript 'eq' denotes the chemical equilibrium value for a given mixture fraction. Transporting the progress variable requires care, since unclosed terms arise when deriving the progress variable transport equation from first principles [82]. Scaled and unscaled progress variable approaches have been tested by Chen et al. [74] , where it is observed that the scaled progress variable approach performs better in capturing local extinction. The transport equations for the Favre-filtered mixture fraction and progress variable, as well as their SGS variances, are written as

$$\overline{\rho}\frac{\mathrm{D}\widetilde{\xi}}{\mathrm{D}t} \simeq \nabla\cdot\left[\left(\overline{\rho\mathcal{D}}+\overline{\rho}\frac{\nu_T}{\mathrm{Sc}_T}\right)\nabla\widetilde{\xi}\,\right], \tag{5}$$

$$\overline{\rho}\frac{\mathrm{D}\sigma^2_{\xi,\mathrm{sgs}}}{\mathrm{D}t} \simeq \nabla\cdot\left[\left(\overline{\rho\mathcal{D}}+\overline{\rho}\frac{\nu_T}{\mathrm{Sc}_T}\right)\nabla\sigma^2_{\xi,\mathrm{sgs}}\right] - 2\,\overline{\rho}\,\widetilde{\chi}_{\xi,\mathrm{sgs}} + 2\,\overline{\rho}\frac{\nu_T}{\mathrm{Sc}_T}\left|\nabla\widetilde{\xi}\,\right|^2, \tag{6}$$

$$\overline{\rho}\frac{\mathrm{D}\widetilde{c}}{\mathrm{D}t} \simeq \nabla\cdot\left[\left(\overline{\rho\mathcal{D}}+\overline{\rho}\frac{\nu_T}{\mathrm{Sc}_T}\right)\nabla\widetilde{c}\,\right] + \overline{\dot{\omega}^*}, \tag{7}$$

$$\overline{\rho}\frac{\mathrm{D}\sigma^2_{c,\mathrm{sgs}}}{\mathrm{D}t} \simeq \nabla\cdot\left[\left(\overline{\rho\mathcal{D}}+\overline{\rho}\frac{\nu_T}{\mathrm{Sc}_T}\right)\nabla\sigma^2_{c,\mathrm{sgs}}\right] - 2\,\overline{\rho}\,\widetilde{\chi}_{c,\mathrm{sgs}} + 2\,\overline{\rho}\frac{\nu_T}{\mathrm{Sc}_T}\left|\nabla\widetilde{c}\,\right|^2 + 2\left(\overline{c\,\dot{\omega}^*} - \widetilde{c}\,\overline{\dot{\omega}^*}\right), \tag{8}$$

where $\mathcal{D}$ is the scalar molecular diffusivity and $\mathrm{Sc}_T = 0.7$ is the turbulent Schmidt number. The sub-grid scalar dissipation rate (SDR) terms $\widetilde{\chi}_{\xi,\mathrm{sgs}}$ and $\widetilde{\chi}_{c,\mathrm{sgs}}$ are unclosed and require modelling. The sub-grid SDR for ξ is modelled using a linear relaxation model $\widetilde{\chi}_{\xi,\mathrm{sgs}} = C_\xi\,(\nu_T/\Delta^2)\sigma^2_{\xi,\mathrm{sgs}}$ [52]. This algebraic expression is suitable for passive scalars, since their gradients are generated through turbulence. However, the effects of chemical reactions, thermal expansion and the multi-scale turbulence–combustion interactions are important for the SDR of reactive scalars [83]. Hence, the term $\widetilde{\chi}_{c,\mathrm{sgs}}$ requires a more sophisticated closure model. The algebraic expression proposed by Dunstan et al. [84], which is part of the LES–FlaRe model development for premixed and partially premixed combustion [72, 73], is adopted here.

The filtered reaction rate term $\overline{\dot{\omega}^*}$ in Eq. (7) takes the form of [82, 85]

$$\overline{\dot{\omega}^*} = \underbrace{\frac{\overline{\dot{\omega}_\psi}}{\psi^{\mathrm{eq}}}}_{\text{Premixed }\overline{\dot{\omega}}_{\mathrm{fp}}} + \underbrace{\overline{\rho\chi_\xi\frac{c}{\psi^{\mathrm{eq}}}\frac{\mathrm{d}^2\psi^{\mathrm{eq}}}{\mathrm{d}\xi^2}}}_{\text{Non-premixed }\overline{\dot{\omega}}_{\mathrm{np}}} + \underbrace{\overline{2\,\rho\chi_{\xi c}\frac{1}{\psi^{\mathrm{eq}}}\frac{\mathrm{d}\psi^{\mathrm{eq}}}{\mathrm{d}\xi}}}_{\text{Cross dissipation term }\overline{\dot{\omega}}_{\mathrm{cdr}}}\;. \tag{9}$$

The three terms on the right-hand side represent the contributions from premixed and non-premixed combustion modes, and their interactions resulting from the cross dissipation rate. These are denoted as $\overline{\dot{\omega}}_{\mathrm{fp}}$, $\overline{\dot{\omega}}_{\mathrm{np}}$ and $\overline{\dot{\omega}}_{\mathrm{cdr}}$ respectively. The cross dissipation term is neglected following previous studies [82, 86].

The first term of Eq. (9) signifies the contribution of premixed combustion mode, where the unscaled reaction rate is $\dot{\omega}_{\psi} = \dot{\omega}_{\mathrm{CO}} + \dot{\omega}_{\mathrm{CO_2}}$. This term is modelled as [86]

$$\overline{\dot{\omega}}_{\mathrm{fp}} = \overline{\rho} \int_0^1 \int_0^1 \frac{\dot{\omega}_{\mathrm{fp}}(\eta, \zeta)}{\rho(\eta, \zeta)} \widetilde{P}(\eta, \zeta)\, \mathrm{d}\eta\, \mathrm{d}\zeta\,, \tag{10}$$

where $\dot{\omega}_{\mathrm{fp}}(\eta, \zeta)$ and $\rho(\eta, \zeta)$ are the flamelet reaction rate and density respectively. They are obtained from one-dimensional unstrained planar laminar premixed flame calculations over the flammability range. The density-weighted joint PDF is approximated using statistically independent beta PDFs as $\widetilde{P}(\eta, \zeta) \approx \widetilde{P}_{\beta}(\eta\,;\,\widetilde{\xi}, \sigma^2_{\xi,\mathrm{sgs}}) \times \widetilde{P}_{\beta}(\zeta\,;\,\widetilde{c}, \sigma^2_{c,\mathrm{sgs}})$, where η and ζ are the sample space variables for the mixture fraction and progress variable respectively. The DNS study by Chen et al. [87] demonstrated that the sub-grid correlation is not influential on the time-averaged statistics because the contribution related to the large-scale fluctuations is resolved in LES. Therefore, the sub-grid correlation is not considered here and the statistical independence assumption is made for simplicity.

The non-premixed contribution of Eq. (9) is modelled using [86]

$$\overline{\dot{\omega}}_{\mathrm{np}} = \overline{\rho}\,\widetilde{c}\,\widetilde{\chi}_{\xi} \int_0^1 \frac{1}{\psi^{\mathrm{eq}}(\eta)} \frac{\mathrm{d}^2\psi^{\mathrm{eq}}(\eta)}{\mathrm{d}\eta^2} \widetilde{P}(\eta)\, \mathrm{d}\eta\,. \tag{11}$$

The non-premixed contribution does not come from counterflow diffusion flamelets and is instead a correction term for the premixed contribution. This term contains the filtered mixture fraction scalar dissipation rate, which is the sum of the resolved and SGS contributions $\widetilde{\chi}_{\xi} = \widetilde{\mathcal{D}}(\nabla\widetilde{\xi} \cdot \nabla\widetilde{\xi}) + \widetilde{\chi}_{\xi,\mathrm{sgs}}$. The non-premixed contribution is typically only significant in the stoichiometric regions, since $\mathrm{d}^2\psi^{\mathrm{eq}}/\mathrm{d}\eta^2$ tends to zero rapidly when moving away from stoichiometry [86].

The final term that requires closure is the reaction source term in Eq. (8) and is written as $(\overline{c\,\dot{\omega}^*} - \widetilde{c}\,\overline{\dot{\omega}^*}) = (\overline{c\,\dot{\omega}_{\mathrm{fp}}} - \widetilde{c}\,\overline{\dot{\omega}}_{\mathrm{fp}}) + (\overline{c\,\dot{\omega}_{\mathrm{np}}} - \widetilde{c}\,\overline{\dot{\omega}}_{\mathrm{np}})$, where $(\overline{c\,\dot{\omega}_{\mathrm{np}}} - \widetilde{c}\,\overline{\dot{\omega}}_{\mathrm{np}}) = 0$ following Eq. (11). The term $\overline{c\,\dot{\omega}_{\mathrm{fp}}}$ is evaluated in a similar manner to Eq. (10) [86].

The Favre-filtered temperature $\widetilde{T}$ is obtained using the filtered enthalpy obtained from its transport equation, the mixture-averaged enthalpy of formation $\widetilde{\Delta h}^0_f$ and an effective specific heat capacity $\widetilde{c}_p$ through the approximation $\widetilde{T} = T_0 + (\widetilde{h} - \widetilde{\Delta h}^0_f)/\widetilde{c}_p$, where '0' refers to a reference temperature, which is 298.15 K in this work. The three thermochemical quantities of the mixture, $\widetilde{\Delta h^0_f}$, $\widetilde{c}_p$ and $\widetilde{\mathcal{M}}$, are calculated in a manner similar to Eq. (10), as described in detail by Ruan et al. [86].

These three terms are included in a look-up table along with the two reaction rate source terms in Eqs. (7) and (8).

Simulations of non-swirling jet flames [73, 86, 88, 89] and swirl-stabilised flames in a gas turbine model combustor [76–80] using the LES–FlaRe modelling framework described here are presented next to shed light on the physical mechanisms for the dynamics and stabilisation of the lifted flame root.

3 Non-swirling Jet Flames

The first configuration to be analysed is the canonical round jet flame configuration (see the schematic in Fig. 1), where several experiments of lifted flames have been undertaken using hydrogen [90–92] and hydrocarbon fuels [10, 17, 93, 94], and the parameters that have been varied include the Reynolds number (fuel jet velocity or nozzle diameter) or the air dilution. These experiments have been used as initial cases for developing the FlaRe modelling approach described in the previous section. The model is assessed on its ability to capture the average lift-off height when the inlet conditions are changed over a wide range. The flame stabilisation mechanisms are explored with the aid of simulation data and the findings are discussed in this section.

3.1 Effects of Inlet Conditions

Hydrogen jet flames [90–92] are investigated first, since the chemical time scales ($\tau_c = \delta_L^0/s_L^0$) are significantly shorter than those for hydrocarbon flames, and fewer species and elementary steps are involved in hydrogen–air combustion. The combustion model was tested *a priori* using DNS data [96, 97] and *a posteriori* using 2D axisymmetric steady-state RANS simulations [86]. The experiment by Cheng et al. [90] was used for model assessment and validation. It was shown that including the non-premixed contribution $\overline{\dot{\omega}}_{\rm np}$ in Eq. (9) improved the predictions of the lift-off height. Neglecting the non-premixed term causes the flame to stabilise closer to the fuel nozzle exit, as observed in the study by Ruan et al. [86]. For RANS simulations, the fluctuations of c and ξ are more significant and the use of statistically independent PDFs may not be suitable because these fluctuations are entirely modelled. It was found that including this correlation within the joint PDF using a *copula* method, as described by Darbyshire and Swaminathan [98], improved the lift-off height prediction. This is shown in Fig. 3a with the comparison against the experiment by Cheng et al. [90]. Additional RANS simulations were undertaken to capture the lift-off height for different jet velocities and compared to the experiment by Brockhinke et al. [92]. The jet velocity was varied between 500 and 850 m/s, which corresponds to pipe Reynolds numbers that are approximately within the range $9000 < \mathrm{Re}_D < 15500$. It is shown in Fig. 3a that the average lift-off heights are well captured by the simulations using the FlaRe approach over the range of jet velocities used and that the

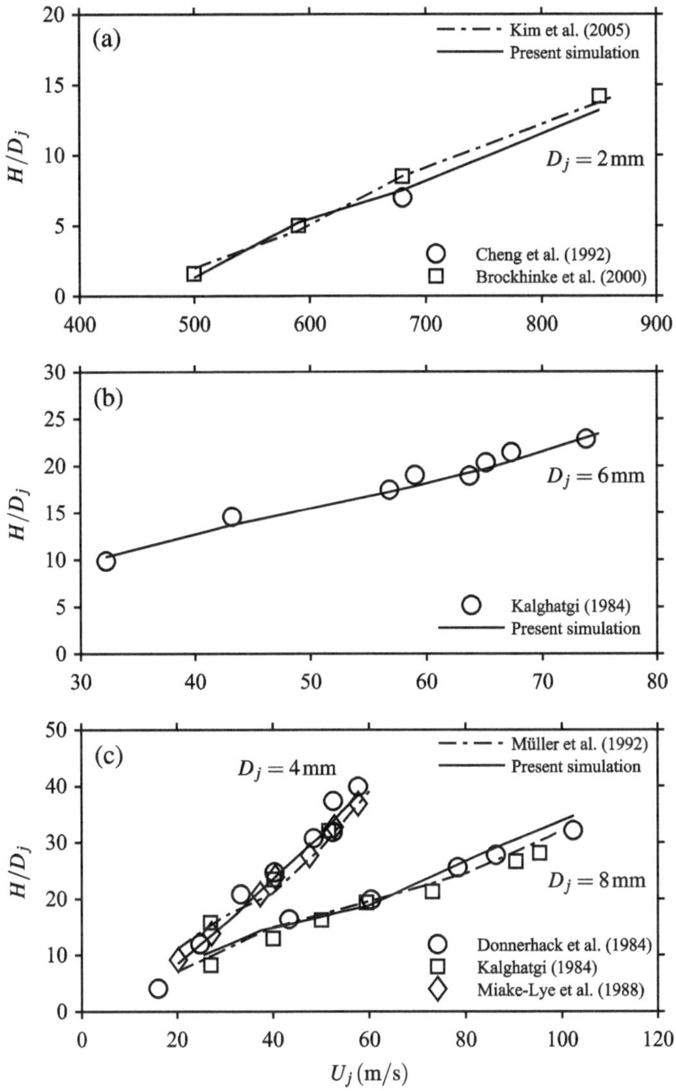

Fig. 3 Comparisons between experimental studies and 2D axisymmetric RANS simulations of the lift-off height for jet flames with different jet velocities for **a** hydrogen [86], **b** propane [95] and **c** methane [95]

lift-off height increases with the jet velocity. The lift-off heights obtained using the CMC model are also good [99] but this approach requires a substantially higher computational cost.

The combustion modelling framework was then applied to jet flame configurations using hydrocarbon fuels [95]. The results of 2D axisymmetric RANS sim-

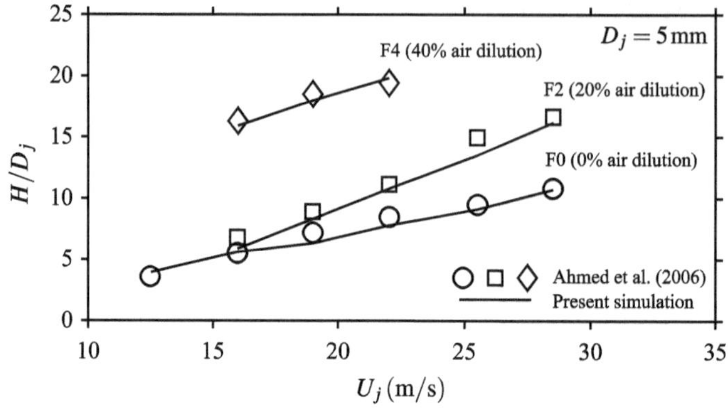

Fig. 4 Comparisons between experimental measurements [93] and 2D axisymmetric RANS simulations [88] of the lift-off height for jet flames with different jet velocities and methane with varying air dilution

ulations for propane and methane jet flames are shown in Fig. 3b, c respectively, where two different nozzle diameters have been used for the methane cases. Previous simulation results using other combustion models are also shown wherever available for comparison. The velocities for the cases with propane are varied between 32 and 74 m/s, which correspond to a Reynolds number range of approximately $41,500 < \mathrm{Re}_D < 96,000$, while the velocities for the cases with methane are varied between 16 and 100 m/s, which correspond to a Reynolds number range of approximately $3600 < \mathrm{Re}_D < 45,000$. It is shown that the lift-off height is well captured for both hydrocarbon fuels over the wide range of jet velocities used. In addition, the lift-off height is more sensitive to the jet velocity when methane is used in comparison with the propane cases. To study the effect of air dilution on the lift-off height, the FlaRe modelling framework was then applied to the jet flame configuration studied experimentally by Ahmed and Mastorakos [93]. The inlet velocity and the amount of air dilution by volume into the fuel jet were varied. These results are shown in Fig. 4 and it is shown that the comparisons between the experimental data and simulation results obtained using 2D RANS [88] are very good. For a given jet velocity, the lift-off height increases with the dilution level, which is due to the stoichiometric mixture fraction contour being shifted towards the centre of the jet, where the local velocities are larger, and the flame stabilises at a location further downstream. For case F4 (40% air dilution), using a velocity that was higher than 22 m/s leads to blow-off in the simulation, which was consistent with the observations in the experimental study [93]. Therefore, it is demonstrated that the FlaRe combustion model is capable of accurately capturing the variations in the average lift-off height over a wide range of velocities and different fuels by using computationally inexpensive 2D axisymmetric steady RANS simulations.

Although the comparisons between the results obtained using the RANS framework and the measured lift-off heights are good, it is also of interest to analyse the transient evolution of the flame leading edge from its ignition location. Experimental data are available in the study by Ahmed and Mastorakos [93] for methane with 30% air dilution for two different jet velocities and ignition locations. Simulations were performed for the transient evolution of the flame leading edge using 2D and 3D URANS simulations [89]. It was demonstrated that the final lift-off height was the same in 2D and 3D simulations for both jet velocity cases, which further justifies the axisymmetric assumption made for the steady RANS calculations presented earlier. However, the interactions between the leading flame point and large-scale structures cannot be captured using the URANS methodology, as the flow field and flame are averaged. Therefore, simulations with the LES framework are required to capture the instantaneous behaviour of the leading flame point and the propagation dynamics. The findings using the LES–FlaRe approach are discussed next.

3.2 *Flame Stabilisation*

In the previous study using the 2D URANS framework [88], various stages of the flame evolution were identified and these compared well with the experimental observations using high-speed movies [93]. The flame kernel is first convected downstream quickly by the incoming flow and the kernel then grows in a spherical shape due to both molecular and turbulent diffusion. The flame then expands in the radial direction and the flame brush upstream edge remains at a location near the ignition location until the flame is able to position itself between the lean and rich flammability mixture fraction contours, where the most upstream point is located along the stoichiometric mixture fraction contour. The flame then propagates upstream until it reaches its final stabilisation height. The previous study by Chen et al. [89] was useful to gain insights into the general flame stabilisation stages, but it did not capture the transient interactions with the large-scale eddies, as the relevant scales associated with turbulence and mixing are modelled using the URANS framework. It was also shown in the study by Chen et al. [89] that compared to the measured position, the computed flame leading edge was always closer to the fuel nozzle during the propagation stage, which suggests that the propagation speed of the flame base was over predicted. Hence, simulations with the LES framework then performed to obtain improvements in the flame leading edge propagation stage and observe the influences on the flame stabilisation due to large-scale eddies.

The LES–FlaRe framework described in Sect. 2 was applied to the same jet burner [93] to study the transient evolution of the flame leading edge by Chen et al. [73]. Two cases were studied using jet velocities of $U_j = 12.5\,\mathrm{m/s}$ and $U_j = 25.5\,\mathrm{m/s}$ of methane with 30% air dilution by volume; these are referred to here respectively as F3-1 and F3-2. The ignition location is $30D_j$ downstream of the fuel nozzle exit for both cases, but measurements have also been taken for the ignition location of $40D_j$ downstream of the fuel nozzle exit. Studies with the LES framework have been undertaken on the evolution of the leading flame edge with a jet velocity of $U_j = 12.5\,\mathrm{m/s}$ using the artificially thickened flame model [100], and with a

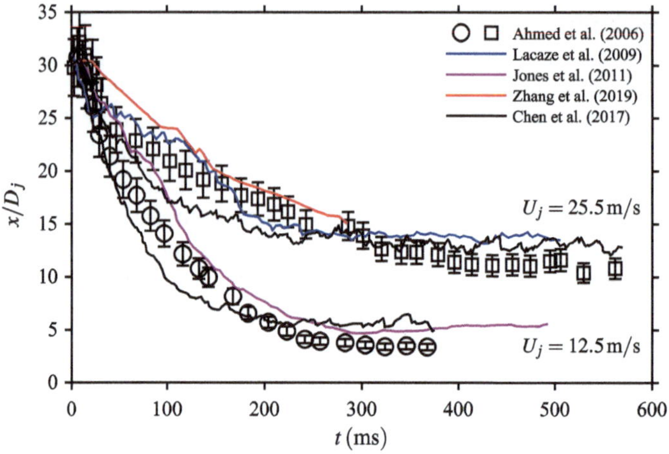

Fig. 5 Comparison of the time series of the flame leading edge location using experimental measurements [93] and from numerical studies [76, 100–102]

jet velocity of $U_j = 25.5\,\mathrm{m/s}$ using the transported PDF approach with Eulerian stochastic fields [101] and the CMC model [102].

The experimental study recorded a time series of the axial position of the most upstream flame point using high-speed line-of-sight imaging and OH-PLIF measurements. In the aforementioned LES studies, the flame leading point was tracked by using $\widetilde{T} = 1200\,\mathrm{K}$ to $1500\,\mathrm{K}$. These simulation results are compared to the experimental measurements for the most upstream flame point in Fig. 5. For the low-velocity case (F3-1), the results are available for the present simulations with the LES–FlaRe approach [73] and the previous transported PDF [101] simulations. It can be seen that the flame leading edge position is under predicted using LES–FlaRe, suggesting that the propagation speed of the flame base is over predicted, which is attributed to over predictions in the local reaction rates. In addition, the flame reaches a final stabilisation height approximately 100 ms earlier than as seen in the experiment. On the other hand, the flame leading edge location is consistently over predicted using the transported PDF model, suggesting the propagation speed of the flame base is under predicted. The final lift-off height for both approaches is approximately $5D_j$ above the fuel nozzle exit, which is $2D_j$ higher than the experimentally observed value of approximately $3D_j$ above the fuel nozzle exit. For the high-velocity case, the LES results obtained using LES–FlaRe [76], thickened flame [100] and CMC [103] approaches are compared. It is shown in Fig. 5 that the flame leading edge position is under predicted using the LES–FlaRe approach during the first 200 ms period, which is similar to the behaviour for case F3-1. The final lift-off height for case F3-2 was approximately $14D_j$, which was approximately $2D_j$ higher than the experimentally observed value of approximately $12D_j$ above the fuel nozzle exit. It had been postulated that the statistical independence assumption for the presumed PDF leads to the over predictions in the final lift-off height for both simulations using the

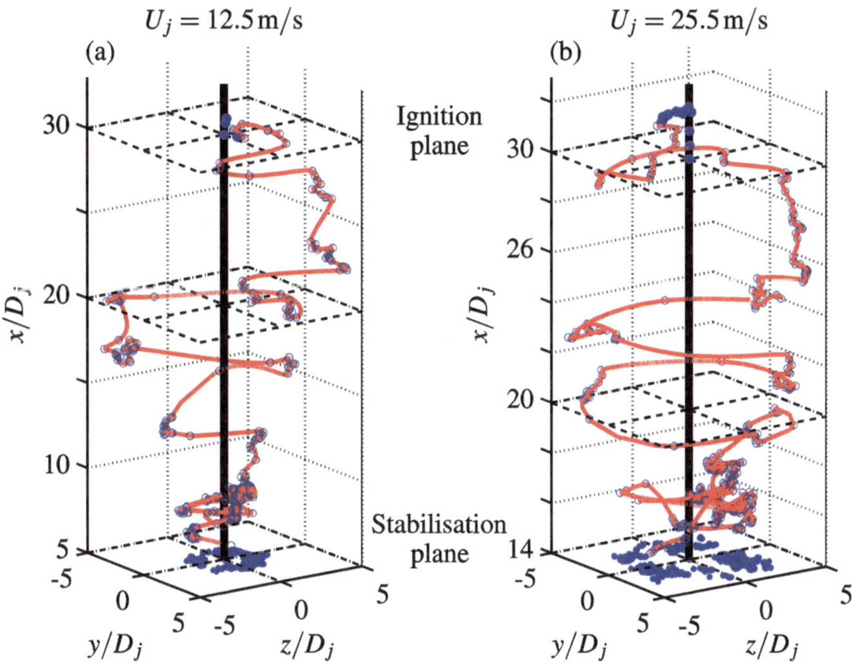

Fig. 6 Evolution of the flame leading edge trajectory for case **a** F3-1 and **b** F3-2 [73]

LES–FlaRe approach [73]. However, it was later shown in the DNS study by Chen et al. [87] that the correlation that represents the interactions between the mixture fraction and progress variables fields was seen to be small for lifted flames. The LES study using the artificially thickened flame approach over predicts the flame leading point position for the first 150 ms, while the final lift-off height was similar to the value obtained using LES–FlaRe. The CMC model consistently over predicts the location of the flame leading edge throughout the entire 300 ms of simulation runtime. Therefore, it is demonstrated that the LES–FlaRe approach performs well in comparison with the other approaches but at a much lower computational cost.

The use of CFD is beneficial in comparison with experimental techniques, since it is possible to extract further information from the simulation results, in order to obtain further insights into the physical mechanisms. However, it should be noted that CFD cannot replace experiments, but both are required to improve the understanding and knowledge of the flame stabilisation mechanisms. An example of an advantage for using CFD is the analysis of the leading flame point trajectory in 3D space for cases F3-1 and F3-2 [73], as shown in Fig. 6a, b respectively, which are obtained using the results from the simulations with LES–FlaRe approach. The ignition plane for both cases is located at $x/D_j = 30$, and it is shown that the flame leading point initially moves up (i.e. downstream) along the central jet axis (solid black line). This downstream movement from the ignition plane is more evident in case F3-2, since the fuel jet velocity is greater. The radial expansion that follows is different between the two cases and also the flame leading edge path, as the local mixing conditions are different for the two LES realisations. After reaching a certain radial distance from the jet axis, which is observed to be approximately $5D_j$, the flame leading point then moves upstream for both cases. The flame edge continues to propagate upstream and follows a path where near-stoichiometric mixtures are present. These mixtures are created due to the presence of large-scale eddies and the induced flame–flow interactions. For case F3-1, it is shown in Fig. 6a at $x/D_j = 20$ that the flame leading edge follows a spiral path towards the stabilisation plane located at $x/D_j = 5$. This motion is similar for case F3-2 from $x/D_j = 24$, but the radius of the spiral remains larger than in case F3-1, as shown in Fig. 6b. It is also seen in Fig. 6b that the trajectory in the region $14 \leq x/D_j \leq 16$ fluctuates more. This is due to the higher incoming velocity, which causes more local extinction to be present, as the mixing time scales as shorter and lead to the creation of mixtures out of the flammability range. When the flame is near the stabilisation plane for both cases, it is shown in Fig. 6 that the flame leading point moves around the jet axis with a circular trajectory. The spread of the scattered points is wider for case F3-2 because the lift-off height is greater and hence, the radial position for a given mixture fraction is larger. It was shown by Chen et al. [73] that the radial location of the leading point is found to be within lean mixtures. However, the presence of mixtures is not the only reason for the flame stabilisation location and its presence may be attributed to other physical processes, such as turbulence–combustion interactions and the presence of large-scale eddies.

It has been discussed thus far that the flame stabilisation processes are controlled by numerous processes, which include local mixing, large flow structures and turbulence–combustion interactions. This causes the flame root's location to change intermittently. These processes are different in swirling flows and the flame stabilisation processes can lead to significant changes in the flame behaviour within a confined flow configuration. These aspects are investigated next for gas turbine combustors.

4 Gas Turbine Combustors

For practical combustion systems, it is possible that unexpected flame behaviour can be present, which can lead to serious consequences. An example is the presence of pressure fluctuations and if their amplitudes are large enough, this leads to increased flow/acoustic oscillations within the engine, flashback of the flame and burning part of the fuel injection system [25, 104]. This behaviour is referred to as thermo-acoustic instabilities, which are the combination of the influences from turbulent flows, heat release from the flame, hydrodynamic stability and acoustics. If the heat release rate is in phase with the pressure oscillations and of similar amplitudes, thermo-acoustic instabilities arise, which can significantly influence the flame stabilisation in practical combustors [105]. Another safety hazard that must be considered for gas turbine combustors is the risk of complete flame blow-off. This issue becomes critical as modern gas turbines operate under very lean conditions and understanding flame stabilisation near the lean blow-off (LBO) limit is of paramount practical importance.

4.1 Model Combustor

The gas turbine model combustor [21, 22] developed by the German Aerospace Centre (DLR) featuring a dual-swirler configuration with detailed measurements for the flow/flame structures is of interest here. Three benchmark flames operating under thermo-acoustically stable (flame A) and unstable (flame B) conditions, and close to the LBO limit (flame C) are listed in Table 1 with their operation details. A schematic of this burner is shown in Fig. 7. Dry air at atmospheric pressure and room temperature enters a single plenum and the flow is split through two radial swirlers. The two co-swirling flows enter the combustion chamber through central and annular nozzles, in order to achieve quick fuel–air mixing. Non-swirling methane is fed through a nozzle ring that contains 72 channels and is located between the two air nozzles. The exit planes of the central air and fuel nozzle ring are 4.5 mm below the exit of the annular air nozzle and the entrance to the combustion chamber. State-of-the-art laser diagnostics techniques have been used to investigate the physical processes in the three flames using stereoscopic particle image velocimetry (stereo-PIV), Raman spectroscopy, laser Doppler velocimetry (LDV), OH/CH/CH_2O/acetone planar laser-induced fluorescence (PLIF) and OH^*/CH^* chemiluminescence with high repetition rates [21, 22, 106–108].

The extensive experimental datasets [21, 22] have served for the further validation of the LES–FlaRe combustion modelling framework, reported in Refs. [76, 77] for flames A and B, and Refs. [79, 80] for flame C. Flame A has also been attempted with the LES framework using other combustion models including FPV and F-TACLES [109], EDC [110] and FGM [111] approaches, and flame C was simulated using the CMC model [103]. To date, the LES–FlaRe approach is the only model to capture the correct behaviours of the unstable flame B undergoing pronounced thermo-acoustic oscillations. Since the statistics for the flow and flame structures have been well captured by the LES–FlaRe approach in previous studies, the focus here is on the transient behaviours of the flame root and its stabilisation mechanisms under stable, unstable and close to LBO conditions. The Raman measurements were

Table 1 Summary of the operating conditions for the three flames in the gas turbine model combustor [21, 22]

Flame	ϕ_{glob}	$\dot{m}_a$ (m/s)	$\dot{m}_f$ (m/s)	S	P_{th} (kW)
A	0.65	18.25	0.697	0.9	34.9
B	0.75	4.68	0.205	0.55	10.3
C	0.55	4.68	0.15	0.55	7.6

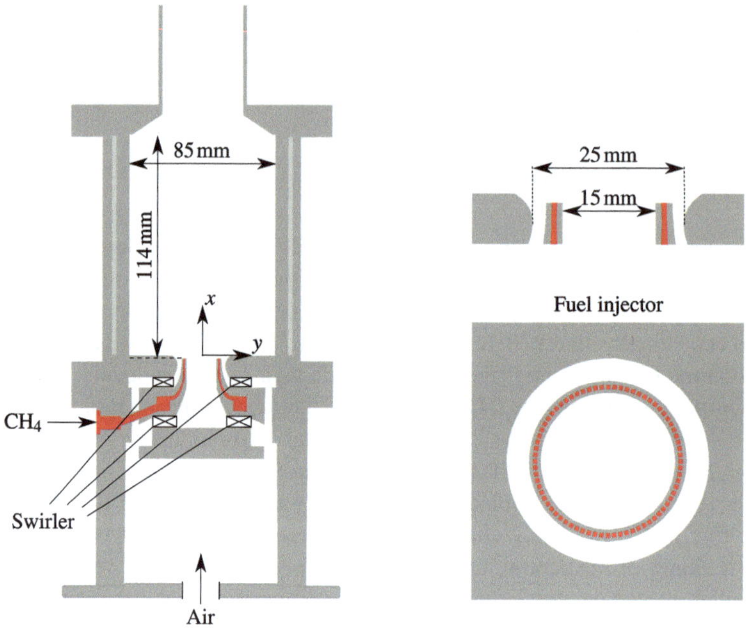

Fig. 7 Schematic of the gas turbine model combustor [21, 22]

used to indicate the mean structure of the flame (temperature and species mass fractions) and identify the mean lift-off height of the flame. The lift-off height H was tracked using the most upstream point of $\widetilde{T} = 1200$ K above the fuel nozzle, as shown in Fig. 8 using H, which is also illustrated in the schematic in Fig. 2.

The measured lift-off heights for flames A, B and C are $H \approx 5$ mm, 4 mm and 6 mm respectively, but the behaviour in the flame root region for the three cases is different. The experimental studies have observed the main flame stabilisation mechanisms, as well as a wide range of transient phenomena. A PVC is observed in isothermal and reacting flows for this configuration, which is a form of hydrodynamic instability that appears in swirling flows usually with high swirl numbers [3]. It is a helical vortex structure residing in the shear layer between the swirling jet and the inner recirculation zone, creating a low pressure region as marked in Fig. 8. It is possible that the PVC can be dampened or disappear when combustion is present [112], but the flame stabilisation mechanisms for all three flames studied here are highly influenced by the PVC, since it is located in the ISL where the flame root is present. The ISL originates from the fuel nozzle (as illustrated in Fig. 2) and is close to the flame and outside of the IRZ, as shown by the pink lines in Fig. 8. The PVC generates regions of a high vorticity magnitude and since the PVC passes through the unburnt reactants, this induces strong mixing processes between the fuel and air. These large-scale structures are convected along the ISL and this leads to flame–vortex interactions, where hot combustion products are entrained into the vortices.

This has been observed in the experiments [107, 113] to be a dominant mechanism for flame stabilisation. To visualise this mechanism, a mid-plane snapshot of the filtered reaction rate and the stoichiometric mixture fraction contour are shown in Fig. 8. This is taken from the flame C simulation when the flame root was stable. In the flame root region where H is marked, it is shown on the right-hand side that the PVC passes through an area of unburnt reactants near the flame root. The stoichiometric mixture fraction is in the vicinity of the PVC, indicating that the fuel–air mixture will become well-mixed and ignition should occur, as the velocity vectors in the inset of Fig. 8 indicate a vortex. The left-hand side of Fig. 8 indicates the next stage, which is PVC induced flame roll-up. The vortex is close to the stoichiometric mixture fraction and the flame root, which is also suggesting that the fuel–air mixture within the vortex is ignited. On the right-hand side of Fig. 8, a large-scale vortex with a high reaction rate has been convected further downstream. The velocity vectors shown in the inset of Fig. 8 suggest that the vorticity is strong, and hot combustion products and radical species are entrained into the vortex region to sustain combustion. The reaction zone is then quenched when vortex breakdown occurs, as shown in the downstream regions of Fig. 8.

The aforementioned PVC induced sequence can be observed continuously as the PVC spins about the centre axis in flame A, which stably retains a 'V' shape flame. Although such a flame stabilisation process is still present in flame B, the flame shape changes significantly with the presence of pressure oscillations, which causes a flat flame to be close to the dump plane, and the PVC cannot be always identified. As observed in the LES studies by Chen et al. [76, 77], the pressure fluctuations caused variations in the air mass flow split between the central and annular air passages during one revolution of the PVC. The different acoustic impedances due to the swirlers cause *fuel jet flapping* behaviour when the PVC is out of plane, i.e. not near the fuel nozzle. This causes enhanced radial mixing of the fuel and leads to a shorter and flattened flame shape, which was also observed in the experiments by Boxx et al. [114] and Steinberg et al. [115]. It has also been observed [77] that the structure of the PVC is influenced by the pressure fluctuations within the combustion chamber. The PVC induced flame roll-up and enhanced mixing effects are present, as also seen in flame A, at the maximum and minimum pressure inside the combustion chamber. The axial distance between the vortices is observed to be higher when the pressure is higher, which suggests that the PVC extends for higher pressures and retracts for lower pressures. This behaviour was also seen in the experimental study by Steinberg et al. [115] and occurred in numerous thermo-acoustic cycles. In addition, there was frequent disappearance of the flame root during large pressure fluctuations, and the root of the PVC contracted and expanded radially for high and low pressures respectively. The flame root was able to reform when the pressure fluctuations become smaller, which was also observed in the experimental study by Boxx et al. [114]. For flame C, the flame stabilisation sequence described for Fig. 8 is also different, since the global equivalence ratio is leaner than in flame B (see Table 1) and both cases have similar inlet velocities. As a result, the stoichiometric mixture fraction contour is further away from the PVC and mixing enhancement is weakened, which occasionally leads to failed ignition during a later flame roll-up

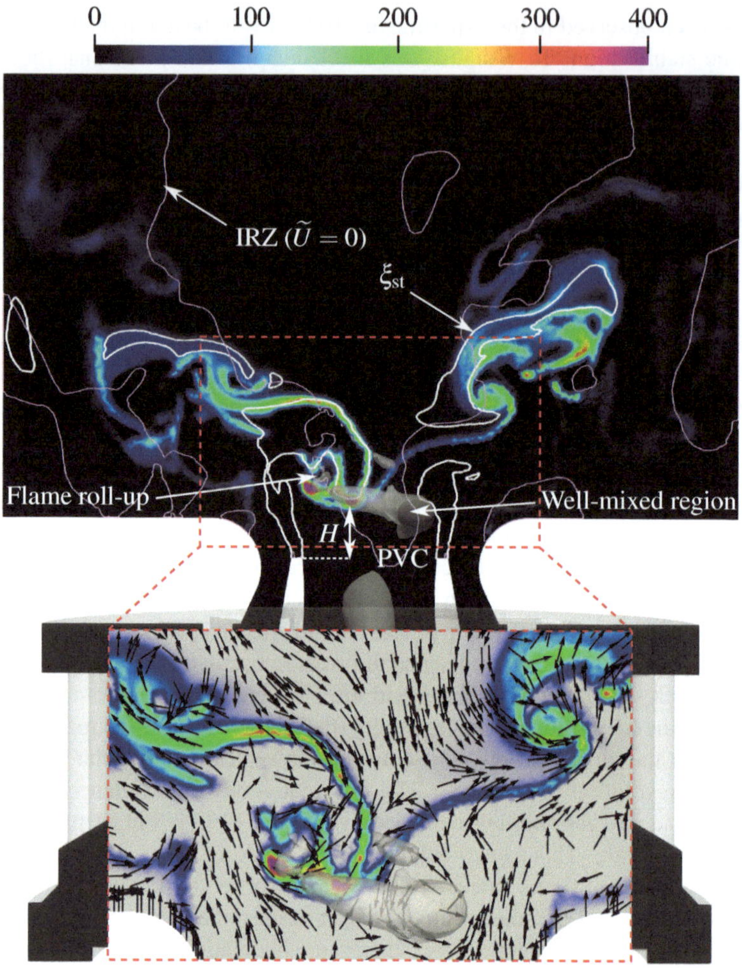

Fig. 8 Snapshot of the filtered reaction rate, the stoichiometric mixture fraction (white lines) and the zero axial velocity contour (pink lines) in the x–y mid-plane for flame C. The inset shows the filtered reaction rate and the velocity vectors

process. Given the different flame base behaviours taking place in the three flames, it is of interest to compare the flame leading edge behaviour, which is discussed next.

As previously mentioned, the flame root and its azimuthal position are highly influenced by the PVC, which has also been seen when analysing time sequences of the LES snapshots for the three flames. The flame leading point, i.e. the most upstream location of the $\widetilde{T} = 1200$ K iso-surface in the IRZ, and the fluctuation of its azimuthal position θ' on the transverse plane are analysed. The fast Fourier transform (FFT) is applied to time series of this fluctuation to obtain the spectral behaviour for the three flames in Fig. 9. For flames A and B, it is shown in Fig. 9a,

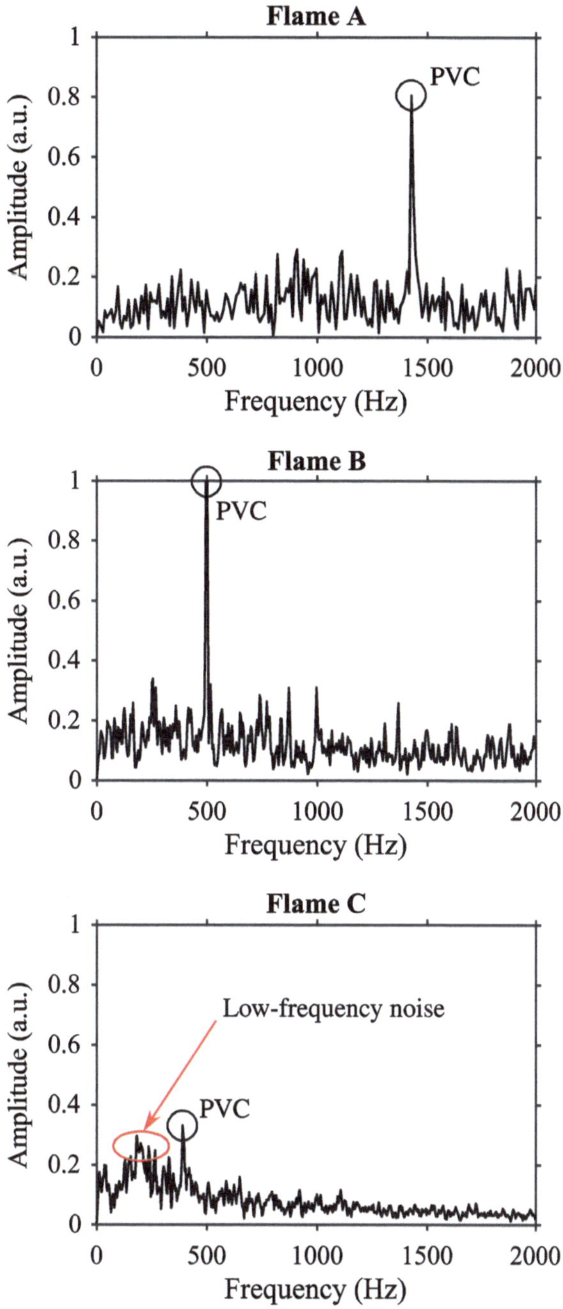

Fig. 9 Spectral behaviour of the azimuthal movement for the flame root using the azimuthal position fluctuations for **a** flame A, **b** flame B and **c** flame C

b respectively that the FFT signals have a single pronounced peak corresponding to the PVC frequency. This suggests that the PVC is the dominant mechanism for the stabilisation of this flame and the azimuthal position of the flame follows the rotation of the PVC. For flame C, the FFT looks significantly different as there is a peak that is at the PVC frequency, but it is much less prominent. Flame C is close to the LBO limit and is highly unstable. This is typical in modern jet engines, since the aim is to operate at fuel-lean conditions in order to reduce pollutant emissions and therefore, there is a risk of flame blow-off. These risks were captured for this configuration in the study by Stöhr et al.[107] using simultaneous stereo-PIV and OH-PLIF techniques. It was observed in the study by Stöhr et al. [107] that flame lift-off occurred 1–2 times per second and lasted between 0.1 and 0.15 s. The lift-off height was observed to reach a height of 30–40 mm. This lift-off event was triggered due to failed ignition at the flame root and within the vortices. This suggests that the PVC induced fuel–air mixing and flame roll-up is weaker and this leads to the flame becoming unstable. This is reflected in Fig. 9c as there is some low-frequency noise below the PVC frequency.

It was also reported by Stöhr et al. [107] that during an unstable blow-off event, the flame root was extinguished and this led to complete flame lift-off, i.e. the flame stabilised at the combustor exit. The ultimate blow-off event was only observed when further reducing the global equivalence ratio. Highly transient flame root dynamics were observed, as also seen for flame B [114], but failed ignition within the convected large-scale vortex structures was also observed before the blow-off event. This occurs if re-ignition does not occur after more than 2 ms, which is the duration for one rotation of the PVC. During the lift-off event, the flame stabilised downstream and the flame then moved back upstream and returned to the location of $H \approx 6$mm. In the LES study by Massey et al. [79], a lift-off event was captured using the LES–FlaRe framework. This was not captured in the CMC study by [103], since the model used was a non-premixed model with single conditioning on the mixture fraction and strong partial premixing is present at the flame root region and vortices. When a lift-off event occurs, different flame shapes were observed leading to significant temporal variations in the the heat release rate [79]. This phenomenon captured in the LES is demonstrated in Fig. 10, which shows two time series of the integrated heat release rate and the lift-off height. The lift-off height is also moving-averaged over a 4.5 ms window and the averaged values are shown using horizontal grey lines. The dash-dotted line in the bottom of Fig. 10 shows the temporal variation of the heat release rate integrated over the entire combustion chamber, denoted as $\overline{\dot{Q}}_{\text{glob}}$, for 45 ms, which is an arbitrarily chosen interval that included a lift-off event in the simulation. The volume integrated heat release rate varies with time, but it is close to the thermal power of $P_{\text{th}} = 7.6$ kW for the experiment. However, it is difficult to identify the lift-off event from this quantity. Therefore, the temporal variation of the heat release rate is integrated over a smaller volume around the fuel nozzle region to gather information near the flame root. This heat release rate, denoted as $\overline{\dot{Q}}_{\text{nozzle}}$, is also shown in Fig. 10, using the solid line, but multiplied by 100 to show on the same scale as the global values. It is evident that the heat release rate in this region

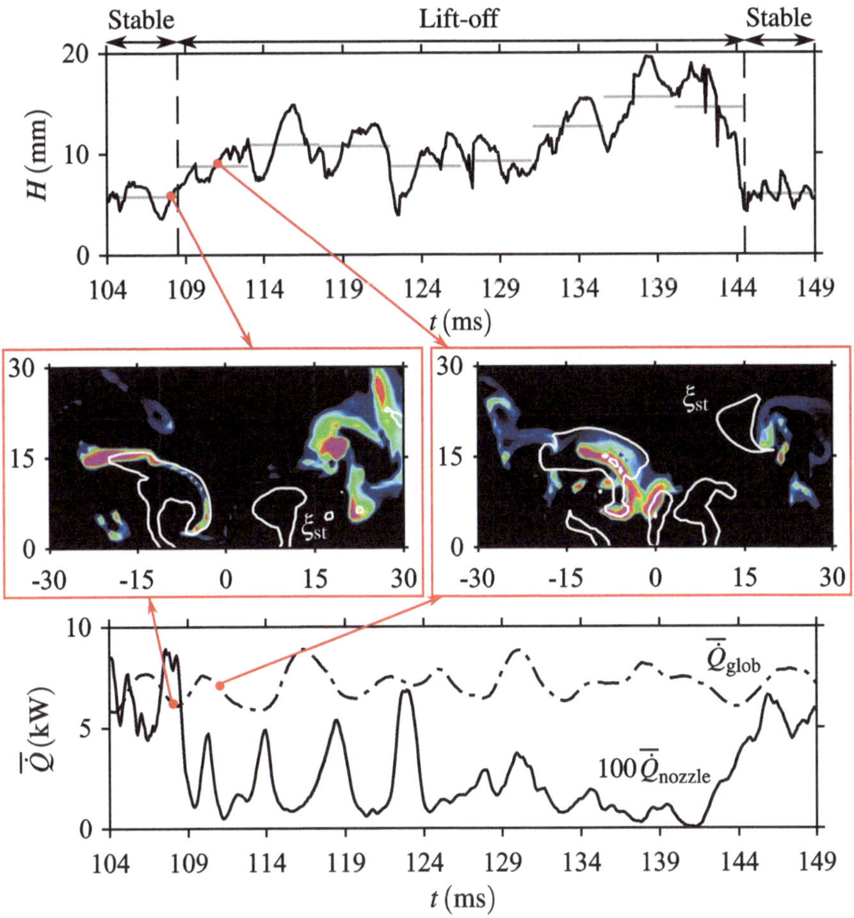

Fig. 10 Time series of the lift-off height and volume integrated heat release rate for a lift-off event in flame C

changes significantly over the time interval shown and it is clear that the flame has lifted off in the range $124 < t < 141$ ms, as the flame is out of the near fuel nozzle region. The trend seen for the lift-off height is directly linked to $\overline{\dot{Q}}_{\text{nozzle}}$. After the first window of 4.5 ms, the lift-off height fluctuates up to $t = 131$ ms, since the unstable behaviour and fluctuating heat release rate are caused by the flame root trying to establish itself. Beyond this time, the averaged lift-off height significantly increases, suggesting that the flame recedes downstream and does not stabilise near the fuel nozzle, which is shown by the low heat release rate seen during this time window. The last window of 4.5 ms shows that the average height is very similar to the first window and hence, the flame leading edge is re-established at its typical location with a more stable flame in its typical 'V' shape.

Two instants are marked with red points in Fig. 10 at $t = 108.05$ ms and 111.05 ms, which represent precursors to the lift-off event. As seen in Fig. 10, the lift-off height increases and the heat release rate in the flame root region decreases after both of these instances. On the right-hand of the first frame, there is no reaction along the stoichiometric mixture fraction line and the reaction rate is very weak in comparison with the reaction rate field in Fig. 8. This is caused by the air entrainment between the stoichiometric reactant mixture and hot products, and this prevents ignition at the flame root region. This causes the flame to move away in the radial direction in the right half of the domain and leads to the sudden drop in the values of $\overline{\dot{Q}}_{\text{nozzle}}$ that is seen in the time series in Fig. 10. Consequently, the flame root recedes downstream, leading to an increase in the flame lift-off height around $t = 108$ ms, as shown in the H time series. As seen in the second frame, a weak vortex is shown on the right-hand side, and a pocket of rich mixture is present along the centreline near the bottom of the combustion chamber in the region where the flame root typically stabilises. This prevented ignition of the vortex on the right-hand side and there was no flame roll-up. The flame then moved to a downstream position and therefore, the flame root is shifted from its typical location leading to an lift-off event.

The analyses presented here of the simulations using the LES–FlaRe framework have demonstrated that the underlying mechanisms that influence the flame root dynamics for flames under thermo-acoustically unstable conditions and near the lean blow-off limit compare well with the experimental observations. This suggests that the modelling framework is suitable for simulating full-scale gas turbine combustors, since the approach is also economical from a computational cost perspective. Combustion models need to have a low computational cost to simulate such configurations, since the grid sizes are typically large and the geometries often include multiple swirlers and complex injection systems. A brief overview of some progress with simulations of industrial burners using the LES–FlaRe framework is presented next.

4.2 Industrial Burners

Simulations of practical configurations are challenging, since experiments which can be used to validate combustion models are scarce. This is because such practical configurations operate at elevated pressures and the reactants are preheated. Therefore, non-intrusive techniques are required, which are very expensive for configurations operating at high pressures. Furthermore, the gases/combusting mixtures can become increasingly translucent with optical depth as pressure increases and thus, laser diagnostics can become challenging. Measurements have been conducted for one configuration, which is the DLR-AT Big Optical Single Sector (BOSS) rig, where OH and spray measurements have been obtained [116–119]. The combustor rig is fitted with a Rolls-Royce developmental lean burn injector having a central pilot surrounded by a main jet. The operating conditions are chosen to represent take-off conditions, and air with a temperature above 700 K and a pressure above 10 bar passes through the swirlers in the burner. Liquid kerosene is injected through fuel atomisers and this produces a partially premixed flame that stabilises in the shear layers formed along the inner and outer recirculation zones. It has been observed that the flame is sensitive to changes in its shape due to turbulence–combustion interactions. The flame is mainly attached in the shear layer of central and outer recirculation zones, and the flame has an 'M' shape. However, it has been observed that the opening angle of the pilot can cause the flame shape to change to a 'V' shape.

The study by Langella et al. [120] used the LES–FlaRe formulation to gain insights into the flame switching behaviour, and it was demonstrated that the flame configurations and spray characteristics were well captured. Analysis of the flames showed that a PVC was not present, but a small vortex formed intermittently close to the pilot and this occurred with a flipping movement of the pilot jet. The change to the 'V' shape occurs when the vorticity is not strong enough, and this caused richer fuel–air mixture pockets, leading to the flame moving upstream and the flame root stabilising closer to the injectors. It was also observed that when the flame has established its 'V' shape, the flame influences the vorticity field and local richer fuel–air mixture pockets were formed. This also causes the flame to change to a 'V' shape. The 'V' shape flame is undesirable, since it is close to the fuel injector system and poses risks of damaging the injector nozzle. Therefore, increasing the vorticity magnitude can create more homogeneous mixtures to prevent 'V' shape flames occurring. Capturing these underlying mechanisms behind flame stabilisation in real-engine combustors with flamelet models demonstrates that such models are promising for modelling full-scale gas turbine combustors and serve as a useful tool for industry.

5 Conclusions

This chapter has presented a broad overview of lifted flames in turbulent flows with and without swirl in laboratory and practical applications used for energy production and propulsion. The stabilisation mechanisms of lifted partially premixed flames are complex and understanding these mechanisms is important for the design of combustion systems. Numerical modelling techniques with the LES paradigm are becoming more widely used as a CFD tool, due to their capabilities of modelling small-scale mixing and turbulence–combustion interactions, and resolving the flow field at the larger scales of turbulence. Different sub-grid combustion models have been briefly discussed with a focus on the LES–FlaRe modelling framework, which offers a reduced computational cost while maintaining physical consistencies for the sub-grid processes and their interactions. The two lifted flame configurations have been investigated: the canonical jet flame and a gas turbine model combustor with co-axial swirlers. The RANS formulation of the FlaRe framework is able to accurately capture the lift-off height for non-swirling jet flames with different fuels and jet velocities. However, the framework lacks the capability of modelling the transient evolution of the flame root and therefore, the LES framework is required. The flame leading edge evolution for two jet velocities obtained using the LES–FlaRe approach compared well against results obtained using more computationally expensive models. In the gas turbine model combustor, the average flame stabilisation height is less sensitive to changes in the inlet velocity and the global equivalence ratio. However, the flame root behaviour and its stabilisation are highly sensitive to changes to these inlet conditions. The simulations undertaken with the LES–FlaRe approach demonstrated that the underlying mechanisms for flames under thermo-acoustically stable and unstable conditions, and close to lean blow-off were well captured in comparison with the experimental observations. The PVC is the dominant mechanism for flame stabilisation even in the presence of self-induced thermo-acoustic instabilities and under lean blow-off conditions. The LES–FlaRe framework has been applied to the industrial burner BOSS rig and it is briefly discussed that the modelling framework is able to produce the switch in flame shape. The underlying flame stabilisation mechanisms for lifted flames can therefore be captured using an LES flamelet based closure in a variety of configurations. The modelling approach can serve as a useful approach for industry, due to the reduced computational cost while maintaining physical consistencies within the framework.

Acknowledgements The authors acknowledge the financial support from Mitsubishi Heavy Industries, Ltd., Japan. This work used the ARCHER UK National Supercomputing Service (https://www.archer.ac.uk) with the resources provided by the UKCTRF (e305).

References

1. Spalding DB (1979) Combustion and mass transfer. Pergamon Press, Oxford, UK
2. Peters N (2000) Turbulent combustion. Cambridge University Press, Cambridge, UK
3. Syred N (2006) Prog Energy Combust Sci 32(2), 93–161
4. Mansour MS (2003) Combust Flame 133(3), 263–274
5. Lyons KM (2007) Prog Energy Combust Sci 33(2), 211–231
6. Wohl K, Kapp NM, Gazley C (1948) Symp Combust Flame Explos Phenom 3(1):3–21
7. Pitts WM (1988) Symp (Int) Combust **22**(1), 809–816
8. Lawn CJ (2009) Prog Energy Combust Sci 35(1), 1–30
9. Vanquickenborne L, van Tiggelen A (1966) Combust Flame 10(1), 59–69
10. Kalghatgi GT (1984) Combust Sci Technol 41(1–2), 17–29
11. Peters N, Williams FA (1983) AIAA J 21(3), 423–429
12. Müller CM, Breitbach H, Peters N (1994) Symp (Int) Combust **25**(1):1099–1106
13. Buckmaster J, Matalon M (1988) Symp (Int) Combust **22**(1):1527–1535
14. Buckmaster J (2002) Prog Energy Combust Sci 28(5), 435–475
15. Domingo P, Vervisch L (1996) Symp (Int) Combust **26**(1):233–240
16. Broadwell JE, Dahm WJA, Mungal MG (1984) Symp (Int) Combust **20**(1):303–310
17. Miake-Lye RC, Hammer JA (1988) Symp (Int) Combust **22**(1):817–824
18. Feikema D, Chen RH, Driscoll JF (1990) Combust Flame 80(2), 183–195
19. Gupta AL, Lilley DG, Syred N (1984) Swirl flows. Abacus Press, Royal Tunbridge Wells, UK
20. Syred N, Beér JM (1974) Combust Flame 23(2), 143–201
21. Weigand P, Meier W, Duan XR, Stricker W, Aigner M (2006) Combust Flame 144(1–2), 205–224

22. Meier W, Duan XR, Weigand P (2006) Combust Flame 144(1–2), 225–236
23. Matalon M (2009) Proc Combust Inst 32(1), 57–82
24. Lieuwen TC (2012) Unsteady Combustor Physics. Cambridge University Press, Cambridge, UK
25. Dowling AP, Mahmoudi Y (2015) Proc Combust Inst 35(1), 65–100
26. Masri AR (2015) Proc Combust Inst 35(2), 1115–1136
27. Pope SB (2000) Turbulent flows. Cambridge University Press, Cambridge, UK
28. Gicquel LYM, Staffelbach G, Poinsot T (2012) Prog Energy Combust Sci 38(6), 782–817
29. Pope SB (2013) Proc Combust Inst 34(1), 1–31
30. Swaminathan N, Bray KNC (eds) (2011) Turbulent premixed flames. Cambridge University Press, Cambridge, UK
31. Chen ZX, Langella I, Swaminathan N (2019) The role of CFD in modern jet engine combustor design. Environ Impact Aviat Sustain Solut IntechOpen
32. Williams FA (1985) Combustion theory, 2nd edn. Benjamin Cummings, Menlo Park, CA, USA
33. Hinze JO (1959) Turbulence. McGraw-Hill, New York, NY, USA
34. Tennekes H, Lumley JL (1972) A first course in turbulence. MIT Press
35. Smagorinsky J (1963) Mon. Weather Rev. 91(3), 99–164
36. Germano M, Piomelli U, Moin P, Cabot WH (1991) Phys Fluids A Fluid Dyn 3(7), 1760–1765
37. Moin P, Squires K, Cabot W, Lee S (1991) Phys Fluids A Fluid Dyn 3(11), 2746–2757
38. Lilly DK (1992) Phys. Fluids A Fluid Dyn 4(3), 633–635
39. Vreman AW (2004) Phys Fluids 16(10), 3670–3681
40. Nicoud F, Ducros F (1999) Flow Turbul Combust 62(3), 183–200
41. Nicoud F, Toda HB, Cabrit O, Bose S, Lee J (2011) Phys Fluids 23(8):085,106
42. Yoshizawa A, Horiuti K (1985) J Phys Soc Japan 54(8), 2834–2839
43. Ghosal S, Lund TS, Moin P, Akselvoll K (1995) J Fluid Mech 286:229–255
44. Kim WW, Menon S (1995) 33rd AIAA aerospace sciences meeting and exhibit. AIAA 95-0356, Reno, NV, USA
45. Butler TD, O'Rourke PJ (1997) Symp (Int) Combust 16(1):1503–1515
46. Colin O, Ducros F, Veynante D, Poinsot T (2000) Phys Fluids 12(7), 1843–1863
47. Kerstein AR, Ashurst WT, Williams FA (1988) Phys Rev A 37(7), 2728–2731
48. McMurtry PA, Menon S, Kerstein AR (1992) Symp (Int) Combust 24(1):271–278
49. Marble FE, Broadwell JE (1977) Project squid technical report TRW-9-PU. Purdue University, West Lafayette, IN, USA
50. Bray KNC, Cant RS (1991) Proc R Soc A 434(1890), 217–240
51. Hawkes ER, Cant RS (2001) Combust Flame 126(3), 1617–1629
52. Pitsch H (2006) Annu Rev Fluid Mech 38(1), 453–482
53. Dopazo C, O'Brien EE (1974) Phys Fluids 17(11), 1968–1975
54. Pope SB (1985) Prog Energy Combust Sci 11(2), 119–192
55. Valiño L (1998) Flow Turbul Combust 60(2), 157–172
56. Rembold B, Jenny P (2006) J Comput Phys 220(1), 59–87
57. Klimenko AY (1990) Fluid Dyn 25:327–331
58. Bilger RW (1993) Phys Fluids A Fluid Dyn 5(2), 436–444
59. Navarro-Martinez S, Kronenburg A, Di Mare F (2005) Flow Turbul Combust 75(1–4), 245–274
60. Klimenko AY, Pope SB (2003) Phys Fluids 15(7), 1907–1925
61. Cleary MJ, Klimenko AY (2009) Flow Turbul Combust 82(4), 477–491
62. Libby PA, Bray KNC (1980) Combust Flame 39(1), 33–41
63. Liew SK, Bray KNC, Moss JB (1981) Combust Sci Technol 27(1–2), 69–73
64. Bradley D, Gaskell PH, Lau AKC (1990) Symp (Int) Combust 23(1):685–692
65. Bray KNC, Libby PA, Masuya G, Moss JB (1981) Combust Sci Technol 25(3–4), 127–140
66. Libby PA, Bray KNC (1981) AIAA J 19(2), 205–213
67. van Oijen JA, de Goey LPH (2000) Combust Sci Technol 161(1), 113–137
68. Gicquel O, Darabiha N, Thévenin D (2000) Proc Combust Inst 28(2), 1901–1908

69. Pierce CD, Moin P (2004) J Fluid Mech 504:73–97
70. Langella I, Swaminathan N, Pitz RW (2016) Combust Flame 173:161–178
71. Langella I, Swaminathan N, Williams FA, Furukawa J (2016) Combust Sci Technol 188(9), 1565–1591
72. Langella I, Swaminathan N (2016) Combust Theory Model 20(3), 410–440
73. Chen Z, Ruan S, Swaminathan N (2017) Proc Combust Inst 36(2), 1645–1652
74. Chen ZX, Langella I, Barlow RS, Swaminathan N (2020) Combust Flame 212:415–432
75. Langella I, Chen ZX, Swaminathan N, Sadasivuni SK (2018) J Propuls Power 34(5), 1269–1284
76. Chen ZX, Swaminathan N, Stöhr M, Meier W (2019) Proc Combust Inst 37(2), 2325–2333
77. Chen ZX, Langella I, Swaminathan N, Stöhr M, Meier W, Kolla H (2019) Combust Flame 203:279–300
78. Chen ZX, Swaminathan N (2020) Fuel 275:117,868
79. Massey JC, Chen ZX, Swaminathan N (2019) Combust Sci Technol 191(5–6), 1019–1042
80. Massey JC, Chen ZX, Swaminathan N (2021) Flow Turbul Combust 106(4):1355–1378
81. Bilger RW (1988) Symp (Int) Combust 22(1):475–488
82. Bray K, Domingo P, Vervisch L (2005) Combust Flame 141(4), 431–437
83. Swaminathan N, Bray KNC (2005) Combust Flame 143(4), 549–565
84. Dunstan TD, Minamoto Y, Chakraborty N, Swaminathan N (2013) Proc Combust Inst 34(1), 1193–1201
85. Domingo P, Vervisch L, Bray K (2002) Combust Theory Model 6(4), 529–551
86. Ruan S, Swaminathan N, Darbyshire O (2014) Combust Theory Model 18(2), 295–329
87. Chen ZX, Doan NAK, Ruan S, Langella I, Swaminathan N (2018) Combust Theory Model 22(5), 862–882
88. Chen Z, Ruan S, Swaminathan N (2015) Combust Flame 162(3), 703–716
89. Chen Z, Ruan S, Swaminathan N (2016) Combust Theory Model 20(4), 592–612
90. Cheng TS, Wehrmeyer JA, Pitz RW (1992) Combust Flame 91(3–4), 323–345
91. Brockhinke A, Andresen P, Kohse-Höinghaus K (1996) Symp (Int) Combust 26(1):153–159
92. Brockhinke A, Haufe S, Kohse-Höinghaus K (2000) Combust Flame 121(1–2), 367–377
93. Ahmed SF, Mastorakos E (2006) Combust Flame 146(1–2), 215–231
94. Røkke NA, Hustad JE, Sønju OK (1994) Combust Flame 97(1), 88–106
95. Ruan S, Chen Z, Swaminathan N (2015) 25th international colloquium on the dynamics of explosions and reactive systems. Leeds, UK
96. Mizobuchi Y, Tachibana S, Shinio J, Ogawa S, Takeno T (2002) Proc Combust Inst 29(2), 2009–2015
97. Mizobuchi Y, Shinjo J, Ogawa S, Takeno T (2005) Proc Combust Inst 30(1), 611–619
98. Darbyshire OR, Swaminathan N (2012) Combust Sci Technol 184(12), 2036–2067
99. Kim IS, Mastorakos E (2005) Proc Combust Inst 30(1), 911–918
100. Lacaze G, Richardson E, Poinsot T (2009) Combust Flame 156(10), 1993–2009
101. Jones WP, Prasad VN (2011) Proc Combust Inst 33(1), 1355–1363
102. Zhang H, Giusti A, Mastorakos E (2019) Proc Combust Inst 37(2), 2125–2132
103. Zhang H, Mastorakos E (2019) Flow Turbul Combust 102(4), 909–930
104. Poinsot T (2017) Proc Combust Inst 36(1), 1–28
105. Huang Y, Yang V (2009) Prog Energy Combust Sci 35(4), 293–364
106. Stöhr M, Sadanandan R, Meier W (2009) Proc Combust Inst 32(2), 2925–2932
107. Stöhr M, Boxx I, Carter C, Meier W (2011) Proc Combust Inst 33(2), 2953–2960
108. Stöhr M, Arndt CM, Meier W (2015) Proc Combust Inst 35(3), 3327–3335
109. See YC, Ihme M (2015) Proc Combust Inst 35(2), 1225–1234
110. Benim AC, Iqbal S, Meier W, Joos F, Wiedermann A (2017) Appl Therm Eng 110:202–212
111. Donini A, Bastiaans RJM, van Oijen JA, de Goey LPH (2017) Flow Turbul Combust 98(3), 887–922
112. Selle L, Lartigue G, Poinsot T, Koch R, Schildmacher KU, Krebs W, Prade B, Kaufmann P, Veynante D (2004) Combust Flame 137(4), 489–505
113. Stöhr M, Boxx I, Carter CD, Meier W (2012) Combust Flame 159(8), 2636–2649

114. Boxx I, Stöhr M, Carter C, Meier W (2010) Combust Flame 157(8), 1510–1525
115. Steinberg AM, Boxx I, Stöhr M, Carter CD, Meier W (2010) Combust Flame 157(12), 2250–2266
116. Schneider D, Meier U, Quade W, Koopman J, Aumeier T, Langfeld A, Behrendt T, Hassa C, Rackwitz L (2010) 27th congress of international council of the aeronautical sciences. Nice, France
117. Freitag S, Behrendt T, Heinze J, Lange L, Meier U, Hassa C, Rackwitz L (2011) 24th european conference on liquid atomization and spray systems. Estoril, Portugal
118. Meier U, Heinze J, Lange L, Hassa C, Rackwitz L, Doerr T (2012) CEAS Aeronaut J 3(1), 45–53
119. Meier U, Lange L, Heinze J, Hassa C, Sadig S, Luff D (2015) J Eng Gas Turbines Power 137:7
120. Langella I, Heinze J, Behrendt T, Voigt L, Swaminathan N, Zedda M (2019) Proceedings of the ASME Turbo Expo 2019: Turbomachinery Technical Conference and Exposition. Volume 4B: Combustion, Fuels, and Emissions, V04BT04A047. ASME GT2019-91879, Phoenix, AZ, USA

Effect of Oxygenation on the Liftoff, Stabilization, and Blowout Characteristics of Laminar Co-flow Jet Flames

Rajesh Kumar, Krishna C. Kalvakala, and Suresh K. Aggarwal

Abstract This chapter summarizes the results of a computational study that examined the effects of a new oxygenation strategy on the flame structure, liftoff height, stabilization, and blowout characteristics in non-premixed laminar co-flow jets. Oxygenated ethylene–air flames were established by employing an innovative strategy of using N_2-diluted fuel jet along with O_2-enriched oxidizer jet such that the stoichiometric mixture fraction (f_{st}) is varied but the adiabatic flame temperature is not materially changed. For each case, the flame liftoff and blowout processes were examined by using CO_2 dilution in the fuel jet. A time accurate, implicit, axisymmetric CFD code that uses a detailed description of transport and chemistry (111 species and 784 reactions) and that includes radiation effects was adopted. The computational model and chemical mechanism were validated using published measurements of the flame length, temperature, and species profiles of ethylene/air jet flames. Simulations indicated burner attached flames for all levels of oxygen enrichment, without CO_2 dilution. The flame stabilization mechanism in the near-field region can be described by a balance between the reaction rate and local scalar dissipation rate. Further, as the oxygen enrichment is increased, it becomes more difficult to get lifted flames. However, once a flame is lifted, its liftoff height increases rapidly with the increase in CO_2 dilution, i.e., a lifted oxygenated flame exhibited a greater sensitivity to CO_2 dilution.

Keywords Oxygenation · Dilution · Liftoff · Blowout

1 Introduction

Flame extinction is one of the fundamental and classical phenomena in combustion science, where both chemistry and transport play a vital role. This phenomenon has high potential to dictate the efficiency and emissions from combustion systems used in transportation and power generation applications. In addition, it has direct relevance

R. Kumar · K. C. Kalvakala · S. K. Aggarwal (✉)
University of Illinois at Chicago, Chicago, IL 60607, USA
e-mail: ska@uic.edu

A. Gupta et al. (eds.), *Advances in Energy and Combustion*, Green Energy and Technology,
https://doi.org/10.1007/978-981-16-2648-7_12

in safe, stable, and reliable operation of these combustion systems. For instance, flame extinction closely correlates with the flame flashback in micro-combustors, flame quenching due to flame-wall interaction in furnaces, and with lean burning limit in gas turbines. This further emphasizes the importance of flame extinction in the design of combustion devices, including furnaces and gas turbine combustors.

In many systems, flame extinction is often described in terms of flammability limits, burning velocity, liftoff, and blowout characteristics. In order to gain fundamental understanding of this mechanism, laminar lifted flames have been commonly investigated [1–7]. In a jet flame configuration, lifted flames can be established in near field or far field by adjusting the flow conditions, dilution of fuel, premixing, or fuel composition. For instance, increase in jet velocity for laminar premixed jets makes the flame to liftoff from the burner [8, 9], which changes the flame structure to a large extent. Increase in liftoff height is accompanied by flame structure transitioning from a non-premixed flame to a double flame and then to a triple flame containing three reaction zones—a lean premixed, a rich premixed, and a non-premixed zone [10, 11]. In addition to jet velocity, several other factors including fuel and air-stream dilution [9, 12–15], temperature of air stream [16–18], and partial premixing [9, 12, 14] have been identified to have a significant impact on stabilization of lifted flames. Lee et al. [9] examined the effects of nozzle diameter and jet velocity on the liftoff characteristics of N_2-diluted, non-premixed propane jet flames. The nozzle diameter and amount of dilution were found to have prominent effect on the liftoff and blowout processes, and three regimes for liftoff and blowout were identified depending on the jet diameter and velocity. Lock et al. [13] examined the effects of fuel- and air-stream dilution on the liftoff and blowout of partially premixed (PPFs) and non-premixed (NPF) methane flames. It was noticed that dilution of either of air or fuel stream with CO_2 tends to liftoff the flame, eventually leading to blowout. However, with fuel-stream dilution, PPFs stabilized at a higher liftoff height and blew out at a lower CO_2 dilution than NPFs. In contrast, with air-stream dilution, NPFs were lifted higher, and blew out at a lower CO_2 dilution than PPFs. In addition, results showed remarkable similarity in the blowout or extinction behavior of co-flow and counterflow flames with respect to the level of partial premixing and air- and fuel-stream dilutions.

In a subsequent study, Lock et al. [14] elaborated on the role of air- and fuel-stream dilution on the flame liftoff and blowout characteristics for three different fuels—methane, acetylene, and ethylene–under PPF and NPF conditions. It was observed that ethylene and acetylene flames exhibit stronger resistance to blowout or in other words are more difficult to extinguish than methane flames. Kim et al. [17] investigated the liftoff characteristics of N_2-diluted propane diffusion flames at elevated temperatures. It was observed that the lifted flames maintain a tribrachial or triple flame structure up to a temperature of 900 K. In addition, while liftoff height was found to decrease with temperature, the lifted flames were stabilized by a balance between local flow velocity and the tribrachial flame propagation speed regardless of temperature. Further, Choi and Chung [18] experimentally examined the liftoff characteristics of methane, ethylene, ethane, and n-butane fuels at elevated temperatures. The study highlighted the importance of auto-ignition temperature on the flame liftoff and blowout behavior. It was observed that at temperatures lower than auto-ignition temperature, the liftoff and blowout characteristics varied with Schmidt number (Sc) of the fuel, while at higher temperatures, regardless of Sc number, ignition delay time of fuel/air mixture had a significant effect on flame liftoff and blowout. The results were consistent with previous studies by Chung and Lee [1, 2], which emphasized the importance of Schmidt number (Sc) in the liftoff process and observed that lifted flames can be stabilized for $Sc < 0.5$ or $Sc > 1$.

In recent times, oxygen-enhanced combustion (OEC) and the use of oxyfuels have gained significant amount of interest for their potential to reduce soot emissions and easier CO_2 capture [19–22]. On the other hand, there are very few studies that highlighted the significance of OEC strategy on liftoff and blowout characteristics [23–26]. Pickett et al. [23] in diesel spray flames observed that the liftoff length was inversely proportional to the ambient oxygen concentration. Xue and Ju [26] through experiments examined the influence of fuel-bound oxygen on the liftoff phenomenon using di-methyl ether (DME), oxygen-diluted propane, and n-butane jet diffusion flames. It was observed that DME flames required higher flow velocities for blowout. Further, an increase in oxygen addition, both propane and n-butane flames changed from direct liftoff regime to direct blowout regime. Chung et al. [27] examined auto-ignited laminar DME flames with high-temperature co-flow air to observe a unique non-monotonic behavior in liftoff height with jet velocity. Chen and Axelbaum [24] investigated the extinction phenomena under counterflow flames through a unique strategy of using oxygenated combustion. The strategy involved simultaneous O_2-enrichment of oxidizer stream and N_2 dilution of fuel steam, such that the stoichiometric mixture fraction was increased without altering the adiabatic flame temperature. It was observed that under counterflow configuration, the oxygenated flames were significantly more resistant to extinction, as their extinction strain rates were an order of magnitude higher than those without oxygenation. In subsequent studies, this oxygenation strategy has been examined under counterflow and co-flow conditions [19, 21, 28–30] for its potential to reduce polycyclic hydrocarbons (PAHs) and soot emissions. However, the effect of such strategy on liftoff and blowout phenomena in jet flames has not been examined in previous studies. Motivated by this, the present study aims to numerically investigate the effect of this oxygenation strategy along with CO_2 dilution on the flame liftoff and blowout behavior. Simulations are performed using a time accurate, implicit algorithm with detailed chemistry and transport models. The oxygenated ethylene flames are established in co-flow jets by removing a specified amount of N_2 from the oxidizer jet and introducing it into the fuel jet. For each oxygenated flame, its liftoff and blowout behavior is characterized by using an increasing amount of CO_2 dilution in the fuel jet. Thus, at a fixed oxygenation level or stoichiometric mixture fraction, the diluent concentration in the fuel jet is increased until the blowout of lifted flame occurs. The detailed flame structure and its propagation characteristics during the liftoff and blowout process are investigated for different oxygenation levels. It is important to note that most of the results in this chapter have been previously reported in an AIAA conference paper [30].

2 Computational Model

Ethylene laminar jet flames in a co-flow configuration are simulated using a time-dependent, axisymmetric algorithm, the UNICORN (Unsteady Ignition and Combustion using Reactions) [31, 32]. The algorithm numerically solves the continuity, axial- and radial-momentum, energy, and species conservation equations on a staggered grid system. The finite difference forms of the momentum equations are obtained using an implicit QUICKEST (Quadratic Upstream Interpolation for Convective Kinematics with Estimated Streaming Terms) scheme and those of the species and energy equations using a hybrid scheme of upwind and central differencing [33]. The pressure field is computed by solving the discrete form of the pressure Poisson equation at all the grid points simultaneously by using the LU (Lower and Upper diagonal) matrix-decomposition technique.

The general form of the conservation equations in cylindrical coordinates can be represented as:

$$\frac{\partial(\rho\Phi)}{\partial t}+\frac{\partial(\rho u\Phi)}{\partial r}+\frac{\partial(\rho v\Phi)}{\partial z}=\frac{\partial}{\partial r}\left(\Gamma^{\Phi}\frac{\partial\Phi}{\partial r}\right)+\frac{\partial}{\partial z}\left(\Gamma^{\Phi}\frac{\partial\Phi}{\partial z}\right)-\frac{\rho u\Phi}{r}+\frac{\Gamma^{\Phi}}{r}\frac{\partial\Phi}{\partial r}+S^{\Phi} \tag{1}$$

where Φ is the variable,ρ is the density, t the time, u and v represent the radial (r) and axial (z) velocity components, respectively. The transport coefficients Γ^{Φ} and source term S^{Φ} in the above equation are provided in Table 1 of Ref. [34], and further details are discussed in Ref. [35]. Further, a sink term based on an optically thin gas assumption is included in the energy equation to account for thermal radiation from the flame [36]. This is expressed as $q_{\text{rad}} = -4\sigma K_p(T^4 - T_0^4)$ [37], where T denotes the local flame temperature, and K_p accounts for the absorption and emission from participating species (CO_2, H_2O, CO, CH_4 and soot particles), and is expressed as $K_p = P\sum X_i K_{p,i}$, where $K_{p,i}$ denotes the mean absorption coefficient of k_{th} species, σ the Stefen–Boltzmann coefficient constant, and T_0 the ambient temperature. The values of $K_{p,i}$ are calculated using a polynomial approximation to the

Table 1 Input values of species mole fractions for various oxygenated flames

CASES—percentage of N_2 removed from air (%)	Fuel stream		Air stream		Stoichiometric mixture fraction (f_{st})
	C_2H_4 mole fraction	N_2 mole fraction	O_2 mole fraction	N_2 mole fraction	
0	1	0.0	0.21	0.79	0.064
25	0.262	0.738	0.262	0.738	0.243
50	0.151	0.849	0.347	0.653	0.423
100	0.081	0.919	1.0	0.0	0.782

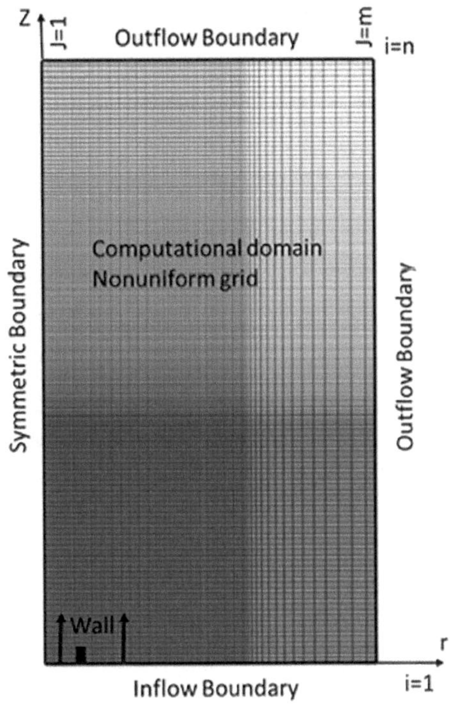

Fig. 1 Schematic of the computational domain and boundary conditions. Adopted from [30]

experimental data provided in Ref. [37]. The thermodynamics and transport properties in these equations are considered as temperature and species dependent. First, the species enthalpies are calculated using the polynomial curve fits developed for the temperature range 300–5000 K. The methodology for calculating the molecular viscosity, thermal conductivity, and the binary diffusion coefficients of the species, and then of the mixture is described in Ref. [15].

Figure 1 illustrates the computational domain simulated in the present study. The domain extends 200 and 50 mm in axial (Z) and radial (r) directions, respectively. In addition, an isothermal insert of 2×0.38 mm is used to simulate the fuel nozzle wall. A grid-independent study has been performed to understand the dependency of grid on computed results. Figure 2 presents a comparison of computed temperature profiles along the center line for three different grid systems and with respect to experimental data. It is observed that the computed temperature profiles are nearly independent of the grid system; thus, a 401 (Z) $\times$ 101(r) grid system will be adopted for rest of the simulations. The outflow boundaries are located sufficiently far from the flame region to minimize the effect of enclosure on the predictions. The inner jet contains fuel and diluent (N_2 and CO_2) mixture, while the outer jet issues oxidizer. The velocity of the inner and outer jets is 35 cm/s and is fixed for all the cases investigated. As mentioned, oxygenated ethylene flames are established by transferring a specified amount of N_2 from the outer jet into the fuel jet. In other words, the oxidizer stream is made richer in oxygen while simultaneously diluting the fuel

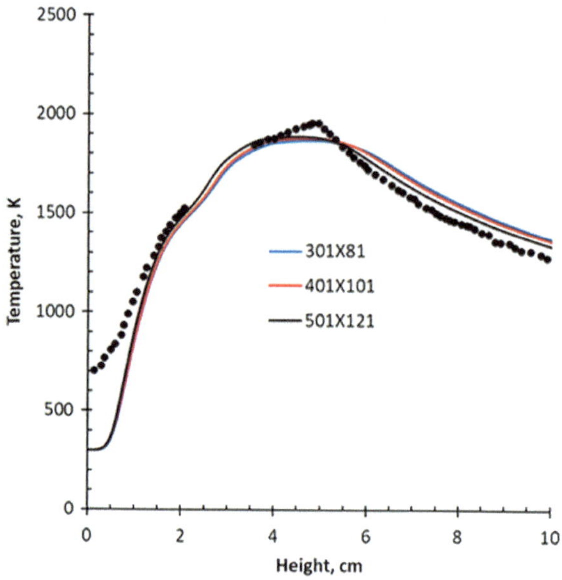

Fig. 2 Comparison of experimental (symbols) and predicted (solid lines) temperature profiles along the central line at different grid sizes. For all the cases, the flame is computed using the same initial conditions. Adopted from [30]

stream with N_2. Consequently, the stoichiometric mixture fraction (f_{st}), (given by Eq. (2)) changes. In addition, for a given f_{st}, the diluent (CO_2) concentration in the fuel jet is slowly increased in order to analyze the flame liftoff, stabilization, and blowout processes. The ethylene flame chemistry is modeled using the USC II mechanism [38] with 111 species and 784 reactions. Table 1 provides details including species mole fraction in the fuel and oxidizer streams for base (or CO_2 undiluted) oxygenation levels, along with the f_{st} values. Table 2 provides the N_2 and CO_2 mole fractions in the fuel stream for all cases at various diluent concentrations simulated.

$$f_{\mathrm{st}} = \left(1 + \left(\frac{Y_F \mathrm{MW}_O \vartheta_O}{Y_O \mathrm{MW}_F \vartheta_F}\right)\right)^{-1} \tag{2}$$

Table 2 Mole fractions of N_2 and CO_2 in the fuel stream for various diluted flames

Cases	0% Oxygen enrichment		25% Oxygen enrichment		50% Oxygen enrichment		100% Oxygen enrichment	
	X_{N_2}	X_{CO_2}	X_{N_2}	X_{CO_2}	X_{N_2}	X_{CO_2}	X_{N_2}	X_{CO_2}
D1	0	0	0.738	0	0.849	0	0.919	0
D2	0	0.5	0.585	0.207	0.738	0.131	0.849	0.075
D3	0	0.67	0.485	0.344	0.653	0.231	0.790	0.140
D4	0	0.75	0.385	0.478	0.556	0.345	0.738	0.196
D5	0	0.80	0.339	0.541	0.506	0.404	0.693	0.246
D6	0	0.83	0.303	0.590	0.485	0.430	0.653	0.289
D7	0	0.85	0.295	0.601	0.465	0.453	0.634	0.309

where Y_f and Y_O are mass fractions of fuel and oxygen at the respective boundaries, ϑ_F and ϑ_O are the stoichiometric coefficients for fuel and oxygen, and MW stands for molecular weight.

3 Results and Discussion

3.1 Validation of Computational Model

The computational algorithm and the reaction mechanism have been extensively validated previously for a variety of steady and unsteady flames including opposed-jet diffusion flames [39], burner-stabilized flames [8, 40], and lifted jet flames [41]. As discussed in cited studies, there is generally good agreement between predictions and measurements in terms of flame topology including the locations of the various reaction zones (i.e., the rich premixed (RP), lean premixed (LP), and non-premixed (NP) zones), and the liftoff heights. We herein provide another validation against the experimental data of nitrogen-diluted ethylene diffusion flames reported by Smooke et al. [42]. Figure 3 compares the measured and predicted temperature and acetylene profiles along the centerline for two N_2-diluted ethylene jet flames. It is observed that the computational model captures the experimental data, both qualitatively and quantitatively to a good extent. The computational model reproduces the rapid rise and subsequent decrease in temperature. However, there are some differences between measured and predicted peak temperatures, which may be attributed to the over prediction of radiation losses by the optically thin model. Figure 3b compares the measured and predicted acetylene profiles along the centerline for the two flames. Again, the computational model reasonably captures the main features of the flame and the changes in the acetylene concentration with N_2 dilution. Thus, overall, the computational model is able to reproduce the experimental flame data in a satisfactory manner.

3.2 Effects of Oxygenation and Dilution on the Flame Structure

Figure 4 presents the computed flame structures for various levels of oxygenation (see Table 1). For each case, the flame structure is shown in terms of heat release contours (q) and velocity vectors (v). The stoichiometric mixture fraction (f_{st}) line for each flame is shown by a black line. As the oxygenation level or f_{st} is increased, the flame height decreases considerably which can be attributed to the increased availability of oxygen combined with the reduced availability of fuel. In other words, at higher values of f_{st}, the required stoichiometric condition for combustion is achieved at a shorter height. In addition, with increase in f_{st}, the heat release rate at the flame

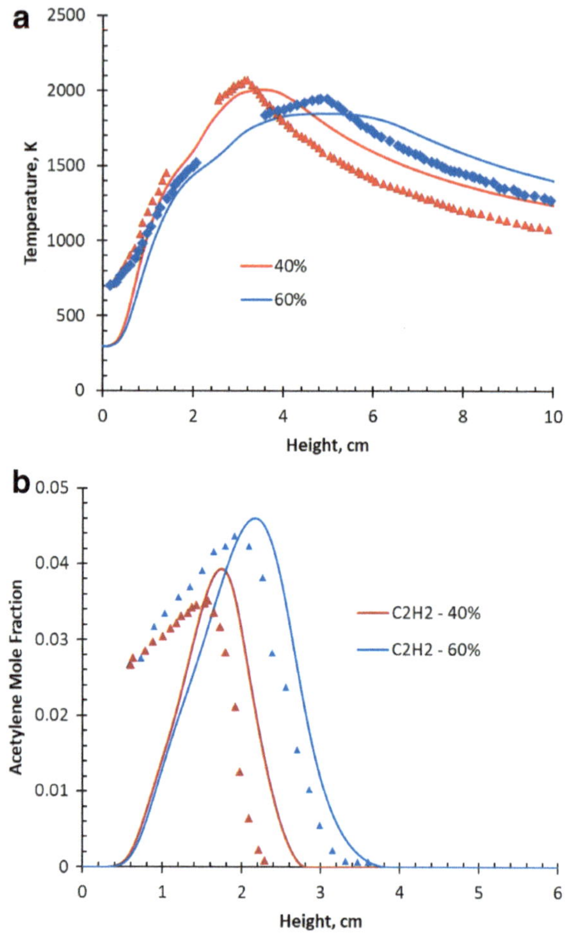

Fig. 3 Comparison of experimental (symbols) [42] and predicted (solid lines) **(a)** temperature and **(b)** acetylene mole fraction profiles along the central line in N_2-diluted ethylene/air co-flow jet flames with 40% and 60% ethylene in the fuel jet. Adopted from [30]

base increases, while for all f_{st} conditions considered, the flame remains attached to the burner. These changes associated with increase in oxygenation on the burner-stabilized flame height and structure, especially near the base region, have significant implications on the flame liftoff behavior, as elaborated in the following sections.

Figure 5 presents the topologies/structures of various oxygenated flames for different CO_2 dilution levels. Again, the flame structure is presented in terms of heat release rate contour, velocity vectors, and stoichiometric mixture fraction line. Figures 5a–d correspond to cases with $f_{st} = 0.064$, 0.243, 0.423, and 0.782, respectively. For any given f_{st}, increase in CO_2 dilution makes the flame to liftoff from the burner and stabilize at a certain height. The liftoff height is observed to be in the low velocity region that extends from the wake of the burner rim. It is also observed that the flame remains attached to the burner until the dilution levels (sum of N_2 and CO_2 mole .fractions) of 66%, 81%, 88%, and 93% for the four oxygenation cases: $f_{st} =$ 0.064, 0.243, 0.423, and 0.782, respectively. Further, any additional increase in CO_2

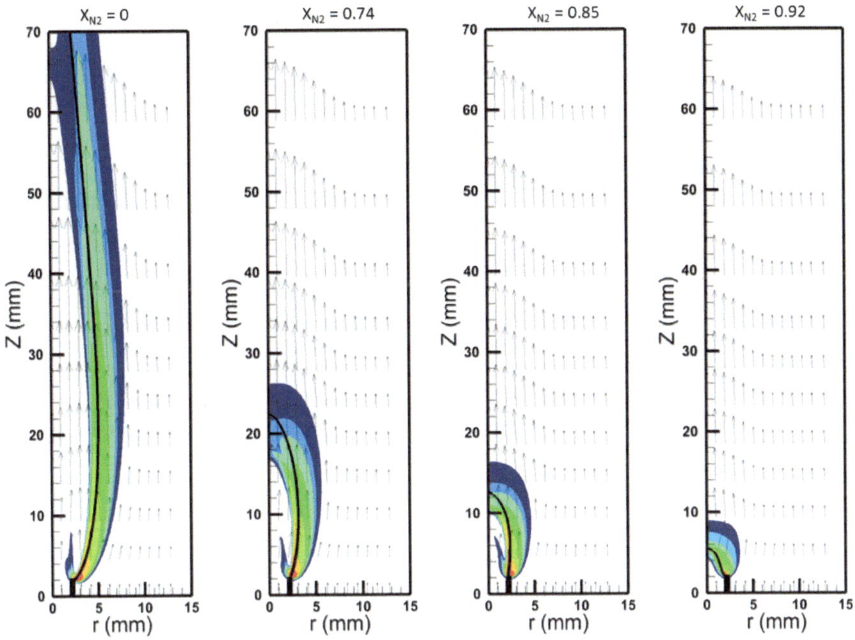

Fig. 4 Predicted heat release rate contours and velocity vectors for various oxygenated flames with f_{st} = 0.064, 0.243, 0.423, and 0.782 (Table 1). The dark black line represents the stoichiometric mixture fraction (f_{st}). Adopted from [30]

dilution, results in a sudden increase in liftoff height, followed by flame blowout. For instance, for the flame (a) with $f_{st} = 0.064$, 75% dilution lifts the flame base to a height of 12 mm which then moves rapidly to a height of 41 mm as the dilution level increases to 83%. The flame blows out for a dilution level of 85%. In addition, for all cases with different f_{st}, the lifted flame at lower heights exhibits a double flame, containing rich premixed (RP) and non-premixed (NP) reaction zones. With increase in the dilution level (which results in higher flame heights), the mixing in the wake region above the burner rim is enhanced, which allows for increased entrainment of oxidizer into the fuel stream. As a consequence, the lifted flame structure changes from double to triple flames. For $f_{st} = 0.782$ flame, the rich premixed zone is weaker and appears as a merged flame due to high dilution rate in the fuel stream. In addition, the lifted flames become increasingly more sensitive to CO_2 dilution as the oxygen enrichment level is increased, which is discussed in the next section.

In order to characterize the effect of oxygenation on the flame structure, we present in Fig. 6 the state relations for two extreme cases ($f_{st} = 0.064$ and 0.782, or 0% and 100% oxygenation) in terms of the profiles of major reactant and product species (C_2H_4, O_2, H_2O, and CO_2) and intermediate fuel species (H_2 and CO) with respect to the mixture fraction (f) at an axial location 2 mm above the burner. Here $f = 1$ and 0 indicate the fuel and oxidizer sides, respectively. Following Bilger [43], the mixture fraction $Z(F)$ is defined as:

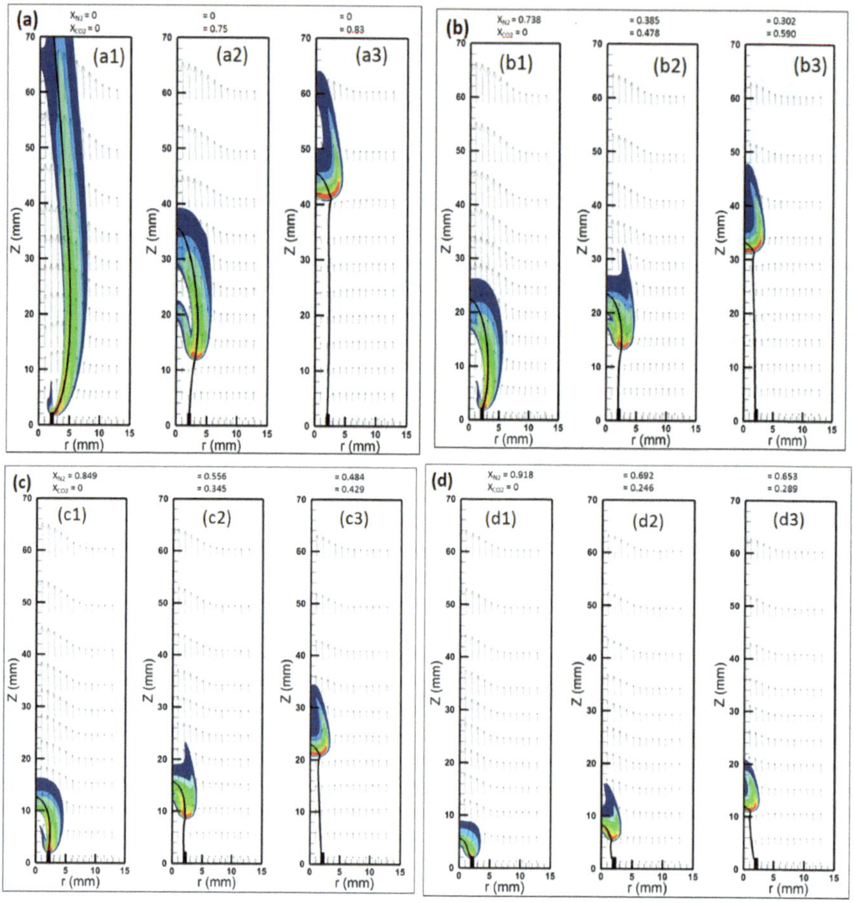

Fig. 5 Predicted heat release rate contours and velocity vectors for various oxygenation levels, f_{st} = 0.064 (**a**), 0.243 (**b**), 0.423 (**c**), and 0.782 (**d**), and various dilution levels$\left(X_{CO_2}\text{ and }X_{N_2}\right)$. The dark black line represents the stoichiometric mixture fraction (f_{st}). Adopted from [30]

$$Z(F) = \sum\left[N_{Ci}Y_i\left\{\frac{M_C}{M_i}\right\}\right] + \sum\left[N_{hi}Y_i\left\{\frac{M_h}{M_i}\right\}\right]$$

Here, N_{Ci} and N_{hi} are the number of carbon and hydrogen atoms, respectively, in species i, Y_i the mass fraction of species i, M the molecular weight, and summation is over all species. Since the fuel or oxidizer stream may also contain diluent, Z can be normalized as follows:

$$f = \frac{[Z(F) - Z(F,l)]}{[Z(F,r) - Z(F,l)]}$$

where subscripts r and l correspond to fuel and oxidizer streams, respectively.

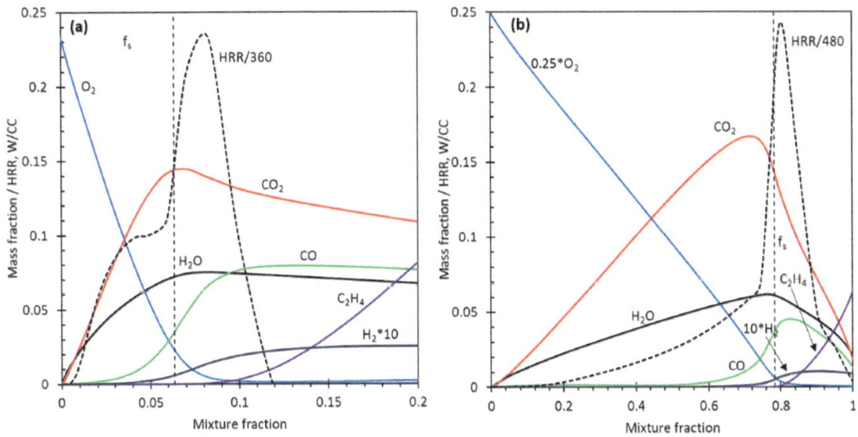

Fig. 6 State relationships in terms of scalar profiles with respect to mixture fraction (f) for two flames with $f_{st} = 0.064$ (**a**) and $f_{st} = 0.782$ (**b**), but without CO_2 dilution. The vertical dashed line represents the stoichiometric mixture fraction (f_{st}) value. Adopted from [30]

Figure 6 depicts a typical non-premixed flame structure in the mixture fraction coordinate for both flames. The reaction zone is located by the peak values of heat release rate (HRR), CO_2, and H_2O mole fractions near $f = f_{st}$. In addition, it is indicated that as the level of oxygenation is increased, the stoichiometric mixture fraction (f_{st}) increases from $f_{st} = 0.0637$ to 0.7817 for the 100% O_2-enriched case, and the flame location moves from the oxidizer to fuel side. The increased f_{st} value also implies that mixing required to establish the flame is reduced compared to the case with lower f_{st} value.

3.3 Effect of Dilution on Flame Liftoff and Blowout

Figure 7 presents the variation in flame liftoff height with respect to diluent mole fraction for all four f_{st} cases. It is observed that, with addition of diluent, the flame liftoff height for 0% oxygenated flame initially increases in a gradual, nonlinear fashion, and then followed by a rapid increase until the blowout occurs (at $X_{CO_2} = 0.85$). Further, these results for the flame without oxygenation ($f_{st} = 0.064$) are consistent with previous experimental data [14] in terms of CO_2 mole fraction required for blowout of ethylene/air flames. In short, this again underlines the capability of the computational algorithm and the chemical mechanism used in this study. In contrast to 0% oxygenated flames, the liftoff height for oxygenated flames (25, 50 and 100%) increases very rapidly after flame liftoff, indicating greater sensitivity of the flame to dilution level. Further as shown in Fig. 7, the critical dilution $\left(X_{CO_2} + X_{N_2}\right)$ levels for extinction are 5%, 89%, 92%, and 94% for 0%, 25%, 50%, and 100% oxygen-enriched flames, respectively. Critical values of CO_2 and N_2 dilutions for blowout

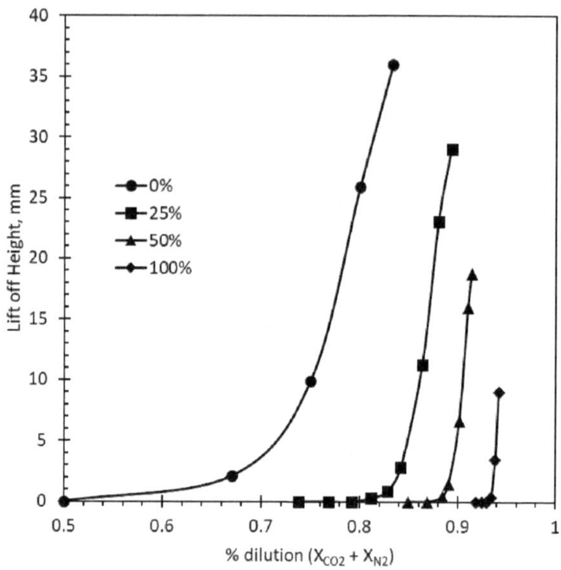

Fig. 7 Liftoff height plotted as a function of fuel-stream dilution ($CO_2 + N_2$) for various oxygenated flames; 0%, 25%, 50%, and 100% correspond to oxygenation level by removing N_2 from air stream and adding it to fuel stream as indicated in Tables 1 and 2. Adopted from [30]

are provided in Table 2 (Case: D7). In addition, with the increase in the level of oxygenation, flames tend to blowout at a lower height and the range of dilution from liftoff to blowout reduces drastically.

The stabilization of lifted non-premixed flames has generally been explained based on the dynamic balance between a triple flame speed and local flow velocity. This stabilization is based on the existence of a triple flame structure at the flame base, which is more appropriate for flames that are stabilized in the far field. For lifted flames stabilized in the near field, Briones and Aggarwal [15] and Takahashi and Katta [44] hypothesized a stabilization mechanism based on a dynamic balance between the characteristic reaction rate and scalar dissipation rate. To examine this hypothesis in the context of present study, we have computed the scalar dissipation rate for various dilution and oxygenation levels. The scalar dissipation rate is defined as $\chi = 2D_{F-\mathrm{mix}}(\nabla f)^2$, where f is the mixture fraction and $D_{f-\mathrm{mix}}$ is the fuel diffusivity with respect to the mixture.

Figure 8 presents the variation of local χ with respect to axial position along the fst line for various oxygenated flames (D6 case in Table 2) near blowout (extinction) conditions. For each oxygenated flame, the flame base is located at the point of minimum χ value. The enhanced blowout characteristics of oxygenated flames can be explained by an increasingly lower value of the scalar dissipation rate (χ), implying stronger reactivity as the amount of oxygenation is increased.

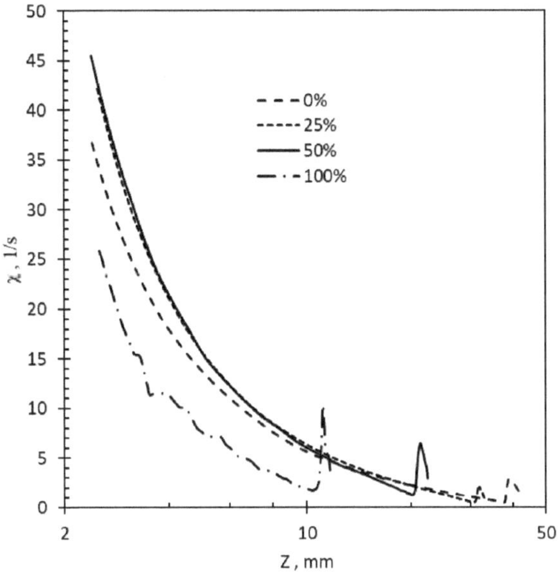

Fig. 8 Scalar dissipation rate (χ) plotted versus the axial position (z) along the stoichiometric mixture fraction line for flames near blowout. 0, 25, 50, and 100% correspond to oxygenation level by removing specified amount of N_2 from oxidizer stream and adding it to fuel stream. Adopted from [30]

3.4 *Blowout (Extinction) Characteristics of Oxygenated Co-flow Flames*

Several studies have highlighted the importance of using the peak flame temperature (T_{max}) as an indicator of flame strength or its resistance to extinction [11, 14]. Figure 9

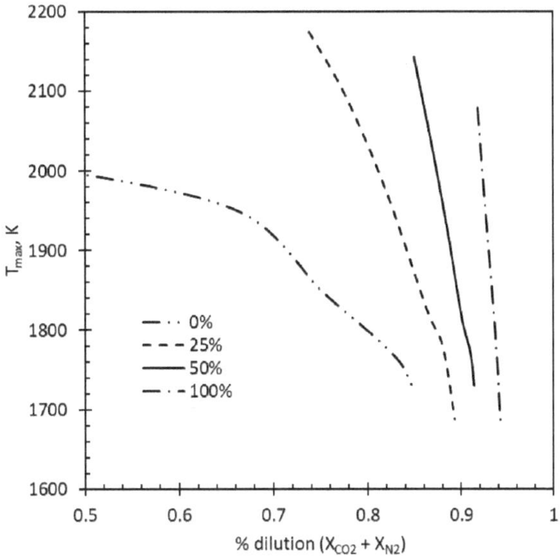

Fig. 9 Peak flame temperature plotted as a function of ($CO_2 + N_2$) dilution for various oxygenated flames. Adopted from [30]

plots T_{max} as a function of fuel-stream dilution for the four oxygenation cases. For each case, T_{max} decreases as the amount of dilution is increased, and the minimum T_{max} reached indicates flame condition close to extinction. For the base case (f_{st} = 0.064), the peak flame temperature decreases more gradually as the amount of dilution is increased. In contrast, at higher oxygenation levels, T_{max} decreases more rapidly with dilution, with the amount of dilution required for flame extinction or blowout increasing with f_{st}. However, for all the four cases, the predicted temperatures near blowout are within a narrow range of 1680–1730 K. Further, for the base case, T_{max} exhibits a nonlinear behavior because the diluent has two effects: the dilution effect and the thermal effect that cools the flame. On the other hand, for oxygenated flames, T_{max} decreases almost linearly, with the thermal effect of CO_2 dilution being more prominent. In oxygenated flames, thermal effect of CO_2 is more prominent than the dilution effect of either CO_2 or N_2. For instance, for the flame with 100% enrichment, as CO_2 dilution is increased from $X_{CO_2} = 0.246$ to 0.289, T_{max} decreases from 1763 to 1682 K. Any further increase in CO_2 causes the flame blowout or extinction.

4 Conclusions

A computational investigation on the flame liftoff, stabilization, and blowout characteristics of oxygenated non-premixed ethylene/air jet flames has been carried out. A time accurate, implicit algorithm that uses a detailed description of the chemistry and includes radiation effects is used for this study. The oxygenated flames are established by simultaneously using an O_2 enriched oxidizer jet along with a N_2 diluted fuel jet in such a way that the adiabatic flame temperature is not materially changed. The flame liftoff and blowout processes are then examined by using CO_2 dilution in the fuel jet. The computational algorithm and chemical mechanism have been validated using the experimental data on flame structure reported in terms of temperature and species profiles for diluted ethylene/air jet flames [42]. Important observations are as follows,

1. Results indicate burner-attached flames for all levels of oxygen enrichment, but without CO_2 dilution. As CO_2 dilution is increased to a critical value, the flame lifts off from the base and stabilizes in the near-field region. With further increase in dilution, the lifted flame gets stabilized further downstream of burner rim in a low velocity region. With still further increase in CO_2 dilution, the liftoff height rapidly increases until the blowout occurs. The amount of critical dilution required for extinction (blowout) increases with the increase in oxygenation level. For instance, the predicted critical dilution $\left(X_{N_2} + X_{CO_2}\right)$ values for oxygenation levels of 0%, 25%, 50%, and 100% are 0.85, 0.89, 0.92, and 0.94, respectively, highlighting the increased flame resistance to extinction. Also, for the 0% enrichment case, there is a good agreement between the measured and predicted values of CO_2 $\left(X_{CO_2} = 0.85\right)$ required for flame blowout.

2. As the oxygen enrichment is increased, it becomes more difficult to establish lifted flames. However, once the flame is lifted, its liftoff height increases rapidly with the increase in CO_2 dilution, i.e., a lifted oxygenated flame exhibits greater sensitivity to CO_2 dilution.
3. The predicted range of dilution for stable lifted flames reduces monotonically with the increase in oxygen enrichment. For the flame with 100% oxygen enrichment, the flame lifts off at dilution level of 93% dilution and blows out at 94% dilution, indicating a stable range of only 1%, as compared to a value of 19% for the non-oxygenated flame.
4. The stabilization mechanism for lifted flames in the near-field region can be described by a balance between the reaction rate and local scalar dissipation rate.

Acknowledgements Computations were performed using the UNICORN code provided by Dr. Viswanath R. Katta. His assistance throughout this study is gratefully acknowledged.

References

1. Chung SH, Lee BJ (1991) On the characteristics of laminar lifted flames in a nonpremixed jet. Combust Flame 86:62–72
2. Lee BJ, Chung SH (1997) Stabilization of lifted flames tribrachial flames in a laminar nonpremixed jet. Combust Flame 109:163–172
3. Chen YC, Bilger RW (2000) Stabilization mechanism of lifted laminar flames in axisymmetric jet flows. Combust Flame 122:377–399
4. Ghosal S, Vervisch L (2001) Stability diagram for lift-off and blowout of a round jet laminar diffusion flame. Combust Flame 123:646–655
5. Takahashi F, Schmoll WJ, Katta VR (1998) Attachment mechanism of diffusion flames. Proc Combust Inst 27:675–684
6. Takahashi F, Katta VR (2000) A reaction kernel hypothesis for the stability limit of methane jet diffusion flames. Proc Combust Inst 28:2071–2078
7. Ruetsch GR, Vervisch L, Linan A (1995) Effects of heat release on triple flames. Physics of Fluid 7:1447–1454
8. Azzoni R, Ratti S, Puri IK, Aggarwal SK (1999) Gravity effects on triple flames: flame structure and flow instability. Phys Fluids 11:3449–3464
9. Lee BJ, Kim JS, Chung SH (1994) Effect of dilution on the liftoff of non-premixed jet flames. Proc Combust Inst 25:1175–1181
10. Chung SH (2007) Stabilization, propagation and instability of tribrachial triple flames. Proc Combust Inst 31:877–892
11. Aggarwal SK (2009) Extinction of laminar partially premixed flames. Prog Energy Combust 35:528–570
12. Qin X, Puri IK, Aggarwal SK (2002) Characteristics of lifted triple flames stabilized in the near field of a partially premixed axisymmetric jet. Proc Combust Inst 29:1565–1572
13. Lock A, Briones AM, Aggarwal SK, Puri IK, Hegde U (2007) Liftoff and extinction characteristics of fuel- and air-stream-diluted methane-air flames. Combust Flame 149:340–352
14. Lock A, Aggarwal SK, Puri IK (2009) Effect of fuel type on the extinctions of fuel and air stream diluted partially premixed flames. Proc Combust Inst 32:2583–2590

15. Briones AM, Aggarwal SK (2006) A numerical investigation of flame liftoff, stabilization, and blowout. Phys Fluids 18:043603–1–11
16. Choi BC, Kim KN, Chung SH (2009) Autoignited laminar lifted flames of propane in coflow jets with tibrachial/edge and mild combustion. Combust Flame 156:396–404
17. Kim KN, Won SH, Chung SH (2007) Characteristics of laminar lifted flames in coflow jets with initial temperature variation. Proc Combust Inst 31:947–954
18. Choi BC, Chung SH (2010) Autoignited laminar lifted flames of methane, ethylene, ethane and n-butane jets in coflow air with elevated temperature. Combust Flame 157:2348–2356
19. Skeen SA, Yablonsky G, Axelbaum RL (2010) Characteristics of non-premixed oxygen-enhanced combustion: II Flame structure effects on soot precursor kinetics resulting in soot-free flames. Combust Flame 157:1745–1752
20. Kalvakala KC, Katta VR, Aggarwal SK (2018) Effects of oxygen-enrichment and fuel unsaturation on soot and NO_x emissions in ethylene, propane and propene flames. Combust Flame 187:217–229
21. Kalvakala KC, Aggarwal SK (2021) Soot emission reduction in oxygenated co-flow jet flames. Proc Combust Inst 38
22. Bhatia P, Katta VR, Krishnan SS, Zheng Y, Sunderland PB, Gor JP (2012) Simulations of normal and inverse laminar diffusion flames under oxygen enhancement and gravity variation, Combut Theory Model 16(5):774–798
23. Siebers D, Higgins B, Pickett L (2002) Flame lift-off on direct injection diesel fuel jets: Oxygen concentration effects, SAE Transactions. J Engines 111(3):1490–1509
24. Chen R, Axelbaum RL (2005) Scalar dissipation rate at extinction and the effect of oxygen-enriched combustion. Combust Flame 142:62–71
25. Zhou J, Ai Y, Kong W (2013) The liftoff properties of dimethyl ether jet diffusion flames with preheating, Proc. of ASME Turbo Expo 2013: Turbine Technical Conference and Exposition, 1B, 2013 V01BT04A020
26. Xue Y, Ju Y (2006) Studies on the liftoff properties of dimethyl ether jet diffusion flames. Combust Sci Tech 178(12):2219–2247
27. Al-Noman SM, Chul CB, Chung SH (2018) Autoignited lifted flames of dimethyl ether in heated coflow air. Combust Flame 195:75–83
28. Kalvakala KC, Aggarwal SK (2019) Polycyclic aromatic hydrocarbons and soot emissions in oxygenated ethylene diffusion flames at elevated pressures. J Eng Gas Turb Power 141:Article 071022
29. Kumfer BM, Skeen SA, Chen R, Axelbaum RL (2006) Measurement and analysis of soot inception limits of oxygen enriched co-flow flames. Combust Flame 147:233–242
30. Kumar R, Katta VR, Aggarwal SK (2018) On the effects of oxygen-enrichment on flame liftoff, stabilization and blowout in ethylene flames. In: 2018 AIAA aerospace sciences meeting, AIAA-2018–1478
31. Katta VR, Goss LP, Roquemore WM (1994) Effect of nonunity Lewis number and finite rate chemistry on the dynamics of a hydrogen air jet diffusion flame. Combust Flame 96:60–74
32. Roquemore WM, Katta VR (2000) Role of flow visualization in the development of UNICORN. J Visualization 2:257–272
33. Leonard BP (1979) A stable and accurate convective modeling procedure based on quadratic upstream interpolation. Comput Methods Appl Mech Engg 19:59–98
34. Shu Z, Aggarwal SK, Puri IK, Katta VR (1997) A numerical investigation of the flame structure of an unsteady inverse partially premixed flame. Combust Flame 111:295–311
35. Shu Z, Aggarwal SK, Puri IK, Katta VR (1997) Flame-Vortex Dynamics In An Inverse Partially Premixed Combustor; The Froude Number Effects. Combust Flame 111:276–295
36. Qin X, Puri IK, Aggarwal SK, Katta VR (2004) Gravity, radiation, and coflow effects on partially premixed flames. Phys Fluids 16:2963–2974
37. Siegel R, Howell JR (1981) Thermal radiation heat transfer. Hemisphere Publishing Corporation, New York
38. Wang H, You X, Joshi AV, Davis SG, Laskin A, Egolfopoulos F, Law CK (2007) USC Mech. version II. High-temperature combustion reaction model of H_2/CO/C_1-C_4 compounds. http://ignis.usc.edu/USC_Mech_II.htm

39. Katta VR, Meyer TR, Brown MS, Gord JR, Roquemore WM (2004) Extinction criterion for unsteady, opposing-jet diffusion flames. Combust Flame 137:198–221
40. Shu Z, Krass BJ, Choi CW, Aggarwal SK, Katta VR, Puri IK (1998) An Experimental and numerical investigation of the structure of steady two-dimensional partially premixed methane-air flames. Proc Combust Inst 27:625–632
41. Katta VR, Takahashi F, Linteris GT (2004) Suppression of cup-burner flames using carbon dioxide in microgravity. Combust Flame 137:506–522
42. Smooke MD, Hall RJ, Colket MB, Fielding J, Long MB, McEnally CS, Pfefferle LD (2004) Investigation of the transition from lightly sooting towards heavily sooting co-flow ethylene diffusion flames. Combust. Theory Model. 8:593–606
43. Bilger RW (1989) The structure of turbulent nonpremixed flames. Proc Combust Inst 22:475–488
44. Takahashi F, Katta VR (2009) A reaction kernel hypothesis for the stability limit of methane jet diffusion flames. Proc Combust Inst 28:2071–2078

Laminar Burning Velocity Measurements at Elevated Pressure and Temperatures and the Challenges in Kinetic Scheme Optimization

Robin John Varghese and Sudarshan Kumar

Abstract Reliable data on fundamental combustion parameters are essential to validate kinetic schemes under a wide range of experimental conditions. Laminar burning velocity is one such fundamental intrinsic property of a flammable mixture. The challenges in determining the effect of simultaneous change in temperature and pressure on the burning velocity of low-calorific value syngas–air mixtures are presented. Comparing the present experiments with available kinetic schemes at elevated temperature and pressure reveals poor prediction capabilities. The major contributing factor to this discrepancy is the high sensitivity of key reactions at high temperatures and pressure conditions. The challenges of direct use of the experimental data to optimize kinetic schemes for the syngas–air mixtures are discussed.

Keywords Burning velocity · Optimization · Elevated temperatures and pressures · Diverging channel

Nomenclature

α	Temperature exponent
α_t	Thermal diffusivity
β	Pressure exponent
δ	Flame thickness
ϕ	Equivalence ratio
ρ	Mixture density
S_u	Laminar burning velocity
χ_{H_2}	Hydrogen content

R. J. Varghese (✉) · S. Kumar
Department of Aerospace Engineering, Indian Institute of Technology, Bombay, India
e-mail: robinjohn@iitb.ac.in

A. Gupta et al. (eds.), *Advances in Energy and Combustion*, Green Energy and Technology,
https://doi.org/10.1007/978-981-16-2648-7_13

Abbreviations

FFCM	Foundational fuel chemistry model
GRI	Gas research institute
IGCC	Integrated gasification and combined cycle
LBV	Laminar burning velocity
LCV	Low-calorific value

1 Introduction

The use of fossil fuels has been on the rise since the industrial revolution. Preservation of resources demands efficient combustion of fossil fuels, and the protection of the environment stresses a restraint on the pollutant emissions. Environment (shrinking oil reserves) and ecology (climate change, air quality) concern over the past decades has improved awareness to shift the dependence on these fuels to alternative sources for various power generation needs and transportation. However, the alternative energy source should primarily (a) decrease fossil energy use, (b) reduce pollutant emissions, (c) sustainable, (d) scalable. Synthesis gas derived from coal or biomass is considered a viable alternative for stationary power generation, and biomass gasification results in a mixture of H_2 and CO gas with a significant quantity of inert diluents. The gasification product is characterized by large compositional variability and low-calorific value (LCV). The primary combustion parameters like burning velocity and ignition delay of these LCV syngas mixtures are not available in the literature at high temperature and pressure conditions. However, they are essential for the integrated gasification and combined cycle (IGCC) power plants to efficiently use the resources.

Laminar-burning velocity (LBV) is a fundamental combustion entity that characterizes combustible mixtures' overall reactivity. LBV impacts the stabilization of flames which directly influences the limits of flashback and helps validate kinetic schemes. LBV is also one of the critical parameters that model the combustion of fuels in the spark-ignition engines.

Various stationary and dynamic flame methods are available for the determination of LBV. LBV data on low-calorific value syngas–air mixtures are limited, and experimental research on simultaneous variation of temperature and pressure for LBV determination is non-existent. Most syngas–air mixtures' investigations focus on the effect of single diluents [1–4] or different concentrations of carbon monoxide [5]. LBV is available for the syngas–air system with a considerable variation in H_2/CO ratios at different equivalence ratios for various N_2/ CO_2 dilution [2, 3, 6–9]. Yan et al. [10] reported LBV of three variants of syngas–air mixtures derived from a gasification plant using the heat-flux method at 298–358 K.

Shang et al. [11] investigated the effect of inert diluents (CO_2/N_2) on LBV of producer gas mixtures using the Bunsen flame method. The effect of N_2 was determined to be milder in comparison with the same volume of CO_2. Similarly, Natarajan et al. [2] reported an LBV of syngas mixtures at various temperatures and pressures. The performance of available reaction mechanisms at high-temperature conditions (>450 K) was comparatively inaccurate. Han et al. [12] reported LBV of syngas mixtures at elevated pressures and temperatures for different hydrogen content (χ_{H_2} = 0.25, 0.5, 0.75), with spherically expanding flames. Natarajan et al. [2] and Han et al. [12] report the only available literature for LBV of syngas–air mixtures, where both the change in temperature and pressure were considered individually. The effect of the simultaneous change in the mixture temperature and pressure on the laminar burning velocities of syngas–air mixtures was not considered [2, 12]. This chapter presents LBV for a simultaneous variation of temperature and pressure for syngas–air mixtures and the significance of this consideration for optimizing reaction chemistry.

The synthesis gas compositions considered in the present study are similar to the output of the biomass gasifiers and are described by large quantity (60–80%) of diluents (CO_2 and N_2) [13], and the remaining volume (40–20%) constitutes of combustible gases. Productive use of these compositions in IGCC combustors merits an accurate knowledge of LBV at high temperature and pressure conditions. The limited research available on LBV of syngas–air mixtures does not account for the compositional variability. The validation and optimization of kinetic schemes based on LBV are limited at high temperatures and non-existent at elevated pressures for the present study's compositions.

The experimental method presented in the chapter is the preheated diverging channel method. The in-house developed and perfected method in the *Combustion Laboratory* of the Indian Institute of Technology Bombay has undergone considerable improvements. The most recent was the extension of the method to include a simultaneous change in the thermodynamic conditions (Pressure and Temperature), and the facility was validated using the LBV measurements of methane–air mixtures. In this chapter, LBV of syngas–air mixtures at elevated temperatures and pressures are presented, along with a modified power–law correlation for representing LBV at high temperatures and pressures. The composition of the syngas–air mixture presented in the chapter is referred to as SG from hereon, is H_2-15%, CO-15%, CO_2-15%, N_2-55%. The kinetic scheme optimization challenges for the LCV syngas mixtures with LBV at high temperature and pressures as the target variable for genetic algorithm (GA)-based optimization is also discussed briefly.

2 Methods

2.1 Experimental Method

The experimental technique employed for the measurement of LBV in the present work is the preheated diverging channel. The experimental section consists of a quartz diverging channel of rectangular cross section, an infrared heater, and a gas supply system controlled using a computer. The diverging quartz channel stabilizes planar flames at elevated temperatures, and the infrared heater placed inside the cylindrical pressure vessel heats the diverging channel to achieve a positive temperature gradient along the flow direction. The uniform distribution of temperature and velocity profiles along the transverse direction at various axial positions are presented in [5]. The infrared heater also aids in compensating the heat loss from the flame to the channel walls [5].

Figure 1 shows the relevant details of the diverging channel experimental section along with the cylindrical pressure vessel. The pressure vessel is initially filled at a specified pressure, and the relief valve mechanism used for regulating the pressure inside the chamber is accurate up to ± 0.1 atm. The infrared heater is placed below the rectangular quartz channel and is switched on at the required heating rate. The temperature inside the channel at various axial locations are measured using a k-type thermocouple for various heating conditions. The fuel–air mixture is fed into the channel at a specified inlet velocity and ignited at the channel's exit using an ignition device. The flame gradually moves upstream and stabilizes where the local flow velocity is equal to the laminar burning velocity. LBV is determined from the flame area and unburnt gas temperature at the flame stabilization location using a rearranged mass conservation equation. The Peclet number and Reynolds number of the flow inside the channel are low, and hence, the assumption that the unburnt gas temperature is the same as the temperature of the quartz wall is reasonable [14, 15].

2.2 Computational Method

LBV and sensitivity of various pre-exponential factors of different elementary reactions were determined using the freely propagating flame code (PREMIX). The conservation equations and the calculation of species and temperature profiles of PREMIX are discussed in detail in [16]. LBV prediction capabilities of Aramco 3 (581 species and 3037 reactions) [17–19], Davis mechanism (14 species and 28 reactions) [20] GRI-Mech 3.0 (53 species and 325 reactions) [21], and foundational fuel chemistry model (FFCM-1) [22] (38 species and 291 reactions) were analyzed initially. Comparing the four kinetic schemes' predictions with the present LBV measurements of syngas–air mixtures revealed FFCM-1 to be in closer agreement until elevated temperatures of 500 K and 1 atm. This observation is consistent

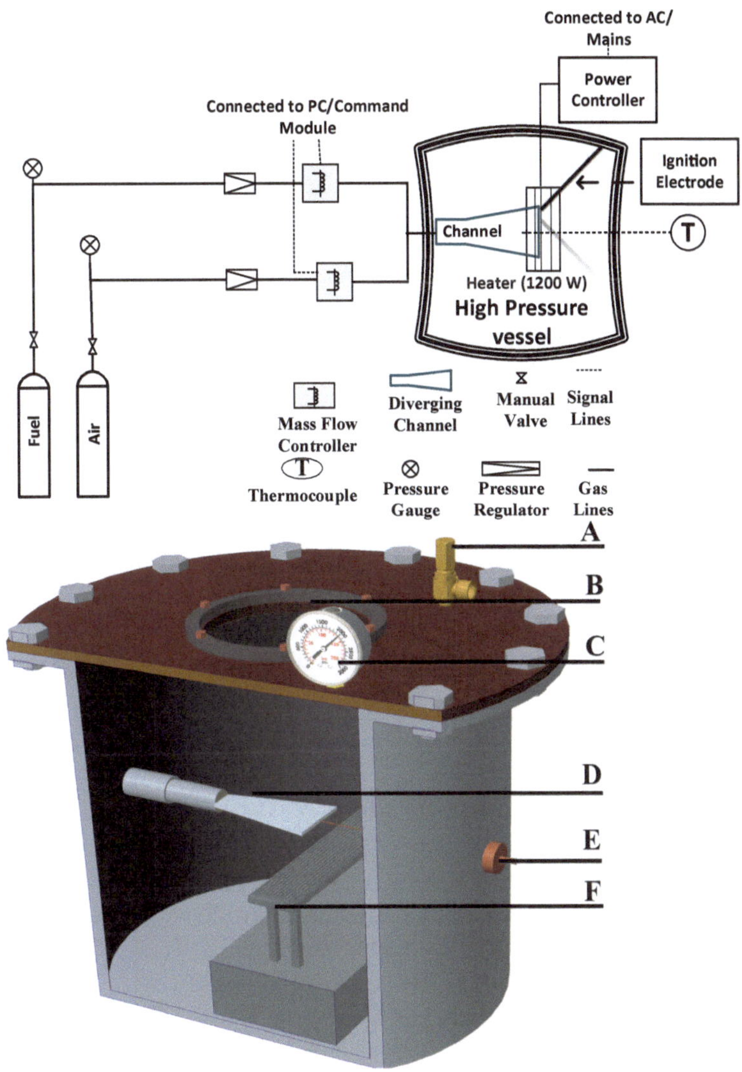

Fig. 1 **a** Schematic of the experimental facility **b** High-pressure vessel 3D model—Sectional view, A-Pressure relief valve, B-Glass window, C-Pressure gauge, D-Diverging channel, E-Thermocouple, F-Infrared heater

with the conclusions of Natarajan et al. [2] for GRI-Mech 3.0 and Davis mechanisms. However, for a simultaneous variation of temperature and pressure, the LBV prediction competency of various kinetic schemes deteriorate. Therefore, a genetic algorithm-based optimization for the best-performing kinetic scheme (FFCM-1) was attempted. The target variables for the optimization and various constraints are discussed in detail in a latter section. The base model for optimization consists of 38 species and 310 reactions. 19 reactions from Aramco 3 were merged to the FFCM-1 model. The Arrhenius parameters for four reactions were specified as logarithmic functions of pressure, and the rate constants were linearly interpolated with respect to the logarithm of pressure in between the specified pressures. This merged kinetic scheme was used as the base model for optimization.

3 Results and Discussion

The SG–air mixture's laminar burning velocities were measured using the diverging channel technique at elevated temperatures (300–650 K) and pressures (1–5 atm). The unique composition of the mixture and unavailability of literature at the thermodynamic conditions presented makes direct comparison challenging. Therefore, the comparison with the best performing kinetic scheme (FFCM-1) and optimized reaction scheme is presented in the following sections. However, the measured LBV for various fuels such as CH_4 [23], LPG [24], C_3H_8 [24], $H_2/CO/CO_2/N_2$ [5, 25–27], C1–C5 alcohols [28, 29], DME [30], and methyl formate [31] using the diverging channel technique were reported to be accurate at various conditions in comparison with other methods. The method's extension to elevated pressures and subsequent validation of the technique using LBV of CH_4-air mixtures were reported recently [32].

3.1 *LBV at High Temperature and Pressure*

The diverging channel facility for the measurement of LBV at elevated temperatures and pressures was used for the syngas–air mixture (SG). Figure 2 shows the measured LBV of stoichiometric SG–air at different temperatures (T_u = 300–650 K) and mixture pressures (P_u = 1, 2, 3, 5 atm).

The measured LBV at various temperatures increases with an increase in the mixture enthalpy at a specified pressure. The increase in LBV with an increase in temperature at various pressures is similar. For instance, the comparisons at T_u = 450, 600 K indicates LBV is doubled for both P_u = 1, 5 atm. However, the extrapolated values of $S_{u,o}$, and power exponents of temperature (α) are distinct. $S_{u,o}$ decreases with an increase in pressure, whereas temperature exponents (α) increase slightly. The decrease in LBV with an increase in pressure may be attributed to the increased density, higher third-body reaction effects, and the reduction in thermal diffusivity.

LBV predictions using the FFCM-1 base model and optimized model are also compared with the measured values in Fig. 2. LBV predictions deviate from the measured values at high temperature and pressure conditions for the base model. The discrepancy in the predictions using the optimized model reduced considerably. The uncertainty in the reaction rate parameters is significantly high, and optimization

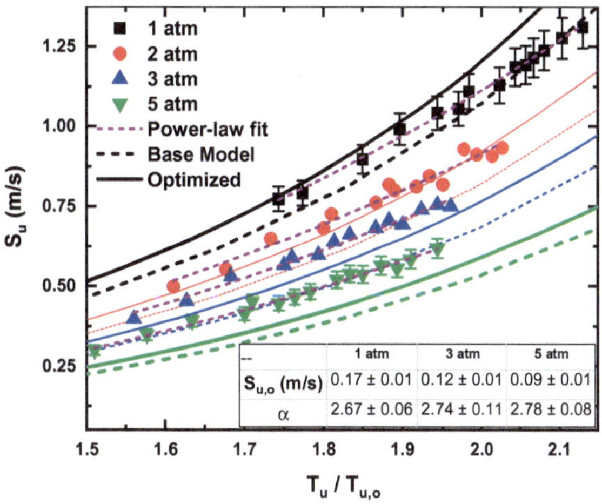

--	1 atm	3 atm	5 atm
$S_{u,o}$ (m/s)	0.17 ± 0.01	0.12 ± 0.01	0.09 ± 0.01
α	2.67 ± 0.06	2.74 ± 0.11	2.78 ± 0.08

Fig. 2 LBV of stoichiometric SG–air mixtures at elevated temperatures and pressures

would be substantially improved if quantum chemistry calculations are considered to develop a base model for the reaction mechanism.

3.2 *Optimization Approach*

A genetic algorithm-based approach to optimize the rate parameters of the mechanism was employed. The target data for optimization was the measured burning velocities of syngas–air mixtures at different equivalence ratios and various temperature and pressure conditions. The overall relative squared error describes the objective function for minimization between the LBV obtained from experiments and simulation. A complete set of the target parameters is provided in the supplementary material. The optimization algorithm's objective is to match the target data with the freely propagating flame speed predictions using each modified rate parameter. The parameters' uncertainty range was defined as the pre-exponential factor (A) (ratio) 1.5, temperature dependence factor, $n = \pm 0.2$, and activation energy Ea $= \pm 21$ kJ/mol. The RMS error of the target data and flame speed predictions using the back-end freely propagating flame code is initially calculated.

The procedure of optimization can be summarized as (1) 15 datasets are generated randomly considering the specified uncertainties of reaction rate parameters as the initial population; (2) The freely propagating flame code is used to calculate the flame speeds using the initial mechanisms; (3) The RMS error of the function evaluation and target data is defined as the objective function and is calculated from (2, 3). (4) The convergence distance and percentage (0.1, 80%) of the objective function are evaluated; (5) New population is generated, and mutation may occur during this step, and flame speeds are calculated for the new population. (6) The objective function is re-evaluated, the populations are merged, and best-fit mechanisms are selected as the

Table 1 Optimized reaction rate constants (Units: cal/mol)

Reaction number	Reaction	A	n	Ea
R4	$H_2 + OH = H + H_2O$	4.4E+13	0	6990
R272	$CH_3 + H(+M) = CH_4(+M)$	1.3E+16	−0.6	383
R280	$CH_3 + OH = CH_2(S) + H_2O$	4.9E+14	−0.7	−446
R281	$CH_3 + OH = CH_2O + H_2$	3.5E+05	1.4	−3244
R282	$CH_3 + OH = CH_2OH + H$	1.6E+10	1.0	3210
R283	$CH_3 + OH = H + CH_3O$	1.2E+09	1.0	11,940
R284	$CH_3OH(+M) = CH_3 + OH(+M)$	2.1E+18	−0.6	92,541
R285	$CH_3OH(+M) = CH_2OH + H(+M)$	7.9E−03	5.0	84,467
R301	$CO + H_2(+M) = CH_2O(+M)$	4.3E+07	1.5	79,600

next generation. (7) The steps are repeated until the predefined maximum generation (2000) is reached. Table 1 shows the optimized rate constants of sensitive reactions.

3.3 Variation of Power Exponents of Temperature and Pressure

The conventional power–law correlation available for laminar burning velocities can be expressed as:

$$S_u = S_{u,o}\left(\frac{T_u}{T_{u,o}}\right)^{\alpha}\left(\frac{P_u}{P_{u,o}}\right)^{\beta} \quad (1)$$

where the exponents of temperature and pressure are considered to be constants.

Figure 3a shows the power exponents of temperature determined from the measured burning velocities at various pressures, along with a comparison with a linear curve fit. The comparisons with the available kinetic schemes show a similar increase in the power exponents with increased pressure. However, the values are slightly underpredicted for both the optimized and base FFCM-1 models. This difference in the predictions after optimization indicates the need for quantum chemistry calculations at high temperature–pressure conditions and expressing the reaction factors in terms of the pressure range. The temperature exponents were considered constant and independent of temperature range and the pressure they were measured. Varghese et al. [32] proposed a modified power–law correlation representing temperature exponents' variation with a change in the mixture pressure. Here, in syngas–air mixtures, a similar power–law correlation is reasonable for an accurate representation of temperature exponents at various pressures.

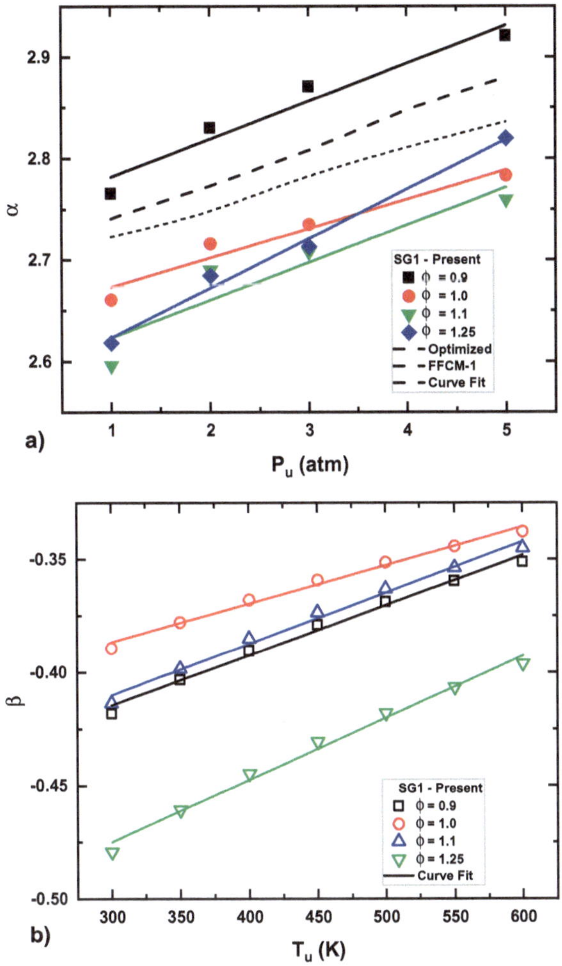

Fig. 3 **a** Power exponents of temperature (α) as a function of pressure and **b** pressure exponents (β) at different temperatures

Similarly, the comparison of power exponents of pressure determined from the measured LBV shows an increasing trend. The predictions of both the kinetic schemes significantly underpredict the pressure exponents. Perhaps measurements at higher pressures (5–20 atm) would be necessary to accurately calculate the pressure exponents' variation.

Figure 3 shows the temperature and pressure exponents at various pressures and temperatures. Power exponents (α) increase linearly with an increase in pressure, and pressure exponent (β) increases with temperature. This clearly shows that the assumption of constant exponents is inaccurate when considering the simultaneous variation of thermodynamic conditions. Therefore, the pressure exponents are expressed in terms of temperature ratios and, similarly, temperature exponents as a linear function of pressure.

$$\alpha = \alpha_o + \alpha_1\left(1 - \frac{P_u}{P_{u,o}}\right); \beta = \beta_o + \beta_1\left(1 - \frac{T_u}{T_{u,o}}\right) \tag{2}$$

The conventional power–law correlation is modified considering the variations of power exponents of temperature and pressure (Eq. 2), and Eq. 1 can be modified as:

$$S_u = S_{u,o}\left(\frac{T_u}{T_{u,o}}\right)^{\alpha_o+\alpha_1\left(1-\frac{P_u}{P_{u,o}}\right)}\left(\frac{P_u}{P_{u,o}}\right)^{\beta_o+\beta_1\left(1-\frac{T_u}{T_{u,o}}\right)} \tag{3}$$

3.4 Comparison of Optimized Model with Available Kinetic Schemes

The diverging channel was used to measure LBV of syngas–air mixtures at elevated temperatures (300–650 K) and pressures (1–5 atm). The LBV determined from the experiments were used to evaluate the performance of available kinetic schemes and the optimized reaction mechanism.

Figure 4 shows the predictions of five different kinetic schemes (GRI-Mech3.0, Aramco3.0, FFCM-1 base model and optimized model) at $T_u = 450$ K and elevated pressures ($P_u = 1 - 5$ atm).

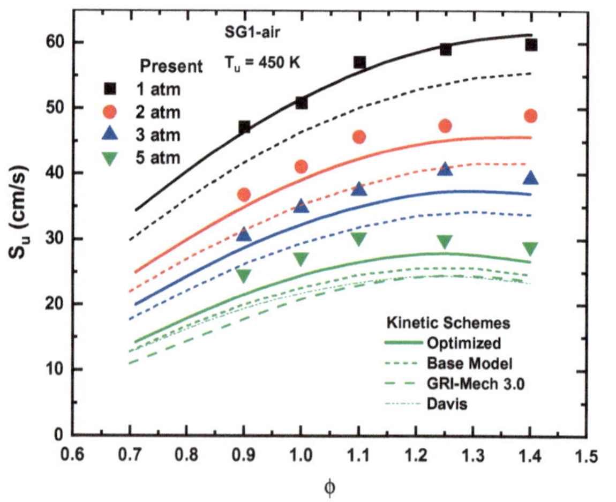

Fig. 4 Comparison of measured burning velocities at $T_u = 450$ K and elevated pressures with various kinetic schemes

LBV decreases with an increase in pressure at $T_u = 450$ K. The mixture density increases with an increase in pressure, reducing the mean free path for molecular collisions and leading to enhanced third-body recombination reactions. The oxidation rate also depends on the change in thermal diffusivity at higher pressures and contributes to the slower flame propagation at elevated pressures. The mixture pressure increase indicates a peak shift of LBV from rich mixtures to near stoichiometric mixtures. For instance, at $P_u = 1$ atm, the peak burning velocity is present for very rich mixtures ($\phi = 1.4$), whereas, at 5 atm, the peak burning velocity occurs for stoichiometric mixtures. The flame structure of stoichiometric mixtures at two different pressures ($P_u = 1, 5$ atm) is compared to provide further insight into the peak shifting of syngas–air mixtures at elevated pressures.

The prediction accuracy of various kinetic schemes has improved from subsequent optimizations. FFCM-1 based model may be considered the best-performing reaction mechanism available for methane–air mixtures when a simultaneous variation of temperature and pressure are considered. Optimization of the base FFCM-1 model has considerably improved the LBV predictions at elevated pressures and temperatures.

3.5 Flame Structure

The PREMIX calculations for freely propagating the SG–air mixture's flames at $T_u = 450$ K, and $P_u = 1, 5$ atm for various equivalence ratios were carried out. Figure 5 shows the comparison of the minor species' mole fractions for different equivalence ratios at 1 atm and 5 atm. For a direct comparison of different equivalence ratios, the abscissa was defined as dimensionless temperature (τ). $\tau = (T - T_u)/(T_{\mathrm{ad}} - T_u)$, where T_u is the unburnt gas temperature and T_{ad} is the adiabatic flame temperature. The mole fractions of H radicals increase with an increase in equivalence

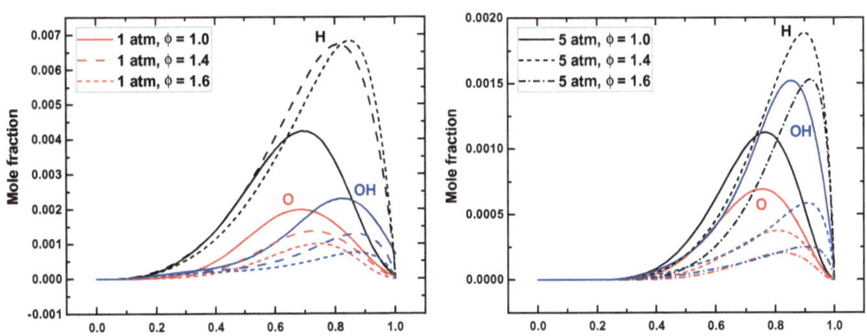

Fig. 5 Flame structure comparison of syngas–air mixtures at $T_u = 450$ K and different pressures (1, 5 atm)

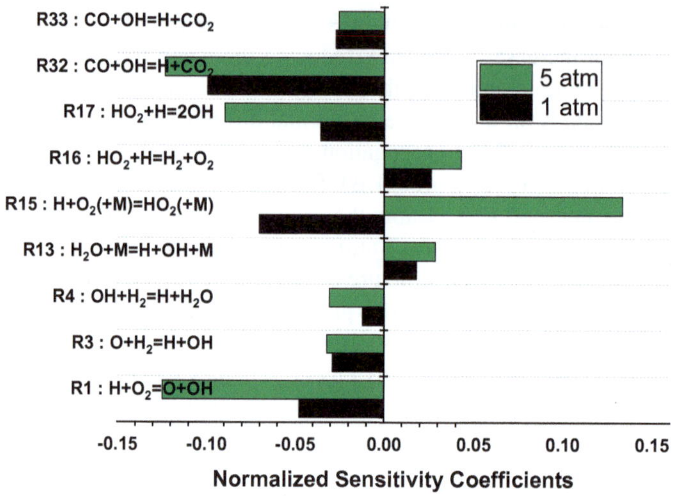

Fig. 6 Normalized sensitivity coefficients of SG–air mixtures at $T_u = 450$ K and different pressures (1, 5 atm)

ratio from stoichiometric until $\phi = 1.4$ and then decreases for $P_u = 1$ atm and $T_u = 450$ K. The values of O and OH radicals decrease continuously as the fuel–air mixture transitions to the rich side. This continuous reduction in mole fractions of O and OH mole fractions is followed for 5 atm case. However, the flame thickness reduces for higher pressures, which can be identified by the peak values closer to dimensionless temperature $\tau = 1$. Turns [33] define flame thickness ($\delta = 2\alpha_t/S_u$) for binary diffusion, unity Lewis number, and linear temperature profile. LBV can be considered to be indirectly proportional to the flame thickness in the case of multi-component diffusion and non-unity Lewis number.

Sensitivity coefficients for gas temperature with respect to the reaction A-factors were calculated for stoichiometric SG–air mixtures at $T_u = 450$ K and $P_u = 1, 5$ atm.

Figure 6 shows the normalized sensitivity coefficients for gas temperature with respect to the reaction pre-exponential factors for stoichiometric SG–air mixtures at an unburnt gas temperature of 450 K and different mixture pressures (1, 5 atm). The sensitive reactions for both the pressures for stoichiometric mixtures remain the same. However, the sensitivity coefficients of the high-pressure mixture are enhanced. The third-body reaction (R15: $H + O_2(+M) = HO_2(+M)$) has positive sensitivity for Pu = 5 atm, whereas the 1 atm case indicates a negative sensitivity. The third-body reactions' enhanced sensitivity is expected at denser mixture conditions due to the enhanced third-body collisions. The amplified negative sensitivity of the reactions at elevated pressures, and the shift in third-body collisions' sensitivity indicates a decrease in burning velocities at elevated pressures may be favored.

4 Conclusions

LBV of practical syngas compositions, which are an output of biomass gasifiers, was measured for various equivalence ratios at elevated temperatures and pressures. The power exponents of temperature and pressure for the measured burning velocities were determined. The temperature exponents were demonstrated to be a strong function of the temperature range at which they are calculated and the mixture pressure. The temperature and pressure exponents' analysis indicated that considering the power exponents to be constant leads to inaccurate extrapolations. The power exponents of temperature and pressure are expressed as pressure and temperature ratios. The dependences were suggested individually earlier by Babkin and Kozachenko [33], Han et al. [34] and a combined modified power–law correlation was proposed by Varghese et al. [32] for methane–air mixtures. The modified power–law correlation and the variation of power exponents of temperature and pressure proposed are accurate for syngas–air mixtures.

The best-performing kinetic scheme (FFCM-1) for the syngas mixtures at ambient conditions performed poorly at high pressure–temperature conditions. The major contributing factor to this discrepancy is the high sensitivity of key reactions at high temperatures and pressure conditions. Genetic algorithm-based optimization of the base model with the current study's experimental results as the target data was attempted. A base kinetic scheme was constructed with FFCM-1 (38 species and 291 reactions) and supplemented by 23 reactions from Aramco 3. The pressure-dependent Arrhenius factors for rate constants were incorporated in the base model. The prediction capabilities for burning velocities at high temperature and pressure conditions have significantly improved with the optimization. The validation datasets of the modified mechanism at the syngas–air mixtures are promising.

Acknowledgements Financial assistance by the Government of India in the form of a fellowship for Senior Research Fellow (SRF) is sincerely acknowledged.

Appendix: Experimental Data

The following tables provide the experimental results obtained using the diverging channel method for various compositions. Table 2 shows the detailed composition of the syngas considered. Tables 3, and 4 show the laminar burning velocity and temperature exponents obtained from the experiments, respectively. Table 5 shows the target parameters used for the optimization.

Table 2 Detailed composition of the syngas used in the present work

Sr. no	% Volume				H_2/CO ratio
	CO	CO_2	H_2	N_2	
SG1	15	15	15	55	1
SG2	20	15	15	50	0.75
SG3	25	15	15	45	0.6

Table 3 Laminar burning velocity of SG1-SG3 with uncertainties at various equivalence ratio at ambient conditions (300 K, 1 atm)

ϕ	SG1	ΔS_u	SG2	ΔS_u	SG3	ΔS_u
Unit	cm/s	cm/s	cm/s	cm/s	cm/s	cm/s
0.71	10.21	±0.49	12.50	±0.58	15.97	±0.80
0.90	16.14	±0.77	20.00	±0.99	23.67	±1.17
1.00	18.23	±0.91	23.00	±1.14	26.10	±1.26
1.10	20.65	±1.00	24.20	±1.45	28.95	±1.45
1.25	21.62	±1.03	26.27	±1.34	30.59	±1.53

Table 4 Temperature exponent (α) of different compositions at various equivalence ratios (ϕ) and $P_u = 1$ atm

ϕ	SG1	$\Delta\alpha$	SG2	$\Delta\alpha$	SG3	$\Delta\alpha$
0.70	3.06	±0.09	2.62	±0.10	2.50	±0.05
0.90	2.77	±0.08	2.56	±0.09	2.37	±0.05
1.00	2.67	±0.08	2.51	±0.09	2.35	±0.05
1.10	2.60	±0.08	2.46	±0.09	2.33	±0.05
1.25	2.63	±0.07	2.43	±0.09	2.30	±0.05

Table 5 Target data for optimization

(ϕ)	P_u(atm)	T_u(K)	LBV (cm/s)	P_u(atm)	LBV (cm/s)	(ϕ)	P_u(atm)	T_u(K)	LBV (cm/s)	P_u(atm)	LBV (cm/s)
0.7	1	300	9.7	3	4.5	1.3	2	300	16.8	5	9.3
0.9	1	300	15.5	3	10.0	1.4	2	300	16.4	5	8.6
1.0	1	300	17.4	3	12.0	0.7	2	450	24.6	5	10.5
1.1	1	300	20.1	3	13.0	0.9	2	450	39.4	5	26.1
1.3	1	300	20.5	3	14.0	1.0	2	450	43.7	5	28.7
1.4	1	300	20.2	3	12.5	1.1	2	450	48.2	5	31.9
0.7	1	450	33.6	3	16.1	1.3	2	450	49.8	5	29.2
0.9	1	450	47.8	3	32.0	1.4	2	450	51.5	5	30.4
1.0	1	450	51.4	3	36.5	0.7	2	600	60.0	5	26.6
1.1	1	450	57.6	3	39.0	0.9	2	600	88.9	5	60.6
1.3	1	450	59.6	3	42.0	1.0	2	600	95.5	5	63.9
1.4	1	450	60.4	3	40.8	1.1	2	600	104.5	5	70.5
0.7	1	600	81.0	3	39.9	1.3	2	600	107.7	5	65.7
0.9	1	600	106.0	3	73.1	1.4	2	600	115.8	5	74.3
1.0	1	600	110.7	3	80.2						
1.1	1	600	121.6	3	85.1						
1.3	1	600	127.0	3	91.6						
1.4	1	600	131.3	3	94.6						
0.7	2	300	7.0	5	2.8						
0.9	2	300	12.5	5	8.0						
1.0	2	300	14.5	5	9.3						
1.1	2	300	16.2	5	10.4						

References

1. Shih H-Y (2009) Computed extinction limits and flame structures of H_2/O_2 counterflow diffusion flames with CO_2 dilution. Int J Hydrogen Energy 34(9):4005–4013
2. Natarajan J, Lieuwen T, Seitzman J (2007) Laminar flame speeds of H_2/CO mixtures: Effect of CO_2 dilution, preheat temperature, and pressure. Combust Flame 151(1–2):104–119
3. Burke MP, Dryer FL, Ju Y (2011) Assessment of kinetic modeling for lean $H_2/CH_4/O_2$/diluent flames at high pressures. Proc Combust Inst 33(1):905–912
4. Wang J, Huang Z, Kobayashi H, Ogami Y (2012) Laminar burning velocities and flame characteristics of CO–H_2–CO_2–O_2 mixtures. Int J Hydrogen Energy 37(24):19158–19167
5. Varghese RJ, Kolekar H, Hariharan V, Kumar S (2018) Effect of CO content on laminar burning velocities of syngas-air premixed flames at elevated temperatures. Fuel 214:144–153
6. Bouvet N, Chauveau C, Gökalp I, Halter F (2011) Experimental studies of the fundamental flame speeds of syngas (H_2/CO)/air mixtures. Proc Combust Inst 33(1):913–920
7. Burbano HJ, Pareja J, Amell AA (2011) Laminar burning velocities and flame stability analysis of H_2/CO/air mixtures with dilution of N_2 and CO_2. Int J Hydrogen Energy 36(4):3232–3242
8. Lee JH, Kim JH, Park JH, Kwon OC (2010) Studies on properties of laminar premixed hydrogen-added ammonia/air flames for hydrogen production. Int J Hydrogen Energy 35(3):1054–1064
9. Prathap C, Ray A, Ravi MR (2012) Effects of dilution with carbon dioxide on the laminar burning velocity and flame stability of H_2–CO mixtures at atmospheric condition. Combust Flame 159(2):482–492
10. Yan B, Wu Y, Liu C, Yu JF, Li B, Li ZS et al (2011) Experimental and modeling study of laminar burning velocity of biomass derived gases/air mixtures. Int J Hydrogen Energy 36(5):3769–3777
11. Shang R, Zhang Y, Zhu M, Zhang Z, Zhang D, Li G (2016) Laminar flame speed of CO_2 and N_2 diluted H_2/CO/air flames. Int J Hydrogen Energy 41:15056–15067
12. Han M, Ai Y, Chen Z, Kong W (2015) Laminar flame speeds of H_2/CO with CO_2 dilution at normal and elevated pressures and temperatures. Fuel 148:32–38
13. He Y, Wang Z, Weng W, Zhu Y, Zhou J, Cen K (2014) Effects of CO content on laminar burning velocity of typical syngas by heat flux method and kinetic modeling. Int J Hydrogen Energy 39(17):9534–9544
14. Maruta K, Kataoka T, Kim NI, Minaev S, Fursenko R (2005) Characteristics of combustion in a narrow channel with a temperature gradient. Proc Combust Inst 30(2):2429–2436
15. Kumar S, Maruta K, Minaev S, Fursenko R (2008) Appearance of target pattern and spiral flames in radial microchannels with CH_4-air mixtures. Physics of Fluids. 2008;20.
16. Smooke DM, Miller JA, Kee RJ (1983) Determination of adiabatic flame speeds by boundary value methods. Combust Sci Technol 34:79–90
17. Zhou C-W, Li Y, O'Connor E, Somers KP, Thion S, Keesee C et al (2016) A comprehensive experimental and modeling study of isobutene oxidation. Combust Flame 167:353–379
18. Metcalfe WK, Burke SM, Ahmed SS, Curran HJ (2013) A hierarchical and comparative kinetic modeling study of C1–C2 hydrocarbon and oxygenated fuels. Int J Chem Kinet 45(10):638–675
19. Li Y, Zhou C-W, Somers KP, Zhang K, Curran HJ (2017) The oxidation of 2-butene: a high pressure ignition delay, kinetic modeling study and reactivity comparison with isobutene and 1-butene. Proc Combust Inst 36(1):403–411
20. Davis SG, Joshi AV, Wang H, Egolfopoulos F (2005) An optimized kinetic model of H_2/CO combustion. Proc Combust Inst 30:1283–1291
21. Smith GP, Golden DM, Frenklach M, Moriarty NW, Eiteneer B, Goldenberg M et al (1999) GRI-Mech 3.0. p. http://www.me.berkeley.edu/gri_mech.
22. Smith G, Tao Y, Wang H (2016) Foundational fuel chemistry model version 1.0 (FFCM-1). https://web.stanford.edu/group/haiwanglab/FFCM1
23. Akram M, Saxena P, Kumar S (2013) Laminar burning velocity of methane–air mixtures at elevated temperatures. Energy Fuels 27(6):3460–3466

24. Akram M, Kumar S (2012) Measurement of laminar burning velocity of liquified petroleum gas air mixtures at elevated temperatures. Energy Fuels 26(6):3267–3274
25. Varghese RJ, Kolekar H, Kumar S (2019) Laminar burning velocities of $H_2/CO/CH_4/CO_2/N_2$ -air mixtures at elevated temperatures. Int J Hydrogen Energy 44(23):12188–12199
26. Varghese RJ, Kolekar H, Kumar S (2019) Demarcation of reaction effects on laminar burning velocities of diluted syngas–air mixtures at elevated temperatures. Int J Chem Kinet 51(2):95–104
27. Varghese RJ, Kumar S (2020) Machine learning model to predict the laminar burning velocities of $H_2/CO/CH_4/CO_2/N_2$/air mixtures at high pressure and temperature conditions. Int J Hydrogen Energy 45(4):3216–3232
28. Katoch A, Asad M, Minaev S, Kumar S (2016) Measurement of laminar burning velocities of methanol–air mixtures at elevated temperatures. Fuel 182:57–63
29. Katoch A, Chauhan A, Kumar S (2018) Laminar burning velocity of n-propanol and air mixtures at elevated mixture temperatures. Energy Fuels 32(5):6363–6370
30. Varghese RJ, Kishore VR, Akram M, Yoon Y, Kumar S (2017) Burning velocities of DME(dimethyl ether)-air premixed flames at elevated temperatures. Energy 126
31. Kumar R, Katoch A, Singhal A, Kumar S (2018) Experimental investigations on laminar burning velocity variation of methyl formate-air mixtures at elevated temperatures. Energy Fuels 32(12):12936–12948
32. Varghese RJ, Kolekar H, Kishore VR, Kumar S (2019) Measurement of laminar burning velocities of methane-air mixtures simultaneously at elevated pressures and elevated temperatures. Fuel 257:116120
33. Turns SR (1996) Introduction to combustion: McGraw-Hill Companies
34. Han P, Checkel MD, Fleck BA, Nowicki NL (2007) Burning velocity of methane/diluent mixture with reformer gas addition. Fuel 86(4):585–596

Combustion Systems

Direct Numerical Simulation of Preignition and Knock in Engine Conditions

Minh Bau Luong and Hong G. Im

Abstract Theoretical framework and high-fidelity direct numerical simulations (DNS) on preignition and knock encountered in internal combustion (IC) engines are briefly reviewed in this chapter. The theoretical framework is presented first, and systematically followed by one-dimensional and multi-dimensional simulations. A number of high-fidelity simulations under realistic IC-engine conditions were used to demonstrate the super-knock mechanism in the presence of the complex chemistry-turbulence interaction. The ratio of ignition delay time to eddy-turnover time, τ_{ig}/τ_t, and the ratio of the most energetic length scale of turbulence and temperature fluctuations, l_t/l_T, affect the knock intensities are discussed. Different statistical metrics extracted from the multi-dimensional simulations to predict knock intensity are also presented. The quantitative determination of strong and weak ignition modes using the predicted Sa-based metrics is also cover in this chapter.

Keywords Direct numerical simulation (DNS) · Ignition criteria · Strong ignition · Weak ignition

1 Introduction

Modern downsized and boosted internal combustion (IC) engines provide higher thermodynamic efficiencies and cleaner combustion. However, these engines operated under extreme conditions of high load suffer from a higher possibility of abnormal preignition process, a premature auto-ignition induced by hot spots leading to ignition front propagation [1–7]. The mechanism and the source of such irregular preignition events are still not well-understood [8]. The early flames developed from preignition kernels increase the temperature and pressure of the end-gas that may induce a stronger end-gas auto-ignition, leading to extremely high-knock intensities, and even super-knock under high-load and/or high-temperature conditions. As represen-

M. B. Luong (✉) · H. G. Im
Clean Combustion Research Center, King Abdullah University of Science and Technology, Thuwal, Saudi Arabia
e-mail: minhbau.luong@kaust.edu.sa

A. Gupta et al. (eds.), *Advances in Energy and Combustion*, Green Energy and Technology,
https://doi.org/10.1007/978-981-16-2648-7_14

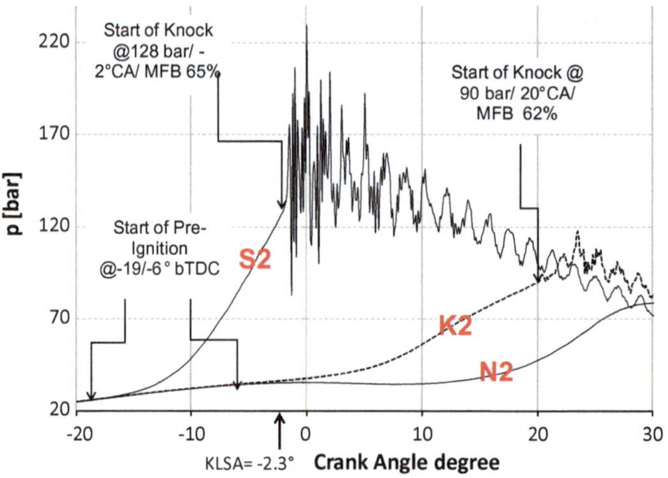

Fig. 1 Examples of in-cylinder pressure traces show preignition and superknock in a turbocharged SI engine: super-knock cycle (S2) with preignition at 19 CAD BTD and KI ~ 14.5 MPa, moderate knock cycle (K2) with preignition at 6 CAD BTDC and KI ~ 2.4 MPa, and normal (N2) cycle without preignition (taken from Refs. [6, 7])

tatively shown in Fig. 1 [6], super-knock cycle (S2) is characterized by high-pressure amplitudes and intensities due to localized detonation development that may lead to severe structural damage to the engine components. Therefore, a reliable criterion to predict irregular combustion phenomena is of critical importance [1, 3, 4, 6, 9–12].

A large spectrum of turbulence length and time scales combined with extremely short time scales of complex combustion process in detonation simulations pose a significant challenge in reproducing the phenomena using computational simulations, which require extremely fine spatial and temporal resolutions to capture the highly intermittent localized characteristics. This chapter reviews our recent efforts to investigate the fundamental issues of the knock events using high-fidelity direct numerical simulations (DNS) with the capability of fully resolving all temporal and spatial scales and the complex interaction of thermochemistry and turbulence [3, 4, 13, 14]. First-principle DNS is capable of unraveling the complex interplay between turbulence and chemical reactions to provide a better understanding of the mechanism of detonation development encountered in modern combustion devices under extreme high-load operating conditions, and to develop a reliable predictive model for real-world industrial applications [15–31].

In the following, a theoretical framework on super-knock development encountered in internal combustion (IC) engines is briefly reviewed, followed by high-fidelity DNSs in the subsequent sections. A systematically parametric set of two-dimensional (2-D) and three-dimensional (3D) DNS simulations are conducted at engine-relevant conditions [1–4, 32]. Effects of various parameters such as length and time scales of turbulence and scalar fluctuations on the occurrence and intensity of detonation are examined and a scaling relation for knock prediction is proposed [33–43].

2 Developing Detonation Regime Diagram

2.1 Theoretical Background

2.2 Zeldovich's Theory on Reactivity Gradient

According to Zeldovich's theory [44], a spontaneous ignition front is defined as a successive ignition events in the presence of spatial reactivity gradients. Its propagation speed, S_{sp}, is determined as

$$S_{sp} = |\nabla \tau_{ig}|^{-1}, \tag{1}$$

where τ_{ig} is the ignition delay time of the local reactant mixture as schematically shown in Fig. 2, defined as the time to reach the maximum heat release as illustrated in Fig. 3.

Based on the relative ratio of S_{sp} to the laminar flame speed, S_{sp}/S_L, and to the sound speed, S_{sp}/a, of the local mixture as illustrated in Figs. 2 and 3, Zeldovich [44] classified four different ignition regimes of a reaction front propagation in a combustible gaseous mixture:

1. *weak ignition*: In the presence of large spatial variations in τ_{ig}, $S_{sp} \sim S_L$, the local ignition establishes a slow-speed thin-front deflagration (see Figs. 2c and 3).
2. *spontaneous ignition*: For a mixture with a smaller level of τ_{ig} variations, $S_{sp} \gg a$, the mixture ignites spontaneously and S_{sp} exhibits a rapid ignition front propagation (see Figs. 2a and 3).
3. *detonation*: If S_{sp} is sufficiently large and comparable to the sound speed, $S_{sp} \sim a$ as visually shown in Fig. 2b, the combustion wave couple with the acoustic wave may lead to detonation development characterized by extremely high-peak amplitudes of pressure oscillations.
4. *homogeneous ignition*: In the asymptotic limit of zero gradient, S_{sp} becomes infinity, resulting in the purely homogeneous ignition mode.

2.3 $\varepsilon - \xi$ Regime Diagram

Gu and Bradley [45] further mapped different propagation modes into a regime diagram characterized by two key non-dimensional parameters, ξ and ε, which are defined as:

$$\xi = a/S_{sp}, \tag{2}$$

$$\varepsilon = (r_{hs}/a)/\tau_e, \tag{3}$$

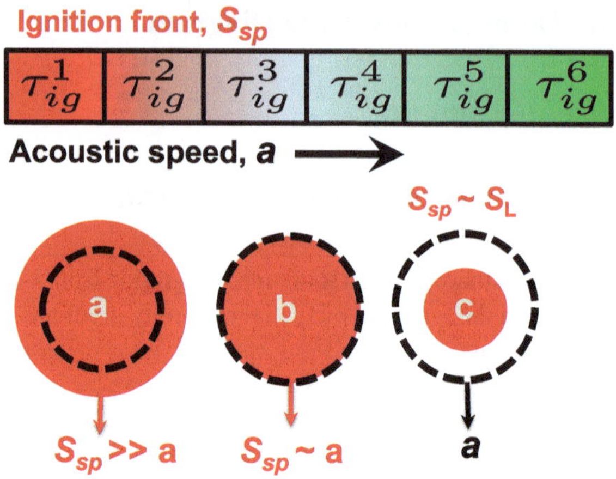

Fig. 2 Schematic of 1-D spatially monotonic distribution of ignition delay times, $\tau_{\mathrm{ig}}^{1} < \tau_{\mathrm{ig}}^{2} \cdots < \tau_{\mathrm{ig}}^{6}$, (first row), and the relative speed between the reaction front and the acoustic wave (second row) with **a** spontaneous ignition, **b** detonation with a perfect coupling between the reaction front and acoustic wave, and **c** deflagration

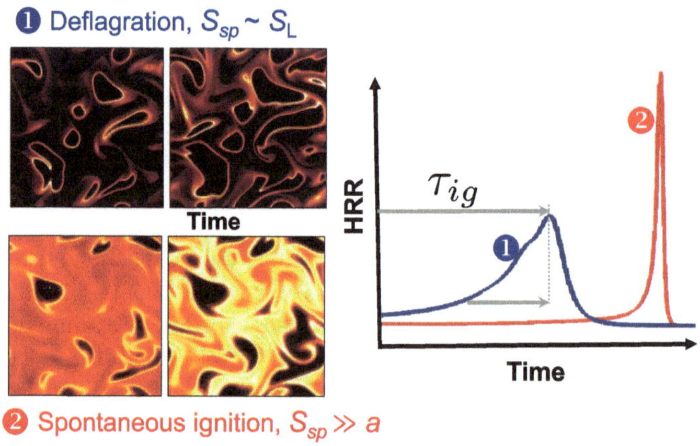

Fig. 3 Representative 2D contours of HRR show deflagrations (first row) and spontaneous ignition fronts (second row) and their temporal evolutions of HRR (right column). The definition of ignition delay time, τ_{ig}, and excitation time, τ_e (the time duration from 5% of the peak HRR to the peak HRR), are shown in the right column

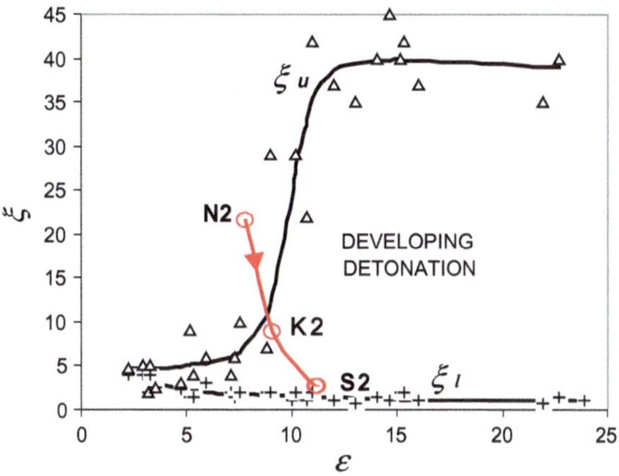

Fig. 4 Conditions for the occurrence of developing detonations in terms of Xi and epsilon for a stoichiometric 50%H_2–50%CO/air mixture at 5 atm [45, 46]. Three points (S2, K2, and N2) in the red line correspond to the three pressure traces shown in Fig. 1 (taken from Refs. [6, 7])

where ε is defined as the ratio of the residence time, r_{hs}/a, of the acoustic wave within the hot spot with a radius, r_{hs}, to the excitation time, τ_e, in which most of the chemical energy is released. τ_e is typically defined as the time interval between 5% of the peak heat release rate (HRR) and the peak HRR as schematically illustrated in Fig. 3.

Based on the range of ξ, five different modes of ignition front propagation within a hot spot were identified [45, 46]:

1. $\xi = 0$ for instant thermal explosion,
2. $0 < \xi < \xi_l$ for spontaneous (supersonic) ignition where the autoignitive wave is ahead of acoustic wave,
3. $\xi_l < \xi < \xi_u$ developing and developed detonation, through the thermoacoustic coupling process referred to as the shock wave amplification by coherent energy release (SWACER) mechanism [47, 48]),
4. $\xi_l \geq \xi < a/S_L$ subsonic autoignitive deflagration where S_L is the laminar burning velocity,
5. $\xi_l \geq \xi < a/S_L$ laminar burning deflagration propagating at the speed comparable to the laminar flame speed.

Figure 4 shows the $\xi - \varepsilon$ peninsula with the upper and lower limit, ξ_u and ξ_l, respectively, between which the developing detonation regime is identified for a stoichiometric 50%H_2–50%CO/air mixture at 5 atm [45, 46]. Note that the upper and lower ξ limits of the C-shaped curve strongly depend on the initial conditions of T_0, P_0, and ϕ_0, fuel types, and the kinetic model used in simulations [49, 50]. In general, the developing detonation regime is expected when ξ ranges between 1 and $O(10)$ quantity [49, 50]. As representatively shown in Fig. 5, due to the NTC behavior of two-stage ignition fuels (e.g., DME), they are much more prone to detonation development (gray region with roughly $1 < \xi < 10$) than single-stage ignition fuels (e.g., ethanol) for the same initial conditions of temperature, pressure, equivalence ratio, temperature gradient.

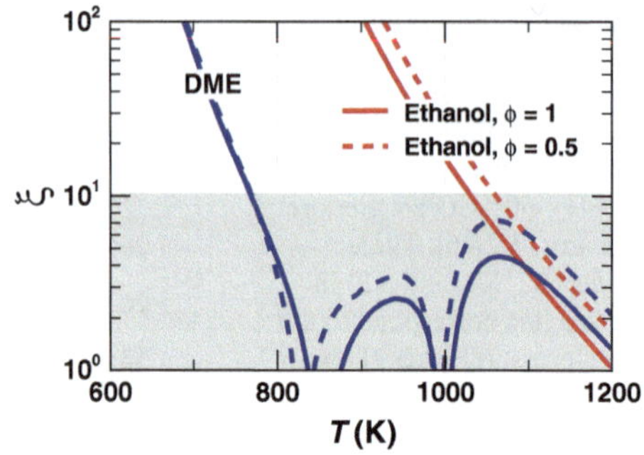

Fig. 5 The ξ values as a function of temperature for ethanol/air mixtures (red) and DME/air mixtures (blue) at ϕ of 0.5 and 1.0, and pressure of 35 atm, and $\partial T/\partial r = -2$ K/mm. The shaded region of $1 \leq \xi \leq 10$ is prone to the developing detonation for a small hot spot, i.e., $\varepsilon < 10$

2.4 *Quantitative Analysis*

Luong et al. [3] proposed three *a posteriori* statistical metrics to quantitatively assess the knock intensity, defined as follows:

$$F_H = \frac{\sum \langle \dot{q} | \, P > 1.1 P_e \rangle}{\sum \dot{q}}, \tag{4}$$

$$F_{\mathrm{CJ}} = \frac{\sum \langle \dot{q} | \, P > P_{\mathrm{CJ}} \rangle}{\sum \dot{q}}, \tag{5}$$

$$F_{\mathrm{VN}} = \frac{\sum \langle \dot{q} | \, P > P_{\mathrm{VN}} \rangle}{\sum \dot{q}}, \tag{6}$$

where $\dot{q}$ is the HRR; P_e, P_{CJ} and P_{VN} correspond to the homogeneous constant-volume equilibrium pressure, the Chapman–Jouguet pressure, and the von Neumann pressure, respectively. The summation is operated over all the cells of the computational domain and integrated over time. As such, F_i $(i = e, CJ, VN)$ measures the fraction of HRR associated with higher pressures exceeding the corresponding P_i, evaluated at the initial conditions of T_0, P_0, and ϕ_0.

According to Fig. 4 [45], the developing detonation regime is bounded by the upper and lower ξ limits of the C-shaped curve, ξ_u and ξ_l, respectively, suggesting that the detonation intensity depends strongly on the statistical quantities extracted from the spatial distribution of ξ such as its volume-average $\overline{\xi}$, probability density function. To this end, F_D, defined as the volume fraction of $\xi_u < \xi < \xi_l$, is proposed to predict knock intensity:

$$F_D = \int_{\xi_l}^{\xi_u} P(\xi) \mathrm{d}\xi, \tag{7}$$

which serves as an *a priori* predictive indicator of the subsequent detonation development while F_H is an *a posteriori* statistical assessment of knock intensity. Note that both F_D and F_H range from zero to unity. F_D was found to have a strong correlation with the fraction of the total heat release associated with detonation extracted from the full DNS data, F_H [3]. The mixture is found to be more prone to strong detonation as F_D approaches unity if $\varepsilon_p \gtrapprox 1$. For each initial condition, F_D together with the volume-average, $\overline{\xi}$, and probability density function, $P(\xi)$, are computed to quantify the knock intensity [3].

In the next sections, the above predictive criteria are discussed using 1D simulations first, followed by selected 2D and 3D simulations with and without turbulence in order to characterize the spontaneous ignition and detonation development.

2.5 One-Dimensional Detonation Development

The developing detonation $\varepsilon - \xi$ diagram [45, 46] has been used in many 1-D studies to investigate the interaction between flame propagation and end-gas auto-ignition, and the effects of boundary conditions, domain size, hot-spot location, hot-spot interaction, and cool flames on subsequent detonation formulation [6, 49–71]. The effects of the bulk-mixture gradient on developing detonation were also studied by Sow et al. [60] and Nogawa and Terashima [72]. In the following, two representative 1-D cases with distinct detonation development characteristics are discussed.

The first case is shown in Fig. 6a features a deflagration to detonation transition (DDT) event where a hot spot is developed into deflagration, gradually consuming the unburned mixture, and finally accelerating to make a transition to detonation. Specifically, the initial temperature profile is designed such that ξ ranges from 0 to 1000 within the hot spot (see Fig. 6). Figure 7 shows the temporal evolution of the temperature and pressure profiles. For such a strong ignition source, autoignition immediately forms a detonation front within the hot spot length, which is subsequently attenuated quickly upon leaving the hot spot (lines 1–4 in Fig. 7). Due to the low reactivity of the unburned mixture at $T_0 = 1000\,\text{K}$, the heat release rate within the reaction zone is not fast enough to sustain the detonation front such that it decays quickly and transits back into deflagration at $x \sim 10\,\text{mm}$. As more fuel/air mixture is gradually consumed by deflagration, the temperature and pressure of the remaining unburned mixture ahead of the flame are increased due to the compression heating effect (lines 5–7 in Fig. 7). Due to the increased reactivity of the unburned mixture together with the flame acceleration, the deflagration front transits into detonation at $x \sim 20$ mm and rapidly consumes the rest of the unburned mixture (lines 8–12 in Fig. 7). Much higher pressure peaks are observed during this period. These processes are typically referred as to deflagration to detonation transition (DDT).

The second case (i.e., T_0 of 1200 K with T_{peak} of 1264 K in Fig. 6) features a direct detonation initiation in which the reaction front and the pressure wave are perfectly coupled to form an extremely high shock wave. The initial hot spot is designed for ξ to range between 3 and 6 (see Fig. 6). Due to the high reactivity of the unburned mixture with T_0 of 1200 K, the hot spot immediately gives rise to a detonation front within the hot spot length, thereby rapidly consuming the unburned mixture (see

Fig. 8). Once the detonation is established, the rapid heat release rate in the reaction zone is able to sustain the steady detonation front that propagates and consumes the rest of the unburned mixture.

The above cases show that the two parameters, ξ and ε, within the hot spot and the bulk mixture determines the nature of the early detonation development. Figure 9 shows the compiled results of a series of 1-D simulations for various combinations of these parameters. For a fixed T_0 of 1200 K and ϕ of 1, the 1-D simulations were performed by varying T_{hs} and r_{hs} such that both ξ and ε are varied accordingly. As seen in Fig. 9, the maximum pressure above the equilibrium pressure (i.e., $> 1.1P_e$) for $0.2 < \xi < 20$. The peak P_{max} occurs at $\xi \sim 6$ where the perfect coupling between the reaction front and the acoustic wave evolving into a stable detonation shock wave. The range of ξ was used to determine the volume fraction of ξ and F_D to predict the amplitude of knock intensity in multi-dimensional cases.

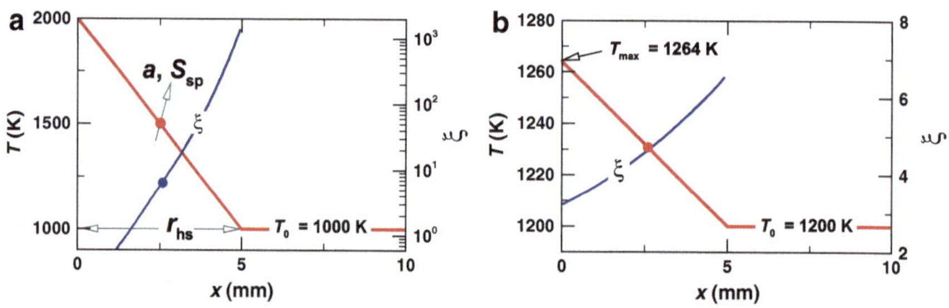

Fig. 6 Two representative initial conditions of a hot spot with a 1-D linear-temperature profile: **a** $T_0 = 1000$ K and $T_{max} = 2000$ K featuring deflagration to detonation transition (DDT) (see Fig. 7), and **b** $T_0 = 1200$ K and $T_{max} = 1264$ K featuring direct detonation development (perfect coupling between the acoustic wave and ignition front, see Fig. 8). The speed of sound, a, and ξ are evaluated at the center of hot spots (marked as dots). The spatial distribution of ξ within the hot spots is also shown. Note that τ_{ig} and τ_e are computed by using a 0-D homogeneous adiabatic constant-volume reactor using the localized temperature and pressure

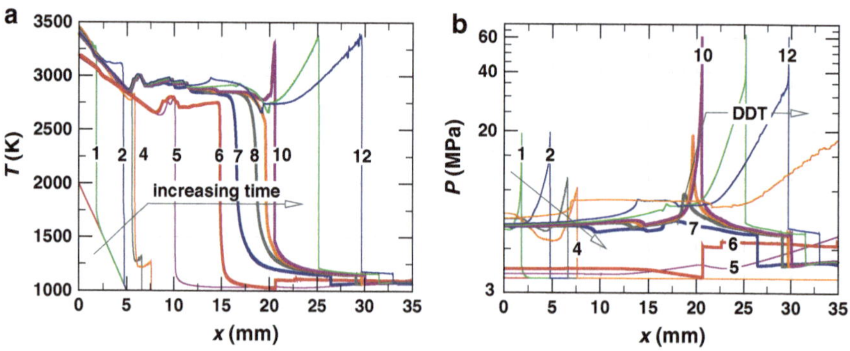

Fig. 7 The temporal evolutions of **a** temperature and **b** pressure at different time. The initial conditions of T_0 of 1000 K and P_0 of 35 atm (see Fig. 6), a stoichiometric ethanol/air mixture

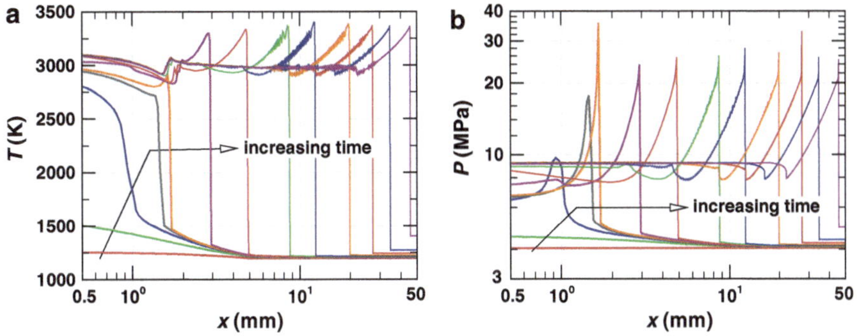

Fig. 8 The temporal evolutions of **a** temperature and **b** pressure at different time.The initial conditions of T_0 of 1000 K and P_0 of 35 atm (see Fig. 6), a stoichiometric ethanol/air mixture

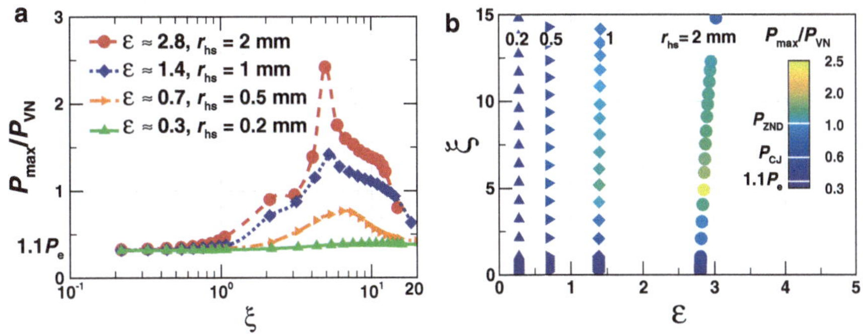

Fig. 9 **a** The normalized maximum pressure as a function of ξ, and **b** the $\xi - \varepsilon$ diagram for different hot spot sizes (taken from Luong et al. [3])

2.6 2-D and 3-D Simulations of Super-Knock Under Realistic IC-engine Conditions

The basic characterization in 1D problem is now extended to multi-dimensional conditions in the presence of spatial fluctuations in temperature and velocity fields. Parametric 2-D and 3-D simulations were systematically performed to validate the criteria proposed in Fig. 2.4. The distribution of ξ and ε were varied by changing T' over a wide range from 0.5 to 20 K, and l_T of 1, 2, 5, and 10 mm. For each initial condition, the full detailed information of spatial distribution of ξ and the corresponding statistical quantities of ξ can be computed. Figure 10 shows the spatial distribution of ξ of four cases with different T' and l_T that result in no-knock, mild-knock and super-knock [3].

Figure 11 shows a representative 2-D contour of temperature field with T_0 of 1200 K, T' of 15 K, and l_T of 5 mm. According to Peters et al. [53, 73], the temperature field is subdivided into the *dissipation elements* over which the temperature profile

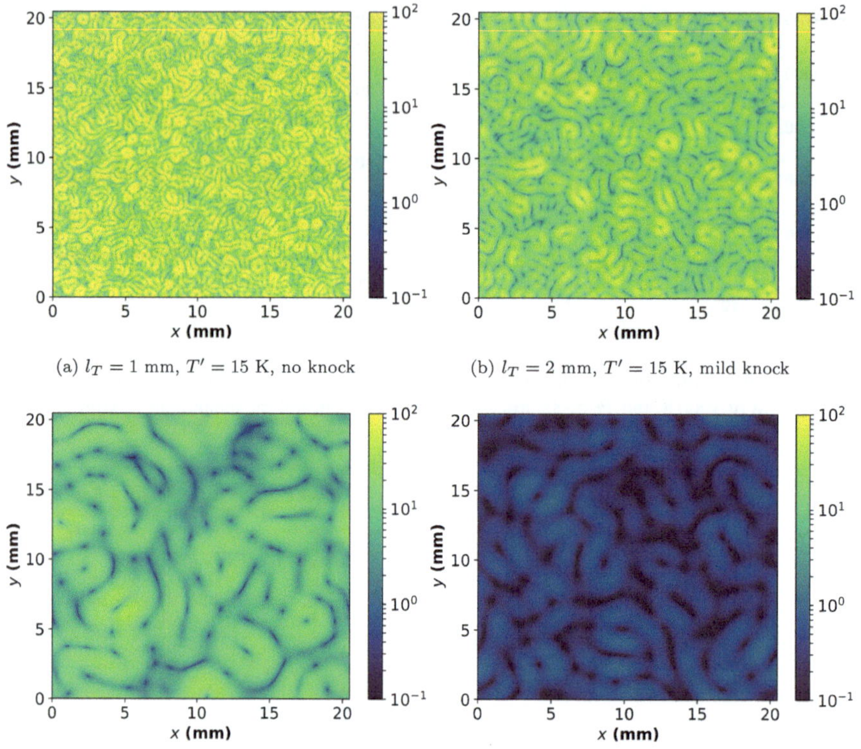

(a) $l_T = 1$ mm, $T' = 15$ K, no knock

(b) $l_T = 2$ mm, $T' = 15$ K, mild knock

(c) $l_T = 5$ mm, $T' = 15$ K, super-knock

(d) $l_T = 5$ mm, $T' = 0.6$ K, mild knock

Fig. 10 Representative 2-D contours of ξ distribution computed from four different initial conditions for three different l_T of 1 mm, 2 mm, 5 mm, with T' of 15 K (**a–c**) and 0.6 K (**d**)

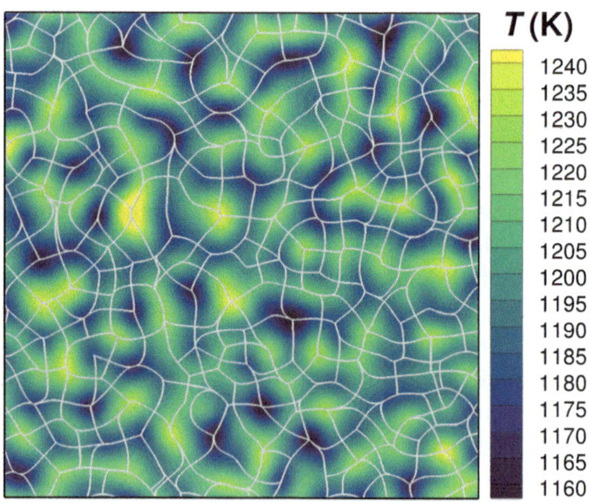

Fig. 11 Representative 2-D contour of temperature field with T_0 of 1200 K, T' of 15 K, and l_T of 5 mm overlaid by the dissipation elements. The white lines mark the boundary between the dissipation elements

is monotonic, ranging from the peak to the trough of each dissipation element. For isotropic temperature field fluctuations, Peters et al. [53] found r_{hs} comparable with the mean distance of the dissipation elements, l_{DE}, and approximated as $l_{DE} \approx 2\lambda_T$, where λ_T is the Taylor mixing scale [53]. In the multi-dimensional problems, l_{DE} is believed to be a minimum size of a hot spot of unburned reactants required so that it allows a sufficient run-up distance of detonation development [12, 53].

Based on the Taylor mixing scale, Towery et al. [12] proposed a statistical model to evaluate the predictive ε_p for highly compressible homogeneous isotropic turbulence fluctuations. ε_p is defined as

$$\varepsilon_p = 2\lambda_T / (a_{rms} \tau_{exo}) \tag{8}$$

where a and τ_{exo} are the mass-weighted sound speed and exothermicity time, respectively (see Ref. [12] for details). In Refs. [3, 4], at the the initial conditions of $T_0 = 1200$ K, $P_0 = 35$ K, and $\phi_0 = 1$, $\varepsilon_p = 1$ corresponds to the initial temperature field of $l_T = 3.3$ mm. The cases with $\varepsilon_p < 1$ (i.e., l_T of 1 and 2 mm with ε_p of 0.3 and 0.6) are less likely to develop detonation while the cases with ε_p of 1.2, 1.4, and 3.8 for l_T of 4 mm, 5 mm, and 10 mm, respectively, are more likely to develop strong detonation if $1 < \xi < 10$.

For a uniform mixture with only thermal fluctuations, the predictive ξ_p is scaled as [11, 12]

$$\xi_p = a | \frac{d\tau_{ig}}{dT} \nabla T | \cong a \left| \frac{d\tau_{ig}}{dT} \right| |\widetilde{\nabla T}| \tag{9}$$

where the statistical mean temperature gradient is further approximated by $|\widetilde{\nabla T}| \cong T'/\lambda_T$, with the Taylor mixing length scale, $\lambda_T \cong (l_T \alpha / T')^{0.5}$ [11]. These predictive parameters, ξ_p, and ε_p, will be validated in the subsequent sections.

2.6.1 Baseline Cases with No Turbulence

In this subsection, three baseline (no turbulence) cases with l_T of 1 mm, 2 mm, and 5 mm denoted as BL1, BL2, and BL5, respectively, are chosen for discussion. The probability density functions (PDF) of ξ, the temporal evolution of the normalized P_{max}, and the contours of pressure at the time of the peak of F_H for three baseline cases are shown in Figs. 12 and 13. According to the scaling of ξ_p above, with a fixed T' of 15 K, increasing l_T from 1 to 5 mm decreases ξ_p which is consistent with the PDF of ξ shifting toward the developing detonation regime (see Fig. 12). Consequently, BL5 is predicted to be more prone to developing detonation as compared to BL1 and BL2.

Consistent with the PDFs of ξ and $\varepsilon_p = 0.3$, $P_{max} \leq P_{VN}$ and the isocontours of pressure in Figs. 12 and 13, respectively, confirm that BL1 features a spontaneous ignition process. BL2 with $F_{VN} \approx 0.07$ is characterized as a mild detonation process as also quantitatively predicted by $\varepsilon_p = 0.6$. Unlike BL1 and BL2, a strong detonation

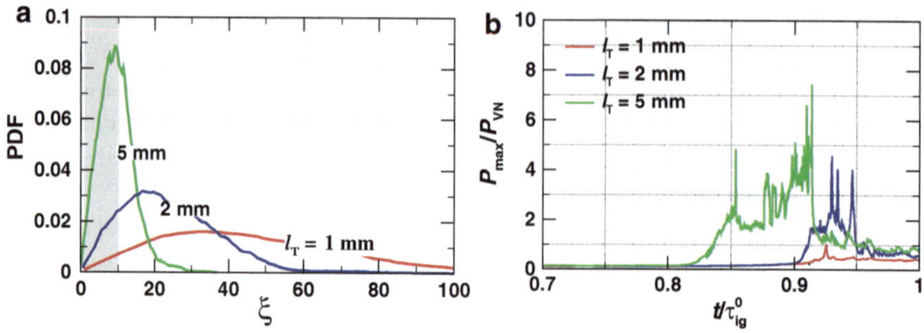

Fig. 12 **a** The probability density functions (PDF) of ξ for three baseline cases (without turbulence) with three different most energetic length scale of temperature, l_T of 1 mm, 2 mm, and 5 mm, featuring no knock, mild knock, and super-knock, respectively, and **b** their temporal evolution of the normalized maximum pressure, P_{max}/P_{VN}, [4]. The gray area in the PDF figure depicts the region of $1 < \xi < 10$ determined by a series of 1-D simulations under the same initial conditions, which is prone to developing detonation (see Fig. 9a) (adapted from Luong et al. [3, 4])

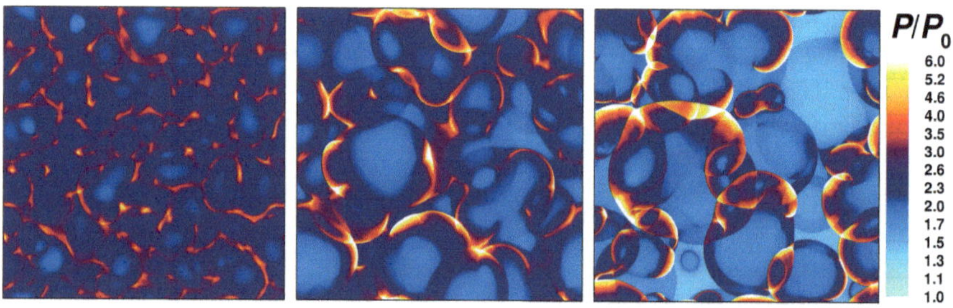

Fig. 13 Representative 2-D contours of pressure for three cases with the same T' of 15 K, and l_T of 1 mm, 2 mm, and 5 mm with no knock, mild knock, and super-knock, from left to right, respectively (adapted from Luong et al. [3])

process is observed for BL5, quantified by $F_{VN} \approx 0.2$ and $P_{max} \gg P_{VN}$, which is accurately predicted by $\varepsilon_p = 1.6$ for BL5.

The statistical quantities, F_H, F_D and $\overline{\xi}$, of all the baseline cases are plotted in Fig. 14. Three main points are noted from Fig. 14. First, F_D has a better agreement with F_H for $\varepsilon_p > 1$ (i.e., cases with $l_T > 4$ mm in Fig. 14b). Second, the magnitude of F_H for 3-D cases is comparable with the corresponding 2-D cases as seen in Fig. 14a despite the peak F_{VN} of the 3-D cases slightly lower than that of the 2-D cases [3]. Third, Fig. 14 also reveals a good correlation between $\overline{\xi}$ and the knock intensity characterized by the F_H magnitude. The correlation indicates that the conditions with $\overline{\xi}$ ranging between 0.5 and 30 experience developing detonation and the strongest detonation intensity (F_H approaches unity) occurs at $\overline{\xi} \sim 6$ with $\varepsilon_p > 1$. Specifically, the maximum values of F_H at $\overline{\xi} \sim 6$ are approximately 0.18, 0.22, 0.55, and 0.75 for l_T of 1 mm, 2 mm, 5 mm, and 10 mm, respectively. These results verify that the prediction of ε_p for knock propensity is consistent with F_H, the detonation intensity is greatly alleviated with decreasing l_T (mild-to-no detonation for $l_T < 3.3$ mm with $\varepsilon_p < 1$).

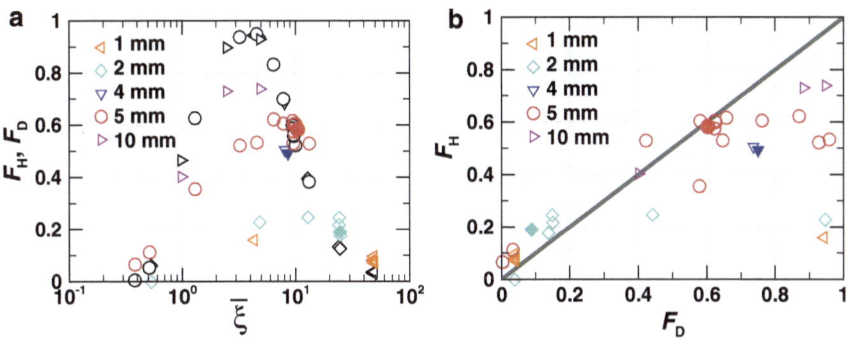

Fig. 14 **a** F_H (color symbols) and F_D (black symbols) as a function of $\bar{\xi}$, with T' ranging from 0.5 K to 20 K for different l_T ranging from 1 mm to 10 mm, **b** the correlation between F_H and F_D. Filled (open) symbols are for 3-D (2-D) cases (taken from Luong et al. [3])

2.6.2 Effect of Turbulence

The previous subsection demonstrated that no detonation was observed for BL1 while BL5 exhibited a strong detonation process because a minimum run-up distance is required for developing detonation as accurately predicted by ε_t. The results suggest that decreasing (increasing) the length scale of temperature fluctuations (hence the dissipation element length) is a key parameter to alleviate (promote) knock intensity. In this subsection, the additional effect of turbulent velocity fields on knock intensity is also investigated [4].

The isocontours of pressure and HRR of two representative 2-D simulations and a 3-D simulation in Figs. 15, 16 and 17 are chosen for discussion. With the same turbulent velocity fluctuation, u' of 83.3 m/s, and root-mean-square temperature fluctuation, T' of 15 K, l_T/l_e and τ_{ig}/τ_t were found to be two key parameters to determine knock propensity.

First, increasing turbulent intensity by increasing u' and fixing $l_e = 5$ mm is not effective in reducing knock intensity since with $\tau_{\mathrm{ig}}/\tau_t$ of 1.0, turbulence is not fast enough to modify the autoignition and the subsequent detonation process as visually seen in Fig. 16. Rather, the main effect of turbulent eddies is to weakly wrinkle and distort the detonation fronts. Second, a finer turbulence structure, l_e of 1 mm $\ll l_T =$ 5 mm, and $\tau_{\mathrm{ig}}/\tau_t$ of 5.0, effectively attenuates the detonation intensity as visually seen in Figs. 16 and 17. High turbulence intensity together with a short mixing-time scale significantly increases the amount of reactants burned by corrugating turbulent flames rather than by the developing detonation phase. Fast and small turbulence structures also induce smaller and finer localized detonation waves during the developing detonation process. Third, for the 3-D cases with a slightly lower F_H, a consistent agreement between the 2-D and 3-D cases was observed in terms of $P_{\max}$, the duration of detonation process, and the level of pressure oscillations, P' [4].

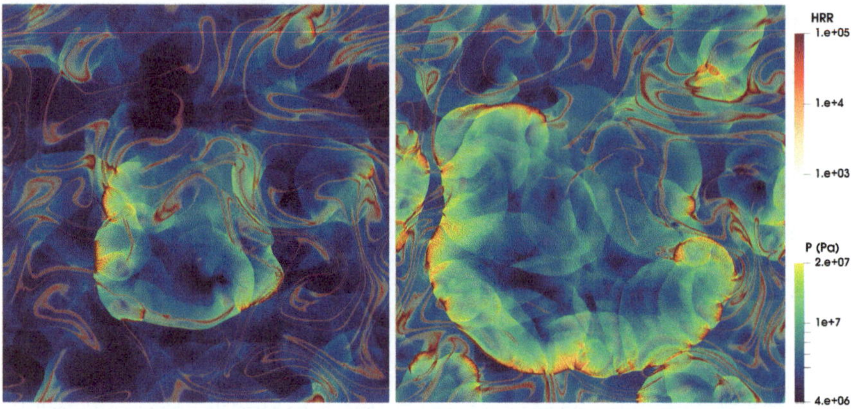

Fig. 15 Two-dimensional contours of pressure and heat release rate (HRR) at the times of onset of detonation development (left) and the peak detonation intensity (right) for a 2-D case with the most energetic length scale of temperature and turbulent field, l_T of 5 mm and l_e of 5 mm, respectively, the turbulent velocity fluctuation, u' of 83.3 m/s, and the ratio of the ignition delay time to turbulent time scale τ_{ig}/τ_t of 1.0 [4] (taken from Desai et al. [14])

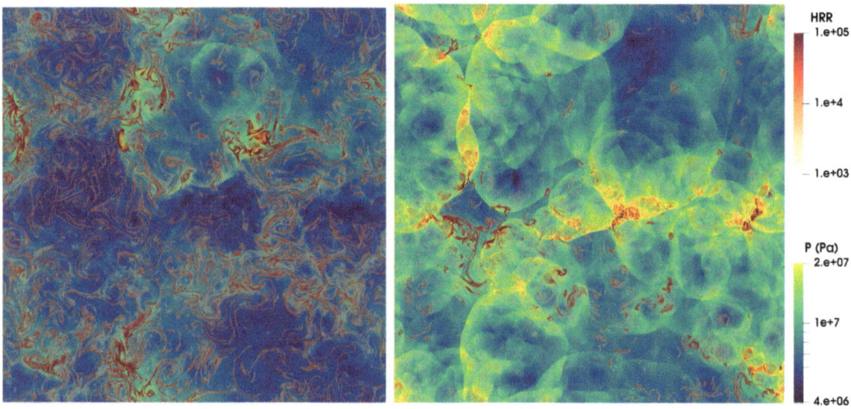

Fig. 16 Two-dimensional contours of pressure and heat release rate (HRR) at the times of onset of detonation development (left) and the peak detonation intensity (right) for a 2-D case with the most energetic length scale of temperature and turbulent field, l_T of 5 mm and l_e of 1 mm, respectively, the turbulent velocity fluctuation, u' of 83.3 m/s, and the ratio of the ignition delay time to turbulent time scale τ_{ig}/τ_t of 5.0 [4] (taken from Desai et al. [14])

Contrary to the case with $l_e = 1$ mm $\ll l_T = 5$ mm being able to attenuate detonation intensity, a fast turbulent mixing, τ_{ig}/τ_t of 5, together with $l_e = 5$ mm $\gg l_T = 1$ mm, promotes detonation development as seen in Fig. 18. Turbulence with a larger eddy size effectively spreads the temperature field over a wider area, stretching out the run-up distance, and thus facilitates the developing detonation [4].

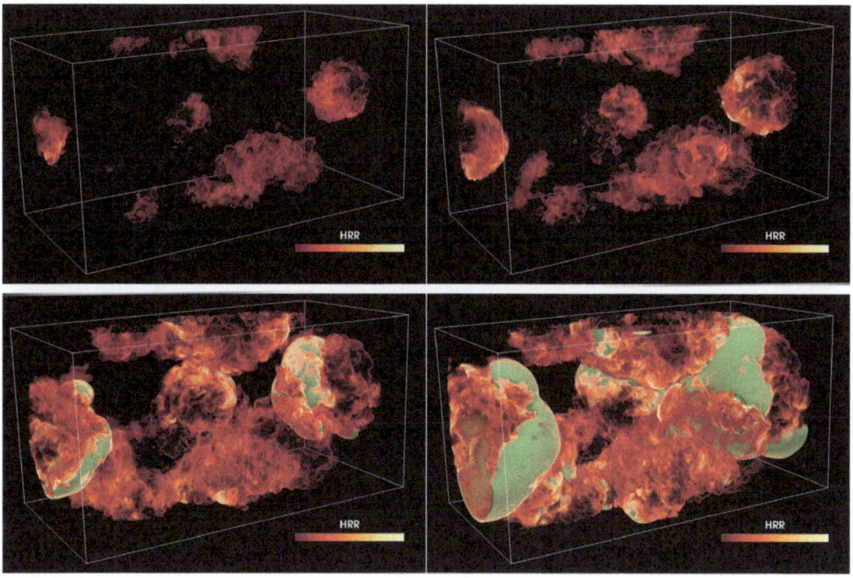

Fig. 17 The volume rendering of HRR and the iso-surface of pressure (green) with $P = P_{CJ}$ at successive times prior to the onset of the detonation formation (top row) until the detonation intensity reaches its peak (bottom row) for a 3-D case with the most energetic length scale of temperature and turbulent field, l_T of 5 mm and l_e of 1 mm, respectively, the turbulent velocity fluctuation, u' of 83.3 m/s, and the ratio of the ignition delay time to turbulent time scale τ_{ig}/τ_t of 5.0 [4] (taken from Desai et al. [14])

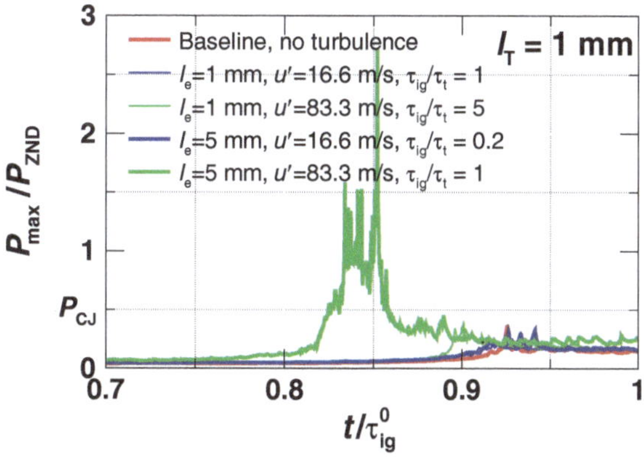

Fig. 18 Temporal evolution of the normalized maximum pressure, P_{max}/P_{VN} for the cases with l_T of 1 mm with different l_e and u' (adapted from Luong et al. [4])

3 Weak and Strong Ignition Modes

3.1 Theoretical Background

Based on Zeldovich's theory and validated by high-fidelity DNS data, Sankaran et al. [15] proposed a criterion to delineate the weak or strong ignition modes, referred to as the Sankaran number (Sa) herein, which is based on the ratio of laminar flame speed, S_L, to the ignition front speed, $S_{sp} = |\nabla \tau_{ig}|^{-1}$ [44].

$$\mathrm{Sa} = \beta \frac{S_L}{S_{sp}}, \tag{10}$$

where β is a weighting factor of 0.5 to account for some delay in developing a deflagration from the ignition kernel due to the interactions of thermal and radial diffusion between the neighboring regions. Sa = 1 serves as the boundary between the strong (Sa < 1) and the weak (Sa > 1) ignition.

In practical combustion devices, the detailed information of the spatially resolved scalar fluctuations is not available, and thus the prediction for in a real engine should be based on the statistical quantities such as the mean and variance of fluctuations in RANS or LES simulations. Im et al. [11] conducted a theoretical scaling analysis by incorporating the turbulent effect into the original Sankaran criterion to derive the turbulent ignition regime diagram as shown in Fig. 19.

For compositionally homogeneous reactant mixtures with turbulent velocity and temperature fluctuations, the diagram classifies ignition modes into three categories: *weak* exhibiting a dominant slow-speed deflagration mode of fuel consumption, *reaction-dominant strong*, and *mixing-dominant strong* exhibiting a dominant volumetric-ignition mode of fuel consumption. Pal et al. [74] and Luong et al. [1] conducted a series of DNS simulations and found consistent results with the pre-

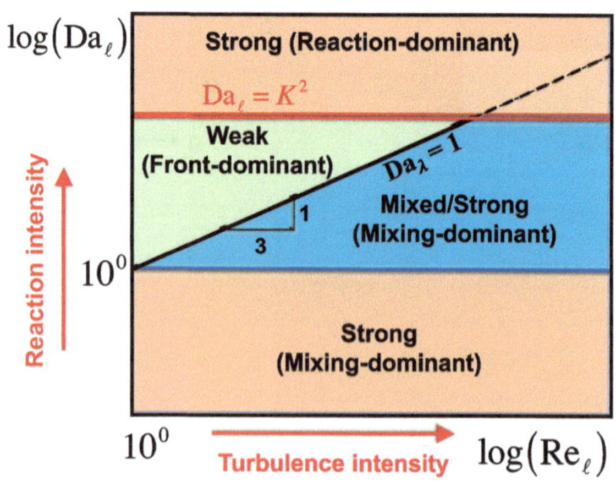

Fig. 19 Regime diagram proposed by Im et al. [11] for strong and weak ignition modes for nearly homogeneous reactant mixture with velocity and temperature fluctuations

dictions by the turbulent ignition regime diagram. The ignition criterion [11] was further extended to accommodate both temperature and concentration fluctuations and validated for the mixtures with NTC and non-NTC behaviors [1].

The gradient of ignition delay time, $\nabla \tau_{\mathrm{ig}}$, for a mixture with a uniform composition and pressure is solely due to temperature variations expanded by the chain rule as [11]:

$$|\nabla \tau_{\mathrm{ig}}| = \left| \frac{\mathrm{d}\tau_{\mathrm{ig}}}{\mathrm{d}T} \nabla T \right| , \tag{11}$$

where $\mathrm{d}\tau_{\mathrm{ig}}/\mathrm{d}T$ represents the ignition delay sensitivity to temperature as representatively shown in Fig. 20b, and ∇T is the temperature gradient of the bulk mixture [11, 74]. In turbulent conditions, the temperature gradient is approximated by the statistical mean temperature gradient, $\widetilde{|\nabla T|}$, which is estimated based on the variance of the temperature, T', fluctuations and the Taylor mixing scale, λ_T, as

$$\widetilde{|\nabla T|} \approx \frac{T'}{\lambda_T} , \tag{12}$$

where the Taylor mixing scale is approximated as

$$\lambda_T = l \mathrm{Re}_l^{-0.5} , \tag{13}$$

with $\mathrm{Re}_l = u'l/\nu$ being the turbulent Reynolds number, ν the kinematic viscosity of the bulk mixture gas, u' the root mean square (RMS) turbulent velocity fluctuation, and l the integral scale of velocity field.

Further scaling analysis yields the expression for the *predicted* Sankaran number, Sa_p, [11] as follows:

$$\mathrm{Sa}_p = K \mathrm{Da}_l^{-1/2}, \quad K = \beta \frac{1}{(\tau_f \tau_{\mathrm{ig}})^{1/2}} \left| \frac{\mathrm{d}\tau_{\mathrm{ig}}}{\mathrm{d}T} T' \right| , \tag{14}$$

where $\mathrm{Da}_l = \tau_t/\tau_{ig}$ is the ignition Damköhler number, defined as the ratio of the turbulence integral time scale, l/u', to the ignition delay time, and $\tau_f = \alpha/S_L^2$ is the characteristic flame time, where α is the thermal diffusivity of the initial bulk mixture. K is referred to as the normalized thermal ignition sensitivity.

In the presence of both temperature and equivalence ratio (concentration) fluctuations, τ_{ig} is a function of T and ϕ such that $|\nabla \tau_{\mathrm{ig}}|$ is expanded by using the chain rule as flows:

$$|\nabla \tau_{\mathrm{ig}}| = \left| \frac{\partial \tau_{\mathrm{ig}}}{\partial T} \nabla T + \frac{\partial \tau_{\mathrm{ig}}}{\partial \phi} \nabla \phi \right| , \tag{15}$$

where $\nabla \phi$ and $\partial \tau_{\mathrm{ig}}/\partial \phi$ are the equivalence ratio gradients, and the ignition delay sensitivity to equivalence ratio, respectively [1]. Assuming that the temperature and equivalence ratio fields are correlated with the turbulent flow field, it results in a comparable integral length scale, $l_T \sim l_\phi \sim l_e$, and $\widetilde{\nabla\phi} \sim \phi'/\lambda_T$ as approximated in the temperature case.

As such K in Fig. 14 is extended as:

$$K = \beta \frac{1}{(\tau_f \tau_{\mathrm{ig}})^{1/2}} \left(\left| \frac{\partial \tau_{\mathrm{ig}}}{\partial T} T' \right| + s \left| \frac{\partial \tau_{\mathrm{ig}}}{\partial \phi} \phi' \right| \right), \tag{16}$$

where the sign of s is to account for the NTC behavior and the correlation of T-ϕ distribution. s is positive if T and ϕ fluctuations are uncorrelated, thereby being additive to $|\nabla \tau_{\mathrm{ig}}|$ that synergistically promotes the deflagration mode regardless of the NTC regime. s adopts a negative (positive) sign for a negative $T - \phi$ correlation if the product of $\partial \tau_{\mathrm{ig}}/\partial T$ and $\partial \tau_{\mathrm{ig}}/\partial \phi$ is positive (negative) that the combined effects of T and ϕ on K become subtractive (additive).

3.2 *Prediction of Weak/strong Combustion Modes*

Many fuels with and without the NTC behavior were chosen to investigate the effect of temperature and equivalence ratio fluctuations on the combustion modes by varying the initial mean temperatures and different fluctuation levels [1, 20, 27, 28, 30, 75–77]. The initial conditions and physical parameters are chosen to be relevant to IC engine conditions [1, 33–43]. The DNS solution were processed to verify the predictive accuracy of the ignition criteria.

Figure 20 shows the homogeneous ignition delay time of DME/air mixtures and ethanol/air mixtures and the sensitivity of ignition delay time to temperature that is needed to compute Sa_p. DME exhibits the negative-temperature coefficient (NTC) regime in which τ_{ig} increases with increasing temperature such that its τ_{ig} varies non-monotonically with temperature. As such, within/near the NTC regime, the variation of $\mathrm{d}\tau_{\mathrm{ig}}/\mathrm{d}T$ for DME is small compared to that of single-stage fuels (e.g., ethanol), which in turn leads to a small $|\nabla \tau_{\mathrm{ig}}|$ for the same temperature fluctuation level. It suggests that within/near the NTC regime, the mixture with T' only is more susceptible to spontaneous ignition as quantitatively predicted by Sa_p in Fig. 21 [2].

Figure 22 shows a typical initial condition used in the simulations. These DNS cases were adopted to perform the statistical analysis of Sa such that the volume-average (mean) Sa, $\overline{\mathrm{Sa}}$, the probability density function of Sa, $P(\mathrm{Sa})$, and the volume fraction of the regions with Sa <1, $F_{\mathrm{Sa},S}$, extracted from the spatial Sa distribution are obtained. In addition, $\mathrm{Sa_p}$ for each simulation was also computed based on the information of the initial mean bulk mixture condition such as T_0, p_0, and ϕ_0, and the RMS values, T' and ϕ' as representatively shown in Fig. 21 as a function of temperature.

The isocontours of HRR for some representative DME cases are shown in Fig. 23. Spontaneous ignition characterized by volumetric ignition is dominant even with T' of 60 K at T_0 of 900 K due to a small variation of τ_{ig} with temperature within/near the

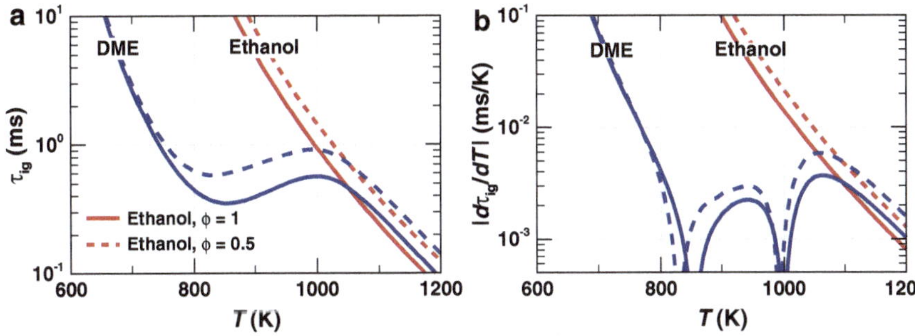

Fig. 20 **a** The 0-D homogeneous ignition delay time, τ_{ig}, and **b** the absolute value of the sensitivity of τ_{ig} to temperature, $|d\tau_{ig}/dT|$, as a function of temperature for ethanol/air mixtures (red) and DME/air mixtures (blue) at ϕ of 0.5 and 1.0, and pressure of 35 atm

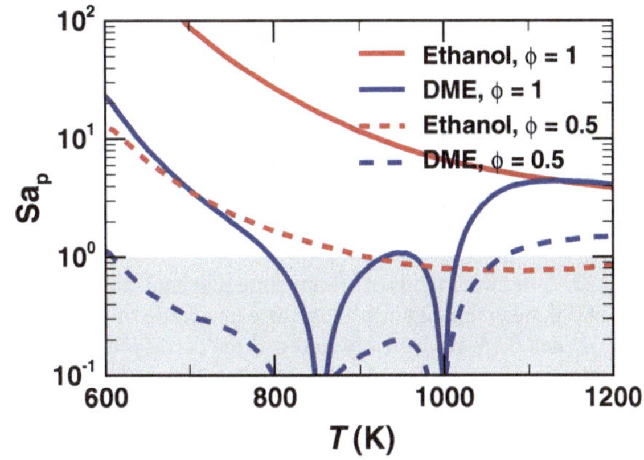

Fig. 21 The predicted Sa number, Sa_p, as a function of temperature for ethanol/air mixtures (red) and DME/air mixtures (blue) at ϕ of 0.5 and 1.0, and pressure of 35 atm, and assuming the identical integral length scale of velocities and temperature field, $l_t = l_T = 1$ mm, $\tau_t/\tau_{ig} = 1$, and $T' = 15$ K. The shaded region of $Sa_p < 1$ is for strong ignition, while the region of $Sa_p > 1$ is for weak ignition

NTC regime as shown in Fig. 20). On the contrary, deflagration propagation becomes dominant for the cases with T_0 of 680 and 1045 K outside the NTC regime. Similar results were found for n-heptane [20, 29–31, 75, 76, 78, 79].

Contrary to the T'-only cases, in the presence of both T and ϕ, especially when they are negatively correlated, T_0 of 900 K results in a much lower HRR (due to the synergistic effect of T' and ϕ') than those of T_0 outside the NTC regime (due to the subtractive effect of T' and ϕ' for the cases with T_0 of 770 and 1045 K), as seen in the bottom row of Fig. 23.

Quantitative assessment of combustion modes for the previous cases was also conducted. For each initial field of a DNS case, the exact distribution of Sa within the domain and its associated statistical quantities can be computed. For direct validations with the predictive quantities such as F_S and $\overline{\text{Sa}}$, the fractional contributions of weak and strong ignition to the total heat release were extracted from the detailed DNS solution.

Figure 22a-b shows representative initial fields of temperature and equivalence ratio distribution. Based on these detailed DNS information, the spatial distribution

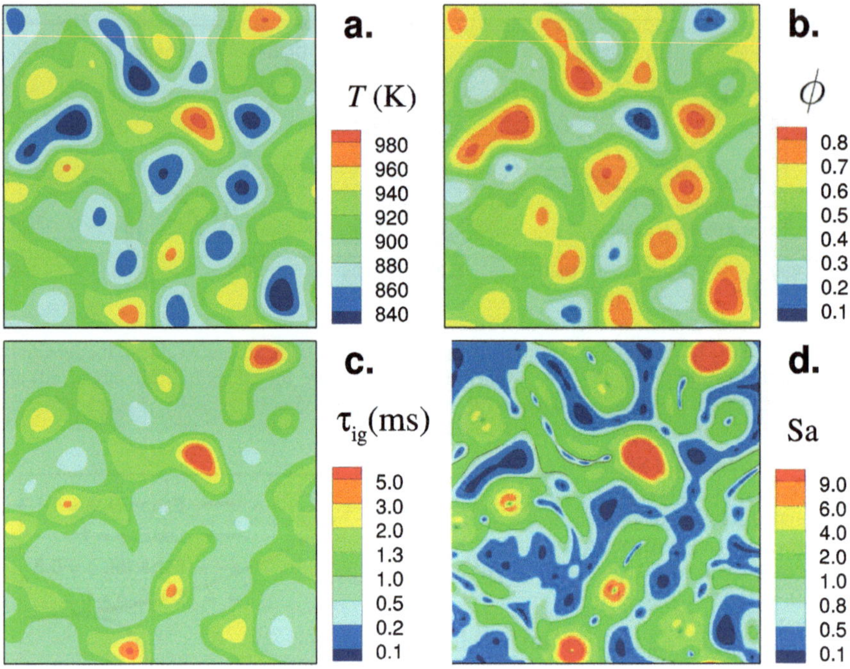

Fig. 22 **a–b** initial fields of temperature and equivalence ratio with a negative T–ϕ distribution at the initial mean temperature and equivalence ratio of 900 K–0.5 with their fluctuation of T' and ϕ' of 30 K and 0.15, respectively, and **c–d** the corresponding ignition delay time and Sa distribution. The black iso-lines of Sa = 1 delineates two distinct combustion modes of strong (Sa $<$ 1) and weak (Sa $>$ 1) ignition (taken from Luong et al. [2])

of τ_{ig} and Sa, and the corresponding probability density function (PDF) of Sa, $P(\text{Sa})$ are computed in Figs. 22c–d and 24, respectively. The volume fraction of the regions attributed to strong ignition was also estimated as

$$F_{\text{Sa},S} = \int_0^1 P(\text{Sa})\text{dSa}\,, \tag{17}$$

which was proposed as a quantitative metric to predict the fractional cumulative heat release rate associated with strong ignition directly computed from DNS solution fields, $F_{\text{Da},S}$ as [20, 27, 28, 30, 75, 76]

$$F_{\text{Da},S} = \frac{\sum \langle \dot{q} |\, \text{Da} > \text{Da}_0 \rangle}{\sum \dot{q}}\,, \tag{18}$$

where the summation is operated over the total number of computational cells in the DNS domain, $\dot{q}$ is the HRR, and Da is Damköhler number defined as the ratio of the

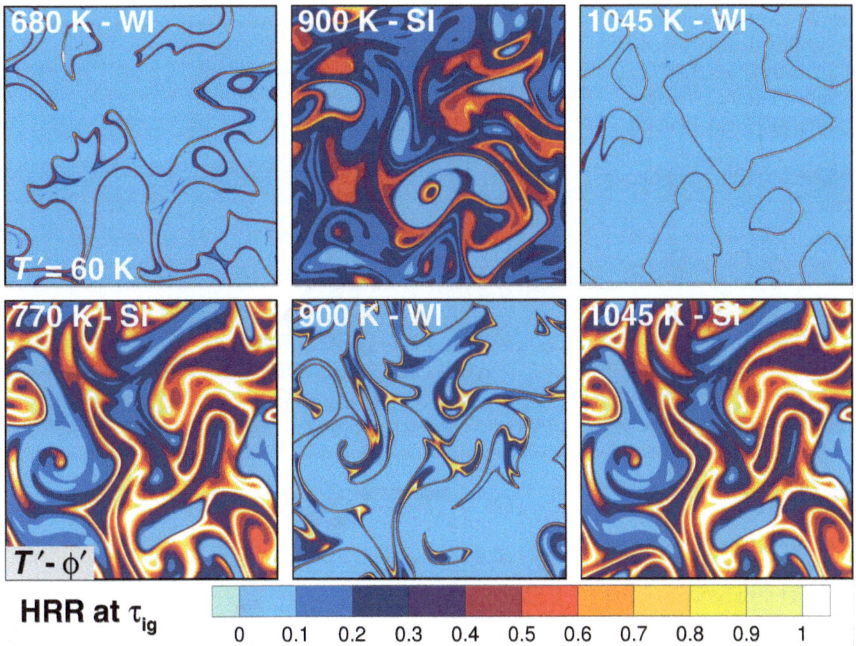

Fig. 23 2-D contours of the normalized HRR at the maximum spatial-averaged HRR occurrence for the DME/air mixtures at the initial pressure of 30 atm. The first (second) row corresponds to the cases with temperature fluctuations only, T' of 60 K, (with both temperature and equivalence ratio fluctuations, T' of 30 K and ϕ of 0.15 [2]) (adapted from Luong et al. [1])

local reaction and diffusion terms for a selected scalar variable [16, 17, 20, 28, 30, 71, 75, 76, 80]. Strong ignition is dominant if $F_{\mathrm{Sa},S} \sim 1$. Strong and weak ignition is delineated by a threshold value, Da_0, of order unity that depends on the chosen initial condition. The significance of the Sa-based metric is that the prediction of combustion modes is based purely on the initial conditions.

Consistent with the qualitative observations in the previous section, $\overline{\mathrm{Sa}}$ and $F_{\mathrm{Sa},S}$ for the DME cases of T_0 at 680 K, 900 K, and 1045 K are found to be, respectively, 4.2 & 0.20, 0.5 & 0.92, and 1.3 & 0.4. The cases with T_0 of 680 K and 1045 K have $\overline{\mathrm{Sa}}$ greater than one, suggesting a weak-ignition dominance, while for the case with T_0 of 900 K, a strong ignition is expected. The predictive accuracy of $\overline{\mathrm{Sa}}$ and $F_{\mathrm{Sa},S}$ is further verified by the magnitude of $F_{\mathrm{Da},S}$ as shown in Fig. 25.

Figure 25 shows a strong correlation between $\overline{\mathrm{Sa}}_{\mathrm{n}} = \overline{\mathrm{Sa}}/(1+\overline{\mathrm{Sa}})$ and $F_{\mathrm{Sa},S}$ regardless of T_0, T', ϕ', and fuel types. A good correlation between Sa_{p} and $F_{\mathrm{Sa},S}$ was also demonstrated, which is attributed to the agreement between the predicted Sa_{p} and the volume-average $\overline{\mathrm{Sa}}$. Based on this universal correlation, the magnitude of either $\overline{\mathrm{Sa}}$ or Sa_{p} can be used to quantitatively infer $F_{\mathrm{Sa},S}$, which characterize the combustion intensity. In general, a higher $\overline{\mathrm{Sa}}$ corresponds to a lower $F_{\mathrm{Sa},S}$, which translates into a lower combustion intensity. In particular, $\overline{\mathrm{Sa}}_{\mathrm{n}} = 0.5$ serves as a transition point from

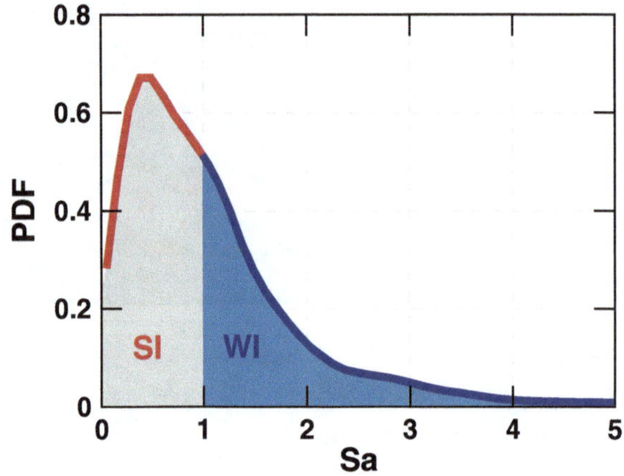

Fig. 24 A representative probability density function of Sa distribution, PDF(Sa). SI and WI denote strong ignition and weak ignition, respectively with a threshold of Sa = 1 (taken from Luong et al. [2])

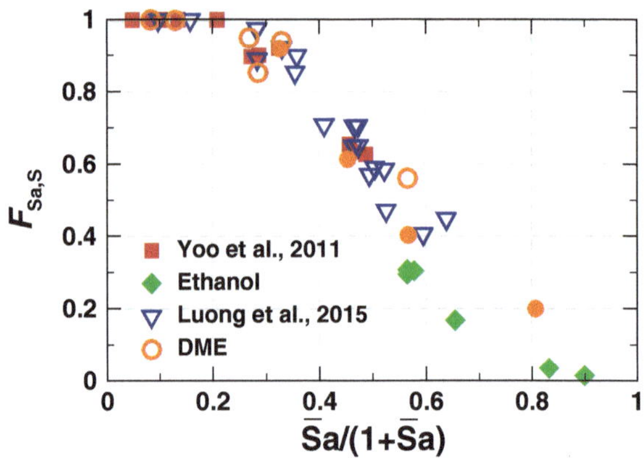

Fig. 25 The volumetric fraction of Sa < 1, $F_{\mathrm{Sa},S}$, as a function of the normalized $\overline{\mathrm{Sa}}$, $\overline{\mathrm{Sa}}_{\mathrm{n}}=\overline{\mathrm{Sa}}/(1+\overline{\mathrm{Sa}})$ to fit within a range between 0 and 1, with 0.5 corresponding to Sa = 1 (taken from Luong et al. [1])

SI to WI dominance with 60% heat released by spontaneous ignition mode ($F_{\mathrm{Sa},S}$ = 0.6). Spontaneous ignition dominance ($F_{\mathrm{Sa},S}$ = 1.0) occurs at $\overline{\mathrm{Sa}} < 0.3$. On the contrary, weak ignition is expected to be dominant at $\overline{\mathrm{Sa}} > 1.0$. In summary, either $\overline{\mathrm{Sa}}$ or Sa_{p} can be used to quantitatively predict the combustion modes over a wide range of initial conditions and different fuel types.

4 Summary

Recent developments in the theoretical study to predict detonation development in IC engine conditions were summarized. A brief review was given on the theoretical framework, starting from the original Zeldovich theory, and the regime diagrams per Bradley [46] and Im [11], and various proposed predictive criteria by Luong et al. [3, 4]. A large collection of DNS simulation data were used for *a priori* and *a posteriori* test for the validation of different predictive criteria based on the initial solution field, and its extension to large turbulent conditions with statistical average and variances has been suggested.

In real combustion engines, additional complexities arise from the front-to-wall and front-to-front interactions which may further enhance or suppress the occurrence and intensity of the detonation event. This is of strong practical significance and is currently being investigated by large-scale simulations [3, 4].

Acknowledgements This work was sponsored by King Abdullah University of Science and Technology and used the resources of the KAUST Supercomputing Laboratory.

References

1. Luong MB, Hernández Pérez FE, Im HG (2020) Prediction of ignition modes of NTC-fuel/air mixtures with temperature and concentration fluctuations. Combust Flame 213:382–393
2. Luong MB, Hernández Pérez FE, Sow A, Im HG (2019) Prediction of ignition regimes in DME/air mixtures with temperature and concentration fluctuations. AIAA SciTech 2019 Forum https://doi.org/10.2514/6.2019-2241
3. Luong MB, Desai S, Hernández Pérez FE, Sankaran R, Johansson B, Im HG (2020) A statistical analysis of developing knock intensity in a mixture with temperature inhomogeneities. Proc Combust Inst 37
4. Luong MB, Desai S, Hernández Pérez FE, Sankaran R, Johansson B, Im HG (2006) Effects of turbulence and temperature fluctuations on knock development in an ethanol/air mixture. Flow Turbul Combust
5. Wang Z, Liu H, Reitz RD (2017) Knocking combustion in spark-ignition engines. Prog Energy Combust Sci 61:78–112
6. Kalghatgi GT, Bradley D (2012) Pre-ignition and 'super-knock' in turbo-charged spark-ignition engines. Int J Engine Res 13:399–414
7. Kalghatgi GT (2015) Developments in internal combustion engines and implications for combustion science and future transport fuels. Proc Combust Inst 35:101–115
8. Figueroa-Labastida M, Badra J, Elbaz AM, Farooq A (2018) Shock tube studies of ethanol preignition. Combust. Flame 198:176–185
9. Meyer JW, Oppenheim AK (1971) On the shock-induced ignition of explosive gases. Symp (Int) Combust 13:1153–1164
10. Lutz AE, Kee RJ, Miller JA, Dwyer HA, Oppenheim AK (1989) Dynamic effects of autoignition centers for hydrogen and c1, 2-hydrocarbon fuels. Symp (Int) Combust 22:1683–1693
11. Im HG, Pal P, Wooldridge MS, Mansfield AB (2015) A regime diagram for autoignition of homogeneous reactant mixtures with turbulent velocity and temperature fluctuations. Combust Sci Technol 187:1263–1275
12. Towery CAZ, Poludnenko AY, Hamlington PE (2020) Detonation initiation by compressible turbulence thermodynamic fluctuations. Combust Flame 213:172–183
13. Hernández Pérez FE, Mukhadiyev N, Xu X, Sow A, Lee BJ, Sankaran R, Im HG (2018) Direct numerical simulations of reacting flows with detailed chemistry using many-core/GPU acceleration. Comput Fluids 173:73–79
14. Desai S, Yu JK, Song W, Luong MB, Hernández Pérez FE, Sankaran R, Im HG (2020) Direct numerical simulations of reacting flows with shock waves and stiff chemistry using many-core/GPU acceleration. Comput Fluids, In revision
15. Sankaran R, Im HG, Hawkes ER, Chen JH (2005) The effects of non-uniform temperature distribution on the ignition of a lean homogeneous hydrogen-air mixture. Proc Combust Inst 30:875–882

16. Chen JH, Hawkes ER, Sankaran R, Mason SD, Im HG (2006) Direct numerical simulation of ignition front propagation in a constant volume with temperature inhomogeneities: I. fundamental analysis and diagnostics. Combust Flame 145:128–144
17. Hawkes ER, Sankaran R, Pébay P, Chen JH (2006) Direct numerical simulation of ignition front propagation in a constant volume with temperature inhomogeneities: II. parametric study. Combust Flame 145:145–159
18. Bansal G, Im HG (2011) Autoignition and front propagation in low temperature combustion engine environments. Combust Flame 158:2105–2112
19. Gupta S, Im HG, Valorani M (2011) Classification of ignition regimes in HCCI combustion using computational singular perturbation. Proc Combust Inst 33:2991–2999
20. Yoo CS, Lu T, Chen JH, Law CK (2011) Direct numerical simulations of ignition of a lean *n*-heptane/air mixture with temperature inhomogeneities at constant volume: parametric study. Combust Flame 158:1727–1741
21. Yu R, Bai X-S (2013) Direct numerical simulation of lean hydrogen/air auto-ignition in a constant volume enclosure. Combust Flame 160:1706–1716
22. El-Asrag HA, Ju Y (2013) Direct numerical simulations of exhaust gas recirculation effect on multistage autoignition in the negative temperature combustion regime for stratified HCCI flow conditions by using H_2O_2 addition. Combust Theory Model 17:316–334
23. El-Asrag HA, Ju Y (2014) Direct numerical simulations of NO_x effect on multistage autoignition of DME/air mixture in the negative temperature coefficient regime for stratified HCCI engine conditions. Combust Flame 161:256–269
24. Bhagatwala A, Lu T, Chen JH (2014) Direct numerical simulations of HCCI/SACI with ethanol. Combust Flame 161:1826–1841
25. Bhagatwala A, Sankaran R, Kokjohn S, Chen JH (2015) Numerical investigation of spontaneous flame propagation under RCCI conditions. Combust Flame 162:3412–3426
26. Bansal G, Mascarenhas A, Chen JH (2015) Direct numerical simulations of autoignition in stratified dimethyl-ether (DME)/air turbulent mixtures. Combust Flame 162:688–702
27. Yoo CS, Luo Z, Lu T, Kim H, Chen JH (2013) A DNS study of ignition characteristics of a lean *iso*-octane/air mixture under and SACI conditions. Proc Combust Inst 34:2985–2993
28. Luong MB, Luo Z, Lu T, Chung SH, Yoo CS (2013) Direct numerical simulations of the ignition of lean primary reference fuel/air mixtures with temperature inhomogeneities. Combust Flame 160:2038–2047
29. Kim SO, Luong MB, Chen JH, Yoo CS (2015) A DNS study of the ignition of lean PRF/air mixtures with temperature inhomogeneities under high pressure and intermediate temperature. Combust Flame 162:717–726
30. Luong MB, Yu GH, Chung SH, Yoo CS (2017) Ignition of a lean PRF/air mixture under RCCI/SCCI conditions: a comparative DNS study. Proc Combust Inst 36:3623–3631
31. Yu GH, Luong MB, Chung SH, Yoo CS (2019) Ignition characteristics of a temporally evolving *n*-heptane jet in an *iso*-octane/air stream under RCCI combustion-relevant conditions. Combust Flame 208:299–312
32. An Y, Jaasim M, Vallinayagam R, Vedharaj S, Im HG, Johansson B (2018) Numerical simulation of combustion and soot under partially premixed combustion of low-octane gasoline. Fuel 211:420–431
33. Mittal G, Sung C-J (2006) Aerodynamics inside a rapid compression machine. Combust Flame 145:160–180
34. Wang Y, Rutland CJ (2005) Effects of temperature and equivalence ratio on the ignition of *n*-heptane fuel spray in turbulent flow. Proc Combust Inst 30:893–900
35. Kokjohn SL, Musculus MPB, Reitz RD (2015) Evaluating temperature and fuel stratification for heat-release rate control in a reactivity-controlled compression-ignition engine using optical diagnostics and chemical kinetics modeling. Combust Flame 162:2729–2742
36. Tang Q, Liu H, Yao M (2017) Simultaneous measurement of natural flame luminosity and emission spectra in a RCCI engine under different fuel stratification degrees. SAE Int J Engines 10:2017–01–0714

37. Schießl R, Maas U (2003) Analysis of endgas temperature fluctuations in an SI engine by laser-induced fluorescence. Combust Flame 133:19–27
38. Kaiser SA, Schild M, Schulz C (2013) Thermal stratification in an internal combustion engine due to wall heat transfer measured by laser-induced fluorescence. Proc Combust Inst 34:2911–2919
39. Söderberg F, Johansson B, Lindoff B (1998) Wavelet analysis of in-cylinder ldv measurements and correlation against heat-release. SAE paper 107:444–456
40. Funk C, Sick V, Reuss DL, Dahm WJA, Turbulence properties of high and low swirl in-cylinder flows, SAE paper (2002) 2002–01–2841
41. Miles PC (2008) Turbulent flow structure in direct-injection, swirl-supported diesel engines. In: Flow and combustion in reciprocating engines, experimental fluid mechanics, Springer, pp 173–256
42. Akkerman V, Ivanov M, Bychkov V (2009) Turbulent flow produced by piston motion in a spark-ignition engine. Flow Turbul Combust 82:317–337
43. Petersen BR, Ghandhi JB (2010) High resolution scalar dissipation and turbulence length scale measurements in an internal combustion engine. SAE Int J Engines 3:65–83
44. Zeldovich YB (1980) Regime classification of an exothermic reaction with nonuniform initial conditions. Combust Flame 39:211–214
45. Gu X, Emerson D, Bradley D (2003) Modes of reaction front propagation from hot spots. Combust Flame 133:63–74
46. Bradley D, Morley C, Gu XJ, Emerson DR (2002) Amplified pressure waves during autoignition: Relevance to CAI engines. SAE Technical Paper (2002) 2002–01–2868
47. Lee JH, Knystautas R, Yoshikawa N (1978) Photochemical initiation of gaseous detonations. Acta Astronautica 5:971–982
48. Lee JHS (2008) The Detonation Phenomenon. Cambridge University Press
49. Su J, Dai P, Chen Z (2020) Detonation development from a hot spot in methane/air mixtures: effects of kinetic models. Int J Engine Res 146808742094461
50. Gao Y, Dai P, Chen Z (2020) Numerical studies on autoignition and detonation development from a hot spot in hydrogen/air mixtures. Combust Theo Modell 24:245–261
51. Bates L, Bradley D, Paczko G, Peters N (2016) Engine hot spots: modes of auto-ignition and reaction propagation. Combust Flame 166:80–85
52. Bates L, Bradley D (2017) Deflagrative, auto-ignitive, and detonative propagation regimes in engines. Combust Flame 175:118–122
53. Peters N, Kerschgens B, Paczko G (2013) Super-knock prediction using a refined theory of turbulence. SAE Int J Engines 6:953–967
54. Dai P, Chen Z, Chen S, Ju Y (2015) Numerical experiments on reaction front propagation in n-heptane/air mixture with temperature gradient. Proc Combust Inst 35:3045–3052
55. Yu H, Chen Z (2015) End-gas autoignition and detonation development in a closed chamber. Combust Flame 162:4102–4111
56. Dai P, Qi C, Chen Z (2017) Effects of initial temperature on autoignition and detonation development in dimethyl ether/air mixtures with temperature gradient. Proc Combust Inst 36:3643–3650
57. Terashima H, Matsugi A, Koshi M (2017) Origin and reactivity of hot-spots in end-gas autoignition with effects of negative temperature coefficients: Relevance to pressure wave developments. Combust Flame 184:324–334
58. Pan J, Wei H, Shu G, Chen R (2017) Effect of pressure wave disturbance on auto-ignition mode transition and knocking intensity under enclosed conditions. Combust Flame 185:63–74
59. Wei H, Chen C, Shu G, Liang X, Zhou L (2018) Pressure wave evolution during two hotspots autoignition within end-gas region under internal combustion engine-relevant conditions. Combust Flame 189:142–154
60. Sow A, Lee BJ, Hernández Pérez FE, Im HG (2019) Detonation onset in a thermally stratified constant volume reactor. Proc Combust Inst 37:3529–3536
61. Pan J, Wei H, Shu G, Chen Z, Zhao P (2016) The role of low temperature chemistry in combustion mode development under elevated pressures. Combust Flame 174:179–193

62. Pan J, Dong S, Wei H, Li T, Shu G, Zhou L (2019) Temperature gradient induced detonation development inside and outside a hotspot for different fuels. Combust Flame 205:269–277
63. Desai S, Sankaran R, Im HG (2019) Unsteady deflagration speed of an auto-ignitive dimethyl-ether (DME)/air mixture at stratified conditions. Proc Combust Inst 37:4717–4727
64. Desai S, Sankaran R, Im HG (2020) Auto-ignitive deflagration speed of methane (CH_4) blended dimethyl-ether (DME)/air mixtures at stratified conditions. Combust Flame 211:377–391
65. Robert A, Richard S, Colin O, Martinez L, De Francqueville L (2015) LES prediction and analysis of knocking combustion in a spark ignition engine. Proc Combust Inst 35:2941–2948
66. Robert A, Richard S, Colin O, Poinsot T (2015) LES study of deflagration to detonation mechanisms in a downsized spark ignition engine. Combust Flame 162:2788–2807
67. Ali MJM, Luong MB, Sow A, Hernández Pérez FE, Im HG (2018) Probabilistic approach to predict abnormal combustion in spark ignition engines. SAE paper 2018–01–1722
68. Wei H, Chen C, Zhou H, Zhao W, Ren Z (2016) Effect of turbulent mixing on the end gas auto-ignition of n-heptane/air mixtures under IC engine-relevant conditions. Combust Flame 174:25–36
69. Chen L, Wei H, Chen C, Feng D, Zhou L, Pan J (2019) Numerical investigations on the effects of turbulence intensity on knocking combustion in a downsized gasoline engine. Energy 166:318–325
70. Zhong L, Liu C (2019) Numerical analysis of end-gas autoignition and pressure oscillation in a downsized SI engine using large eddy simulation. Energies 12:3909
71. Zhang T, Sun W, Wang L, Ju Y (2019) Effects of low-temperature chemistry and turbulent transport on knocking formation for stratified dimethyl ether/air mixtures. Combust Flame 200:342–353
72. Nogawa T, Terashima H (2020) Effects of globally stratified temperature distributions and NTC characteristics on end-gas combustion modes. Combust Sci Technol 1–25
73. Wang L, Peters N (2006) The length-scale distribution function of the distance between extremal points in passive scalar turbulence. J Fluid Mech 554:457–475
74. Pal P, Valorani M, Arias PG, Im HG, Wooldridge MS, Ciottoli PP, Galassi RM (2017) Computational characterization of ignition regimes in a syngas/air mixture with temperature fluctuations. Proc Combust Inst 36:3705–3716
75. Luong MB, Lu T, Chung SH, Yoo CS (2014) Direct numerical simulations of the ignition of a lean biodiesel/air mixture with temperature and composition inhomogeneities at high pressure and intermediate temperature. Combust Flame 161:2878–2889
76. Luong MB, Yu GH, Lu T, Chung SH, Yoo CS (2015) Direct numerical simulations of ignition of a lean *n*-heptane/air mixture with temperature and composition inhomogeneities relevant to HCCI and SCCI combustion. Combust Flame 162:4566–4585
77. Zhang J, Luong MB, Hernández Pérez FE, Han D, Im HG, Huang Z (2019) Exergy loss of dme/air mixtures and ethanol/air mixtures with temperature and concentration fluctuations under HCCI/SCCI conditions: a DNS study, submitted to proceedings of the combustion institute (2019)
78. Luong MB, Yu GH, Chung SH, Yoo CS (2017) Ignition of a lean PRF/air mixture under RCCI/SCCI conditions: chemical aspects. Proc Combust Inst 36:3587–3596
79. Luong MB, Sankaran R, Yu GH, Chung SH, Yoo CS (2017) On the effect of injection timing on the ignition of lean PRF/air/EGR mixtures under direct dual fuel stratification conditions. Combust Flame 183:309–321
80. Echekki T, Chen JH (2003) Direct numerical simulation of autoignition in non-homogeneous hydrogen-air mixtures. Combust Flame 134:169–191

Numerical Investigation of Flow-Acoustics Coupling in a Half-Dump Combustor Using Hybrid CFD-CAA Methodology

Bharat Bhatia and Ashoke De

Abstract The present work investigates the flow-acoustics coupling in a half-dump combustor emulating methane-air non-premixed combustion for varying Reynolds numbers (Re) and fuel flow rates. The study uses hybrid computational aero-acoustics (CAA) simulation methodology, which is based on Lighthill's acoustic analogy, to compute the acoustic flow field. The underlying turbulent flow is modeled using a hybrid Reynolds Averaged Navier Stokes/Large Eddy Simulation (RANS/LES) approach, while the turbulence-chemistry interaction is handled using steady laminar flamelet model (SFLM). Chemical kinetics for methane-air combustion is represented using the GRI-3.0 mechanism. In simulations, time-resolved data is analyzed to understand the flow physics in greater detail, while the comparison is made with the measurements to benchmark the numerical code used herein. In particular, the information of non-dimensional numbers, i.e., Helmholtz and Strouhal numbers, allows to characterize the instability modes and distinguishes between hydrodynamic (vortex shedding modes) and acoustics instabilities (acoustic modes). The observation yields that unsteady pressure fluctuations either come from the natural acoustics mode of the channel or vortex shedding process and thereby exhibiting multiple frequencies in the domain for lower Re case, while a single dominant frequency for higher Re case. Moreover, the computed Rayleigh index using the unsteady heat release and pressure oscillations from reacting conditions confirms the mode of instability, whether it is driving (positive net value) or damping (negative net value), at different frequency levels.

Keywords Hybrid RANS/LES · CFD-CAA · Flow-acoustics coupling

B. Bhatia · A. De (✉)
Department of Aerospace Engineering, Indian Institute of Technology Kanpur, Kanpur 208016, India
e-mail: ashoke@iitk.ac.in

A. Gupta et al. (eds.), *Advances in Energy and Combustion*, Green Energy and Technology,
https://doi.org/10.1007/978-981-16-2648-7_15

1 Introduction

Flow-acoustics couplings or combustion instabilities in real-life combustion systems such as gas turbines, rocket engines, burners, and furnaces are of paramount importance in the last couple of decades [1–11]. The reason is the generated noise and exciting acoustic instabilities, which have a significant impact on the structural loading in confined reacting systems. Vortex shedding reported being one of the essential driving mechanisms of flow-acoustics coupling [5], where the interactions between large-scale and small-scale turbulence structures lead to the generation of periodic heat lease. The shear layer vorticity in dump combustor is primarily formed due to the interaction of high and low-speed streams at the dump plane. Many studies, including both experimental and computational approaches, have already reported the importance of vortex structures on flow-acoustics coupling [1–11]. Also, some studies reported combustion instability in practical combustor using a high fidelity turbulence model, i.e., large eddy simulation (LES) [2, 3, 11].

In this work, we utilize a hybrid computational aero-acoustics (CAA) methodology, based on Lighthill's analogy, for predictions of the acoustic flow field. Mainly, this approach uses two-step procedures to compute the acoustic flow field such as (a) firstly, the scale resolved turbulence calculations are done using high fidelity turbulence model, and then, (b) the acoustic flow field is computed using the CAA formulation invoking the results of the first step. In general, this is the most affordable approach in the current scenario as the direct computation of both the sources and resulting acoustic fields is not a viable solution for practical systems [9, 12]. Moreover, we have invoked hybrid RANS/LES [9, 13] approach to model the underlying turbulent flow field. The hybrid approach utilizes the advantages of both RANS and LES at reasonable computational cost, where the LES model resolves the separated flow regions, and the RANS model becomes active close to the boundary layer (attached to a solid surface). Hybrid RANS/LES has gained a lot of attention or an excellent alternative option to LES for simulating flows with large-scale separation where one can obtain better results with a reasonable computational cost.

Therefore, the primary focus of the present work is two-fold: (a) to access the predictive capability of hybrid RANS/LES model in conjunction with CAA methodology and (ii) to analyze and shed light on the physics of flow-acoustics coupling and the associated unsteady vortex dynamics in a half-dump combustor. The primary aim is to differentiate between hydrodynamic mode of instability (low Re) with the acoustic mode of instability (high Re), which is quite often mixed-up in the open literature and termed as acoustic instability [9]. Also, the emphasis is put to establish the theory behind low Re hydrodynamic instability (primarily due to unsteady vortex dynamics) and high Re acoustic instability (mostly due to strong coupling between heat release and pressure fluctuations). Another goal is to access the applicability of a reduced-order model for practical applications. To achieve the said objective, we have investigated a laboratory scale dump combustor for which measurements are available for comparison and assessment [7]. Furthermore, we have analyzed the unsteady flow features (heat release and pressure fluctuations) to estimate the

Rayleigh index, which confirms the onset of instabilities. Also, the detailed analysis of non-dimensional numbers (Helmholtz and Strouhal numbers) for varying Reynolds number and fuel flow rates differentiates between two instability modes, i.e., natural acoustic modes and hydrodynamics modes (vortex shedding modes) in the combustor.

2 Numerical Details

In the hybrid approach, the hybrid RANS/LES model employs to compute the turbulent flow field, while the sound pressure level is calculated using CAA formulation at the next step. CFD simulations provide the required acoustics source information to the CAA model to compute this far-field sound pressure level. The general form of the Favre-filtered governing equations for the transport of mass, momentum, and turbulence can be expressed as:

$$\frac{\partial}{\partial t}\left(\bar{\rho}\tilde{\phi}\right)+\frac{\partial}{\partial x_j}\left(\bar{\rho}\,\tilde{u}_j\tilde{\phi}\right)=\frac{\partial}{\partial x_j}\left(\bar{\rho}D\frac{\partial\tilde{\phi}}{\partial x_j}\right)+\langle S_\phi\rangle \tag{1}$$

where $\tilde{\phi}$ represents the flow field variable solved for, and $\langle S_\phi\rangle$ represents the source of $\tilde{\phi}$. The scalar diffusion coefficient and mean density are represented by D and ρ, respectively. The turbulence model provides turbulence scale information. In the context of the present work, we have invoked the SST k-ω-based hybrid RANS/LES model in conjunction with a steady laminar flamelet model (SLFM) [13]. The inherent advantage of this hybrid method lies in the calculation of eddy viscosity using zonal information. It automatically switches from RANS region to the LES region through the following modification of dissipation term of the turbulent kinetic energy:

$$Y_k = \rho\beta * k\omega F_{\text{hybrid}} \tag{2}$$

where F_{hybrid} is defined as:

$$F_{\text{hybrid}} = \max\left(\frac{l_t}{C_d\Delta_{\max}}, 1\right) \tag{3}$$

where C_d is the model constant with a value of 0.61, $\Delta_{\max}$ is the maximum local grid spacing (Δ_x, Δ_y, Δ_z), and the turbulent length scale definition $\left(l_t = \frac{\sqrt{k}}{0.09*\omega}\right)$ comes from RANS model. However, the grid-induced separation in some portion of attached boundary layer (where the local surface grid spacing < the boundary layer thickness, i.e., $\Delta_s < d\delta$, with d of the order one) remains a significant problem in hybrid formulation, and this requires some special treatment to avoid such undesirable separation, or instead, the boundary layer needs to be protected. In the present scope of the work, this is achieved through a blending function of the SST model as:

$$F_{\text{hybrid}} = \max\left(\frac{l_t}{C_d \Delta \max}(1 - F_S), 1\right) \tag{4}$$

With $F_S = 0, F_1, F_2$, where F_1, F_2 are the blending functions of the SST model, while $F_S = F_2$ provides the maximum level of protection against grid-induced separation [13].

2.1 Turbulence-Chemistry Interaction Model

The local flame structure of the turbulent flames can be described as an ensemble of laminar and one-dimensional local structures. With unitary Lewis number, the steady equations can be expressed concerning mixture fraction Z and scalar dissipation rate χ as follows:

$$0 = \frac{\chi}{2}\frac{\partial^2 Y_i}{\partial Z^2} + \frac{\dot{\omega}_i}{\rho} \tag{5}$$

$$0 = \frac{\chi}{2}\frac{1}{C_p}\frac{\partial^2 h}{\partial Z^2} - \frac{\chi}{2}\frac{1}{C_p}\sum_{k=1}^{N} h_k \frac{\partial^2 Y_k}{\partial Z^2} - \frac{1}{\rho C_p}\sum_{k=1}^{N} h_k \dot{\omega}_k \tag{6}$$

The connection between the physical space and mixture fraction space is achieved through the scalar dissipation rate, which quantifies the deviation from equilibrium and is defined as:

$$\chi = 2D\left|\frac{\partial Z}{\partial x_j}\right|^2 \tag{7}$$

The scalar dissipation rate varies along the flamelet and is modeled as Eq. (7) as depicted below:

$$\chi(Z) = \chi_{\text{st}} \exp\left[2\left(\text{erfc}^{-1}(2Z_{st})\right)^2 - 2\left(\text{erfc}^{-1}(2Z)\right)^2\right] \tag{8}$$

where erfc^{-1} stands for the inverse of the complementary error function. The turbulent flame brush is represented as the ensemble of diffusion flamelets, where the Favre averaged temperature and species mass fraction of a turbulent flame can be determined as:

$$\tilde{\phi} = \iint \phi(Z, \chi_{\text{st}}) p(Z, \chi_{\text{st}}) dZ d\chi_{\text{st}} \tag{9}$$

where a presumed β-PDF is used to define the probability of the mixture fraction. The temperature and mean density have an extra dimension of mean enthalpy $\widetilde{H}$ to consider the non-adiabatic steady diffusion flamelets. The species mass fraction is assumed to have an eligible effect by the heat loss or gain by the system. The evolution of mixture fraction in the physical space is represented by the transport equation of Z and Z'' is given as:

$$\frac{D}{Dt}(\rho\tilde{Z}) = \nabla.\left(\frac{\mu_{\text{eff}}}{\sigma_t}\nabla\tilde{Z}\right) \tag{10}$$

$$\frac{D}{Dt}(\rho\widetilde{Z''^2}) = \nabla.\left(\frac{\mu_{\text{eff}}}{\sigma_t}\nabla\widetilde{Z''^2}\right) + C_g\mu_t.\left(\nabla\tilde{Z}\right)^2 - C_d\rho\frac{\varepsilon}{k}\widetilde{Z''^2} \tag{11}$$

where $Z'' = Z - \bar{Z}$, $\mu_{\text{eff}} = \mu_l + \mu_t\sigma_t C_g$, and C_d are 0.85, 2.86, and 2.0, respectively. More details regarding flamelet modeling can be found in the literature [13–16].

2.2 *Acoustics Modeling Using Lighthill's Analogy (CFD-CAA)*

Lighthill's [17] analogy separates the analysis of acoustics into two steps. The first step is sound generation induced by fluid flow in any real continuous medium. The second step is sound propagation in an acoustic medium at rest, exerted by external fluctuating sources, which are calculated from the first step. In Lighthill's analogy, the momentum equation is rewritten in the conservative form while leaving out the viscous stresses as:

$$\rho\left[\frac{\partial u_i}{\partial t} + u_j\frac{\partial u_i}{\partial x_j}\right] = -\frac{\partial P}{\partial x_i} \tag{12}$$

Eliminating ρu_i and subtracting $c_o^2\left(\frac{\partial^2\rho'}{\partial x_i\partial x_j}\right)c_0^2(\delta^2\rho'/\delta x_i\delta x_j)$ gives the Lighthill's equation

$$\frac{\partial^2\rho'}{\partial t^2} - c_o^2\frac{\partial^2\rho'}{\partial x_i\partial x_i} = -\frac{\partial^2 T_{ij}}{\partial x_i\partial x_j} \tag{13}$$

where $T_{ij} = \rho u_i u_j + (P - c_o^2\rho)\delta_{ij}$. More details about the Lighthill's analogy can be found in the literature [13, 17].

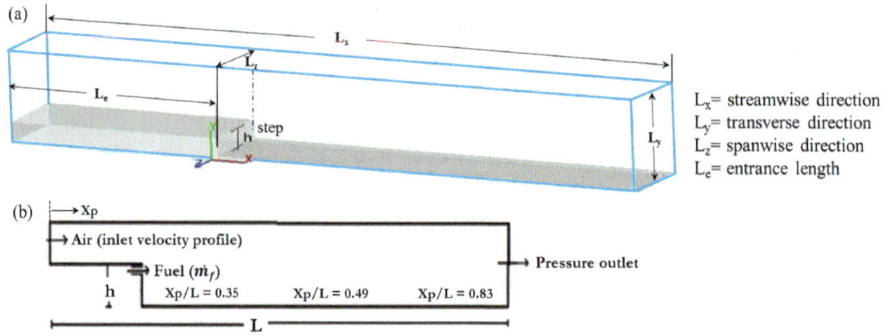

Fig. 1 Computational domain: **a** 3D view with dimensions, **b** BCs and probing locations

2.3 Setup of the CFD Calculation

2.3.1 Computational Domain and Grid

Because of the nature of the combustor geometry, the simulations are performed in a 3D configuration of the given combustor geometry [7], as depicted in Fig. 1a. Figure 1b projects the 2D sectional view where different lengths are identified as streamwise length $L = L_x = 48$ h (includes the entrance/inlet section of $L_e = 14.67$ h), transverse length $L_y = 2$ h, and spanwise length of $L_z = 2$ h. Here, h represents the step height, and the value of 30 mm has been used in the present calculations. The computational domain consists of a structured mesh of hexahedral cells, where the streamwise and spanwise directions use uniform grid spacing, and the transverse direction invokes non-uniform distribution along with near-wall grid refinement. The used grid (fine mesh) consists of 936 × 91 × 14 grid points in all three directions. Also, two other meshes, having 624 × 61 × 9 grid points (medium mesh) and 312 × 31 × 5 grid points (coarse mesh) , have been considered for grid independence analysis.

2.3.2 Boundary Conditions

The boundary of the computational domain at the highest axial distance is set to the pressure outlet (Fig. 1b). For the current configuration, the Reynolds number based on step height is defined as $\mathrm{Re} = U_o h/\nu \mathrm{Re} = U_0 h/\nu$, and the expansion ratio as $ER = L_y/(L_y - h) = 2$. At the left inlet boundary ($X = -14.6$ h), the turbulent inlet velocity profile is set using the information of turbulent boundary layer thickness at the step ($\delta_{99}/h = 1.2$). All the walls are no-slip walls, and the periodic boundary condition is utilized along the spanwise directions. For stability, the Courant–Freidrich–Lewy (CFL) condition determines the time step chosen for the simulations.

The time-resolved unsteady computations provide statistical data. To achieve useful statistical information, we have collected large numbers of samples (~20 flow-through times) after completion of the initial transient phase (~10 flow-through times). Figure 1b shows the fuel injection location, which is located 1 mm below the step corner. The desired fuel flow ($\dot{m}_f$) rate is obtained through a square hole of 0.13 h, which is centered along the spanwise direction. The unsteady signals of pressure and heat release fluctuations are also monitored in consistency with the measurement's locations [7]. Figure 1b depicts three such points, which are located at $X_p/L = 0.35, 0.49,$ *and* 0.83, where X_p is the distance from the inlet of the computational domain, and L is the length of the combustor to maintain consistency with the measurements in reference [7].

2.3.3 Computational Algorithm

The flow simulations are performed ANSYS FLUENT [13], where the pressure and velocity coupling is achieved through the Semi-Implicit Method for Pressure Linked Equations (SIMPLE) algorithm. The second-order discretization scheme is consistently used for all the terms. Particularly, the convective fluxes of all the equations excluding momentum are discretized using the second-order upwind scheme, while the bounded second-order central difference scheme is invoked for momentum equations; while the diffusion term discretization involves second-order central differencing. Time discretization employs a second-order implicit scheme, and the methane-air combustion is represented using detailed chemistry, i.e., GRI 3.0 mechanism containing 52 species and 325 reactions [18]. The second-order discretization scheme is consistently used for all the terms.

3 Results

In this section, we initially present non-reacting simulation results along with grid independence and compare them with the measurements to have confidence in the grid and turbulence model. Finally, we analyze the coupled reacting flow simulations for varying fuel flow rates and Reynolds numbers to examine flow-acoustics coupling in the chosen half-dump combustor.

3.1 Grid Independence and Non-reacting Flow Results

Here, we report two different cases for validation of the solver.

(a) Re = 5100

The backward facing step (BFS) geometry is schematically shown in Fig. 1. This case was experimentally investigated by Jovic and Driver [19]. The Re based on step

Table 1 Recirculation bubble length (X_r/h) comparison for Re = 5100

Re	Experimental	Coarse mesh (0.22 M)	Fine mesh (0.62 M)
5100	6.1	6.32	6.15

height is 5100. We have considered two different sets of grids: One is with coarser gird of 0.22 M (million) cells and another one with fine resolution of 0.6 M (million) cells. The computational domain has streamwise length $L = L_x = 30$ h (includes the entrance/inlet section of $L_e = 10$ h), transverse length $L_y = 6$ h, and spanwise length of $L_z = 4$ h. The expansion ratio of this case is $ER = 1.2$. At the left-most boundary, the turbulent inlet velocity profile is imposed ($\delta_{99}/h = 1.2$), while the rightmost boundary is set to pressure outlet. No-slip boundary condition is used for the walls, the periodic boundary condition along the spanwise directions, and the free-stream condition is utilized at the top-most boundary along the transverse direction. The statistics are collected over ~30 flow-through times after completion of the initial transient phase (Table 1).

Figure 2 shows the predicted results of mean and rms of axial velocity at two different axial locations using a fine grid only. Mean and rms velocity profiles are in good agreement with the measurements, and the size of the recirculation bubble is well predicted.

(b) Re = 18,000

The considered configuration is already illustrated in Fig. 1. The domain dimension and grid details are already reported in Sect. 2.3. Here, we have analyzed three sets of grids to achieve grid independence. Figure 3 depicts the mean and rms profiles of axial velocity at different locations. The mean and rms profiles using medium and fine grids are in excellent agreement with each other alongside the measurements of [7]. However, the rms profiles using the coarse grid show some discrepancies, and these discrepancies are more visible in Fig. 4 which illustrates the recirculation

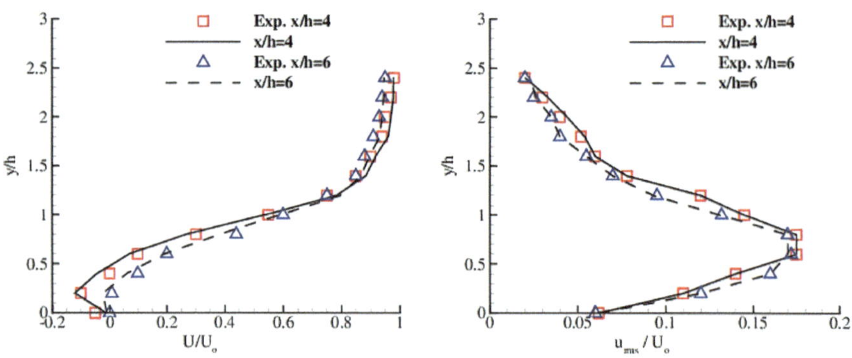

Fig. 2 Mean and rms of axial velocity for Re = 5100. Symbols are measurements [19], and lines are predictions

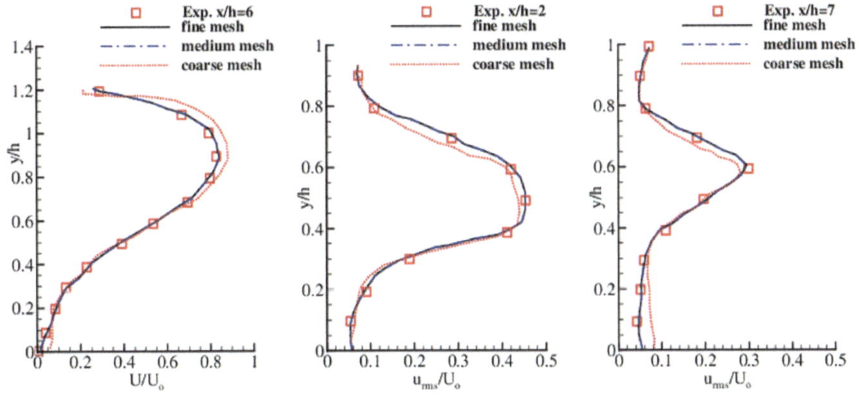

Fig. 3 Mean and rms of axial velocity for Re = 180,00. Symbols are measurements [7], and lines are predictions

Fig. 4 Recirculation bubble length (X_r/h) for Re = 18,000. Lines represent the streamwise velocity at first cell above the wall

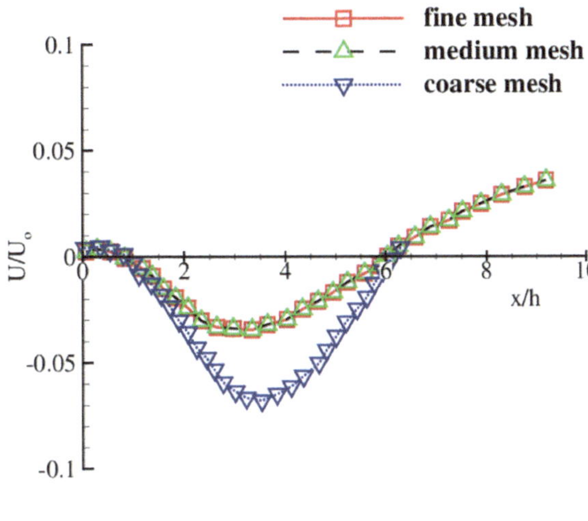

bubble predictions.

The velocity profiles are well captured in both the recirculation and recovery regions. Necessarily, the turbulent activities are dominant in the recirculation regions and decay gradually in the downstream locations. At downstream, the inner layer of Reynolds stresses behaves like an ordinary turbulent boundary layer, whereas in the outer layer, it decays slowly due to the presence of large eddies that are generated in the upstream shear layer. Coarse mesh shows some over predictions in the recovery region. Moreover, the length of the recirculation bubble using coarse mesh is X_r/h = 6.16, while the same using the other two meshes is $X_r/h = 6.03$ and is in good agreement with the experiments ($X_r/h = 6.01$). Although the overall data predictions using medium and fine meshes are in god agreement with measurements, the fine mesh has been considered for details CAA calculations due to its better grid resolution in wall units. The grid resolution of the fine mesh in wall units is $x^{+}=2$, $y^{+}_{min} = 0.14$, $y^{+}_{max} = 3$, and $z^{+} = 6$, where '+' superscript denotes near-wall units in terms of friction velocity u_τ and kinematic viscosity ν.

3.2 *Reacting Flow Results Using CFD-CAA*

The study on the reacting cases is conducted on various fuel and air mass flow rates with detailed chemistry. In the detailed discussion, the results of reacting flow simulations are compared with the results of non-reacting flow simulations (that is cold flow simulations), which helps to segregate the frequency contribution due to hydrodynamic factors. This result along with the instantaneous pressure and heat release signal helps in the determination of the driving mechanism of combustion instability. Moreover, the Rayleigh index is also plotted to support the data. The simulation processes, for both reacting flow and non-reacting flow, are done for two extreme cases with Reynolds number Re = 47,000 and Re = 18,000 while for the intermediate Reynolds number (i.e., Re = 10,000, 26,000, 30,000, 32,000, 36,000, and 45,000), only reacting flow simulations are carried out.

The numerical simulations are carried for eight different Reynolds numbers and fuel flow, as shown in Table 2. First, the analysis of all the three fuel mass flow rate cases is shown, which highlights the mass flow effect on the combustor characteristics. Later, the detailed investigation is carried out for the fuel mass flow rate of 142 mg/s. For analyzing the acoustic signal at various locations across the domain, three receiver probe points are located in a similar position as in the experiments. The dominant frequency is noted at each probe. This frequency is used in the calculations and analysis carried out for the simulations, including the calculation of Helmholtz

Table 2 Various cases for flow-acoustic analysis. Here, quantities marked with (*) are calculated according to the given equation

Reynolds no. x 10^4	Fuel flow rate					
	142 mg/s		229 mg/s		316 mg/s	
	ϕ*	MR*	ϕ*	MR*	ϕ*	MR*
1.0	0.04762	0.275882				
1.8	0.027603	0.085149	0.044515	0.137318	0.061426	0.189486
2.6	0.019308	0.040811	0.031138	0.065815	0.042968	0.090819
3.0	0.017438	0.030654				
3.2	0.015142	0.026942				
3.6	0.013297	0.021287	0.021443	0.034329	0.02959	0.047372
4.5	0.011114	0.013624				
4.7	0.010661	0.012489	0.017192	0.020141	0.023724	0.027792

(*He*) number and Strouhal (*St*) number. The definitions of He and St are recasted as:

$$
\begin{aligned}
He &= f * L/c \\
St &= f * h/U
\end{aligned}
\tag{14}
$$

where f is observed frequency, L is the length of the combustor, c is the speed of the sound at the reference condition (inlet in this case), h is the height of the step at dump plane, and U is the average airflow velocity at the inlet. The advantage of using these non-dimensional numbers lies in the fact that the acoustic and the hydrodynamic contributions to the frequency become discernible. Helmholtz number takes care of the acoustics contribution while the Strouhal number looks into the hydrodynamic factors.

(*)—The calculation of the equivalence ratio and the momentum ratio is carried out by the following formula:

$$
\begin{aligned}
\phi &= s * (F/A) \\
\mathrm{MR} &= \frac{(\rho u^2)_F}{(\rho u^2)_{ox}}
\end{aligned}
\tag{15}
$$

3.2.1 Fuel Flow Rate of 142 Mg/S ($\dot{m}_f$ = 142 mg/s)

Figure 5 illustrates two non-dimensional numbers, i.e., Helmholtz number (*He*) and Strouhal number (*St*), at different locations for varying Reynolds number (Re). Notably, when the *He* corresponding to the dominant frequency at the probing point is constant with the changing Reynolds number (Re), the frequency f also remains constant, while the change of *He* with respect to Re indicates the role of hydrodynamics as a contributor to the frequency f. The results indicate that for the lower Reynolds number range, if the observed frequency of the flow is equivalent to the vortex shedding frequency in the recirculation zone, then *St* decreases hyperbolically, and *He* linearly increases, while *St* approaches a constant value at higher Reynolds number. The results here show four data clusters in Helmholtz number plot in Fig. 5a. The corresponding data can also be observed showing Strouhal number variation against the Reynolds number range (Fig. 5b). The data matches the experimental data quite well.

Figure 5c illustrates the schematic of the whole process of lock-on (represented from [7]). The named clusters—I, II, III, IV represent the pattern of variation of the dominant frequency with the increasing Re. Here, f_a denotes the acoustic frequency while f_v denotes the vortex shedding frequency. Cluster I denotes the complete hydrodynamic dominance, while Cluster II shows acoustic dominance. Cluster III and IV exhibit the regime when the flow field gets lock-on the combustor acoustic modes. Hence, the data points (Fig. 5a) in the nearly constant *He* (Cluster II) appear to be transition region dominated by the acoustic modes, where the transitioning takes

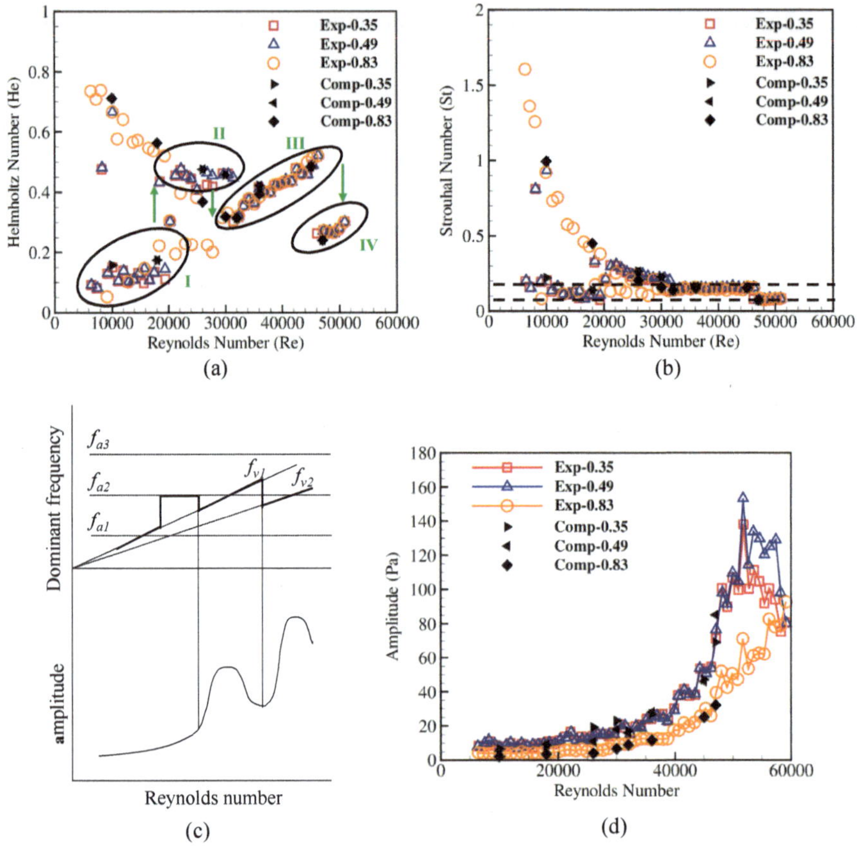

Fig. 5 **a** Helmholtz number, **b** Strouhal number at three points for $\dot{m}_f$ = 142 mg/s, **c** schematic of the dominant frequency measured at the probe with varying Re showing flow-acoustics lock-on during combustion, and **d** amplitude versus Re. Experimental data are taken from [7]

place from linearly increasing trend at lower Re (cluster I) to a linearly increasing trend at higher Re (cluster III & IV). The 1st transition (Cluster I→II) is primarily attributed to the vortex shedding at the dump plane, and it is similar to the case of typical vortex shedding phenomena in the wake of oscillating bodies. With the increase of Reynolds number in the lock-on regime, this vortex shedding frequency shift into the natural frequency of the acoustic mode of the duct, as reported earlier [7]. The corresponding data points in *St* plot exhibit an asymptotic behavior to a constant value (after decreasing hyperbolically) with increasing Re that signifies the nonlinear rise in amplitude due to flow-acoustics lock-on. The dominant frequency for this floc-acoustics lock-on happens to be the vortex shedding frequency at the step/dump plane. Mostly, the data points in the linearly increasing portion of the plot (cluster I, III, IV) are between successive vortex shedding modes, and the linear variation with different slopes indicates different modes of vortex shedding, suggesting the possibility of vortex pairing or merging in a lower shedding frequency (lower Strouhal number) and consequently a lower slope. The observed dominant frequency corresponds to the vortex shedding, as the vortex shedding frequency at the step gives rise to the *St* = 0.2, which happens to be the frequency at which flow-acoustic lock-on takes place [7, 20]. The decrease of Strouhal number (Cluster IV) to half of its value in cluster III is indicative of the dominance of the vortex merging process in high Reynolds number cases.

This interesting phenomenon is represented by the cluster (III & IV), which after the initial linear growth of *He* (shown as cluster III), suddenly drops to cluster IV. This is a feature of flow-acoustic lock-on (Fig. 5a). This difference of lock-on phenomena is visible for just the clusters III and IV, and not clusters I. This means that although the linearly increasing clusters III and IV show hydrodynamic behavior (just like the cluster I) but they are controlled by the combustor acoustics (through the lock-on phenomena) and accompanied by a steep rise in the amplitude (Fig. 5d).

The abrupt increase in the amplitude of noise corresponds to the flow-acoustics lock-on at higher Re. Along with this amplitude plot (Fig. 5c), the schematic of flow-acoustics lock-on (Fig. 5c) suggests the actual process of shifting the natural acoustic mode of the duct into vortex shedding variation with increasing Re. The present predictions using hybrid RANS/LES approach seem to be much better compared to previous URANS predictions [21], where authors reported all the modes to be acoustics mode at all Re. However, the present simulations capture two distinct modes at lower Re due to hydrodynamic instability, and a single mode at higher Re when flow-acoustics lock-on takes place/hydrodynamic instability merges with acoustics instability as reported in the literature [20].

3.2.2 Fuel Flow Rate of 229 Mg/S ($\dot{m}_f = 229\ \mathrm{mg/s}$)

A similar trend is observed in the higher fuel flow rate of 229 mg/s, where the four Reynolds number cases are simulated (Re = 18,000, 26,000, 36,000, and 47,000). Again, in this case, the results here indicate four data clusters in Helmholtz number plot in Fig. 6a, like the previous case. The corresponding data can also be observed in Fig. 6b showing Strouhal number variation against the Reynolds number range. The results match the experimental data quite well. A notable difference is that the slope of the linearly increasing cluster I, III, and IV has increased on the Helmholtz plot, and the transition difference has decreased. This shows the higher sensitivity of the vortex shedding phenomenon to the increasing Reynolds number. Further, a delay of 3rd transition is also evident from both the figures, especially while looking at the measurements [7]. However, the present simulations with 4 Re do not completely cover the 3rd transition, as we have done for the previous case (Fig. 5). The frequency of the vortex shedding can be known from the Strouhal number plot in Fig. 6b, whereas the cluster II corresponding to the constant portion of the plot indicates the dominance of acoustic modes in Fig. 6a.

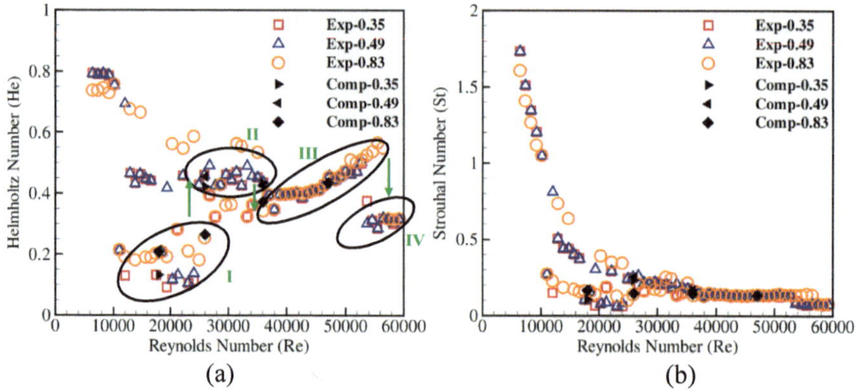

Fig. 6 **a** Helmholtz number and **b** Strouhal number at three points for $\dot{m}_f = 229$ mg/s. Experimental data are taken from [7]

3.2.3 Fuel Flow Rate of 316 Mg/S ($\dot{m}_f = 316$ mg/s)

For this case also, four Reynolds number cases are simulated as specified in Table 2. The data agree with the experimental results quite well. In this case, the second transition almost diminishes. Moreover, the third transition is delayed significantly, also evident from both the figures. The frequency of the vortex shedding can be known from the Strouhal number plot in Fig. 7b, while the largely flat region in Fig. 7a indicates the domination of acoustic modes.

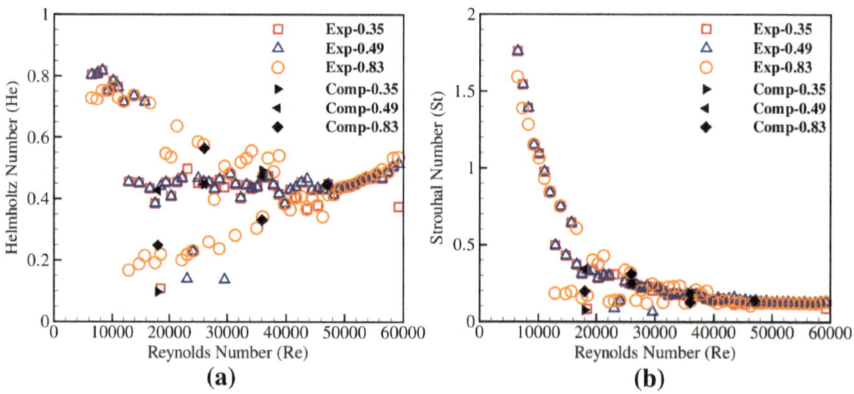

Fig. 7 **a** Helmholtz number and **b** Strouhal number at three points for $\dot{m}_f = 316$ mg/s. Experimental data are taken from [7]

3.2.4 Detailed Discussion for $\dot{m}_f = 142$ mg/s

The literature strongly suggests that the movement of the recirculation bubble at the dump plane is primarily the driving mechanism for flow-acoustics coupling in the present case [8]. Primarily, the shear layer oscillations at the downstream of the dump plane also modulate the heat release fluctuations and lead to the generation of acoustics oscillations. Also, the frequency of this shear layer oscillations matches with the natural acoustics modes of the combustor. These two processes are in close resonance with each other through the symptomatic rise and fall of the amplitudes. When the periodic pressure and heat release fluctuations are in phase, it excites the high-amplitude pressure oscillations. The presence of such high-amplitude pressure oscillations leads to the generation of acoustics oscillations, and sustenance becomes stronger in the presence of heat generation, as reported by Rayleigh [22]. Thus, this acoustic excitation becomes stronger when the unsteady pressure and heat release fluctuations get into the same phase. The most widely used criterion for such acoustics excitation is known as Rayleigh's criterion and is expressed using local Rayleigh Index (*RI*) over one cycle of instability as [23]:

$$RI = \frac{1}{T}\int\limits_{t}^{t+(2\pi/\omega)} p'q'dt \tag{16}$$

where p' and q' are the pressure and heat release oscillations, T is the time for one cycle of instability, and ω is the frequency in rad/s. In general, the *RI* is a good marker to identify the growth or damping of instabilities under reacting conditions. The positive value of *RI* indicates the coupling of pressure and heat release fluctuations to give rise to amplified pressure fluctuations, while the negative value of *RI* represents the damping of the same.

Before moving the detailed identification of the frequencies, let us examine how the natural acoustic modes of the present combustor can be estimated. Here, one can use the simple relationship to evaluate the natural acoustic modes of the combustor as $f = \frac{c}{2}\left[\left(\frac{n_x}{l_x}\right)^2 + \left(\frac{n_y}{l_y}\right)^2 + \left(\frac{n_z}{l_z}\right)^2\right]^{1/2}$, where n_x, n_y, n_z are the order of modes, c is the speed of sound, and l_x, l_y, l_z are the length of the domain in three directions. Due to the size of the combustor geometry, there exists no dominant modes along the transverse or spanwise directions as the first mode in 'y/z' direction is ~2900 Hz under cold condition, i.e., $T = 300$ K. Hence, it is understood that the only longitudinal modes are dominant in this combustor and can approximately be estimated as $f = \frac{c}{2}\left(\frac{n_x}{l_x}\right)$. The first few natural acoustics modes along the longitudinal direction are approximately ~120.5, 241, 361.5, 602.75, 723.31, 843.8, 964.4, 1085, 1205.51 Hz, and so on.

Here, we restrict our discussion two Re cases, i.e., Re = 47,000 and 18,000. For both cases, we will present reacting flow results followed by the non-reacting results to establish the resemblance between the instability mode and driving mechanism.

All these cases, the sampling rate is 50 kHz ($\Delta t = 2 \times 10^{-5}$ s), while a large number of sampling size and block size are considered for calculations.

Figure 8 illustrates the predictions for Re = 47,000 at two extreme locations in the domain. As observed, multiple dominant frequencies (58, 148, 293, 567, 713, 860, 1007, 1152, 1302, 1573, 1702, 1866 Hz, and so on in the range of 2000 Hz) are observed from the PSD plots while the observed frequencies are in similar range at two locations. Since there exists feedback between pressure and heat release fluctuations at the natural acoustic mode of the combustor, the spectra exhibit multiple peaks at these locations, which are either harmonics or superposition of two different

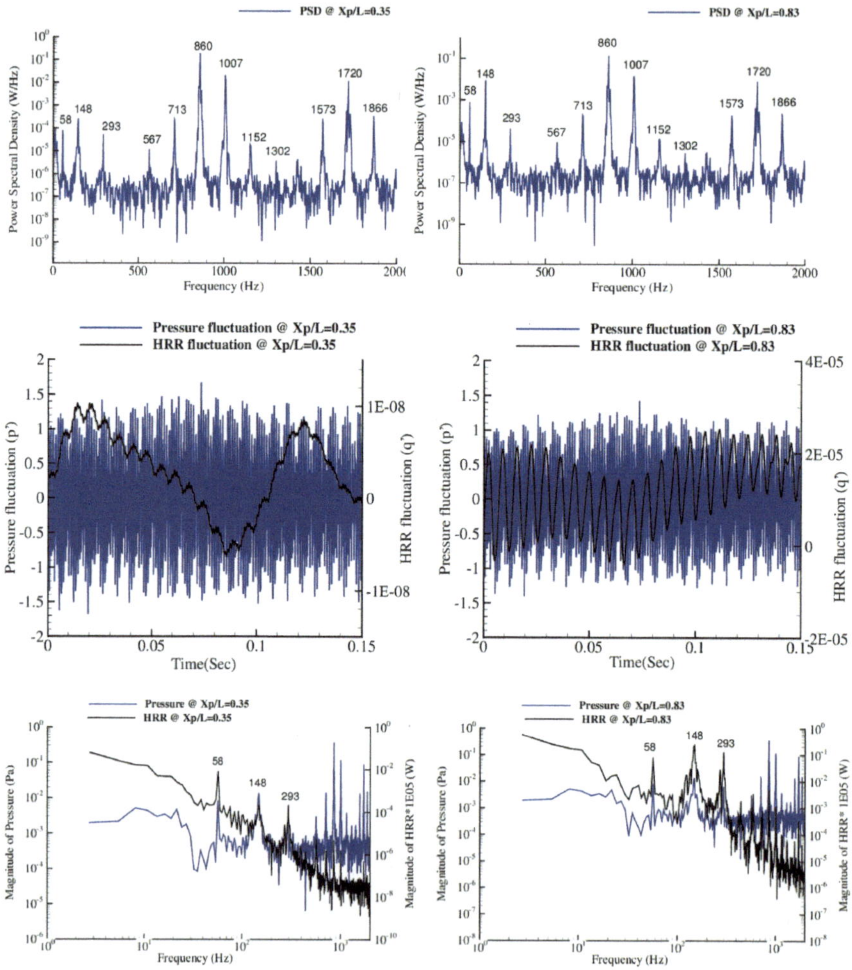

Fig. 8 Reacting CFD-CAA results for Re = 47,000 ($\dot{m}_f$ = 142 mg/s) at two locations: PSD versus frequency, HRR and pressure fluctuations versus time, PSD of pressure and HRR versus frequency

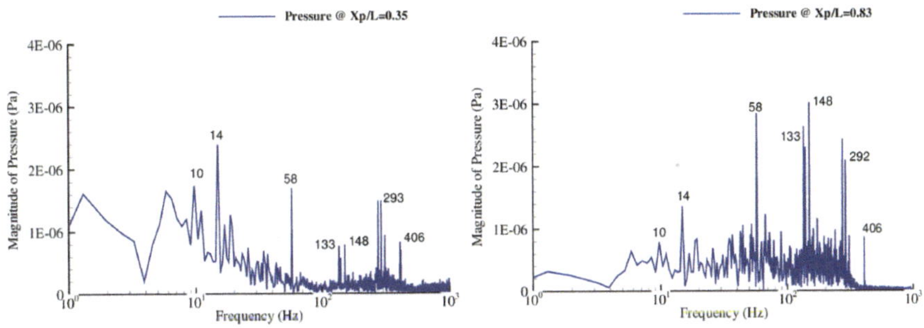

Fig. 9 Non-reacting CFD-CAA results for Re = 47,000 at two locations: PSD of pressure versus frequency

modes. To identify the driving force, one has to look at the complete spectra of both pressure and heat release, as depicted in Fig. 8 at both locations. Two points, i.e., $X_p/L = 0.35$ belongs to the vortex shedding region, and $X_p/L = 0.83$ is far away from the vortex shedding region, show similar behavior. While looking at pressure and heat release fluctuation together, one can notice the most correlated frequency (i.e., beat frequency) is ~58 Hz. In particular, this frequency gets generated primarily due to the upstream shear layer oscillations, i.e., movement of the recirculation bubble. Analyzing the other observed peaks in spectra, one can notice that they either belong to the natural acoustic mode of the combustor or to a superposition of natural acoustic mode and lock-on mode (58 Hz). For example, 293 Hz is the superposition of second natural mode + lock-on mode (~241 + 58 Hz), 567 Hz is the superposition of 4th natural mode + lock-on mode (~482 + 58 Hz), 713 Hz and 860 Hz are approximately sixth and seventh natural mode, 1007 Hz is the superposition of 8th natural mode + lock-on mode (~964 + 58 Hz), and similarly, other modes can also be identified. While correlating these spectra with the cold flow results in Fig. 9, noticeably, the reacting flow results strongly resemble the cold flow predictions. As usual, multiple peaks, including some low amplitude peaks in the vortex shedding region, are also present in cold flow results (Fig. 9), but those mostly include the ones which are dominant in reacting flows (Fig. 8). Interesting to note here is that the cold flow predictions also exhibit the lock-on frequency of 58 Hz which confirms the excitation of natural acoustic mode with lock-on to vortex shedding mode in the shear layer. The presence of low amplitude peaks supports the cyclic behavior of shear layer vortex (at step) due to recirculation and flow reversal.

To further support the above observation, the net value of the local Rayleigh index is estimated to confirm that heat release and pressure oscillations are in phase. Figure 10 illustrates the local Rayleigh index (*RI*), which appears to be positive for most of the dominant frequencies obtained for this Reynolds number, which in turn indicates the feedback between pressure and heat release fluctuations. The positive value of RI at the most correlated frequency (58 Hz) strongly suggests the excitation of acoustic mode with lock-on to the vortex shedding mode in the shear layer region [7, 8]. Thus, this correlated frequency is termed as acoustic locked-on frequency.

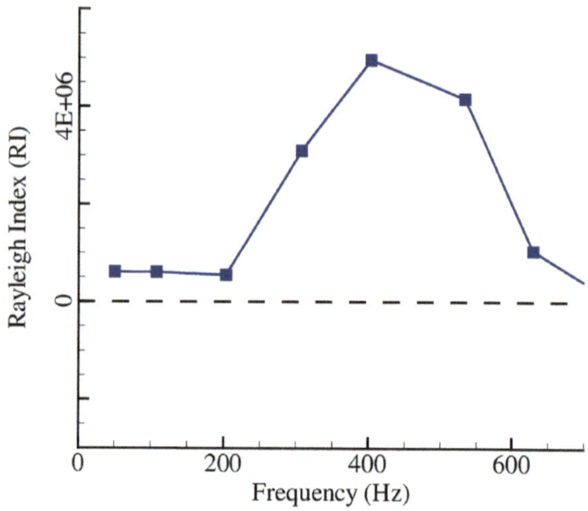

Fig. 10 Reacting CFD-CAA results: rayleigh index for Re = 47,000 for $X_p/L = 0.83$

Moreover, the positive value of *RI* indicates that instability is driving (i.e., positive net value) at most dominant frequencies, and the presence of acoustic locked-on frequency and acoustic modes shows that acoustics is the driving mechanism of combustion instability at this Reynolds number. This finding indicates that the instability or flow-acoustics coupling in this combustor primary takes place due to shear layer oscillations (vortex shedding) coupled with unsteady heat release fluctuations at the dump plane.

While analyzing the data for Re = 18,000, one can notice completely different behavior, as illustrated in Fig. 11. The observed dominant frequencies are also different at different locations, i.e., 42 Hz and 136 Hz, at $X_p/L = 0.35$ and 0.83, respectively. The different dominant frequencies at different locations confirm the excitation of hydrodynamic instabilities without locking on to the acoustics mode. The hydrodynamic mode appears primarily due to shear layer oscillations at the dump plane (motion of recirculation bubble). The first frequency (42 Hz) arises due to the motion of the recirculation bubble, while the second one (136 Hz) is approximately the first bulk mode frequency of the combustor duct. Hence, unlike the higher Re case, the pressure and heat release fluctuations are not in phase and do not attribute to the resonant acoustic mode (Fig. 11). The plot local RI supports the argument also, as illustrated in Fig. 12, as no feedback exists between pressure and heat release rate at this Re. It means the combustion at this Re is entirely stable, and the instabilities are hydrodynamic in nature. Moreover, the non-reacting spectra (Fig. 13) further bolsters this argument as the observed dominant frequencies are the same as of reacting cases (Fig. 11). Hence, the mode of instability is purely hydrodynamic in nature at lower Re, while flow-acoustics lock-on takes place at higher Re.

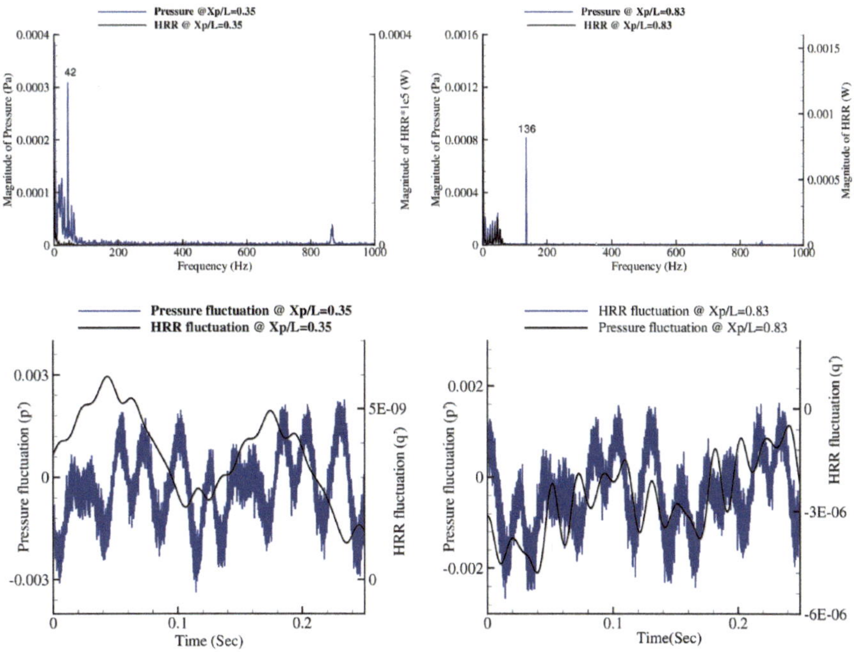

Fig. 11 Reacting CFD-CAA results for Re = 18,000 ($\dot{m}_f = 142$ mg/s) at two locations: PSD of pressure and HRR versus frequency, HRR and pressure fluctuations versus time

Fig. 12 Reacting CFD-CAA results: Rayleigh Index for Re = 18,000 for $X_p/L = 0.83$

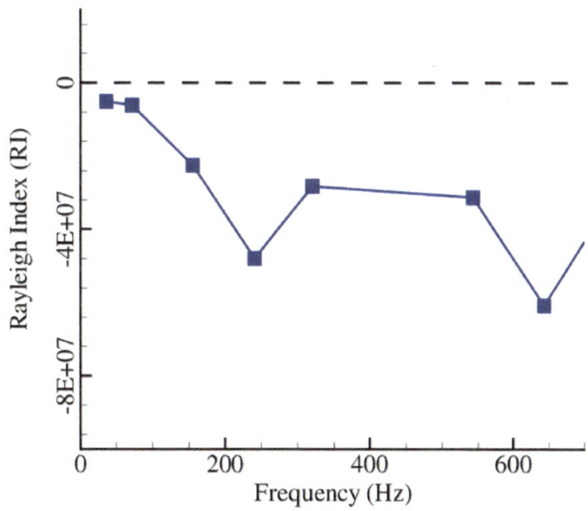

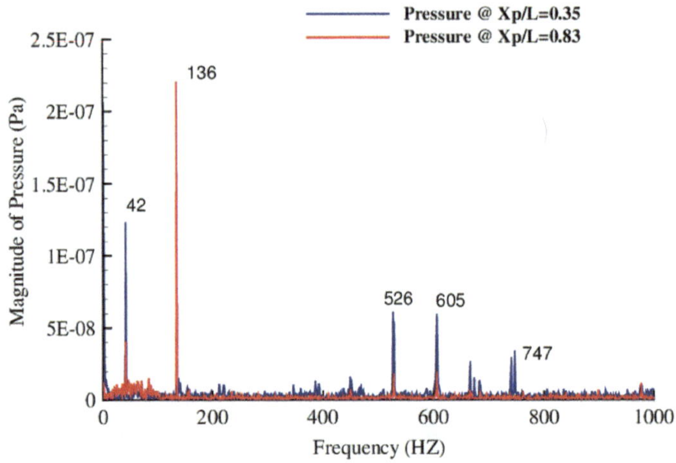

Fig. 13 Non-reacting CFD-CAA results for Re = 18,000 at two locations: PSD of pressure versus frequency

3.2.5 Instability Map

Further, the ratio of flow timescale to the acoustic timescale is calculated for this combustor. This ratio is equal to $\chi = St/He = \left(\frac{h}{L}\right)\left(\frac{1}{M}\right)$, where M is the Mach number of the airflow at the combustor inlet (Table 2). The characteristic length for the acoustic oscillations is the length of the combustor, while the characteristic length for the shear flow or heat release process is the step height. The data are taken from all the three receiver probe points in a numerical domain for the three mass flow rates at different Reynolds number. The plot χ against the equivalence ratio ϕ indicates a map of the combustor's performance with respect to its acoustic stability behavior, as shown in Fig. 14. The flow-acoustics interaction is found to occur when the order of χ is equal to unity or *1/n*, where *n* is the integer multiple of natural acoustic mode. This suggests the ratio of the hydrodynamic timescale to the acoustics timescale of the chamber.

The map explains the dominance of acoustics in higher Re or lower ϕ flows. It is similar to what has been discussed earlier on *St* and *He* plots. For all the three fuel flow rates, *St* is found to depend hyperbolically on flow velocity, in other words, directly proportional to *1/u*. This observation for all the fuel flow rates is clearly visible together in this single instability map (Fig. 14), because as ϕ (depends hyperbolically on flow velocity) decreases, χ (= *St/He*) also decreases.

As discussed earlier, the data corresponding to two lower most Re cases (Re = 10,000 and Re = 18,000) lying in cluster I of Fig. 5a show the complete dominance of the hydrodynamics. These two cases, hence, have a value of χ nearer to or greater than 1.0 and are far apart from the rest of the higher Re data points. The predicted results clearly indicate that the shear layer oscillation is the primary reason behind the instabilities associated with this half-dump combustor. At this lower Re, the acoustic

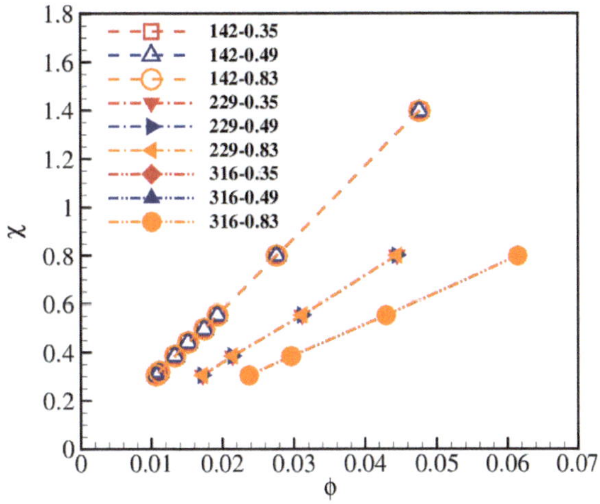

Fig. 14 Map for acoustic instability for three locations for different fuel flow rates

pressure does not get coupled with the shear layer oscillation to create a feedback loop between flow and acoustics. Hence, the hydrodynamic mode of instability is the dominant one without flow-acoustics coupling at lower Re, whereas at higher Re cases (of cluster III and IV in Fig. 5a), the value of χ decreases to less than 0.5, and the acoustics pressure couples with the shear layer oscillation and provides a necessary feedback loop between flow and acoustics. Hence, flow-acoustic lock-on happens and acoustic instability becomes the dominant mode of instability at higher Re.

As expected, the curve is found to shift rightwards with an increase in the fuel rate or equivalence ratio, but with a gentler slope. It indicates a lesser differences in the domination of the acoustics or the hydrodynamic factors for their respective cases of Re. This is evident in the previous plots of *He* (Fig. 5a, 6a and 7a), where the 2nd transition second difference vanishes as the fuel flow rate is increased, and at the highest case of fuel flow rate in this study, it is only acoustics dominated.

It is understood that under the turbulent reacting condition, the heat release (shear layer burning) becomes periodic due to the energy cascading process (vortex breaks down to small scales) and becomes higher with increasing turbulent conditions (higher Re cases). Henceforth, this periodic heat release, in particular, feeds back the necessary energy to acoustics pressure to have flow-acoustics lock-on situation at higher Re cases as compared to lower Re cases. This also stands for the acoustics-dominated cases of high-fuel flow rate where the heat release is also high.

4 Conclusions

The present work reports the flame-acoustics coupling behavior in a half-dump combustor using numerical simulations. Hybrid RANS/LES methodology is used for the simulation of the turbulent flow to increase computational efficiency. Further, the steady laminar flamelet model (SLFM) is used for turbulence-chemistry interaction for the reacting flow case. Using CAA methodology, the unsteady pressure and heat oscillations are monitored at specific locations to get the information of the flow features and the calculation of Helmholtz and Strouhal numbers, Rayleigh Index, at different frequencies. From the analysis of Helmholtz and Strouhal numbers, it is concluded that the hydrodynamic frequency dominates at smaller Reynolds numbers, while the flow-acoustics lock-on takes place at higher Reynolds numbers. The increase in fuel flow rate delays the transition and decreases the abrupt difference in each transition. Further, the detailed analysis confirms the existence of a correlation between pressure fluctuations and heat release at different Reynolds numbers. The observation suggests that acoustics is the driving mechanism for the combustion instability at high Reynolds number, while hydrodynamics is the driving mechanism for combustion instability at low Reynolds.

Acknowledgements Financial support for this research is provided through Science and Engineering Research Board (SERB), India. Simulations are carried out on the computers provided by the Indian Institute of Technology Kanpur (IITK) (www.iitk.ac.in/cc), and the manuscript preparations, as well as data analysis, have been carried out using the resources available at IITK. This support is gratefully acknowledged.

References

1. Candel S (2002) Combustion dynamics and control: progress and challenges. Proc Combust Inst 29(1):1–28
2. Yu KH, Trouve AC, Daily JW (1991) Low frequency pressure oscillations in a model ramjet combustor. J Fluid Mech 232:47–72
3. Menon S (1992) Active combustion control in a ramjet using large-eddy simulations. Combust Sci Technol 84:51–79
4. Poinsot T, Veynante D (2005) Theoretical and Numerical Combustion. R. T. Edwards, Philadelphia
5. Schadow SC, Gutmark EJ (1992) Combustion instability related to vortex shedding in dump combustors and their passive control. Prog Energy Combust Sci 18:117–132
6. Sivasegaram S, Whitelaw JH (1987) Oscillations in axi-symmteric dump combustors. Combust Sci Technol 52:413–426
7. Shreenivasan OJ (2009) Onset of acoustic instability by vortex shedding in a non-premixed combustor, Ph.D. dissertation, Indian Institute of Technology, Chennai, India
8. De A, Kushari A, Sudharsan K, Kondreddy G (2016) CFD-CAA simulation of flow acoustic coupling in a half-dump combustor. J Propul Power 32(1):95–107
9. David P, An overview of computational aero acoustic modeling at NASA Langley, Aerodynamics, Aerothermo-dynamics, and Acoustics Competency, NASA Langley Research Center, Hampton, VA 23681-2199, U.S.A
10. Kailasanath K, Gardner JH, Boris JP, Oran ES (1987) Numerical simulations of acoustic vortex interactions in a central dump ramjet combustor. J Propul Power 3:525–533
11. Angelberger C, Egolfopoulos F, Veynante D (2000) Large eddy simulations of chemical and acoustic effects on combustion instabilities. Flow Turbul Combust 65(2):205–220
12. Christopher KW, Tam (2006) Recent advances in computational aeroacoustics. Fluid Dyn Res 38(9):591–615

13. ANSYS Fluent 18.0 User's guide, Canonsburg, PA, USA
14. Saini R, De A (2017) Assessment of soot formation models in lifted ethylene/air turbulent diffusion flame. Therm Sci Eng Prog 3:49–61
15. Saini R, Prakash S, De A, Yadav R (2018) Investigation of NOx in piloted stabilized methane-air diffusion flames using finite-rate and infinitely-fast chemistry based combustion models. Therm Sci Eng Prog 5:144–157
16. Reddy M, De A, Yadav R (2015) Effect of precursors and radiation on soot formation in turbulent diffusion flame. Fuel 148:58–72
17. Lighthill J (1978) Acoustic streaming. J Sound Vib 61(3):391–418
18. Smith GP, Golden DM, Frenklach M, Moriarty NW, Eiteneer B, Goldenberg M, Bowman CT, Hanson RK, Song S, Gardiner WJ, Lissianski VV (2000) GRI 3.0, Gas Research Institute, Chicago, IL. http://www.me.berkeley.edu/gri_mech
19. Jovic S, Driver D (1994) Backward-facing step measurements at low Reynolds number, $Re_h = 5100$, NASA Ames Research Center
20. Chakravarthy SR, Obla J (2007) Experimental characterization of onset of acoustic instability in a non-premixed half dump combustor. J Acoust Soc Am 122(1):120–127
21. Nath A (2012) Unsteady reynolds-averaged navier-stokes simulation of flow-acoustic coupling in a half-dump combustor. Master's dissertation, Indian Institute of Technology, Chennai, India
22. Rayleigh JWS (1945) The theory of sound. Dover, Newyork
23. Putnam AA (1971) Combustion driven oscillations in industry. Elsevier Publishing Co, New York

Emissions from FEHA Fueled RQL Gas Turbine Combustor Under High-Pressure and High-Temperature Simulated in-Flight Conditions

Hitoshi Fujiwara and Keiichi Okai

Abstract Combustion rig tests of a rich burn–quick quench–lean burn (RQL) gas turbine combustor and of a single concentric lean-burn combustor were performed using kerosene (Jet A1)100% hydro-treated ester and fatty acid (HEFA) alternative jet fuel. The results showed a clear tendency on the suppression of non-volatile PM (nvPM) emission with increase in blending ratio of HEFA. A flight mission analysis utilizing the present experimental data gave emission profiles over the flight path.

Keywords Hydro-treated ester and fatty acid (HEFA) · Fuel property · Combustion test · Particulate matter · Flight mission analysis

Nomenclature

ATM	Air Traffic Management
ASTM	American Society for Testing and Materials
CO	Carbon monoxide
CO_2	Carbon dioxide
EI	Emission Index
FAME	Fatty and methyl-ester fuel
FT-SPK	Fischer–Tropsch Synthetic Paraffin Kerosene
HEFA	Hydro-treated ester and fatty acid
JAXA	Japan Aerospace Exploration Agency
LTO	Landing and takeoff
NEDO	New Energy Industrial Technology Development Organization
NO_x	Nitrogen oxides
nvPM	Non-volatile particulate matter
ROC	Rate of climb
RQL	Rich burn Quick quench and Lean burn

H. Fujiwara (✉) · K. Okai
Propulsion Research Unit, Aeronautical Research Directorate, Japan Aerospace Exploration Agency, 7-44-1 Jindaiji-Higashi, Chofu 182-8522, Tokyo, Japan
e-mail: fujiwara.hitoshi@jaxa.jp

A. Gupta et al. (eds.), *Advances in Energy and Combustion*, Green Energy and Technology,
https://doi.org/10.1007/978-981-16-2648-7_16

SPK	Synthetic paraffin kerosene
TEC	Toyo Engineering Cooperation
THC	Total hydrocarbons
TRA	Technology Reference Aircraft
UOP	Universal oil products

1 Introduction

Global climate change due to the rapid increase in CO_2 emissions especially caused by aviation is one of the critical issues that can be solved through international collaborations. Although the amount of CO_2 emissions from aviation consists of only around 2% of the total CO_2 emissions, it is of crucial and important to start suppressing emissions caused by air traffic as early as possible. Recent rapid growth of aviation transportation has caused many activities that could curtail CO_2 reduction form aviation. In other fields, much efforts are in progress, such as electric and hybrid vehicles in the automotive sector, and carbon dioxide capture and storage (CCS) installations in power generation plants. It should be noted that more than 90% of CO_2 emissions from commercial aircraft operations are generated by large aircraft, indicating areas of pursuing future research to reduce commercial aircraft emissions with a focus on technology applicable to large commercial aircrafts [1].

In aviation, alternative fuels are considered one of the important options to suppress CO_2 emission. Their specifications are strictly defined in ASTM D7566 "Standard Specification for Aviation Turbine Fuel Containing Synthesized Hydrocarbons" [2]. Its Annexes define not only the chemical and physical properties but also the manufacturing process of those fuels, which is of prime importance from the aviation safety point of view.

The JAXA investigations started in 2010 on the bio-derived aviation alternative fuels. At JAXA, the impact of introducing bio-fuels in turbofan jet engines was investigated [3] and combustion testing initiated in 2011 with an available bio-fuel (that was not certified) called FAME fuel (Fatty acid methyl-ester) [4]. As expected, the fuel had higher CO emission at low load conditions due to low flame temperatures [5].

In 2017, Mitsubishi Power along with their partners (Toyo Engineering Cooperation (TEC), JERA and JAXA) started a project to conduct a pilot-scale plant testing on a Fischer–Tropsch synthetic paraffin kerosene (FT-SPK) fuel produced from lignocellulosic biomass. This project was supported by NEDO [6]. For this project, JAXA participated in conducting the final combustion tests of the (bio-fuel) product.

This chapter presents the results of our investigations on the aviation-certified bio-fuels to understand the limitations and potentials of bio-derived aviation fuels as a contribution toward more environmentally friendly aviation. It further focuses on emission characteristics with Jet A1 and HEFA fuels using same high-temperature and high-pressure that simulated in-flight test conditions. A simplified flight mission analysis was conducted along with data provided on the emission profiles over the flight path.

The alternative turbine fuel used in this study was hydro-treated ester and fatty acid (HEFA) made from tallow fat. The fuel was provided by Honeywell UOP/Nikki Universal. The manufacturing process for this fuel is specified in Annex 2 of ASTM D7566. The HEFA fuel was available for use in commercial flights, if the blending ratio did not exceed 50 vol.% as per ASTM D7566.

The specifications for HEFA fuel are provided in ASTM D7566 Annex 2 [7]. To understand the similarities and differences between HEFA fuel and the baseline Jet A1 fuel, the properties of HEFA fuel were analyzed. Table 1 summarizes the results. Several considerations on the analysis are summarized below:

1. HEFA showed a very low freezing point, which is considered to be caused by branched paraffins.
2. Dynamic viscosity and surface tension of HEFA and baseline jet fuel differed only a little from each other. The slight differences might affect the atomization behavior of HEFA fuels as compared to the baseline fuel.
3. The density of HEFA is lower than that of baseline jet fuel.
4. The lubricity of HEFA is lower than that of baseline jet fuel due to lack of sulfur content.
5. The most remarkable difference between the two fuels lie in its aromatics content. HEFA contains fewer aromatic compounds, while the baseline jet fuel contains about 20% of aromatic compounds on volume basis, which consists of many benzene structure compounds and its derivatives that are low polycyclic aromatics hydrocarbons (PAHs).

2 Experimental Apparatus and Conditions

A schematic diagram of the single burner combustion rig test is shown in Fig. 1. Photograph of the JAXA AP7 high-pressure test facility is shown in Fig. 2. The facility allows combustion testing at maximum inlet temperature, pressure, and air mass flow of 1000 K, 40 bar, and 4 kg/s, respectively. This test rig is usually used for the demonstration of a new combustor concept and for the development of some innovative measurement technologies. An exhaust gas sampling probe having eight ($\varphi = 0.8$) sampling holes was located at the exit of the combustor liner.

Exhaust sample gas was introduced to the measurement instruments using a stainless-steel pipe that had a valve to control the mass flow rate and temperature of the sample gas prior to introducing the gas sample to the instrument. NOx was measured using a chemiluminescence detector (CLD), CO and CO_2 were measured using nondispersive infrared (NDIR) detectors, and THC was measured using a flame ionization detector (FID). They were incorporated in the Horiba MEXA ONE analyzer. Particulate matter (PM) mass was measured using a photo-acoustic soot sensor (PASS) in AVL MSS 483. PM number was measured through condensation particle counter (CPC).

Table 1 Chemical and physical analysis of the fuels

Properties	HEFA	D7566 Table A2.1-2	Jet A1	Jet A1: HEFA = 50:50	D7566 Table 1	Test method
Freezing point (°C)	−58.5	<−40	−48.0	−51.5	<−47	JIS K 2276
Dynamic viscosity [mm^2/s] @15 °C	1.963	N/A	1.655	1.786	<8 (−20 °C)	JIS K2283
Surface tension [mN/m]	22.5	N/A	23.6	23.1	N/A	JIS K 2241
Density [g/cm^3] @15 °C	0.7554	0.73-0.77	0.7886	**0.7732**	0.775-0.840	JIS K 2249-1
Net heat of combustion [MJ/kg]	44.14	N/A	43.38	43.73	>42.8	JIS K 2279
Physical distillation						JIS K 2254
Initial boiling point [°C]	146.5	N/A	148.5	148.5	N/A	
10% recovered temp. [°C]	165.0	<205	164.5	165.0	<205	
50% recovered temp. [°C]	208.0	Report	193.5	198.5	Report	
90% recovered temp. [°C]	253.5	Report	237.5	247.0	Report	
Final boiling point [°C]	269.0	<300	259.0	262.0	<300	
Lubricity [mm]	1.04	N/A	0.83	**0.87**	<0.85	ASTM D5001
Carbon [mass%]	84.7	N/A	86.1	85.4	N/A	JPI-5S-65
Hydrogen [mass%]	15.2	N/A	13.8	14.5	N/A	
Sulfur [mass%]	<0.0001	<0.0015	0.0006	0.0003	<0.3	JIS K 2541-6
Composition						ASTM D1319
Paraffin [vol. %]	98.4	Report	79.1	89.1	N/A	

(continued)

Table 1 (continued)

Properties	HEFA	D7566 Table A2.1-2	Jet A1	Jet A1: HEFA = 50:50	D7566 Table 1	Test method
Olefin [vol. %]	0.9	N/A	3.0	1.6	N/A	
Total aromatics [vol. %]	0.7	N/A	17.9	9.3	8-25	
Aromatics						ASTM D6379
Benzene [vol. %]	<0.1	N/A	19.1	10.2	N/A	
[mass%]	<0.1	N/A	21.3	11.6	N/A	
Naphthalene [vol. %]	<0.1	N/A	0.1	0.1	N/A	
[mass%]	<0.1	N/A	0.2	0.1	N/A	
Total aromatics [vol. %]	<0.1	N/A	19.2	10.3	8.4-26.5	
[mass%]	0.1	N/A	21.5	11.8	N/A	

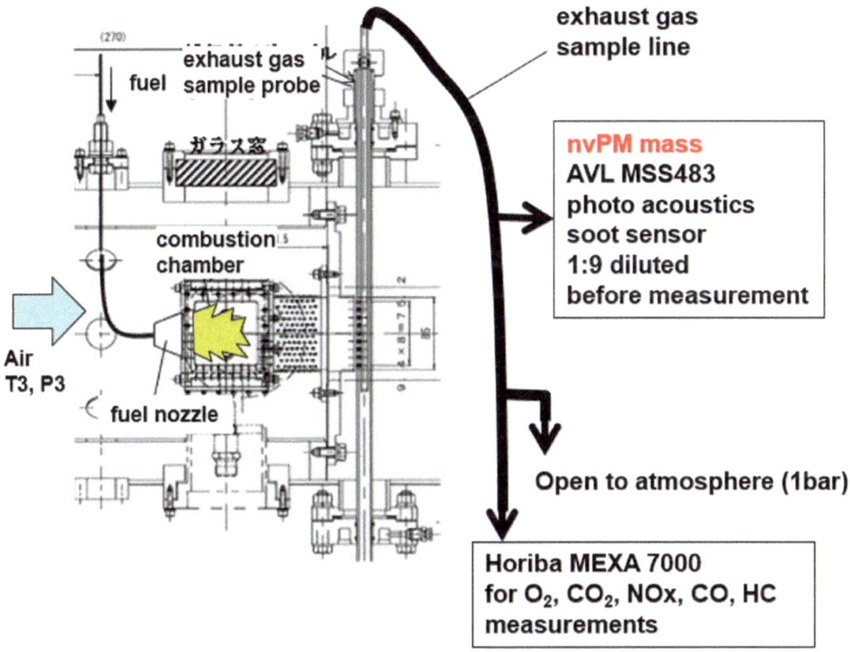

Fig. 1 A schematic diagram of the combustion test rig

Fig. 2 Photograph of the JAXA AP7 high-pressure combustion test rig

Fig. 3 RQL combustor

Figure 3 shows the RQL combustor [8] examined in this study. A Parker-Hannifin type air blast fuel nozzle, as shown in Fig. 4, was used for the injection of fuel into the combustor. In tests, 10% of total air flow entered through the upstream while the remaining 90% of total air entered through the air holes on the liner (as shown in

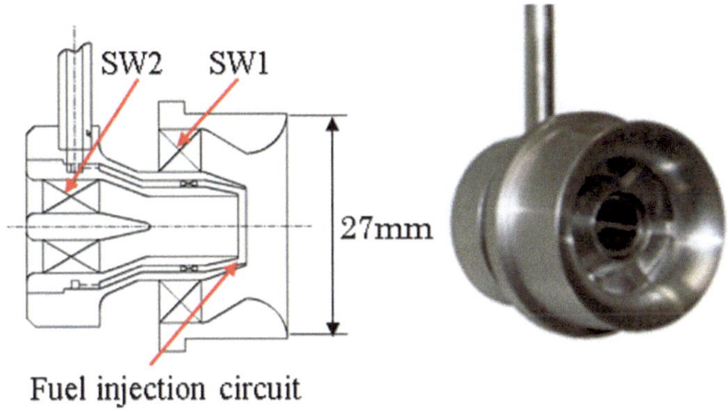

Fig. 4 Parker-Hannifin type air blast fuel nozzle

Fig. 3) that served as combustion/dilution air. The fuel nozzle had only one fuel inlet and no fuel staging were performed.

Inlet air conditions, shown in Table 2, were tested, wherein the total air/fuel ratio (AFR) ranged from approximately 38–120. Inlet air conditions were determined, so that the emissions close to in-flight conditions can be obtained. The present efforts at JAXA had been in force since 2013 with focus on engine technology development and is called "Research and Development for Green Engine Technology." Figure 5 shows the present condition in reference to the target of "Research and Development for Green Engine Technology." The target engine is a 20,000 lbf class turbofan engine with bypass ratio of 13.

Table 2 Experimental condition

Case #	Representation	Inlet temperature (K)	Inlet pressure (kPa)	Pressure loss ratio (%)
Case 1	Idle (7% MTO)	524	548	2.5
Case 2	Approach (30%MTO)	644	1350	4.6
Case 3	Cruise	739	1350	4.1
Case 4	Climb temp. 13.5 bar	803	1350	4.2
Case 5	Climb temp. 20 bar	803	2000	4.2
Case 6	MTO temp. 20 bar	837	2000	4.1

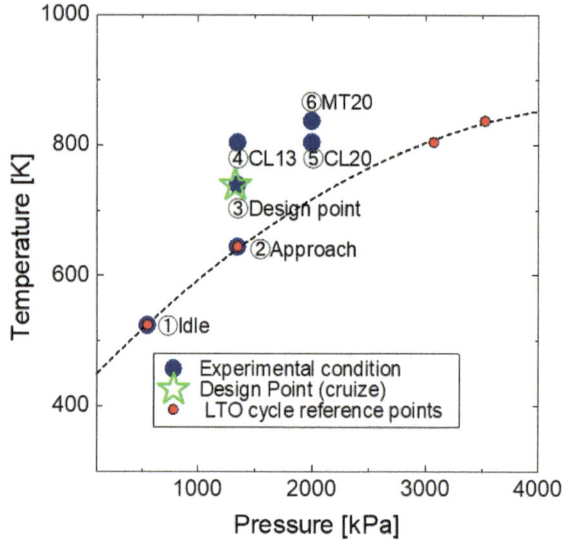

Fig. 5 Experimental conditions and referenced R&D target

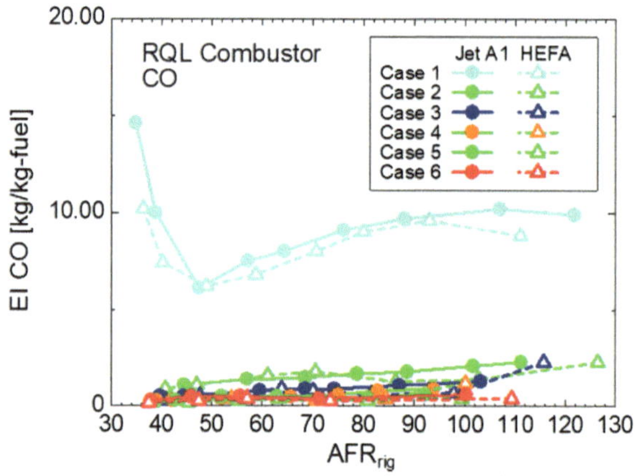

Fig. 6 Emission index (EI) of CO emission characteristics for Jet A1 and HEFA fuels

3 Experimental Results[1]

Combustion tests with fuel supply of 100% HEFA or 100% jet fuel were performed to investigate the effects of fuel change from jet fuel to HEFA. Figure 6 shows emission index (EI) for CO as a function of the air/fuel ratio. The EI CO decreased with increase in of thermal load intensity for both the fuels. A remarkable difference can be seen for the idle condition as compared to the other conditions. The results show that for each condition, the emission characteristics are similar from the two fuels.

[1] From AIAA 2020–1220 [9]; reprinted by permission of the American Institute of Aeronautics and Astronautics, Inc.

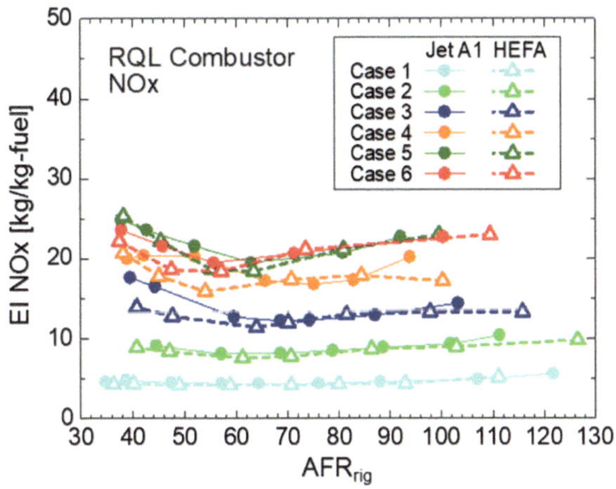

Fig. 7 Emission index (EI) of NO_X emission characteristics for Jet A1 and HEFA fuels

Figure 7 shows the emission index (EI) of NO_X as a function of air/fuel ratio. The EI NO_X increased with the increase of pressure and temperature for both the fuels. For each condition, the emission characteristics were observed to be similar for the two fuel cases.

Figure 8 shows the emission index (EI) of THC as a function of air/fuel ratio. The EI THC decreased with increase of thermal load intensity for both the fuels, similar to that observed for EI CO. For each condition, the emission characteristics were similar for the two fuel cases.

Figures 9 and 10 show mass (mg/m^3) and number of nvPM (particles/cm^3) emission, respectively. NvPM mass for each fuel generally increased with the increase in thermal load intensity, but this tendency is rather complex. The reduction of nvPM mass with HEFA was more critical than the number of nvPM. These results are in good agreement with the previous study by the authors [10].

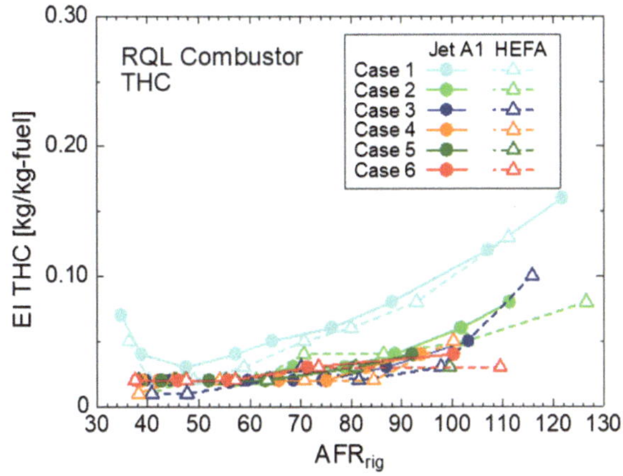

Fig. 8 Emission index (EI) of THC emission characteristics for Jet A1 and HEFA fuels

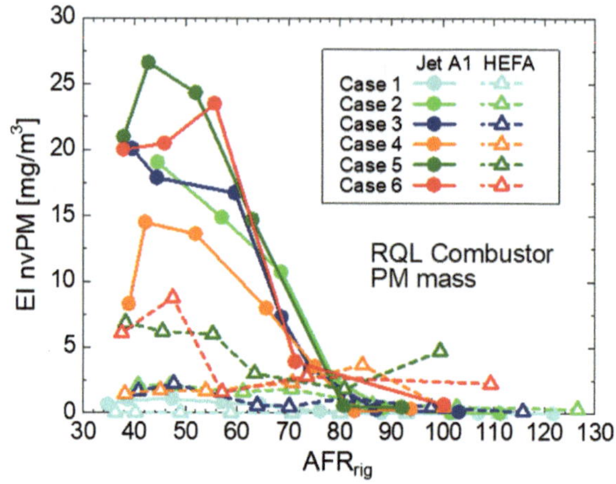

Fig. 9 PM mass emission (mg/m^3) characteristics for Jet A1 and HEFA fuels

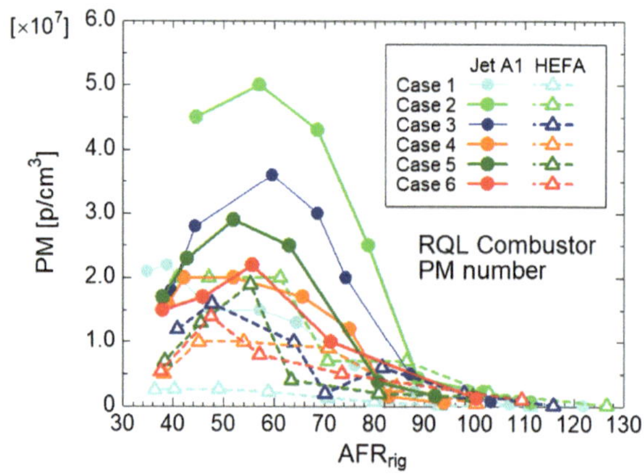

Fig. 10 PM number emission ($particles/cm^3$) characteristics for Jet A1 and HEFA fuels

In order to have a better understanding of the PM emission characteristics, pressure and temperature effects were investigated. Figure 11 shows the effect of pressure on PM emission characteristics at a constant temperature condition of 803 K. These figures show that as the pressure increased both mass and number of particle emission decreased. However, the pressure effect is less eminent in the PM number case.

Figure 12 shows the effect of temperature on PM emission characteristics at constant pressure condition of 1350 kPa. In the case, temperature effect on PM mass was not as clear as that for the constant temperature case. However, the results clearly showed that the PM number decreased with increase in temperature.

There are a number of reports available on the soot-emission characteristics for both PM mass and number of particles in combustion of conventional jet fuel and certified bio-fuel (blend). However, not many deals with the actual in-flight and/or rating conditions. Moore and co-workers [11] demonstrated reduction of soot emission from alternative fuels compared to conventional jet fuel at cruise condition,

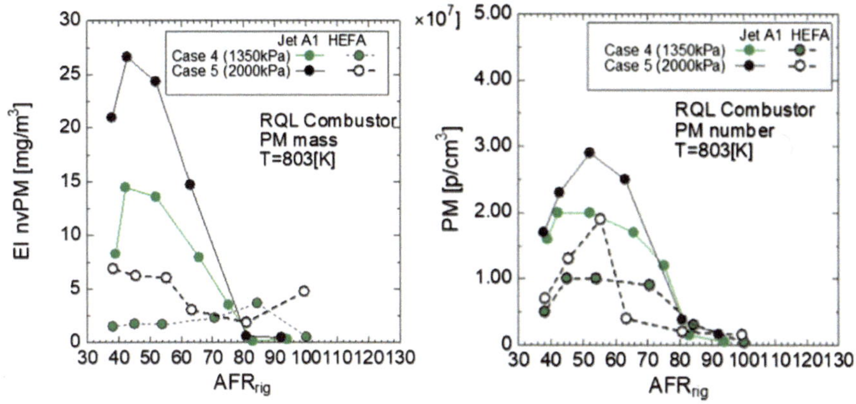

(a) PM mass density (b) PM number concentration

Fig. 11 Effect of pressure on PM mass density and particle number concentration emission characteristics for Jet A1 and HEFA at constant temperature condition

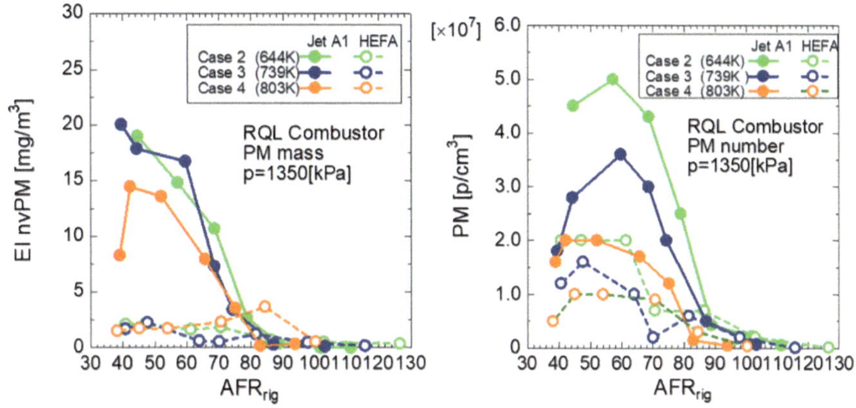

(a) PM mass density (b) PM number concentration

Fig. 12 Effect of temperature on PM mass density and particle number concentration emission characteristics for Jet A1 and HEFA at constant pressure condition

suggesting a reduction of climate effects (contrail formation). The general tendency seen in the present study is in consistent with the reference. Anderson and co-workers [12] demonstrated suppression of soot number and mass for various part load conditions. They showed that the mass reduction remained unchanged with the change of thermal load and that particle number density decreased with the increase of engine power. This tendency is consistent with the findings reported in the present study. Brem and co-workers [13] reported on their measurement of both particulate mass and number and showed that effect of mass and number is comparable and that the improvement of emission characteristics by blending HEFA fuel is marginal when investigated at the defined landing–takeoff (LTO) cycle condition. This is slightly different from the present results, but it should be noted that the reference work is HEFA/Jet A blend that had a relatively smaller fraction of HEFA content in the blend. If the ratio of HEFA in the fuel blend increases, the discrepancy in tendency for mass and number might become apparent. Liati and co-workers examined in-detail the emission of soot under both idle and climb-out condition with due attention paid to soot reactivity [14]. They suggested that at climb-out (high-load) condition, the HEFA blend showed higher reactivity potential to bear higher health risk compared to soot produced from Jet A1 fuel. Their study points out the importance of evaluation for the HEFA fuel in a number of different projects.

A correlation for soot formation is given by Döpelhheuer [15]. From their equation, the soot density (in mg/cm^3) is reported to be proportional to $p^{1.35}$. The data from Fig. 11a is re-plotted in Fig. 13a with correlated values for pressure as given in ref. [32]. The results showed a good general agreement. Also, previous report by the authors [10] showed that at relatively low pressure and temperature condition (350 kPa and 450 K), the variation of soot density (in mg/m^3) with HEFA concentration (%) in the mixture fuel showed quasi-linear rapid decay, although variation of number concentration with fuel mixture ratio was linear, as seen in Fig. 13b. Assuming this as a valid tendency at elevated pressures, the correlation above was replotted and the results are shown in Fig. 14. The results obtained showed that, with the fact that the mass density curve with respect to fuel mixture ratio was a rapid decay, while the value of soot concentration kept to a low value with increase of HEFA in the fuel mixture due to the rapidly decaying tendency.

The effect of fuel change from baseline jet fuel to HEFA is remarkable, especially on soot formation and soot emission. The reduction of emission from HEFA to

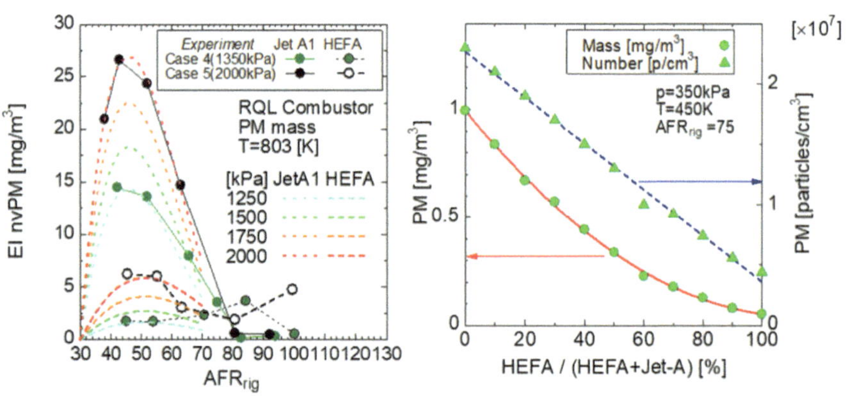

(a) Experimental data and correlation of PM mass concentration profile.

(b) Tendency of change of PM mass concentration with fuel mixture ratio.

Fig. 13 Data for correlation of soot concentration in the fuel mixture under elevated pressure condition

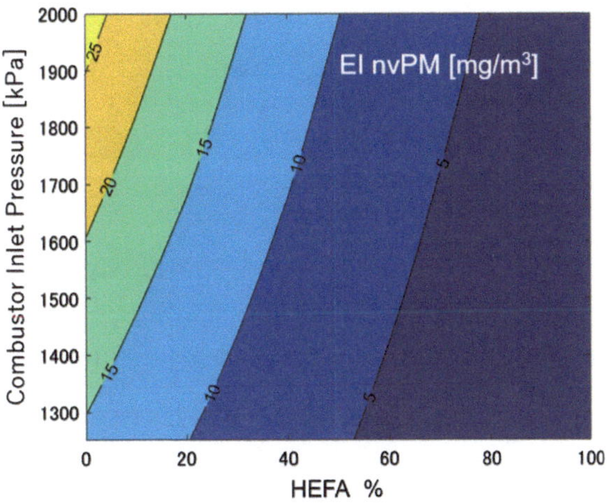

Fig. 14 Correlation of soot concentration in the fuel mixture under elevated pressure condition

baseline jet fuel is attributed largely to the reduced aromatic concentration in the HEFA fuel. Other combustion properties tested included the emission of CO, NOx, and THC. They all showed almost no effect from the change of fuel. In the following sections, based on the emission data analyzed above, a simplified flight mission analysis was conducted and is discussed below.

4 Simplified in-Flight Emission Analysis

The present experimental condition was determined based on an engine operation condition for the conceptual engine configuration studied in "Research and Development for Green Engine Technology." The concept engine was defined under reference JAXA's concept technology reference aircraft TRA 2022A described by Nomura [16]. In the following, investigation on emission profiles of a hypothetical flight mission is presented.

The TRA2022A is one of the options defined in the study of conceptual design study of 120-seat class future passenger aircraft, that is anticipated to enter service in early 2020s. One of the goals of TRA2022s is to reduce fuel consumption by more than 30% compared to baseline reference aircraft. Figure 15 shows a conceptual drawing of the vehicle and its mission profile. Table 3 shows specifications of TRA 2022A. The description of mission flight profile in Nomura [16] provides in-detail aerodynamic characteristics with the empirical data. In the present chapter, simplified version of vehicle definition was used and the lift-to-drag coefficient relation was depicted from an empirical expression of the aerodynamic characteristics given by Sun and co-workers [17], see Fig. 16. The engine performance data is given as maps of lapse rate and other expressions as seen in Fig. 17. Basic performance description of the turbofan engine was determined from the previous work by Fukuyama and co-workers [18, 19]. For the descent performance, description by Jenkinson and co-workers [20] was referenced.

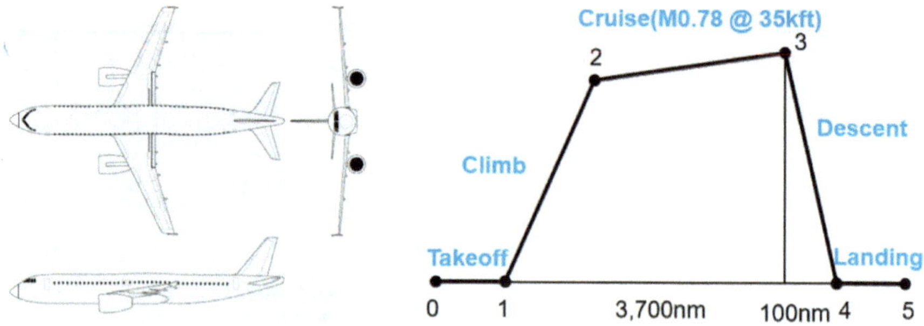

(a) TRA 2022A three-view drawing (b) Mission profile of TRA 2022A

Fig. 15 Reference Aircraft model TRA-2022A [16]

Table 3 Main characteristics

Cruise speed	M0.78
Range	360 nm
Number of seats	126
Main wing width	98.1 ft
Main wing area	1014.0 ft^2
Main wings aspect ratio	9.5
Fuselage length	123.3 ft
Fuselage diameter	12.2 ft
Takeoff thrust	20,348 × 2 lbf
Maximum takeoff weight	127,980 lb

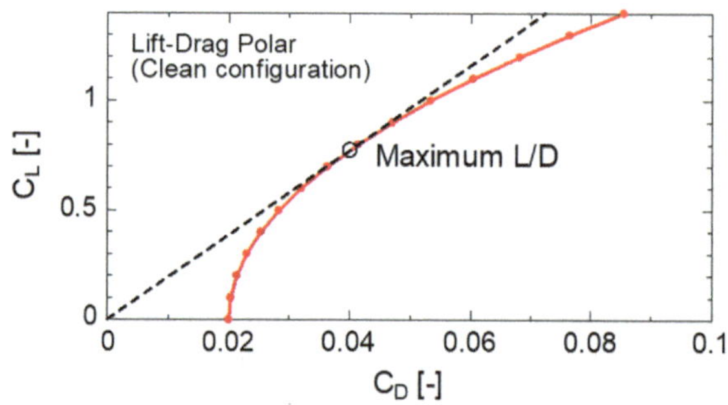

Fig. 16 Lift–Drag polar for flight mission analysis

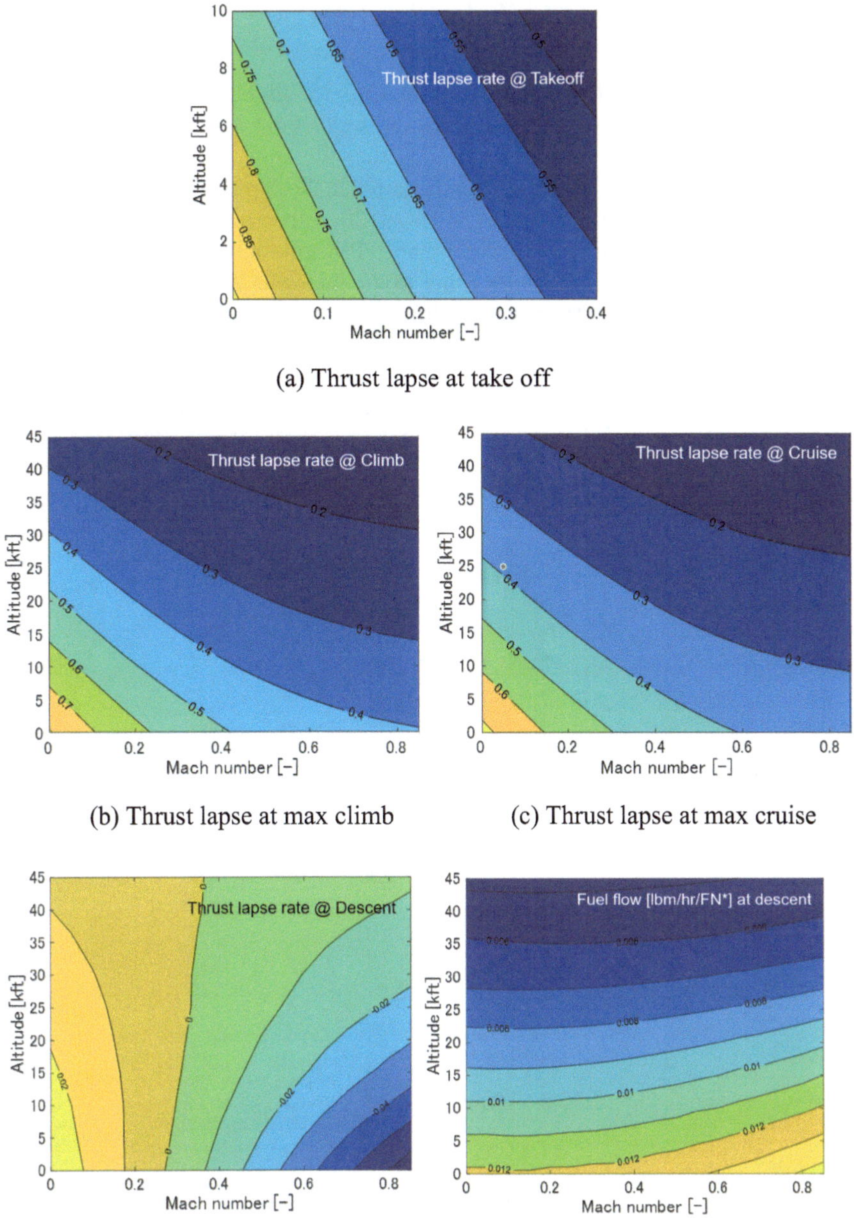

(a) Thrust lapse at take off

(b) Thrust lapse at max climb

(c) Thrust lapse at max cruise

(d) Thrust lapse at descent

(e) Fuel flow [lb/hr/F_N*] at descent

Fig. 17 Information for engine characteristics calculation

5 Results of in-Flight Emission Analysis

Based on the mission definition shown in Fig. 15, flight mission analysis was conducted. The methods used were based on the method described by Fukuyama and co-workers [18, 19] as well as in the other literature [20]. Figure 18 shows calculated 2D-flight path. The results on variation in the corresponding fuel flow rate and mission fuel burn are shown in Figs. 19 and 20, respectively. As seen previously, emission data remained unchanged other than nvPM emissions. In the following, fuel flow and emissions are representing either of the fuels (Jet A1, HEFA, and mixture of the two). For soot (nvPM) emissions, differences between the fuels are examined.

By referencing to the flight profile given above, several emission profiles were investigated by utilizing the experimental data shown in this chapter.

There are many empirical formulas proposed for describing EINOx, and the method described by Chandrasekaran and Guha was utilized here [21].

Fig. 18 Flight path

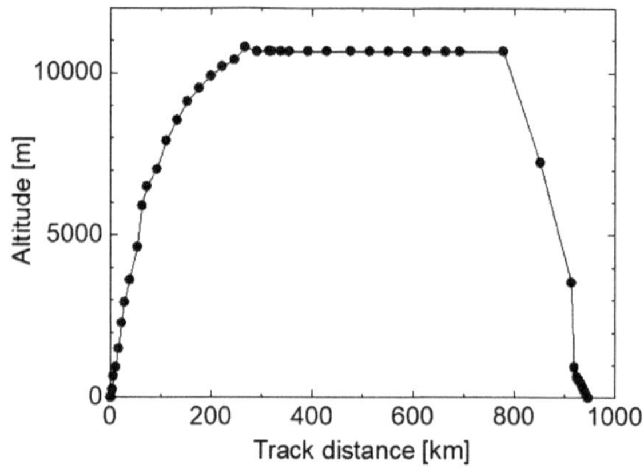

Fig. 19 Fuel flow variation

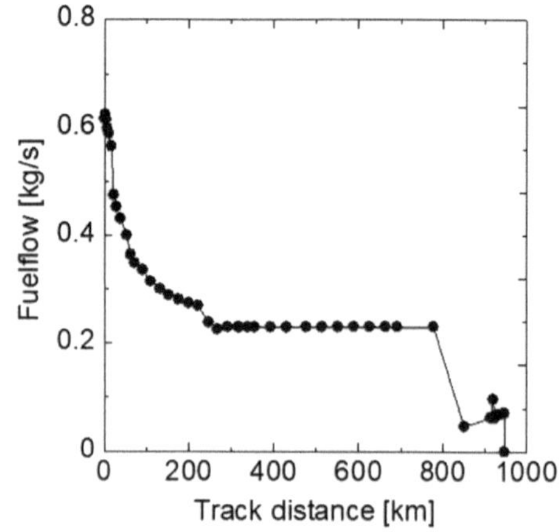

Fig. 20 History of fuel burn

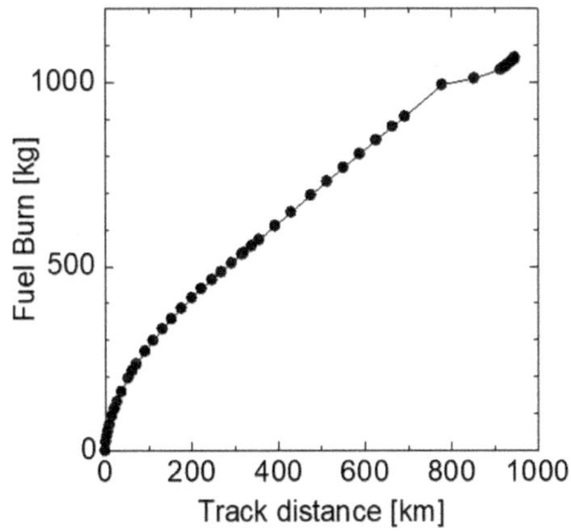

$$\mathrm{EINOx} = 0.00036\sqrt{p_{tc}e^{\frac{T_{tc}}{200}}}$$

where p_{tc} and T_{tc} are combustor inlet pressure and temperature, respectively. The expression was adjusted to the experimental data and correlated as shown in Fig. 21a. The NOx emission history in relation to flight distance is presented in Fig. 21b. The figure shows that the NO_x outflow rate to be initially large, then decreases and then kept to a relatively low value over the cruise phase. This suggests that emissions under takeoff and climb phase conditions are quite large, and that the total amount of emission should be relatively small but stand out in summation under cruise altitude conditions. In each flight altitude and path, these data can be utilized to investigate the effect of emission on the atmosphere and climate.

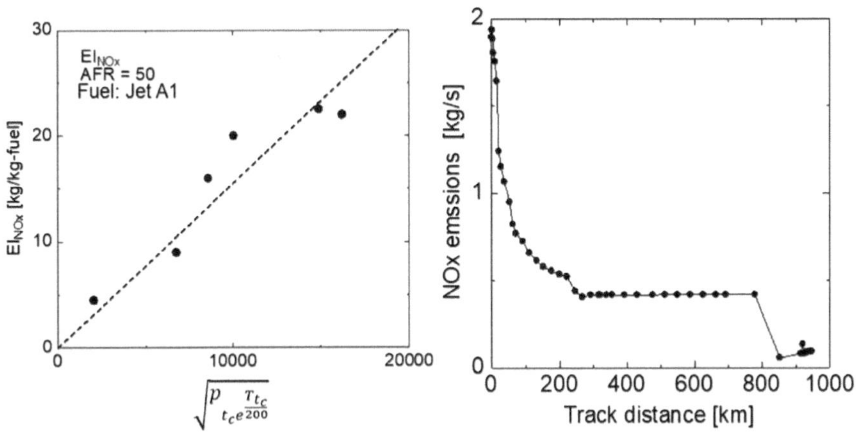

(a) NO_x emission from Correlation (b) NO_x emission rate over flight

Fig. 21 NO_x emission consideration

Similarly, soot-emission characteristics are described by utilizing the expression provided in [22].

The expression for conventional jet fuel is written as:

$$c_{\mathrm{soot}} \propto p_3 e^{\left(-\frac{20{,}000}{T_{\mathrm{FL}}}\right)}$$

where p_3 is combustor inlet pressure and T_{FL} is defined as the flame temperature. The results obtained from this correlation are shown in Fig. 22a. Using the expression, the results on soot-emission profile are shown in Fig. 22b. The emission profile is similar to the one for NO_X, as seen from Fig. 21b, but one remarkable point is that the ratio of the emission value of cruise phase to take off for the jet fuel case (0% HEFA) is relatively large compared to the NO_x case. Soot emission is now recognized as increasingly important since it serves as kernel to generate contrails, which is considered as a potential of source on global warming. Based on the previous investigation as seen in Fig. 14, HEFA and mixture fuel cases are also plotted, see Fig. 22b. Increase in HEFA fuel in the mixture resulted in decreased soot emission over the flight path. Inclusion of relatively small fraction (20%) of HEFA decreased large amount of soot emission due to the tendency seen in Fig. 14.

Finally, emission of water vapor and CO_2 were investigated. The expression used was as given in [23].

When fuels consumption is given as:

$$f = t \cdot \mathrm{sfc} \cdot \mathrm{Th}$$

where, t, sfc, and Th represent the time, specific fuel consumption, and thrust, respectively. The emission of CO_2 an H_2O can be written as:

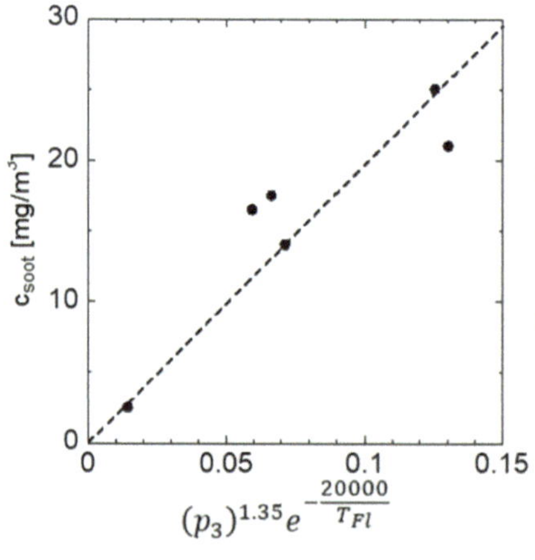

(a) Soot emission from Correlation

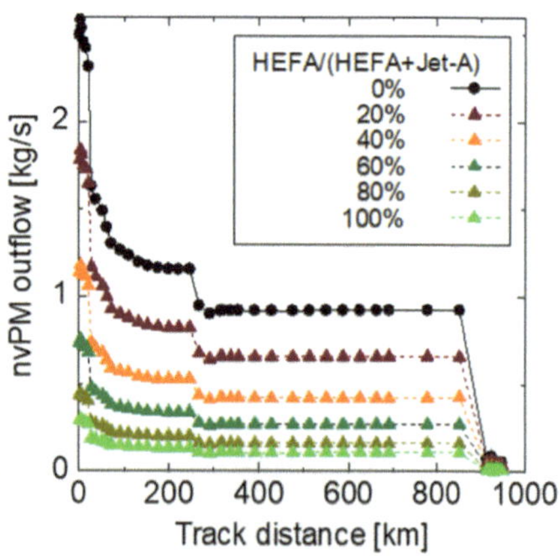

(b) Soot outflow rate over flight

Fig. 22 nvPM emission consideration

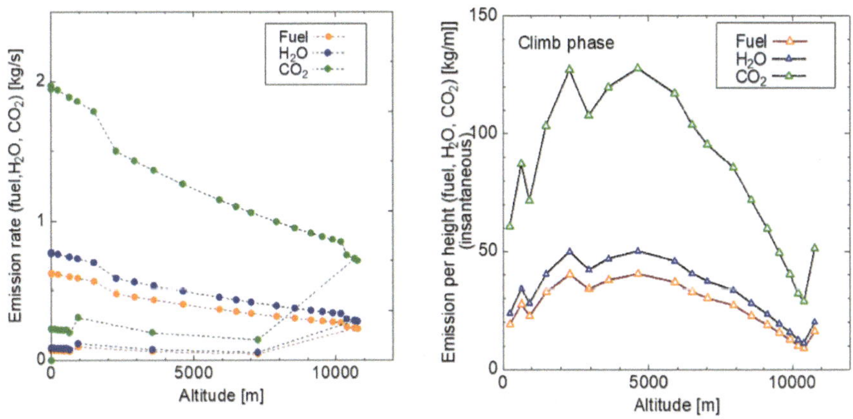

(a) Emission rate over the flight profile. (b) Emission per height at climb phase.

Fig. 23 CO_2 and H_2O emission consideration

$$E_{CO_2} = EI_{CO_2} \cdot f = 3.155 \cdot f$$
$$E_{H_2O} = EI_{H_2O} \cdot f = 1.237 \cdot f$$

Figure 23a shows the emission profiles of fuel, CO_2, and H_2O with respect to flight altitude. As the expression implies, the emission of CO_2 and H_2O are simply related to a constant factor (i.e., fuel consumption). Therefore, the reduction of these emissions is realized by decreasing the fuel consumption. However, more importantly, emission amount at each altitude may affect differently to the atmosphere. Contrail's formation mechanism is complex and related to the amount of water (supercooled) vapor, soot or aerosol kernels, and other factors. These emission profiles with respect to the altitude would be important. Figure 23b shows the emission scale indicating amount per height, which is emission outflow rate in kg/s divided by rate of climb (ROC). This implies that in the middle of climb phase, one can have a relatively large amount of emission.

To examine the effect of those emissions on climate change, many flight histories should be referred and integrated over many years. But as a first attempt, these profiles provide insights into the reason and effects of each emission with respect to a specified flight mission. This chapter shows an example to interpret the experimental data to understand the direct effect of emission characteristics to the atmosphere.

6 Conclusions

Combustion tests with jet fuel and HEFA up to 20 bar and 837 K were performed. The results showed that mass and number density of PM emission was suppressed with HEFA, while the emissions of NOx, CO, and THC with HEFA fuel were similar to those with Jet A1 fuel. At higher pressure and temperature, the suppression effect of PM emission with HEFA fuel was apparent. However, the hypothetical diameter determined from the PM mass and number density data suggested that the particle diameter becomes closer between the Jet A1 and HEFA fuels. This study clearly showed that pressure and temperature effect on the PM emission to be large and that further investigations are necessary with actual conditions in order to closely evaluate the soot-emission characteristics under flight conditions including the impact of climate.

The present study demonstrated a way to utilize emission data at ground test facility for the simulations on in-flight evaluation, thus to be used for the investigation of direct effect on emissions to the atmosphere. The study clearly defined a hypothetical turbofan engine and aircraft that flies. The corresponding combustion and emission data from experiments with simulated combustor entry condition (pressure and temperature) can easily provide prediction map data for the emissions during flight. By calculating a mission profile, some emission characteristics were obtained as histories of time and/or flight path (location). To further investigate the effect of emissions on climate change, assumptions and information on Air Traffic Management (ATM) should be incorporated. For those purpose, the present table-style emission data would be beneficial.

Acknowledgements This work was in part supported by JSPS KAKENHI Grant Number 19H02341.

The authors thank Dr. Yoshitaka Fukuyama of JAXA for his guidance and comments on engine systems analysis.

The authors also thank Professor Ashwani K. Gupta of the University of Maryland for his comments and extensive revisions to the chapter.

References

1. National Academies of Sciences, Engineering, and Medicine (2016) Commercial aircraft propulsion and energy systems research: reducing global carbon emissions. The National Academies Press, Washington DC. 10.17226/23490
2. ASTM D7566-18 (2018) Standard specification for aviation turbine fuels containing synthesized hydrocarbons, ASTM International, West Conshohocken, PA
3. Fukuyama Y, Fujiwara H, Okai K (2009) Alternative fuels and their impact on the turbofan engine design and performance under the realistic flight conditions, ACGT paper No. 51, asian joint congress on gas turbines 2009 (ACGT 2009), Tokyo, Japan (ACGT2009-051)
4. Okai K, Fujiwara H, Hongoh M, Shimodaira K (2012) Application of a Bio-fuel to a single sector combustor for an experimental small aero-engine, asian joint congress on gas turbines 2012 (ACGT 2012), Shanghai, China, (ACGT 2012-2121)
5. Lefebvre AH, Ballal DR (2010) Gas Turbine Combustion: alternative fuels and emissions, 3rd edn, CRC press, Taylor & Francis
6. Okai K, Mizuno T, Shinoda K, Fujii A, Kojima Y, Sakai K, Shibahara Y (2020) Development and testing of integrated process of woody biomass gasification and fischer-tropsch synthesis for bio-derived aviation fuel, AIAA 2020-3670, AIAA Propulsion and Energy Forum

7. Abe Y, Toba M, Mochizuki T, Yoshimura Y, Oxidative degradation behaviour of fatty acid methyl ester in fish oil biodiesel and improvement of oxidation stability by partial hydrogeneration, J Japan Petroleum Inst 52(6):307–315 (in Japanese)
8. Makida M, Yamada H, Shimodaira K (2014) Detailed research on rich-lean type single sector and full annular combustor for small aircraft engine. In: 29th Congress of the international council of the aeronautical sciences (ICAS 2014), St. Petersburg, Russia, 2014 (ICAS2014-0628)
9. Okai K, Fujiwara H, Makida M, Shimodaira K (2020) The effect of the fuel change from petroleum kerosene to HEFA alternative jet fuel on the emission of an RQL gas turbine combustor under high pressure and high temperature conditions, AIAA 2020-1220, AIAA Sci-Tech 2020
10. Okai K, Fujiwara H, Makida M, Shimodaira K (2019) The effect of the fuel change from petroleum kerosene to HEFA alternative jet fuel on the number of nvPM emission of an RQL gas turbine combustor, AIAA 2019-1772, AIAA Sci-Tech 2019
11. Moore RH, Thornhill KL, Weinzierl B, Sauer D, D'Ascoli E, Kim J, Lichtenstern M, Scheibe M, Beaton B, Beyersdorf AJ, Barrick J, Bulzan D, Corr CA, Crosbie E, Jurkat T, Martin R, Riddick D, Shook M, Slover G, Voigt C, White R, Winsead E, Yasky R, Ziemba LD, Brown A, Schlager H, Anderson BE (2017) Biofuel blending reduces particle emissions from aircraft engines at cruise conditions, nature. https://doi.org/10.1038/nature21420
12. Anderson B (2020) The ACCESS-II science and implementation teams, alternative-fuel effects on contrails and cruise emission (ACCESS-2) flight experiment.https://science.larc.nasa.gov/large/data/ACCESS-2/presentations/4%20RichMoore_FuelEffectsOnBC.pdf. Accessed 20 Dec. 2020
13. Brem B, Durdina L, Elser M, Schoenenberger D, Setyan A, Wyss S, Munoz M, Schreiber D, Liati A, Haag R, Rentsch D, Fischer A, Mohn J, Heeb NV (2018) Non-volatile particle matter mass and number emissions of an aero gas turbine fuelled with alternative fuel blends. In: 22nd ETH-Conference on combustion generated nanoparticles, June 18th 2018
14. Liati A, Schreiber D, Alpert PA, Liao Y, Brem BT, Corral Arroyo P, Hu J, Jonsdottir HR, Ammann M, Dimopoulos P, Eggenschwiler PD (2019) Aircraft soot from conventional fuels and biofuels during ground idle and climb-out conditions: Electron microscopy and X-ray micro-spectroscopy. Environ Pollut 247:658–667
15. Döpelhheuer A (2001) Quantities, characteristics and reduction potential of aircraft engine emissions. SAE 2001-01-0308
16. Nomura T (2013) Conceptual design of future passenger aircraft aimed at reducing fuel consumption, JAXA-RR-13-007, Dec 2013
17. Sun J, Hoekstra JM, Elebroek J (2018) Aircraft drag polar estimation based on a stochastic hierarchical model, presented n 8th SEAR innovation days, Dec 2018
18. Fukuyama Y, Yamane T (2018) Flight mission fuel burn evaluation for aero-engine related technology variation, ACGT 2018-TS45. In: Proceedings of the asian congress on gas turbines, Aug 2018
19. Fukuyama Y, Yamane T, Nishizawa T (2019) Overview of aero-engine, airframe and flight characteristics influence on the flight mission energy consumptions, IGC-2019-234. In: Proceedings of the international gas turbine congress 2019 Tokyo, Nov 2019
20. Jenkinson LR, Simpkin P, Rhodes D (1999) Cibil Jet aircraft design. AIAA Education Series, American Institute of Aeronautics and Astronautics, Inc.
21. Vera-Morales M, Hall CA (2009) Modelling performance and emissions from aircraft for the aviation integrated modelling project. In: AIAA 2009-1263, 47th AIAA aerospace sciences meeting and exhibit, Jan 2009
22. Chandrasekaran N, Guha A (2012) Study of prediction methods for NOx emission from turbofan engine. J Propul Power 28(1)
23. Kugele A (2005) Aircraft particulate matter emission estimation through all phases of flight. EEC/SEE/2005/0014, EUROCONTROL Experimental Centre
24. Srichar B, Ng HK, Chen NY (2012) Integration of linear dynamic emission and climate models with air traffic simulations. In: AIAA 2012-4756, AIAA guidance, navigation and control conference, Aug 2012

Performance Evaluation of Upward Swirl Combustor with Reverse Fuel Injector and Hydrogen Blending

Parag Rajpara, Rupesh Shah, and Jyotirmay Banerjee

Abstract Modern gas turbine combustors are required to provide high combustion efficiency, stability over wide range of operating conditions and fuel flexibility while meeting strict low emission norms, with other requirements such as low heat loads on combustor liner wall. One of the promising ways to meet such requirements is the provision of swirler. In conventional Can combustor, the swirler is positioned at upstream end of the primary region. This arrangement is termed as backward swirl arrangement. When swirler is placed between primary and secondary region, it is termed as upward swirl arrangement. Influence of key parameters on the performance of a Can-type upward swirl gas turbine combustor is reported in this chapter. First, a relative assessment of thermal and emission characteristics of conventional Can-type combustor and upward swirl Can-type combustor is presented. Test results of exit temperature profile, liner wall temperature and emission characteristics at exit of the combustor are compared for both the combustors. Results of large eddy simulation (LES) are compared to identify and understand the flow and combustion regimes using methane as a fuel. Temperature distribution obtained in conventional combustor indicates that backward swirl flow arrangement provided for flame stability and air–fuel mixing has drawback that it produces localized hot spots of high temperature gradients in the primary zone resulting in the formation of thermal NO_x emissions. In addition, the thermal pattern created due to swirling of combustion products exerts high heat loads on the liner wall of combustor. When air is allowed to enter in the primary zone through upward swirl arrangement, it helps to confine the flame near the combustor axis. This peculiar flow regime set the liner wall at a low temperature compared to conventional combustion chamber. Comparison of emission characteristics demonstrates that NO_x emissions are significantly low in upward swirl combustor compared to conventional combustor. However, low NO_x levels are achieved by incurring unacceptable penalties in combustion chamber performance in terms of low combustion efficiency and high CO emission levels.

P. Rajpara
Department of Mechanical Engineering, Marwadi University, Rajkot, Gujarat, India

R. Shah (✉) · J. Banerjee
Department of Mechanical Engineering, Sardar Vallabhbhai National Institute of Technology, Surat, Gujarat, India
e-mail: rds@med.svnit.ac.in

A. Gupta et al. (eds.), *Advances in Energy and Combustion*, Green Energy and Technology,
https://doi.org/10.1007/978-981-16-2648-7_17

To address the difficulties of upward swirl combustor, a modification to the fuel injection strategy is proposed. A new concept of reverse fuel injection is introduced. The existing upward swirl combustor has conical-shaped injector where fuel jets are arranged on the periphery of 90° cone. In reverse fuel injection strategy, the fuel exits the fuelling device in reverse axial direction towards the wall of hemispherical dome. The dome wall decelerates the fuel jet first, and then fuel flows in all directions. The injector length is chosen as variable for parametric investigation. Research indicates that the burned gas recirculation and highly turbulent shear flow caused by reverse fuel flow pattern improves fuel–air mixing and produces larger energy density during combustion. This improves the combustion efficiency and reduces CO emission levels remarkably. The length of injector has significant effect on increase in NO_x levels and wall temperature. The reverse fuel injector with 5 mm injector length is identified as optimum configuration for achieving noticeable reduction in CO emission without any adverse effect on NO_x levels and wall temperature. Chemistry between fuel and air is further augmented by blending methane with hydrogen. Results of experiments conducted for constant energy input condition are presented and discussed. For the same energy input, hydrogen addition to methane increases the mole fractions of free radicals H, O and OH, which promotes chemical reactions and produces intensive combustion. From flame visualization, it can be said that flames are relatively broader and shorter at higher hydrogen concentrations. Shorter flame obtained at higher hydrogen percentage indicates more rapid combustion and improved flame stability. Addition of hydrogen up to 5% in fuel blend contributes to additional reduction in CO emissions with marginal increase in NO_x emissions.

Keywords Upward swirl combustor · Reverse fuel injector · Hydrogen blending · CO and NO_x emissions · Combustion efficiency · Flame visualization

1 Introduction

Combustion chambers are required to burn fuel completely, create little pressure drop, produce an acceptable temperature profile, maintain stable operation over a wide range of conditions and satisfy stringent emission norms. Another important requirement is to maintain low heat loads on liner wall of gas turbine combustor even at higher engine pressure ratio. All these requirements are difficult to be satisfied simultaneously. Enrichment of mixing between fuel and air by means of swirl is one of the promising way to meet such performance requirements. Swirling of inlet air in combustion enhances the rate of mixing by creating strong turbulence at the mixing layers. This helps in achieving higher fuel burning rate, flame stability and controlled emission. Several researchers have studied the relation between swirl flow and NO_x emissions in conventional gas turbine combustors [1–8]. In general, conventional combustors have swirler positioned at the upstream end of the combustor dome around the fuel injector (as shown in Fig. 1). This arrangement is termed as backward swirl arrangement. Studies have shown that backward swirl arrangement provides

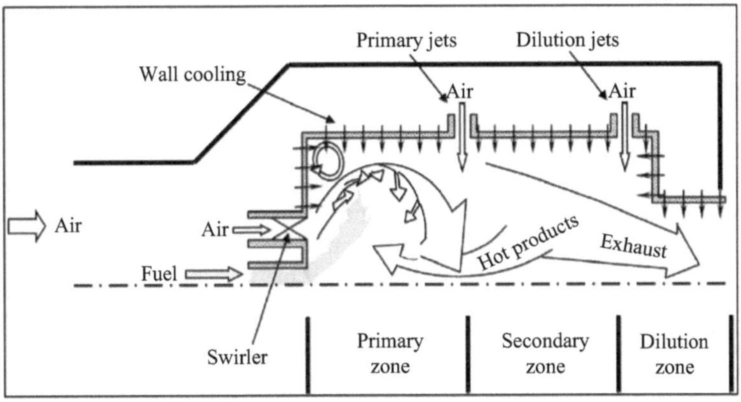

Fig. 1 Conventional gas turbine combustor features

flame stability and air–fuel mixing in conventional combustor. This arrangement has drawback of localized hot spots of high temperature gradients in the primary zone. This facilitates formation of thermal NO_x emissions. In addition, the thermal pattern created due to swirling of combustion products exerts high heat loads on the liner wall of combustor. There are many practical methods and approaches available to alleviate liner wall temperature and NO_x levels in gas turbine combustion chambers. Traditional mechanisms available for wall-cooling and emission control in the field of gas turbine combustion are discussed first followed by philosophy of upward swirl.

2 Wall-Cooling Techniques

Extensive efforts have been put by the researchers towards reducing wall-cooling requirements from conventional gas turbine combustors. Wassell and Bhangu [9] suggested that the composite liner wall structure made out of Transply[1] is capable of reducing cooling air requirement significantly as compared to the traditional film cooling in combustor. This large saving in cooling air benefits in terms of better improvement in pattern factor and control of emissions. Wadia [10] recommended the use of composite matrix liner, as it demands least amount of cooling air if cost is not an issue. Bailey et al. [11] studied the cooling characteristics of combustor liner using jet impingement method along with rib turbulators[2] for heat transfer augmentation. They reported that rib turbulators enhance the heat transfer and keep the liner relatively cool. However ribbed structure results in increase of overall engine weight. Various

[1] **Transply** is one kind of composite liner structure developed by Rolls-Royce Company. It is produced by brazing of two or more metal alloys having high temperature strength.

[2] **Turbulators** are designed to develop turbulent flow and help to augment heat transfer on tube side of fire tube boilers, heat exchangers and other heat transfer devices.

other methods such as film cooling, jet impingement cooling, effusion cooling[3], transpiration cooling[4], combined slot injection and effusion cooling and many other are also preferred by researchers [12–18] to keep low liner wall temperature in gas turbine combustor. However, at higher engine pressure ratio, the cooling effectiveness of air used for liner cooling reduces due to higher inlet air temperature. Increase in the pressure ratio also leads to an increase of turbine inlet temperature. This is beneficial in improving the pattern factor in order to maintain uniform temperature distribution at the exit of combustor. However, higher turbine inlet temperature has adverse effect on liner wall temperature, specifically at the tail end of combustor. Supply of excess air for combustor liner cooling means lesser air for combustion. This in turn results into incomplete combustion and formation of carbon monoxide and unburned hydrocarbons. Therefore, the only option left with combustor engineers and researchers is to use the available cooling air most efficiently to minimize the wall-cooling requirements in gas turbine combustor with optimum emission of pollutants.

2.1 Emission Control Mechanisms

The principal pollutants emitted by gas turbines are carbon dioxide (CO_2), carbon monoxide (CO), unburned hydrocarbons (UHC), water vapour (H_2O), NO_x, SO_x, O_2, N_2 and particulates. When emission control methods were originally introduced, NO_x was the primary regulated gas turbine emission. Various methods such as water/steam injection [19], lean premixed prevaporized combustion [20], catalytic combustion [21], rich-burn/quick-mix/lean-burn combustor [22], staged combustion [23], selective catalytic reduction [24] and exhaust gas recirculation [25] worked well in alleviating NO_x emissions from gas turbine combustor. However, the fulfilment of these techniques depends on reducing the temperature of flame, which tends to increase the formation of carbon monoxide (CO) and unburned hydrocarbons (UHC). Other concerns with these techniques are those of durability, maintainability and safety.

[3] **Effusion cooling** is the simplest approach of wall cooling defined by large number of holes drilled on liner wall normal to the wall. It can be applied to specific portion or entire liner wall. The advanced effusion cooling technique is "Angled Effusion Cooling (AEC)" in which the cooling holes are drilled at an angle to the liner wall. AEC improves cooling effectiveness of the cool air film but increases the liner weight.

[4] **Transpiration cooling** technique is one in which the cooling air passes through pores uniformly dispersed over the liner surface. The cooling air jets emerging from each pore quickly settle to form a protective layer of cool air over the entire surface of liner wall. This technique is most efficient one, but its practical application is hampered due to limitation of available porous material.

2.2 *Philosophy of Upward Swirl*

Philosophy of upward swirl is new to the field of gas turbine combustion technology. It has potential to meet conventional combustor requirements. However, this technology has not received much attention. For the very first time, Arai et al. [26] and Furuhata et al. [27] have coined the concept of upward swirl to achieve low NO_x emissions in micro gas turbine combustors. The concept is presented in Fig. 2.

The combustor consists of primary zone and the secondary zone, and both are connected by the throat zone. In order to enhance burned gas recirculation in the primary zone, swirler was set at the end of the primary zone. The combustion air was introduced through swirler and forced to flow upward towards the centrally positioned fuel injector. It was reported that burned gas recirculation and highly turbulent shear flow caused by upward swirl promote mixing and provide a premixed like combustion resulting in low NO_x emissions from combustor. Combustion tests showed that emissions of CO and NO_x were very low in the wide range of flow conditions.

In another study, Shah and Banerjee [6] made the prototype of Can-type upward swirl combustor. Combustion experiments were carried out using conventional Can-type combustor, backward swirl combustor and upward swirl Can-type combustor. Obtained results showed that the annular entry of primary air in upward swirl configuration helps to confine the flame near the combustor axis. Thus, the liner wall remains at a very low temperature in upward swirl combustor compared to backward swirl Can combustor. Experimental results demonstrated that upward swirl combustor has substantially low NO_x emissions compared to backward swirl combustor. Unfortunately, the poor combustion efficiency and increase in formation of CO emissions are

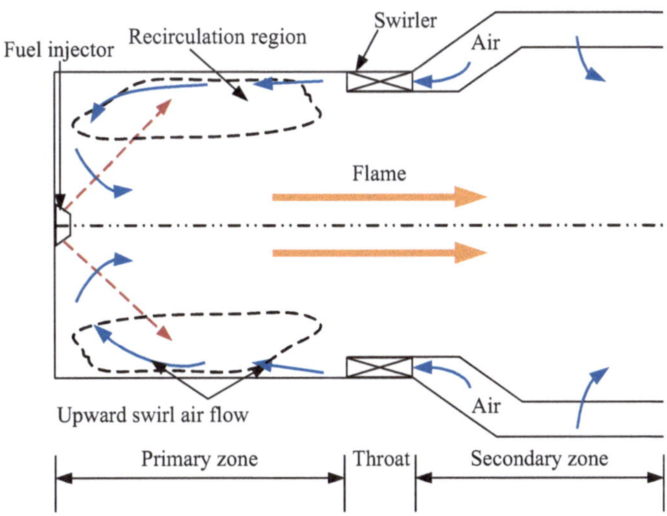

Fig. 2 Schematic diagram of upward swirl combustor [27]

formidable challenges that must be overcome to secure implementation of upward swirl combustion in operational gas turbine.

Motivated by the desire to improve the performance of a Can-type upward swirl combustor and to address above-mentioned challenges, parametric investigations are performed. First, the performance of backward swirl Can-type combustor and upward swirl Can-type combustor is assessed. Thermal and emission characteristics are compared. Strategies adopted to improve the performance of upward swirl Can combustor are tested, and results are discussed.

3 Backward Swirl and Upward Swirl Can-Type Combustor

The conventional combustor under consideration is shown in Fig. 3. It is a roughly three/fourth scale model of a Can-type gas turbine combustor. It is representative of the Rolls-Royce Tay gas turbine combustor [6]. The conventional combustor presented in Fig. 3 comprises of fuel injector, swirler followed by hemispherical head (dome), one row of primary holes and one row of secondary holes on cylindrical barrel and circular to rectangular nozzle. Fuelling device is a conical-shaped fuel injector where fuel jets are arranged on 90° cone. It injects gaseous fuel through ten 1.7 mm diameter holes circumferentially placed on pitch circle diameter of a 90° cone located centrally at the upstream end of hemispherical head. A swirler with 10 flat vanes at 45° angle to the flow is placed circumferentially around the fuel injector on the head of the combustor. In conventional combustor, one row of primary holes and one row of secondary holes are located in the intermediate barrel section. There are 6 primary holes and 12 dilution holes having diameter of 10 mm pierced through the liner surface in conventional combustor.

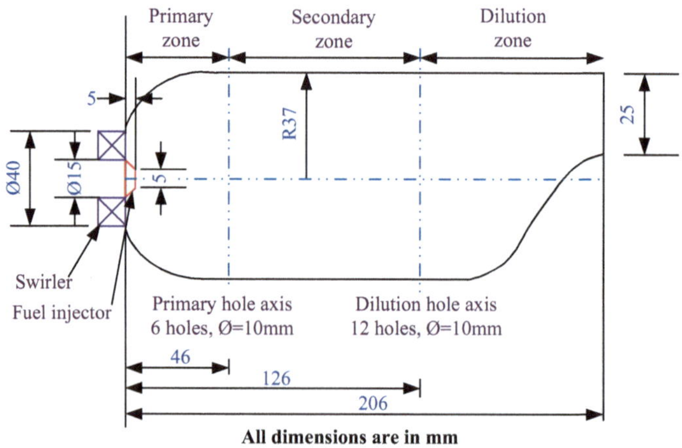

Fig. 3 Schematic diagram of backward swirl Can combustor

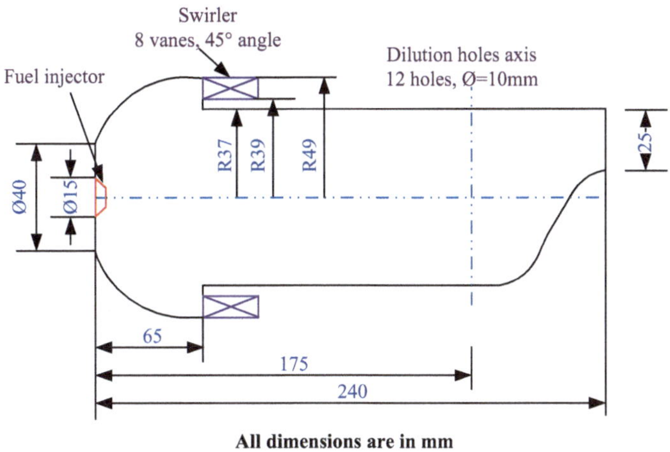

Fig. 4 Schematic diagram of upward swirl Can combustor

The Can combustor geometry developed with upward swirl arrangement is shown in Fig. 4. In upward swirl configuration, swirler is set between primary and secondary zones. The primary air required for combustion enters into the combustion chamber through swirler and flows in upstream direction towards the centrally located fuel injector. Fuel is injected into combustion chamber through a conical-shaped fuel injector having 10 number of fuel holes of diameter 1.7 mm arranged on 90° cone located at the centre and at the upstream end of the dome region. One row of dilution air holes is provided on the liner wall of the combustor barrel in the dilution zone.

4 Experimental Procedure and Test Conditions

Figure 5 represents the schematic diagram of experimental test setup developed for reactive experimentation on Can-type gas turbine combustor. The test setup is equipped with fuel line, airlines and the measurement system. Measurements of temperature and emission at the exit of combustor are carried out. The temperature distribution at the exit of combustor is measured by L-shaped K-type thermocouples. Emissions are measured at the exit of combustor using "Testo" made flue gas analyser (Testo 340). Thermal imaging of outer wall of combustor is performed using "Testo" made thermal imaging camera (Testo 875-1i). Details of various equipments and measurement accuracy have already been explained elsewhere [28].

All experiments are conducted at atmospheric conditions. In present study, the total air flow rate to the combustor is kept constant at 0.07 kg/s. Normally, the gas turbine engines operate at an air to fuel ratio (AFR) in the range of 40–70. Experiments are performed for different values of air–fuel ratio (AFR) ranging between 50 and 70.

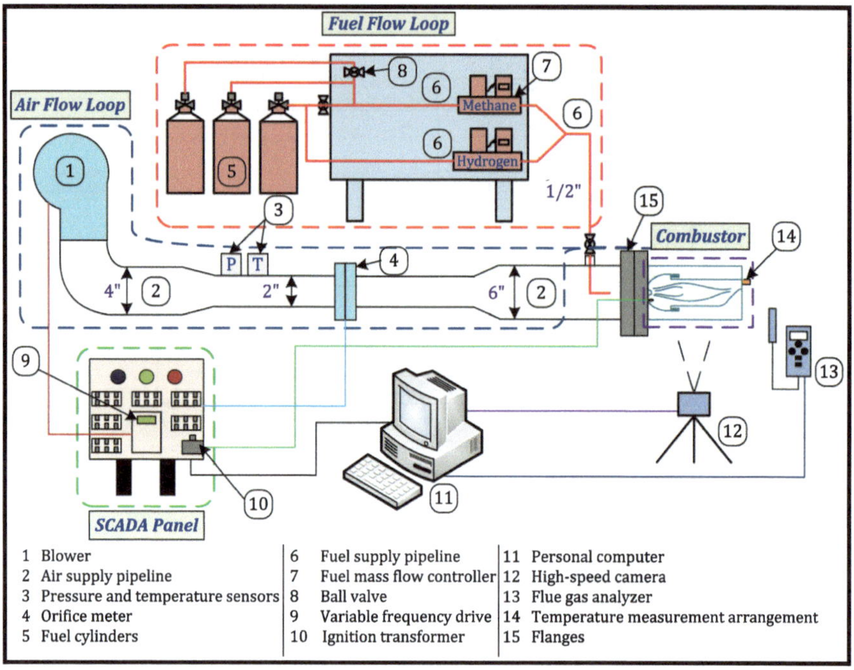

Fig. 5 Schematic diagram of upward swirl combustor test rig

5 Numerical Methodology

Computations are performed to understand and analyse flow and combustion physics inside the combustor. Computational fluid dynamics (CFD) tool Ansys Fluent is used to model flow, turbulence, heat transfer and reactions for the combustor. Large eddy simulation (LES) is used as the main approach to describe the turbulent flow. The effect of bulk radiation is included using discrete ordinates (DO) model. The interaction between turbulence and chemistry in a non-premixed flow regime is discussed by means of the well-known PDF model. The detailed description about governing equations and physical models may be found elsewhere [28].

In present study, computations are carried out with initial air and fuel temperature as 300 K. Combustor inlet (air and fuel) surfaces are defined as mass flow inlet. Pressure boundary condition is assigned at combustor outlet. Total mass flow rate of air at combustor inlet is kept constant at 0.07 kg/s. The fuel mass flow rate is varied to achieve desired value of air-fuel ratio. In present work, computational results obtained for AFR equal to 50 are compared with the test results using methane as fuel. Further details about solution methods and convergence criteria for different governing equations are available in [28].

6 Conventional Can Combustor Versus Upward Swirl Can Combustor

Assessment of thermal and emission characteristics of conventional and upward swirl Can combustor is carried out through evaluation of exit temperature variation, flame characteristics, emission level and thermal imaging of combustor outer wall.

6.1 Exit Temperature Variation

To measure temperature variation at the combustor exit, thermocouples are placed at the mid of the rectangular exit section as shown in Fig. 6. Test results obtained at AFR equal to 50 for the conventional and the upward swirl combustor are compared as illustrated in Fig. 7.

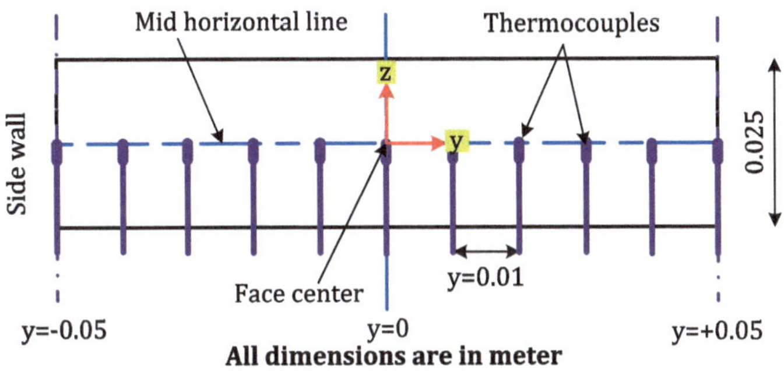

Fig. 6 Sketch diagram of temperature measurement arrangement at combustor exit

Fig. 7 Exit temperature variation for conventional (—▲—) and upward swirl (—▲—)

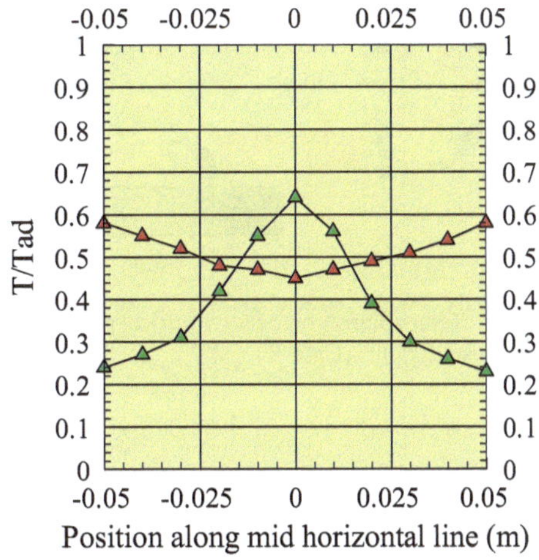

In Fig. 7, the ordinate represents the position along the mid-horizontal line while the abscissa represents the temperature value normalized by the adiabatic flame temperature value of methane with air as oxidizer. In Fig. 7, zero position represents the face centre location of the exit section and ±0.05 represents the sidewalls of combustor on mid horizontal line at combustor exit. The comparison reveals that temperature variation at the exit is opposite in both combustors. In conventional combustor, the lowest temperature is in the vicinity of face centre of exit plane while the highest temperature is observed at the sidewalls of exit section. On the other hand, the sidewalls of upward swirl combustor are at relatively low temperature compared to core region of the exit section. This depicts that alteration in swirl flow arrangement significantly influences the thermal flow patterns inside the Can combustor. In the following section, the flame characteristics of two combustors are studied and discussed.

6.2 Flame Characteristics

For AFR 50, simulation results for both the swirl flow configurations are compared in terms of temperature contours on vertical plane of combustor and is shown in Fig. 8. It implies from Fig. 8 that the swirling flow and primary jet impingement utilized for flame stability and air–fuel mixing in conventional combustor forces the combustion products to flow near the combustor liner wall. This sets the combustor liner wall to a very high temperature. When air is allowed to enter in the primary zone through upward swirl arrangement, it helps flame to confine near the combustor axis. This peculiar flow regime set the liner wall at a low temperature compared to the conventional Can combustor. Same can be observed from temperature contours shown in Fig. 8.

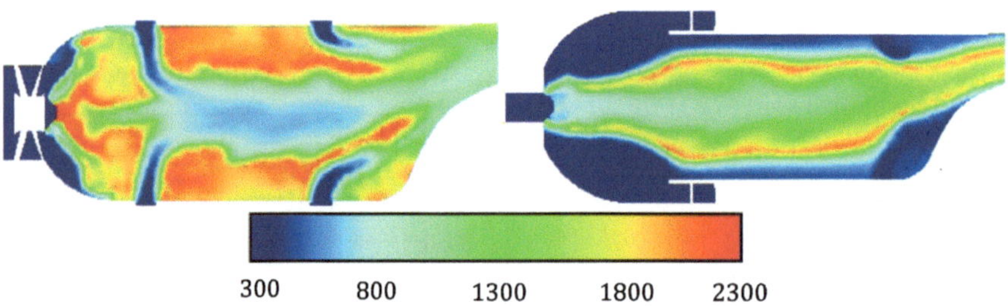

Fig. 8 Temperature contours on vertical plane of backward swirl combustor (left) and upward swirl combustor (right) (Unit: K)

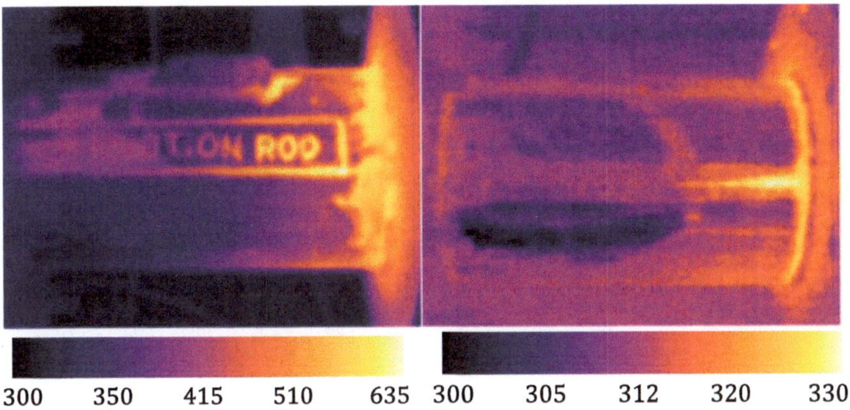

Fig. 9 Thermal images of outer wall surface for conventional combustor (left) and upward swirl combustor (right) (Unit: K)

6.3 *Outer Wall Temperature*

The change in flue gas flow regime within the combustion chamber caused by different swirl flow arrangements result into significant change in the wall temperature distribution. To estimate the effect of swirl flow arrangement on wall temperature, the combustor outer wall temperature is measured with the help of infrared thermography. Figure 9 compares the thermal images of outer cylindrical wall surface for conventional and upward swirl combustors. Test results show that outer wall surface is relatively cool in case of upward swirl combustor compared to conventional combustor. This is mainly attributed to different thermal flow patterns within the combustion chamber of two configurations. Since high temperature gases dominate the liner wall region of backward swirl combustor, the radiative intensity in this region is maximum. Therefore, the radiative heat transfer from liner to outer wall is larger resulting in higher temperature at the outer wall surface. On the other hand, in upward swirl combustor, the main flow resulting from burning of fuel–air mixture flows in the vicinity of combustor axis. Therefore, the liner wall region remains at relatively low temperature. Since the relative distance between the liner wall and the outer wall and the nature of wall surfaces are same in both combustors, the radiative heat transfer from liner to outer wall is relatively small in upward swirl combustor compared to backward swirl.

6.4 *Emission Characteristics*

The emission performance of two combustors is evaluated by carrying out experiments to measure pollutant species such as NO_x and CO at different values of AFR

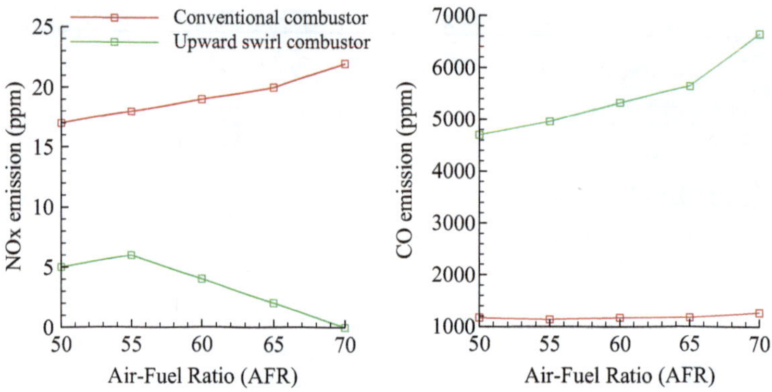

Fig. 10 Measured emission values (at 15% O_2)

as illustrated in Fig. 10. Test results show that NO_x levels are very low in upward swirl compared to conventional backward swirl at all AFRs. The mechanism of NO_x formation in combustor is correlated to its flame characteristics. In backward swirl combustor, the fuel injector and the dome wall are exposed to high temperature gases in the primary region (as seen in Fig. 8). The high temperature wall and fuel injector act as high-energy inert reactants, which accelerates the formation of two active molecules of atomic oxygen (O). This situation favours the generation of nitrogen oxides (NO_x) in conventional combustor through reactions $O + N_2 \rightarrow NO + N$ and $N + O_2 \rightarrow NO + O$ mainly governed by Zeldovich mechanism [28]. On the other hand, in upward swirl configuration, fuel burns in the vicinity of combustor centreline and the combustor core remains at higher temperature compared to combustor wall. Hot spots of high temperature gradients observed in the primary zone of conventional combustor are eliminated in upward swirl combustor. This result into significant reduction in NO_x levels from upward swirl combustor compared to backward swirl combustor.

Although the upward swirl method is very effective in reducing wall temperature and NO_x levels, but CO levels increase significantly with AFR as can be seen from Fig. 10. The increase in CO emission is an indication of poor fuel–air mixing and slow fuel burning rate. Inefficient fuel burning in upward swirl is reflected as lower exit gas temperature and poor combustion efficiency as shown in Table 1. To meet the goals of low CO emission and high combustion efficiency while maintaining

Table 1 Performance parameters at the exit (AFR = 50)

Parameters	Conventional combustor	Upward swirl combustor
Average temperature (K)	1183	991
Combustion efficiency (%)	92.45	72.3

NO_x levels and wall temperature within an acceptable range, the modification in fuel injection strategy of upward swirl is required. The performance of upward swirl combustor with modified fuel injection system is evaluated.

7 Assessment of Upward Swirl Combustor with Reverse Fuel Injector

A new concept of reverse fuel injection is introduced to improve combustion efficiency and reduce CO emission level for upward swirl combustor. The concept is shown in Fig. 11. In reverse fuel injection, fuel is supplied into combustor through injection holes in reverse axial direction towards the wall of hemispherical dome in primary zone. The wall of hemispherical dome decelerates the fuel jet first, and then, the fuel spreads in all directions in the primary zone.

To arrive at optimum dimensions of the reverse fuel injector geometry, the injector length (distance between injection holes and the dome wall) is considered as a variable. Parametric study is carried out with three variants of reverse fuel injector having injector length equal to 5 mm, 10 mm and 15 mm. Figure 12 presents schematic diagram of three variants of reverse fuel injector geometry. Results obtained for methane fuel are presented here, which shows the influence of fuel injection strategy on flow field, combustion and emission characteristics of upward swirl combustor.

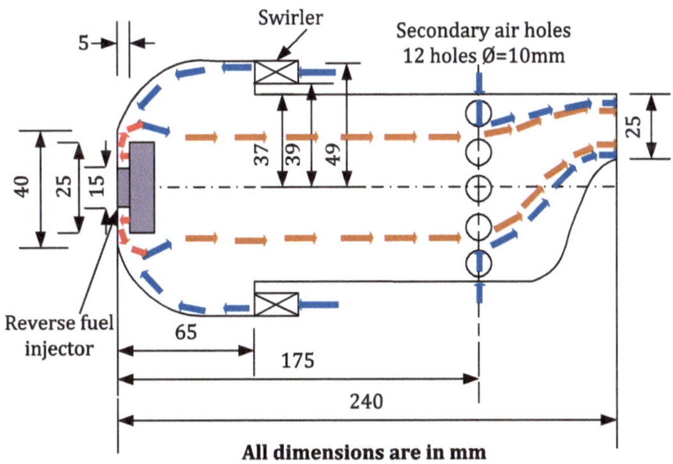

Fig. 11 Schematic flow diagram of upward swirl combustor with reverse fuel injector

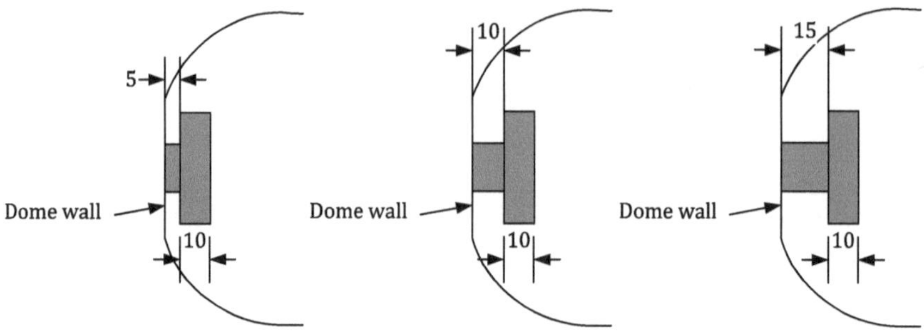

Fig. 12 Schematic diagram of different reverse fuel injector variants

7.1 *Influence of Reverse Fuel Injection on Flow Structure*

Effect of fuel injection strategy on flow structure is carried out by means of plotting streamlines along the longitudinal plane of combustor. Figure 13a–d demonstrates streamline plots on longitudinal plane of combustor for existing conical fuel injector and three variants of reverse fuel injector. For upward swirl combustor with conical fuel injector, streamline pattern depicts that fuel jets are not able to penetrate deep into swirling primary air and mixing take place close to the injector in the vicinity of combustor axis (refer Fig. 13a). Streamlines show that fluid expands and accelerates in the secondary zone. Streamlines are straight and parallel to the combustor axis in the secondary zone. The nature of streamlines in dilution zone depicts that flow converges as it passes through a circular to rectangular nozzle section.

Flow structure of upward swirl combustor appears to be significantly altered in primary and secondary regions when reverse fuel injection strategy is used. For the

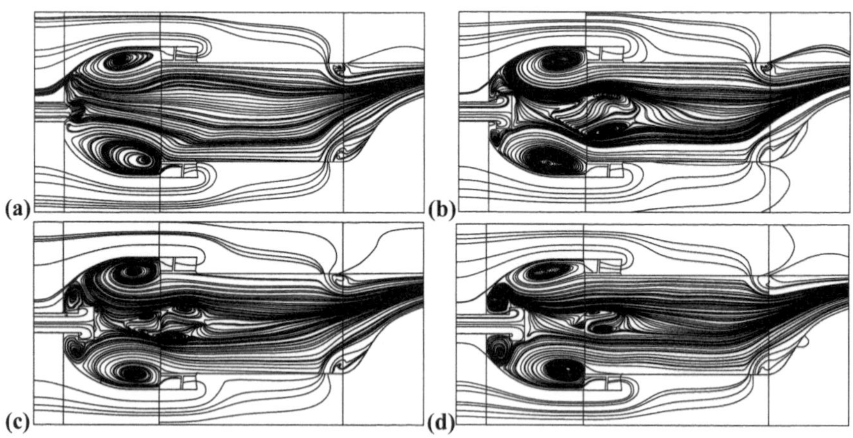

Fig. 13 Streamline on vertical plane for **a** conical fuel injector, **b** 5 mm reverse fuel injector, **c** 10 mm reverse fuel injector, **d** 15 mm reverse fuel injector

first model of reverse fuel injector with 5 mm injector length, streamline pattern show that fuel jet strikes at the dome wall and then spreads in all directions in the primary zone (refer Fig. 13b). The primary air from swirler flows towards the fuel injector and interacts with the fuel jet near the injector wall. The air–fuel mixture prepared in this region propagates in the downstream direction. The downstream propagation of gaseous mixture is retarded by the flow of swirling primary air as can be seen from a bunch of air flow streamlines near the injector wall in Fig. 13b. This reduces the fluid velocity in the centre of the flow resulting in the formation of wake in the downstream of the primary zone as demonstrated in Fig. 13b. The formation of wake is beneficial in terms of achieving better mixing.

With an increase in injector length from 5 mm to 10 mm, the fuel jet strikes at the dome wall from a longer distance. The increased size of horseshoe vortex formed due to fuel–air interaction near the injector wall reveals that fuel jet spreads more in radial direction with higher strength (as seen from Fig. 13c). The strength of central recirculating region decreases, which reveals that combustion gases produced from fuel–air burning moves faster.

With further increase in injector length from 10 mm to 15 mm, the size of horseshoe vortex formed due to fuel–air interaction increases revealing increased fuel jet penetration into the primary air (refer Fig. 13d). However, the size and strength of the central recirculating region decreases, which indicates that overall residence time of combustion gases reduces in combustor due to faster mixing between fuel and air in primary region.

7.2 Influence of Fuel Injection on Combustion

Influence of fuel injection on combustion is determined in the form of energy density and fuel mass fraction variation for different fuel injector geometries. The energy density quantifies the amount of heat released during combustion. The energy density of flow is defined as

$$E = \rho \times C_p \times T \tag{1}$$

where ρ is density, C_p is specific heat and T is temperature.

Figure 14a–d illustrates the LES results of energy density contours for different types of fuel injection system at AFR 50. The energy density is higher in regions where fuel–air mixture burns. In upward swirl combustor, fuel–air mixture burns in the vicinity of combustor centreline, and as a consequence, the energy density is seen higher in this region.

With reverse fuel injector of 5 mm injector length, the rapid near-field mixing and formation of central recirculating region causes better mixing of combustion products, which produces more amount of energy density in combustor. This is revealed from Fig. 14b where the region of higher energy density is wider in 5 mm reverse fuel injector combustor. As a result of intensified reactions, fuel burning rate

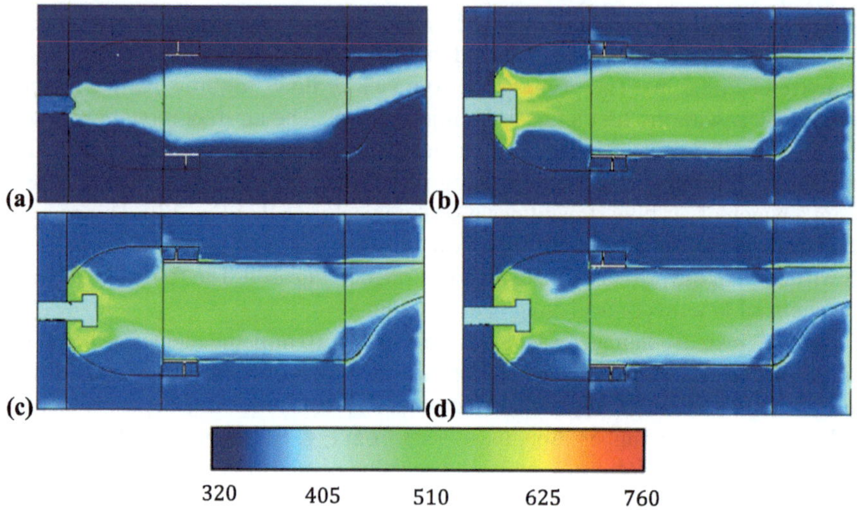

Fig. 14 Energy density contours for **a** conical fuel injector, **b** 5 mm reverse fuel injector, **c** 10 mm reverse fuel injector, **d** 15 mm reverse fuel injector (unit: kJ/m^3)

is enhanced, which in turn decreases the amount of unburned fuel in combustor as can be seen from Fig. 15a.

Injection of fuel from 10 mm distance from the dome wall produces wider reaction zone in the primary region. This can be observed from increased volume of high-energy density region in the primary zone as shown in Fig. 14c. The high-energy

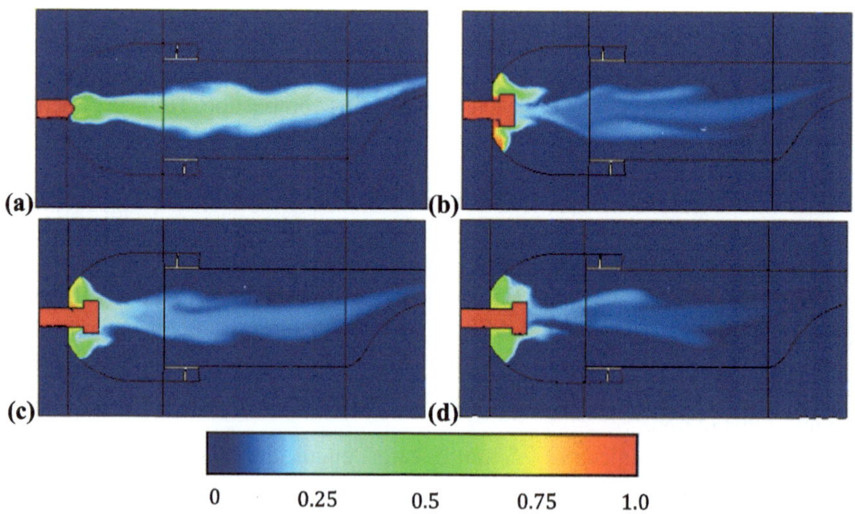

Fig. 15 Fuel mass fraction contours for **a** conical fuel injector, **b** 5 mm reverse fuel injector, **c** 10 mm reverse fuel injector, **d** 15 mm reverse fuel injector

density region expands more when fuel is injected from 15 mm distance from the dome wall. Wider reaction zone produced at higher injector length is associated with the rapid fuel–air mixing caused by increased fuel jet penetration. However, an increase of reverse fuel injector length produces little effect on augmentation in energy density and reduction in unburned fuel in combustor.

7.3 *Influence of Fuel Injection on Wall Temperature*

Figure 16a–d illustrates thermal images of combustor outer wall for all type of fuel injector configurations considered here. It is found that increasing the reverse fuel injector length (distance between fuel injection holes and the dome wall) elevated the combustor outer wall temperature. As discusses earlier, the combustion gases expand radially towards the liner wall by increasing the length of the injector, resulting in an increase in the rate of heat transfer from the inner to the outer wall, thereby increasing the surface temperature of the outer wall. For 5 mm injector length, the elevation in wall temperature is negligible, whereas the maximum wall temperature is obtained when injector length is increased to 15 mm.

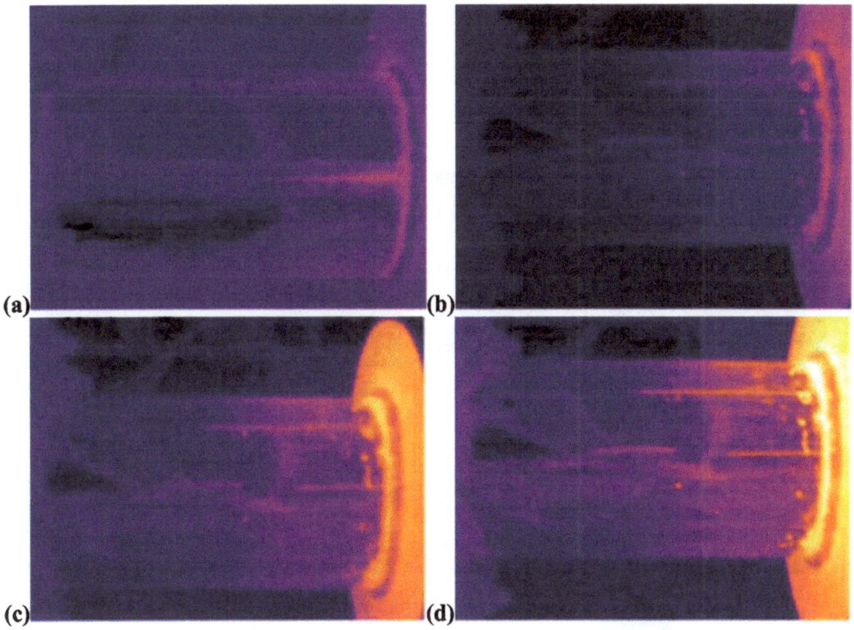

Fig. 16 Thermal images of outer wall for **a** conical fuel injector, **b** 5 mm reverse fuel injector, **c** 10 mm reverse fuel injector, **d** 15 mm reverse fuel injector (unit: K)

7.4 Influence of Fuel Injection on Emission Characteristics

Emission measurement in terms of NO_x and CO is obtained for different fuel injector geometry. Measured values of NO_x emission for all cases at different AFR values are illustrated in Fig. 17. The upward swirl combustor with conical fuel injector emits very low amount of NO_x emission at all AFRs. NO_x emission increases when reverse fuel distribution method is used. The mechanism of NO_x emission formation is correlated with the flame characteristics, which is presented in Fig. 18a–d.

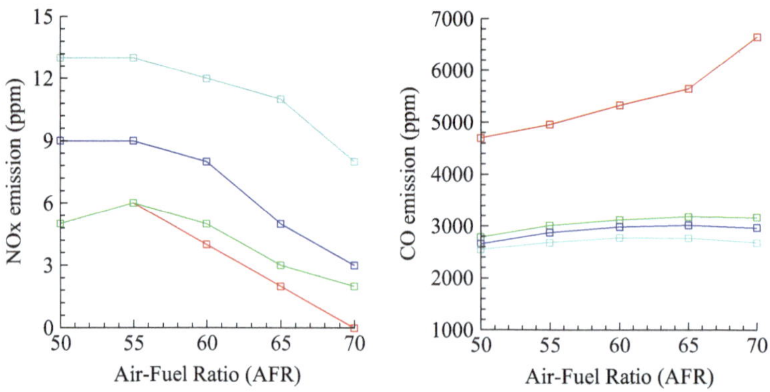

Fig. 17 Emission characteristics for different fuel injection systems conical fuel injector (—□—), 5 mm (—□—), 10 mm (—□—) and 15 mm reverse fuel injector (—□—)

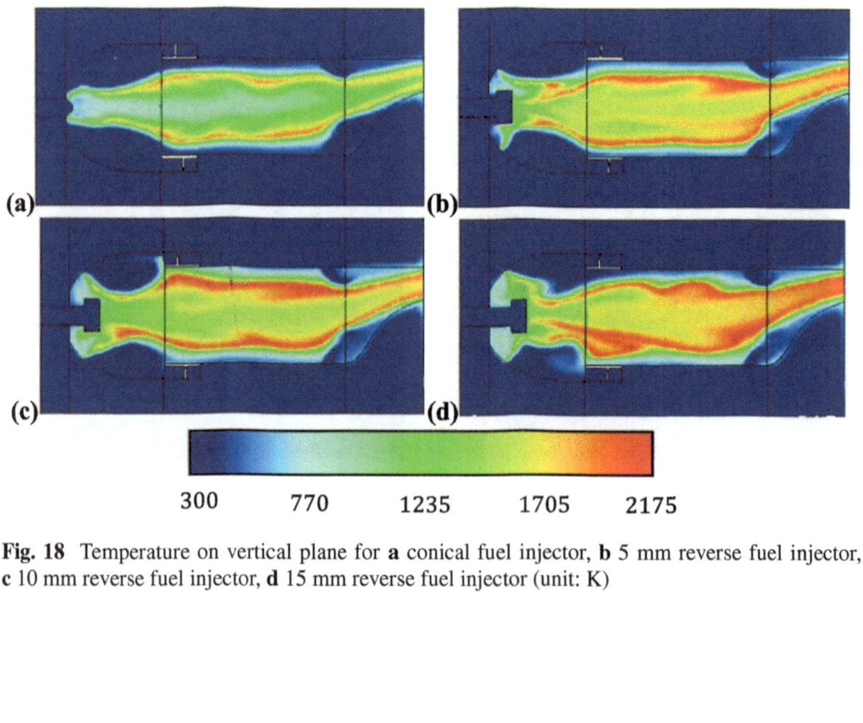

Fig. 18 Temperature on vertical plane for **a** conical fuel injector, **b** 5 mm reverse fuel injector, **c** 10 mm reverse fuel injector, **d** 15 mm reverse fuel injector (unit: K)

For upward swirl combustor with conical fuel injector, slower burning rate of fuel results into less amount of energy produced during combustion, which in turn result into lower primary zone temperature and corresponding value of NO_x emission is low. Fuel delivery through reverse fuel injector improves fuel–air mixing, which leads to better combustion and larger energy release. This leads to an increase in gas temperature, which accelerates the formation of thermal NO_x according to the Zeldovich mechanism [28]. It is found that at low AFR values the concentration of NO_x emission remains same for conical fuel injector and the first model of 5 mm reverse fuel injector. Reverse fuel injector with 5 mm length produces little effect on NO_x emission elevation at higher AFR value. Increasing the distance of fuel injection holes from the dome wall (injector length) promotes the formation of NO_x emission from combustor. This is because as the length of injector increases, reaction zone gets wider in the primary zone, which in turn result into higher flame temperature in primary zone and thereby produces large amount of NO_x emission. Among three variants of reverse fuel injector, lowest NO_x emission is achieved in 5 mm injector, whereas highest value of NO_x emission is produced in 15 mm injector.

The effect of fuel injection on CO levels is illustrated in Fig. 17. Using reverse fuel injection strategy, efficient mixing is achieved between fuel and air. This results in enhanced rate of fuel burning and better combustion of CO into CO_2. When compared for the three variants of reverse fuel injector, it is found that increasing the injector length produces little effect on CO reduction and performance improvement. This is because as the injector length increases, the depth of fuel jet penetration in primary zone increases while the size and strength of central recirculation zone in combustor decreases (refer Fig. 13b–d). Larger fuel jet penetration in radial direction helps in achieving better fuel–air mixing and intensive combustion in primary zone, while reduced size and strength of central recirculation zone decreases the residence time of combustion products. This slows down the rate of combustion of CO into CO_2. The performance parameters such as average exit gas temperature and combustion efficiency are improved due to better combustion of fuel in reverse fuel injection system as shown in Table 2.

From the measured results, it is observed that reverse fuel injection system works well in terms of achieving reduction in CO and maintaining low NO_x and low wall temperature when injector length is kept at 5 mm. Increasing the length of fuel

Table 2 Performance of combustor for various fuel injection systems (AFR = 50)

Fuel injector models	Average exit temperature (K)	Combustion efficiency (%)
Conical fuel injector	991	72.35
5 mm reverse fuel injector	1099	83.65
10 mm reverse fuel injector	1102	83.97
15 mm reverse fuel injector	1112	85.02

injector produces little effect on CO reduction but has remarkable effect on increase of NO_x emission and wall temperature. There is scope for further reduction in CO emission level using 5 mm reverse fuel injector geometry. The CO emission can further be alleviated by blending hydrogen with methane, which augments combustion chemistry between fuel and air. Analysis of chemical reaction, flame structure and emission characteristic is reported for different methane–hydrogen fuels used in upward swirl combustor for the optimum geometry of 5 mm reverse fuel injector. Results are presented and discussed in the following sections.

8 Evaluation of Upward Swirl Combustor with Reverse Fuel Injector and Hydrogen Blending

Hydrogen is a carbon-free, high-energy-density substance that has been widely used as a supplement with methane/natural gas in gas turbines. The strategy of adding hydrogen fraction to methane has advantage that it makes the system partially carbon-free, preserves a clean environment, extends flammability limits and reduces the total cost of fuel mixture [29–34].

The upward swirl combustor with reverse fuel injector is evaluated for hydrogen blended methane fuel under constant energy input condition. The methane–air combustion case (0% H_2) is considered as a reference case. A variable fraction of hydrogen, between 0 and 10% in mass, is added at constant energy input condition. Various parameters such as methane and hydrogen flow rates, total fuel mixture flow rate, total airflow rate and equivalence ratio for same energy input condition are listed in Table 3.

Table 3 Parameters for constant energy input condition

Cases	H_2 % (by mass)	Energy input (kW)	Fuel mixture flow rate (kg/s)	CH_4	H_2	Total air flow rate (kg/s)	Equivalence ratio (Ø)
				Volume flow rate (SLPM)			
1	0	70	0.0014	129	0	0.07	0.345
2	1	70	0.00138	126	10	0.07	0.342
3	2	70	0.00136	123	20	0.07	0.340
4	3	70	0.00134	120	31	0.07	0.338
5	4	70	0.00132	117	41	0.07	0.337
6	5	70	0.001301	114	51	0.07	0.336
7	7	70	0.001261	107	72	0.07	0.333
8	10	70	0.001201	98	102	0.07	0.328

8.1 Influence of Hydrogen Addition on Chemical Kinetics

The chemical kinetic effect of hydrogen addition is analysed by means of mole fraction profiles of various flame species derived from LES results. Some chemical species including major species CH_4 and CO_2 and the free radicals O, H and OH are selected for analysis. Figure 19a–e illustrates the mole fraction profiles of various chemical species for pure methane–air combustion (reference case) and the fuel mixture containing 10% hydrogen.

As illustrated in Fig. 19a–b, with hydrogen addition, the mole fraction of CH_4 reduces while the mole fraction of CO_2 increases across the combustor as compared to the reference case of pure methane–air combustion. The decrease/increase in carbon-related species in the flame is due to decrease in the molar fraction of CH_4 reactant and the chemical kinetic effect. The dominant chemical reactions contributing to CH_4 are identified and presented as follows [35],

$$OH + CH_4 \leftrightarrow CH_3 + H_2O \tag{2}$$

$$H + CH_4 \leftrightarrow CH_3 + H_2 \tag{3}$$

$$O + CH_4 \leftrightarrow OH + CH_3 \tag{4}$$

$$OH + CO \leftrightarrow H + CO_2 \tag{5}$$

From Eqs. (2–4), it is implied that main CH_4 consumption reactions are the reactions of reactant CH_4 with intermediate species H, O and OH. These free radicals H, O and OH are highly active during combustion and play major role in the chemical reactions [35, 36]. The main reactions forming free radicals H, O and OH are

$$OH + H_2 \leftrightarrow H + H_2O \tag{6}$$

$$H + O_2 \leftrightarrow O + OH \tag{7}$$

As illustrated in Fig. 19c–e, the mole fraction of intermediate radicals H, O and OH increases as hydrogen is added. Increase of H_2 mol fraction due to hydrogen addition increases the rate of reaction of Eq. (6) and thus forms more H and promotes the reaction rate of Eq. (7). The rate of consumption of CH_4, therefore, increases due to promotion of Eq. (7) as hydrogen is added. This implies that combustion reactions are intensified and combustion of methane–air flame is promoted. The increase of CO_2 mol fraction due to hydrogen addition is attributed to the increase of OH radicals, which promotes the oxidation of CO into CO_2 as per Eq. (5).

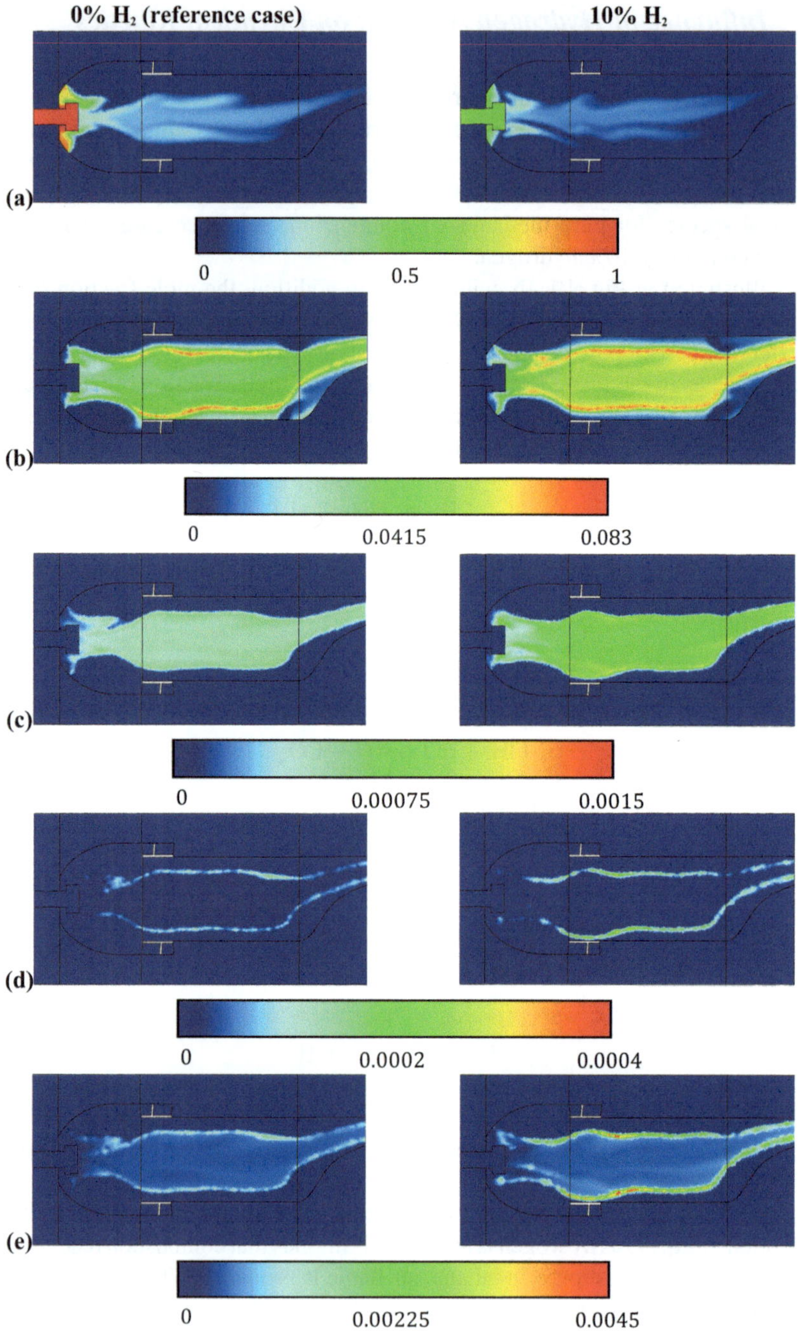

Fig. 19 Mole fraction profiles of various species **a** CH_4, **b** CO_2, **c** H, **d** O and **e** OH

8.2 *Influence of Hydrogen Addition on Flame Structure*

Flame visualization study is carried out to understand the behaviour of different hydrogen–methane flames. Luminous photographs of hydrogen–methane flames captured in a quartz chamber are presented in Fig. 20a–h. Flame visualization reveals that when methane alone is burned in combustor, a much wider and longer flame is produced, which is demonstrated by wider blue region in Fig. 20a for reference case (0% H_2). Methane–air flame is observed to extend beyond the downstream end of combustor as reaction continues to occur at the downstream. This is due to poor mixing in the reaction zone when methane alone is burned.

A little change is observed in the wider and longer blue flame region produced for methane–air combustion when the hydrogen mass concentration is increased from 0 to 2% in fuel (as shown in Fig. 20b–c). This might be because of little change in volumetric flow rate of fuel (with fixed air flow rate) at combustor inlet at same energy input. Significant change in structure and appearance of methane–air flame

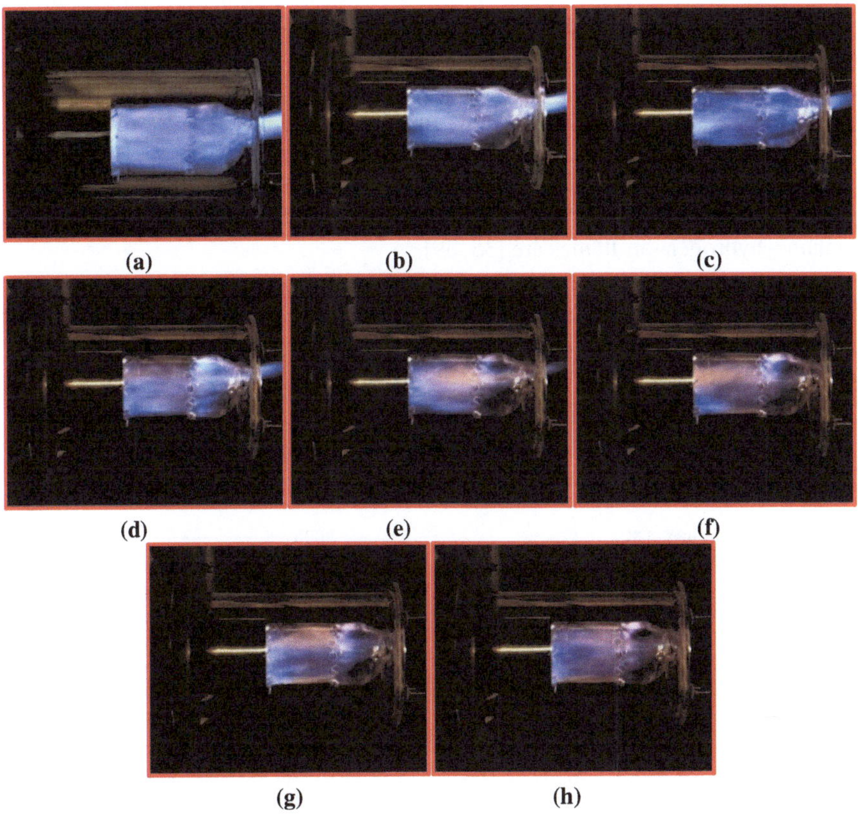

Fig. 20 Luminous photographs of different methane–hydrogen flames **a** 0% H_2, **b** 1% H_2, **c** 2% H_2, **d** 3% H_2, **e** 4% H_2, **f** 5% H_2, **g** 7% H_2, **h** 10% H_2

is observed when hydrogen fuel mass concentration is increased from 2 to 10% in fuel. Figure 20d–h shows that the flames are wider but shorter at higher hydrogen percentages. In addition, the brightness of visible blue flame region is decreased and flame appearance is changed from typical blue flame to pale orange-blue. Shorter flames at higher hydrogen percentages are mainly attributed to high reactivity of hydrogen, which increases the burning rates and shifts the reaction zone upstream towards the burner head [37, 38].

8.3 Influence of Hydrogen Addition on Emission Characteristics

Figure 21 shows the influence of hydrogen addition over the emission levels of CO and NO_x for the same energy input. As illustrated in Fig. 21, hydrogen addition to methane produces remarkable reduction in CO emission level. Reaction described in Eq. (5) is the main reaction involved in consumption of CO. This reaction is identified as the most important reaction for hydrocarbon combustion and nearly all of the heat released during combustion is expected to occur in this reaction [35]. As discussed earlier, the addition of hydrogen at same energy input increases the mole fraction of free radicals H, O and OH in the flame. Increase of OH concentration in the flame promotes the reaction rate and thereby converts CO into CO_2.

The NO_x concentration level, on the other hand, show opposite trends with hydrogen enrichment. The dominant reactions contributing to NO formation in methane–hydrogen–air flames are [35, 39],

$$N + OH \leftrightarrow NO + H \tag{8}$$

$$N + O_2 \leftrightarrow NO + O \tag{9}$$

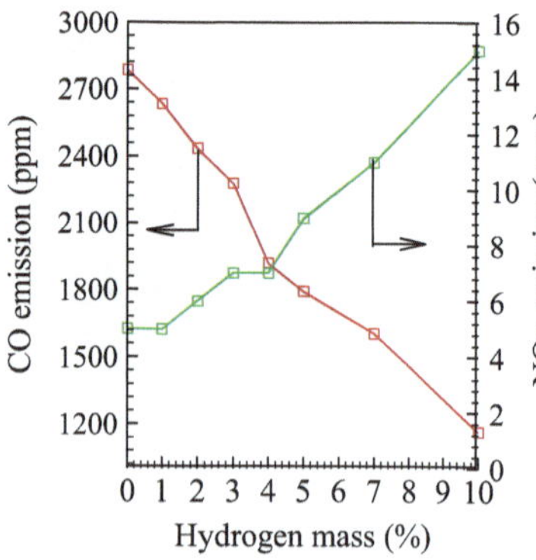

Fig. 21 Emission levels at different hydrogen percentages

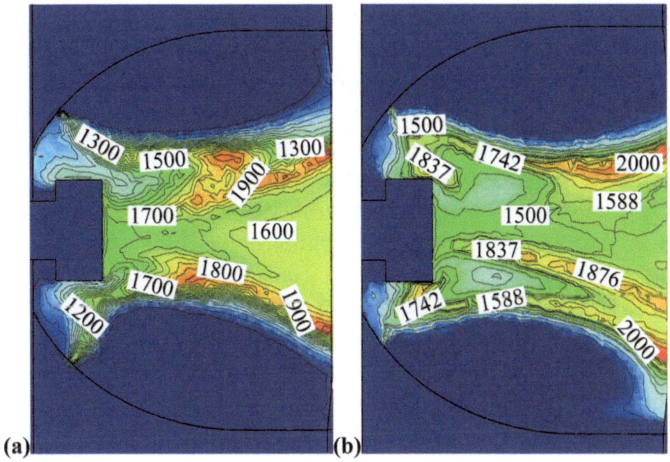

Fig. 22 Flame temperature in primary zone **a** reference case (0% H_2), **b** 10% H_2

$$N_2 + O \leftrightarrow NO + N \tag{10}$$

$$HNO + H \leftrightarrow H_2 + NO \tag{11}$$

The reactions described in Eqs. (8–11) occur in high temperature regions [35]. The NO_x formation mechanism is correlated with flame temperature in the primary zone. Figure 22 presents LES temperature contours on vertical plane of combustor obtained at AFR 50 for reference case and combustion with 10% hydrogen fuel blend. It is observed that the flame temperature in primary zone increases with hydrogen blending. This is because hydrogen has higher combustibility than methane, which results into intensified reactions and thus higher reaction zone temperature. This favours the formation of thermal NO_x at higher hydrogen concentration.

It is interpreted that injection of hydrogen by 5% in fuel blend produces another 40% reduction in CO emission level with marginal increase in level of NO_x emission from 5 ppm to 8 ppm. Further increase in hydrogen concentration from 5% to up to 10% in fuel mixture reduces the CO level, but results in higher combustion temperature and higher levels of NO_x. Therefore, limiting the hydrogen concentration to 5% in the fuel mixture can contribute to a noticeable reduction in CO emission level with marginal increase in the NO_x emission level in upward swirl combustor.

9 Conclusions and Future Endeavours

Low NO_x levels and low wall-cooling requirements are achieved in a Can-type combustor by adopting upward swirl philosophy. Upward swirl however suffers from poor combustion efficiency and large CO emissions. The development of reverse fuel injection strategy appears to be promising for improvement in combustion efficiency and reduction in CO emission levels from upward swirl combustor. The length of injector has significant influence on increase in wall temperature and the NO_x level. The length of reverse fuel injector can be kept around 5 mm to achieve better performance in terms of reduced CO level with negligible effect on the wall temperature and the NO_x level. The CO emission can further be alleviated by blending methane with hydrogen. For the same energy contribution at the combustor inlet, injection of hydrogen up to 5% in the fuel mixture reduce CO emissions by another 40% with little rise in NO_x emission value from 5 to 8 ppm.

In future, the performance evaluation of upward swirl combustor with reverse fuel injector needs to be carried out under different operating conditions. To achieve high level of combustion efficiency with low level of CO and NO_x emissions, the air supply into upward swirl combustor may be redistributed by suitable mechanism.

Acknowledgements The authors wish to acknowledge the funding of this current research by the Aeronautics Research and Development Board (ARDB), New Delhi, Government of India, under the Grant-In-Aid Scheme (Sanction letter no. ARDB/01/1041742/M/I dated 7 October 2014).

References

1. Goldstein RJ, Lau KY, Leung CC (1983) Velocity and turbulence measurements in combustion systems. Experiments in Fluids 1. https://doi.org/10.1007/bf00266261
2. Bicen AF, Jones WP (1986) Velocity characteristics of isothermal and combusting flows in a model combustor. Combust Sci Technol 49. https://doi.org/10.1080/00102208608923900
3. Bicen AF, McGuirk JJ, Palma JMLM (1989) Modelling gas turbine combustor flow fields in isothermal flow experiments. Proc Inst Mech Eng Part A: J Power Eng 203. https://doi.org/10.1243%2FPIME_PROC_1989_203_015_02
4. Jones WP, Wilhelmi J (1989) Velocity, temperature and composition measurements in a confined swirl driven recirculating flow. Combust Sci Technol 63. https://doi.org/10.1080/00102208908947115
5. Koutmos P, McGuirk JJ (1989) Investigation of swirler/dilution jet flow split on primary zone flow patterns in a water model can-type combustor. J Eng Gas Turbines Power 111. https://doi.org/10.1115/1.3240253
6. Shah RD, Banerjee J (2014) Development of an upward swirl CAN type combustor. Ph. D. thesis. National Institute of Technology Surat; 2014
7. Shah RD, Banerjee J (2016) Thermal and emission characteristics of a CAN combustor. Heat Mass Transfer 52. https://doi.org/10.1007/s00231-015-1572-9
8. Zhang K, Ghobadian A, Nouri JM (2017) Comparative study of non-premixed and partially-premixed combustion simulations in a realistic Tay model combustor. Appl Therm Eng 110. https://doi.org/10.1016/j.applthermaleng.2016.08.223
9. Wassell AB, Bhangu JK (1980) The development and application of improved combustor wall cooling techniques.In: ASME international gas turbine conference and products show, Mar 1980
10. Wadia AR (1988) Advanced combustor liner cooling technology for gas turbines. Defence Sci J 38

11. Bailey JC, Intile J, Fric TF, Tolpadi AK, Nirmalan NV, Bunker RS (2003) Experimental and numerical study of heat transfer on turbine combustor liner. J Eng Gas Turbine Power 125. https://doi.org/10.1115/1.1615256
12. Gao L, Ekkad SV, Bunker RS (2003) Heat transfer under stretched arrays of impinging jets. In: ASME IGTI conference, Atlanta, GA 2003
13. Hebert R, Gao L, Ekkad SV, Bunker RS (2004) Effect of streamwise pressure gradient on jet impingement heat transfer. In: 6th ASME ISHMT joint conference, India, Jan 2004
14. Scrittore JJ, Thole KA, Burd SW (2005) Experimental characterization of film-cooling effectiveness near combustor dilution holes. In: ASME turbo expo 2005: power for land, sea, and air, Jan 2005
15. Li L, Liu T, Peng XF (2005) Flow characteristics in an annular burner with fully film cooling. Appl Therm Eng 25. https://doi.org/10.1016/j.applthermaleng.2005.03.009
16. Li L, Peng XF, Liu T (2006) Combustion and cooling performance in an aero-engine annular combustor. Appl Therm Eng 26. https://doi.org/10.1016/j.applthermaleng.2005.11.023
17. Para SR, Li X, Subbuswamy G (2009) Numerical study of mist film cooling in combustor at operating conditions. ASME International Mechanical Engineering Congress and Exposition, Jan 2009
18. Ali ABS, Kriaa W, Mhiri H, Bournot P (2012) Numerical investigations of cooling holes system role in the protection of the walls of a gas turbine combustion chamber. Heat Mass Transfer 48. https://doi.org/10.1007/s00231-011-0932-3
19. Touchton GL, September (1984) Influence of gas turbine combustor design and operating parameters on effectiveness of NO_x suppression by injected steam or water. In: ASME joint power generation conference
20. Leonard G, Stegmaier J (1994) Development of an aeroderivative gas turbine dry low emissions combustion system. J Eng Gas Turbines Power 116. https://doi.org/10.1115/1.2906853
21. Enga BE, Thompson DT (1979) Catalytic combustion applied to gas turbine technology. Platinum Metals Rev 23
22. Mosier SA, Pierce RM (1980) Advanced combustion systems for stationary gas turbine engines. EPA Contract 68
23. Adachi S, Iwamoto A, Hayashi S, Yamada H, Kaneko S (2007) Emissions in combustion of lean methane-air and biomass-air mixtures supported by primary hot burned gas in a multi-stage gas turbine combustor. Proc Combust Inst 31. https://doi.org/10.1016/j.proci.2006.07.239
24. Davis LB, Washam RM (1989) Development of a dry low NO_x combustor. In: ASME international gas turbine and aeroengine congress and exposition, June 1989
25. Wilkes C, Gerhold B (1980) NO_x reduction from a gas turbine using exhaust gas recirculation. ASME Paper No. 80-JPGC/GT-5
26. Arai M, Amano S, Furuhata T (2007) Combustion characteristics in a micro gas turbine combustor with a recirculation zone induced by an upward swirl. J Environ Eng 2. https://doi.org/10.1299/jee.2.124
27. Furuhata T, Amano S, Yotoriyama K, Arai M (2007) Development of can-type low NO_x combustor for micro gas turbine (fundamental characteristics in a primary combustion zone with upward swirl). Fuel 86. https://doi.org/10.1016/j.fuel.2007.01.005
28. Rajpara P, Dekhatawala A, Shah R, Banerjee J (2018) Influence of fuel injection method on performance of upward swirl can-type combustor. Appl Therm Eng 130. https://doi.org/10.1016/j.applthermaleng.2017.11.017
29. Phillips JN, Roby RJ (1999) Enhanced gas turbine combustor performance using H_2-enriched natural gas. In: ASME international gas turbine and aeroengine congress and exhibition, June 1999
30. Schefer RW, Oefelein J (2002) Reduced turbine emissions using hydrogen-enriched fuels. In: Proceedings of the 2002 US DOE hydrogen program review-NREL/CP-610-32405. June 2002
31. Schefer RW, Wicksall DM, Agrawal AK (2002) Combustion of hydrogen-enriched methane in a lean premixed swirl-stabilized burner. Proc Combust Inst 29. https://doi.org/10.1016/s1540-7489(02)80108-0

32. Juste GL (2006) Hydrogen injection as additional fuel in gas turbine combustor. Evaluation of effects. Int J Hydrogen Energy 31. https://doi.org/10.1016/j.ijhydene.2006.02.006
33. Cozzi F, Coghe ALDO (2006) Behavior of hydrogen-enriched non-premixed swirled natural gas flames. Int J Hydrogen Energy 31. https://doi.org/10.1016/j.ijhydene.2005.05.013
34. Emadi M, Karkow D, Salameh T, Gohil A, Ratner A (2012) Flame structure changes resulting from hydrogen-enrichment and pressurization for low-swirl premixed methane–air flames. Int J Hydrogen Energy 37. https://doi.org/10.1016/j.ijhydene.2012.04.017
35. Wang J, Huang Z, Tang C, Miao H, Wang X (2009) Numerical study of the effect of hydrogen addition on methane–air mixtures combustion. Int J Hydrogen Energy 34. https://doi.org/10.1016/j.ijhydene.2008.11.010
36. Burbano HJ, Amell AA, Garcia JM (2008) Effects of hydrogen addition to methane on the flame structure and CO emissions in atmospheric burners. Int J Hydrogen Energy 33. https://doi.org/10.1016/j.ijhydene.2008.04.020
37. Kim HS, Arghode VK, Linck MB, Gupta AK (2009) Hydrogen addition effects in a confined swirl-stabilized methane-air flame. Int J Hydrogen Energy 34. https://doi.org/10.1016/j.ijhydene.2008.10.034
38. Shih HY, Liu CR (2014) A computational study on the combustion of hydrogen/methane blended fuels for a micro gas turbines. Int J Hydrogen Energy 39. https://doi.org/10.1016/j.ijhydene.2014.07.046
39. Kahraman N, Tangöz S, Akansu SO (2018) Numerical analysis of a gas turbine combustor fueled by hydrogen in comparison with jet-A fuel. Fuel 217. https://doi.org/10.1016/j.fuel.2017.12.071

Fires

Quantification of Enhanced Fire Severity in Modern Buildings

Dharmit Nakrani and Gaurav Srivastava

Abstract Modern materials used in the interior as well as exterior of a building pose new fire safety challenges. From an interior perspective, the traditional materials such as natural wood and fabrics are being increasingly replaced by modern synthetic alternatives, which are typically petroleum-based and are hence, more susceptible to fire. Similarly, from an exterior perspective, current building façade systems are predominantly based on combustible materials such as aluminum composite panels as opposed to the traditional non-combustible materials such as terracotta, concrete, and brick masonry. Modern materials offer significantly improved performance in certain aspects compared to their traditional alternatives—for instance, most petroleum-based polymers are much better thermal insulators compared to concrete or masonry and hence provide significantly better energy efficiency to a building. However, these modern materials are also significantly more susceptible to being vehicles of fire spread and hence pose challenges to the designers and engineers, where they have to find a balance in this *pareto* optimal choice of modern versus traditional. The current work examines this issue and presents methods to quantify fire severity of modern buildings both from interior and exterior perspectives. It is clear that the downside of modern materials is too prominent to be ignored, and some of the estimation methods must be factored in the modern building design philosophy.

Keywords Fire risk · Building fire · Façade · ACP · MDF · Furnishing

Nomenclature

P Probability
T Temperature

D. Nakrani · G. Srivastava (✉)
Department of Civil Engineering, Indian Institute of Technology Gandhinagar, Palaj, Gandhinagar, India
e-mail: gauravs@iitgn.ac.in

A. Gupta et al. (eds.), *Advances in Energy and Combustion*, Green Energy and Technology,
https://doi.org/10.1007/978-981-16-2648-7_18

1 Introduction

In recent decades, interior and exterior of buildings have witnessed major changes in the types of materials being used. In most modern habitats, there has been a noticeable enhancement in the use of petroleum-based polymeric materials such as plastics, foams, and adhesives. Consequently, modern building fires have also become more devastating than earlier. From 1993 to 2017, a total of 92.6 million fire incidents were reported, which caused more than one million deaths and significant economic losses; the total annual financial loss due to fires is estimated to be 1% of the world GDP [1, 2]. Developing countries such as India and Pakistan recorded the highest number of fire-related casualties and the second-highest number of fires [2].

The modern built environment possesses higher internal fuel load (room furnishings that can catch fire) primarily due to the use of synthetic materials in contemporary furnishings as compared to traditionally available alternatives. Traditional furnishings were typically made from natural materials such as solid wood, cotton, and wool. In modern furnishings, engineered wood products such as plywood, particle boards (made from wood chips, sawdust, glue, formaldehyde-based resins, and wax) have replaced natural solid wood, and polyurethane, nylon, polyester, polyethylene (PE), acrylic, and other petroleum-based polymers have replaced cotton and wool. These synthetic materials usually have greater calorific values, lower ignition temperatures, faster burning times, and produce much more toxic gases during combustion when compared to natural materials.

Another recent change in the building landscape, especially in developing nations, is the increase in the number of high rise buildings in cities for both commercial and residential purposes as a result of the increasing population density. Rapid urbanization, expanding cities, and rising heights of buildings intensified discussions on sustainability of the modern built environment [3]. Consequently, sustainability and energy performance became the main parameters in building design and led to changes in the building envelope materials and their installation techniques. These building envelopes, typically known as façade or curtain wall systems, render the building sustainable, make them more energy efficient, and enhance the esthetics. However, many materials used in building envelops (such as insulation boards and aluminum composite panels) are also petroleum-based and significantly enhance the susceptibility of such buildings to fire. In the 2017 fire at the Grenfell Tower, London, its façade system was found to be the main contributor toward the immense severity of the incident.

The present work intends to provide a bird's eye view toward quantification of the enhanced fire severity of modern buildings due to changes in their internal furnishings and external façades.

2 Fire Severity Due to Internal Furnishings

Changes in the interior of buildings primarily include (a) the increased use of more flammable synthetic materials such as plastics and textiles, (b) the increased quantity of combustible materials, and (c) the use of goods with the unknown composition and uncertain flammability behavior. Over time, home and office contents have transitioned from being comprised of natural materials to dominated by synthetic materials [4].

Modern office and residential buildings use corrugated plastic roofs, partition walls, plastic doors, false ceiling panels, core panels for walls, interior finish materials, etc. Table 1 lists the time to ignition and peak heat release rates (HRR) observed through cone calorimeter tests (at incident heat flux level of 50 kW/m^2) for some materials which were used traditionally in residential and office occupancies and have been widely replaced by their modern alternatives.

It can be observed from Table 1 that time to ignition is considerably shorter for most of the polymeric materials being used in modern occupancies as compared to the traditional materials. Also, peak HRR is greater for modern materials made from plastics, acrylic fabric as compared to traditional materials. Further, polymeric materials have greater calorific values compared to traditional materials [5]. A study led by Babrauskas [6] compared specimens of upholstered chairs made with different materials. The cotton-padded chair covered in a cotton fabric produced a peak HRR of 370 kW at 910 s after ignition. The foam-padded chair covered in polyolefin fabric produced a peak HRR of 1990 kW at 260 s after ignition. While both chairs exhibited similar total heat released of about 420 MJ, the synthetic chair synthetic chair yielded almost five times greater peak HRR than that of the natural chair and this peak was attained four times faster.

Table 1 Combustion characteristics of various interior materials used in residential and office buildings

	Material	Time to ignition (s)	Peak HRR (kW/m^2)	Calorific value (MJ/kg)
Traditional materials	Ghana teak wood	25	298	19
	Pine wood	30	160	20
	Heavy cotton fabric	8	189	15
Modern alternatives	Particle board	24	249	21
	Plywood	23	328	20
	Polyester	6	210	22
	Acrylic fabric	12	386	29
	Polyurethane	5	227	24
	Polyethylene (PE)	31	725	40

The survey by Zalok [7] for office buildings showed a significant decrease in the contribution of cellulosic material (70%) and a greater proportion of plastics (22%) over the past decades. Further, the survey conducted by Issen [8] for single-family residential homes in the US found the contribution of cellulosic materials to be 65% and plastics around (25%). The decrease in the proportion of cellulosic materials is compensated with an increase in the proportion of polymeric materials. The introduction of appliances such as computers, printers, air-conditioners, and refrigerators in modern occupancies has primarily led to an increase in plastic-based combustible contents. This high intensity of fuel load in modern occupancies plays a key role in faster fire propagation, shorter flashover times, and rapid changes in fire dynamics. A full-scale experimental study conducted by Kerber [4] aimed at characterizing fire development within modern and legacy rooms, concluded that flashover can occur within 5 min of ignition in modern rooms, whereas legacy rooms took almost 30 min to flashover (flashover is the point when a local fire suddenly engulfs an entire room; any evacuation or first-responder firefighting operations are effective only before the flashover stage). The development of room temperatures in legacy and modern rooms is shown in Fig. 1 [1, 4].

The overall increase in the total amount of combustible materials within a typical room, measured through fire load energy density (FLED)—the amount of combustibles per unit area of a compartment—has been clearly observed through several fire load surveys across the world. Table 2 summarizes the FLED values in office-type occupancies (of about 30 m^2 floor area) over the last five decades reported by various surveys and studies. It can be observed that the FLED has significantly increased due to the use of more flammable materials.

A study by Khan and Srivastava [15] carried out for office and dormitory buildings observed that, on an average, fire load present in modern buildings is about three times greater than what is prescribed by building codes (used for design). Based on the FLED and the compartment characteristics (area, size, windows, doors, etc.), one can predict the time–temperature curve to be expected during a potential fire through

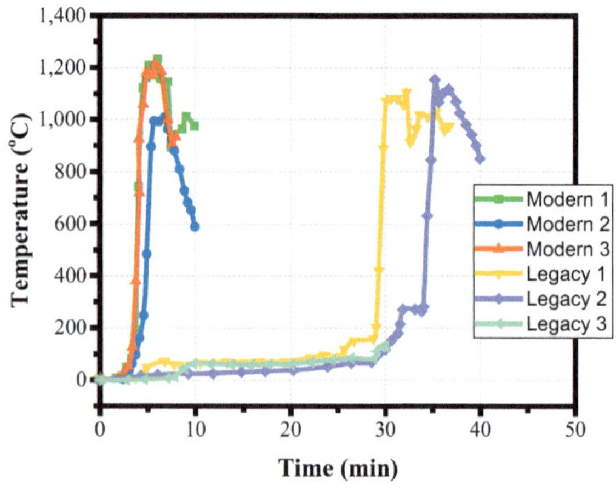

Fig. 1 Development of room temperature in legacy and modern building rooms. [1, 4]

Table 2 FLED of office buildings over the past five decades around various geographical locations

Timeframe	Location	Mean FLED (MJ/m^2)	References
1970–1980	USA and UK	365	[9, 10]
1980–1990	New Zealand	496	[11]
1990–2000	New Zealand and India	590	[12–14]
2000–2010	Canada	850	[7]
2010–2020	India	1205	[15]

the parametric relations available in the Eurocodes [16]. Figure 2 shows some such parametric fire curves for a typical modern compartment considering the increased FLED along with the standard fire curves (the ISO 834 and the ASTM E119 fire curves). The standard fire curves are the ones which are used to prescribe design requirements for building components. Many building components are required to have a one-hour fire rating, indicating that the components should not fail when subjected to a standard fire for a period of one hour. However, if one observes the parametric fire curves of Fig. 2, it is evident that a component tested with the standard fire will not sustain one hour of the parametric fire (since the temperatures attained by the parametric fire are greater than that of the standard fire).

The enhancement in the fire risk can be characterized probabilistically through Monte Carlo simulations by considering appropriate probability distribution of FLED (typically, generalized extreme value or gamma distributions [15]) based on the observed data. This allows the consideration of the uncertainties associated with the actual distribution of FLED and a robust quantification of the risk. Given that standard fire temperatures play a key role in defining fire ratings of different building

Fig. 2 Parametric fire curves compared with standard fire curves [15]

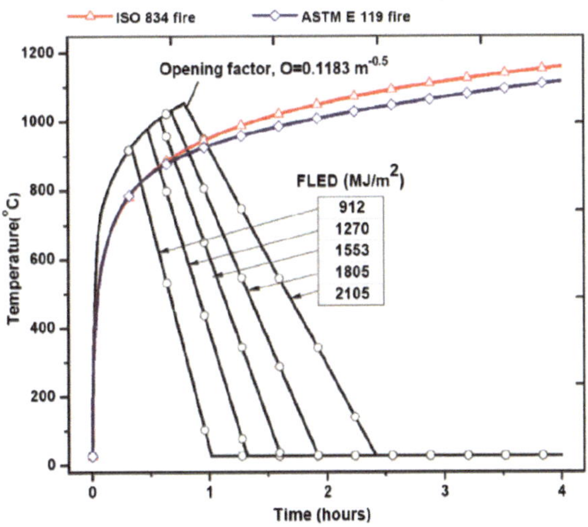

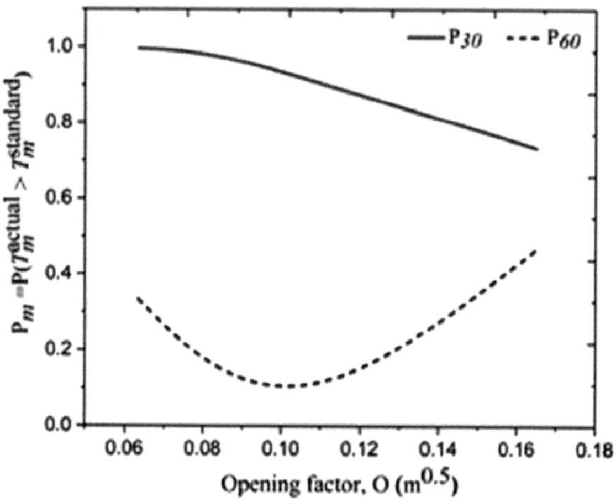

Fig. 3 P_{30} and P_{30} versus opening factor for dormitory buildings [15]

components, one may define the fire risk as the probability of the real fire temperature exceeding the standard fire temperature during different times of a fire. Specifically, it can be defined as

$$P_m = P(T_m^{\text{actual}} > T_m^{\text{stdandard}})$$

where m denotes time in minutes, T_m^{actual} is the actual fire temperature after m minutes, and $T_m^{\text{stdandard}}$ is the standard fire temperature after m minutes. Figure 3 shows P_{30} and P_{60} computed for dormitory building configurations using Monte Carlo simulations with a sample size of 1000 [15]. The opening factor quantifies the extent of openings in the room in the form of doors and windows (that will allow fresh air ingress and aid in combustion). It is interesting to note that the opening factor of 0.10 $m^{0.5}$ comes out as the 'optimal' opening factor from a P_{60} point of view. Many such insights can be obtained through such fire risk quantification studies.

Modern polymeric materials also impact the amount of smoke and hazardous gases generated during their combustion which ultimately reduces the egress time for occupants and increases health and life risks as compared to traditional materials. It can be seen from Fig. 4 that CO and CO_2 yield obtained through cone calorimeter results of cellulosic materials (wood, cotton fabric, etc.) is significantly lower as compared to modern polymeric materials (polyester fabric, nylon fabric, PE, polyurethane foam, etc.).

3 Fire Severity Due to External Façades

Façade is an important component of modern buildings which is typically non-load bearing, and its purpose is to separate the indoor environment from external

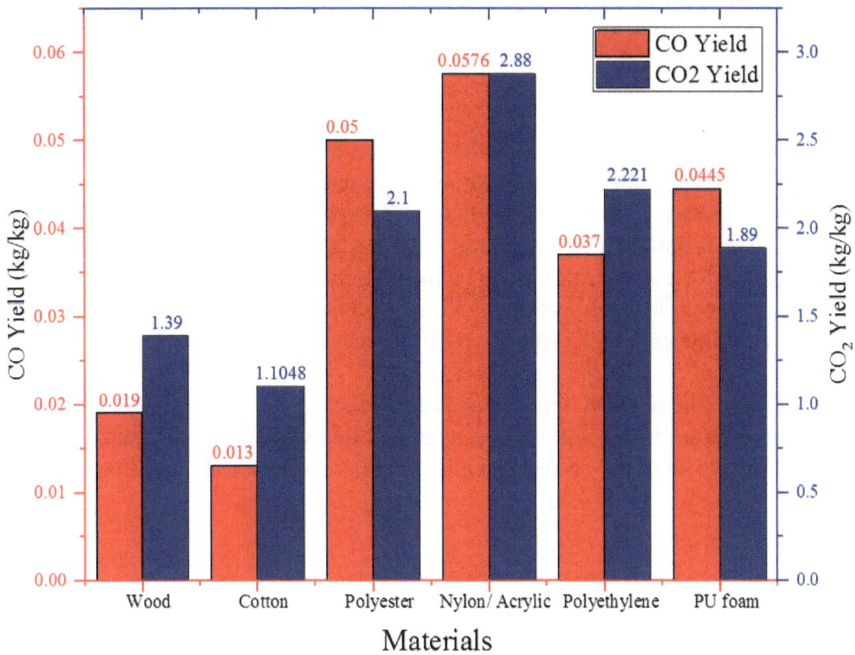

Fig. 4 CO and CO_2 yield of different materials used in residential and office buildings [17]

factors. It also caters to aesthetic, economic, energy, and security considerations. Earlier cladding systems comprised of materials such as terracotta stone, cement tiles, masonry veneer, pre-cast concrete panels, and non-structural glass. These cladding materials were non-combustible and hence possessed negligible fire risk to the building occupants.

Modern commercial and residential buildings increasingly utilize a façade system based on glass or a combination of glass and other materials such as aluminum composite panels (ACP), medium/high-density fiberboards (MDF/HDF), and high-pressure laminates (HPL). Further, refurbishing of tall concrete buildings often involves covering the exterior with a rain-screen façade system, consisting of an outer screen, a cavity, and an inner layer of insulation. This outer screen is usually ACP, HPL, or mineral fiberboard. The increased use of ACP in the construction sector is confirmed by the global sales data. Construction application accounted for around 52% of the global ACP market in 2018. Global ACP market was valued at USD 5.33 billion in 2018 and is projected to reach USD 8.71 billion in 2026 [18]. The newer materials usually possess superior insulating properties as compared to traditional concrete- or masonry-based systems and hence positively impact the ambient thermal insulation of the buildings, which leads to lower costs for maintaining the inside temperatures during regular operations.

While such façade systems render building energy-efficient, reduce air and water infiltration, and enhance esthetics, they can become easy vehicles for movement of fire along the building [19, 20]. Bonner and Rein [21] found that the number of worldwide fires in tall buildings with spread via façade is currently, on average, 4.8 per year, and the total has increased by seven times in the last 30 years. Nyugen et.al. [22] summarized recent major fire accidents from around the world. Investigations carried out after these fire accidents have shown that the presence of combustible materials in the cladding, typically a polymer core sandwiched between aluminum panels, has been a major contributor in the rapid spread of fire.

Before the installation in actual buildings, façade systems are usually required to be certified for their fire performance using applicable testing standards ranging from small (bench) to large-scale tests. Results from one such bench-scale test, using the cone calorimeter, which provides HRR for several façade materials (at incident heat flux level of 50 kW/m^2) are shown in Fig. 5.

Table 3 lists the ignition time and peak HRR (at incident heat flux of 50 kW/m^2) of commonly used façade materials. The table highlights the short time to ignition and high HRR values for modern façade materials, whereas the traditional façade

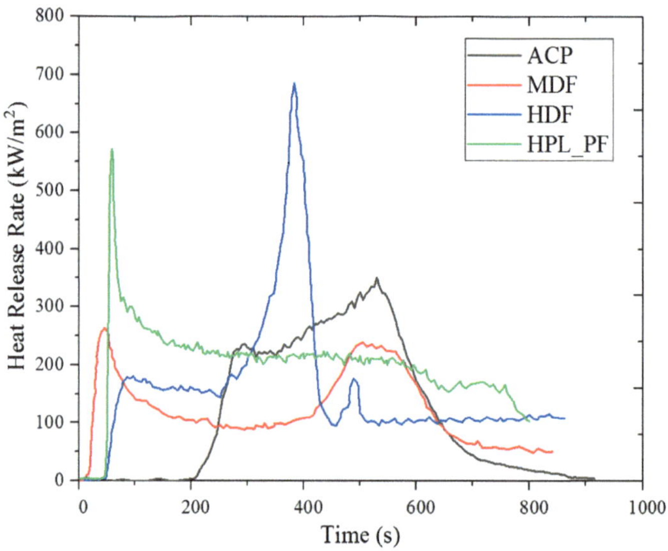

Fig. 5 Heat release rate of various façade materials

Table 3 Combustion characteristics of different façade materials

Material	Thickness (mm)	Time to ignition (s)	Peak heat release rate (kW/m^2)	References
ACP	4	207	350.96	[23]
MDF	12	19	263.52	[24]
HDF	8	44	685.96	[24]
HPL-PF	10	48	572.12	[25]

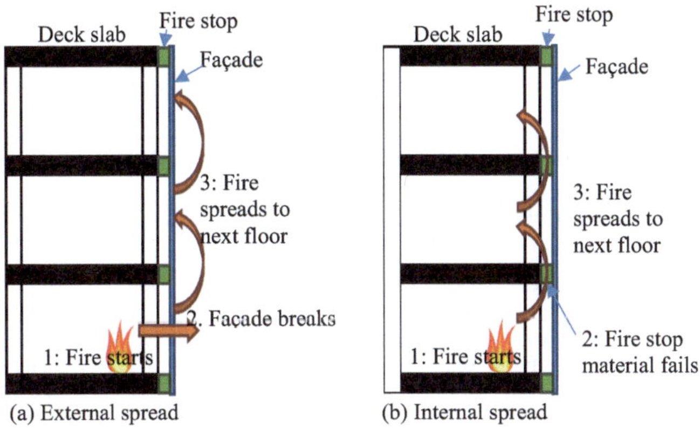

Fig. 6 Mechanisms of fire spread in the façade systems

systems are non-ignitable. The time to ignition of ACP and HPL is higher than that of MDF and HDF as the former comprise of a combustible core of polymeric material sandwiched between metal sheets thereby reducing the direct exposure, whereas the latter are wood-based panels.

White and Delichatsios [26] studied various small-scale tests and observed that these tests are applied only to individual material components and represented specialized exposure conditions. Moreover, bench-scale tests also did not help to ascertain the fire spread mechanism over the façade systems in real-fire scenarios. Fire spread in façade systems typically takes place through two mechanisms: external spread through the leap-frog mechanism, and internal spread [27], as shown in Fig. 6.

It is therefore essential to understand how the whole façade system will perform and which fire spread mechanism will contribute to fire spread over the cladding systems. Srivastava et al. [23] highlighted the need for considering the system in its entirety rather than performing component-level testing. Despite many years of use of façade systems, full-scale testing remains the only robust option to gain knowledge about the performance of these systems under real fire scenarios.

A summary of full-scale façade tests of various building standards is given in Table 4. It can be observed from Table 4 that the geometry, fire source, specimen support details, test duration, measurements, and acceptance criteria vary greatly for different testing standards. Also, all full-scale façade tests simulate an internal post-flashover fire with flames ejecting from windows. However, these tests do not give an idea about the performance of these systems under the impact of exterior fire sources or the interaction of façade fire with room furnishings of upper floors.

The full-scale standard façade tests typically consider a single compartment setup, i.e., the fire compartment is at the ground floor with a standard wood crib as the fire source and no compartment (and hence, no fire load) at upper floors of the experimental setup. Thus, the interaction between the furniture of the compartment and the façade system cannot be studied for upper floors.

Table 4 Summary of different full-scale façade fire tests

Test	Arrangement	Façade dimensions	Standard fire source	Test duration (min)	Failure criteria
ISO 13,785:2002	Re-entrant corner "L" arrangement (wing wall)	H: 4 m above window lintel W: 3 m Wing width: 1.2 m	5.5 MW gas burner in enclosure	23–27	None specified
BS 8414 part 1 and Part 2	Re-entrant corner "L" arrangement (wing wall)	H: 6 m above window soffit W:2.6 m Wing Width: 1.5 m	3 MW timber crib in opening	30	Exterior or interior fire spread 5 m above window within 15 min
DIN 4102—20	Re-entrant corner "L" arrangement (wing wall)	H: 5.5 m W: 2 m wing width: 1.4 m	320 kW gas burner in opening	20	Exterior or interior fire spread 3.5 m above the window
NFPA 285	Single wall surface	Two-story test frame; H: 5.3 m W: 4.1 m	900 kW gas burner in bottom enclosure and 400 kW gas burner in window (ignited 5 min after room burner)	30	Exterior fire spread > 3.05 m above the window
SP FIRE 105	Single wall surface	H: 6 m W: 4 m includes a 500 mm eave at top	2.5 MW Heptane fuel tray in an enclosure	15	Exterior fire spread > 3.2 m above the window
CAN/ULC S134	Single wall surface	H:7.25 m above window W: 6 m	5.5 MW propane burner or timber crib in an enclosure	25	Exterior fire spread > 5 m above the window
GB/T 29,416	Re-entrant corner "L" arrangement (wing wall)	H:9 m W: 2.6 m Wing width: 1.5 m	Timber crib or gas burner in an enclosure	–	–

Srivastava et al. [27] developed a full-scale façade testing facility at Indian Institute of Technology Gandhinagar, India, where full-scale façade systems can be studied under real fire scenarios. The testing facility is a three-story steel structure with three independent compartments at each level, as shown in Fig. 7. The plan dimensions of the compartments at each story are $10' \times 20'$ (approximately 3 m $\times$ 6 m) with a story height of $10'$. The facility has been designed to simulate both fire spread mechanisms

Fig. 7 Full-scale façade fire experimental facility at IIT Gandhinagar [27]

(external and internal). Two faces (Faces 1 and 2) of the test structure can simulate external spread mechanism, while one face (Face 3) has been designed to simulate the internal spread mechanism. This facility allows (i) to test façade systems including curtain wall and perimeter firestop as per actual installation practices, (ii) use of real furniture to simulate real fire scenario, and (iii) provision of fully furnished compartments at all floor levels to enable the study of interaction between façade fire and compartment components.

Several full-scale tests with different combinations of façade assemblies such as glass-ACP and glass-MDF have been performed at this facility. Before the onset of fire in experiments, fire scenarios are developed using a combination of real office/residential furniture and rated wooden cribs with the desired FLED based on the considered design parameters. The test structure is instrumented with K-type thermocouples for temperature measurements, internal video cameras to monitor smoke movement, an elaborate smoke collection system, and external infrared cameras to measure surface temperatures.

Through numerous experiments, it has been observed that façade fire in assemblies with glass and ACP can reach from one story to the next through the leap-frog mechanism in about 3 min and hence can severely limit the response and egress time for higher floors of a building. Once the fire breaches the compartment of origin and combustible façade catches fire, the vertical spread over the external surface between story is less than 3 min. Further, the experiments showed that the façade system can be exposed to heat flux levels in excess of 100 kW/m^2 with maximum temperatures reaching 1000 °C. The existing testing standards consider lower incident flux levels, and the gas burners used in most existing test methods can reach maximum temperatures of around 700 °C. These findings indicate a significant gap between current testing standards and real fire performance of façade systems [23].

Glass-MDF façade assembly tested under real fire scenarios has been found to be better than glass-ACP systems as no flame spread was observed over the surface once

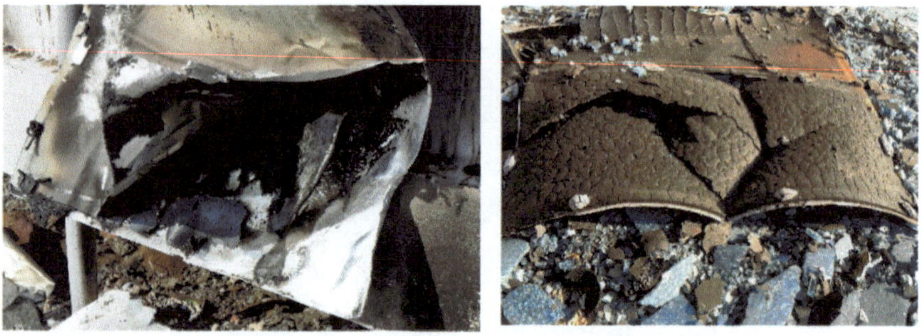

(a) ACP with exposed PE core residue (b) MDF after getting charred

Fig. 8 Behavior of different façade materials after being exposed to real fire scenarios

the fire breached the compartment of origin. External vertical spread over the MDF surface was not observed as opposed to rapid flame spread over the ACP surface. Failure of MDF panels wherein they fall out from their positions occurred in the compartment of origin within 10 min of the ignition, whereas ACP typically falls out within 4 min of the ignition.

ACP comprises of a PE infill placed between two aluminum sheets. Aluminum melts around 660 °C, while PE can catch fire in a matter of seconds (see Table 1). During a fire, flame temperatures typically range from 600 to 800 °C, and thus, aluminum begins to melt early. The combustible core of PE aids the melting of aluminum as the polymer at the periphery catches fire and begins to drip and vaporize thereby increasing the local fuel concentration. However, MDF is a wood-based panel and hence does not melt. During combustion, it usually chars and retains its strength for a relatively longer time when compared to ACP. The combustion of PE leaves residue in the aluminum sheets, while MDF fails after considerable charring, as shown in Fig. 8. Due to these inherent material-level differences in ACP and MDF, an ACP-based façade tends to fail much faster than that of an MDF-based façade system.

When façade panels begin to fail (and fall), they create more and more openings on the façade side of the building. This allows more access for fresh air inside the fire compartment and makes it easier for the fire to sustain. The extent of openings, measured through the opening factor (or ventilation factor), can be used as an indicator of the spread of fire within a compartment as well as the inter-story spread. Figure 9 shows the cumulative ventilation factors for two full-scale experiments: one with glass-ACP façade system and other with glass-MDF system with the same initial conditions and fire scenario. It can be observed that in the ACP-based system, the ground and first floors become fully open from the façade-side within 10 min of ignition, whereas in the MDF-based system, it takes about 20 min for the ground floor to become fully open and the fire does not propagate to first or second floors (they remain completely closed from the façade-side).

Overall, the degree to which a façade system might become involved in a fire is dependent on its overall flammability. By flammability, it is referred to the ease

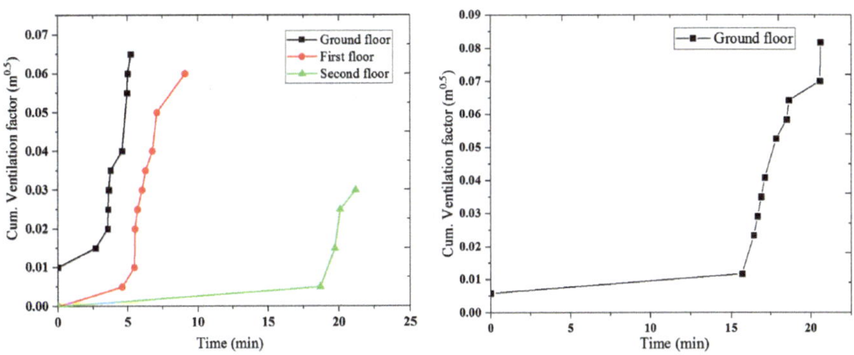

(a) Glass-ACP façade assembly (b) Glass-MDF façade assembly

Fig. 9 Ventilation factor w.r.t time plots for different façade assemblies

of fire to ignite and spread through the façade. Flammability is a broader concept that depends on how different materials are combined, oriented, and connected in the façade system, and how they interact during a fire. While the traditional non-combustible façade systems cannot become vehicles of fire like their modern combustible counterparts, they are also less energy-efficient and increase the overall heating/cooling costs of the building during day to day operations.

4 Conclusions

This disposition aimed to highlight the dilemma of current designers and engineers dealing with modern buildings. The advantages related to esthetics and enhanced functionality of modern materials is on one side, and the disadvantage of their greater fire susceptibility is on the other. The engineers (and perhaps occupants) need to be cognizant of the need to balance the two sides in order to achieve a reasonably safe design. Quantification of the enhanced fire severity due to modern materials is an essential exercise for anyone to understand the flip side of modern materials.

Cone calorimeter results of various traditional materials and modern engineered alternatives have shown that modern materials are easier to ignite and have higher HRR values. Various fuel load surveys from across the globe have shown the consistent rise in the overall fire load within buildings. Specifically, for India, the fire loads have become three times greater than those reported in previous studies conducted in the 1990s. The proportion of natural cellulosic materials has decreased considerably in recent times, whereas the proportion of plastics and synthetic materials, which have much higher calorific values than cellulose, has increased. Parametric design fire curves provide further quantification of the fire severity by predicting time–temperature curves of potential fires within building compartments. Parametric curves computed through actual fire loads observed in recent surveys tend to be more severe than the standard fire curves which are typically employed for characterizing fire ratings of different components of a building. A probabilistic characterization of the enhanced fire risk due to the changes in the materials used for room furnishings can also be done through Monte Carlo simulations based on the developed probability models for fire loads to assess optimal design parameters for building compartments.

Modern exterior façade materials such as ACP, HDF, MDF, and HPL are also highly susceptible to ignition and flame spread, while the traditional façade materials such as terracotta stone, cement tiles, masonry veneer, pre-cast concrete panels, and non-structural glass are all non-combustible. The inability of bench-scale tests in assessing the performance of modern combustible façade systems under real fire scenarios and full-scale standardized facade tests available have been discussed along with their limitations. This highlights the need for full-scale testing in actual site conditions under real scenarios. A comparison of the performance of dominantly used glass-ACP façade system and a glass-MDF system under real fire scenarios showed significant insights. It was observed in glass-ACP assembly that the time of vertical spread of fire between stories is less than 3 min indicating an extremely short window of time available for egress and rescue operations. The tests also showed that during real fires, these combustibles façade systems can be exposed to heat flux levels above 100 kW/m^2 which the standardized full-scale tests do not account for.

Overall, this work provided insights toward the quantification of enhanced fire severity of modern buildings due to changes in internal furnishings and external façade systems. It is expected that more research in this area with some of the discussed factors will lead to the improvement of building standards and eventually produce safer designs in the future.

Acknowledgements The support from the Centre for Safety Engineering, IIT Gandhinagar, is gratefully acknowledged. In the development of the full-scale fire testing facility, the support from Underwriters Laboratories LLC is acknowledged. The support from Ministry of Human Resource Development (Government of India) under the Uchchatar Avishkar Yojana (UAY) is also acknowledged.

References

1. Kodur V, Kumar P, Rafi MM (2019) Fire hazard in buildings: review, assessment and strategies for improving fire safety. PSU Res Rev 4(1):1–23
2. Brushlinsky NN, Ahrens M, Sokolpv S, Wagner P (2019) World fire statistics
3. Zeng C, Deng X, Dong J, Hu P (2016) Urbanization and sustainability: comparison of the processes in 'BIC' countries. Sustain 8(4):1–18
4. Kerber S (2012) Analysis of changing residential fire dynamics and its implications on firefighter operational timeframes. Fire Technol 48(4):865–891
5. Ljungberg LY (2007) Materials selection and design for development of sustainable products. Mater Des 28(2):466–479
6. Babrauskas V, Lawson RJ, Walton DW, Twilley HW (1982) Upholstered furniture heat release rates measured with a furniture calorimeter
7. Zalok E (2011) Validation of methodologies to determine fire load for use in structural fire protection. Fire Prot Res Found, 1–65

8. Issen LA (1980) Single-family residential fire and live loads survey. NIST Interagency/Internal Rep
9. Baldwin R, Law M, Allen G (1970) Survey of fire loads in modern office buildings—some preliminary results. Fire Saf Sci 808:1–1
10. Culver CG (1976) Survey results for fire loads and live loads in office buildings. Natl Bur Stand Build Sci Ser 85
11. Barnett CR (1984) Pilot fire load survey carried out for the New Zealand Fire Protection Association. MacDonald Barnett Partners, Auckl
12. Buchanan AH, Abu AK (2017) Structural design for fire safety, 2nd edn. Wiley, West Sussex
13. Kumar S, Rao CVSK (1997) Fire loads in office buildings. J Struct Eng 123(March):365–368
14. Kumar S, Rao CVSK (1995) Fire load in residential buildings. Build Environ 30(2):299–305
15. Khan NA, Srivastava G (2018) Enhanced fire severity in modern Indian dwellings. Curr Sci 115(2):320–325
16. Eurocode 1 (2004) Actions on structures—part 1–2: general actions -actions on structures exposed to fire, London
17. Wg Cdr GA Sachdev VS, Kotresh TM, Vyawahare MK (2004) Heat release characteristics of the basic materials used for flying clothing. Ind J Aerosp Med 48(1):53–58
18. "Aluminium Composite Panels (2019) Global market analysis, insights and forecast, 2019–2026
19. Gandhi P et al (2017) Performance of glass-ACP façade system in a full-scale real fire test in a G+2 structure. Procedia Eng. 210:512–519
20. Peng L, Ni Z, Huang X (2013) Review on the fire safety of exterior wall claddings in high-rise buildings in China. Procedia Eng. 62:663–670
21. Bonner M, Rein G (2018) Flammability and multi-objective performance of building façades: towards optimum design. Int. J. High-Rise Build. 7(4):363–374
22. Nguyen K, Weerasinghe P, Mendis P, Ngo T, Barnett J (2016) Performance of modern building façades in fire: a comprehensive review. Electron J Struct Eng 16(1):69–86
23. Srivastava G, Nakrani D, Ghoroi C (2020) Performance of combustible facade systems with glass, ACP and firestops in full-scale, real fire experiments. Fire Technol 56(4):1575–1598
24. Lee BH et al (2011) Evaluating the flammability of wood-based panels and gypsum particleboard using a cone calorimeter. Constr Build Mater 25(7):3044–3050
25. McKenna et al (ST) Fire behaviour of modern façade materials—understanding the Grenfell Tower fire. J Hazard Mater 368(2018):115–123 (2019)
26. White N, Delichatsios M (2015) Fire hazards of exterior wall assemblies containing combustible components, 1st edn. Springer, New York
27. Srivastava G et al (2018) Development of a unique full-scale real-fire façade testing facility at IIT Gandhinagar. Curr Sci 115(9):1782–1787

The Tea Light Candle and the Global Waste Problem

Ali S. Rangwala and Kemal S. Arsava

Abstract In 2012, the World Bank estimated that each person living on planet Earth produces approximately 1.2 kg of waste per day, amounting to 1.2 billion tons per year. By 2025, this number is expected to reach a staggering 2.2 billion tons, which raises the fundamental question: What is and what happens to waste? Domestic burning of waste is a frequent disposal technique in many parts of the world. Further, even when waste is moved to dumpsites in many developing countries, it is not uncommon for the material to be burned by open, uncontrolled fires. In short, *fire* is the primary mode of waste disposal. Fire provides volume reduction and prevents disease, but inefficient combustion brings its hazards in the form of toxic/noxious gasses. This paper will discuss the work we have been doing at WPI related to burning hazardous waste cleanly and efficiently by observing a tea light candle's behavior.

Keywords Fire · Flame refluxer · Candle · Waste burning

1 Background

The world population generates 1–2 billion metric tons of municipal solid waste per year globally[1]. In the last decade, this waste's nature has shifted to include increasing amounts of plastics and electronic waste as shown in Table 1. The "waste" can be considered a reasonably good combustible with an approximate heat of combustion slightly higher than wood and equal to 20 kJ/g.

Fire is the most efficient means of reducing the volume of solid waste. This has led developed areas to invest in waste incineration facilities, which require a large amount of infrastructure and capital investment. In less-developed areas, fires for waste management are typically open-air and present both health and environmental hazards. The purpose of this paper is to demonstrate new techniques in burning waste outdoors, which can reduce the environmental impact by increased combustion temperatures and added capability to burn waste streams that are moist and have lower

A. S. Rangwala (✉) · K. S. Arsava
Worcester Polytechnic Institute, Worcester, MA 01609, USA
e-mail: rangwala@wpi.edu

A. Gupta et al. (eds.), *Advances in Energy and Combustion*, Green Energy and Technology,
https://doi.org/10.1007/978-981-16-2648-7_19

Table 1 Typical composition of municipal waste

	%	kJ/g
Paper	41	10.4
Yard waste	15.2	9.3
Plastics	10	46
Food waste	16.8	7.6
Wood	3.8	16
Textiles	2.2	15
Rubber and leather	2.4	19.5
Glass	7.6	
Metals	9	
Miscellaneous	2	
		kJ/g
Coal		29
Natural gas		45
Propane		50
Methanol		22
Wood		16.5
Waste		20

heat of combustion. A simple example of a tea light candle is used to demonstrate the problem and the solution.

When the waste is burned in the open, the fire takes the form of a diffusion flame as shown in Fig. 1. Typically, fires of this type burn at a regression rate of 0.1 mm/s. This rate is slow, mainly because fuel vaporization is slow because of the highly endothermic nature of the pyrolysis process. If we imagine 1 cubic foot (0.31 × 0.31 × 0.31 m) of solid waste burning as a pool in one plane, it will burn out in (0.31 m)/(0.1 mm/s) ~ 52 min. However, if the same solid waste can be burned in a premixed mode, like in a Bunsen burner, for example, then its regression rate increases

Fig. 1 Comparison between rates of burning between a turbulent pool fire and a Bunsen burner flame. The fuel consumption rate is prolonged when burning as a pool, which results in incomplete combustion and an overall slow-burning process. For waste burning, this translates to a more significant strain on the environment

to ~ 1 m/s and will burn out in (0.31 m)/(1 m/s) = 0.3 s! This drastic reduction in burn time has advantages. The increased temperatures cause the flame to be less sooty as shown in Fig. 1, which compares a hydrocarbon pool fire (diffusion) and a Bunsen burner premixed flame. The question then is, can we increase the burning rate or regression rate of the free diffusion flame? We will proceed to answer this question by using a common tea light candle and show how we can increase its regression rate by a factor of 10 times.

2 The Tea Light Candle

2.1 *How Long Does It Take for a Tea Light Candle to Burn?*

Let us consider the example of a candle flame shown in Fig. 2. If paraffin is ignited with a match, it is challenging to burn. The candle's paraffin melts because of heat from the flame; it travels by capillary forces through the wick, where it then evaporates to become paraffin vapor, a gaseous fuel. The process is illustrated in Fig. 2.

The slow-burning candle flame takes around 4 h, 30 min to be fully consumed. The long burning time makes it amenable for decorations at home. The wick ensures that burning is complete, resulting in a bright yellow flame with no visible soot. Solid paraffin in a tea light represents an ideal solid fuel because it is homogeneous and melts at low temperatures to form a low viscosity liquid. Polyethylene and PVC, typical plastics in solid waste, will melt at low temperatures, but they have different flow properties from paraffin in the molten state. It will burn slowly by first using the flame's heat to form a melt pool, gasifying to create fuel vapor. The entire process is inherently slow because of the conversion of condensed fuel to vaporized fuel. An obvious way of burning the candle faster is to increase the number of wicks discussed in the next section.

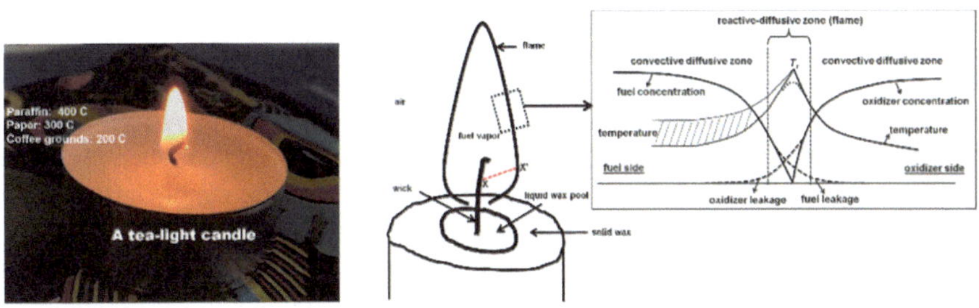

Fig. 2 Candle flame [1]

Fig. 3 Addition of wicks to increase the burning rate of a tea light candle

2.2 *How Can the Burning Rate Be Increased?*

The burning rate of the tea light candle can be increased by adding more wicks as shown in Fig. 3. With two wicks, the candle is consumed in 2 h, with three wicks in 1 h 20 min, and four wicks in 1 h as shown in Fig. 3. Adding more wicks does not improve the burning further because the wicks increase the paraffin's melting rate, and at some point, when the paraffin is completely melted in the container, it burns like a small pool fire. At this stage of burning, the purpose of wicks is lost.

2.3 *How Can the Burning Rate Be Increased Even Further?*

Figure 4 shows the tea light candle, with the addition of four copper rods of 1 mm in diameter. The height of the copper rods was approximately two times the base's height as shown in Fig. 4. As the candle burns, it heats the copper rods, which then capture and conduct ("reflux") the flame heat back to the paraffin. The wax is observed to melt in 2 min, and in 22 min, a pool fire is observed and the wicks drop off. The burning rate continues to accelerate because of the copper. At ~30 min, the formation of bubbles is observed on the copper rod's surface in the molten pool, and consequently, the burning rate increases because of the improvement in heat transfer rates because of nucleate boiling. The flame height at around 33 min is ~10 times a normal candle flame, and the candle is consumed in 37 min. The enhanced flame height after 30 min is shown in Fig. 4. It would have been impossible to generate an increase in flame height of this magnitude without the onset of nucleate boiling that occurs because of the rods' presence.

Figure 5 shows a plot of the burning time of the tea light candle. Adding wicks improves the burning rate to some extent. However, the addition of the copper rods improves the burning rate much more. Further, the molten pool is wholly consumed

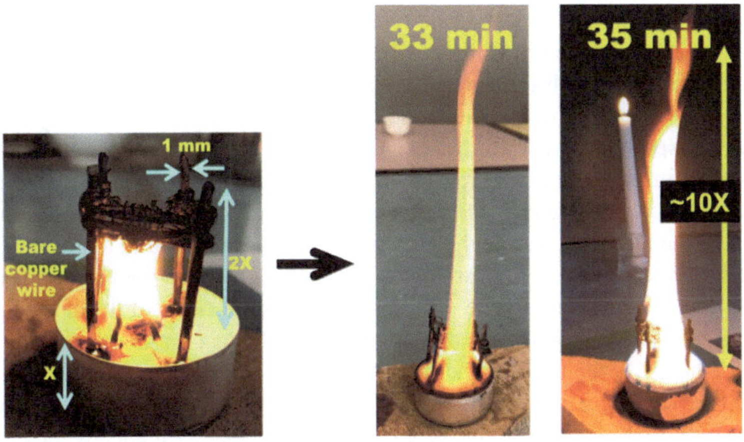

Fig. 4 Increase in burning rate by a factor of 10 by addition of four copper rods—the Flame Refluxer™ concept

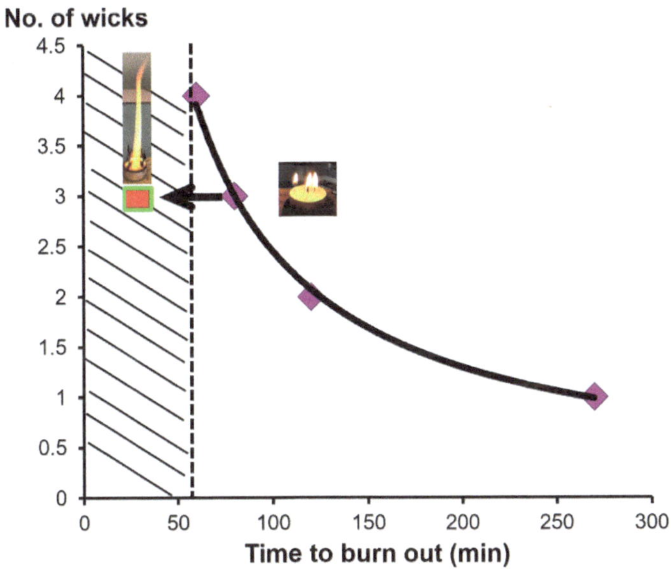

Fig. 5 Burning of a tea light candle. The copper rods can burn out the tea light candle in ~35 min by creating an additional heat transfer pathway via conduction through the copper

from 30 to 37 min. Thus, if additional wax were to be supplied at this stage of burning, then it would be consumed at a significantly higher rate (~10 times).

The slower burning during normal usage is the resistance to heat transfer from the flame to the surface. Torero et al. [2] characterized the theoretical efficiency of heat transfer for burning condensed fuels showing it could be as low as 0.3%. With such a small fraction of the total heat released actually returned to the fuel surface, the vaporization and consequently combustion is slow as most of the heat is lost to the environment through the buoyant convection plume and thermal radiation [2–6]. The addition of copper rods to the tea light candle creates a secondary pathway for heat transport to the paraffin wax. This pathway is coupled to the flame, and as heat is transferred, the burning rate improves, which enables further heat transport and so on. Once nucleate boiling starts on the copper rod's surface, the burning rate increases by a factor of 10, and the paraffin candle is consumed in a few minutes!

The baseline (unenhanced) burning rate of a tea light-sized pool (or diameter) of paraffin wax was measured to be ~2 g/m^2 s. What is the cause of the enhancement, especially the role of the bubble dynamics during the last 7 min of the burn when the flame height is ten times higher than the baseline? The controlling parameters' role can be further explained, assuming that the heated copper rods will produce bubbles because of subcooled nucleate boiling. Recent experiments by the research group have demonstrated that the bubbles generation rate per unit area $\dot{b}''_g \sim 4 \times 10^5$ bubbles/m^2-s [7, 8]. Note that the generation rate will vary with several parameters, like the thermofluid properties of molten wax and the surface roughness of the copper. However, the ballpark number allows us to get an order of magnitude analysis on bubbles' influence. Assuming that the bubbles are spherical and the average bubble size is $r \sim 1$ mm [9],

$$\dot{m}''_{\text{bubble-vapor}} = \frac{4}{3}\pi r^3 \rho_g \dot{b}''_g, \tag{1}$$

where $\dot{m}''_{\text{bubble-vapor}}$ is the mass flux of fuel vapor generated by bubbles reaching the surface and consequently releasing the vapor of density ρ_g. Substituting $r = 1$ mm, $\rho_g = 7300$ g/m^3, and $\dot{b}''_g = 4 \times 10^5$ bubbles/m^2-s leads to $\dot{m}''_{\text{bubble-vapor}} = 1.5$ g/m^2-s. Further, the mass flux of fuel vapor because of a micron droplet spray by bubble bursts on the surface, $\dot{m}''_{\text{bubble-spray}}$, is given by

$$\dot{m}''_{\text{bubble-spray}} = \frac{4}{3}\pi r^3_{\text{drop}} \rho_\ell n_{\text{drop}} \dot{b}''_g, \tag{2}$$

where n_{drop} is the number of droplets produced per bubble burst and r_{drop} equals the average drop radius. Spiel [10] experimentally observed that a 1 mm bubble bursting on the water surface creates anywhere from 50 to 200 drops of 10–100 μm size. Assuming that each bubble of paraffin vapor upon bursting produces ~100 drops of 100 μm size and substituting $\rho_\ell = 749{,}000$ g/m^3, $r_{\text{drop}} = 10^{-4}$ m, $n_{\text{drop}} = 100$, and $\dot{b}''_g = 4 \times 10^5$ bubbles/m^2-s into Eq. (2) give $\dot{m}''_{\text{bubble-spray}} = 15.65$ g/m^2-s.

The above example shows that a significant enhancement in the burning rate over the baseline, increasing from 2 to 17.2 g/m^2-s (~9×), is achieved mainly because of bubble bursts on the surface. With an increase in the rod surface area, for example, by changing surface conditions [11, 12], the burning rate will increase further. Also, by optimizing the rod geometry and the number of rods, the time to melt the pool and create the nucleate boiling can be reduced even further. In short, the tea light candle can be burned in an even shorter time. One can hypothesize that the upper limit will be controlled by the critical heat flux (CHF) condition during subcooled nucleate boiling. Besides, it will depend on the film thickness of the bubble at the time of bursting. The latter is coupled with flame heat flux by evaporation, a parameter not explored in the literature [9]. Elucidation of these mechanisms has significant implications and has never been addressed in fire research.

3 The Flame Refluxer Concept and Applications

We have seen that conductive nonflammable objects, such as metal rods, can be placed into a condensed fuel to enhance the burning rate. This idea led to the development of a patented technology named the Flame Refluxer™ [14–16]. The radiative and convective heat generated by the combustion is directed back to condensed fuel to create a feedback loop that effectively enhances the burning rate. The increased burning rate also results in a reduction of smoke [16]. The simple but effective design of the Flame Refluxer™ requires no moving parts, atomizing nozzles, or compressed air, which makes it more reliable than existing burner technologies. The fact that it uses heat generated by combustion itself to improve burning makes it amenable to rapid waste burning. Figure 6 shows a timeline of the project's evolution from concept to a tested technology at both laboratory and field. The idea has been successfully applied to burn hydrocarbon liquids, emulsions, waste liquids trapped in the soil, human/animal waste, and solid fuel slurries mixed in water. Details of the Flame Refluxer™ in different applications are available in reports [17, 18] and papers [13, 16, 19–22].

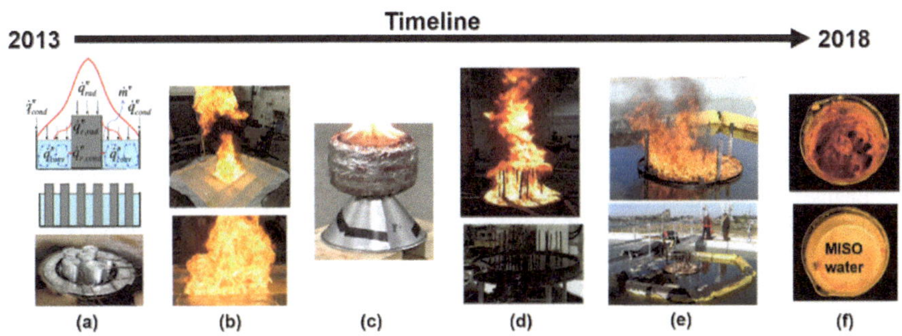

Fig. 6 Timeline of the Flame Refluxer™ from concept to application in different areas, **a** transformer fires [13], **b** arctic oil spill burning [5], **c** waste burner (solid fuel), **d** waste oil burner (liquid) [14], **e** offshore in situ burning [15], **f** biowaste burning

4 Conclusions

The paper demonstrates how a tea light candle's burning rate can be increased by a factor of 10 by using four copper rods immersed in the wax. Nucleate boiling (formation of bubbles on the metal's surface immersed in the liquid) is the main reason for significant enhancement of mass burning rate. The metal object immersed in the flame zone creates a high-efficiency pathway for heat transfer from the hot flame zone (~800–1000 °C) to the cooler fuel layer (~200–400 °C). This is because the portion of the object immersed in the fuel can easily reach temperatures above the fuel boiling point resulting in subcooled nucleate boiling, which increases the heat transfer rates by order of magnitude ($\times 100$). This is the main reason for the fuel's faster evaporation and the corresponding increase in the burning rate.

The idea of immersing a metal object in fuel to enhance its burning rate led to the development of a patented technology named Flame Refluxer ™. This technique can be extended to burn a variety of wastes quickly and efficiently, creating a pathway for efficient waste disposal in specific applications related to waste burning.

Acknowledgements The Flame Refluxer is funded by the Bureau of Safety and Environmental Enforcement, U.S. Department of the Interior, Washington, DC, under contract number E14PC00043. The contents do not necessarily reflect the BSEE's views and policies, nor does mention of the trade names, or commercial products constitute endorsement or recommendation for use. Many thanks to people, who made contributions to this study, including Karen Stone, Kristi McKinney, Glenn Mahnkenn, Professor Raghavan Vasudevan (IIT Madras), Professor Jose Torero (University College, London), and students at WPI (in alphabetical order) Apoorv Walavalkar, Aravind Krishnan, Dave Petrow, Drew Egger, Francisco Kang, Hamed Farmahini, Hayri Sezer, Honggang Wang, John Ramos, Karl T. Sundberg, Kathryn Hall, KiJana M. Haney, Li Chang, Lydia Shi, Mahesh S. Kottalgi, Nathaniel G. Sauer, Nathan S. Ghion, Nicolas Tobar, Panyawat Tukaew, Raymond Ranellone, Shijin Kozhumal, Shivaprasad Arava, Steve Stephansky, Trevor Borth, Veronica Kimmerly, Xiaoyue Pi, Zhaozhao Meng.

References

1. SFPE (2008) SFPE handbook of fire protection engineering, 4th edn. Society of Fire Protection Engineering, Bethesda
2. Torero JL, Olenick SM, Garo JP, Vantelon JP (2003) Determination of the burning characteristics of a slick of oil on water. Spill Sci Technol Bull 8(4):379–390
3. Brosmer M, Tien C (1987) Radiative energy blockage in large pool fires. Combust Sci Technol 51(1–3):21–37
4. Gritzo L, Nicolette V, Tieszen S, Moya J, Holen J (1995) Heat transfer to the fuel surface in large pool fires. In: Chan SH (ed) Transport phenomenon in combustion. Taylor and Francis
5. Shi X, Bellino P, Simeoni A, Rangwala A (2016) Experimental study of burning behavior of large-scale crude oil fires in ice cavities. Fire Saf J 79:91–99
6. Jiang F, De Ris J, Qi H, Khan M (2011) Radiation blockage in small scale PMMA combustion. Proc Combust Inst 33(2):2657–2664
7. Pi X, Rangwala AS (2019) The influence of wire orientation during nucleate pool boiling in subcooled dodecane. Int J Heat Mass Transf 137:1247–1257

8. Sauer N, Pi X, Arsava KS, Rangwala AS (2020) Controlling mechanisms of burning rate enhancement while using Flame RefluxerTM technology during in situ burning of crude oil spills. In: Proceedings of international oil spill conference
9. Poulain S, Villermaux E, Bourouiba L (2018) Ageing and burst of surface bubbles. J Fluid Mech 851:636–671
10. Spiel DE (1995) On the births of jet drops from bubbles bursting on water surfaces. J Geophys Res Oceans 100(C3):4995–5006
11. Jones BJ, McHale JP, Garimella SV (2009) The influence of surface roughness on nucleate pool boiling heat transfer. J Heat Trans 131(12)
12. Chung SH, Law CK (1984) Importance of dissociation equilibrium and variable transport properties on estimation of flame temperature and droplet burning rate. Combust Flame 55:225–235
13. Shi X, Rangwala AS (2013) Influence of immersed objects on liquid pool boilover. AICHE Annual Meeting, San Francisco
14. Rangwala AS, Shi X, Arsava KS, Mahnken G (2017) Methods and systems for clean-up of hazardous spills. Google Patents
15. Arsava KS, Rangwala AS, Mahnken G (2019) Systems and methods for in-situ clean up of burnable materials. Google Patents
16. Tukaew P (2017) Outdoor gas emission sampling system: a novel method for quantification of fires in outdoor conditions
17. Rangwala AS, Arsava KS, Mahnken G, Shi X (2015) A novel experimental approach to enhance burning of oil-water emulsions by immersed objects. In: B.o.S.a.E. Enforcement (ed) Worcester Polytechnic Institute
18. Rangwala AS (2017) An offshore oil burn enhanced by floating immersed objects. In: B.o.S.a.E. Enforcement (ed) Worcester Polytechnic Institute
19. Arsava KS, Raghavan V, Rangwala AS (2018) Enhanced oil spill clean-up using immersed thermally conductive objects. Fire Technol 54(6):1745–1758
20. Arsava KS, Borth TJ, Fields SL, Hansen K, Kozhumal S, Mahnken G, Petrow D, Rangwala AS, Sezer H, Stone K (2019) Application of flame Refluxer^{TM} concept to ISB–experimental results of 5 field trials in mobile, alabama. In: Proceedings-42nd AMOP technical seminar on environmental contamination and response. Emergencies Science and Technology Section
21. Sezer H, Arsava KS, Kozhumal SP, Rangwala AS (2017) The effect of embedded objects on pool fire burning behavior. Int J Heat Mass Transf 108:537–548
22. Sezer H, Arsava KS, Rangwala AS (2017) Oil spill clean-up using immersed metal wool. J Environ Chem Eng 5(5):5196–5206

Sustainable Energy Technologies

Approach to Solve Renewable Energy Problems

Sourabh Kumar and Ryoichi S. Amano

Abstract Computational fluid dynamics had shown significant development in software capability and hardware during the last two decades. It has become an essential tool in various product development stages and is used actively in advanced research technology. Renewable energy varies from traditional to new resources such as biomass, fuel cells, water, wind, and solar. These renewable sources are playing a significant role in electric generation, and both energy producers and consumers are demanding more deployment of these sources. There is a great advantage in using numerical simulation methods for innovative renewable technologies such as wind energy, biomass, solar power, hydropower, gas and steam turbines, fuel cells, tidal, wave, and many others. CFD analysis provides an additional resource for these innovative renewable energy sources to predict results that can save time and money and reduce physical prototyping and testing resources. This paper intends to outline and discuss all CFD applications in renewable energy sources, helping understand these methods' advantages and usage.

Keywords Computational fluid dynamics · Renewable energy · Wind turbine · Hydropower · Biomass · Tidal · Solar and geothermal

Abbreviations

A	Area
Cp	Pressure coefficient
CFD	Computational fluid dynamics
DES	Detached Eddy simulation
k	Velocity

S. Kumar
Jacobs Engineering Inc, Bingham Farm, USA

R. S. Amano (✉)
University of Wisconsin, Milwaukee, WI, USA
e-mail: amano@uwm.edu

A. Gupta et al. (eds.), *Advances in Energy and Combustion*, Green Energy and Technology,
https://doi.org/10.1007/978-981-16-2648-7_20

K	Kelvin
LES	Large eddy simulation
m	Meter
OWSC	Oscillating wave surge converter
OWC	Oscillating wave converter
P	Power output
PA	Point absorber
P_{ave}	Power output
Q	Volume of water flow
RANS	Reynolds averaged Navier–Stokes
SST	Shear stress transport
T	Time
U	Velocity
VOF	Volume of fluid
WEC	Wave energy converter
W	Watt

Nomenclature

η	Efficiency
ρ	Density
g	Acceleration due to gravity
h	Acceleration due to gravity
ω	Specific rate of dissipation

1 Introduction

Fluid dynamics involves a partial differential equation for physical laws. CFD solvers transform these partial differential equations into algebraic equations and solve using computers numerically. Computational fluid dynamics is a method of solving numerically the laws that govern the movement of fluid. The complete system can be transformed into a virtual product and can be visualized virtually for analysis. One of the advantages of using CFD over the experimental method is that the flow analysis can be achieved without interfering with the flow and can access the flow variable's details where conventional experimental techniques fail. CFD can be implemented to save time and resources during design iterations for faster and economical than tests. CFD analysis can be used for measuring all quantities with higher resolution in space and time. CFD process can be a handy tool for minimizing the test if the methodology has been verified and validated with tests. With the advancement in the numerical algorithm and computing power, there is the widespread use of CFD techniques in various fields such as gas turbine, heat transfer, multiphase, combustion, and several other areas.

Overall, the energy demand increases over the years, and dependence on fossil fuels causes environmental problems. Increase the use of renewable energy sources will solve the environmental concern along with solving power demand. Renewable energy is the energy source that replenishes naturally, and these resources are inexhaustible but are limited by the amount of energy per unit of time [1]. Over the last century, a significant energy generation source has been fossil fuel, which results in greenhouse gas emissions. Reduction in emissions and air pollution can be made by increasing renewable energy sources such as water, wind, bioenergy, solar, tidal, hydro, geothermal, and ocean energy.

This chapter is mainly explaining the CFD application and methods in the field of renewable energy sources. CFD simulations are appropriate and of great importance for virtual investigation and prediction of renewable energy systems. These CFD simulations can provide several design iterations for flow around wind turbine, waves, hydro, solar, and biomass. These CFD simulations can reduce the prototype test and increase trust in the new renewable sources. The main objective of this chapter is to describe the various renewable energy sources briefly. CFD simulation methods are discussed to provide insight into the process used for modeling. There is an attempt to introduce and review the methods used to address renewable energy sources problem using CFD simulations.

2 Renewable Energy

Renewable energy is a natural source that can be replenished over time. Renewable energy sources are more uniformly distributed geographically than fossil fuels. These systems are independent of fuel supply with clean output. As environmental control is becoming more stringent, there is an increase in demand for renewable sources. Renewable energy generation over the years has been shown in Fig. 1 [2]. Over the years, the major obstacle to the increase in renewable energy use was higher cost than conventional sources. Due to the increase in innovative methods, the cost is coming down, shown in Fig. 2 [3].

Some of the primary renewable energy sources currently being used are:

1. Hydropower
2. Biomass power
3. Geothermal power
4. Ocean power (tidal/wave/thermal)
5. Wind power
6. Solar power (solar heating and thermal electric/ photovoltaic).

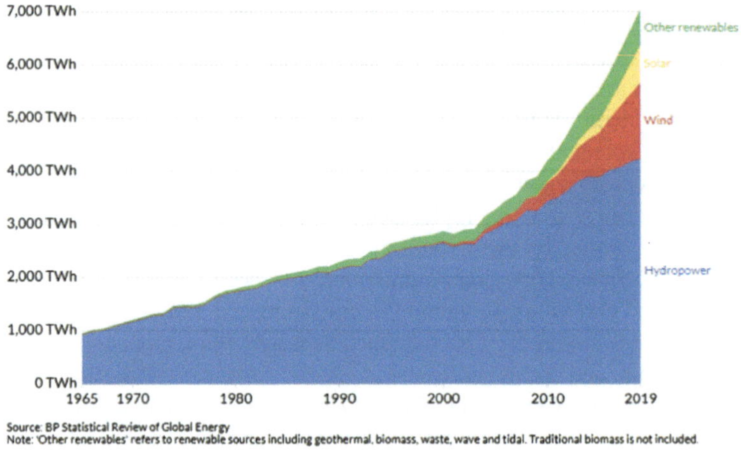

Other renewables refer to geothermal, biomass, waste and tidal. Traditional biomass is not included

Fig. 1 Renewable energy generation worldwide [2]

	GLOBAL WEIGHTED-AVERAGE COST OF ELECTRICITY (USD/KWH) 2018	COST OF ELECTRICITY: 5TH AND 95TH PERCENTILES (USD/KWH) 2018	CHANGE IN THE COST OF ELECTRICITY 2017-2018
Bioenergy	0.062	0.048-0.243	-14%
Geothermal	0.072	0.060-0.143	-1%
Hydro	0.047	0.030-0.136	-11%
Solar photovoltaics	0.085	0.058-0.219	-13%
Concentrating solar power	0.185	0.109-0.272	-26%
Offshore wind	0.127	0.102-0.198	-1%
Onshore wind	0.056	0.044-0.100	-13%

Fig. 2 Global electricity costs [3]

2.1 Hydropower

Hydropower is the oldest source of energy used for electricity generation. The hydropower plant's mechanism works on flowing water. Change in water elevation called head determines the available energy of water flows through the penstock and pushes the turbine for electricity generation. Due to water being the primary source of hydroelectric power, the plants are made near the source location. Figure 3 [4] shows a traditional hydropower plant setup. Hydropower is a well-established technology that accounts for global electricity generation. This water energy is either

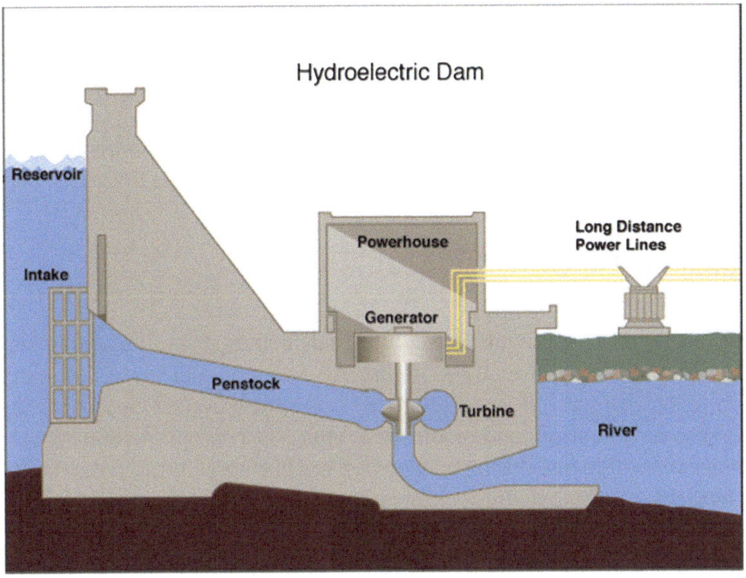

Fig. 3 Hydroelectric power generator [4]

in the form of potential energy (reservoirs) or kinetic energy (rivers). In both cases, electricity is generated by passing the water through large water turbines.

Most of the conventional hydroelectric facilities include:

1 Run of the river system—the water current's force applies pressure on the turbine, including a weir to divert water flow to hydro turbines.
2 Storage system—water is accumulated in the reservoirs created by dams on streams and rivers and is released through hydro turbines to generate electricity [4].

Power output from a dam is a product of efficiency η, the potential energy per unit volume, and the volume of water flowing per second Q

$$P = n\rho g h Q \tag{1}$$

Power output is dependent on hQ.

Hydro turbines are rotating equipment that converts high-pressure water to mechanical power. For low head and high flow rate, Kaplan turbines, which are axial flow, are used. For higher heads, the Francis turbines, which are radial inflow reaction turbines, are used. Pelton wheels are used for small flow rates where a nozzle is used to accelerate the high-pressure water stream to the high-velocity jet, impinges on the wheel turbine's blades. The overall efficiency of converting waterpower to electricity is being 80% [5].

The significant advantage of hydropower is low operating costs, a low impact on the atmosphere, long plant life, and quick response to electricity demands [6]. Some additional benefits are water for irrigation and flood control. The return of investment for these plants is extended. Some severe social and environmental issues need to be considered, such as population displacement, sedimentation, water quantity changes, flooding, and aquatic life. Overall developing hydropower sites can be economically competitive with fossil fuel plants and being renewable and pollution-free.

2.2 Biomass Power

Biomass is a plant- and animal-derived material such as straw, logs, dung, and crop residue, which are used directly or indirectly as fuels, and these fuels are often called biofuels. Biomass is generally used by combustion or by converting it to biofuels such as bioethanol and biodiesel. Biomass is a carbon-neutral source as the amount of carbon released during combustion has been previously removed from the atmosphere during photosynthesis into making plant material.

These are agricultural crop, or they are byproduct which can replace some fossil fuel consumption. Another form of biomass is animal waste or human waste, where digesters can produce methane, and the residue of this process can be used as farm fertilizers.

The essential processes that use the energy of biomass are as follows.

1. Combustion: Woody plants and grasses can be burnt directly to provide process heat or electric power.
2. Gasification: The process that converts biomass into gaseous fuel with hydrogen, carbon dioxide, and carbon monoxide is known as gasification. Gasification is done at high temperatures without combustion with controlled amounts of oxygen or steam, which conserves the heating value with fuel combustible in the boiler furnace.
3. Pyrolysis: Thermal decomposition of biomass produces combustible products in solid, liquid, and gaseous forms.
4. Fermentation: Distillation of carbohydrates produces ethanol liquid fuel used as a blender in gasoline for motor fuel.
5. Anaerobic digestion: Produces a gaseous mixture of carbon dioxide and methane.

The heating value of biomass is lost due to easy use, which increases the production cost. The overall price to the consumer is higher for biofuels, and hence, biomass energy supply must use unutilized byproduct and low capitalization cost for biomass conversion to higher-value fuel.

For biomass energy crops except for wood, a year-round steady supply of feedstock for energy needs storage post-harvesting, which increases the cost of biomass fuels. Kumar and Reisel [7] presented a mathematical model for calculating energy usage and corn ethanol production cost. The model investigated solar and wind energy as potential energy sources to reduce carbon footprints during ethanol production.

2.3 Geothermal Power

Geothermal energy is the earth's natural heat, tapping hot temperature via steam at great depth or moderate temperature at shallow depth. The different temperature ranges of water or steam at different locations can be effectively used for various applications such as

(a) Hot water or steam from high-temperature geothermal reservoirs can be used for electric generation
(b) Moderate and low can be used for direct applications such as heating buildings.

There is a high temperature below the earth level. This hot interior has a high magnitude of energy to be exploited by drilling wells to the depth of 10 km and pumping high-temperature steam and water effectively for geothermal energy plants.

The most practical and economic sites for geothermal energy are surface temperature, higher and underground water, and steam close to the surface. Figure 4 has a typical geothermal power plant. Production well supplies warm water and steam mixture and fed to the separator for separating to steam from liquid—the liquid being pumped back through the injection well. Steam is provided to a turbine for electricity generation. The condenser cools the exhaust steam and reinjects it to the ground.

Some of the challenges of geothermal resources can be locating a reservoir of enough high temperature, which are assessable by drilling. In addition to this, supply should sustain for a considerable duration, and the resistance to the pump should be less to make the plant economical. Lately, geothermal heat pumps are utilizing

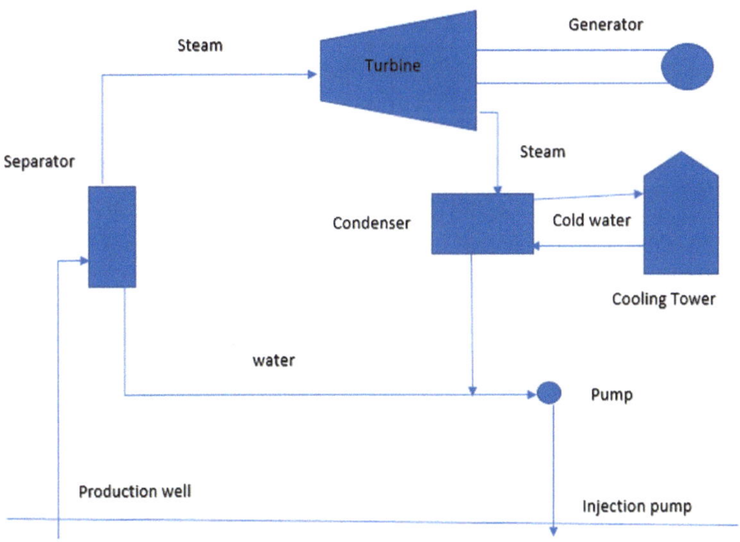

Fig. 4 Flow of fluid in geothermal power plant

the heat storage capacity of underground water or steam. Water from the reservoir supplies heat to the heat pump and then releases a high amount of heat during the cold weather.

2.4 *Ocean Power (Tidal/wave/thermal)*

Tidal

The regular tidal rise and fall of the ocean level can be used to generate electricity. The tidal rise and fall are due to the gravitational force on the opposite side of the earth by moon and sun, which produces higher elevation toward and away from the moon and lower elevation in the middle. Spring tides are due to the full moon, which is maximum elevation, and the time when solar effects are minimum is called neap waves.

Tidal power system uses ocean current to drive turbines to generate electricity. Tidal barrage for energy generation uses a low head turbine and can work most of the day between high and low tides by using barrage. Sluice gates located along the barrage control the level of water across the turbine. Sluice gates are closed when the tides turn in or start to come in, and barrage holds back the rising sea level, creating a water level difference, which can be used for power generation. A similar reverse phenomenon is used during low tides to generate electricity called tidal barrage ebb generation—Fig. 5 [8] shows the tidal barrage power system.

Power from the tidal barrage

$$P_{\text{ave}} = \frac{\rho g A h^2}{2T} \tag{2}$$

The advantage of this system is carbon-free energy, and it improves flood protection. Large tidal barrage construction has some drawbacks, including enormous capital cost, long construction times and intermittent operations, and long-term ecological impact.

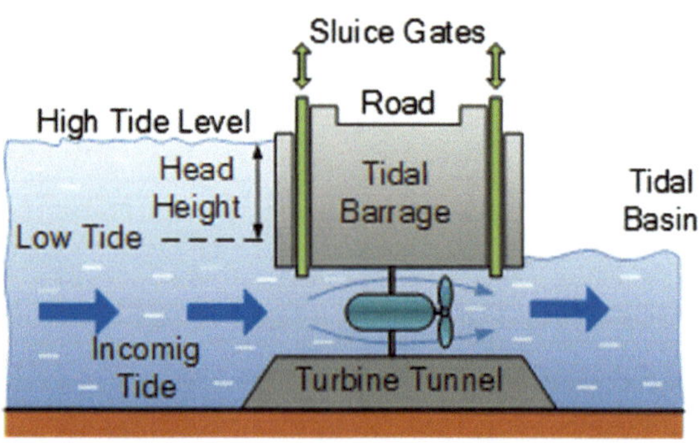

Fig. 5 Tidal barrage energy generation [8]

Wave: The effect of wind mainly causes the waves on the surface of the sea. The wave crest experiences air closer and moves faster. From Bernoulli's theorem, air pressure is reduced; hence amplitude increases, and waves are generated. As wave crest collapses, fluid is displaced and forced to rise above the equilibrium level. The motion of fluid below the water level decays exponentially with depth. The wave energy moves with the wave, and by using suitable mechanical devices, the energy in a wave can be converted to mechanical power. The wavelength of waves is the superposition of different wavelengths; hence, the tool should be robust enough to withstand random fluctuations of waves.

Incident power per unit width of the wavefront is [9]

$$P = 1/4\,\rho g a^2 \sqrt{\frac{g\lambda}{2\pi}} \tag{3}$$

Most of the wave power is available when the wind exceeds its average value. Wave energy at the location is available for a small fraction of time with 15–20% capacity. There are two types of devices primarily used for extracting wave energy.

(a) Floating bodies anchored in place
(b) structure fixed to the seabed.

By the flow of the waves, parts of the structure will have relative displacement producing power. These are the power produced by a device that intercepts a fraction of waves. Some of the wave power devices being used are oscillating water column, which uses air turbine in a duct above the water surface. The base is being open to the sea so that water oscillates in a vertical direction. That results in the air above the water surface to be in phase with the water column-free surface and drives the air turbine. The turbine used is called the Wells turbine. Figure 6 [10] shows an oscillating water column.

Submerged devices use the change in pressure below the surface when waves pass, with wave crest having high pressure and low pressure for wave trough.

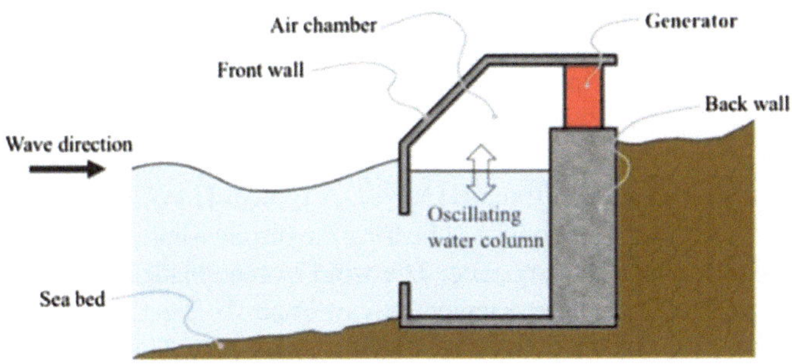

Fig. 6 Oscillating water column [10]

Floating devices float on the water surface with incident waves. The shapes are designed such that most wave energy is absorbed with minimum reflection and transmission.

Thermal

This is the concept of generating mechanical power using the differential temperature of warm surface water and cold deep water. The warm water is for the heat source, and cold deep water for cooling to condense working fluid. These are termed ocean thermal energy conversion (OTEC) plants with small thermodynamics efficiency as the temperature difference is low. There are mainly two types of plants utilized.

(A) Open cycle: This utilizes water evaporated from the warm stream under low pressure to supply a turbine discharging to a lower pressure condenser cooled by cold water.
(B) Closed cycle: This cycle uses working fluid as ammonia for more effective turbine output use. Evaporator and condenser are more efficient and expensive for this cycle.

Figure 7 shows the fluid flow in both cycles. A large volume of warm and cold water is needed with large diameter ducts. Some of the issues with these power systems are fouling water supply pipe and heat exchanger by a marine organism and a heat exchanger's corrosion. The power generated should be higher than the pumping power. The whole system cost should be kept economical for making this renewable energy to be competitive.

2.5 *Wind Power*

Since ancient times, wind energy has been in use for mechanical power. The present-day wind turbine has the latest technology in design and manufacturing and is mainly used for electrical power generation. The wind generation source is the sun, which is absorbed by water and land and heats the surrounding air. A high amount of power is in the wind, which can solve the power issue, but as the wind source is diffuse, it is tough to harness this. Wind availability is variable with time and location.

Wind Turbines

(A) Horizontal axis wind turbine (HAWT): A typical HAWT is shown in Fig. 8. The main components consist of bearing the turbine shaft, gearbox, and generator enclosed inside the nacelle. The wind turbine blades are mounted to the shaft and nacelle and are oriented by yaw control. The rotor is upwind of the tower to avoid blades not getting enough wind—this kind of wind turbine is predominantly used for power generation.

(B) Vertical axis wind turbine (VAWT): This type of wind turbine's main rotor shaft is a traverse of wind with all significant components at the base. One

Fig. 7 **a** Open cycle and, **b** closed cycle ocean thermal power plant

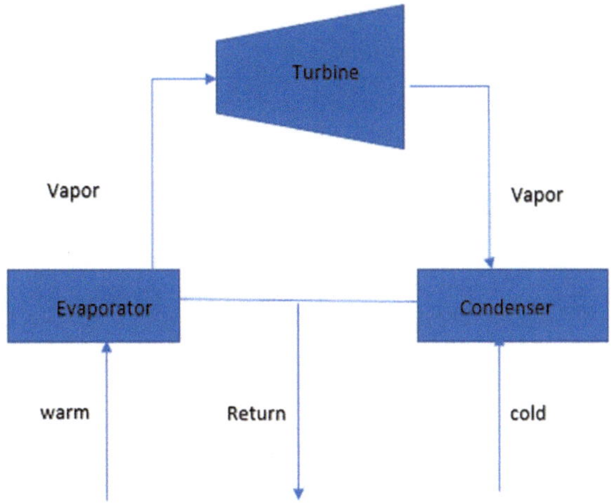

(a)

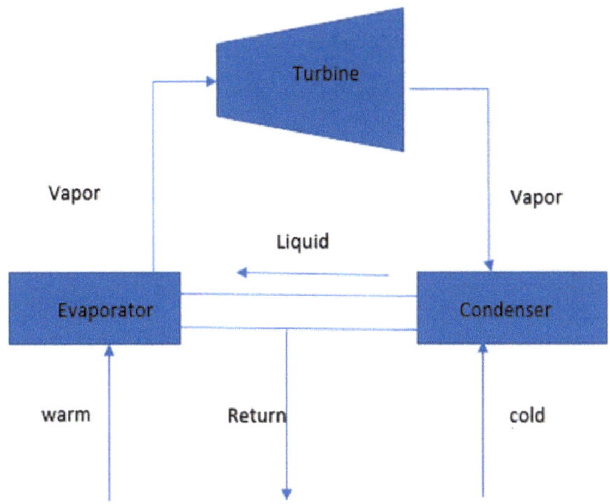

(b)

popular design is the Darrieus design, driven by lift forces generated by wind flow over the aerofoil shaped blades. VAWT is easier to maintain since it does not need a yaw mechanism with a gearbox and generator at the ground level. Overall, these are less cost-effective compared to HAWT.

Power of HAWT: The kinetic energy of the wind provides wind power flowing over a cross-sectional area. An average direction of the wind is

$$P = 1/2A\rho U^3 \tag{4}$$

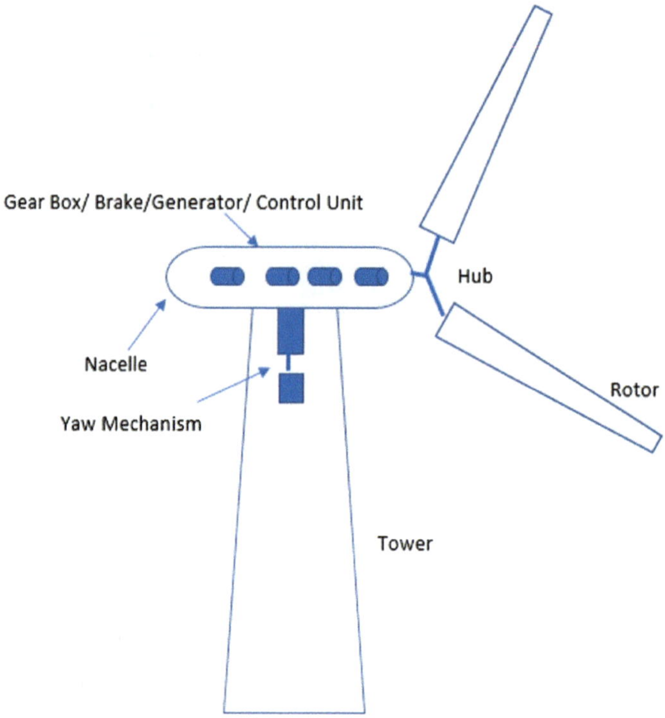

Fig. 8 Horizontal axis wind turbine

However, not all power can be extracted by a wind turbine because some fraction of the kinetic energy moves downstream to maintain airflow. This effect of a theoretical maximum of 59% for extracting power from wind is known as the Betz limit.

Hence, power extracted by the wind turbine is

$$P = 1/2 A\rho U^3 Cp \tag{5}$$

where Cp is power coefficient.

The blade's design is such that the thrust on the turbine is generated and translated to rotational energy by shaping the wind blades as the aerofoil. The turbine blades are designed with a twist, which increases going toward the root to maintain the angle of attack optimum.

Blade width also increases going toward the root of the blade. For turbine blades, several factors are considered, such as blade strength, material, and fatigue. For higher power efficiency, a large tip speed ratio of 6–10 is needed.

Wind site investigation for wind turbine installation plays a significant role. Over a period, excellent and regular measurements are needed to investigate a turbine or number of turbines, wind farm.

Lately, the acceptance of wind warmth is positive because they are a renewable energy source with zero carbon emission. There are some concerns, such as a visual threat to birds and space. Overall, the wind power return of investment depends on capital cost, an ongoing cost, maintenance, operational cost, and the revenue from the power generated.

2.6 Solar Power

Sun emits electromagnetic radiation, and energy flux per unit area is called irradiation. Solar irradiance is of two types of beam irradiance, which retains solar direction and diffuses irradiance, which does not control the direction. The solar radiation reaching the earth has a wide band of wavelength. The fraction of incoming beam that is scattered or absorbed depends on the wavelength of incoming light. Scattering is mostly for low wavelengths, and absorption is for long wavelengths. The average solar power incident on earth is 1000 Wm^{-2}, far greater than the current world power consumption.

Solar energy can be used as a source of power in the form of electricity and thermal. Photovoltaic solar panels can do the conversion to electricity, and other ways can be used as thermal applications such as indoor heating of spaces and fluids.

Solar energy is a clean and renewable power source that can be harnessed anywhere with proper conversion methods. There are several efficient conversion methods available and will briefly be discussed.

A. Flat plate collector: The most economical form of solar thermal energy collector is a flat plate collector. It has a thermally conducting collector plate with a passage that consists of fluid for heat transfer. There is a need for optimum design to minimize the ratio of cost to heat collection.
B. Focusing collector: The main reason to use focusing solar collector systems is to increase solar radiation's intensity on the collector and get solar energy at high temperatures with a small collector area. Solar irradiation increases by concentration ratio, which is mirror dimension to image dimension. The focusing system collects solar energy at high temperatures and can be used in the power generation's heat engine cycle.
C. Photovoltaic cells: These are solid-state devices used to generate electricity from solar energy. The photovoltaic cells are made of n-type and p-type semiconductor materials doped with another element. This creates donor and receptor electrons and at the junction of two layers creates an electric potential difference. Photovoltaic cell efficiencies are rated for a solar irradiance of 1 KW/m^2 under maximum power conditions. Photovoltaic power is direct current (DC) power and needs to be converted to alternating current. The maximum efficiency of photovoltaic cells falls in the range of 25–30%.

Some of the other developing technologies based on solar energy are electrochemical cells, solar-driven Stirling engines, and solar chimney.

3 Computational Fluid Dynamics

Computational fluid dynamics are a set of methods given to the computer to perform fluid flow simulations. It involves analyzing various systems, including fluid flow, heat transfer, multiphase flow, and chemical reaction. High-fidelity CFD models can be visualized in a computer by using a virtual environment. CFD is presently used as simulation tools in various industries to influence design at an earlier stage, saving a lot of time and resources. For any CFD analysis, the most important thing is to identify and formulate the flow problem in terms of the physics such as 2D or 3D, symmetrical or not, turbulent or laminar, compressible or incompressible, and transient or steady. This section is a brief outline of the CFD analysis process, and the attempt is to address the CFD analysis outline.

For all significant CFD solvers, three major elements are

1. Preprocessor
2. Solver
3. Post-processor.

The activities involved in preprocessor ranges from definition of the geometry of the computational domain, grid generation, selection of physical systems to simulated, fluid property definition, and specifying the boundary conditions. The solution of all physical quantities is based on the grid generated to capture the physics accurately. The accuracy can be captured by having a finer grid in a high gradient area compared to a low gradient area. The flow problems are defined at nodes inside each grid, and hence, the accuracy is highly dependent on grids.

Material properties of the fluid and selecting a proper turbulence model for physics are essential parameters for analysis.

Solvers provide a solution to the underlying equations by approximating flow variables by simple functions, discretization by substituting the approximate function into governing flow equation and solving the algebraic equations. The major CFD solvers use a finite volume approach for solving the problem. There is a relationship between the numerical algorithm and physical conservation in the finite volume approach.

Post-processing is understanding and visualizing the physical properties of the simulation results obtained from numerical data. Post-processing of solution involves data visualization of various flow fields such as vector plots, domain geometry, grid visualization, surface plots, and streamlines. There are options for creating animation for dynamic results.

4 Application of CFD in Renewable Energy Source

4.1 Hydropower CFD

Hydropower generation consists of a turbine, intake valve, spillway, dam, and the entire unit. CFD tools can be applied to an individual component of a power generation unit or a whole unit for rapid and cost-effective improvements. This simulation can help understand the complex working mechanism to design new facilities, upgrade and maintain the existing one.

Water turbine applications are mainly divided into two.

1. Impulse turbines which are for low flow rate and high heads such as Pelton turbine. These turbines convert the head of water to kinetic energy, which is discharged to a runner's curved bucket to get useful work.
2. Reaction turbines such as Francis, propeller, Kaplan, and Deriaz, which are medium to high flow rates which are below 450 m head. These get energy by converting accelerating water flow in a runner with pressure drop. These turbines can be arranged horizontally or vertically.

CFD is being implemented in predicting complex flow in hydraulic machines for continuous improvement in all aspects of design such as efficiency, power output, and head. It helps to understand the detailed flow pattern for customized projects, reducing the cavitation and improving efficiency. The main reason for discussing different types of water turbines is to identify the CFD method. The first step is to identify the type of turbine being used for power generation.

Steady-State Reynolds Averaged Navier–Stokes (RANS) model

The steady flow RANS model application is one of the dominant tools for water turbine flow design and analysis. This analysis takes into account for loss effect, boundary layer effects, and overall component flows effect. The analysis of the draft tube, diffuser, and pump turbine gives a more realistic flow analysis.

Keck et al. [11] discussed the basic principle of hydraulic turbines with particular emphasis on using CFD as a tool. They used 3D Euler and 3D Navier–Stokes for Francis runner simulation, and the CFD domain included the whole Francis turbine. Figure 9 shows the Francis turbine grid used for simulation. They compared measured and numerical results for circumferential and meridional velocity components at the runner outlet and demonstrated computational results' reliability. They concluded that details of flow separation, loss sources, and loss distribution in components, matching components at design and off the system, and low-pressure levels with risk of cavitation can be modeled with CFD. The significant challenges were grid generation and grid modification. Simulation of vertical Kaplan turbine was done, which had optimized guide vane and runner blade design for maximum efficiency [12]. Based on these simulations, flow and pressure coefficient efficiency shell have been calculated with similar specific speeds (Fig. 10).

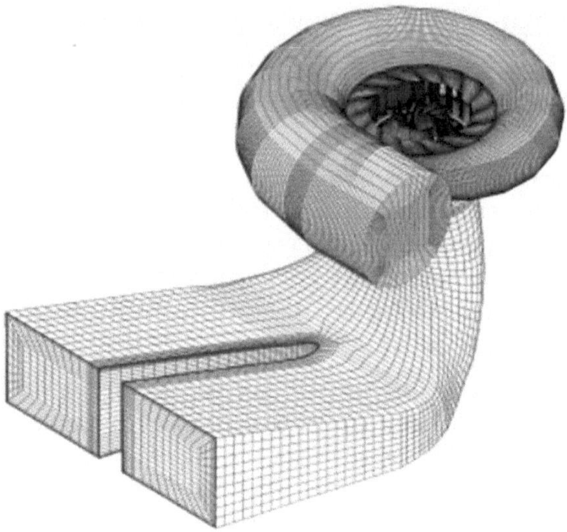

Fig. 9 Grid used in Francis turbine [11]

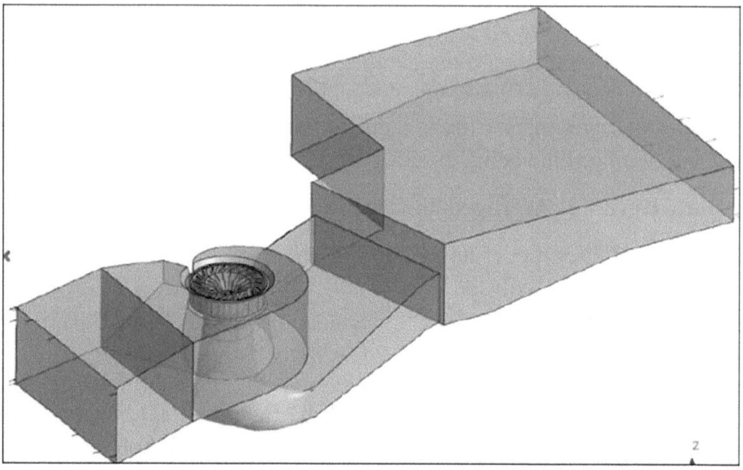

Fig. 10 Vertical Kaplan turbine [12]

Manitoba Hydro used CFD solver to get water flow behavior around construction dams and permanent structures. They used it to guide spillway design, intake and tailrace channel design, and river management during construction. The results' main output is discharge capacity, water levels, velocities, and flow patterns at various points on the construction timeline and different spillway gates [13].

Moving reference frame

The rotating domain's inclusion and development enhanced the hydraulic machines' CFD prediction capability. In this technique, an interface between the rotating and stationary domain was implemented, effectively analyzing all turbine components. This method can be used with steady-state simulations as well. Keck et al. [14, 15] used the full-stage simulation method to calculate the first numerically predicted hill chart, which used rotating and non-rotating components. These simulation results are validated with experiment and have a good agreement. They concluded that an unsteady rotor–stator interface has a more realistic flow and disturbance transported.

Additional development in simulation technique: The advancement in simulation computing resources has given the luxury of implementing other physics to accurately predict the flow. The simulation physics includes multiphase, cavitation, turbulence, fluid–structure coupling, and unsteady flow. Tilahun et al. [16] discussed in detail the design, modeling, and performance analysis of a Pelton turbine. They studied how the size of the turbine affects the efficiency of the turbine. Some of the physical assumptions were considered for simulation for reducing computational expense, such as no casing, symmetrical case, single jet, no hub, periodic torque, and scaling down. For simulation, the multiphysics model of liquid and air, domain interface for rotating and stationary as fluid-fluid and K-ω SST turbulence model was used. The flow visualization study provides insights into the reasons for efficiency and low efficiency in reduced size of the turbine are caused by a large amount of water leaving bucket through the lip. Akinyemi et al. [17] studied electrical generation capacity by evaluating paddle wheel with different speed, and the effect of side and the bottom fin was evaluated. Curved paddle wheels had higher efficiency than straight paddle wheels.

Some of the critical physical phenomenon to be modeled with the unsteady flow is dynamic loading on structural components, which produces.

1. Draft tube vortex: In the draft tube, a rotating pressure vortex is created with low pressure in the center, which can create damage, and hence some additional measures such as aeration through hub and fins are used. High-fidelity CFD simulations can capture these complex phenomena of rolling up vortices at the shear layer between the outer and inner regions. Cavitation is caused when the static pressure of the vortex core drops below vapor pressure. For these types of simulations, DES and LES turbulence model approach needs to be adopted.
2. Von Karman vortex at trailing edge of stay vane/guide vane and runner blade: Lockey et al. [18] investigated CFD analysis of von Karman vortex shedding. In the first step, the steady-state analysis was carried out to understand velocity distribution at the stay vane's (stationary) trailing edge to confirm shredding frequency. In the second step, advanced time-dependent CFD analysis of vortex shedding was analyzed. Von Karman shedding can be used as an additional numerical parameter for accuracy.
3. Guide vane and runner blade interactions-Fluid structure Interaction: One of the essential aspects of turbine design is to analyze the structure's response to the dynamic load of the flow field. Pressure amplitude is determined from the CFD

flow field for the structure's natural frequency, and then FEA analysis is done to determine the structure's response. Sick et al. [19] investigated the pressure pulsation from CFD to the structure of the turbine. They presented the natural frequency of a Francis runner, including added mass effect due to surrounding water. This coupling of fluid mechanics with structure mechanics is becoming more relevant for robust and efficient design.

Application of CFD in the hydraulic turbine is becoming more critical, instrumental in designing and understanding complex flow patterns. The use of CFD will enable us to study highly complex multiphysics and accurately predict the turbine's flow. Overall, CFD usage improves and reduces the time for designing turbines.

4.2 *Biomass Power*

Computational fluid dynamics is becoming an essential tool for developing and designing new processes and equipment to convert biofuel into fuels and improve component performance. Biomass is becoming cost-effective and is increasingly becoming an important renewable source with improved performance using CFD.

Biomass thermochemical conversion refers to gasification of fuel, pyrolysis, carbonization for heat energy. The main difference among these methods is feeding properties, oxidizer, temperature, and time. In this section, the CFD application for biomass thermochemical process is discussed to optimize design and operation. Wang et al. [20] examined the fundamentals involved in developing a CFD solution. This section briefly outlines the approach taken by Wang et al.

Gasification and pyrolysis

This process is the thermal process of producing solid (charcoal), liquid (bio-oil), and gas in the absence of air. The CFD process could provide flow and temperature patterns and concentration details to improve the process.

Fletcher et al. [21] discussed a CFD model to simulate an entrained flow biomass gasifier's flow and reaction. Biomass in this model is simulated using the Lagrangian approach as it enters the gasifier, and it releases it is volatile and undergoes gasification. Figure 11 shows the geometry used for the CFD solution. Lower inlets were used for the inlet of biomass and air and upper inlet for the steam inlet. The initial condition lower chamber was with 1000 K air to start the devolatilization for particles. The upper chamber has 30% steam at 400 K. The biomass was assumed to be dry when it enters the gasifier, with water initially present in the biomass being injected as steam with heated air. Equations for the concentration of CH_4, H_2, CO, CO_2, H_2O, O_2, and reaction with CO, CO_2, and H_2O are modeled. Gas composition details with temperature and velocity at different scenarios can be evaluated. They suggested that simulation can examine the effect of gasifier height and steam flux in the upper inlet.

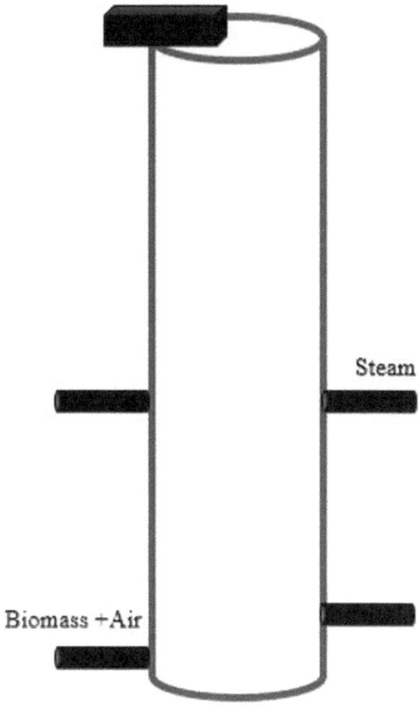

Fig. 11 Geometry of gasifier [21]

Sharma [22] developed a model for the reactive, porous bed of biomass gasifier to simulate pressure drop, the temperature profile in the bed, and flow rates. He concluded temperature profile being susceptible to gas flow rate and heat generation in the oxidation zone. In contrast, high bed temperature, gas flow rate, and reduction in feedstock particle size increase in pressure drop through the gasifier. The downstream zones' temperature is more sensitive to any change in heat generation in the bed compared to the upstream zone. Figure 12 shows the schematic and sectional view of the gasifier used for modeling. The packed bed in the gasifier was modeled based on a saturated porous medium in which fluid flow increases inflow due to the particles' gasification constituting the beds. This simulation predicted temperature profile and pressure drop for varying operating parameters such as gas flow rate, heat generation in oxidation zone, the moisture content in the feedstock, particle size, and input air temperature. Luo et al. [23] modeled wood pyrolysis in a fluidized bed reactor, and they simulated the main operation parameters for product distribution.

Biomass combustion

One of the most extensive applications of CFD models has been to power station boiler and furnaces. These simulations can be used for qualitative and quantitative analysis of the biomass system to improve the current process.

Kaer et al. [24] discussed CFD analysis of straw-fired grate boiler. Based on CFD analysis, they suggested poor mixing in the furnace is the main reason for high amounts of unburnt carbon in the fly ash. Initial boundary conditions were incorporated using stand-alone in-house code. The boiler is comprised of 395 air nozzles

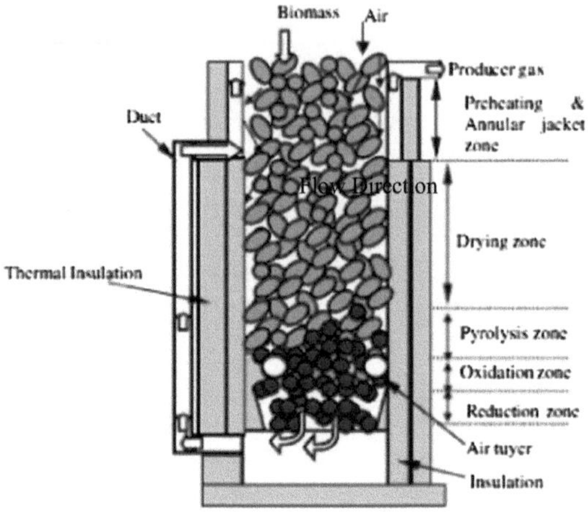

Sectional view of gasifier (22)

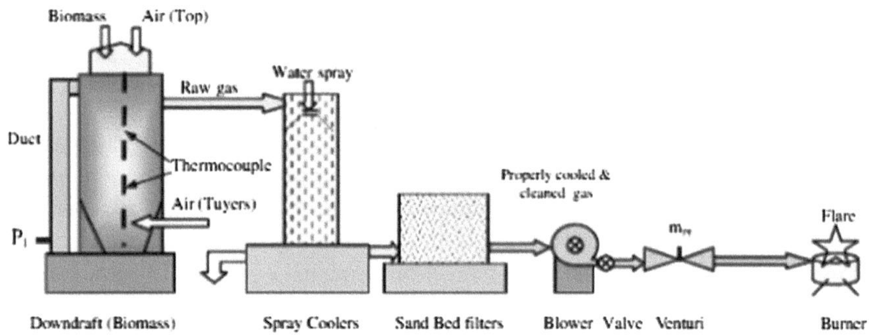

Fig. 12 Schematic diagram of a gasifier system [22]

that influence flow and combustion pattern. The platen secondary and tertiary superheater in the first and second passes were modeled as slabs, and primary superheater tube banks in the third pass were accounted for using source terms in the third pass were accounted for using source terms in momentum and energy equations to simulate heat transfer and pressure drop. Figure 13 shows the boiler section used for simulation. In these kinds of boiler, thermochemical fuel simulation takes place on the grate. The heat released by heterogeneous combustion of entrained particles is insignificant, and only gas combustion is considered. Gas combustion was modeled using the eddy breakup model and two-step reaction mechanism with CO as the intermediate species.

Kaer et al. [25] discussed the biomass-fired ash problem. They investigated deposits on furnace walls and superheater tubes, which includes impacts on heat transfer rates. The model was validated with straw-fired boiler and predicted the

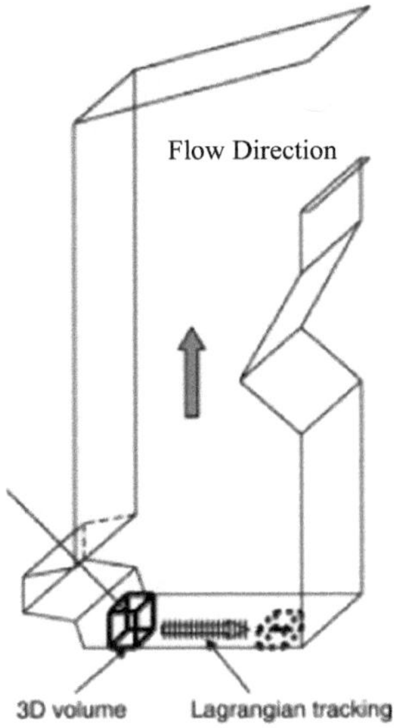

Schematic of fuel bed used for numerical model (24)

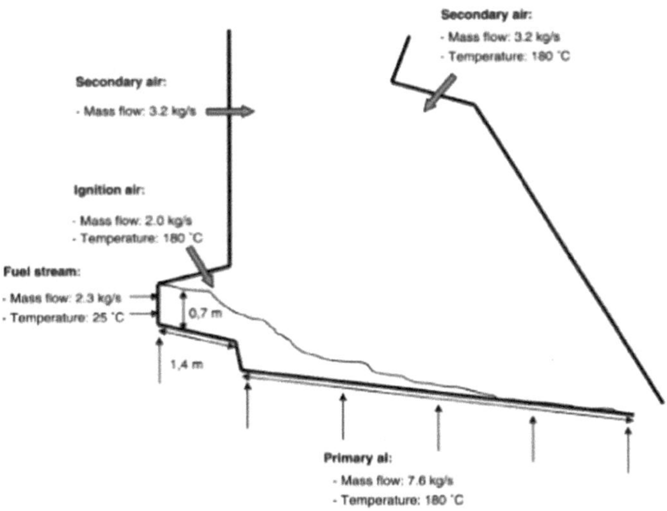

Fig. 13 Schematic of lower section of grate-fired boiler with operating conditions [24]

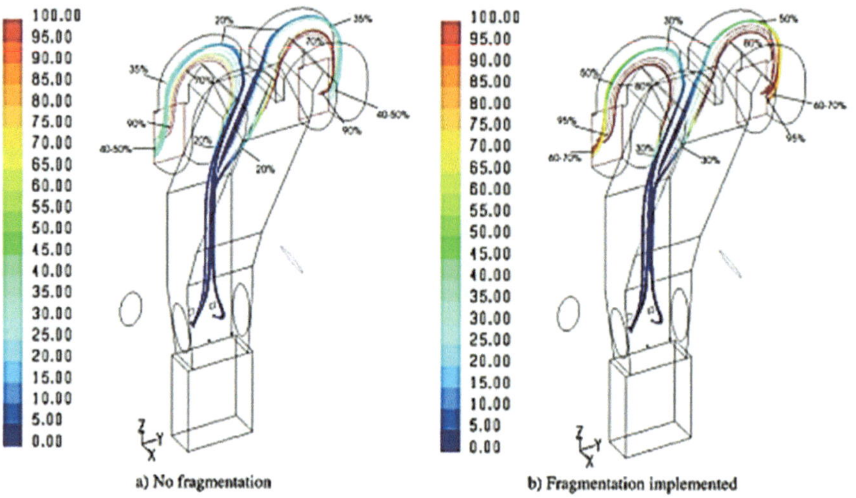

Fig. 14 Particle trajectories of charcoal burnout [26]

correct operational trends. In this analysis deposit, formation models were included got a comprehensive investigation.

Syred et al. [26] worked on fragmentation models for solid fuel combustion using CFD. They suggested that existing models for solid fuel combustion are robust for initial flow field aerodynamics, temperature but are not accurate when followed through the complete system. Using enhanced retention of burning particles in centrifugal force fields can be resolved by changes in particle size distribution and fragmentation as particle gasify or burn. In this study, single-step fragmentation was used, which needs to be refined for more accurate analysis. A proper understanding of the path of a burning particle in coal-burning plants is required. Typical temperatures are 1300 K with a 10-step reaction scheme using finite rate eddy dissipation models. Figure 14 shows the fragmentation application and particle transport.

Biogas Fermentation

By using CFD, improving fermentation material fluid mixing for biogas in anaerobic digesters can provide distribution and visualization of flow patterns. This virtual method is gaining importance because it allows for an optimal design of digester shape, mixing pattern, and different fluid parameters and methods.

The biogas simulation can be having low Reynolds number turbulence. Multiphase flow simulations are commonly used for biogas, and mostly Eulerian–Lagrangian, and Euler–Euler are employed. The biogas reforming reactor was modeled in CFD, which included reaction kinetics for exhaust gas reforming biogas to synthetic gas [27]. The biogas reactor was modeled as an isothermal heterogeneous reactor consisting of two zones connected at the interface, heat zone, and monolithic zone. Inside monolithic zone reforming surface reaction occurs with the presence of rhodium catalyst.

Dapelo et al. [28] developed the Euler–Lagrangian model to simulate gas mixing of sludge for anaerobic digestion. Momentum transfer from bubbles to liquid drives the fluid motion, and they are modeled together. The diameter of vertical columns is small, which reduces the bubble rise. The focus was to resolve flow patterns away from bubble plume.

Wang et al. [29] discussed in detail the CFD applications in biogas fermentation to concentrate mainly on analyzing flow patterns during mixing. The mixing schemes of biogas are mixing medium, mixing parameters, and digester shape. The mixing parameters are kinetic energy diffusion, phase, and temperature distribution. Mixing mediums are solid, liquid, and gas divided into mechanical mixing, hydraulic mixing, and pneumatic mixing. The shape of the digester has a critical influence on the flow pattern. Many CFD shows that these parameters can change the flow pattern significantly.

Biofuels: CFD has been utilized for the modeling and improvement of equipment for the biofuel production process. Biodiesel is produced through the transesterification reaction, where triglyceride feedstock reacts with alcohol to produce fatty acid methyl esters (FAME) with a catalyst that generates biodiesel glycerol. For homogeneous catalysis, stirred tank reactor along with tubular reactors is used. For heterogeneous catalysis, the trickle bed reactor is used. [30].

Adeyemi et al. [31] developed numerical modeling of alkali catalyzed waste cooking oil transesterification in a stirred tank reactor. They implemented the Reynolds average Navier–Stokes equation, a 3D model with rate constant and eddy dissipation model for turbulence. Multiple reference frame model was chosen to describe the rotational motion of impeller.

Boer et al. [32] used CFD to encapsulate the liquid–liquid flow phenomenon in reaction. They investigated the visualization results and studied different simulation settings. Figure 15 shows the simulation setup along with the boundary condition for the CFD of biodiesel. CFD has been applied to modeling and improvement of equipment for the biofuels production process. For current CFD models, flow, turbulence, energy, mass transfer, and chemical reactions can be implemented.

4.3 Geothermal Power

Geothermal wells' potential is being explored all around for new techniques and systems. The most preferred means of extracting the thermal energy contained in geothermal wells are electrical and mechanical. Experiments are costly for the geothermal power production process. CFD modeling can be used effectively to provide an initial preliminary understanding of the whole process. Predicting behavior and production performance of geothermal wells can be done for optimized power production. This section outlines some of the research CFD methods for various aspects of the geothermal system.

Huang et al. [33] studied the wellbore phenomenon by analyzing flow and heat transfer conditions in the well. Wellbores in enhanced geothermal systems extend

Variable/Setting	Typical Value
Dispersed droplet diameter	0.055 mm
Free surface model	None
Homogeneity	Nonhomogeneous
Turbulence (Fluid dependent)	Continuous: k-Epsilon Dispersed: Dispersed phase zero equation
Surface tension coefficient	0.0292 J/m2
Drag force	Ishii-Zuber
Boundary	**Value**
Inlet (Normal speed)	0.41 ms-1
Outlet (Average Static Pressure)	20 psi
Tube wall	Smooth, no-slip wall

Fig. 15 Simulation setting and boundary conditions [32]

from the ground surface, providing large heat transfer areas between the flowing fluid and the surrounding formations. They simulated unsteady fluid flow within a vertical injection and production well with heat transfer. Figure 16 shows the domain used for simulation. The fractured reservoir was located at some distance from the ground surface, with some distance kept between two wellbores with a specified diameter. Details of the CFD setting include rock formation enclosing the reservoir set to be the impermeable solid phase for heat transfer conduction. The fractured reservoir is a fluid region, and velocity inlet and pressure outlet boundary conditions are applied to the inlet of injection and production well. Initial temperature increases with depth with a constant gradient. The time-evolving simulation shows that almost the entire pipe is hydro fully developed but not thermally fully developed in the initial phase. The upper section's heat transfer direction is opposite to that in the lower section due to the opposite gradient vector.

Renaud et al. [34] studied 2D axisymmetric, single-phase, and steady state of the reservoir and wellbore heat extraction from a geothermal well near a magmatic chamber using a simplified geological assumption. The model included many assumptions, and with current computing power 3D, multiphysics simulation can be done. They concluded that a decrease in mass flow rate production drives more heat from the reservoir but cools down faster when flowing up to the surface.

Jiang et al. [35] studied the heat exchange process in a geothermal reservoir. They considered heat reservoirs as porous medium and domain, including an open channel for injection and production wells. Figure 17 shows the three different region schematic diagram. In the simulation, a single domain with multiple subregions is

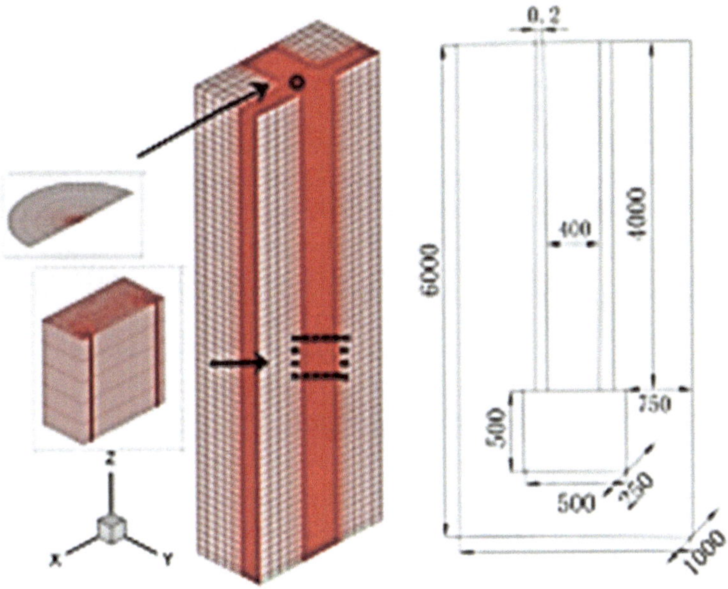

Fig. 16 Schematic of simulated domain [33]

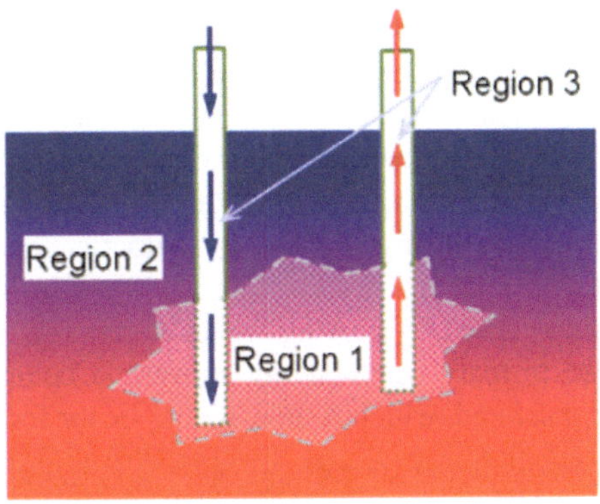

Fig. 17 Schematic of multiple regions; region 1: porous heat reservoir; region 2: doublet wall boreholes; region 3: solid rock enclosing heat reservoir [35]

considered. Region 1 is viewed as the heat reservoir, region 2 the rock enclosing the heat reservoir, and region 3 the injection and production wells.

In addition to CFD for the geothermal system, some component level CFD is also essential for increasing the plant's efficiency. Mubarok et al. [36] studied numerical simulation of two-phase geothermal fluid flows in the production pipeline using a different flow meter. These CFD models were instrumental in developing the predicting capability of fluid flow behavior, including net pressure drop, velocity, turbulent kinetic energy, mass flow rate, temperature, enthalpy, and entropy. These results are useful for selecting the appropriate pressure flow meter to measure geothermal wells' output and power potential.

Yousefi et al. [37] studied a geothermal reservoir that was modeled as a bottom heated square box filled with the water-CO_2 mixture. Different parameter effect on overall performance was investigated. They concluded that if CO_2 percentage increases, then there is a substantial increase in convective heat transfer. They suggested including multiphase analysis for the accurate prediction of heat transfer.

Overall, the increased use of CFD in the geothermal plant for system and component level can increase plant efficiency. More research in this field is needed to enhance CFD usage.

4.4 Ocean Power (Tidal/wave/thermal)

Ocean power can be a significant contributor to a satisfying increase in energy demands over the years. As for other sources, CFD can be a useful tool for early stage optimization and fine-tuning the device design before the final physical prototype.

Windt et al. [38] discussed in detail wave energy converters (WECs) and the use of high-fidelity numerical modeling, and the best practices guidelines for applying CFD in the field of wave energy.

For numerical simulation of wave energy converters, device type, device motion, and the number of devices are important aspects. Based on the operating principle, wave energy converters are classified as point absorber (PA), absorbs power from the wavefield, terminators, aligned parallel to incident wavefronts, being geometrically wide to intercept and stop the incoming wave. Attenuators are placed perpendicular to the incoming tide and derive power from the wave's phase difference. Oscillating wave surge converters extract energy from horizontal wave motion. Pressure differential devices are placed below the ocean surface and utilize the change in pressure caused by passing waves to pump fluid through the system. Figure 18 shows the primary wave energy converters being used. Device motion also needs to be considered, which is an essential factor for model preparation.

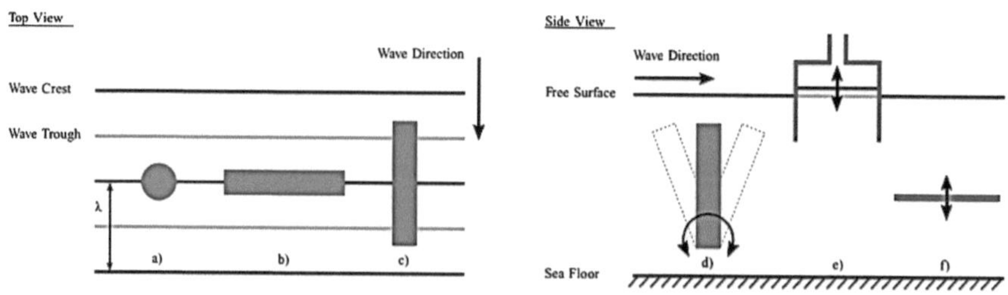

Fig. 18 Different WEC categories, **a** PA, **b** terminator, **c** attenuator, **d** OWSC, **e** OWC, and **f** pressure differential [35]

Tatum et al. [39] modeled tidal steam turbine, which included profiled flow and gravity waves. A full-scale tidal stream turbine was transiently modeled in and operating at maximum power point with both in and out phase waves with its period of rotation. They concluded that a naturally varying sea state might yield a soothing effect in this turbine response. For this analysis, a rectangular cross section was modeled with a turbine situated below the free surface. The velocity profile and the wave characteristic were included in the CFD model. In this analysis, k-ω SST Reynolds averaged Navier–Stokes (RANS) equation was used with transient and free surface modeling.

Karthikeyan et al. [40] compared marine current turbines with different pitch angles. They suggested 200 hub pitch angles for simulation due to the wider operating and peak value of Cp and CFD for various free stream velocities. Jafaria et al. [41] worked on a wave energy generator CFD analysis. In this research wave, an energy converter was simulated inside a wave tank, including water. This analysis included RANS equations coupled with the volume of fluid (VOF) model to generate three-dimensional numerical linear propagating waves and solve the fluid flow. In this simulation, the converter's upper buoy was designed to move upward by ocean waves developed.

Li et al. [42] studied a wave energy generating system. A 6 DOF analysis technique is applied to CFD analysis on a floating body with incoming waves. A spring constant is adopted to control multifloating bodies' motion and calculate the total average power absorption. The results showed average power absorption varies with both positions of the floating body and wavelength.

Some of the major points to consider during the ocean wave CFD analysis are its definition and applications, including domain definition, load, and performance studies for results. Fidelity of the subsystem needs to define for accuracy of the results. The free surface interface needs to be defined as the most accurate and efficient way. Depending on the computational cost and applications, a turbulence model needs to be chosen for 3D fluctuation of flow qualities. During these simulations, one of the important approaches is to consider wave generation and absorption, such as relaxation method, static boundary method, dynamic method, mass source, impulse source, and numerical beach method. The body representation method, which uses boundary fitted mesh with mesh distortion, can be implemented. For more large amplitude displacement, multi-DOF body motion methods such as overset mesh can be utilized. Validation of such CFD methods is vital for confidence in the data.

Schmitt et al. [43] discussed challenges and advantages in the application of RANS CFD methods in the design process of wave energy converters. They concluded that the accuracy and repeatability of CFD simulations are compared to experimental testing. An essential aspect of the CFD method is mesh adaptation of surface motion.

4.5 Wind Power

Wind energy is one of the most important renewable sources of energy. The computational method has been extensively used over the years to improve wind turbine performance. To reduce the uncertain behavior of forces and to reduce the design risk application of CFD methods and its validation is crucial. Extensive research and effort have been made to understand the aerodynamics of wind turbine and the pressure dominated flow in the surrounding. The present wind energy technology has been advanced with a significant contribution of aerodynamic research.

To accurately numerical prediction of the wind turbine, there is a need to understand wind turbines' aerodynamics. Vermeer et al. [44] discussed the aerodynamics of wind turbines in detail and regarding wakes made a distinct division into the near and far region. Near the trail, regions are area one rotor diameter behind the rotor. In this region, the rotor's presence is apparent by many blades, blade aerodynamics, stalled flow, tip vortices, and 3-D effects. Vortices generated at tip and root have sharp gradients in the velocity and turbulence intensity. Momentum and energy from flow provide power to the turbine, which has an axial pressure gradient and a decrease in axial velocity.

For this region, performance and physical processes of power extraction are crucial. This region is considered uniform, steady, and parallel flow. The important aerodynamics aspect being wind shear, rotor-tower interaction, yawed conditions, and stall.

For the far wake region, the focus is more on wind farm situation and wake models, wake interference and turbulence models. These kinds of analyses are more focused on mutual influence when placed in clusters. The following wind turbine will experience lower velocity with high turbulence intensity. This unsteady effect and lower velocity decrease power.

Sanderse et al. [45] discussed six different methods used for wind turbine analysis. These methods are

1. Kinematic method: This method is based on an analytical approach that uses the far wake's self-similar nature to obtain expressions for velocity deficit and turbulence intensity.
2. Blade element momentum theory (BEM): This method uses momentum balance on rotating stream tube through turbine and forces generated by aerofoil lift and drags at various sections along the blade. This can be used to get the aerodynamics blade characteristic.
3. The vortex lattice method is a numerical method used for modeling the thin lifting surface of discrete vortices and is used at the preliminary stage of design. The influence of thickness and viscosity is neglected.
4. The panel method is a numerical scheme for solving inviscid, incompressible, and irrotational flow. Blade geometry and viscous effects can be included with boundary layer code.
5. Generalized actuator disk method: In this method, rotor is represented as an actuator disk or actuator line. The disk is represented by a porous disk that

allows the flow to pass through the rotor. The actuator disk exerts a force on the flow, which is added to the momentum equations, and turbulence sources are also included, which are actuator disk, actuator line, and actuator surface model.

6. Direct method: This method includes all the CFD methods used for the full modeling of the rotor. With the development of computational resources, a new approach to predict wind turbine wake using high-fidelity turbulent model is made.

The fully resolved rotor or direct method includes the rotor's exact geometry and the sliding mesh method for the rotor region. For accurate prediction of the boundary layer, fine resolution of mesh is required near the blade surface and wake region. Initially, steady-state simulation can be run using a moving reference frame model, which uses motion sources from the rotating part, but the mesh is static. This can be used for accurate transient solutions.

Inlet boundary conditions need to be accurate for wind turbine wakes, and hence sheared velocity profile, the anisotropy of turbulence, and the inflow's stationary nature are essential. Some of the methods are for RANS simulations prescribing velocity components and turbulence quantity. For LES, unsteady inflow data are necessary, two different types of methods exist: synthesized inlet methods and precursor simulation methods. A turbulent inflow profile is used close to the rotor to prevent turbulent fluctuations decay. The standard method for the LES and RANS approach is to use a wall function to use wall friction [45]. For upper boundary symmetry condition can be used, the streamwise direction can use periodic condition.

Kalvig et al. [46] studied a wind turbine's performance with an actuator disk, actuator line, and a fully resolved rotor. The CFD results were validated with wind turbine experimental results and concluded that a fully resolved rotor mesh for direct method rotor produces the most accurate results. Figure 18 is the wind tunnel test section. Figure 19 shows the mesh used in the simulation with a turbine rotor in focus (Fig. 20).

Sumner et al. [47] reviewed RANS-based models and LES-based models for complex flow. They studied by dividing the simulation based on a length scale smaller than the rotor diameter and the other at a larger one. Tachos et al. [48] discussed aerodynamics characteristic of the NERL phase II rotor, a horizontal axis downwind wind turbine rotor, used the RANS turbulence model, and concluded that the CFD could predict the aerodynamics of wind turbine rotor. Figure 21 shows the periodic domain considered for simulation.

For wind turbine aerodynamics, CFD simulations can reproduce results comparable to test effects and are currently being actively used for aerodynamic analysis. The major challenge is the modeling of turbulence and flows separation. For all practical purposes, RANS-based CFD methods are performing better than other analytical methods. LES-based CFD methods are eventually going to be actively used for wind aerodynamical use.

With the current success and more confidence with simulation tools, there has been an increase in using it in the design stage. Siemens's general-purpose analysis

Fig. 19 Model wind turbine in the wind tunnel [46]

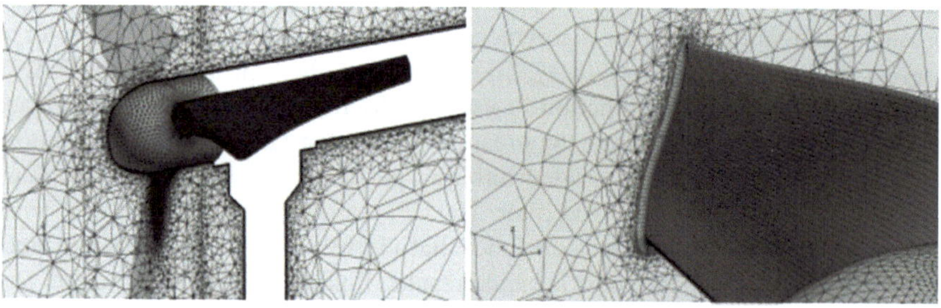

Fig. 20 Mesh around the wind turbine and blade [46]

tools provide airflow around the turbine blade and give a fluid and structural link. This kind of analysis can be effective for a quick change in geometry with automatic CFD. German company Windgiant used CFD to design a multibladed fan for a residential area. Green Energy Technologies used CFD for the development of the Wind Cube turbine system for an urban area. Vestas wind turbine manufacturers used automated methodology to predict the aerodynamics of the blade [49].

4.6 Solar Power

Solar energy is the cleanest and accessible electricity sources in the area with high solar radiation. There is a need for developing new processes and methods to get innovative and cost-effective solar panels. Simulation methods can be used effectively

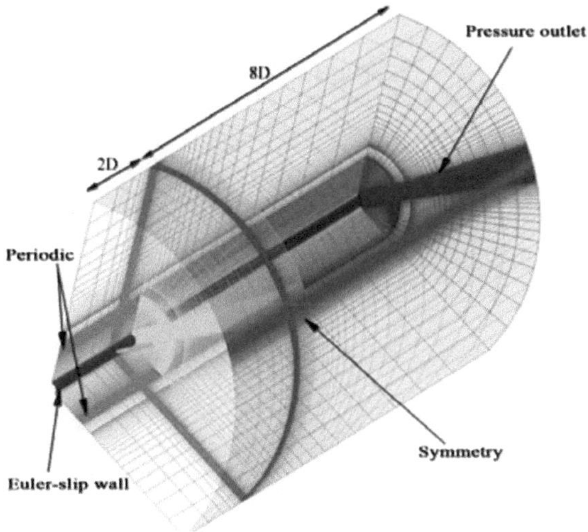

Fig. 21 Periodic domain used for simulation [48]

throughout the solar energy industry. In this section, some of the CFD applications in the solar industry are discussed.

Solar still and desalination: Desalination of water is done with thermal and membrane methods. One of the thermal desalination methods uses solar energy as a heat source, and evaporated water, after colliding with cold glass, is converted to pure water.

Kharea et al. [50] studied a multiphase 3D CFD model of simple solar stills. Solar stills by Kharea et al. It was a transient simulation with results validated with experimental data. Figure 22 shows a solar still schematic diagram, and a similar geometry was used for simulation. They concluded that high solar intensity results in quick evaporation–condensation rate and hence higher distillate, and the productivity decreases with an increase in water depth.

In this simulation, the volume of fraction (VOF) model can be used for air and water with long-run hours with higher time steps. The boundary condition for glass, bottom, and collecting surface can be assumed constant temperature. Solar intensity can be based on absorption factor and emissivity of glass, water, and bottom. The sidewall's heat transfer coefficient needs to maintain constant and no-slip wall condition for the liquid phase and free slip for the gas phase—some simulation assumptions like constant thermophysical properties of aluminum, glass, and air (Fig. 22).

This simulation with modified physics to capture physical phenomenon can be used as a powerful tool for designing parametric analysis of solar stills.

Solar collector: Solar electric heating systems use heat collector, which uses solar energy to generate electricity. In this kind of system, a parabolic linear collector is used to heat and increase the working fluid temperature. A working fluid pile lines pass and directly receive heat from solar radiation and are reflected from parabolic plates. Figure 23 shows a solar collector.

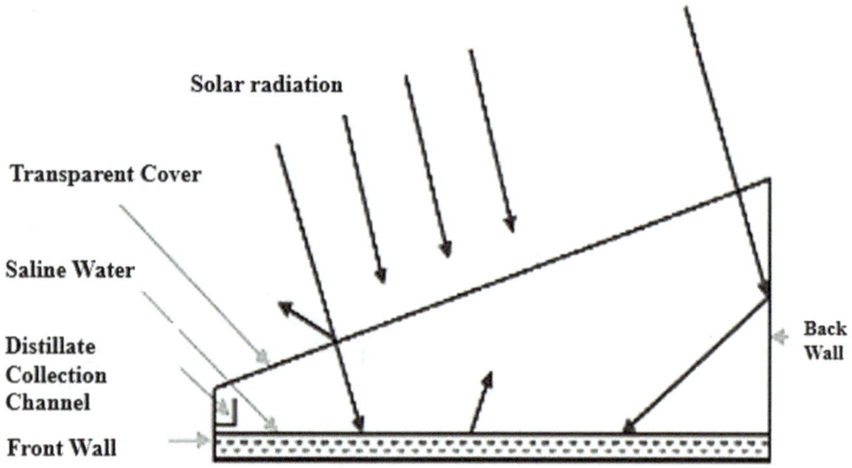

Fig. 22 Energy flow in solar still [50]

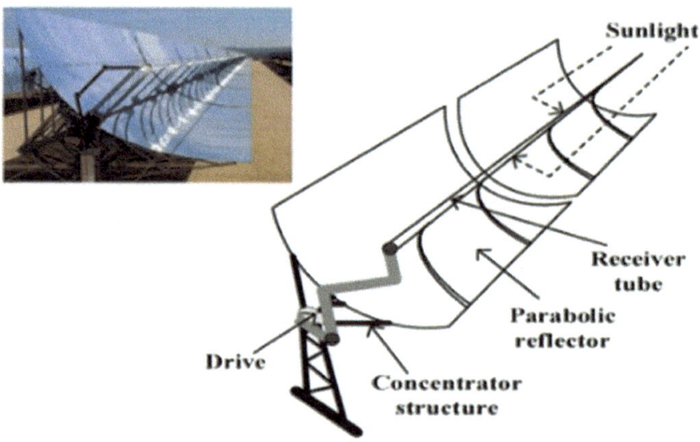

Fig. 23 Energy flow in solar still [51]

Manjunath et al. [51] studied solar air heater CFD analysis. These air heaters have a lower thermal performance level due to high heat losses and poor air heat transfer characteristics. They reviewed thermal performance enhancer by using turbulence enhancers such as corrugations, vortex generators. Gawande et al. [52] studied solar air heaters used different approaches to investigate a solar air heater's thermal performance. They reviewed different articles for the type of computational domain (2d/3d), turbulence model used, type of commercial software, and range of operating and geometrical parameters used. Figure 24 shows a conventional solar air heater.

For CFD analysis, defining material properties, selecting an appropriate physical model and turbulence model, and proper boundary conditions at all boundary zones are essential for this kind of analysis.

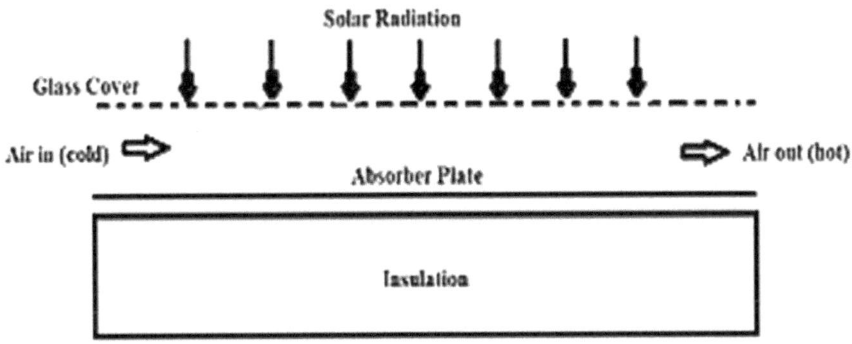

Fig. 24 Conventional solar air heater [53]

Avila-Marin et al. [53] studied modeling strategies for porous structure as a solar receiver in a central receiver system. They discussed central receiver systems, volumetric absorbers, which work at a higher temperature and concentration factor. Volumetric absorbers are porous structures where the air is heated directly by a porous matrix. They concluded that the porous absorber simulation's most accurate strategy is direct simulation or discrete ordinate or Monte Carlo methods for the radiative heat transfer.

Once the solar panel is mounted on a roof, it will have to withstand wind loading from all directions. To understand the sunroof panel's force to withstand the wind load, a safe, reliable CFD simulation can help. When CFD results are available, then solar panel mounts can be designed to withstand those loads.

5 Conclusions

This chapter's main objective was to discuss the importance of renewable energy sources and CFD application in those fields. All renewable sources are briefly discussed to provide background information. CFD is becoming an essential tool in all aspects of all energy sectors' processes and equipment. With the increase in computing resources and CFD algorithms, complex phenomena are achieved by simulation. Over the years, the larger spatial discretization and increased resolution in high physics gradient can be achieved. Turbulence modeling is the center of most CFD simulations; with the current trend RANS modeling, it is used predominantly, but LES is also significantly implemented.

This chapter briefly discussed the CFD methodology for accurate flow and physics prediction to reduce the design cycle and prototype costs. Global warming has put renewable energy sources in the forefront, and a sooner energy system transition is on the horizon. CFD-supported research in this field will continue to grow and expand for an innovative solution. The critical research and reviews were discussed in this chapter to enhance the understanding of CFD application. Different methodologies were discussed, which will be helpful for CFD applications.

Fuel cells are one of the renewable sources and were not discussed in this chapter. In fuel cells, fluid flow, heat transfer, electrochemistry, reactants oxidation, and reduction rates are computed. CFD will be instrumental in future developments and analysis of fuel cells and all other renewable energy sources.

References

1. https://www.eia.gov/energyexplained/renewable-sources
2. https://ourworldindata.org/renewable-energy
3. https://www.utilitydive.com/news/renewable-energy-prices-keep-falling-when-do-they-bottom-out/555822/
4. https://www.eia.gov/energyexplained/hydropower/
5. Bay JA, Golomb DA, Energy and the environment, 1st Edition, Oxford University Press, New York, 2002
6. https://www.mr-cfd.com/services/cfd-consulting/renewable-energy-mr-cfd-cfd-analysis-consultation-training-simulation/
7. Kumar S, Reisel JR (2010) Modelling of energy usage for the refining of ethanol from corn. Int J Sustain Energy
8. https://www.alternative-energy-tutorials.com/tidal-energy/tidal-barrage.html
9. Andrew J, Jelley N (2007) Energy science: principle, technology, and impacts 1st Edition, Oxford University Press, New York, 2007
10. Vertechy R, Fontana M, Rosati Papini GP, Bergamasco M (2016) Oscillating-water-column wave-energy-converter based on dielectric elastomer generator. In: Proceedings of SPIE-the international society of optical engineering
11. Drtina P, Sallaberger M (1999) Hydraulic turbine- basic principles and state-of-the-art computational fluid dynamics applications. J Mech Eng Sci 213:85–102
12. http://www.hfm.tugraz.at/en/references/turbine/development-of-hydraulics-fors-a-vertical-kaplan-turbine.html
13. Sydor KM, Pamela J (2010) Waterman engineering and design: the value of CFD modeling in designing a hydro plant. Renew Energy World
14. Keck H, Drtina P, Sick M (1996) Numerical hill chart prediction by means of CFD stage simulation for a complete Francis turbine. In: Proceedings of the 18 IAHR symposium, Valencia
15. Keck H, Sick M (2008) Thirty years of numerical flow simulation in hydraulic turbomachines. Acta Mech 201:211–229
16. Nigussie T, Engeda A, Dribssa E (2017) Design, modeling, and CFD analysis of a micro hydro pelton turbine runner: for the case of selected site in Ethiopia. Int J Rotating Mach 2017:17. (Article ID 3030217)
17. Akinyemi OS, Liu Y (2015) CFD modeling and simulation of a hydropower system in generating clean electricity from water flow. Int J Energy Environ Eng 6:357–366
18. Lockey K, Keller M, Sick M, Gehrer A (2006) Flow induced vibrations at stay vanes: experience at site and CFD simulation of von Kármán vortex shedding. Hydro2006, Porto Carras, Greece
19. Sick M, Lais S, Weiss TH (2007) Numerical prediction of flow induced dynamic load in water turbines: recent developments and results. In: Proceedings of Hydro2007, Granada, Spain
20. Wang Y, Yan L (2008) CFD studies on biomass thermochemical conversion. Int J Mol Sci 9:1108–1130
21. Fletcher DF, Haynes BS, Christo FC, Joseph SD (2000) A CFD based combustion model of an entrained flow biomass gasifier. Appl Math Model 24(3):165–182

22. Sharma AK (2007) Modeling fluid and heat transport in the reactive, porous bed of downdraft (biomass) gasifier. Int J Heat Fluid Flow 28:1518–1530
23. Luo ZY, Wang SR, Cen KF (2005) A model of wood flash pyrolysis in fluidized bed reactor. Renew Energy 30(3):377–392
24. Kaer SK (2004) Numerical modelling of a straw-fired grate boiler. Fuel 83(9):1183–1190
25. Kaer SK, Rosendahl LA, Baxter LL (2004) Extending the capability of CFD codes to assess ash related problems in biomass fired boilers. In: Division of fuel chemistry, no 12, Proceedings of 227th ACS annual meeting¸ Anaheim California, Mar 28–Apr 1, 2004
26. Syred N, Kurniawan K, Grifths T, Gralton T, Ray R (2007) Development of fragmentation models for solid fuel combustion and gasification as subroutines for inclusion in CFD codes. Fuel 86(14):2221–2231
27. Hamedi MR, Tsolakis A, Lau CS (2014) Biogas upgrading for on-board hydrogen production: reforming process CFD modelling. Int J Hydrog Energy 39(24):12532–12540
28. Dapelo D, Alberini F Bridgeman J (2015) Euler-Lagrange CFD modelling of unconfined gas mixing in anaerobic digestion. Water Res 85(15):497–511
29. Wang J, Xue Q, Guo T, Mei Z, Long E, Wen Q, Huang W, Luo T, Huang R (2018) A review on CFD simulating method for biogas fermentation material fluid. Renew Sustain Energy Rev 97:64–73
30. Quiroz-Pérez E, Gutiérrez-Antonio C, Vázquez-Román R (2019) Modelling of production processes for liquid biofuels through CFD: a review of conventional and intensified technologies. Chem Eng Process Intensif 143:107629
31. Adeyemi NA, Mohiuddin A, Nor RM (2013) CFD modeling of waste cooking oil transesterification in a stirred tank reactor. World Appl Sci J 21:151–158. https://doi.org/10.4028/www.scientific.net/AMM.390.86
32. de Boer K, Bahri PA (2009) Investigation of liquid- liquid two phase flow in biodiesel production. In: Proceedings of 7th international conference on CFD in the minerals and process industries
33. Huang X, Zhu J, Li J (2015a) Analysis of wellbore heat transfer in enhanced geothermal system using CFD modeling. In: World geothermal congress, April 2015, p 6
34. Renaud T, Stebel M, Verdin PG, Falcone G (2018) CFD modeling of a high enthalpy geothermal context geo-energy engineering centre. In: 43rd workshop on geothermal reservoir engineering, Stanford University
35. Jiang F, Luo L, Chen J (2013) A novel three-dimensional transient model for subsurface heat exchange in enhanced geothermal systems. Int Commun Heat Mass Transfer 41:57–62
36. Mubarok MH, Cater JE, Zarrouk SJ (2020) Comparative CFD modelling of pressure differential flow meters for measuring two-phase geothermal fluid flow. Geothermics 86
37. Yousefi S, Atrens AD, Sauret E, Dahari M, Hooman K (2015) CFD convective flow simulation of the varying properties of CO_2–H_2O mixtures in geothermal systems. Sci World J 2015:8. https://doi.org/10.1155/2015/843068
38. Windt C, Davidson J, Ringwood JV (2018) High-fidelity numerical modelling of ocean wave energy systems: a review of computational fluid dynamics-based numerical wave tanks. Renew Sustain Energy Rev 93:610–630
39. Tatum S, Allmark M, Frost C, O'Doherty D, Mason-Jones A, O'Doherty T (2016) CFD modelling of a tidal stream turbine subjected to profiled flow and surface gravity waves. Int J Marine Energy 15:156–174
40. Karthikeyan T, Ezhilsabareesh K, Samad A (2016) Parametric analysis of a tidal current turbine using CFD techniques. Progress Renew Energies Offshore
41. Jafari M, Babajani A, Hafezisefat P, Mirhosseini M, Rezania A, Rosendahl L (2018) Numerical simulation of a novel ocean wave energy converter. In: International scientific conference "environmental and climate technologies", CONECT
42. Li KM, Parthasarathy N, Choi YH, Lee YW (2013) Study on wave energy generation of multi floating bodies for energy absorption by CFD. J Korean Soc Power Syst Eng 17(5)
43. Schmitt P, Doherty K, Clabby D, Whittaker T (2012) The opportunities and limitations of using CFD in the development of wave energy converters. In: Proceedings of the international conference on marine and offshore renewable energy

44. Vermeer LJ, Sorensen JN, Crespo A (2003) Wind turbine wake aerodynamics. Prog Aerosp Sci 39:467–510
45. Sanderse B, Van Der Pijl SP, Koren B (2011) Review of computational fluid dynamics for wind turbine wake aerodynamics. Wind Energy 14:799–819
46. Kalvig S, Manger E, Hjertager B (2014) Comparing different CFD wind turbine modeling approaches with wind tunnel measurements. J Phys Conf Ser 555:012056
47. Sumner J, Watters CS, Masson C (2010) CFD in wind energy: the virtual, multiscale wind tunnel. Energies 3:989–1013
48. Tachos NS, Filios AE, Margaris DP, Kaldellis JK (2009) A computational aerodynamics simulation of the NREL phase II rotor. Open Mech Eng J 3:9–16
49. https://www.digitalengineering247.com/article/renewable-energy-options-multiply-with-cfd-solutions/
50. Khare VR, Singh AP, Kumar H, Khatri R (2016) modelling and performance enhancement of single slope solar still using CFD. In: International conference on recent advancement in air conditioning and refrigeration, RAAR
51. Manjunath MS, Madhwesh N (2019) A review of CFD investigations on heat transfer augmentation of forced convection solar air heater through enhanced fluid turbulence levels. Int J Mech Prod Eng Res Dev 9:1309–1322
52. Gawande VB, Dhoble AS, Zodpe DB, Chamoli S (2016) A review of CFD methodology used in literature for predicting thermo-hydraulic performance of a roughened solar air heater. Renew Sustain Energy Rev 54:550–605
53. Avila-Marin AL, Fernandez-Reche J, Martinez-Tarifa A (2019) Modelling strategies for porous structures as solar receivers in central receiver systems: a review. Renew Sustain Energy Rev 111:15–33

Energy Recovery from Waste Tires Via Thermochemical Pathways

Kiran R. G. Burra, Zhiwei Wang, Matteo Policella, and Ashwani K. Gupta

Abstract Demand for automotive tires has been increasing at a significant rate to reach production rate of 2.9 billion in 2017 that contributed to the generation of almost 1 billion waste tires. The pathways for handling of these waste tires have been limited to aggregate construction applications and direct incineration as fuel. The rate of recycling feasibility was limited by the availability of waste tires with structure intact for their upgrading and reuse in the economy. In the USA, although 40% of the waste tires are disposed by combustion for energy recovery in cement kilns, pulp and paper mills, and electricity generation, the presence of sulfur from vulcanized rubber, the production of hazardous pollutants, and low temperatures caused due to slow heterogenous combustion leads to significant energy and resource losses. Additionally, this pathway is limited to thermal energy production and thus lacks versatility. Alternative thermochemical pathways such as pyrolysis and gasification offer better pathways for the utilization of these wastes as they provide uniform products such as synthetic gas, bio-oil and char. Versatility is also achieved via syngas production as it is a precursor to liquid fuels and various other essential petrochemicals. These pathways provide improved energy efficiency, feasibility, and scalability for increased waste tire utilization and value outcome compared to the current application pathways. Results are reported from the investigations on high-temperature pyrolysis and CO_2-assisted gasification of waste tires with focus on the evolutionary behavior of syngas production, its constituents, and energy yield in lab-scale fixed-bed reactor. The impact of CO_2 addition, temperature, and the addition of biomass feedstock to the waste tire was investigated to understand the feasibility of waste tire disposal via this pathway while also utilizing the CO_2 pollutant and maintaining high energy efficiency. Lack of inhibitive effects observed when tire was co-processed with biomass, makes the disposal of waste tires along with other

K. R. G. Burra · Z. Wang · M. Policella · A. K. Gupta (✉)
Department of Mechanical Engineering, The Combustion Laboratory, University of Maryland, College Park, MD 20742, USA
e-mail: akgupta@umd.edu

Z. Wang
College of Environmental Engineering, Henan University of Technology, Zhengzhou 450001, PR, China

A. Gupta et al. (eds.), *Advances in Energy and Combustion*, Green Energy and Technology,
https://doi.org/10.1007/978-981-16-2648-7_21

wastes easier along with established feasibility regime for efficient and economical disposal of waste tires while recovering energy and utilizing CO_2.

Keywords Waste tires · Syngas · Gasification · Pyrolysis · Recycling · Energy recovery

Nomenclature

α	Extent of mass-loss conversion
BET	Brunauer–Emmett–Teller
BR	Butadiene rubber
$\text{Energy}_{\text{input}}$	Electrical energy consumed by furnaces
GC-TCD	Gas chromatography-thermal conductivity detector
LHV	Low-heating value
M_i	Mass flow rate of species '*i*'
$\text{m}_{\text{feedstock}}$	Initial mass of the feedstock in the lab-scale reactor
m_{syngas}	Cumulated mass of syngas
$M(T)$	Mass of sample left in the TGA at temperature '*T*'
NR	Natural rubber
OEE	Overall energy efficiency
PET	Polyethylene terephthalate
ρ_i	Density of species '*i*' at GC inlet
SBR	Styrene–butadiene rubber
T	Temperature
TDF	Tire-derived fuel
TGA	Thermogravimetric analysis
V_i	Volumetric flow rate
X_i	Mole fraction of species '*i*' recorded by the GC

1 Introduction

With the growing penetration of automobiles in the marketplace, their sustainability is dependent on the availability and management of the fuel, and its materials. The increased utilization of fossil fuels in transportation sector and industry is leading to lack of fuel availability and increase in global warming from the imbalance between carbon emissions and its sequestration. This can be seen by the trend of CO_2 concentration in the atmosphere which reached 415 ppm in 2020 from ~280 ppm in the pre-industrial era [1]. Several efforts are being made here to support CO_2 management and control in the transportation sector with increased penetration of electric vehicles and light-duty trucks. California has promulgated 100% electric vehicle sales by 2035 and phase out of gasoline powered vehicles by 2045. While renewable alternatives and carbon-free transportation are still in developing phase, their scalability and penetration to completely replace the carbon-based infrastructure are questionable. These resources need to be supplemented with carbonaceous fossil fuel resources for the foreseeable future to achieve sustainable energy while seeking carbon neutral footprint. All electric vehicles are justifiable only with the electricity produced from renewable sources.

In addition to the fuel crisis, the handling and disposal of waste tires created by increases in the automotive sector poses significant challenge to sustainably continue producing or recovering these tires. An estimated 2.9 billion automobile tires were produced in 2017, while almost 1 billion waste tires were generated [3]. Predictions based on the current demand growth in automotive sector of 4.1% are expected to result in production of almost 3.2 billion tires by 2022 [4]. Such a rapid growth with the current trends of recovery will be unsustainable. In the USA alone, 255 million waste tires were disposed in 2017, wherein 15% were landfilled, and 41% were combusted for energy recovery [2]. Figure 1 reveals the statistics in 2017 and the current modes of utilization of waste tires which account for 81.4% going to the market while 15% landfilled and the rest remained unaccounted [2]. Tire-derived fuel (TDF) accounts for 41% of the utilization. Other sectors include cement kilns (19%), pulp and paper mills (12%), and electric utilities (10%) for the energy recovered. Almost 30% of this waste tires are ground for use in applications that included asphalt for roads, sport surfaces and mulch, and extrusion/molded products. Utilization in construction offers lower-value applications compared to other applications. The

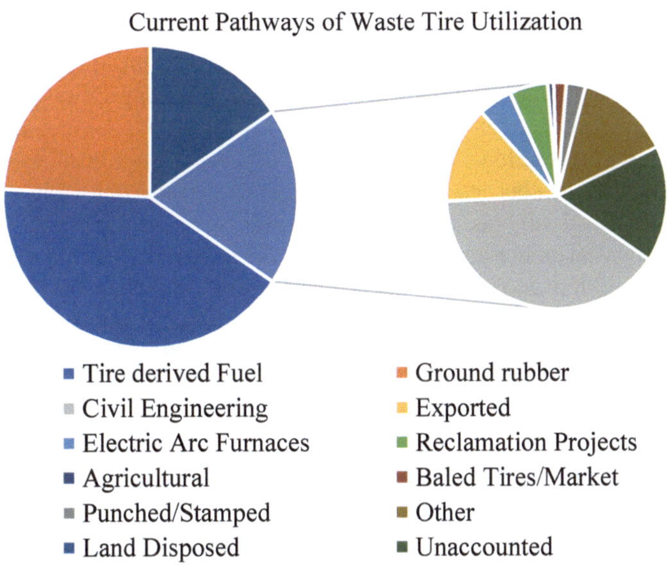

Fig. 1 Fate of waste tires in the USA in 2017 [2]. [Color codes are does not account for all the colors used]

infrastructure that accounted for in TDF includes fossil fuel equivalent combustion units where the waste tire is possibly mixed with other fuels such as coal, and biomass for energy recovery. While these pathways operate with solid fuel efficiencies which is lower than gas-phase combustion due to lower temperatures attained, they also pose environmental hazard in terms of emissions of polychlorinated dibenzodioxins, Zn, SO_x from high S content in vulcanized rubber, polyaromatic hydrocarbons, particulates, odors and other toxic emissions [5]. Thus, to operate in low-environmental impact mode regulated by the government, it requires multi-stage flue-gas cleaning, which significantly lowers the net efficiency of the waste tire utility and recuperated energy.

These statistics reveal the need for alternative pathways for waste tire disposal, especially for energy purposes with the consideration of the imminent fuel crisis and rapidly increasing CO_2 concentration in the environment. Thermochemical pathways such as pyrolysis and gasification for oil and syngas yields provide significant advantages in this aspect of feedstock utilization. Pyrolysis involves thermal decomposition of carbonaceous feedstocks at high temperature to form syngas, oil and char products [6]. Depending on the desired product phase, the operating conditions are modified. High heating rates and low vapor residence times are used when oil is desired, while low heating rates and high residence times are used for increased char residue, and high temperatures above 700 °C when syngas is desired. In the case of gasification, the sample is thermally decomposed at temperatures above 750 °C in the presence of a mild oxidizing/gasifying agent such as H_2O, CO_2, diluted O_2/air to yield significantly high syngas compared to pyrolysis [6]. While temperature has a crucial role in the yield of syngas, its composition is significantly controlled by the gasifying agent where steam (H_2O) yields H_2-rich syngas, CO_2 yields CO-rich syngas, and air/O_2 is used for autothermal operation. Although these processes are endothermic, the reformed uniform syngas, which is the focus of this chapter, is of high quality in terms of its compositional uniformity, heating value, yield, energy recovery and capability to modify composition with minimal downstream resources, via water–gas shift reaction. Additionally, the versatility in applications offered by syngas is unparalleled, because various pathways have been extensively studied and implemented industrially such as Fischer–Tropsch synthesis for liquid fuels from syngas [6]. Significant literature is available on the utilization of these pathways for waste tire conversion, and some of it will be discussed in the context of evolutionary behavior and energetic feasibility later in this chapter to establish the knowledge and its gap compared to the available studies [3–5, 7–29]. This will be preceded by understanding of waste tire constituents, composition and their behavior in the context of thermochemical conversion.

2 Characterization of the Waste Tires

To characterize waste tires' feedstock, we need to initially investigate the composition and its variability in the waste tires collected. Typically, tires consist of a blend of natural (NR) and synthetic rubber (SR) wherein polybutadiene rubber (BR) and styrene–butadiene rubber (SBR) are the most common examples of SR [30]. Carbon black is added to this blend for enhanced resistance against abrasion, in addition to extracting heat from the tread and the belt. For reinforcement, steel wires/textile fabrics (polyester, rayon and nylon cords) are incorporated into this rubber mix for enhancing strength and durability of tires [23]. During the vulcanization of the rubber constituents, ZnO is added to control and improve its physical properties, while sulfur is added as cross-linking agent between the elastomer chains while hardening the final product and improving its resistance against thermal deformation at high temperatures. While these are the major components, additional trace compounds are also added to improve their quality and manufacturing, such as a mix of aromatic hydrocarbons called extender oils and resins as plasticizers for improved shaping and molding, antioxidants and antiozonants to combat oxidization and ozone influence [5]. Different sources, i.e., different automobiles classes, have their specific needs and thus, the characteristics of the tires utilized are also modified in terms of the elastomers' composition, reinforcement and its material, treading, geometry, carbon black and other additives. For example, while passenger car tires are composed of 35% natural rubber (NR) and 65% polybutadiene rubber (BR), truck tires contain 51% NR, 10% BR and 39% styrene–butadiene rubber (SBR) [2]. Table 1 provides further differentiation on the composition of typical passenger car and truck tires. Literature also provides proximate and ultimate analysis of waste tires varying in their sources and properties, and they are summarized in Table 2 [5]. This table reveals the significance of waste tires in terms of the calorific value of this feedstock that is seen more than 30–35 MJ/kg with some as high as 40 MJ/kg. These heating values

Table 1 Typical constituents and composition of waste tire and their utility [14, 30]

Component	Passenger tire (wt. %)	Truck tire (wt. %)	Comments/examples
Rubber	47	45	Synthetic and natural rubbers, examples: styrene–butadiene rubber, natural rubber (polyisoprene), nitrile rubber, chloroprene rubber, polybutadiene rubber
Carbon black	21.5	22	Added to strengthen the rubber and help in abrasive resistance
Metal	16.5	21.5	Steel belts and chords for strength
Textile	5.5	–	Reinforcement
Zinc oxide	1	2	Used with stearic acid to control vulcanization and enhance physical properties of rubber
Sulfur	1	1	Used to control rubber polymer chains, harden and prevent excessive deformation at high temperatures
Additives	7.5	5	Example: Clay/silica as a partial replacement to carbon black

Table 2 Proximate and ultimate analysis of various waste tires and their calorific value [5]

Elemental analysis (wt. % dry basis)						Proximate analysis (wt. % as received basis)				Calorific value (MJ/kg)	References
C	H	N	S	O	Ash	Ash	Volatiles	Fixed carbon	Moisture		
81.72	6.54	0.55	1.87	2.68	6.64	6.64	62.58	30.07	0.71	nr	[31]
83.8	6.9	0.6	2	2.3	4.4	4.3	63.4	30.4	1.9	nr	[32]
83.8	7.6	0.4	1.4	3.1	3.7	3.7	67.3	28.5	0.5	36.45	[33]
82.36	6.92	2.3	1.4	2.03	5	4.95	73.74	20.22	1.09	37.06	[25]
75.5	6.75	0.81	1.44	15.5	+	20.1	57.5	20.85	1.53	29.18	[21]
85.9	8	0.4	1	2.3	2.4	2.4	66.5	30.3	0.8	40	[19]
75.4	7.03	0.21	1.62	5.09	10.21	10.21	62.32	26.26	1.31	33.29	[34]
84.33	7.81	0.49	1.66	3.32	2.4	7.1	62.2	29.4	1.3	40	[12]
82.8	7.6	0.5	1.3	4.5	3.3	3.3	68.7	27.2	0.8	36.46	[35]
86.7	8.1	0.4	1.4	1.3	2.1	8	61.9	29.5	0.7	36.2	[8]
80.29	7.25	0.31	1.84	4.9	5.41	5.3	67.5	25.2	2.1	37.3	[36]
85.05	6.79	0.5	1.53	1.75	4.4	4.35	62.24	32.28	1.14	34.9	[37]
81.5	7.1	0.5	1.4	3.4	6.1	6.07	64.87	28.56	0.5	36.8	[38]
86.09	6.74	0.19	1.93	1.35	3.7	3.7	65.5	29.4	0.9	nr	[26]
86.7	6.9	0.3	1.9	0.9	3.3	4.4	64	30.7	0.9	31.8	[39]
83.92	6.83	0.78	0.92	3.39	4.16	4.16	64.97	30.08	0.75	38.6	[40]
85.25	7.94	0.41	1.38	1.19	3.83	3.83	64.09	31.14	0.94	nr	[41]
83	6.79	0.32	1.37	3.46	5.06	5	64.1	29.7	1.2	35	[42]
67.08	6.12	0.17	2.05	24.58	+	19.1	59.69	19.45	1.72	27.37	[11]
81.79	7.99	0.48	1.81	3.04	4.9	4.88	65.74	28.98	0.4	38.3	[43]

(continued)

Table 2 (continued)

Elemental analysis (wt. % dry basis)						Proximate analysis (wt. % as received basis)				Calorific value (MJ/kg)	References
C	H	N	S	O	Ash	Ash	Volatiles	Fixed carbon	Moisture		
84	7.19	0.49	1.42	3.3	3.6	3.6	65.6	30	0.8	38.8	[44]
83.15	6.78	0.28	1.77	0.84	7.1	7.1	61.9	29.9	1.1	37.35	[45]
74.3	7.2	0.9	1.71	15.89	+	18.9	58.2	21.3	1.6	30.5	[17]

are comparable to some of the high-quality fuels such as anthracite and bituminous coals, and biofuels such as alcohol fuels, but lower than domestic and transportation fuels such as gasoline, kerosene and methane which are higher than 42–45 MJ/kg. Table 2 also reveals significantly high carbon content and fixed carbon in the waste tires which reveals high quality of the feedstock and emphasizes the importance of carbon black's presence in the context of its conversion as this means high levels of char residue from pyrolysis. By extension, it also means that if gasified, major reactions responsible for mass conversion are heterogenous reaction of the fixed carbon components with the gasifying agent. This reveals the potential of waste tires for use as an ideal source of energy and for coal replacement applications.

So, for the context of understanding the feasibility of waste tires, we restrict our discussions in this chapter to specific types of tires in terms of the vehicle class. From the statistics of 2017 given in Table 3, significant portions (almost 73%) of waste tires are from passenger car, while commercial truck and bus tires account for up to 11.4% of the total waste tires collected [2]. With this table as the basis, we restricted our experimental investigation to passenger vehicle waste tire. Further details of our materials and methods and observations will be preceded by various results available in the literature regarding pyrolyzing and gasifying waste tires for syngas production.

Table 3 Types of tires disposed in the USA in 2017 [2]

Tire class	Millions of Tires	Market %	Average weight (kg)	Weight (1000 × tons)
Light-duty tires	**254.6**	**88.6**	**10.2**	**2864.3**
Passenger tire replacements	209.7	73.0		
Light truck tire replacements	31.3	10.9		
Tires from scrapped vehicles	13.6	4.7		
Commercial tires	**32.7**	**11.4**	**54.4**	**1962.0**
Medium, wide-base, heavy truck replacement tires	19.7	6.9		
Tires from scrapped trucks and buses	13.0	4.5		
Total tires hauled	**287.3**	**100.0**	**15.24**	**4826.3**
Used tires culled	**37.9**	**13.2**	**15.24**	**637.1**
Net scrap tires	**249.4**			**4189.2**

3 Thermochemical Conversion of Waste Tires

Various types of processes were proposed and investigated to convert waste tires for energy, fuels or chemicals recovery. Significant portion of the waste tire pyrolysis had been vested in the production of oil, possibly via fast pyrolysis. Literature also provides us with a compiled comparison of the state of the art of pyrolysis of waste tires from different studies in several review articles [5, 14, 16]. Thus, we limit the scope of this chapter, in the context of pyrolysis results reported in the literature, to the results compiled from these review articles and their findings. Table 4 provides us with a comparison of various pyrolysis studies reported and reveals the relative yield of oil, char and gas products [14]. From this table, one can see that the char yield was around 30–50% (by wt.), depending on the reactor conditions used. Such a yield of char is significantly high that is limited to the processes designed for oil production. This is a result of the high carbon black and thus fixed carbon content, which is a characteristic of the waste tire feedstock. Additionally, Table 5 provides information on the quality of oil produced in comparison with petrochemical fuels. This reveals the requirement of downstream catalytic refining of the oil, possibly novel hydrogenation, to make it compatible for fuel applications, especially in terms modifying the flash point, calorific value enhancement, particulate and carbon residue removal, and improving the boiling point characteristics [14]. The requirement of significant refining adds to the issue of high char yields which leads to an end-result of relatively low yields of the final fuel products. The high char yield, unwanted high gas yield and catalytic refining requirements currently limit the capability of pyrolysis as a pathway for the purposes of oil/liquid fuel production.

To alleviate these issues, various strategies have been proposed. As the gas yield from this pyrolysis for liquid process is in the range of around 2–25% (by wt.), valuable utilization of this by-product is essential to support the overall process. The high char yield from pyrolysis still makes this process less attractive and calls for another strategy to target both the issues on high gas and less char yields. Gasification is proposed to avoid direct formation of liquid yield and reform all the intermediates from waste tires into syngas rich in H_2, CO, CH_4, impurities of C_2 and C_3 hydrocarbons and CO_2. This not only improves the effective yield of the desired product to above 70% (by wt.), but also provides uniform and versatile syngas, which can be used for various applications. Even modifying pyrolysis to focus on gas yield, such as high temperature mesh reactor tests shown in Table 4 reveals the gas yield to be higher than 70% (by wt.) and minimized char yield to fixed carbon content levels.

Significant knowledge is available on gasification of waste tires involving various reactor systems, operating conditions and gasifying agents, and they are summarized in some of the well-compiled review articles in the literature [16]. To obtain a complete picture on the gasification of waste tires, one needs to first investigate pyrolysis under similar conditions as this is the first inherent step in gasification. Conesa et al. found from pyrolysis of waste with high vapor residence times (>10 min) that the increase in temperature from 450 to 1000 °C, changed the char yield from 35 to 37% due to the high residence time, while the gas yield increased from 27 to 62%

Table 4 Summary of different pyrolysis tests on waste tires and their relative product yields [14]

Reactor	Experimental conditions	Maximum oil yield				References
		T (°C)	Oil (wt.%)	Char (wt.%)	Gas (wt.%)	
Fixed bed, batch	400–700 °C	500	40.26	47.88	11.86	[46]
Fixed bed, batch	500–1000 °C, 1200 °C/min	500	58	37	5	[36]
Fixed bed, batch	300–720 °C, 5–80 °C/min, 50 g	720	58.8	26.4	14.8	[19]
Fixed bed, batch	450–600 °C, 5 °C/min, 3 kg	475	58.2	37.3	4.5	[12]
Fixed bed, batch	950 °C, ~2 °C/min, 1000 kg	950	20.9	40.7	23.9	[47]
Fixed bed, batch	350–600 °C, 5 & 35 °C/min,	400	38.8	34	27.2	[10]
Fixed bed, batch	300–700 °C, 15 °C/min, 175 g	700	38.5	43.7	17.8	[48]
Fixed bed, batch	375–500 °C, 10 °C/min, 10 g	425	60	~30	~10	[11]
Fixed bed, batch, internal fire tubes	375–575 °C, 750 g	475	55	36	9	[49]
Fixed, wire mesh, fast reactor	390–890 °C, 70–90 °C/s	860	~5	~22	~73	[29]
Moving screw bed	600–800 °C, 3.5–8 kg/h throughput	600	48.4	39.9	11.7	[31]
Rotary kiln	550–680 °C, 4.8 kg/h	550	38.12	49.09	2.39	[50]
Rotary kiln	450–650 °C, 12–15 kg/h	500	45.1	41.3	13.6	[37]
Fluidized bed	740 °C, 1 kg/h	740	30.2	48.5	20.9	[51]
Fluidized bed	750–780 °C, 30 kg/h	750	31.9	38	28.5	[51]
Fluidized bed	700 °C, 200 kg/h	700	26.8	35.8	19	[51]
Fluidized bed	450–600 °C, ~220 kg/h	450	55	42.5	2.5	[52]
Circulating fluidized bed	360–810 °C, 5 kg/h	450	~52	~28	~15	[35]
Conical spouted bed	425 and 500 °C	500	~62	~35	~3	[53]

(continued)

Table 4 (continued)

Reactor	Experimental conditions	Maximum oil yield				References
		T (°C)	Oil (wt.%)	Char (wt.%)	Gas (wt.%)	
Vacuum, conical spouted bed	425 and 500 °C, 25 & 50 kPa	500	~60	~34	~4	[54]
Vacuum	485–550 °C, batch (80–180 kg) and continuous	520	45	36	6	[55]
Vacuum	500 °C, pilot-scale semi-continuous	500	56.5	33.4	10.1	[56]
Vacuum	450–600 °C, 100 g batch	550	47.1	36.9	16	[40]
Closed batch reactor	350–450 °C, 30 °C/min	450	~63	~30	~7	[57]
Drop tube reactor	450–1000 °C, 30 g/h	450	37.8	35.3	26.9	[26]

at the expense of oil yield that decreased from 38% to negligible levels [26]. This is because of high-temperature operation leads to enhanced secondary reactions such as thermal cracking of the hydrocarbon intermediates to form light hydrocarbons, H_2, CO and CO_2 at significant vapor residence times. These results are also evident from the data shown in Tables 6 [14]. From Table 4, one can see that high heating rates (70–90 K/s), high temperature and vapor residence times (15 min) in a wire-mesh microreactor led to significantly high gas yield of 73% [29]. But at temperatures of 450–600 °C with short residence times (~2 min), heating rate of 5 °C/min led to maximum oil yield of 58%. These results also showed that temperature had minimal effect on the relative yields. Although significant literature is available on pyrolytic breakdown of waste tires, focus on syngas yield and evolutionary behavior during gasification is still lacking. Our studies into high-temperature pyrolysis of waste tires and its mixtures with biomass provide comprehensive information (given in the following sections) that establishes the feasibility of syngas recovery from waste tires.

Various reactor systems have been utilized for the gasification of waste tires, which included fixed bed, bubbling and circulating fluidized bed reactors of both lab-scale and pilot-scale, rotary-kiln reactors and plasma gasifiers [5, 16]. Knowledge on the gasification at pilot scale has been limited to using air and steam as the gasifying agent. The addition of air oxidized some of the inflammable components evolved from waste tire decomposition while the exothermic reaction from this oxidation supports the endothermic decomposition reactions responsible for volatile evolution. This leads to syngas with relatively high CO_2 compared to other gasifying agents but provided with the advantage of autothermal operation, i.e., alleviation of the need of external heat/energy source for continuous operation and maintaining the desired temperature. Due to the ease of energy transfer and its similarity to incineration,

Table 5 Properties of oil products obtained from waste tire pyrolysis in the literature [14]

Property	Ref. [47]	[37]	[10]	[58]	Kerosene [59]	Gas oil	Light fuel oil
Flash point (°C)	20	17	65	43	40	75	79
Carbon residue (%)	2.2	1.78	–	–	–	< 0.35	–
Density (kg.l^{-1})	0.91	0.962	0.833	0.924	0.84	0.78	0.89
Viscosity (cSt at 40 °C)	6.3	–	–	3.77	1.2	3.3	21
Viscosity (cSt at 50 °C)	–	2.44	1.01	0.924	–	–	–
Viscosity (cSt at 60 °C)	2.38	–	–	–	–	–	–
Carbon (wt.%)	88.0	84.26	79.61	–	–	87.1	85.5
Hydrogen (wt.%)	9.4	10.39	10.04	–	13.6	12.8	12.4
Nitrogen (wt.%)	0.45	0.42	0.94	–	–	0.05	0.15
Sulfur (wt.%)	1.5	1.54	0.11	0.72	0.1	0.9	1.4
Oxygen (wt.%)	0.5	3.39	9.3	–	–	–	–
Initial B.Pt (°C)	100	–	38.5	70	140	180	200
10% B.Pt (°C)	140	–	58.2	114.5	–	–	–
50% B.Pt (°C)	264	–	174.8	296.1	200	300	347
90% B.Pt (°C)	355	–	–	386.4	315	–	–
Calorific value (MJ/kg)	42	41.7	42.66	38	46.6	46	44.8
Ash (wt.%)	0.002	Trace	–	0.31	–	0.01	0.02
Moisture (vol.%)	4.6	0.88 (wt.%)	–	–	–	0.05	0.1

autothermal mode is the most preferred way of gasification currently used at large scale at the expense of syngas quality [3]. A novel pathway of pilot-scale solar air gasification was reported in Wieckert et al. concerning the conversion of waste tires in comparison with various other feedstocks including coal that reported sustainable energy efficiency of 27% in waste tire conversion [60]. While solar concentrator supported gasification has significant potential in waste tire conversion, the lack of knowledge and the challenges associated with energy losses in the current solar

Table 6 Proximate and ultimate analysis, and calorific value of the tested feedstocks [4]

Sample	Volatile (wt.%)	Fixed carbon (wt.%)	Ash (wt.%)	Moisture (wt.%)	LHV (MJ kg^{-1})	Carbon (wt.%)	Hydrogen (wt.%)	Nitrogen (wt.%)	Oxygen (wt.%)	Sulfur (wt.%)
Waste tire	62.5	27.9	8.9	0.7	33.3	81.85	6.66	1.7	9.8	1.37
Pine bark	63.9	26.2	4.8	5.2	18.8	50.37	4.20	1.61	43.81	0.03

concentrator reactors limit this pathway but is a pursuable challenge for carbon neutral production of energy/chemicals from waste tires and other wastes.

Increase in temperature was found to increase the syngas yield and thus the energy recovery ratio (syngas energy/waste tire energy) as found by Raman et al. who reported the energy recovery ratio improved linearly from 0.1 to 0.4 with increase in temperature from 900 to 1060 K [61]. Other studies have found 0.4 for the highest energy recovery obtained under these conditions of air gasification. In the case of steam gasification, Donatelli et al. found that increase in steam/feedstock ratio decreased the calorific content of the syngas (possibly by reforming all the hydrocarbons into H_2 and CO) to result in decreased energy recovery ratio [20, 62]. Various gasifying agents were investigated to convert waste tires including air, steam, and CO_2 but studies on CO_2 gasification are visibly limited [16]. While air gasification provides autothermal operation and steam gasification provides high H_2 yield, the advantages of CO_2 gasification are multi-fold and more eco-friendly. Incorporation of CO_2 into waste tire gasification leads to utilization of CO_2 for char gasification and volatiles reforming which results in syngas enhancement and improved carbon conversion. It also assists in net conversion of CO_2 into valuable syngas which leads toward making this process more carbon neutral. High-temperature CO_2 is available as flue gas in some industries, such as coal combustion plants, which can be synergistically used for waste tire conversion for effective carbon utilization. Further studies into the scope of CO_2 gasification with respect to waste tire conversion are necessary, and lab-scale studies have been carried out by our group in this direction that support effective and efficient conversion of waste tires using CO_2. Following sections provide detailed experimental setup and result that established feasibility of CO_2-assisted gasification of waste tires as a sustainable pathway for waste tire conversion.

The influence of catalyst in waste tire conversion was conducted using dolomite addition that resulted in lowered calorific value of syngas, and thus the energy recovery ratio. However, the utilization of transition metal (Ni, Co, Fe, Cu) supported on Al_2O_3 carried out by Zhang et al. reported the formation of carbon nanotubes in the char yield that constituted to 8–12 wt.% of the char [63]. Studies on the comparison of commercial Ni-based catalysts, with mineral catalysts such as dolomite and olivine, were also reported in the literature using bubbling fluidized bed gasifier. These studies showed improved H_2 and syngas yield via reforming/cracking reactions at the expense of calorific output. The issues of catalyst deactivation from coke deposition were as expected due to high levels of char formation tendency and hydrocarbon presence with Ni/Al_2O_3 leading to H-abstraction and enhanced H_2 yield from 24 to 57%. Utilization of this carbon deposition for the production of carbon nanotubes was also investigated [63]. Further studies into such a pathway of H_2 and carbon nanotubes are essential to establish its feasibility and operating conditions for optimal products as control over the quality of carbon nanotubes, and their extraction is critical to establish their production. Syngas with H_2 content as high as 99% was also produced from the gasification of waste tires in the presence of CaO [64]. Although the review article claims CaO to be catalyst, H_2 enhancement was carried out via well-known sorption enhanced reforming to remove CO_2 via carbonization of

CaO into $CaCO_3$ which can drive the product equilibrium to high H_2 partial pressures [16]. But the reusability of CaO was low, and thus the state-of-the-art feasibility is questionable as reported in gasification and sorption-enhanced reforming studies in the literature.

The gasification of waste tires with other feedstocks including coal products such as lignite, biomass/biowaste such as pinewood, palm kernel shells, municipal sludge, olive husk, almond shells, palm fruit and plastic wastes such as polyethylene terephthalate (PET) from plastic bottles, and their mixtures were also carried out for various selected goals [16]. Almond shells and palm fruits were added to improve the char reactivity from an otherwise stable low-reactive tire char and found significant improvement in carbon conversion levels via co-conversion. Even in the cases of co-processing, although the values of syngas yield, quality and energy with respect to temperature, catalyst and feedstock ratio were carried out, systematic studies into the impact of feedstock composition and the feedstock ratio on the syngas components and conversion behavior, especially the understanding of possible nonlinear synergy is still lacking in the literature. Enhanced syngas evolution via nonlinear interaction between polymer wastes and biomass/coal were reported with selected plastics but such studies involving waste tires are still lacking [65–70]. Studies to understand the impact of co-processing waste tire with biomass via pyrolysis and gasification have also been carried out and reported in the following sections [4, 15, 18].

4 Methods and Materials

4.1 Feedstock Materials

For studies to establish the feasibility of waste tire conversion via thermochemical pathways, we utilized used-passenger car tire as such waste constitutes to almost 90% of the total waste tires collected in the USA [2]. The waste tire used was a Goodyear winter radial tire reinforced with textile fibers instead of metal wires. For lab-scale pyrolysis and gasification studies, this tire was cut into small cubes having each side of approx. 1.5–2 cm. Thermogravimetric studies were also carried out to characterize this feedstock, and for such microscale studies, the waste tire was frozen using liquid N_2 and then grounded with rubber particles of approx. ~100 μmin size. For each case, the total tested feedstock of waste tire included a mix of 60 wt.% from the tire tread and 40 wt.% from the side wall. Due to reinforcement and tire construction, the composition of the waste tire varied from the side wall to tread. This proportional mixture was utilized to accurately represent a real waste tire.

For co-pyrolysis and co-gasification studies, pine bark was chosen as a representative forestry residue biomass feedstock that was added to the waste tire. The pine bark feedstock was obtained from a local nursery, and the acquired feedstock was dried at 105 °C for 24 h to minimize moisture content before blending with the waste tire. The samples acquired were characterized using proximate and ultimate analysis to establish the fixed carbon and volatile content along with carbon, hydrogen, nitrogen and sulfur (CHNS) content of the chosen feedstocks. Thermogravimetric analysis (TGA) was utilized following modified ASTM 7582–15 standard for proximate analysis, while low heating value (LHV) of the samples and the pyrolysis of char blends was obtained using rapid screening device (5E-KCIII, China); see Table 6 for these results.

4.2 Thermogravimetric Analysis (TGA)

Thermogravimetric analysis was carried out using TA Instruments SDT Q600 with each sample weighing approximately 1–3 mg. The TGA was equipped with horizontal beam balance in an electric furnace and was capable of operating at temperatures of up to 1500 °C. The samples of waste tire and pine bark were ground to 140 mesh particle size separately and as blends for these analyses. The samples were heated from room temperature to 100 °C and maintained for 10 min to remove moisture, followed by heating the sample at 10 or 20 °C/min to reach to temperatures of 950 °C in the presence of high-purity Ar at 100 ml/min. During this heating, the sample mass was recorded to understand the kinetics of their thermal decomposition. At 950 °C, the sample was maintained for 10 min, while the purge gas was changed to dry air. This step was carried out to combust the solid-char residue thus leaving behind only the ash contents. The mass-loss data from these tests allowed us to not only obtain the pyrolysis kinetics but also proximate analysis.

4.3 Lab-Scale Fixed Bed Reactor

Pyrolysis and CO_2-assisted gasification experiments were conducted using a lab-scale fixed-bed reactor system that was operated in semi-batch mode at different set temperatures of up to 1000 °C. The reactant gases flowed through the reactor continuously along with the product gases while for any given test, the sample feedstock was introduced as a batch of fixed mass. This reactor system was powered by two electric furnaces—Lindberg/Blue M Mini-Mite for preheating the incoming gases and Lindberg/Blue M 1200 °C Split-hinge tube furnace for sample heating at any defined temperature. The input gases included N_2 as the tracer gas, CO_2 as gasifying agent, and Ar to purge residual gases from the system. The flow rates of these gases were controlled separately using orifice flow meters. The evolved product gases were collected in sampling bottles (for short sample interval collection) or transported directly to a gas chromatograph for detailed analysis of the gases evolved. The feedstock sample holder was made of stainless-steel wire mesh, and this was inserted into the uniform heated zone of the reactor system at appropriate time via a quick-disconnect coupling at back end of the reactor. A schematic diagram of the

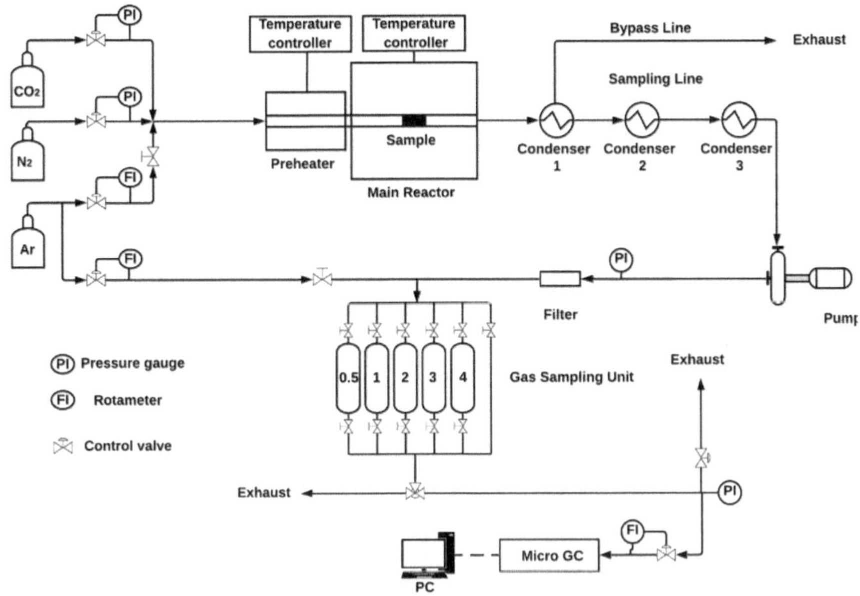

Fig. 2 Schematic of the experimental setup used for pyrolysis and gasification [15]

experimental setup details is provided in Fig. 2. The setup was equipped with ice-bath condenser and particulate filters to cool the product gases and remove moisture, tar and other particulate content in the product gases prior to their transport for gas analysis. Online product gas analysis was carried out using gas chromatography-thermal conductivity detector (GC-TCD) (Agilent Micro GC 3000A) which was calibrated for the quantification of mole fractions of H_2, CO, CO_2, CH_4, C_2H_4, C_2H_2, C_2H_6, O_2 and N_2 (using Restek refinery gas standard #1). The time required for each sample for gas analysis was almost 2.6 min. Additionally, the online gas sampling system was also equipped with gas sampling bottles in case of requirement of sampling times faster than 2.6 min. The electrical furnaces were monitored for their energy consumption (EML-2000, Canada) to estimate the energy requirements of each of the pyrolysis and gasification tests to obtain accurate values of overall energy efficiency beyond typically reported on energy recovery data. The char samples were characterized for their morphology using scanning electron microscopy (JSM-6510, Japan) at beam voltage of 20 kV after these samples were sputtered with Pt for better scanning characteristics. The char samples were also characterized for their surface area and pore volume using surface area and porosity analyzer (Tristar II 3020, USA).

Each test began by first allowing the input gases to go through the reactor system while the two furnaces reached the desired set point temperatures. The temperatures of interest were from 700 to 1000 °C. The input gases were chosen depending on the test conducted. For pyrolysis and co-pyrolysis tests, 2.1 sccm of N_2 flowed while avoiding entry of any other gas to maintain inert conditions during purging and analyzing the evolved gases. In the case of CO_2 gasification, a gas mixture of 75 vol.% CO_2 and 25 vol.% N_2 with a total flow rate of 2.1 sccm was used to allow for CO_2-assisted gasification. Here the vapor residence times were comparable to that in pyrolysis tests. Once the setpoint temperature was attained and the gas flow rates fixed, the sample holder containing 35 g of feedstock sample was inserted into the main reactor via quick-disconnect coupling. Tested feedstock samples included waste tire, pine bark and their mixtures at varying relative mass fractions of 1:3, 1:1 and 3:1 to not only analyze the influence of the mass ratio but also compare the results with the weighted results from separate processing of pure feedstocks to examine any possible interaction. While most of the product gases evolved from the main reactor were vented out, a fraction of it passed through the ice-bath condenser system that removed moisture and tar prior to analyzing the product gases. This dry and tar-free syngas were collected at 0.5, 1, 2, 3, 4 min from the start of the reaction using sampling bottles as such short sampling times of 1 min were not possible for online gas analysis using micro-GC and that our previous tests using different feedstocks had revealed that during this time, most of the syngas was released. From 5 min, this syngas was sent directly to the micro-GC for direct online gas analysis where the gas was collected at time intervals of 2.6 min and analyzed for the mole fractions of different gases against the calibrated gases.

$$M_i = \frac{X_i}{X_{N_2}} * V_{N_2} * \rho_i \tag{1}$$

Using the mole fraction of each gas species with respect to mole fraction of N_2, the gas flow rate of each of the gas species was calculated as the flow rate of N_2 was constant and it did not change during the reaction. From the known flow rate of N_2, it can be used as internal calibration reference to measure the flow rate of each of the gas species using the Eq. (1). In this equation, X_i is the measured mole fraction of species '*i*' at given time, V_{N_2} is the volumetric flow rate of N_2 at the reactor inlet, ρ_i is the standard density of species '*i*' and M_i is the mass flow rate of species '*i*' at given time. This method of obtaining gas flow rates assumes ideal gas behavior of the species, the inertness of N_2 toward these reactions, and that the syngas sample composition is representative of the composition of the evolved syngas prior to venting. Additionally, since each gas sample was collected over span of 10 s, this data represented an averaged value over that timespan at each time duration. This was carried out until the syngas at the GC was negligible. Ar was used to flush the reactor system and sampling lines of residual gases between each test. After this, the gas samples from sampling bottles were analyzed individually using micro-GC while Ar flushing was carried between the analyses for each of these bottles. The operating conditions of this experimental setup are summarized in Table 7.

Table 7 Summary of the operating reactor conditions for pyrolysis and gasification tests [15]

Pyrolysis	
Reactor temperature	673, 773, 873, 973, 1073, and 1173 K
Operating pressure	Atmospheric
Tracer gas and inert medium	2.1 sccm of N_2 (at standard 294 K and 1 atm.)
Sample	35 g of waste tire 2 × 2 cm pieces (60% tread and 40% sidewall)
Apparent vapor residence time	~21 s
Isothermal reaction time	2.5 h
Gasification	
Reactor temperature	973, 1073, 1173, and 1273 K
Operating pressure	Atmospheric
Tracer gas and inert medium	2.1 sccm of 75% vol. CO_2 and 25% vol. N_2 gas mixture (at standard 293 K and 1 atm.)
Sample	35 g of waste tire 2 × 2 cm pieces (60% tread and 40% sidewall)
Apparent vapor residence time	~21 s
Isothermal reaction time	53 min

5 Results and Discussion

5.1 *Thermogravimetric Analysis*

5.1.1 Waste Tire Decomposition

This analysis was carried out for gaining insights into the pyrolytic kinetics and its associated mass-loss behavior during the thermal decomposition of waste tire. This is essential as it provides us with the characteristic temperatures of decomposition and possibly also the compositional details of the rubbers used in the examined waste tire sample. With inert operation of TGA tests, one could quantify the volatile content sourced from the rubbers, accelerators and other additives present in the tire. The combustion step using air at high temperature was carried out to quantify the inorganic content and the fixed carbon content (mostly carbon black) in the case of waste tire. TGA tests in CO_2 environment were avoided since the decomposition here is slow and at their characteristic temperatures (due to low heating rate) which were significantly low for CO_2 to be active. Additionally, to normalize the mass-loss data obtained from TGA into non-dimensional format, conversion (α) was obtained, and the results are reported using this data and its derivative with respective to instantaneous temperature (DTG = $d\alpha/dT$). Normalization was carried out according to Eq. (2), where $M(T)$ is the mass of the remaining sample in the TGA at temperature T, T_o is the initial temperature (100 °C) and T_f is the final temperature (950 °C).

$$\alpha(T) = \frac{M(T_o) - M(T)}{M(T_o) - M(T_f)} \tag{2}$$

This conversion data for thermal decomposition of waste tire is presented in Fig. 3 along with DTG (derivative with respect to temperature). These results show similarity compared to those reported in the literature. The DTG data is useful to represent the behavior of waste tire decomposition and reveals the volatile evolution to be a combination of three stages (i.e., showing three peaks) at temperatures of 450–600 K, 600–680 K, and 680–800 K. Each of these regions could be a combination of decomposition patterns from different constituents, such as decomposition of unstable additives such as extender oils and other low-boiling point additives that contribute to the first stage of decomposition while the second stage could be from natural rubber (NR) decomposition and initiation of decomposition of polybutadiene rubber (BR), and styrene-butadiene rubber (SBR). The third stage could be significantly from the decomposition of BR and SBR based on the straight elastomer and is based on the TGA studies reported in the literature [41]. Most of the literature on such analysis of tires has reported the latter two stages corresponding to elastomers decomposition and in some cases, the second stage contributed more to the mass-loss than the third due to different proportions of the rubber mix. The high mass-loss from the third peak revealed that the specific waste tire examined here contained significantly high BR and SBR proportions compared to NR and that the ratio of BR to SBR could be predicted from the position and contribution of the third peak based on the literature reported on varying ratios.

$$\begin{aligned}\mathrm{DTG}_{\text{waste tire}} = w_{\mathrm{NR}} * \mathrm{DTG}_{\mathrm{NR}} + w_{\mathrm{BR}} * \mathrm{DTG}_{\mathrm{BR}} \\ + w_{\mathrm{SBR}} * \mathrm{DTG}_{\mathrm{SBR}}\end{aligned} \tag{3}$$

To quantify the relative composition of individual rubbers in the waste tire examined, we employed a three-component curve-fit to the DTG data based on the straight

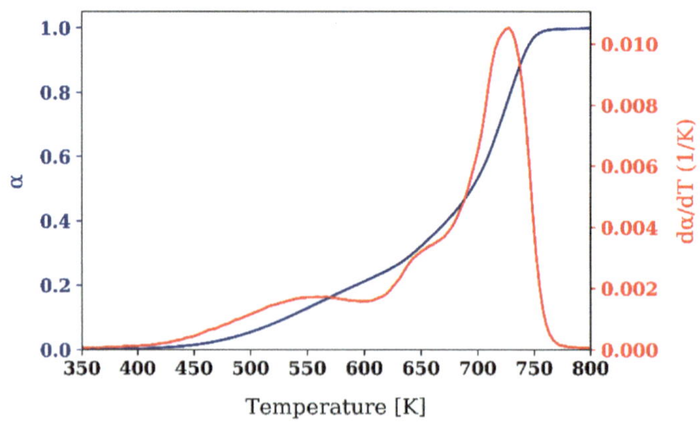

Fig. 3 Extent of mass-loss (α) and its derivative with temperature during waste tire pyrolysis [15]

Table 8 Estimated composition of elastomers in the tested waste tire [15]

Polymers	Composition (wt.%)
Natural rubber (NR)	5
Butadiene rubber (BR)	40
Styrene–butadiene rubber (BR)	55

elastomer DTG data from the literature, neglecting the presence of any interaction's contribution to the mass distribution and additives contribution to the mass-loss in the latter two stages. Equation (3) revealed that the equation used to represent such a model was favorable, where w_i represent the mass fraction of the respective elastomer in the waste tire, and the DTG_{NR}, DTG_{BR} and DTG_{SBR} were DTG of the respective straight-chain elastomers obtained from the literature. The $DTG_{waste\ tire}$ obtained from our data and the fit obtained had a coefficient of correlation of $R^2 = 0.996$. The estimated compositional results from this analysis are given in Table 8 that reveals significant presence of BR and SBR.

5.1.2 Decomposition of Waste Tire-Pine Bark Blend

Investigations involving TGA on blends of waste tire with other feedstocks are essential to understand the presence of any interaction and establish the feasibility of a feed-flexible gasifier where waste tires can be converted in the presence of any other feedstocks for sustainable operation and avoid over dependence on any individual feedstock. For this, blends of waste tire and pine bark were mixed in defined fraction and ground into similar size powder form. The relative fractions of each in the blends were 1:1, 1:3 and 3:1. These samples will be represented by *WxPy* where *x* represents waste tire mass fraction and y represents pine bark mass fraction; for example, W1P3 represents a sample with 25 wt.% waste tire and 75 wt.% pine bark. The results from these blends will be compared with the calculated weighted results from the individual components. Figure 4 reveals the conversion and DTG behavior

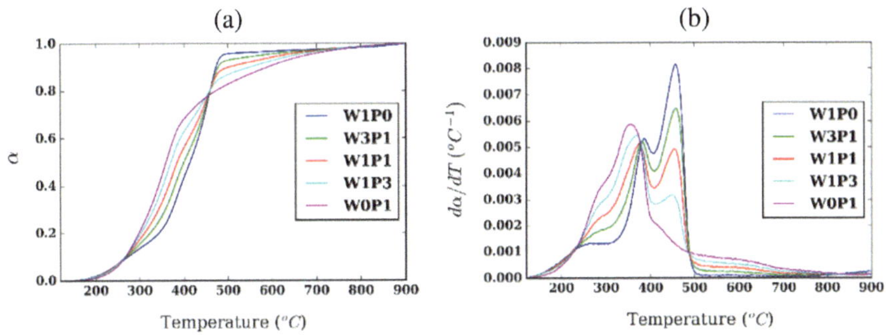

Fig. 4 Effect of waste tire-pine bark mixture ratio on **a** the extent of mass-loss, α, and **b** its derivative [18]

of the blends in comparison with pure components—waste tire (W1P0) and pine bark (W0P1). Pine bark also exhibited the presence of multiple peaks corresponding to the decomposition of lignocellulosic components such as hemicellulose, cellulose and lignin, respectively, with increase in temperature. Note that lignin partially contributed to all the stages of decomposition. Pine bark decomposition started early from 220 °C while waste tire pyrolysis was concentrated between 250 and 550 °C. Comparison of blended DTG with weighted results (not shown here), i.e., comparing DTG_{WxPy} with $(x^*DTG_{\text{waste tire}} + y^*DTG_{\text{pine bark}})/(x + y)$, revealed the lack of any difference in behavior. Although this reveals the lack of any synergistic interaction between waste tire and pine bark, it is important to note the lack of any inhibitive behavior either. This means that the decomposition of either components was possibly occurring independent of each other. This also means that conversion of waste tire in the presence of biomass feedstocks such as pine bark can be carried out in the same reactor without the loss of any efficiency and easily predict the decomposition behavior without any issues of interaction. Although any interaction seems to be missing according to the DTG behavior, further lab-scale studies were also carried out via pyrolysis and CO_2 gasification to confirm such behavior and observe for any high-temperature, heating rate effects and volatile–volatile interaction behaviors that cannot be observed from TGA results. Such a behavior on the lack of interaction from TGA results but presence of significant synergistic interaction during lab-scale high temperature gasification and pyrolysis was observed in various biomass–plastic mixtures such as pinewood–polyethylene. The lab-scale studies described here in the later sections revealed the observation of such a synergistic behavior.

5.2 *Lab-Scale Pyrolysis and Gasification*

5.2.1 Product Gas and Char Yield

Waste Tires

Online gas chromatography analysis was used to analyze and quantify dry, tar-free and particulate-free product gases for H_2, CO, CO_2, CH_4, C_2H_4, C_2H_6 and C_2H_2 as they represent the major compounds evolved from pyrolysis and gasification of tires. In this chapter, we refer to the combination of H_2, CO, CH_4, C_2H_4, C_2H_2 and C_2H_6 as syngas since these components are important due to their calorific value, while the combination of this syngas with CO_2 will be referred to as product gas since one can consider CO_2 as a by-product. From the mass flow rates obtained with time for different temperatures, cumulative yield of these components was calculated over a 50-min time duration for both pyrolysis and gasification to understand the influence of temperature and gasifying agent on the syngas evolution and thus the carbon conversion. Char yields were also measured at the end of reaction for these conditions to gain insights into the tendency of charring and the loss of yield by forming char as by-product.

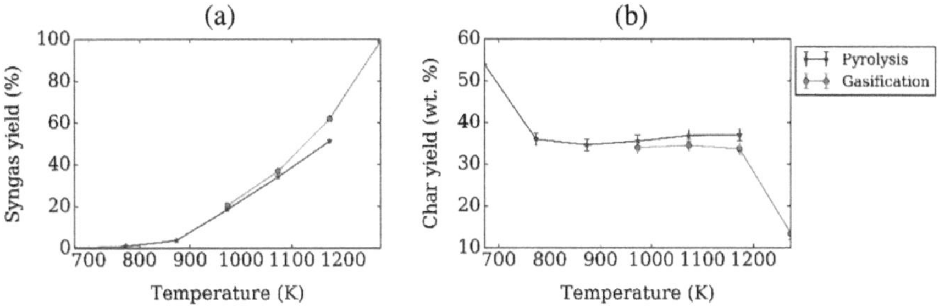

Fig. 5 Effect of temperature on **a** syngas and **b** char yield during pyrolysis and CO_2-assisted gasification of waste tires [15]

Figure 5 reveals the variation of syngas and char yield (as wt. % of waste tire mass after 2.5 h) with temperature and gasifying agent. At lower temperatures of 700–800 K, insignificant amounts of syngas evolved due to the inability of conversion of the volatiles evolved from waste tire to further crack or interact with CO_2 to form any of the evolved syngas components examined. But, at these temperatures char yield was more than 50% that decreased to ~38% as the temperature was increased to beyond the maximum temperature of complete conversion (~770 K) as seen from the DTG results presented in Fig. 3. From 800 K, syngas yield increased strongly with temperature because at high operational temperatures of the lab-scale reactor, the sample experienced high heating rates immediately after placing it in the reactor, leading to increased volatiles evolution. These volatiles undergo secondary reactions such as thermal cracking, and CO_2-reforming in the gas phase leading to the formation of syngas components. These gas-phase reactions leading to the formation of H_2, CO and light hydrocarbons are endothermic and thus with increase in temperature, the syngas yield is enhanced. In the case of gasification, temperatures above 973 K are required for CO_2 to actively react with the intermediates from the decomposing waste tire. Syngas yield as high as 50% from pyrolysis and 100% from gasification was obtained from waste tire conversion. The literature also reported high gaseous yields and were dependent on the reactor used to obtain high temperatures and heating rates, high vapor residence times and slow quenching that provided the necessary conditions for enhanced secondary reactions of cracking and reforming in the gas phase. Comparing the yields from pyrolysis and gasification, increase in temperature increased the differences in the gas yield as the char reforming reactions also contributed leading to improved CO yields.

In the case of char yield, beyond the temperature of complete devolatilization from TGA results (770 K), the char yields did not change significantly with temperature during pyrolysis as the char obtained was stable, similar to the char residue from TGA results. But in the case of gasification, beyond 1173 K, the char yield decreased significantly by about 50% due to Boudouard reaction of the char with CO_2 which was only active at high temperatures beyond 1100 K based on our previous studies on CO_2-assisted gasification of various other kinds of feedstocks.

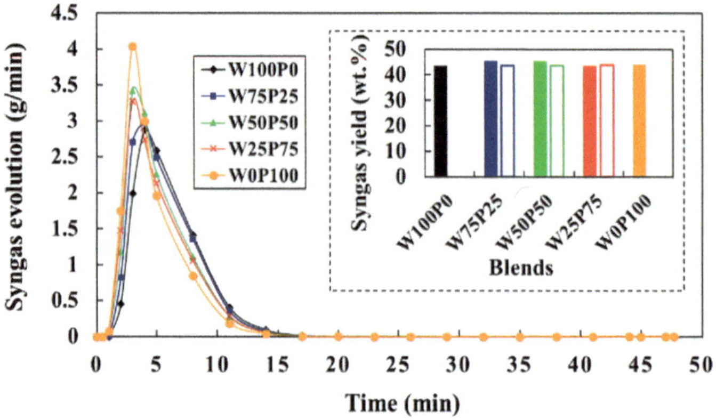

Fig. 6 Effect of waste tire-pine bark blend ratio on the evolution behavior and yield of syngas during co-pyrolysis at 900 °C (solid bars: experimental; hollow bars: weighted yields) [4]

Waste Tires with Pine Bark

Investigations into co-pyrolysis and co-gasification of waste tires with pine bark at 900 °C were conducted using lab-scale reactor system to examine any possible interaction. Figure 6 shows the influence of feedstock mixture fraction on the syngas yield during co-pyrolysis in comparison with estimated yields calculated by weighted aggregate of the individual feedstocks processed separately. The syngas yields revealed no significant interaction between waste tire and pine bark pyrolysis when processed together as the blend co-pyrolysis yields were similar to the weighted yields. At the examined temperature, the syngas yield was found to be ~40 wt.% for all the blend samples and pure samples. This was probably due to the high thermal stability of fixed carbon in waste tire leading to its low reactivity toward pine bark pyrolysis. Variation of char yield after 48 min with blend ratio during co-pyrolysis was also measured, and the results are shown in Fig. 7. While no significant interaction was observable, as the differences between the yields from co-pyrolysis and the weighted yields were negligible, the yield decreased with increase in pine bark content due to relatively high charring tendency from waste tire from its fixed carbon compared to pine bark.

Co-gasification was carried out at both 800 and 900 °C under similar conditions as in our other gasification studies but at different blend ratios. Figure 8 reports the influence of feedstock mixture on the syngas yield at these temperatures. The results obtained showed a decrease in syngas yield from co-gasification at 800 °C compared to separate feedstock gasification, but no such inhibition was observed at 900 °C. Further investigation into the composition of the syngas can provide us with information about any such interaction.

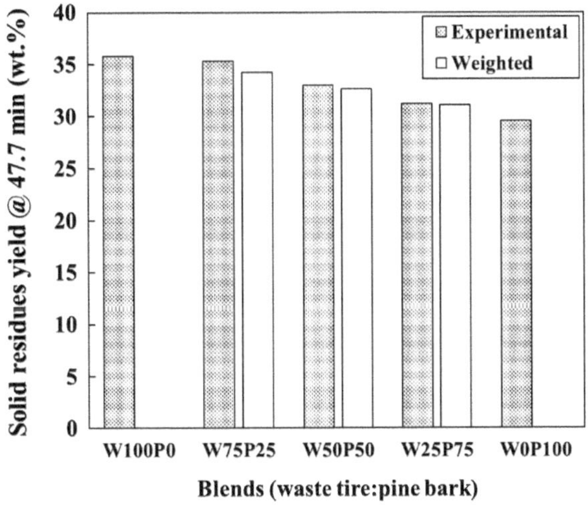

Fig. 7 Effect of feedstock blend ratio on the char yield from co-pyrolysis of waste tire and pine bark at 900 °C [4]

5.2.2 Syngas Components' Evolution and Yield

Waste Tires

From online product gas analysis capabilities using our lab-scale reactor system, we were able to obtain the temporal evolutionary behavior of flow rates of syngas and its components. Figure 9 reveals the impact of temperature on the evolutionary behavior of syngas during pyrolysis and gasification from the waste tire sample. The flow rate curves showed peak, and with increase in temperature, the peak value increased, but its position shifted toward earlier (shorter) times. This is because increase in temperature not only drives the equilibrium toward syngas product due to the global endothermicity of both pyrolysis and CO_2-assisted gasification, but also improves the kinetics of these reactions. This behavior was observed in the evolution of all the major components of syngas; see Fig. 10. Hydrogen (H_2) and CH_4 were the dominant components impacted by the temperature change during pyrolysis while CO and H_2 during CO_2-assisted gasification. While the evolution of syngas flowrate completed in 25 min during pyrolysis, it decreased to a constant nonzero value during CO_2-assisted gasification and then stayed at that flow rate for extended periods of time; see Fig. 10. CO evolution showed similar behavior that extended for a long period of time during CO_2-assisted gasification. During this process, the contribution of CO to syngas yield was very high. This was due to low reactivity of tire-char residual with CO_2 after 25 min, possibly due to high crystallinity, low imperfections and O, H content, and low surface area compared to other carbonaceous materials such as biomass. Long gasification times were reported to be required even in the case of steam gasification despite relatively higher reactivity of tire char with steam compared to CO_2.

Figure 11 reveals the cumulative yields of the individual syngas components and their variation with temperature during pyrolysis and CO_2-assisted gasification, while

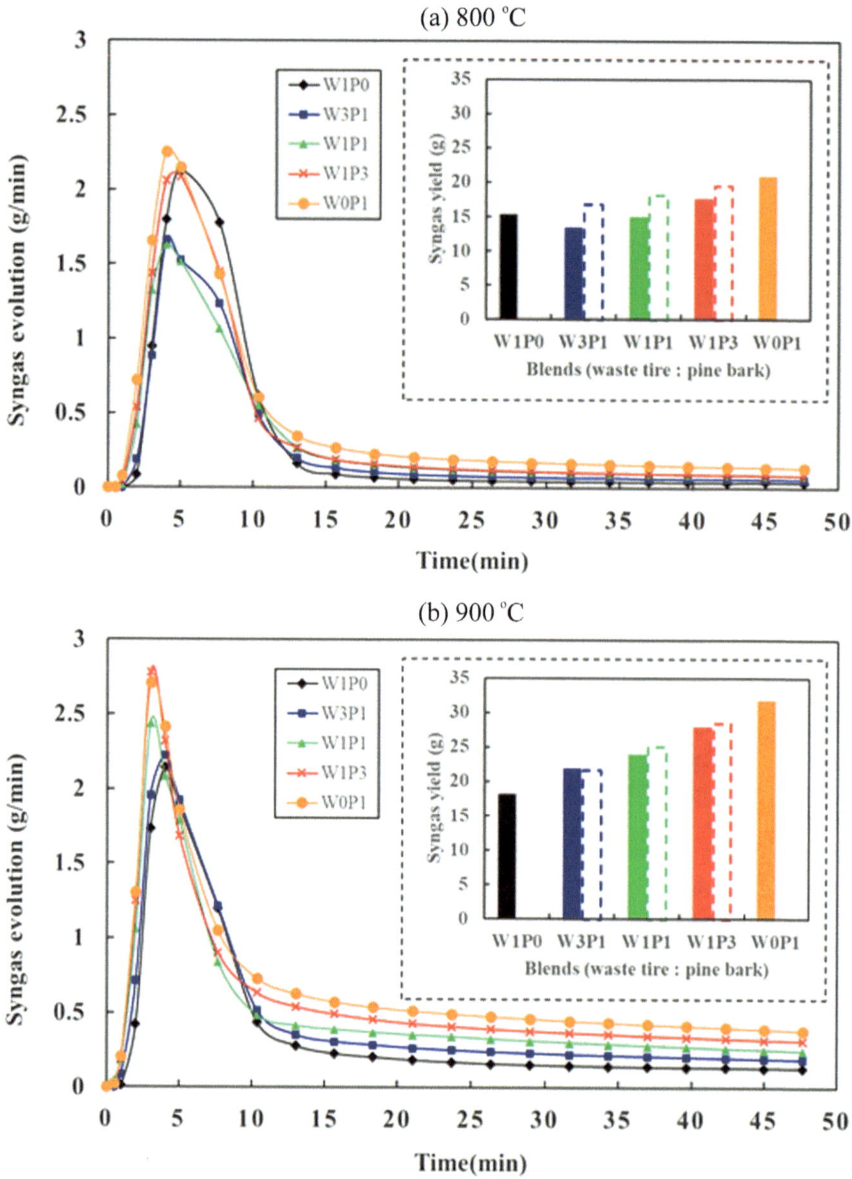

Fig. 8 Effect of feedstock blend ratio on the syngas yield from co-gasification of waste tire and pine bark at **a** 800 °C and **b** 900 °C [18]

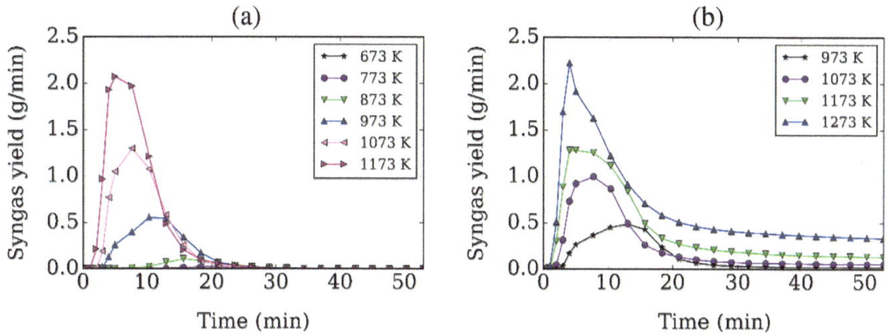

Fig. 9 Effect of temperature on the evolution of syngas from waste tires under **a** pyrolysis and **b** gasification [15]

Fig. 12 shows their mole fraction in cumulative product gas yield (including CO_2). In pyrolysis, increase in temperature increased the yields of H_2, CO, CH_4 and C_2 hydrocarbons. During pyrolysis, decarboxylation and decarbonylation occur of the oxygenates in the waste tire such as extender oils, secondary reactions with char, along with decarboxylation of inorganic content such as $CaCO_3$, $CaSiO_4$ and other metal carbonates. Low molecular mass hydrocarbons (CH_4, and C_2) were generated by gas-phase cracking of the volatiles released from the decomposition of BR and SBR polymers in addition to H_2. Hydrogen is also released from aromatization and cyclization of the volatiles in the gas phase, while these reactions are significantly enhanced with temperature. Comparing the H atomic content in waste tire (from ultimate analysis shown in Table 6) with the yields of H_2 and CH_4 revealed that significant portion of H content in the solid sample got converted to syngas.

Figure 12 reveals that while the net CO_2 yield enhanced with temperature, its mole fraction in product gas decreased. At low-temperature pyrolysis, while decarboxylation reaction is favored to release CO_2, insignificant conversion of heavy hydrocarbon intermediate volatiles into syngas components resulted in high CO_2 mole fraction. As the temperature increased, other components enhanced due to the contribution of enhanced secondary gas-phase reactions at a higher rate leading to the observed net reduction in CO_2 mole fraction. This is also because of enhanced equilibrium drive toward higher CO content at high temperature compared to CO_2. This enhanced thermal cracking of heavy intermediates from rubber components can be seen from enhanced CH_4 mole fraction in product gas; see Fig. 12. However, in the case of CO_2-assisted gasification, this increase in mole fractions of CH_4 and C_2 with temperature is met with opposing reactions such as dry hydrocarbon reforming to form H_2 and CO which results a decrease in these hydrocarbon mole fractions beyond 1173 K. Equations (4)–(11) represent the major reactions during pyrolysis and gasification of waste tires and biomass. Improved rate of increase in CO yield due to Boudouard reaction also contributed to the lowered hydrocarbon yields at high temperature; see Eq. (5).

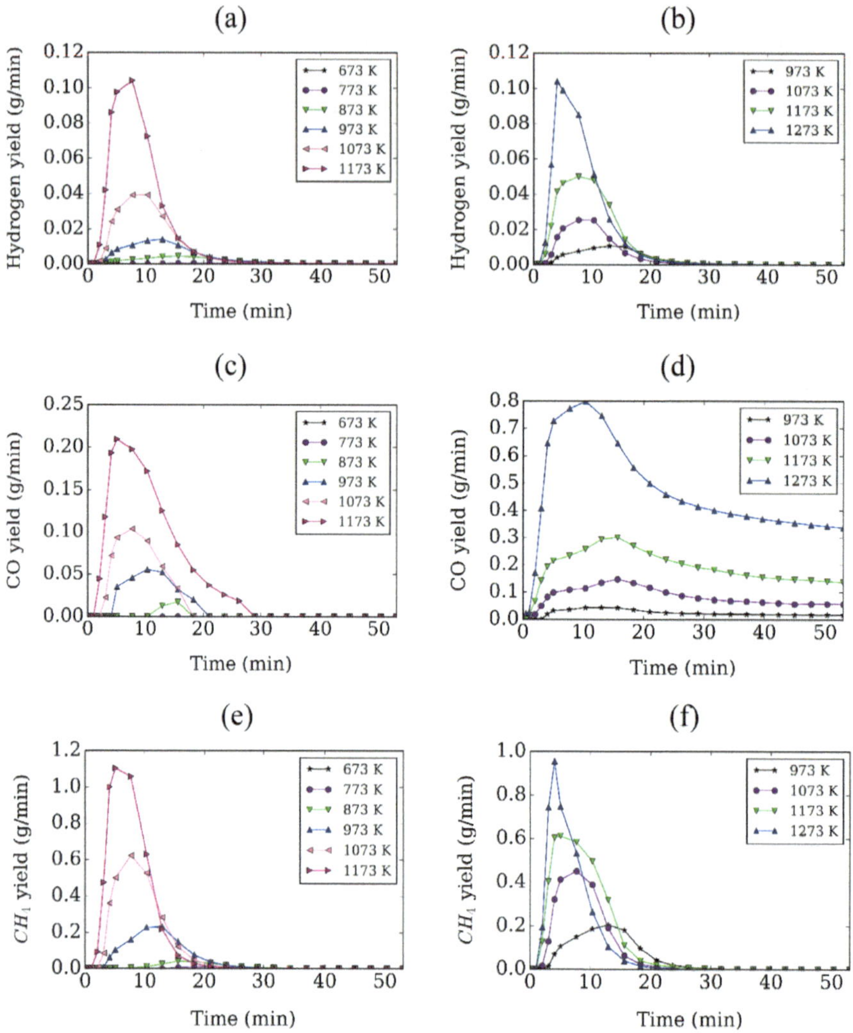

Fig. 10 Effect of temperature on the evolution of H_2, **a, b**, CO, **c, d** and CH_4, **e, f** during pyrolysis (left) and CO_2-assisted gasification (right) of waste tires [15]

$$\begin{aligned} &\text{Waste tire} + \text{pine bark} + CO_2 \\ &\quad \rightarrow H_2 + CO + CO_2 + CH_4 + C_mH_n \\ &\quad (C_2H_2 + C_2H_4 + C_2H_6 + C_3H_8 + \ldots) \\ &\quad + H_2O + \text{char} + \text{tar}\ \Delta H > 0 \end{aligned} \tag{4}$$

$$C + CO_2 \rightleftharpoons 2CO \quad \Delta H = +172\,\text{kJ/mol} \tag{5}$$

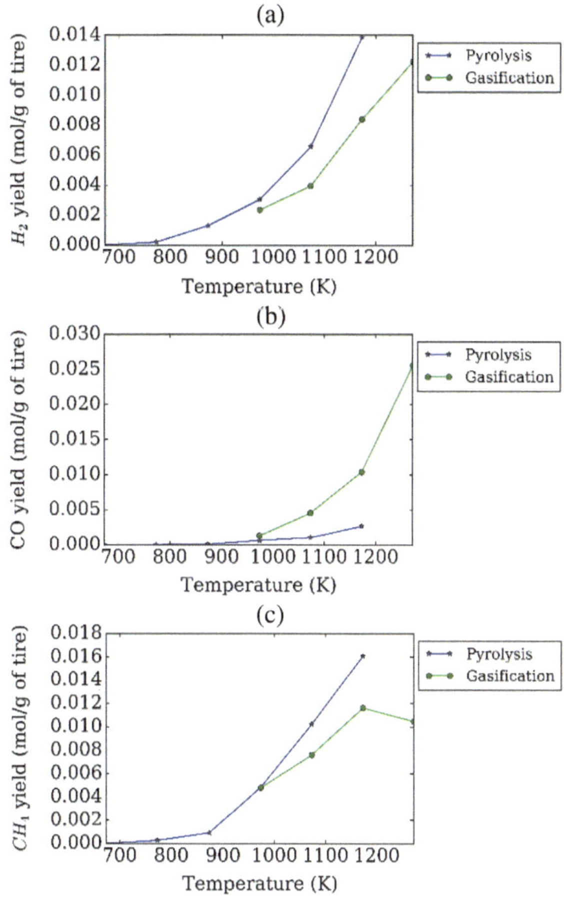

Fig. 11 Effect of temperature on the cumulative yields of **a** H_2, **b** CO and **c** CH_4 from pyrolysis and CO_2-assisted gasification of waste tires [15]

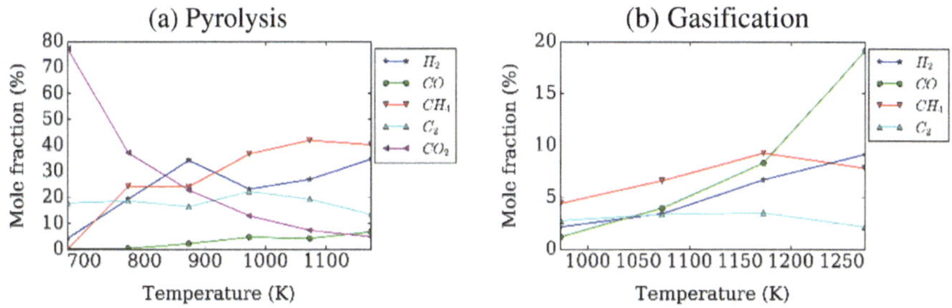

Fig. 12 Effect of temperature on the mole fractions of individual species from cumulative syngas yield from **a** pyrolysis and **b** CO_2-assisted gasification of waste tires [15]

$$C + 2H_2 \rightarrow CH_4 \quad \Delta H = -75\,\mathrm{kJ/mol} \tag{6}$$

$$C + H_2O \rightarrow CO + H_2 \quad \Delta H = +131\,\mathrm{kJ/mol} \tag{7}$$

$$CO + 3H_2 \rightleftharpoons CH_4 + H_2O \quad \Delta H = -206\,\mathrm{kJ/mol} \tag{8}$$

$$CO_2 + H_2 \rightleftharpoons CO + H_2O \quad \Delta H = 41\,\mathrm{kJ/mol} \tag{9}$$

$$\begin{aligned} &\mathrm{Tar}\left(C_xH_y\right) + xCO_2 \rightarrow 2xCO + (y/2)H_2\Delta H \\ &\quad = 0.9\,\mathrm{to}\,0.3\,\mathrm{MJ/mol} \end{aligned} \tag{10}$$

$$\mathrm{Tar} \rightarrow CH_4 + H_2 + H_2O + C_mH_n \quad \Delta H > 0 \tag{11}$$

Equations (5), (7), (9) and (10) contribute to provide significant enhancement of CO yield with increase in temperature, and the rate of increase was higher at high temperature. At temperatures beyond 1073 K, CO yield increased with temperature by 1.5 times/100 K to provide yields as high as 25.6 mmol/g which accounted for 0.72 g per gram of waste tire feedstock. Below this temperature, the yield of all the syngas components from CO_2-assisted gasification was almost the same as that from pyrolysis due to low activity of CO_2 at these temperatures. TGA results and literature also reveal this to be the case while this also supports CO_2-assisted gasification to be an extension of pyrolysis to include CO_2 reforming of the evolving volatiles and char into CO and H_2; see Figs. 5 and 10 that reveal lowered hydrocarbon and char yields with increase in CO yield compared to pyrolysis at high temperatures.

Waste Tires with Pine Bark

To investigate further into any possible synergistic or inhibitive interaction between waste tire and pine bark during their co-processing via pyrolysis and CO_2-assisted gasification, we examined the evolution of syngas components and yield and compared the results with the weighted data calculated from separate conversion of the feedstock. The weighted yields were calculated with the initial mass fraction of the feedstock components to be their respective weights in the aggregate.

Figure 13 reveals the influence of feedstock mass fraction on the variation of major syngas components' mass flow rate with time along with their cumulative yield compared with weighted yields during co-pyrolysis. The results reveal the yields from blended feedstock to be almost same as the weighted yields from mono-conversion. This means that in pyrolysis to obtain gaseous yields, waste tire and pine bark feedstock show no interaction with each other and the behavior of the reaction products was a superimposition of their behavior when pyrolyzed separately. This

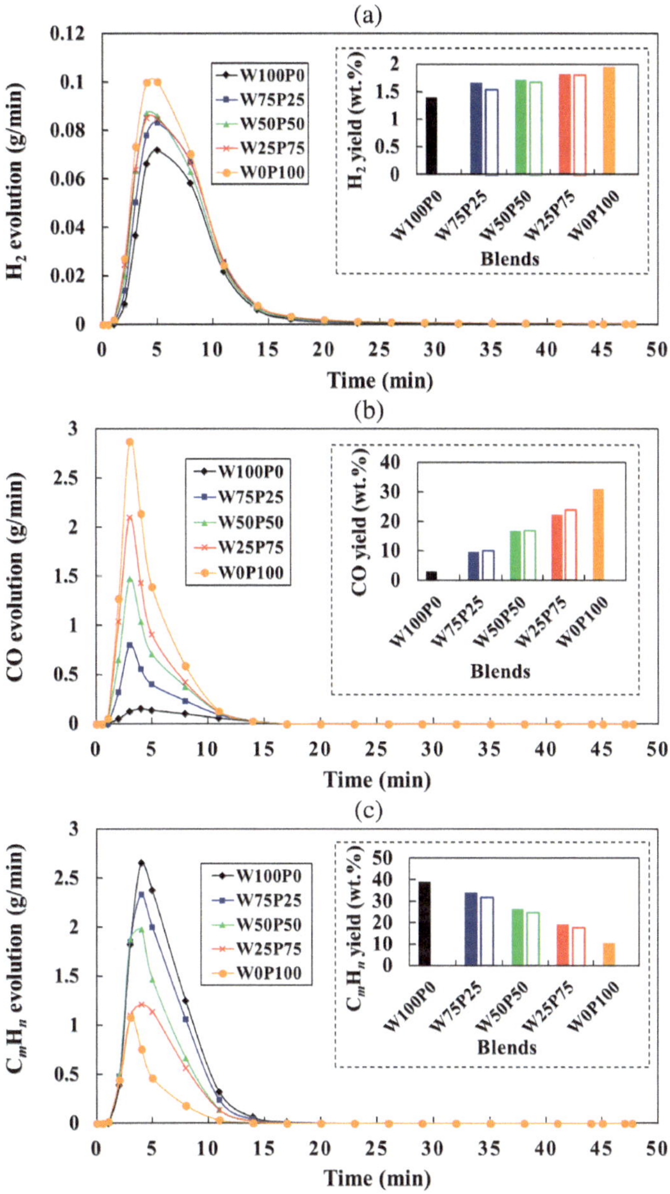

Fig. 13 Effect of feedstock blend fraction on the evolution and yield of **a** H_2, **b** CO and **c** C_nH_m from co-pyrolysis of waste tire and pine bark [4]

simplifies the modeling and design of feed-flexible reactors as it alleviates the need for any further studies into nonlinear behavior. Additionally, this also provides the design option to modify the syngas composition and their respective component yield by modifying the mixture composition of the feedstock. While the syngas yield did not change significantly with the feedstock in the case of co-pyrolysis, hydrocarbon yield was favored with high waste tire content, while CO yield was favored with high pine bark content, due to oxygenates present in the biomass. H_2 yield also increased with increase in pine bark content but to a lesser extent compared to other species. As the net syngas yield was almost the same for all the feedstock mass fractions, it suggests that the decreased mass of C_nH_m with increase in pine bark was appropriately compensated by the increase in CO mass yield. The variation in CO yield varied from less than 5 wt.% from pure waste tire pyrolysis to almost 30 wt.% from pure pine bark, while C_nH_m yield varied between 40 wt.% from pure waste tire to ~10 wt.% from pure pine bark.

Similar data showing the evolution of syngas components and their yield is reported in Fig. 14 for both 800 and 900 °C that not only provide information about the variability on the evolutionary behavior with feedstock variation, but also provide insights into possible interaction that could lead to non-additive yields. Increase in temperature led to significant increase in the peak H_2 mass flow rate from ~0.035 g/min to as much as ~0.075 g/min due to enhanced cracking and CO_2-reforming. Increase in pine bark content also increased the H_2 flow rate and yield due to the higher reactivity of the biomass to react with CO_2 to form H_2 and CO. At 800 °C, some variation between the H_2 yields from the blends and the weighted results was found. However, at 900 °C, these differences disappeared to finally provide blended yields same as the weighted yields. CO flow rate and yield also increased with increase in pine bark content not only due to its higher reactivity compared to waste tire but also due to inherently higher oxygenate content in the biomass. The CO yield also increased with temperature by almost twice from 800 to 900 °C due to enhancement of CO from pyrolysis along with char gasification and CO_2-reforming. The behavior of CO flow rate with time showed not only a peak but also a steady CO yield from ~10-min time due to Boudouard reaction involving char-CO_2 gasification to provide enhanced CO yields. Such a constant flow rate was due to slow reaction rate of the char with CO_2, possibly due to low porosity. Synergistic reduction of C_nH_m was also found at 800 °C, which also led to lower syngas yield, as seen from Figs. 8 and 14. But this was not observed at 900 °C, possibly due to higher influence of gas-phase reactions with CO_2 leading to similar yields as the weighted yields.

From these results, we can see that at operational temperatures of 900 °C required for CO_2-assisted gasification, the syngas component yields from co-gasification were superimposed sum of the yields from individual components to prove the compatibility of waste tire with biomass gasifiers. So, subject to the net product output required, the feedstock can be modified to include waste tires without the loss of conversion efficiency.

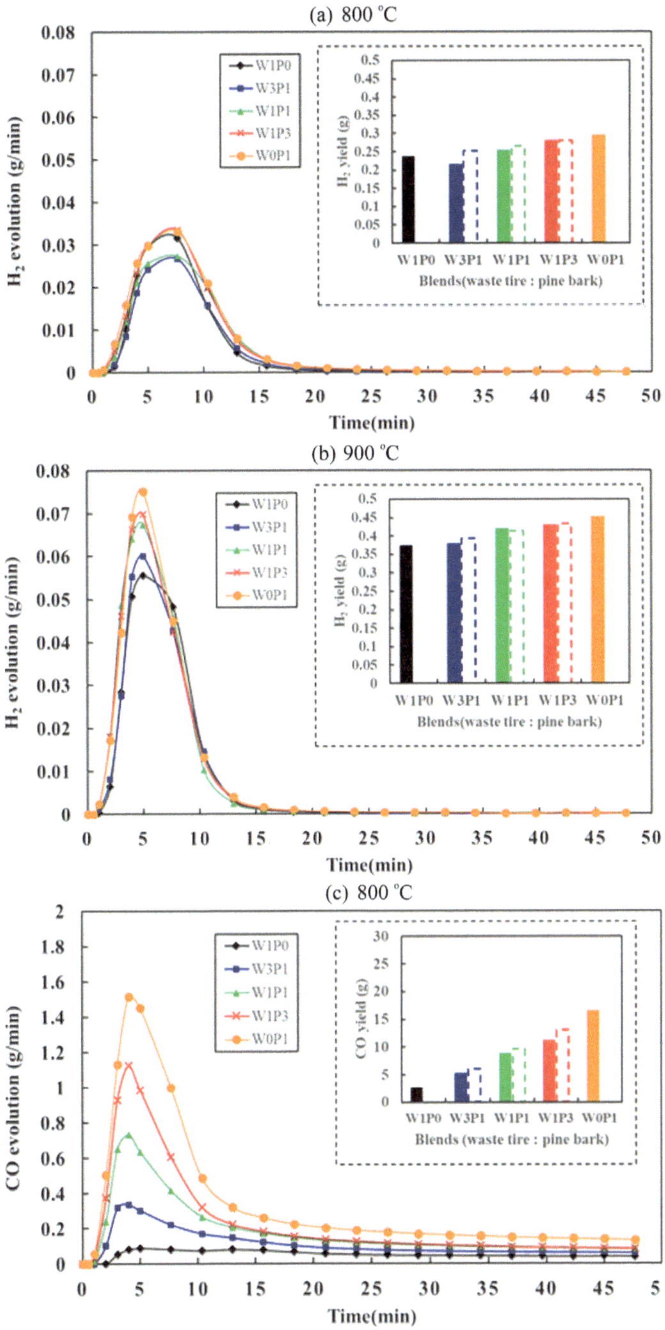

Fig. 14 Effect of temperature and feedstock blend fraction on the evolution and yield of **a, b** H_2, **c, d** CO and **e, f** C_nH_m from co-gasification of waste tire and pine bark at 800 °C and 900 °C, respectively [18]

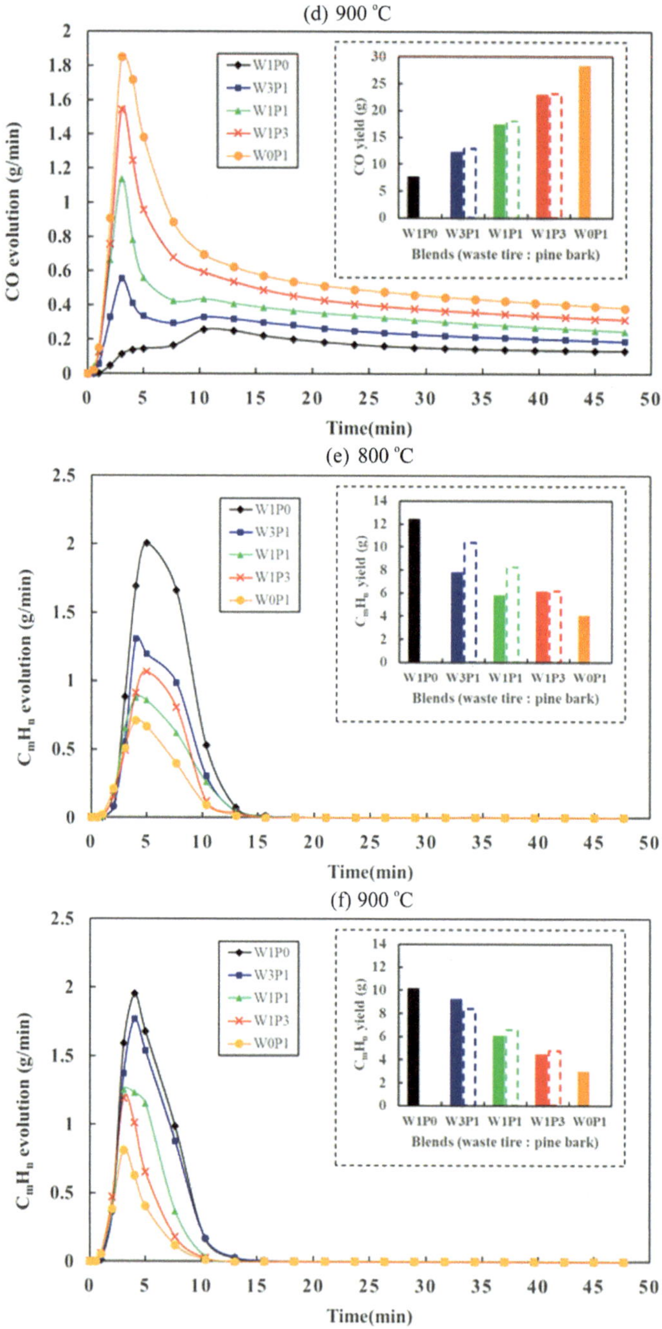

Fig. 14 (continued)

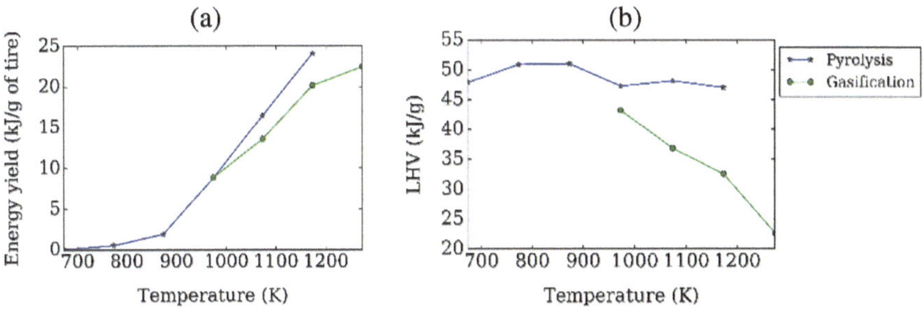

Fig. 15 Effect of temperature on **a** energy yield and **b** LHV of the syngas yield from pyrolysis and CO_2-assisted gasification of waste tire [15]

Fig. 16 Effect of temperature on the overall energy efficiency of pyrolysis and CO_2-assisted gasification of waste tire [15]

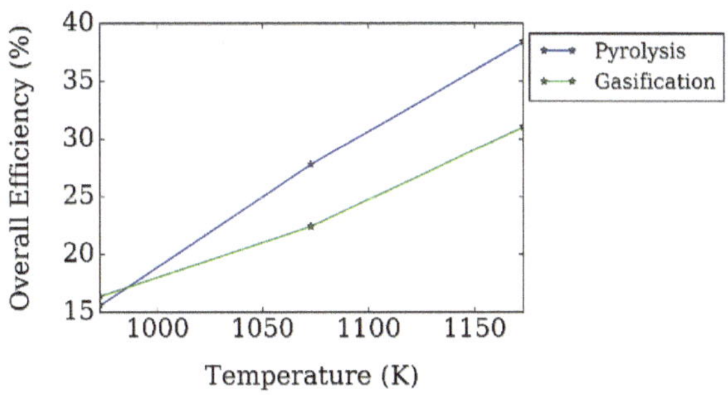

5.2.3 Syngas Energy Recovery and Overall Energy Efficiency

Waste Tires

The energy yield from pyrolysis and gasification in the form of calorific content of syngas was obtained by aggregating the yield of individual syngas components' yield multiplied with their respective lower heating value (LHV). Similarly, LHV of the CO_2-free syngas was also obtained by dividing the respective energy yield with the syngas yield to provide a perspective of the quality of flammable gases that can be obtained. From this energy yield, the overall energy efficiency was also calculated using the Eq. (12) which also considers the electrical energy consumed by the furnaces.

$$\text{Overall energy efficiency} = \frac{m_{\text{syngas}} * \text{LHV}_{\text{syngas}}}{\text{Energy}_{\text{input}} + m_{\text{feedstock}} * \text{LHV}_{\text{feedstock}}} \quad (12)$$

Figure 15 reveals the energy yield and LHV of the syngas, while Fig. 16 reveals the overall energy efficiency of both pyrolysis and CO_2-assisted gasification of waste tire. The energy yield increased with increase in temperature for both gasification and pyrolysis although higher content of hydrocarbons in the syngas from pyrolysis led to higher energy yield from pyrolysis compared to gasification, which was dominated by CO yield due to higher LHV of hydrocarbons compared to CO. So, the energy yield and efficiency reported here are from the perspective of utilizing the products for energy recovery purposes. Pyrolysis yielded overall energy efficiency of almost 38%, while gasification yielded 30% to support the feasibility of waste tire conversion for energy recovery. Although this may seem relatively low, it only accounts for syngas energy and significant portions of energy were also transformed into char residue that offers its own value.

Waste Tires with Pine Bark

To further understand the energy yield and overall energy efficiency, variations obtained from co-pyrolysis and co-gasification of waste tire with pine bark were also examined. Additionally, to obtain accurate representation of the overall energy recovery feasibility, heating value of char residue was also measured and the calorific content of this char was also incorporated into the overall energy efficiency as the product energy output to observe the capability of our lab-scale reactor experiments in establishing the feasibility of this process. Additionally, studies incorporating pine bark also revealed the pathways of modifying energy input and output for better design of conversion reactors.

Figure 17 reveals the overall energy efficiency variation with time along with inset cumulative results of efficiency at different times that included consideration on char energy from co-pyrolysis. The variation with time reveals the efficiency to reach maximum at 10 min while beyond that temperature resulted in a slow decrease in the efficiency. At 15 min, the efficiency reached as high as 45% while considering only syngas energy, and by 47.7 min the efficiency reached to 35%. Note that reactors need to be designed by considering both the extent of conversion and the process efficiency. So, these results will provide with the required understanding to design reactors for high yield and efficiency along with optimal syngas composition. Incorporation of the char energy into efficiency after 47.7 min led to almost constant efficiency of 48% for all the waste tire-pine bark blends examined. This makes the overall process to be feasible and sustainable. Comparison of the energy efficiency from the blends with the weighted results, we can also see a slight non-additive synergistic enhancement in the efficiency when the feedstocks used were blends compared to separate pyrolysis. Such synergistic enhancement could form cumulative enhancement of all the syngas yields to a lesser extent contributing to an overall significance.

From these results, we can see that significant portions of the feedstock were converted into char leading to the limitations of efficiency. While the efficiency is acceptable, the char needs to be either converted for energy purposes or some other utility for better economic output from waste tire conversion. A combination of syngas production with extraction of char including carbon black for utilization in material applications is necessary for sustainable operation of this pathway. Figure 18 shows surface area of the char produced from co-pyrolysis of waste tire and pine bark.

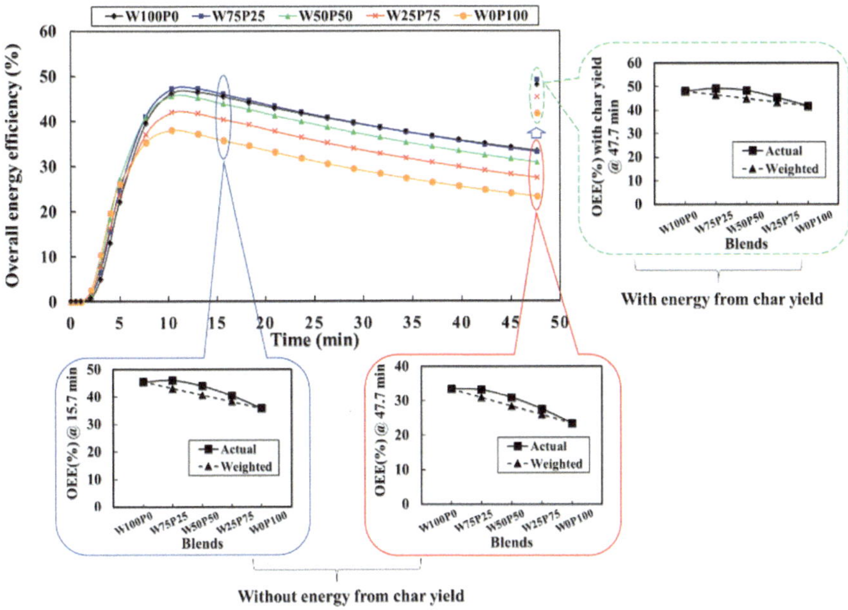

Fig. 17 Effect of feedstock blend fraction on the overall energy efficiency (OEE) from co-pyrolysis of waste tire and pine bark [4]

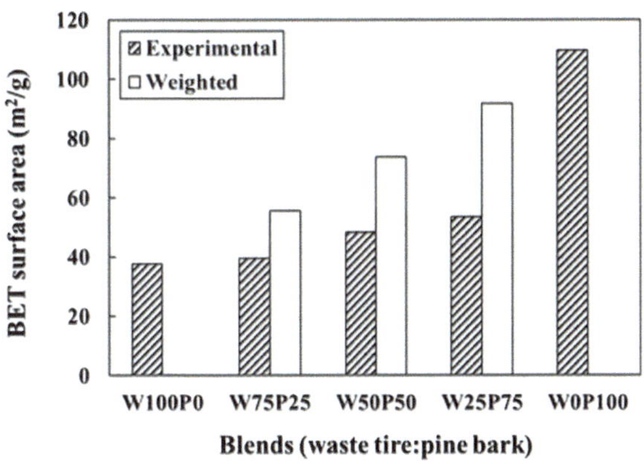

Fig. 18 Effect of feedstock blend fraction on the surface area of char residue from co-pyrolysis of waste tire and pine bark [4]

The results show relatively low-surface area from waste tire compared to pine bark, and their co-pyrolysis led to inhibitive lowered char surface area compared to separate pyrolysis. Figure 19 shows the difference in the morphology of chars from waste char and pine bark and their variation with co-pyrolysis. Further studies are essential to explore future direction toward the applicability of these chars for high-value purposes, such as carbon catalysts, activated carbon and other valuable carbon solid products to make thermochemical conversion of waste tires economically attractive.

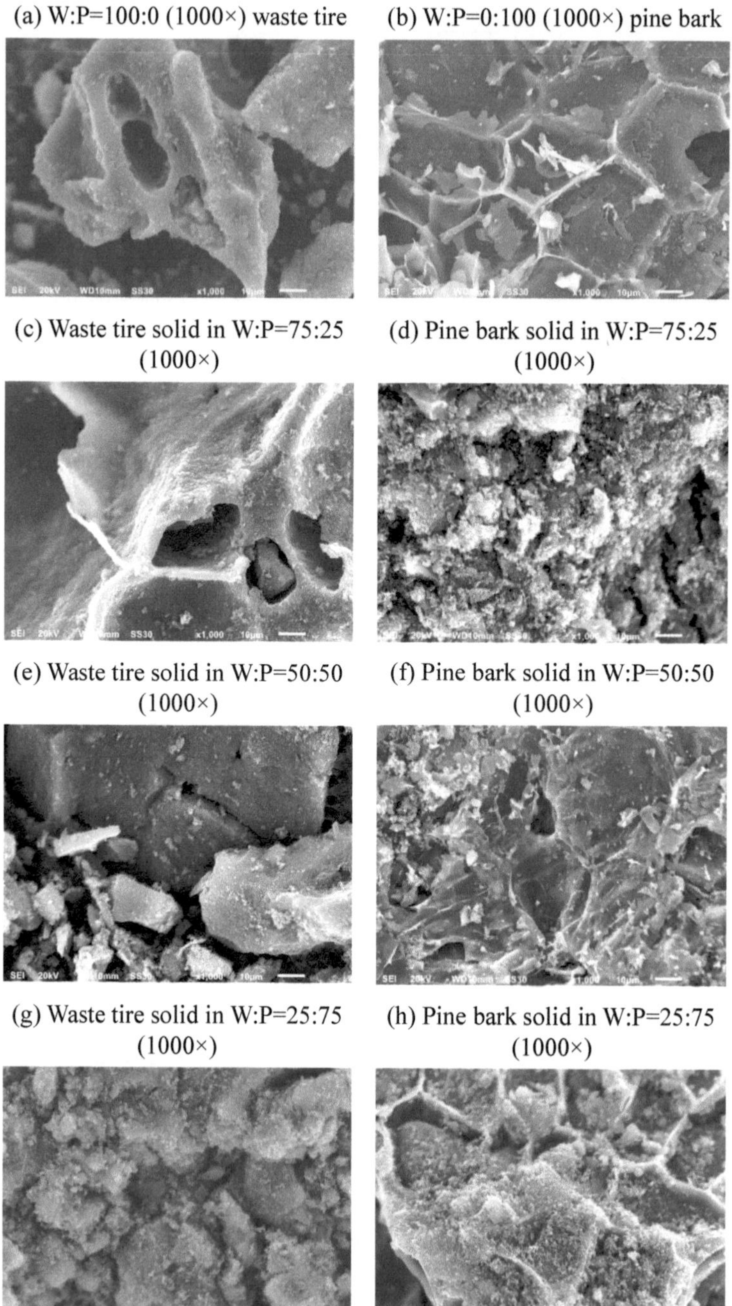

Fig. 19 Morphology of char residue from waste tire and pine bark during their co-pyrolysis at different blend ratios [4]

5.2.4 CO_2 Consumption During CO_2 Gasification

In addition to providing waste tire disposal and energy recovery, CO_2-assisted gasification also provided with positive utility to high-temperature CO_2 utilization and thus reduced net carbon emissions from this disposal process. In our studies, we characterized the capability of CO_2 gasification in CO_2 utilization by examining the net CO_2 consumption during the overall gasification process. This was carried out for both waste tire gasification and its co-gasification with different ratios of pine bark. Figures 20 and 21 reveal the influence of temperature and feedstock composition on the cumulative CO_2 consumption. Increase in temperature led to increase in CO_2 consumption with values as high as 0.7 g of CO_2 consumed per each gram of waste tire at 1000 °C. This consumption rises from the Boudouard reaction and gas-phase CO_2 reforming reactions which enhanced with temperature. Comparison revealed CO_2 consumption during pine bark conversion was higher than waste tire conversion.

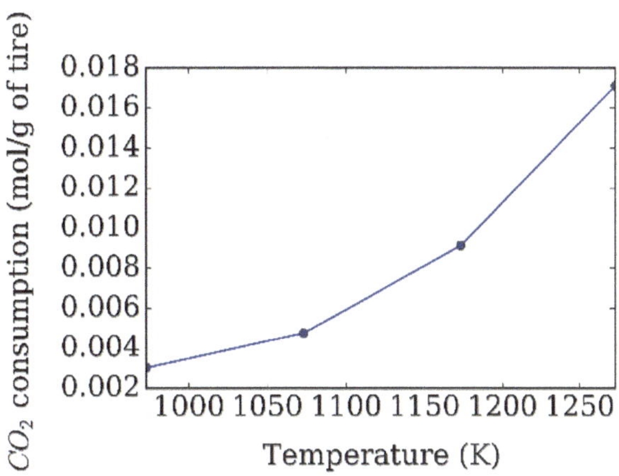

Fig. 20 Effect of temperature on CO_2 consumption during gasification of waste tires [15]

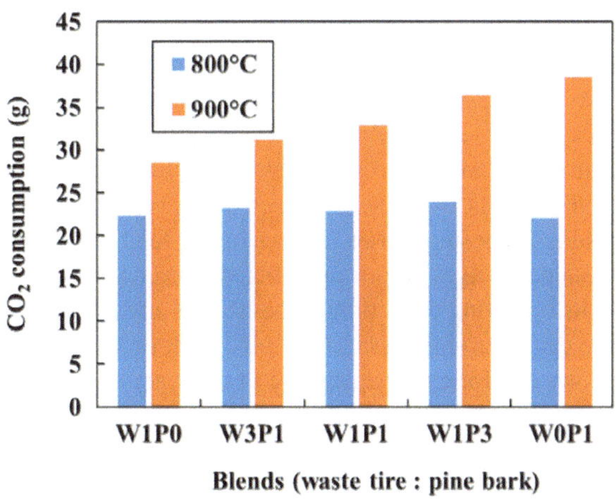

Fig. 21 Effect of feedstock blend ratio on CO_2 consumption during co-gasification of waste tires with pine bark [18]

Additionally, co-gasification also seemed to not provide any non-additive behavior and was similarly superimposable as the syngas yields from co-gasification.

These studies provide important pathways that establish the capability of thermochemical pathways of pyrolysis and CO_2-assisted gasification for efficient, sustainable and economic recovery of waste tires to energy.

6 Conclusions

This chapter provides improved understanding for efficient disposal of real waste tires via thermochemical pathways such as pyrolysis and CO_2-assisted gasification in return for syngas which provides with versatility in utilization and uniformity in handling. Syngas obtained from this pathway is useful for various applications including use as fuel for energy recovery and value-added chemicals production such as methanol, and Fischer–Tropsch transportation fuels. Thermogravimetric analysis revealed the waste tire samples to thermally decompose between temperatures of 120–550 °C and included mass-loss rate peaks of different elastomer content and the additives such as extender oils. Curve-fitting this data with TGA results from straight-chain elastomers of similar nature to those present in the examined waste tire revealed the composition to be dominated by butadiene and styrene–butadiene rubbers, while natural rubber's contribution was only about 5%. Pyrolysis and gasification were carried out using a lab-scale fixed bed reactor to understand the evolution, yield and composition of the product gases extracted. Temperature had significant influence on the syngas yield and composition, where high temperatures led to increased syngas. Endothermic reactions such as thermal cracking of volatile intermediates released from waste tire pyrolysis, CO_2-reforming of the hydrocarbons into H_2 and CO, and Boudouard reaction of CO_2 with char residue were enhanced on both equilibrium and kinetic front to increase the yield and lead to faster release of syngas. Waste tire's char poses a difficulty in gasification as the CO yield from its conversion was low and continued for extended periods of time that concluded low reactivity of its chars. BET surface area value of char from waste tire was only about a half of that from pine bark char. Co-pyrolysis and co-gasification of waste tire and pine bark biomass were also conducted in various proportions of their blends. The results revealed almost no synergistic or inhibitive interaction between waste tire conversion and pine bark conversion, and the results obtained were found to be a weighted superimposition of the results from their respective conversions when conducted separately. These results are a step closer to developing feed-flexible gasifiers that can provide with sustained syngas output and efficiency from various feedstocks with minimal changes to its operating conditions. The lack of any inhibitive interactions means that waste tire can be co-processed with pine bark or vice versa, and the predictable behavior allows one for easier designing of the reactor, while the feedstock can be adjusted to maintain the energy input and other design parameters.

Overall energy efficiency was also calculated for these processes based on the syngas's calorific content, the measured electrical energy requirements for the lab-scale reactors and the calorific content of the feedstock. The results revealed efficiency to improve with increase in temperature that reached as high as 38% in pyrolysis and 30% in CO_2-assisted gasification wherein higher unreformed hydrocarbons in pyrolysis products and their high heating value compared to reformed CO from gasification led to this differences. Incorporating char energy from the measured heating values of the chars to the overall energy efficiency resulted in efficiency to reach almost 50%. Additionally, the co-processing of waste tire with pine bark was able to provide a cumulative synergistic enhancement in efficiency beyond the weighted efficiency that could be achieved by separate conversion of the two components. These results establish the feasibility of pyrolysis and CO_2-assisted gasification of waste tire and its mixtures with biomass as sustainable pathway for their disposal while providing usable syngas.

A major challenge for future studies into these pathways is the waste tire char that was found to be of low reactivity and surface area while being a significant by-product that accounted for almost 30% of the sample mass. Developing pathway for conversion and utilization of this char is essential to not only make these pathways sustainable and efficient in material management, but also to provide economic support. Future research needs to focus on producing high-value carbon products from these char residues, such as carbon catalysts, activated carbon for sorption, electrochemical and energy storage applications to replace expensive materials (such as synthetic graphite and fossil fuel-derived activated carbon), high-quality solid fuel, and carbon dots for biomedical and other imaging applications. These activities will help provide improved value to economically support the thermochemical pathways so that waste tires can be disposed in an eco-friendly fashion while sustainably recovering energy and producing value-added chemicals.

Acknowledgements This work was supported by Office of Naval Research (ONR), and it is gratefully acknowledged. K. G. Burra received Ann. G Wylie fellowship, and this support is gratefully acknowledged.

References

1. Keeling R (2020) Scripps institute of oceanography 2020. https://scrippsco2.ucsd.edu/. Accessed 26 Sep 2020
2. Association U tire rubber. 2017 U. S. scrap tire management summary about the U. S. Tire Manufacturers Association 2018
3. Machin EB, Pedroso DT, de Carvalho JA (2017) Energetic valorization of waste tires. Renew Sustain Energy Rev 68:306–315. https://doi.org/10.1016/j.rser.2016.09.110
4. Wang Z, Burra KG, Zhang M, Li X, Policella M, Lei T et al (2020) Co-pyrolysis of waste tire and pine bark for syngas and char production. Fuel 274:117878. https://doi.org/10.1016/j.fuel.2020.117878

5. Martínez JD, Puy N, Murillo R, García T, Navarro MV, Mastral AM (2013) Waste tyre pyrolysis—a review. Renew Sustain Energy Rev 23:179–213. https://doi.org/10.1016/j.rser.2013.02.038
6. Burra KG, Gupta AK (2018) Thermochemical reforming of wastes to renewable fuels. In: Runchal AK, Gupta AK, Kushari A, De A, Aggarwal SK (eds) Energy propuls. A sustain. Technol. Approach, Springer, Singapore, pp 395–428. https://doi.org/10.1007/978-981-10-7473-8_17
7. Hasan A, Dincer I (2019) Comparative assessment of various gasification fuels with waste tires for hydrogen production. Int J Hydrogen Energy 44:18818–18826. https://doi.org/10.1016/j.ijhydene.2018.11.150
8. González JF, Encinar JM, Canito JL, Rodríguez JJ (2001) Pyrolysis of automobile tyre waste. Influence of operating variables and kinetics study. J Anal Appl Pyrolysis 58–59:667–83. https://doi.org/10.1016/S0165-2370(00)00201-1
9. Colom X, Faliq A, Formela K, Cañavate J (2016) FTIR spectroscopic and thermogravimetric characterization of ground tyre rubber devulcanized by microwave treatment. Polym Test 52:200–208. https://doi.org/10.1016/j.polymertesting.2016.04.020
10. Banar M, Akyıldız V, Özkan A, Çokaygil Z, Onay Ö (2012) Characterization of pyrolytic oil obtained from pyrolysis of TDF (Tire Derived Fuel). Energy Convers Manag 62:22–30. https://doi.org/10.1016/j.enconman.2012.03.019
11. Kar Y (2011) Catalytic pyrolysis of car tire waste using expanded perlite. Waste Manag 31:1772–1782. https://doi.org/10.1016/j.wasman.2011.04.005
12. Cunliffe AM, Williams PT (1998) Composition of oils derived from the batch pyrolysis of tyres. J Anal Appl Pyrolysis 44:131–152. https://doi.org/10.1016/S0165-2370(97)00085-5
13. Karatas H, Olgun H, Engin B, Akgun F (2013) Experimental results of gasification of waste tire with air in a bubbling fluidized bed gasifier. Fuel 105:566–571. https://doi.org/10.1016/j.fuel.2012.08.038
14. Williams PT (2013) Pyrolysis of waste tyres: a review. Waste Manag 33:1714–1728. https://doi.org/10.1016/j.wasman.2013.05.003
15. Policella M, Wang Z, Burra KG, Gupta AK (2019) Characteristics of syngas from pyrolysis and CO_2-assisted gasification of waste tires. Appl Energy 254:113678. https://doi.org/10.1016/j.apenergy.2019.113678
16. Oboirien BO, North BC (2017) A review of waste tyre gasification. J Environ Chem Eng 5:5169–5178. https://doi.org/10.1016/j.jece.2017.09.057
17. Ucar S, Karagoz S, Ozkan AR, Yanik J (2005) Evaluation of two different scrap tires as hydrocarbon source by pyrolysis. Fuel 84:1884–1892. https://doi.org/10.1016/j.fuel.2005.04.002
18. Wang Z, Burra KG, Lei T, Gupta AK (2019) Co-gasification characteristics of waste tire and pine bark mixtures in CO_2 atmosphere. Fuel 257:116025. https://doi.org/10.1016/j.fuel.2019.116025
19. Williams PT, Besler S, Taylor DT (1990) The pyrolysis of scrap automotive tyres. Fuel 69:1474–1482. https://doi.org/10.1016/0016-2361(90)90193-T
20. Portofino S, Donatelli A, Iovane P, Innella C, Civita R, Martino M et al (2013) Steam gasification of waste tyre: influence of process temperature on yield and product composition. Waste Manag 33:672–678. https://doi.org/10.1016/j.wasman.2012.05.041
21. RofiqulIslam M, Haniu H, Rafiqulalambeg M (2008) Liquid fuels and chemicals from pyrolysis of motorcycle tire waste: product yields, compositions and related properties. Fuel 87:3112–3122. https://doi.org/10.1016/j.fuel.2008.04.036
22. Chen DT, Perman CA, Riechert ME, Hoven J (1995) Depolymerization of tire and natural rubber using supercritical fluids. J Hazard Mater 44:53–60. https://doi.org/10.1016/0304-3894(95)00047-X
23. Lah B, Klinar D, Likozar B (2013) Pyrolysis of natural, butadiene, styrene-butadiene rubber and tyre components: modelling kinetics and transport phenomena at different heating rates and formulations. Chem Eng Sci 87:1–13. https://doi.org/10.1016/j.ces.2012.10.003

24. Berrueco C, Esperanza E, Mastral FJ, Ceamanos J, García-Bacaicoa P (2005) Pyrolysis of waste tyres in an atmospheric static-bed batch reactor: analysis of the gases obtained. J Anal Appl Pyrolysis 74:245–253. https://doi.org/10.1016/j.jaap.2004.10.007
25. Xiao G, Ni M-J, Chi Y, Cen K-F (2008) Low-temperature gasification of waste tire in a fluidized bed. Energy Convers Manag 49:2078–2082. https://doi.org/10.1016/j.enconman.2008.02.016
26. Conesa JA, Martín-Gullón I, Font R, Jauhiainen J (2004) Complete study of the pyrolysis and gasification of scrap tires in a pilot plant reactor. Environ Sci Technol 38:3189–3194. https://doi.org/10.1021/es034608u
27. de Marco RI, Laresgoiti M, Cabrero M, Torres A, Chomón M, Caballero B (2001) Pyrolysis of scrap tyres. Fuel Process Technol 72:9–22. https://doi.org/10.1016/S0378-3820(01)00174-6
28. Kim S, Park JK, Chun H-D (1995) Pyrolysis kinetics of scrap tire rubbers. I: Using DTG and TGA. J Environ Eng 121:507–14. https://doi.org/10.1061/(ASCE)0733-9372(1995)121:7(507)
29. Zabaniotou A, Stavropoulos G (2003) Pyrolysis of used automobile tires and residual char utilization. J Anal Appl Pyrolysis 70:711–722. https://doi.org/10.1016/S0165-2370(03)00042-1
30. Evans A, Evans R (2006) The composition of a tyre: typical components. Waste Resour Action Program 2006. http://www.wrap.org.uk/sites/files/wrap/2 - Composition of a Tyre - May 2006.pdf=https://doi.org/10.3233/HAB-1993-4306
31. Aylón E, Fernández-Colino A, Murillo R, Navarro MV, García T, Mastral AM (2010) Valorisation of waste tyre by pyrolysis in a moving bed reactor. Waste Manag 30:1220–1224. https://doi.org/10.1016/j.wasman.2009.10.001
32. Scala F, Chirone R, Salatino P (2003) Fluidized bed combustion of tyre derived fuel. Exp Therm Fluid Sci 27:465–471. https://doi.org/10.1016/S0894-1777(02)00249-2
33. Lee JM, Lee JS, Kim JR, Kim SD (1995) Pyrolysis of waste tires with partial oxidation in a fluidized-bed reactor. Energy 20:969–976. https://doi.org/10.1016/0360-5442(95)00049-M
34. Chang YM (1996) On pyrolysis of waste tire: degradation rate and product yields. Resour Conserv Recycl 17:125–139. https://doi.org/10.1016/0921-3449(96)01059-2
35. Dai X, Yin X, Wu C, Zhang W, Chen Y (2001) Pyrolysis of waste tires in a circulating fluidized-bed reactor. Energy 26:385–399. https://doi.org/10.1016/S0360-5442(01)00003-2
36. Leung DYC, Yin XL, Zhao ZL, Xu BY, Chen Y (2002) Pyrolysis of tire powder: influence of operation variables on the composition and yields of gaseous product. Fuel Process Technol 79:141–155. https://doi.org/10.1016/S0378-3820(02)00109-1
37. Li S-Q, Yao Q, Chi Y, Yan J-H, Cen K-F (2004) Pilot-scale pyrolysis of scrap tires in a continuous rotary kiln reactor. Ind Eng Chem Res 43:5133–5145. https://doi.org/10.1021/ie030115m
38. Roy C, Labrecque B, de Caumia B (1990) Recycling of scrap tires to oil and carbon black by vacuum pyrolysis. Resour Conserv Recycl 4:203–213. https://doi.org/10.1016/0921-3449(90)90002-L
39. Senneca O, Salatino P, Chirone R (2000) A fast heating-rate thermogravimetric study of the pyrolysis of scrap tyres. Fuel Energy Abstr 41:104. https://doi.org/10.1016/S0140-6701(00)90920-2
40. Zhang X, Wang T, Ma L, Chang J (2008) Vacuum pyrolysis of waste tires with basic additives. Waste Manag 28:2301–2310. https://doi.org/10.1016/j.wasman.2007.10.009
41. Murillo R, Aylón E, Navarro M, Callén M, Aranda A, Mastral A (2006) The application of thermal processes to valorise waste tyre. Fuel Process Technol 87:143–147. https://doi.org/10.1016/j.fuproc.2005.07.005
42. Larsen MB, Schultz L, Glarborg P, Skaarup-Jensen L, Dam-Johansen K, Frandsen F et al (2006) Devolatilization characteristics of large particles of tyre rubber under combustion conditions. Fuel 85:1335–1345. https://doi.org/10.1016/j.fuel.2005.12.014
43. López FA, Centeno TA, Alguacil FJ, Lobato B (2011) Distillation of granulated scrap tires in a pilot plant. J Hazard Mater 190:285–292. https://doi.org/10.1016/j.jhazmat.2011.03.039
44. Betancur M, Daniel Martínez J, Murillo R (2009) Production of activated carbon by waste tire thermochemical degradation with CO_2. J Hazard Mater 168:882–887. https://doi.org/10.1016/j.jhazmat.2009.02.167

45. Díez C, Martínez O, Calvo LF, Cara J, Morán A (2004) Pyrolysis of tyres. Influence of the final temperature of the process on emissions and the calorific value of the products recovered. Waste Manag 24:463–9. https://doi.org/10.1016/j.wasman.2003.11.006
46. Aydın H, İlkılıç C (2012) Optimization of fuel production from waste vehicle tires by pyrolysis and resembling to diesel fuel by various desulfurization methods. Fuel 102:605–612. https://doi.org/10.1016/j.fuel.2012.06.067
47. Williams PT, Bottrill RP, Cunliffe AM (1998) Combustion of tyre pyrolysis oil. Process Saf Environ Prot 76:291–301. https://doi.org/10.1205/095758298529650
48. Laresgoiti M, Caballero B, de Marco I, Torres A, Cabrero M, Chomón M (2004) Characterization of the liquid products obtained in tyre pyrolysis. J Anal Appl Pyrolysis 71:917–934. https://doi.org/10.1016/j.jaap.2003.12.003
49. Islam RM, Hossain Joardder MU, Kader MA, Islam Sarker MR (2011) Valorization of solid tire wastes available in Bangladesh by thermal treatment. In: Proceedings of the WasteSafe 2011—2nd international conference on solid waste management in the developing countries, Khulna, Bangladesh, pp 1–9
50. Galvagno S, Casu S, Casabianca T, Calabrese A, Cornacchia G (2002) Pyrolysis process for the treatment of scrap tyres: preliminary experimental results. Waste Manag 22:917–923. https://doi.org/10.1016/S0956-053X(02)00083-1
51. Kaminsky W, Mennerich C, Zhang Z (2009) Feedstock recycling of synthetic and natural rubber by pyrolysis in a fluidized bed. J Anal Appl Pyrolysis 85:334–337. https://doi.org/10.1016/j.jaap.2008.11.012
52. Williams PT, Brindle AJ (2003) Fluidised bed pyrolysis and catalytic pyrolysis of scrap tyres. Environ Technol 24:921–929. https://doi.org/10.1080/09593330309385629
53. Olazar M, Aguado R, Arabiourrutia M, Lopez G, Barona A, Bilbao J (2008) Catalyst effect on the composition of tire pyrolysis products. Energy Fuels 22:2909–2916. https://doi.org/10.1021/ef8002153
54. Lopez G, Olazar M, Aguado R, Elordi G, Amutio M, Artetxe M et al (2010) Vacuum pyrolysis of waste tires by continuously feeding into a conical spouted bed reactor. Ind Eng Chem Res 49:8990–8997. https://doi.org/10.1021/ie1000604
55. Roy C, Chaala A, Darmstadt H (1999) The vacuum pyrolysis of used tires. J Anal Appl Pyrolysis 51:201–221. https://doi.org/10.1016/S0165-2370(99)00017-0
56. Pakdel H, Pantea DM, Roy C (2001) Production of dl-limonene by vacuum pyrolysis of used tires. J Anal Appl Pyrolysis 57:91–107. https://doi.org/10.1016/S0165-2370(00)00136-4
57. Miranda M, Pinto F, Gulyurtlu I, Cabrita I (2013) Pyrolysis of rubber tyre wastes: a kinetic study. Fuel 103:542–552. https://doi.org/10.1016/j.fuel.2012.06.114
58. Murugan S, Ramaswamy MC, Nagarajan G (2008) The use of tyre pyrolysis oil in diesel engines. Waste Manag 28:2743–2749. https://doi.org/10.1016/j.wasman.2008.03.007
59. Harker JH, Backhurst JR (1981) Fuel and energy
60. Wieckert C, Obrist A, Von ZP, Maag G, Steinfeld A (2013) Syngas production by thermochemical gasification of carbonaceous waste materials in a 150 kW th packed-bed solar reactor. Energy Fuels 27:4770–4776. https://doi.org/10.1021/ef4008399
61. Pattabhi Raman K, Walawender WP, Fan LT (1981) Gasification of waste tires in a fluid bed reactor. Conserv Recycl 4:79–88. https://doi.org/10.1016/0361-3658(81)90036-9
62. Donatelli A, Iovane P, Molino A (2010) High energy syngas production by waste tyres steam gasification in a rotary kiln pilot plant. Experimental and numerical investigations. Fuel 89:2721–2728. https://doi.org/10.1016/j.fuel.2010.03.040
63. Zhang Y, Wu C, Nahil MA, Williams P (2015) Pyrolysis-catalytic reforming/gasification of waste tires for production of carbon nanotubes and hydrogen. Energy Fuels 29:3328–3334. https://doi.org/10.1021/acs.energyfuels.5b00408
64. Lerner AS, Bratsev AN, Popov VE, Kuznetsov VA, Ufimtsev AA, Shengel' SV et al (2012) Production of hydrogen-containing gas using the process of steam-plasma gasification of used tires. Glas Phys Chem 38:511–6. https://doi.org/10.1134/S1087659612060041
65. Burra KG, Gupta AK (2018) Synergistic effects in steam gasification of combined biomass and plastic waste mixtures. Appl Energy 211:230–236. https://doi.org/10.1016/j.apenergy.2017.10.130

66. Wu Z, Yang W, Tian X, Yang B (2017) Synergistic effects from co-pyrolysis of low-rank coal and model components of microalgae biomass. doi:https://doi.org/10.1016/j.enconman.2016.12.060
67. Zhang Y, Geng P, Liu R (2018) Synergistic combination of biomass torrefaction and co-gasification: Reactivity studies. Bioresour Technol 245:225–233. https://doi.org/10.1016/j.biortech.2017.08.197
68. Chen W, Shi S, Zhang J, Chen M, Zhou X (2016) Co-pyrolysis of waste newspaper with high-density polyethylene: synergistic effect and oil characterization. Energy Convers Manag 112:41–48. https://doi.org/10.1016/j.enconman.2016.01.005
69. Déparrois N, Singh P, Burra KG, Gupta AK (2019) Syngas production from co-pyrolysis and co-gasification of polystyrene and paper with CO_2. Appl Energy 246:1–10. https://doi.org/10.1016/j.apenergy.2019.04.013
70. Ahmed II, Nipattummakul N, Gupta AK (2011) Characteristics of syngas from co-gasification of polyethylene and woodchips. Appl Energy 88:165–174. https://doi.org/10.1016/j.apenergy.2010.07.007

Biomass Conversion to Power in Southeast Asian Countries: Current Situation and Perspectives

Somrat Kerdsuwan and Krongkaew Laohalidanond

Abstract As fossil fuels are still the main primary energy source used among the Association of Southeast Asian Nation countries (ASEAN), ASEAN member states have the ambition to reduce their dependence on fossil fuel and to encourage the utilization of renewable energy, which can be integrated into the energy mix. Hence, alternative energy efficiency and renewable energy plans have been issued in ASEAN. As almost all countries are agricultural-based economies, there are abundant agricultural and forestry by-products or residues that can be used as lingo-cellulosic biomass feedstock for heat and power generation, e.g., rice husks and straw, cassava rhizome, palm oil shell, fronds and empty fruit bunches, corn stalks and empty cobs, as well as sugarcanes tops, leaves, and bagasse. This lingo-cellulosic biomass can be converted to power through direct combustion and indirect combustion technology. The boilers generally used in the direct combustion of biomass power plants are stoker-fired, fluidized bed and pulverized boilers, and equipped with cyclone, electrostatic precipitator, or bag filters to control emissions. In contrast, indirect combustion using gasification technology, e.g., a downdraft, updraft, or cross-draft gasifier, is used to generate syngas which can run a gas or diesel engine to generate power of less than 1 MW_e. A suitable business model of a power plant is a small power producer (SPP), where the installed power capacities are lower than 90 MW_e. There are many key drivers to achieve renewable energy (RE) goals, e.g., tax incentives and feed-in-tariff power buying schemes. It is to believed that the increase in the portion of RE in ASEAN can contribute to the depletion of fossil fuel and the emission of greenhouse gas, finally to achieve sustainable development goals (SDGs).

Keywords Biomass · Power plant · Renewable energy · ASEAN countries

S. Kerdsuwan (✉) · K. Laohalidanond
Department of Mechanical and Aerospace Engineering, Faculty of Engineering, The Waste Incineration Research Center, Science and Technology Research Institute, King Mongkut's University of Technology North Bangkok, Bangkok, Thailand

A. Gupta et al. (eds.), *Advances in Energy and Combustion*, Green Energy and Technology,
https://doi.org/10.1007/978-981-16-2648-7_22

1 Introduction

According to the increase in the world population, more comfortable living standards, and industrialization and globalization, it is inevitable that a great amount of energy is consumed [1, 2]. There is a strong link between the global population and global energy consumption, as graphically shown in Fig. 1, where it is evident that the growth in the global population leads to an increase in global energy consumption. In 2020, the world population is showing an increase of 46.15% from 1990, and the world energy consumption has risen to up to 84.18% [3, 4]. It is predicted that the global population and global energy consumption will increase in the future [3, 5].

The population growth and the final energy consumption in the ASEAN countries, comprising Brunei Darussalam, Cambodia, Indonesia, the Lao People's Democratic Republic (Lao PDR, Laos), Malaysia, Myanmar, the Philippines, Singapore, Thailand, and Vietnam (Fig. 2) behave as other countries in the world, as mentioned above. ASEAN's population has been estimated to be 667.3 million in 2020, which corresponds to a share of 8.56% of the world population in 2020. Regarding the economical point of view, the GDP of ASEAN member states was 3 trillion US dollars, with a share of 3.5% in the global GDP, in 2018 [6, 7]. With the total area of 4,522,945 m^2, the population density accounts for 147 inhabitants per square kilometer, as listed in Table 1 [3, 8]. The United Nations [8] forecasted that ASEAN's population will continuously increase to 791.98 million by 2050, as shown in Fig. 3. According to economic expansion, urbanization, and industrialization in ASEAN, the final energy consumption has sharply increased over the past few decades, from 178.33 metric tons of oil equivalent (Mtoe) in 1990 to 625.07 Mtoe in 2020, and it is projected based on the business-as-usual model (BAU) to be 1,152.71 Mtoe in 2040 [9], as illustrated in Fig. 4.

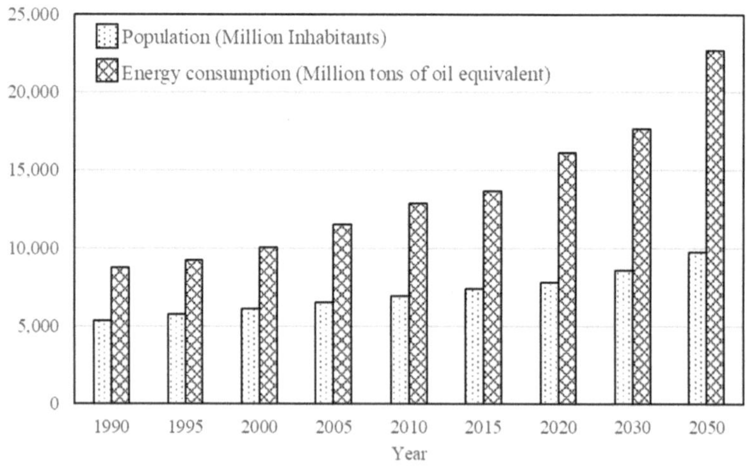

Fig. 1 Relationship between population and energy consumption from 1990 to 2050 [3–5]

Fig. 2 ASEAN members

Table 1 Area, population, and density of each country in ASEAN in 2020 [6, 10]

Country	Area (sq km)	Population (Million, %)	Density (Capita/sq km)	GDP[a] (Billion US, %)
Brunei	5765	0.44 (0.07)	75.80	13.56 (0.45)
Cambodia	181,035	16.72 (2.51)	92.36	24.63 (0.82)
Indonesia	1,904,569	273.52 (40.99)	143.61	1.04 (34.88)
Laos	236,800	7.28 (1.09)	30.74	18.10 (0.61)
Malaysia	330,803	32.37 (4.85)	97.85	358.41 (12.00)
Myanmar	676,578	54.41 (8.15)	80.42	77.26 (2.59)
The Philippines	342,353	109.58 (16.42)	320.08	342.69 (11.48)
Singapore	710	5.85 (0.88)	8,239.44	364.08 (12.19)
Thailand	513,120	69.80 (10.46)	136.03	505.06 (16.91)
Vietnam	331,212	97.34 (14.39)	293.89	241.04 (8.07)
Total	4,522,945	667.31 (100.00)	147.54	2.99 (100.00)

[a]GDP in 2018

The main energy sources in ASEAN have changed from time to time. For example, biomass and waste were a dominant energy source during the 1990s; however, due to the rapid development in terms of industrialization, urbanization, and motorization, the share of fossil fuel, e.g., oil, coal, and natural gas in the ASEAN energy mix increased from 56.4% in 1990 to 75% in 2015, while the share of biomass and waste decreased from 40% in 1990 to 19% in 2015. It is predicted that the share of fossil

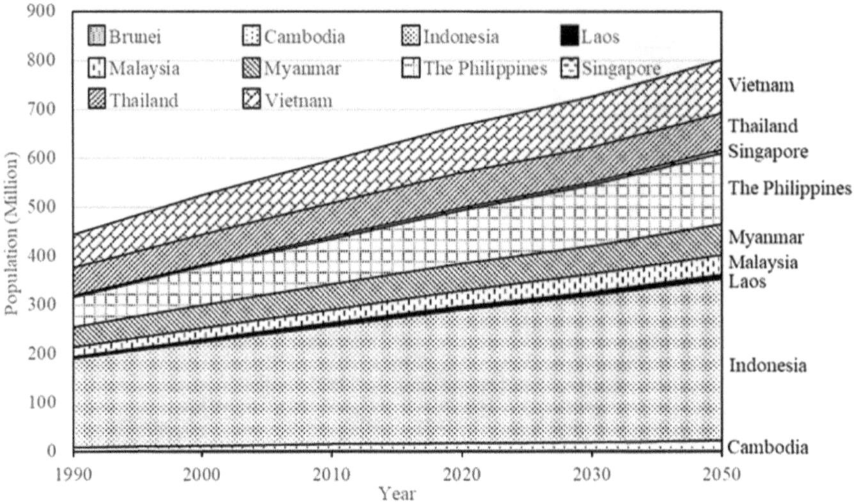

Fig. 3 ASEAN's population and its projection to 2050 [8]

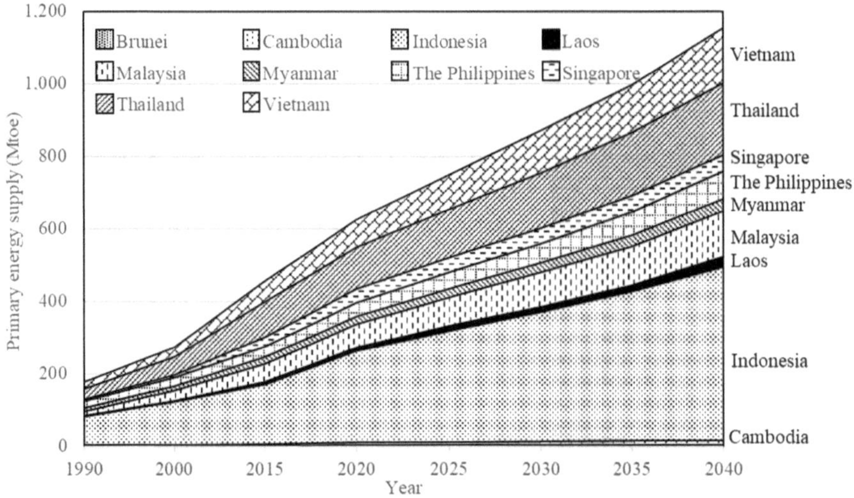

Fig. 4 ASEAN's final energy consumption and its projection to 2040 [9]

fuel in primary energy supply will rise to 80% in 2050 and the share of biomass and waste will decline to 9.2% by 2050 [11]. Figure 5 presents the primary energy supply in ASEAN from 1990 to 2050 [11]. Among the energy sources, coal, oil, and natural gas are the major energy sources in ASEAN.

According to the aforementioned, ASEAN countries will face a challenge in energy security, as they rely on fossil fuel energy sources, which have limited world reserves. In order to tackle this problem, the increase in the share of renewable

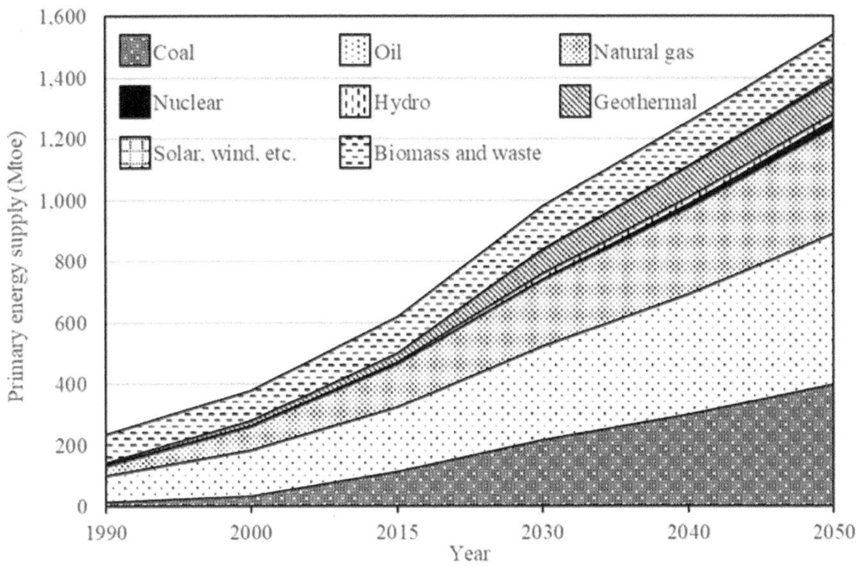

Fig. 5 ASEAN's primary energy supply and its projection to 2040 [11]

energy, e.g., solar, wind, geothermal, hydro, biomass, etc., in the energy mix has been encouraged. This chapter will deal with the using of biomass as a part of renewable energy in ASEAN to reduce the fossil fuel by setting up the goal to achieve. The technologies to convert biomass to power as well as business model used will be discussed, together with the key drivers and perspectives to approach the setting up target.

2 Renewable Energy in ASEAN

Not only the limitation of fossil fuel reserves but their use as the main energy source leads to unfavorable impacts on the environment, particularly regarding greenhouse gas (GHG) emissions, which is the key contributor to the global warming effect. ASEAN has recognized the problem of both energy security and problems with the environment, and hence, the ASEAN plan of action for energy cooperation (APAEC) 2016–2025 has been issued, by which the share of renewable energy of 23% in the power sector, the transportation sector, and the industrial sector should be achieved by 2025 [12]. Although some ASEAN countries produce electricity from renewable energy sources, e.g., the countries along the Mekong River (a trans-boundary river in East Asia and Southeast Asia and runs through China, Myanmar, Laos, Thailand, Cambodia, and Vietnam) have a 40–80% share of hydro energy for electricity production and some countries produce electricity from geothermal, solar PV, and wind; almost all countries have strongly relied on fossil fuel for electricity production, as seen in Fig. 6 [13, 14]. The current energy situation is far behind the target of reaching a 23% share of renewable energy in energy mix by 2025. In order to reach this target, renewable energy has been promoted at the national level though specific key drivers. The renewable energy target compared to the achievement in 2017 and the policy of each country are tabulated in Table 2.

In Table 2, the national policy is differentiated from country to country. The countries on the Mekong riverside promote the utilization of hydropower, while the use of geothermal energy is partly encouraged in Indonesia and the Philippines, where the countries sit on the "ring of fire." Solar, wind, and biomass, including waste, and biogas are mostly fostered in Thailand and Vietnam.

The key drivers for the achievement of renewable energy targets can be categorized according to national policy, fiscal incentives, regulatory instruments, and grid access [16, 17]. National policy is announced by the government in order to reach renewable energy targets and related laws. Fiscal incentives encourage plant developers to invest in renewable energy through the reduction of upfront costs, e.g., income tax exemptions, VAT exemptions, import/export fiscal benefits, accelerated

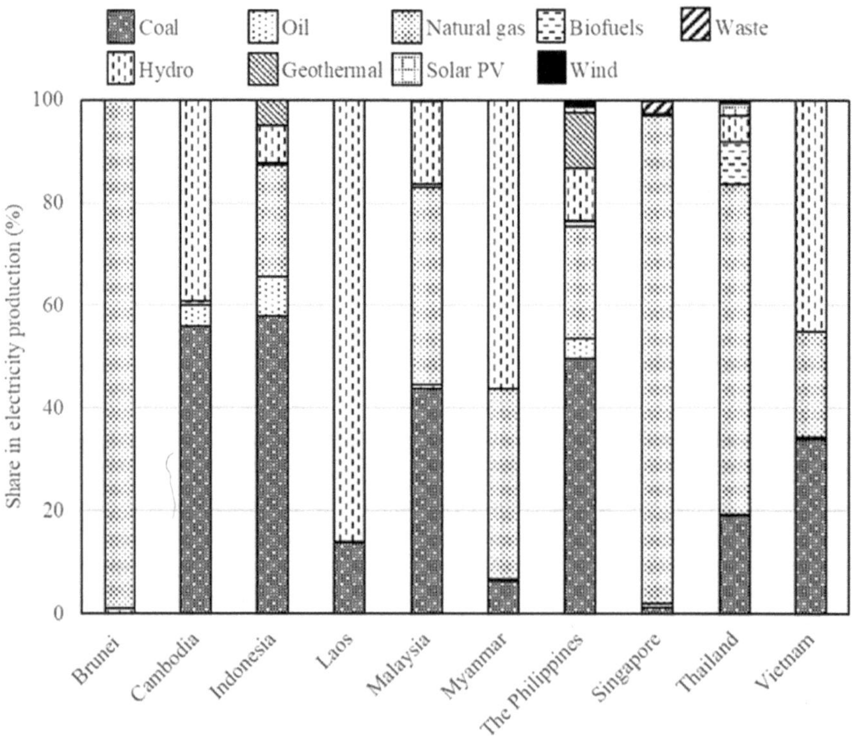

Fig. 6 Share of energy source for electricity production in 2017, exception for Laos, where the information is based on the year 2015 [13, 14]

Table 2 National policy for promoting renewable energy in power generation [15–18]

Country	Target	Current achievement	Key drivers
Brunei	• 10% RE share in electricity production in 2035	0.05%	Need to be developed
Cambodia	• 2.2 GW (80%) hydropower in 2020	n.a.	Tax incentives
Indonesia	• 16.7 GW RE planed new capacity during 2019–2028 (hydro 9.5 GW, geothermal 4.6 GW, solar 0.9 GW, wind 0.9 GW and biomass 0.8 GW • 23% RE in 2025 • 31% RE in 2030	• 9.5 GW RE in 2018 (hydro 5.6 GW, geothermal 1.9 GW, solar 0.1 GW, wind 0.1 GW and biomass 1.8 GW	Feed-in-tariff
Laos	• 0.7 GW RE in 2025 (small hydro 0.4 GW, biomass 0.1 GW, solar 0.1 GW and wind 0.1 GW • 30% RE in total energy consumption in 2025 • >15 MW hydropower in 2015	n.a.	Tax incentives
Malaysia	• 2.1 GW RE in 2020[a] • 20% RE in 2025[a]	4% in 2017	Feed-in-tariff and capital subsidies
Myanmar	• Hydro 8.9 GW, other RE 2 GW (total 46% RE) by 2030–2031	• Hydro 3.3 GW, other RE 0.1 GW in 2018	Need to be developed
The Philippines	• 9.9 GW RE planed new capacity during 2010–2030 (hydro 5.4 GW, geothermal 1.5 GW, solar 0.3 GW, wind 2.3 GW, biomass 0.3 GW, and marine 0.1 GW	• 1.7 GW RE in 2018 (hydro 0.1 GW, geothermal 0.1 GW, solar 0.9 GW, wind 0.4 GW, and biomass 0.2 GW)	Tax incentives feed-in-tariff and capital subsidies
Singapore	• Solar 0.35 GW in 2020 and 1 GW beyond 2020	• Solar 0.2 GW in 2018	Tax incentives and feed-in-tariff
Thailand	• 30% RE in 2036 (solar 6 GW, hydro 3.3 GW, wind 3 GW, biomass 6.78 GW, biogas 0.6 GW) • 20.8 GW RE planed new capacity during 2018–2037[a]	• 10.4 GW RE in 2018	Tax incentives and feed-in-tariff

(continued)

Table 2 (continued)

Country	Target	Current achievement	Key drivers
Vietnam	• 21% RE of 60 GW installed in 2020 • 13% RE of 96 GW in 2025 • 23% RE of 130 GW in 2030 (small hydro 20.2 GW, Solar 4.3 GW, wind 2.7 GW, and biomass 2.7 GW)	• 18.5 GW RE in 2018 (hydro 18 GW, solar 0.1 GW, wind 0.2 GW, and biomass 0.2 GW)	Tax incentives and feed-in-tariff

[a]Excluding large hydro

depreciation, etc. Regulatory instruments are the incentives provided to developers in terms of feed-in-tariffs (FiT), grid access quotas, net metering, certificates, etc. The grid access quota is to ensure that developers obtain access to the power distribution line for renewable-based electricity [16, 17]. Additionally, the facilitation of permission and license procedures can also motivate the investment in renewable energy [16, 17, 19]. Table 2 shows the key drivers for the achievement of renewable energy targets at the nation level.

Although there are many types of renewable energy sources to be executed in the future, only the cost of electricity generation from biomass and hydro is competitive against conventional fuel since energy conversion technologies have fully been developed and commercialized [20]. Besides cost-effective power production, biomass plays an important role in global renewable energy and it is environmental friendly. Since 2000, the biomass energy supply has continuously increased from 42.8 EJ in 2000 to 55.6 EJ in 2017, which corresponds to the share of 69–79% in total renewable energy supply [21], as presented in Fig. 7. Considering the biomass energy supply in Asia, it can be found that the share of bioenergy in renewable energy supply in 2017 accounted for 65% of 33.2 EJ, Fig. 8 [21]. A similar trend of bioenergy supply in total renewable energy supply in ASEAN was reported with its share of 87% [22]. It can be implied that ASEAN owns an abundant amount of bioenergy, both in terms of agricultural residue and forestry residue, since agriculture products are the core driver in economic growth in most ASEAN countries [6, 22], whose agriculture land has a share ranging from 10% in Laos to more than 40% in Thailand and the Philippines [23, 24]. According to the high bioenergy supply and cost-competitive power generation, this chapter discusses biomass energy in ASEAN and its conversion technology, especially regarding power generation.

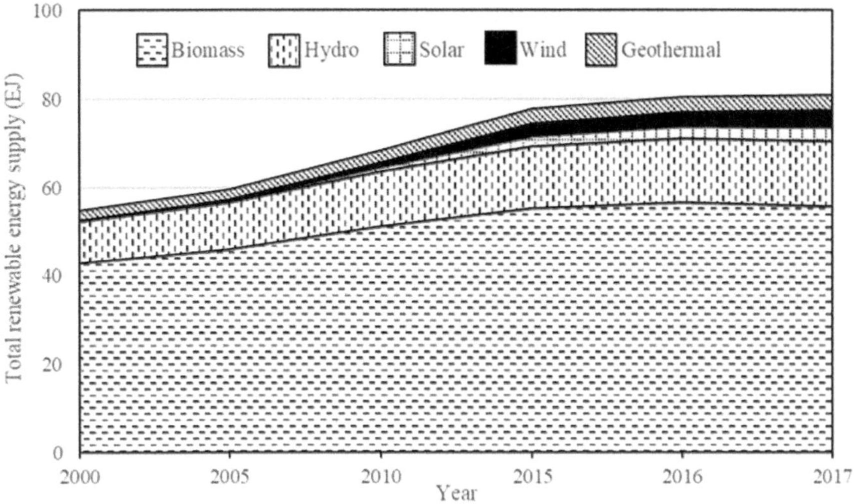

Fig. 7 Global renewable energy supply in EJ [21]

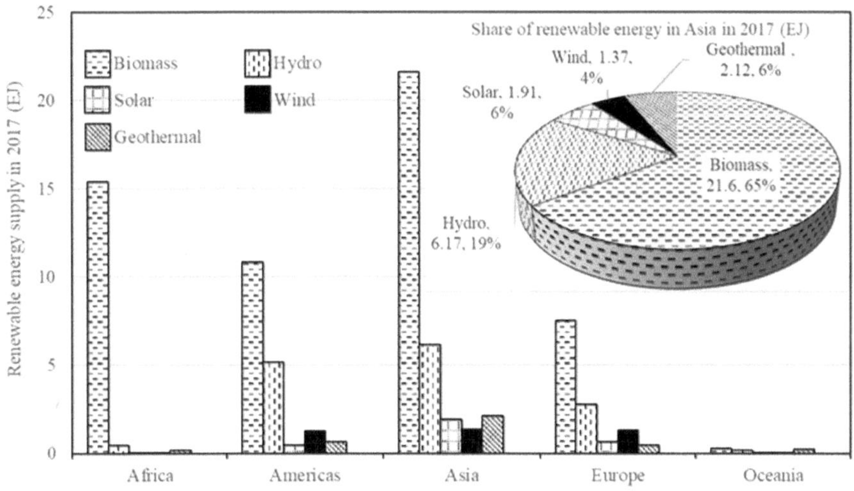

Fig. 8 Renewable energy supply in EJ by continents and share of renewable energy in Asia [21]

3 Biomass as Feedstock for Energy Conversion

3.1 Biomass Residue

Since most ASEAN countries are agro-economic, there are enormous agricultural land and forestry land, as can be noticeably seen in Table 3, for the cultivation of agriculture crops and forestry products.

The top-six ASEAN agricultural products are rice, cassava, palm oil, coconut, corn, and sugarcane, with the annual production capacity of 115.4, 71.2, 60.1, 35.8, 33.7, and 18.5 million tons, respectively. Table 4 shows the average production

Table 3 Land use in each ASEAN country in 2011 (%) [24]

Country	Agricultural land	Forest land	Other land[a]
Brunei	2.5	71.8	23.7
Cambodia	32.1	56.5	11.4
Indonesia	31.2	51.7	17.1
Laos	10.6	67.9	21.5
Malaysia	23.2	62	14.8
Myanmar	19.2	48.2	32.6
The Philippines	41.0	25.9	33.1
Singapore	1.0	3.3	95.7
Thailand	41.2	37.2	21.6
Vietnam	34.8	45.0	20.2

[a]Built-up areas, roads, and other transportation features, barren land, or wasteland

Table 4 Annual amount of agricultural crops in each ASEAN country during 2015–2019 (1000 t) [25–28]

Country	Rice	Cassava	Palm oil	Coconut	Corn	Sugar cane
Brunei	1	3	0	0	0	0
Cambodia	5,445	7,646	0	68	915	223
Indonesia	35,552	16,119	38,300	17,970	11,440	2,095
Laos	1,901	2,279	0	0	1149	214
Malaysia	1,818	53	19,108	512	58	4
Myanmar	12,782	416	0	538	2,228	380
The Philippines	11,612	2,749	88	14,334	7,769	2,217
Singapore	0	0	0	0	0	0
Thailand	18,783	31,509	2577	896	5,145	11,974
Vietnam	27,471	10,441	0	1,495	4,951	1,440
Total	115,365	71,215	60,073	35,813	33,655	18,547

capacity of each agricultural crop by country during 2015–2019. It can be remarkably noticed that Brunei and Singapore have no agricultural products because their economies do not rely on agriculture but on crude oil and natural gas for Brunei and on the business and service sectors for Singapore [24].

As agricultural countries, there are substantial amounts of residue derived from the above-mentioned crops, which will be discussed in greater detail.

Rice straw and rice husks are the residue from rice planting. Rice straw is the stems and leaves left over in rice fields after the harvesting process. Although rice straw can be used as fodder, fertilizer, mushroom growth carriers, etc., the utilization of rice straw in ASEAN can hardly be found, with only 20%, 5%, and 1% of total rice straw in Thailand, the Philippines, and Malaysia [29], respectively. In practical terms, many countries in ASEAN directly return rice straw to the field as a fertilizer after harvesting in order to clear the land for new planting [29, 30], e.g., 29.7–40.2% of rice straw is burnt in the Philippines [31]. Rice husks are the outer rice shell that occurs during the rice milling process. Rice husks have been considered as an energy source for decades [32], since they are easily collected in rice mills and have a uniform and homogeneous shape, which can be conveniently handled. Rice husks have served as solid fuel for power generation in most ASEAN, such as Thailand, Malaysia, and Cambodia [33–35]. Examples of rice straw and rice husks are illustrated in Fig. 9.

Cassava is the second-ranked agricultural crop mostly planted in ASEAN. Its root is the raw material for the starch-derived industry, e.g., human food, animal feed, and chemicals [36]. Additionally, the cassava root is recognized as an energy-containing plant for the production of liquid biofuel via fermentation [36, 37]. The residue from cassava is cassava rhizome (Fig. 10), which is unused and discharged in the field. Some research has focused on the utilization of cassava rhizome for the production of liquid biofuel through the pyrolysis process and solid biofuel through the carbonization or torrefaction process [37, 38]. The use of cassava rhizome as fuel in co-firing power plants can be found in Thailand [39–41].

(a) (b)

Fig. 9 Paddy rice residue (**a**) straw and (**b**) husks

Fig. 10 Cassava rhizome

Palm oil trees are widely cultivated in Indonesia, Malaysia, and Thailand. Palm oil fruit bunches are harvested and delivered to the palm oil industry for crude palm oil production. Not only used as raw material for the food industry, palm oil can be used as raw material in the transesterification process for the production of fatty acid methyl esters, which can be blended in diesel oil and served as bio-diesel for compression ignition engines [42, 43]. The by-products from palm oil processing are trunks, fronds, empty fruit bunches (EFBs), and shells [44, 45]. Palm oil shells are completely used as fuel for steam and power generation in the palm oil manufacturing process [46], while other by-products are fibrous materials, which are generally utilized as raw material for pulp, paper, and particle board production, as well as for energy production [43, 46]. Figure 11 shows the residue from palm oil tree and palm oil mill.

Coconut is a multipurpose plant with a variety of uses, from high-grade food ingredients, cosmetics, and chemicals to furniture and home decor [47, 48]. Coconut fruit is harvested and processed for food production, e.g., coconut milk, coconut oil, and coconut meat, as well as for some chemical manufacturing. Coconut wood is used as tables, chairs, parts of houses, etc. The coconut residue comprises shell, fiber, and frond. Shell and fiber can occur during the manufacturing process and can be used for many purposes, e.g., as aggregate in concrete and road construction material, as agricultural substrate, etc. On the other hand, coconut fronds are left over in the field after harvesting. Therefore, there is an abundant amount of coconut fronds available [49, 50]. A picture of coconut fronds is presented in Fig. 12.

Corn or maize is mostly planted for use as a primary source of animal fodder production and bioethanol production [51]. The residue from corn is the stem, peel, and empty corn cob [51, 52], as shown in Fig. 13. Similar to rice straw, corn stalks, including stems and leaves, are left over in the fields and directly return to the field as a fertilizer for the next planting [31]. Corn peels and empty corn cobs are by-products

Fig. 11 Palm oil residues (a) shells, (b) EFBs, (c) fronds, and (d) trunks

Fig. 12 Coconut fronds

Fig. 13 Corn residue (a) stems and leaves, (b) peels, and (c) empty corn cobs

from the corn seed removal process. Corn stems, peels, and empty corn cobs can be utilized as feedstock for energy production [51].

Sugarcane is a sugar-containing crop that is used in the sugar mill industry. The residue from sugarcane is the tops, leaves, and bagasse (Fig. 14). Like rice straw and corn stems, sugarcane tops and leaves (TLs) are burnt and directly returned to the field [31]. Bagasse is a dry effluent from the sugar mill process after extracting the juice. Bagasse originating in sugar mills is mainly used as fuel in situ for steam and power generation [53, 54].

Table 5 summarizes the agricultural crop residue considered as biomass available highlighted in this chapter. Palm oil shells and bagasse are not included in this chapter, since almost all of them are completely used in situ in palm oil extraction plants and sugar mills.

Forestry products also play a significant role as solid woody biomass, since they can be used as fuel for traditional cooking and heating in the rural areas where electricity cannot be accessed. The main forestry products considered in this chapter

Fig. 14 Sugarcane residue (a) tops and leaves and (b) bagasse

Table 5 Main agricultural residue considered as biomass available in ASEAN

Agricultural crop	Residue
Paddy rice	Rice straw, rice husks
Cassava	Cassava rhizome (CR)
Palm oil	Empty fruit bunches (EFBs), fronds, stems
Coconut	Fronds
Corn	Stems, peels, empty cobs (ECs)
Sugarcane	Tops and leaves (TLs)

(a) (b)

Fig. 15 Forestry biomass (**a**) rubber trees and (**b**) eucalyptus trees

are rubber trees, eucalyptus trees, and white popinac trees [55], as shown in Fig. 15. Apart from agricultural crops and woody biomass, there are some fast-growing, cellulose-rich crops, e.g., cattail and napier grass, which can be used for energy production [55].

3.2 Crop-to-Residual Ratio (CRR)

In order to estimate the amount of available biomass, it is essential to investigate the crop-to-residual ratio (CRR), which is defined as the amount of unused residue per one-unit mass of agricultural crop. The CRR has been investigated, based on a case study of Thailand, by harvesting the agricultural crop from the planting area of 3 × 3 square meters and by weighing the amount of agricultural crop and its residue. Finally, the CRR is calculated based on one kilogram of agricultural crop. Figure 16 shows the example of planting area for the investigation of CRR and its values are listed in Table 6.

The amount of biomass can be estimated using Eq. (1) and the amount of biomass from the top-six agricultural crops in ASEAN is shown in Table 7.

Fig. 16 Example planting area for the investigation of CRR

Table 6 CRR: a case study of Thailand

Agricultural crop	Residue	CRR
Paddy rice	Straw	0.38
	Husks	0.22
Cassava	CR	0.12
Palm oil	EFBs	0.22
	Fronds	0.38
	Stems	0.72
Coconut	Fronds	1.00
Corn	Stems	1.10
	Peels	0.24
	ECs	0.13
Sugarcane	TLs	0.23

Table 7 Estimated annual amount of biomass from the top-six agricultural crops in ASEAN (average from 2015 to 2019) (1000 t)

Agricultural crop	Residue	Amount (1000 t/year)
Paddy rice	Straw	43,838.70
	Husks	25,380.30
Cassava	CR	8,545.80
Palm oil	EFBs	13,216.06
	Fronds	22,827.74
	Stems	43,252.56
Coconut	Fronds	35,813.00
Corn	Stems	37,020.50
	Peels	8,077.20
	ECs	4,375.15
Sugarcane	TLs	4,266.04

Table 8 Annual amount of forestry products in each ASEAN country during 2015–2019 (1000 t) [28]

Country	Wood charcoal	Wood pellets	Wood chips and particles	Wood residue
Brunei	0	0	0	0
Cambodia	37	0	0	0
Indonesia	664	134	1,788	388
Laos	93	0	0	0
Malaysia	29	396	788	1966
Myanmar	332	0	0	0
The Philippines	407	0	0	0
Singapore	5	0	0	0
Thailand	1,466	128	3,142	6,060
Vietnam	414	1,459	13,900	560

$$\text{Amount of biomass} = \text{CRR} \times \text{amount of agricultural product} \tag{1}$$

Considering woody biomass from forestry, the amount of some forestry products by country was statistically recorded by the Food and Agriculture Organization of the United Nations [28]. The average amount of some forestry products that can be utilized as fuel is given in Table 8.

3.3 *Biomass Properties*

The Waste Incineration Research Center (WIRC) has investigated biomass properties in Thailand [56]. Biomass has been collected and ground to the size of less than 1 mm in diameter for fuel property determination. The fuel properties to be investigated in this study are proximate analysis, ultimate analysis, and heating value and density, which are shown in Tables 9, 10 and 11.

Proximate analysis and ultimate analysis as well as heating value were investigated according to ASTM [57]. Bulk density was calculated using Eq. (2).

$$\text{Bulk density} = \text{mass of biomass in container}/\text{volume of container} \tag{2}$$

Table 9 Proximate analysis (% wt. dry basis)

	Residue	Moisture[a]	VM	FC	Ash
Paddy rice	Straw	44.06	67.31	16.23	16.46
	Husks	14.99	63.52	18.28	18.20
Cassava	CR	60.66	73.87	19.04	7.09
Palm oil	EFBs	22.34	80.38	16.70	2.92
	Fronds	56.65	72.04	20.13	7.83
	Stems	65.54	76.36	18.22	5.42
	Shells[b]	12.00	77.50	18.52	3.98
Coconut	Fronds	68.77	73.17	19.09	7.74
Corn	Stems	53.64	72.65	16.62	10.73
	Peels	37.28	75.82	21.17	3.01
	ECs	47.79	77.40	20.23	2.37
Sugarcane	TLs	48.91	73.58	19.86	6.56
	Bagasse[b]	50.73	85.21	11.89	2.90
Rubber timber		22.99	77.57	21.11	1.32
Eucalyptus timber		38.32	76.48	22.71	0.81
White popinac timber		35.57	76.54	21.43	2.03
Cattail		72.39	74.97	16.19	8.84
Napier grass		86.92	64.94	20.82	14.24

[a]As received basis, *VM* Volatile matter, *FC* Fixed carbon
[b]From literature [58]

4 Biomass Conversion Technology

4.1 Technology

Biomass can be converted to heat and power through either the thermo-chemical, biological, or mechanical process. The selection of a suitable process depends on the biomass type and end product required, as shown in Fig. 17.

Since biomass available in ASEAN, listed in Table 7 is lingo-cellulosic material, which mostly contains cellulose, hemicellulose, and lignin [59, 60]. The conversion of lingo-cellulosic material to power is a thermo-chemical process, including carbonization, gasification, and pyrolysis for intermediate fuel production and combustion for heat and power production [22, 59, 61–63].

Rice husks, palm oil shells, and bagasse have successfully been utilized in commercial-scale power plants in ASEAN for decades [64]. According to the experience of commercial-scale rice husk power plants [65], 1 ton of paddy rice can produce approximately 650–700 kg of white rice and 220 kg of rice husks as by-product. White rice is final product for sale as food and rice husks are used as fuel for heat and power generation. With respect to energy consumption, a rice mill consumes

Table 10 Ultimate analysis (% wt. dry basis)

	Residue	C	H	N	S	O	Ash
Paddy rice	Straw	39.34	5.52	1.05	0.05	37.58	16.46
	Husks	39.62	5.25	1.27	0.01	35.65	18.20
Cassava	CR	45.11	5.98	1.38	0.02	40.42	7.09
Palm oil	EFBs	49.29	6.97	0.48	0.00	40.34	2.92
	Fronds	47.42	5.85	0.18	0.00	38.72	7.83
	Stems	45.77	5.80	0.21	0.24	42.56	5.42
	Shells[a]	50.57	5.70	0.32	0.02	39.50	3.98
Coconut	Fronds	43.75	6.05	0.00	0.00	42.46	7.74
Corn	Stems	42.86	5.85	1.56	0.05	38.95	10.73
	Peels	46.02	6.23	0.99	0.01	43.74	3.01
	ECs	47.82	6.27	1.13	0.01	42.40	2.37
Sugarcane	TLs	45.77	6.06	0.83	0.08	40.70	6.56
	Bagasse[a]	43.33	6.16	0.24	0.06	47.31	2.90
Rubber timber		49.64	6.22	0.11	0.00	42.71	1.32
Eucalyptus timber		50.35	6.34	0.40	0.00	42.10	0.81
White popinac timber		49.31	6.12	0.46	0.02	42.06	2.03
Cattail		45.78	5.96	0.63	0.14	38.65	8.84
Napier grass		42.64	4.46	0.78	0.03	37.85	14.24

[a]From literature [58]

30–60 kWh/t of paddy rice for the milling and drying process, where rice husks of 220 kg can produce approximately 90–125 kWh of electricity, as shown in Fig. 18.

Figure 19 illustrates the amount of palm oil and residue from the palm oil industry, as well as the energy required for palm oil production. Approximately, 140–200 kg of palm oil can be extracted from one ton of fresh fruit bunches. The by-products from the processing are 600–700 kg of palm oil mill effluent, which can be fermented to produce 20 cubic meters of biogas, 190 kg of fiber and shells, as well as 230 kg of empty fruit bunches, which can be used as fuel to generate approximately 120 kWh of electricity.

Likewise, bagasse as a by-product from sugar mills has proven to be capable of being used as feedstock for commercial power plants. One ton of sugarcane can generate approximately 100–121 kg of sugar and 290 kg of bagasse, which can be utilized as fuel for the production of 100 kWh of electricity, while a sugar mill plant needs 25–30 kWh of electricity and 0.4 t of steam for the production process, as presented in Fig. 20.

The biomass power plant technology can be categorized according to direct combustion and indirect combustion (gasification), depending on the plant scale. A large-scale power plant with a power generation capacity greater than 1 MW_e uses direct combustion technology, while indirect combustion technology (gasification)

Table 11 Bulk density and heating value

	Residue	Bulk density (kg/m^3)	HHV (kcal/kg)	LHV (kcal/kg)
Paddy rice	Straw	63.97	3,696.36	3,395.46
	Husks	90.00	3,696.67	3,393.33
Cassava	CR	193.00	4,090.00	3,764.17
Palm oil	EFBs	80.00	4,910.00	4,550.00
	Fronds	213.90	4,250.00	3,940.00
	Stems	530.00	4,120.00	3,810.00
	Shells[a]	400.00	4,572.51	4,942.37
Coconut	Fronds	288.40	3,900.00	3,570.00
Corn	Stems	46.25	3,975.00	3,647.50
	Peels	20.13	4,146.00	3,776.00
	ECs	120.32	4,315.00	3,975.00
Sugarcane	TLs	295.00	4,209.29	3,879.29
	Bagasse[a]	120.00	3,560.56	4,466.64
Rubber timber		306.37	4,390.00	4,060.00
Eucalyptus timber		357.00	4,460.00	4,113.33
White popinac timber		253.00	4,383.33	4,053.33
Cattail		169.00	4,280.00	3,970.00
Napier grass		166.00	4,240.00	3,440.00

[a]From literature [58]

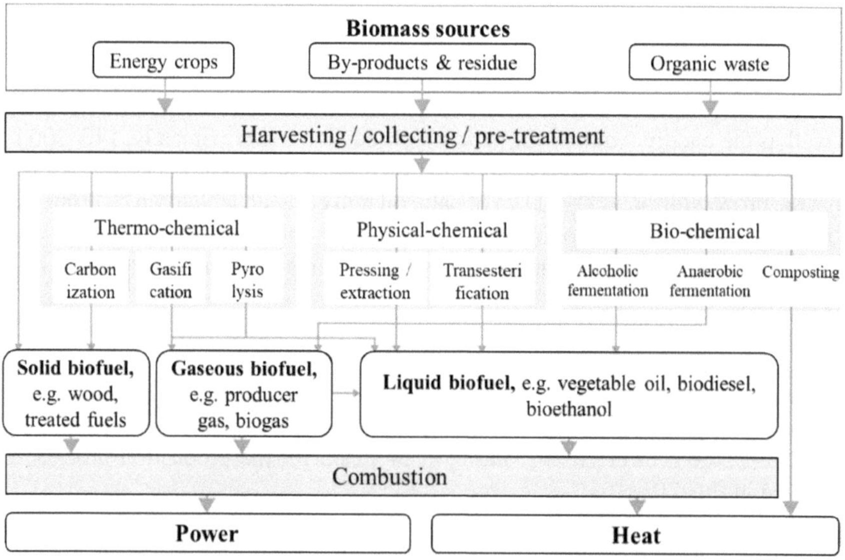

Fig. 17 Biomass conversion pathways for power production. Adapted from [59]

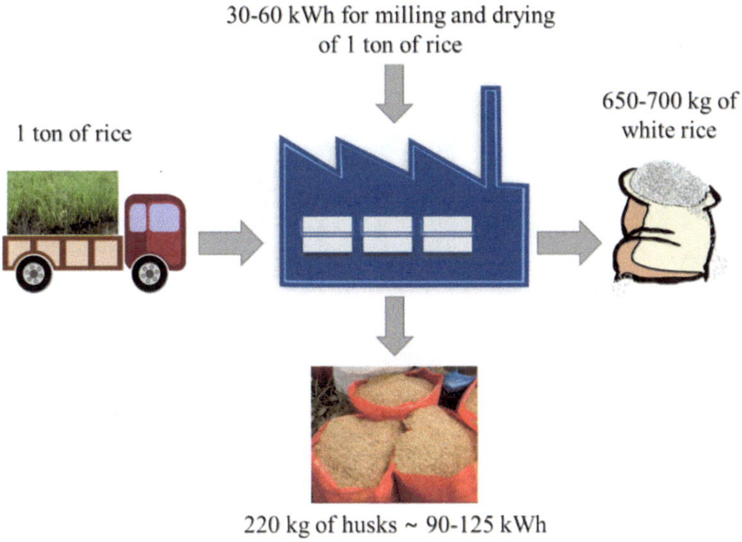

Fig. 18 Power production capacity from rice industry. Adapted from [65]

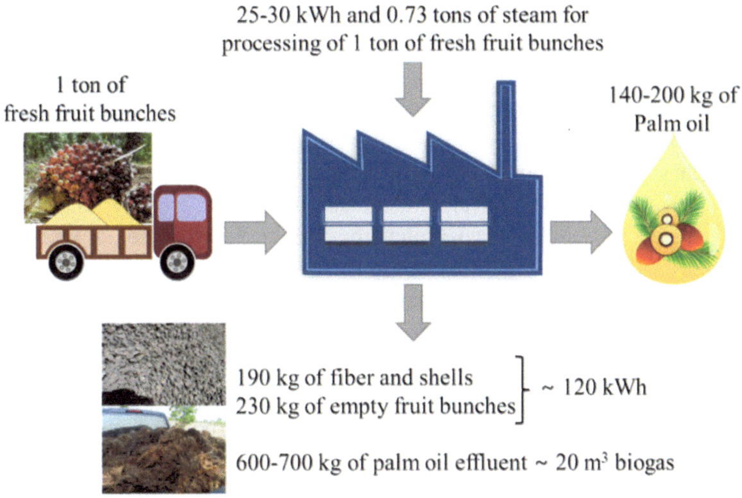

Fig. 19 Power production capacity from palm oil industry. Adapted from [65]

is suitable for a small-scale biomass power plant with the capacity of less than 1 MW of electricity.

In direct combustion technology, biomass is completely burnt with an excess amount of air to generate heat, which can further be used for steam generation in the boiler. The steam from the boiler drives the steam turbine, coupled with the generator, to produce electricity. The flue gas from the process must be cleaned for the removal of

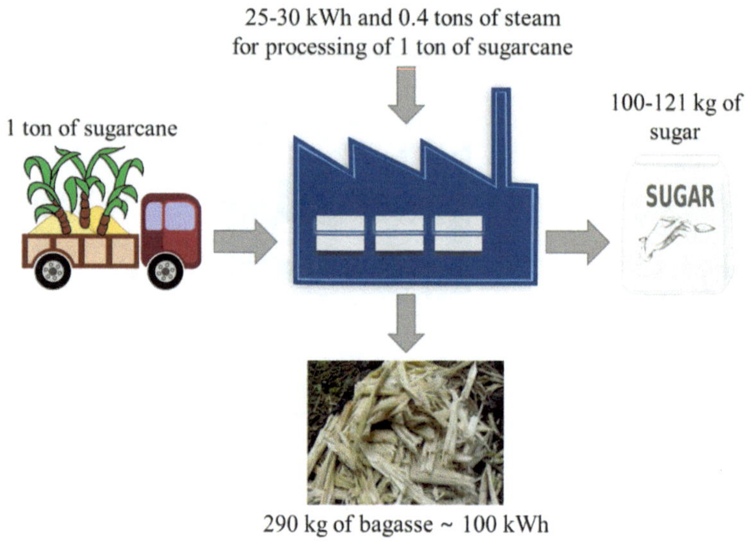

Fig. 20 Power production capacity from sugar industry. Adapted from [65]

particulates, oxide of nitrogen, and oxide of sulfur using an electrostatic precipitator (ESP) or bag filter in order to comply with emission regulations, which are required to control only the particulates, SO_2 and NO_x [66, 67]. Boilers generally used in biomass power plants are stoker-fired boilers (<50 MW of electricity), fluidized bed boilers (30–300 MW of electricity), and pulverized boilers (>600 MW of electricity) [68, 69]. The stoker-fired boiler is a conventional boiler for biomass and is still in use today because it is reliable, convenient, and suitable for different types of feedstock [68]. Compared to a stoker-fired boiler, the fluidized bed boiler provides higher combustion efficiency because the heat-carrier medium, e.g., sand, inside the boiler contributes to the continuous heat transfer; however, a fluidized bed boiler is suitable for small-size and homogenous feedstock, and therefore the fuel preparation process is required [68, 69]. The pulverized boiler is used in practical terms for coal-fired power plants. When using biomass as feedstock in a pulverized boiler, the biomass must be dried and finely ground [69]. Figure 21 is a schematic diagram of direct combustion technology.

In contrast to the direct combustion process, biomass is incompletely burnt in the sub-stoichiometric condition (inadequate amount of air) in the gasifier to produce

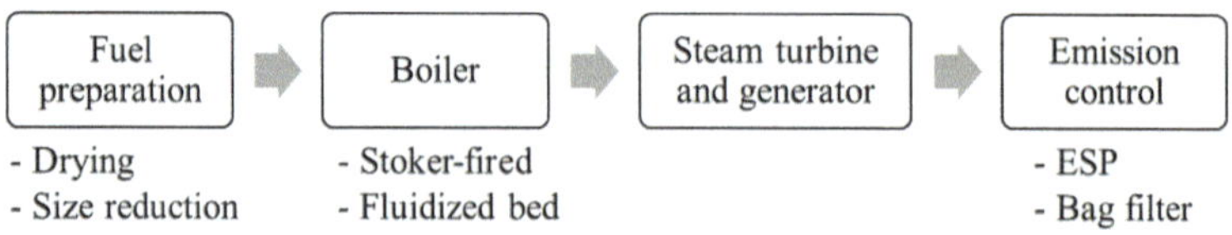

Fig. 21 Direct combustion technology

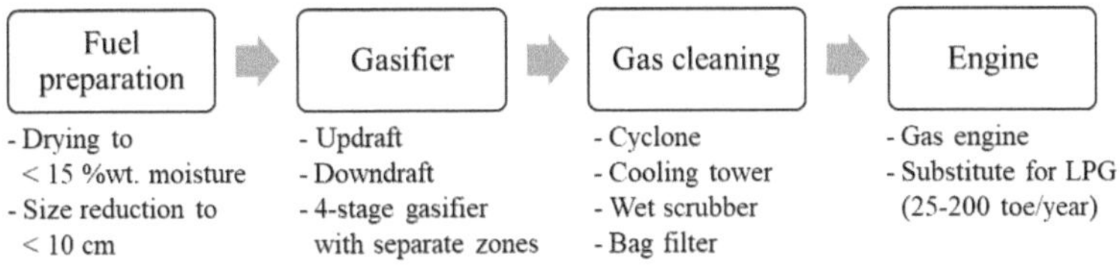

Fig. 22 Indirect combustion (gasification) technology

combustible gas (syngas) in the indirect combustion technology. The syngas must be cleaned to meet the requirements of the engine before being used as fuel in an internal combustion engine connected with a generator for electricity production. There are normally three types of gasifiers used for power production from biomass: an updraft gasifier, a downdraft gasifier, and a four-stage gasifier with a separate reaction zone [70]. Compared to the updraft gasifier, the downdraft gasifier produces lower tar- and particulate-containing syngas [58, 70]. Both the downdraft gasifier and the updraft gasifier can more easily be operated and have lower investment costs than the four-stage gasifier, in which four reaction zones—drying, pyrolysis, oxidation, and reduction—take place in separate reactors. Resulting from the separation of the reaction zones in different reactors, tar and volatile matter are oxidized in the gas phase without contacting the solid material, and consequently, the tar can efficiently be decreased [70]. The syngas cleaning process includes cyclone, cooling tower, wet scrubber, and bag filter, as illustrated in Fig. 22.

4.2 *Examples of Direct Combustion Technology*

An example of a direct combustion biomass power plant is illustrated in Fig. 23. This biomass power plant is located in Petchaboon province, northern Thailand, and requires 954,025 t/year of bagasse as feedstock in a stoker-fired boiler to produce 27 MW of electricity. 15 and 4 MW_e thereof are used in a sugar mill and in a power plant, respectively, and the remaining 8 MW_e is connected to a power distribution grid and sold to the Provincial Electricity Authority (PEA). Particulates and any pollutants are removed from the flue gas by an electrostatic precipitator before being discharged to the atmosphere [71].

The next example is a biomass power plant that uses multi-fuel, including rice husks and wood chips as feedstock. This biomass power plant is located in Nakhon Ratchasima province in northeastern Thailand and purposes to produce 55 MW of electricity (as designed). Rice husks are used as the main feedstock with the amount of 310,100 t/year and wood chips serve as supplementary fuel with the amount of 47,800 t annually. Feedstock is introduced into the circulating fluidized bed (CFB) boiler using sand as a heat carrier in order to produce 45 MW of electricity, 25 MW_e of which is distributed in a grid and sold to the Electricity Generating Authority of Thailand (EGAT), 15 MW_e of which is delivered to a rice mill and the rest, 5 MW_e,

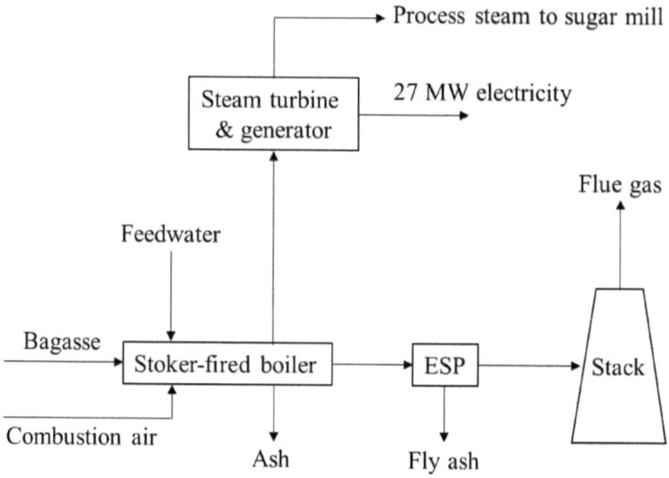

Fig. 23 Direct combustion bagasse power plant. Adapted from [71]

is used in a power plant. The flue gas from the boiler is cleaned in 2 steps: a hot cyclone separator and ESP. A hot cyclone separator is connected next to the boiler, where large particulates are trapped and re-circulated in the boiler for re-combustion. ESP is used to remove fine particulates and any pollutants from the flue gas before the flue gas is released through the stack. Figure 24 is a process flow diagram of a multi-fuel power plant [71].

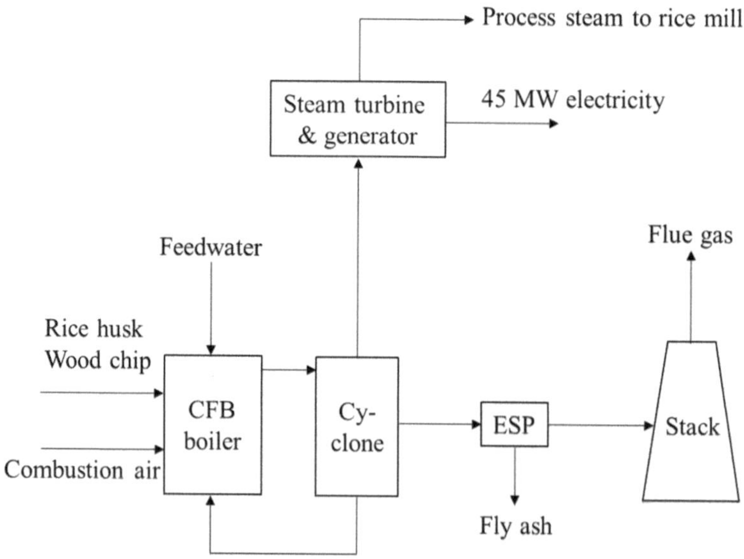

Fig. 24 Rice husk and wood chip in a multi-fuel power plant. Adapted from [71]

4.3 Examples of Indirect Combustion Technology

Unlike direct combustion technology, the indirect combustion (gasification) of biomass has not been widely used on a commercial scale yet. Twelve pilot biomass gasification plants for heat and power production in Thailand have been introduced to demonstrate the development and the feasibility of this technology. Among the 12 gasification plants, 7 plants use a downdraft gasifier, 4 plants apply an updraft gasifier, and only 1 plant uses 4-stage gasification: drying, pyrolysis, oxidation, and reduction, with separate reactors. Regarding the biomass feedstock, forestry residue, e.g., wood chips and wood waste, is used as feedstock in 7 of the 12 plants, 4 of them utilize multi-fuel from forestry residue and agricultural residue, and only one plant uses empty corn cobs as single feedstock. The end-use of 8 pilot plants is heat as a substitute to liquefied petroleum gas (LPG) for application in the production process and only 4 pilot plants aim to produce electricity with the small capacity of 250–350 kW_e [70]. The details of the 12 biomass gasification pilot plants are given in Table 12.

5 Biomass for Heat and Power in ASEAN: The Perspective

As already discussed, nowadays fossil fuel, e.g., coal, oil, and natural gas, is still the main primary energy source in ASEAN. In order to reduce the dependence on fossil fuel and to encourage the utilization of renewable energy, alternative energy efficiency and renewable energy plans have been issued in ASEAN member states. In Sect. 1, the final energy consumption was presented in a business-as-usual (BUA) model. In this section, the final energy consumption is shown based on the alternative-policy-scenario (APS) model, by which the total energy consumption is predicted using available alternative energy efficiency and renewable energy plans as an assumption. The final energy consumption BAU model in 2025, 2030, 2035, and 2040 is compared to APS, as illustrated in Fig. 25. It can be seen that the final energy consumption based on the APS model can decrease by 12–15% compared to the BAU model.

Considering the final energy consumption by source, the share of renewable energy in final energy consumption will increase from 11.6% for the BAU model to 13.8% for the APS model by 2040. This means that after implementing alternative energy efficiency and renewable energy plans, the utilization of renewable energy is predicted to increase, as shown in Fig. 26.

As almost all countries in ASEAN, excluding Brunei and Singapore, are agricultural-based economies, there are abundant agricultural and forestry by-products or residue that can be used as lingo-cellulosic biomass feedstock for power generation, e.g., rice husks and straw, cassava rhizome, palm oil shells, fronds and empty fruit bunches, corn stalks and empty cobs, as well as sugarcane tops, leaves, and bagasse. This lingo-cellulosic biomass can be converted to power through direct combustion and indirect combustion technology. The installed capacity and projected

Table 12 Twelve biomass gasification pilot plants in Thailand [70]

Feedstock	Preparation	Gasification	Gas cleaning	Application
Mixed wood chips	Size reduction	Downdraft with air/steam	Cyclone Cooling tower Venturi scrubber Bio packed bed filter	250 kW_e power production by gas engine
Empty corn cobs	Size reduction	Downdraft with air	Cyclone Spray tower Wet scrubber Fabric filter	350 kW_e power production by gas engine
Wood chips	Size reduction	4-stage gasification: drying, pyrolysis, oxidation, and reduction	Hot gas filtration Gas cooler Scrubber	250 kW_e power production by combined heat and power
Eucalyptus wood	Size reduction	Downdraft with air	Cyclone Filter	300 kW_e power production by gas engine
Wood waste from furniture	Size reduction	Downdraft with air	Cyclone Scrubber, filter	Heat used in hotel (25 toe/year)
Wood waste	Size reduction	Downdraft gasifier with air Using plasma torch	Cyclone Wet scrubber	Heat for fertilizer drying (84 toe/year)
Wood chips	Size reduction	Updraft gasifier with air	Gas cooler Cyclone Fabric filter	Heat for palm oil industry (60 toe/year)
Wood chips, empty corn cobs, bagasse, straw	Size reduction	Updraft gasifier with air	Gas cooler Cyclone Fabric filter	Heat for rubber drying (60 toe/year)
Palm empty fruit bunches, husks, chips	Size reduction	Updraft gasifier with air	Gas cooler Cyclone Fabric filter	Heat for drying palm fruit (132 toe/year)
Wood waste from furniture	Size reduction	Downdraft gasifier with air Double throat	Cyclone Filter	Heat for ceramic process (205 toe/year)
Corncob, wood chips	Size reduction	Cross-draft and updraft with air		Heat for corn mill industry (115 toe/year)
Empty corn cobs, shells, coconut fiber	Size reduction	Downdraft with air	Cooling tower Scrubber, filter	Heat for steam generation (192 toe/year)

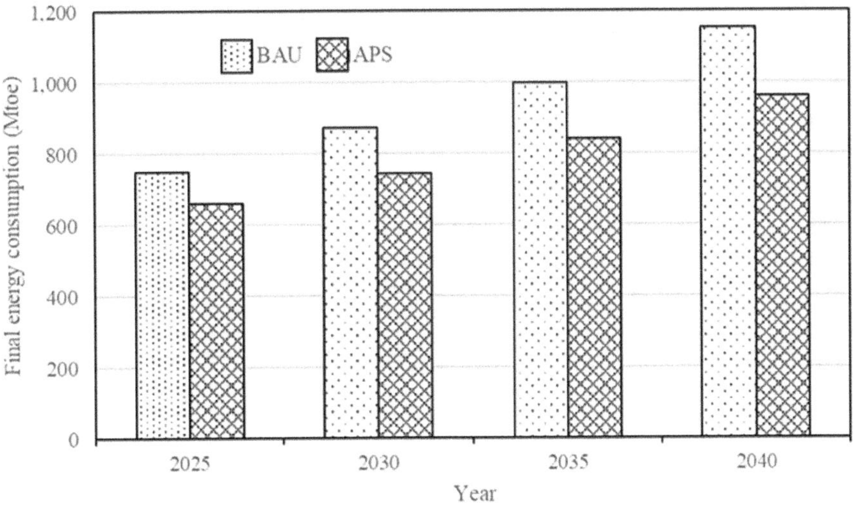

Fig. 25 Final energy consumption BAU and APS. Adapted from [9]

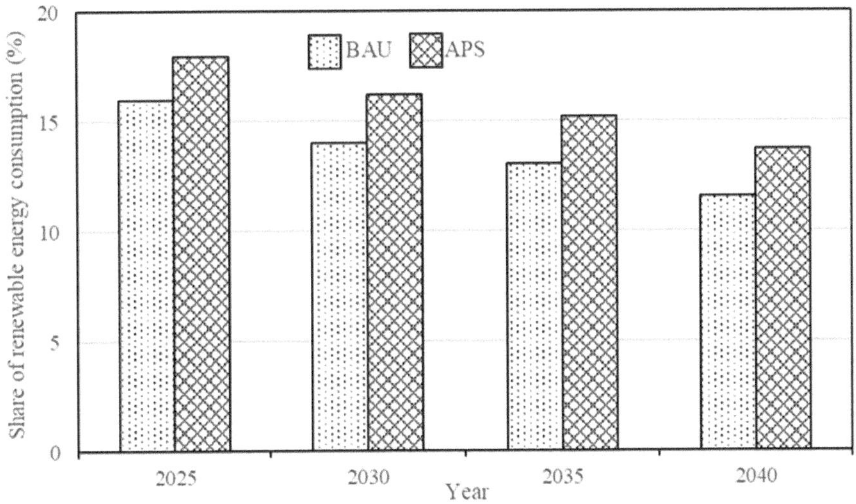

Fig. 26 Share of renewable energy consumption BAU and APS. Adapted from [9]

target of power generation from biomass according to country policy are listed in Table 13.

Brunei and Singapore have no biomass in the country, and therefore there is no target on biomass power plant. However, other countries have the ambitious target to replace their energy from fossil fuel with biomass, especially Thailand, Indonesia, Malaysia, and the Philippines. Almost of biomass power plants use residue

Table 13 Installed capacity and projected target of power generation from biomass according to the country's policy [9, 17, 22, 71–78]

Country	Installed capacity	Projected target
Brunei	0	No target
Cambodia	23 MW	74 MW by 2040
Indonesia	1856 MW in 2018	5500 MW by 2025
Laos	39.7 MW in 2016	58 MW by 2025
Malaysia	330 MW in 2015	800 MW by 2020 1340 MW by 2030
Myanmar	Very few	147 MW in 2035 17% share in electricity production in 2050
The Philippines	220 MW in 2015	192.31 MW additional capacity in 2018 313.42 MW additional capacity in 2019-2022 500 MW additional capacity in 2023-2040
Singapore	0	No target
Thailand	2451.82 MW in 2014	5570 MW in 2036
Vietnam		1% share in electricity production in 2020 ~1.2% share in electricity production in 2025 ~2.1% share in electricity production in 2030

from the production line as feedstock in order to produce heat and power for in-house utilization, and the excess power is sold to the national grid line. These are considered as small power producers (SPP), where the installed capacities are lower than 90 MW_e. With key drivers such as tax incentives and the feed-in-tariff power buying scheme, it is believed that ASEAN member states can achieve the target of power generation from biomass in order to increase the portion of RE in ASEAN, as well as to reduce the utilization of fossil fuel and the emission of greenhouse gas to achieve the sustainable development goals (SDGs) as suggested by the United Nations.

References

1. Ram A, Saurabh T, Ashwani KA, Pankaj K (2019) Population-urbanization-energy Nexus: a review. Resources 136(8)
2. Yoshihisa S (2007) Energy consumption: an environmental problem. IEEJ Trans 1:12–16
3. Worldometer (2020) World population by year. https://www.worldometers.info/world-population/world-population-by-year/
4. Enerdata (2020) Global energy statistical yearbook 2019. https://yearbook.enerdata.net/
5. US Energy Information Administration (2020) EIA projects nearly 50% increase in world energy usage by 2050, led by growth in Asia. https://www.eia.gov/todayinenergy/detail.php?id=41433
6. ASEAN Secretariat (2019) ASEAN Key Figures 2019
7. The World Bank GDP (Current US). https://data.worldbank.org/indicator/NY.GDP.MKTP.CD

8. United Nation (2020) World population prospects 2019. https://population.un.org/wpp/Download/Standard/CSV/
9. ERIA Energy outlook and energy saving potential in East Asia 2019
10. ASEANStatsDataPortal Gross domestic product (GDP) in ASEAN, at current prices (nominal), in US dollars (Annually; Quarterly). https://data.aseanstats.org/
11. Endang JMS, Tan CS, Goh CH (2019) Energy security in ASEAN region: its challenges. In: IOP conference series: earth and environment science, vol 268
12. Visal V, Beni S, Alysius DP, Nadhilah S (2020) A review of renewable energy development and its policy under nationally determined contrinutions in ASEAN. Int J Smart Grid Clean Energy 9(1):149–161
13. International Energy Agency (2020) Data and statistics, data table. https://www.iea.org/data-and-statistics/data-tables?country=VIETNAM&energy=Electricity&year=2017
14. ERIA (2018) Lao PDR energy statistics
15. Renewable Energy Institutes (2019) Renewable energy to replace coal power in Southeast Asia, Pragmatism to deliver a sustainable bright future
16. IRENA (2018) Renewable energy market analysis, Southeast Asia
17. ERIA (2018) Distributed energy system in Southeast Asia
18. Ministry of Energy (2020) Thailand power development plan 2015–1036 (PDP 2015). https://www.egat.co.th/en/images/about-egat/PDP2015_Eng.pdf
19. ASEAN Center for Energy (2016) ASEAN renewable energy policies
20. ASEAN Center for Energy (2019) Levelised cost of electricity (LCOE) for selected renewable energy technologies in the ASEAN member state II
21. World Energy Association (2019) Global bioenergy statistics
22. Maw M, Dagmar J, Myo MW, Aung MT, Tomáš P (2019) Biomass energy: an overview of biomass sources, energy potential, and management in Southeast Asian Countries. Resource 81(8)
23. OECD/FAO (2017) OECD-FAO agricultural outlook 2017–2026
24. Central Intelligence Agency (2010) The world factbook. https://www.cia.gov/library/publications/the-world-factbook/fields/288.html
25. United States Department of Agriculture (2020) Foreign agricultural service. https://ipad.fas.usda.gov/countrysummary/Default.aspx?id=VMSjdd
26. World Data Atlas (2020) Agriculture, crop production. https://knoema.com/atlas/topics/Agriculture
27. Statista (2020) Farming. https://www.statista.com/statistics/658221/asia-pacific-cassava-production-by-country/
28. (2020) Food and Agriculture Organization of the United Nation, FAOSTAT. http://www.fao.org/faostat/en/#data/FO
29. Rosmiza MZ, Davies WP, Rosniza ACR, Mazdi M, Jabil MJ (2014) Farmers' knowledge on potential uses of rice straw: an assessment in MADA and Sekinchan, Malaysia. Malays J Soc Space 10(5):30–43
30. Danutawat T, Nguyen TKO (2007) Effects from open rice straw burning emission on air quality in the Bangkok Metropolitan region. ScienceAsia 33:339–345
31. ESCAP-CSAM (2018) Status of straw management in Asia-Pacific and options for integrated straw management. http://www.un-csam.org/Publication/StatusOfStrawMgrAP_final_31July2018.pdf
32. Kuen SL, Paul HW, Lin CJ, Ching IJ A process development for gasification of rice husk. Fuel Process Technol
33. Jittima P, Shabbir HG (2017) Sustainable utilization of rice husk ash from power plants: a review. J Clean Prod 167:1020–1028
34. N.a. (2020) Resource efficient and cleaner production for sustainable rice milling in Cambodia. https://prospernet.ias.unu.edu/wp-content/uploads/2012/09/SPC-learning-case-10_teaching-note_final.pdf
35. Shafie SM, Mahlia TMI, Masjuki HH, Rismanchi B (2012) Life cycle assessment (LCA) of electricity generation from rice husk in Malaysia. Energy Procedia 14:499–504

36. Reinhardt H, Keith F (2008) Cassava production and utilization in Asia and its potential as a bio-fuel. In: NexGen bio-ethanol conference, Bangkok, Thailand, April 22–24
37. Nakorn T, Prasert R, Pruk A, Adisak P (2017) Biochar production from cassava rhizome in a semi-continuous carbonization system. In: 4th International Conference on Power and Energy Systems Engineering, CPESE 2017, Berlin, Germany, September 25–29
38. Suntorn S, Keartisak S, Adisak P (2012) Production of bio-oil via fast pyrolysis of cassava rhizome in a fluidised-bed reactor. Energy Procedia 14:668–673
39. The Consulting Engineer Association of Thailand (2020) Study and design of power plant using cassava rhizome fired. http://www.ceat.or.th/2010/index.php/energy/288-study-and-design-of-power-plant-using-cassava-rhizome-fired-.html
40. Power Solution Technologies Public Company Limited (2020). http://www.pst.co.th/productGroup-78-view.html
41. Consultants of Technology Company Limited (2020). https://www.cot.co.th/th/project/study-and-design-of-cassava-rhizome-fired-power-plant-eastern-and-northeastern-regions-of-thailand/
42. Ogan IM, Marie-Josée D, Michael N (2015) Palm oil: processing, characterization and utilization in the food industry—a review. Food Bioscience 10:26–41
43. Sheil D, Casson A, Meijaard E, Van Nordwijk M, Gaskell J, Sunderland-Groves J, Wertz K, Kanninen M (2009) The impacts and opportunities of oil palm in Southeast Asia: what do we know and what do we need to know? https://www.cifor.org/publications/pdf_files/OccPapers/OP-51.pdf
44. Muhammad JI, Azizah B, Farah HA, Rizafizah O (2018) Palm oil industry in South East Asia and the effluent treatment technology—a review. Environ Technol Innov 9:169–185
45. Kerdsuwan S, Laohalidanond K (2011) Renewable energy from palm oil empty fruit bunch. In: Renewable energy—trends and applications, IntechOpen Publisher, pp 123–150. ISBN: 978-953-307-939-4
46. Shuit SH, Tan KT, Lee KT, Kamaruddin AH (2009) Oil palm biomass as a sustainable energy source: a Malaysian case study. Energy 34:1225–1235
47. Ahuja SC, Siddharth A, Uma A (2014) Coconut—history, uses, and folklore. Asian Agri-History 18(3):221–248
48. Romulo NA (2009) The situation and prospects for the utilization of coconut wood in Asia and Pacific. http://www.fao.org/3/a-am610e.pdf
49. Neetesh K, Dilip K (2014) Utilization of coconut shell in different forms in concrete. Int J Sci Res Dev 2(7):158–160
50. Tay LT, Ramadhansyah PJ, Norhidayah AH, Haryati Y, Dewi SJ (2015) A review of utilization o coconut shell and coconut fiber in road construction. Jurnal Teknologi (Sci Eng) 76(14):121–125
51. Elpiniki S, Alexandra S, Georgios C, Nicholaos D (2020) Maize as energy crop. https://www.intechopen.com/books/maize-production-and-use/maize-as-energy-crop
52. Otchere AG, Hagan EB (2014) Potential for electricity generation from maize residues in rural Ghana: a case study of Brong Ahafo Region. Int J Renew Energy Technol Res 3(5):1–10
53. Yasujiro W (2003) Utilization of bagasse energy in Thailand. Mitig Adapt Strat Glob Change 8:253–260
54. Mitr Phol Bio Power (2020) Case study: bagasse cogeneration development in Thailand's sugar industry. http://www.fao.org/fileadmin/templates/rap/files/meetings/2014/140723-d1s3.Bagasse.pdf
55. Saranya Y (2004) Bioconversion of alpha cellulose from weeds to ethanol. https://cuir.car.chula.ac.th/xmlui/handle/123456789/27065 (in Thai)
56. Kerdsuwan S, Loahalidanond K (2016) Availability and characteristic studies of using 23 biomass residues for developing of distributed community power plants. In: European biomass conference and exhibition proceedings, pp 185–188
57. ASTM International
58. Loahalidanond K (2008) Theoretische Untersuchungen und thermodynamische Modellierungen der Biomasse vergasung und der Fischer-Tropsch-Synthese zur Herstellung von

Dieselkohlenwasserstoffen aus thailändischen Biomassen. Dissertation. RWTH Aachen Univerity
59. Daniela T, Martin D, Volker L, Jan L, Andreas O (2015) Flexible bioenergy supply for balancing fluctuating renewables in the heat and power sector—a review of technologies and concepts. Energy Sustain Soc 5(35)
60. Department of Alternative Energy Development and Efficiency, Ministry of Energy Thailand (2020) Feasibility study on cellulosic ethanol production in commercial scale. (in Thai)
61. Kusch S, Maria VM (2009) Integration of lignocellulosic biomass into renewable energy generation concepts. ProEnvironment 2:32–37
62. Kerdsuwan S, Laohalidanond K (2015) Advancements in biomass and waste gasification for power generation. In: Handbook of clean energy systems. Wiley
63. Gupta AK, Kerdsuwan S (2014) Efficient energy conversion of wastes and fuels in power systems. Appl Sci Eng Prog
64. (2014) The working group for bioenergy science technology and innovation policy for Thailand in the context of AEC, ASEAN bioenergy technology status report 2014
65. EC-ASEAN COGEN Programme (COGEN 3) (2004) Applicability of European Technologies in ASEAN market
66. (2020) Emission standard, Stationary engines-World Bank guidelines. https://dieselnet.com/standards/inter/wbank.php#emstd
67. International Finance Cooperation, World Bank Group (2008) Environmental, health, and safety guidelines-Thermal power plants. (2020) https://www.ifc.org/wps/wcm/connect/f82a5f06-f3f7-4033-8ea6-b767523cda8e/FINAL_Thermal%2BPower.pdf?MOD=AJPERES&CVID=jqeD9Eg&id=1323162579734
68. Arnie I (2020) Biomass equipment options for steam and power. http://biomassmagazine.com/articles/2470/biomass-equipment-options-for-steam-and-power
69. Wolf JP, Dong N (2013) Biomass combustion for power generation: an introduction: biomass combustion science, technology and engineering. Elsevier Inc.
70. Krongkaew L, Palita C, Somrat K (2015) Status of using biomass gasification for heat and power in Thailand. Energy Procedia 79:385–390
71. (2020) Report on environmental impact assessment (EIA) and monitoring. (in Thai) http://eia.onep.go.th/
72. Trois D (2019) Bioenergy power plant in Indonesia: rules and implementation. German-Indonesia Business Forum, Jakarta
73. Phonepasong S (2017) Bioenergy use in Lao PDR. In: Regional workshop on sustainable bioenergy production and use in Southeast Asia: lessons learnt and future opportunities, 16–17 November 2017, Hanoi, Vietnam
74. (2013) National biomass strategy 2020: new wealth creation for Malaysia's biomass industry. https://www.cmtevents.com/MediaLibrary/BStgy2013RptAIM.pdf
75. Asian Development Bank (2016) Myanmar energy sector, assessment, strategy and road map. https://www.adb.org/sites/default/files/institutional-document/218286/mya-energy-sector-assessment.pdf
76. Ruby BDG Biomass sector roadmap. https://www.doe.gov.ph/sites/default/files/pdf/announcements/iloilo-a-bs3-02-biomass-sector-roadmap.pdf
77. Ministry of Energy Thailand (2015) Alternative energy development plan: AEDP2015. http://www.eppo.go.th/images/POLICY/ENG/AEDP2015ENG.pdf
78. Approval of the revised national power development master plan for the 2011–2020 period with the vision to 2030. Hanoi, 18 March 2016. https://policy.asiapacificenergy.org/sites/default/files/PDP%207%20revised%20Decision%20428-QD-TTg%20dated%2018%20March%202016-ENG.pdf

Role of Artificial Roughness in the Performance Improvement of Solar Air Heaters

Navneet Arya, Rajneesh Kumar, and Varun Goel

Abstract Solar radiation is a free, abundant, and clean source of thermal energy. A solar air heater (SAH) is a simple device which acts as a heat exchanger in a way that it extracts thermal energy from the solar radiation. There are various applications in which SAH is used such as space heating and crop drying. As the performance of SAH is low due to less heat transfer coefficient of air which acts as a working fluid in SAHs. Among different techniques, artificial roughness is considered to be the best technique for improving the performance of SAH. Providing roughness element in the direction of flow will help in creating local turbulence which helps in augmenting the heat transfer. Due to increase in turbulence, friction also increases in the flow channel. So, roughness may be provided in such a manner that no significant rise in friction takes place. A lot of roughness geometries were investigated experimentally, numerically to see the impact of roughness elements on heat transfer and friction. The impact of roughness elements on the performance of SAH has been studied for roughness geometry using computational fluid dynamics (CFD). It is a powerful numerical tool which can be utilized to study the flow structure near roughness elements in solar air heaters. This chapter provides the comprehensive information about the different cross-sectional roughness elements, and the effect of these roughness elements on the performance of SAH has also been discussed.

Keywords Roughness elements · Solar air heater · Flow structure · CFD · Heat transfer

Nomenclature

h_{avg} Convective heat transfer coefficient

N. Arya · V. Goel
Mechanical Engineering Department, National Institute of Technology Hamirpur, Hamirpur 177005, India

R. Kumar (✉)
Mechanical Engineering Department, CGC Chandigarh Engineering College, Landran (Mohali), Ajitgarh, Punjab, India

A. Gupta et al. (eds.), *Advances in Energy and Combustion*, Green Energy and Technology,
https://doi.org/10.1007/978-981-16-2648-7_23

k	Thermal conductivity
W	Width of absorber plate
H	Duct height
v	Velocity

Abbreviations

CFD	Computational Fluid Dynamics
f_{penalty}	Frictional penalty
Nu_{enh}	Nusselt number enhancement
SAH	Solar air heater
TPP	Thermohydraulic performance parameter

1 Introduction

1.1 Energy Scenario

Energy is the one of the essential component for living comfortable life. Most of our day-to-day activities are accomplished with the use of energy. Population growth is one of the main reason which significantly effects the amount of energy consumption [1].

The International Energy Agency (IEA) published a report tilted, 'World Energy Outlook' in 2012 and in which the total energy supply was estimated to 532.5 EJ, and among this, 53.8% is harvested from fossil fuels. Prediction shows that the global annual energy demand going to be increased to 778 EJ by 2035 [2]. It is also being expected that over the next two decades, global energy demand is increased by 50% [3], and for meeting the expected energy demand, rate of consumption of conventional fuels (i.e., fossil) will increase drastically. Beside this, the bigger challenge is faced by the energy sector, specially oil and gas companies. It is estimated that these companies consume one-tenth of the produced energy [2].

The biggest problem that is associated with the fossil fuels is the degradation of environment because of the emission of greenhouse gases (GHG) which leads to global warming. In the next two decades, it is expected that there is 40% increase in the CO_2 emissions, and it would increase to 44.1 Gt by 2035 from 31.2 Gt in 2011. The CO_2 emission is also accompanied with the other greenhouse gases, and therefore, it is estimated that other greenhouse gases increased up to 1000 ppm equivalent of CO_2 with the increase of CO_2 emission. With this, the average global temperature is expected to be increased by 6 °C [2].

1.2 Importance of Solar Energy

The sun is the power source of the energy in our solar system, and the radiation comes from the sun supports the life on the earth. The solar radiations formed due to 5.5×10^3 K temperature on the sun surface which is resulted due to fusion of hydrogen [4, 5].

Solar energy is a clean and sustainable source of energy which is freely available in on the Earth during sunshine. Approximately, 1.6×10^5 terawatts (TW) of energy are received from the sun in terms of solar radiation. Among them, 5×10^4 TW are reflected back to the atmosphere and rest is utilized for performing various activities on Earth such as photosynthesis, wind and wave conversion, and solar heating or ocean thermal energy [6] as shown in Fig. 1.

As per the report published by IEA in 2011, it is claimed that India receives 5×10^3 trillion kWh of solar energy, and it is interesting to see that it is much higher than the annual global consumption of energy [7, 8]. Understanding the importance of solar energy, Government of India had taken an initiative and in 1992 established an exclusive ministry, i.e., Ministry of New and Renewable Energy (MNRE), which devoted to promote use of renewable energy and technology associated with it [9]. Beside this, a non-banking financial institution has been established under MNRE to provide financial assistance for projects which are dedicated for development of sustainable energy-efficient components.

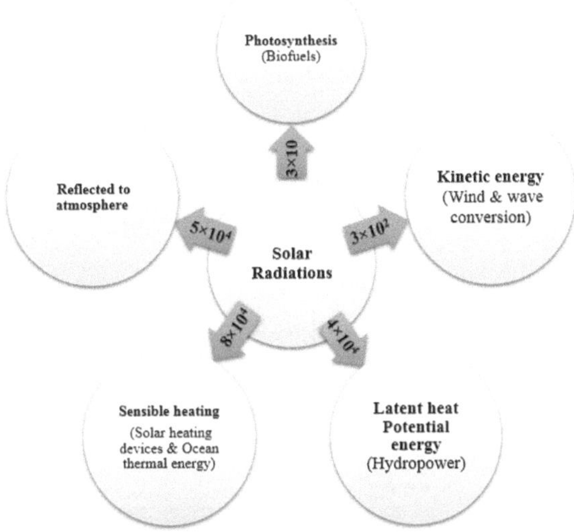

Fig. 1 Occurrence of various activities from the solar energy (units 10^{12} W)

1.3 Introduction to Solar Air Heater

Solar air heater (SAH) is a type of heat exchanger which helps in extracting heat from the solar radiations and utilized same to heat the air. A SAH consists of a glass plate, absorber plate (usually painted black to increase heat absorbing capacity), and both of them are encased in a wooden box in such a way that the glass plate is exposed to the direct sun light, while the absorber plate faces the glass plate. The air is guided to flow either between the glass and absorber plate or at the back of absorber plate depending on the design. Figure 2 shows the schematic of SAH. There is various application in which SAH can be used such as drying corps or dry fruits or herbs, heating living space, and drying timbers. [10–15].

There are many factor on which it can be concluded that the SAH is advantageous over the other solar energy harvesting devices such as solar water heater (SWH), and those factors are:

1. The fabrication of system is simple and compact.
2. Only one working medium is required, i.e., air.
3. As working fluid is air, there is no corrosion.
4. Leakage of working fluid does not cause serious effect on the environment, apart from low performance of SAH.
5. There is marginal variation in the pressure in the system and the surrounding.
6. In comparison with the SWH, maintenance of SAH is cheaper.

Among these advantages, there are some demerits are also associated with the SAH, and they are highlighted below:

1. The main problem observed in the SAH is associated with its performance because of low heat transfer coefficient of air.

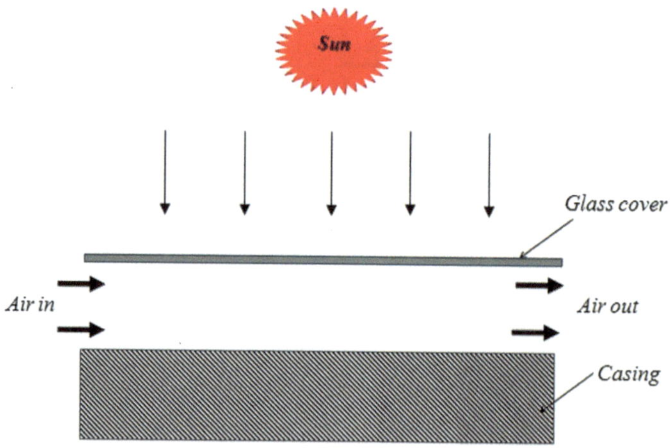

Fig. 2 Schematic of conventional SAH

2. As there is marginal change in the air density, the large amount of working fluid needs to be handled.
3. Due to working medium, the use of SAH is limited to temperature values.

2 Artificial Roughness and Its Importance in SAH

The working medium used in SAH is air, and the heat transfer coefficient obtained between the absorber plate and air is low which leads its poor performance. There are different techniques available in the literature which can be used to increase heat transfer coefficient in SAH as presented in Table 1. All these techniques improves heat transfer either by increasing heat transfer area or by increasing heat transfer coefficient of air.

The artificial roughness is one of the simplest and easiest ways of improving heat transfer through the SAH. In this technique, the roughness elements are either pasted or directly fabricated over absorber plate, and it is assumed that the area of roughness element is negligible in comparison with the area of absorber plate. Therefore, it can be concluded that the heat transfer is improved by providing roughness over the absorber plate, and this improvement in heat transfer is achieved because of creating local turbulence which results in local heat transfer coefficient increment, and hence, heat transfer through the SAH increases. Initially, this technique was used in cooling of turbine blade passages in which roughness is placed on both sides of the passage. But, in case of SAH, the roughness is provided only on single side where absorber plate is placed. With the time, this technique emerged as most effective method for increasing the performance of SAH.

The formation of viscous sub-layer is one of the main reasons that causes low heat transfer in SAH. These layers oppose transfer of heat from air to plate (absorber), and it leads to higher temperature zone near the absorber plate. For improving the heat transfer through the SAH, it becomes important to break these layers. The main

Table 1 Different techniques for improving performance of SAH

Method	Principle
Multi-glazing	By reducing losses from glass cover
Multi-passing	By providing more time to air for the heat transfer
Recycle	Re-heating of air by re-directing the conventional air outlet again into SAH and forced to flow at the back of absorber plate
Jet impingement	Making air jet to cause higher local turbulence in the SAH for improving SAH performance
Winglets	Creates vortices in the flow domain for increasing interaction with the absorber plate
Fins	By increasing the area for heat transfer
Artificial roughness	Break-up of viscous sub-layers which resulted flow separation along with secondary flow generation

purpose of providing artificial roughness over absorber plate is to break viscous sub-layer in such a way that the effective heat transfer takes place within the SAH.

2.1 General Roughness Parameters

The size and arrangement of the roughness over the absorber plate are decided on the basis of roughness parameters. As effective design of SAH is concerned, selection of roughness parameter value is very important. Generally, relative rib height (e/D) and relative rib pitch (P/e) are two important roughness parameters, and these roughness parameters can be non-dimensionalized as relative rib height (defined as the ratio between rib height and hydraulic diameter; e/D) and relative rib pitch (defined as ratio between rib pitch and rib height; P/e). However, there are other roughness parameters as well such as rib aspect ratio, rib attach angle, relative chamfered height, etc. But, the impact of roughness also depends upon the orientation/arrangement of the rib over the absorber plate.

With the variation of roughness parameters, there is noticeable changes which can be seen in heat transfer and flow characteristics. Actually, there is a formation of viscous sub-layer over the absorber plate under turbulent flow which acts as barrier in heat transfer from absorber plate to air flowing through the SAH. The e/D parameter helps in breaking of viscous sub-layers, and for this, it is provided to have rib height greater than the viscous sub-layer thickness. The formation of viscous sub-layers under turbulent flow condition on the absorber plate can be significantly affected with the change in relative rib height (e/D). While the flow separation and occurrence of reattachment points vary with the variation of P/e value for the rib element, there is tremendous variation in heat transfer which is observed along with the flow characteristics in SAH. Therefore, it can be concluded that both these parameters are very important for designing of energy-efficient SAH.

3 Literature Review

In recent years, a lot of research has been carried out for improving the thermal and hydraulic performance of SAH. Though, the air has low heat transfer coefficient, but beside this, there is development of laminar sub-layers over the absorber plate which acts as a barrier in transfer of heat from absorber plate to the air flowing through the SAH [16–18], and therefore, it poses poor performance. There are different techniques which can be used for improving SAH performance, and these techniques are presented in Table 1. Among all the techniques, artificial roughness is more promising for heat transfer improvement [16].

In present section, the literature review is carried out on artificial roughness depending on the shaped rib roughness geometries which can be employed over the absorber plate for improving the performance of (SAH). The schematic view of

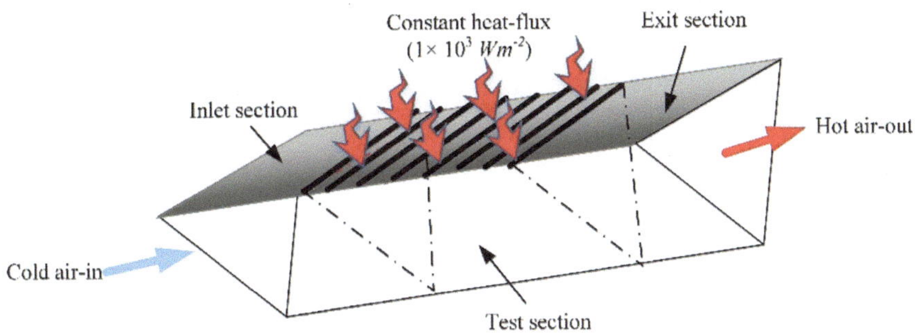

Fig. 3 Schematic view of the SAH with rib elements

SAH with rib roughness elements is shown in Fig. 3. The roughness element can be of various shapes, and these rib elements can be arranged in repeated or periodic pattern over the absorber plate. Figure 4 shows the schematic view of basic geometries of the roughness elements.

In this current chapter, the literature review has been carried out using different cross-sectional roughness elements at a time, and different possible patterns considered by the researchers have been presented in Figs. 5, 6, 7, and 8 for circular, square, rectangular, and chamfered ribs, respectively. It is seen that transverse rib pattern is very popular in which the direction of air is perpendicular to the rib element, while some complex patterns such as arc shape, V-shape, and S-shape were also investigated by the researchers. Under the following headings, the literature review has been performed:

1. Circular cross-sectional ribs and their different patterns
2. Square and rectangular cross-sectional ribs and their different patterns
3. Chamfered al ribs and their different patterns.

In conventional SAH, there is occurrence of low temperature gradient over the absorber plate which leads to low heat transfer coefficient or heat transfer. The heat transfer coefficient can be increased by creating local turbulence in the flow field, and therefore, the repeated rib roughness is provided over the absorber plate. These repeated rib roughness helps in developing secondary flow (caused due to flow separation, vortices and reattachment of viscous sub-layers) around the rib elements along with the main flow, and these associated flows play vital role in the flow dynamics and heat transfer within the SAH and thus the performance of SAH. Furthermore, these flows around the roughness has strong dependency on the shape and pattern of rib elements which is provided over the absorber plate, and it resonates in augmentation of heat transfer along with pressure drop.

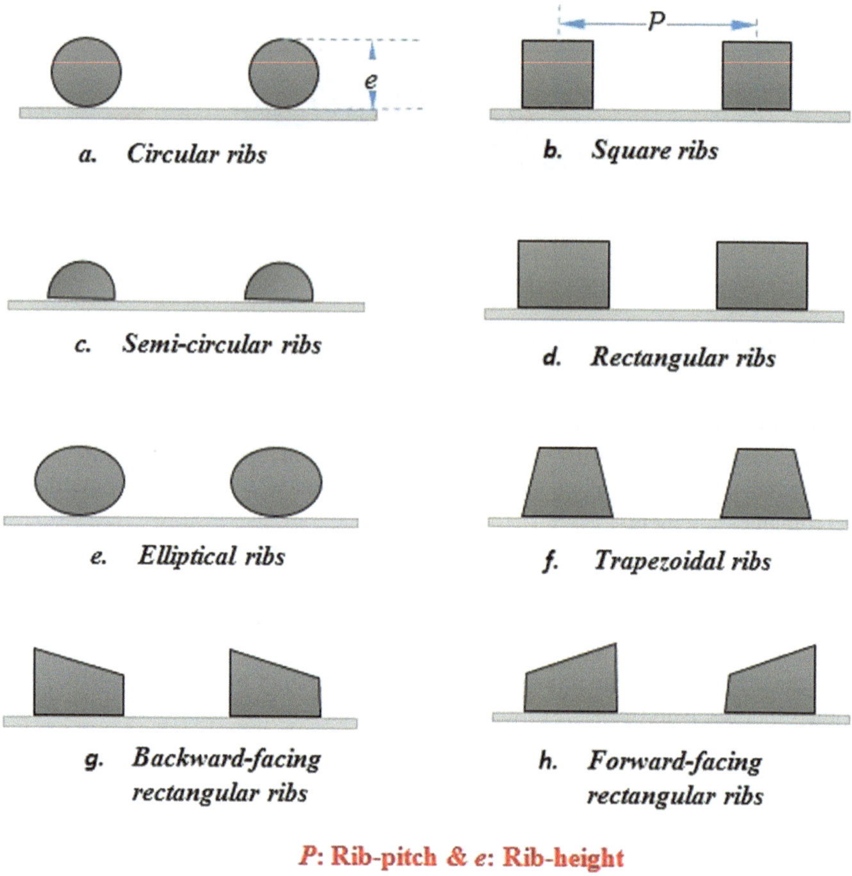

Fig. 4 Basic shape of rib roughness

3.1 Circular Cross-Sectional Ribs and Their Different Patterns

The circular cross-sectional rib element is widely used by the researchers in SAH for improving performance because it can be easily fabricated. Different roughness patterns were proposed and studied by the researcher over the absorber plate as it is shown in Fig. 5. The transverse arrangement of circular cross-sectional rib element is most common [19–27], while in some studies rib element are arranged in complex patterns as well such as single or multiple V-pattern with or without gaps [28, 29], single or multiple arc-shaped with or without gaps [30–32], innovative arc-shaped [33], etc. The details of parameters considered by different researcher for various rib patterns is provided in Table 2.

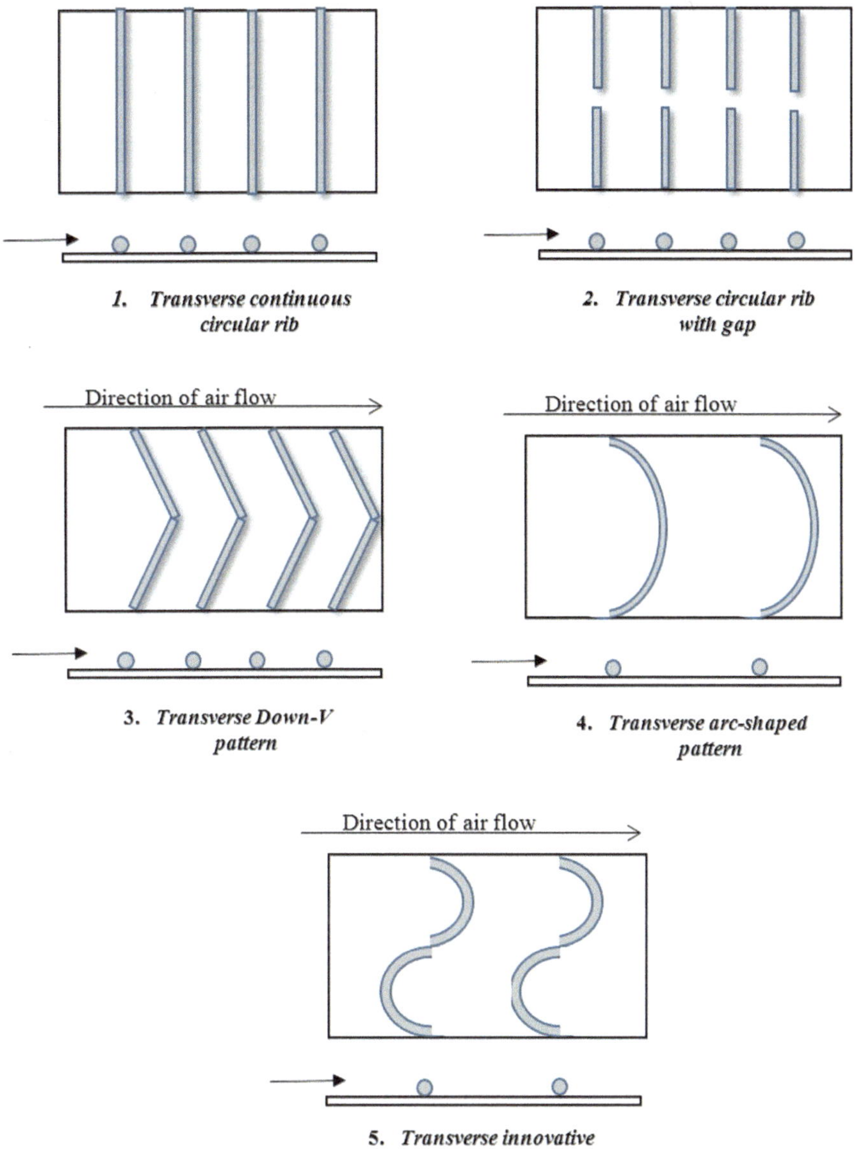

Fig. 5 Different patterns which are studied by the researchers using circular cross-sectional roughness

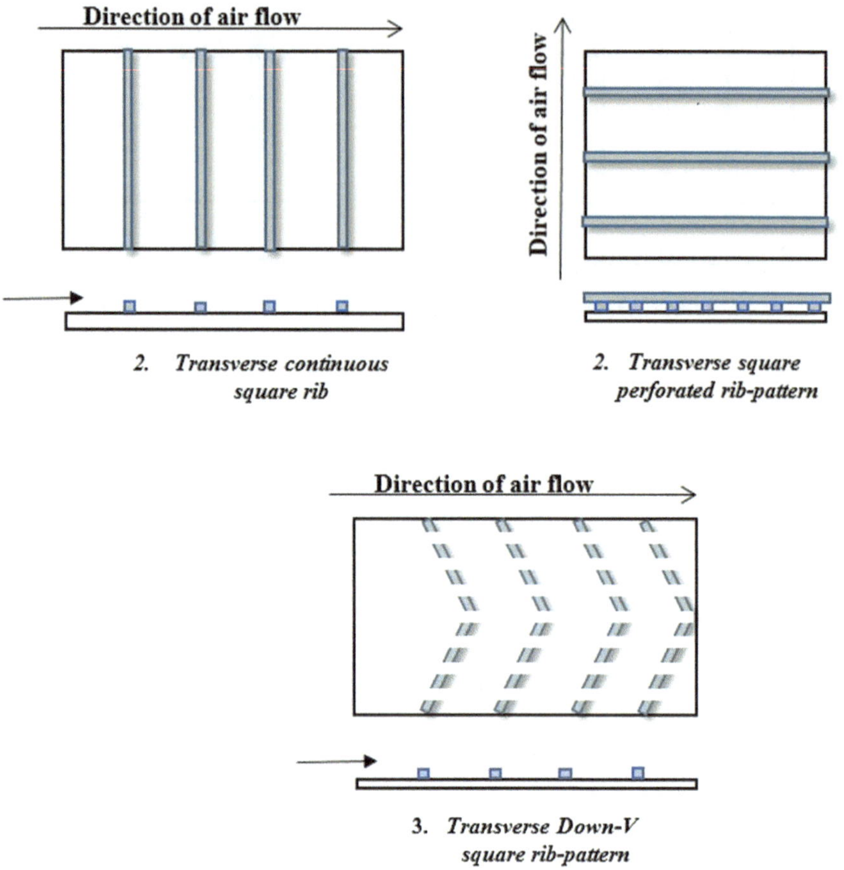

Fig. 6 Different patterns which are studied by the researchers using square cross-sectional roughness

The rib top surface plays an important role in the separation of free-shear layer which occurred by placing rib elements in the flow direction. The separated free-shear layer glides over the rib-top surface and results improvement in performance of SAH. In case of circular cross-sectional rib elements, the rib-top surface area is negligible, and thus, the augmentation of heat transfer is poor. Furthermore, these ribs provide low pressure drop as well which leads to low friction factor (f) value. In comparison with smooth SAH, the highest improvement in heat transfer and friction penalty (f_{penalty}) is claimed by Prasad and Saini [20], i.e., 2.38 and 3.39, respectively, for $e/D = 0.033$, $P/e = 10$. While, the highest thermohydraulic performance parameter (TPP) value is claimed by Verma and Prasad [19], i.e., 1.71.

As there is no effective separation of shear layer takes place due to circular rib element, the researchers have adopted different rib patterns over the absorber plate to enhance flow separation. Bhardwaj et al. [27] had used the inclined rib element

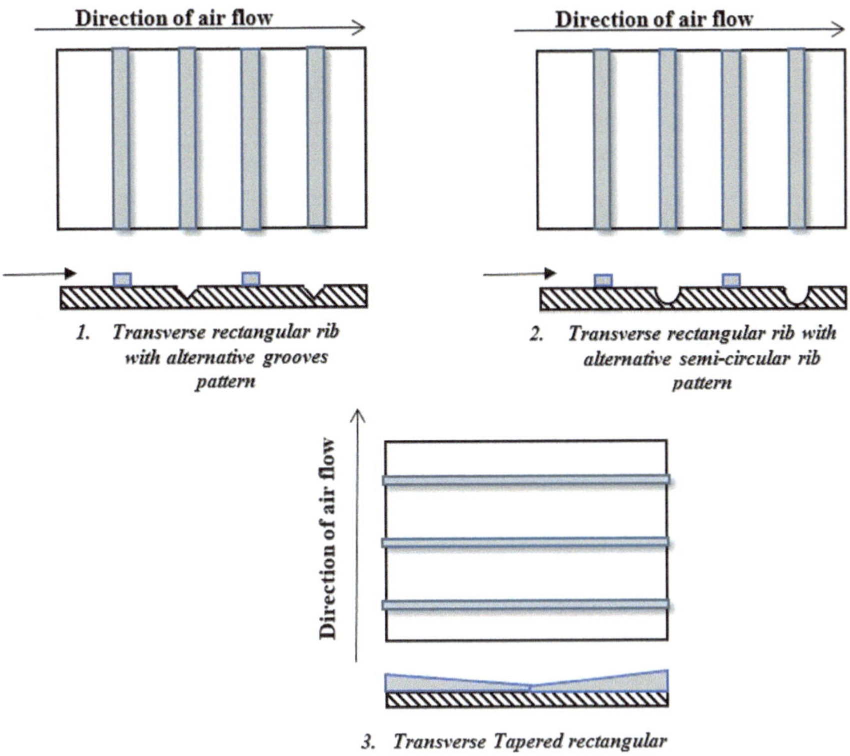

Fig. 7 Different patterns which are studied by the researchers using rectangular cross-sectional roughness

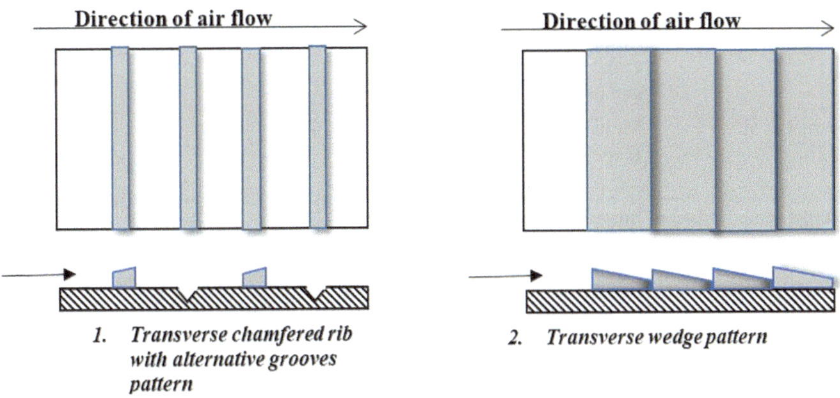

Fig. 8 Different patterns which are studied by the researchers using chamfered roughness

Table 2 Different studies published on circular cross-sectional roughness elements

Authors	Year	Arrangement of rib element	Cross section of flow passage	Range of key parameters		
				Re ($\times 10^{-3}$)	*e*/*D*	*P*/*e*
Verma and Prasad [19]	2000	Transverse continuous	Rectangular	5–20	0.01–0.03	10–40
Prasad and Saini [20]	1988	Transverse continuous	Rectangular	5–50	0.02–0.033	10–20
Yadav and Bhagoria [21]	2013	Transverse continuous	Rectangular	3.8–18	0.021–0.042	7.14–35.71
Chaube et al. [22]	2006	Transverse continuous	Rectangular	3–20	0.09	13.3
Aharwal et al. [23]	2008	Transverse continuous	Rectangular	3–18	0.037	10
Kumar et al. [24]	2017	Semi-circular	Triangular	4–24	0.0167–0.067	7.5–100
Kumar et al. [25]	2018	Transverse continuous	Triangular	4–18	0.015–0.06	4–20
Aharwal et al. [26]	2009	Transverse with gap	Rectangular	3–18	0.018–0.037	4–10
Bhardwaj et al. [27]	2017	Transverse inclined	Rectangular	4.5–28	0.21–0.043	4–16
Singh et al. [28]	2011	Down V-pattern	Rectangular	3–15	0.015–0.043	4–12
Singh et al. [29]	2015	Down V-pattern	Rectangular	3–15	0.043	4–12
Sahu and Prasad [30]	2016	Arc-shaped pattern	Rectangular	2–39	0.023–0.042	10
Pandey et al. [31]	2016	Multi-arc shape with gap pattern	Rectangular	2.1–21	0.016–0.044	4–16
Kumar et al. [32]	2019	Multi-arc shape with gap pattern	Rectangular	11–19	0.043	10
Kumar et al. [33]	2020	Innovative arc-shaped pattern	Rectangular	2.1–23	0.018–0.045	4–16

pattern over the absorber plate, and using this pattern, Nu and f increased by 2.96 and 3.13 times the smooth duct. But, the inclined rib element provides easy path to the secondary flow and guide it toward the side walls of the SAH passage. During this process, the increased temperature lowers the heat carrying capacity of the air and leads to reduction in heat transfer. This reason motivated the researcher toward V-rib pattern in which the secondary flow is directed toward the centerline of the flow passage, and beside this, the distance covered by the secondary flow is reduced which increases heat carrying capacity [28, 29]. The TPP obtained in case of V-rib pattern is 2.06 [29]. Another pattern, i.e., arc-shaped rib, is also investigated by the researchers [30–32]. In case of arc-shaped rib element, the gradual slop is provided to the secondary flow (in comparison with V-pattern) which leads comparatively higher TPP, i.e., 3.45 [31]. Another innovative arc-shaped rib pattern was introduced by Kumar et al. [33] in which the arc-shaped rib elements are arranged in the form of S shape along the width of SAH. The secondary flow associated from these types of rib pattern results in higher local turbulence in between the two successive rib elements and gives best THP value, i.e., 3.61, in comparison with transverse, inclined, V-shaped and arc-shaped rib elements. Therefore, it can be concluded that the pattern of rib element over plate (absorber) has significant effect on the performance of SAH.

3.2 Square and Rectangular Cross-Sectional Ribs and Their Different Patterns

The square and rectangular cross-sectional roughness provides more area over the rib top, and thus, the separation of free-shear takes place much effectively than the circular cross-sectional roughness. Beside this, the local eddies formed at the front of the roughness while flow strikes the rib element is comparatively low in rectangular rib element due flat surface of the roughness element. There are different patterns of rib elements which have been studied by the researcher for square and rectangular cross-sectional roughness, and they are presented in Figs. 6 and 7, respectively. Different studies are performed using different ranges of roughness parameters which are shown in Tables 3 and 4 for square and rectangular cross-sectional rib elements, respectively.

In literature, different patterns are studied using square rib element such as transverse [22, 34–37], perforated transverse [38], and V-shaped [39, 40]. In case of transverse square rib elements, the maximum THP is claimed of the order of 1.88 which is estimated by Singh and Bhagoria [35] by investigating the rectangular duct SAH with the help of commercial ANSYS software using finite volume method. Similar study is also performed by Kumar et al. [37] using triangular cross-sectional duct SAH, and it is seen that under similar range of roughness parameter the maximum THP obtained is 1.44 which is comparatively lower as that obtained for rectangular duct SAH. Thus, an important conclusion can be made that the SAH performance is not only affected by the shape or pattern of the rib roughness, but it varies significantly with the cross section of flow passage through which air flows.

The effect of perforated rib element has been investigated by the Singh and Singh [38]. It is concluded from the study that the for low Reynolds number, i.e., Re < 3000, the flow reattachment zone is limited to small area and which restrict the increment in Nu. But, for higher Re (i.e., Re> 9000), the flow reattachment zone is stretched to successive rib element, which helps better augmentation of heat from the absorber plate, and due to this, the thermal efficiency of the SAH is found to be 68.55%. In

Table 3 Different studies published on square cross-sectional roughness elements

Authors	Year	Arrangement of rib element	Cross section of flow passage	Range of key parameters		
				Re ($\times 10^{-3}$)	*e*/*D*	*P*/*e*
Taslim et al. [34]	1997	Transverse continuous	Square	2–18	0.133–0.25	5–10
Chaube et al. [22]	2006	Transverse continuous	Rectangular	3–20	0.09	13.3
Singh and Bhagoria [35]	2014	Transverse continuous	Rectangular	3.8–18	0.042	10.71
Abhay et al. [36]	2018	Transverse continuous	Rectangular	3.8–18	0.042	7.14–17.86
Kumar et al. [37]	2019	Transverse continuous	Triangular	4–18	0.02–0.04	5–15
Singh and Singh [38]	2018	Transverse wave profiled	Rectangular	3–15	0.042	4–30
Karwa and Chauhan [39]	2009	Discrete V-down	Rectangular	1.07–26.35	0.07	6.6–53.3
Kumar et al. [40]	2012	Discrete V-down	Rectangular	2–20	0.043	10

Table 4 Different studies published on rectangular cross-sectional roughness elements

Authors	Year	Arrangement of rib element	Cross section of flow passage	Range of key parameters		
				Re ($\times 10^{-3}$)	*e*/*D*	*P*/*e*
Jaurker et al. [41]	2006	Rectangular	Rectangular	3–21	0.0181–0.0363	4.5–10
Aghie et al. [42]	2015	Rectangular	Rectangular	10	0.018–0.037	10
Kumar et al. [16]	2017	Rectangular	Triangular	4–18	0.02–0.04	5–15
Layek et al. [43]	2007	Rectangular with alternative grooved	Rectangular	3–21	0.022–0. 04	4.5–10
Alfarawl et al. [44]	2017	Rectangular with alternative semi-circular	Rectangular	12.5–86.5	0.0468	6.6–53.3
Gupta and Varshney [45]	2017	Tapered rectangular	Rectangular	3.8–18	0.042–0.084	3.6–17.9

another pattern, the rib elements are arranged in discrete V-shaped, and the effect of providing gaps has been investigated [40]. By discretion of the rib elements, the local velocity of secondary flow increased drastically at the location of gap which produces highly intense local turbulent in the flow field and gives considerably higher augmentation in heat transfer. Thus, the TPP found in discrete V-shaped pattern is 3.7.

In case of rectangular cross-sectional rib elements, three different patterns have been investigated by the researcher, i.e., transverse [16, 41, 42], combination of rectangular rib with alternative groove [43, 44], and transverse tapper rib [45]. As the rib-top area is higher in case of rectangular rib element in comparison with circular and square rib, the highest augmentation of heat is observed. Two different studies were found in the literature in which rectangular [41, 42] and triangular [16] flow passage has been used by the researchers. The highest TPP obtained is 1.85 [41] and 1.89 [16] in rectangular [41] and triangular [16] duct SAH, respectively. It is found that the THP obtained in the triangular duct is comparatively higher as that in rectangular duct SAH. In other two studies, Layek et al. [43] and Alfarawl et al. [44] used the rectangular rib element in combination with alternative triangular and circular groove, respectively. It is claimed that with this combination (i.e., rib with triangular groove), the significantly higher local turbulence occurred in the flow passage due to vortices developed around and in the groove with its break-up with the alternative rib element. Due to this reason, the heat transfer increased by 170% in comparison with smooth SAH. While in case of rectangular rib with successive circular groove, the TPP found is 1.39. In comparison with the above-discussed rectangular rib patterns, the best TPP is found in case of transverse tapered rectangular rib elements, i.e., 1.91. Due to rib tapered rib element, the turbulent kinetic energy increased, and it is guided toward the centerline (where the rib height is minimum), and thus, the vortex formation takes place which results higher thermal performance in comparison with other discussed patterns.

3.3 Chamfered Ribs and Their Different Patterns

The different configuration of chamfered rib roughness is presented in Fig. 8, and the range of parameter considered is presented in Table 5. The transverse rib roughness is commonly used by the researcher [46–51], and most of the studies were carried out using rectangular flow passage [46–49]. The highest THP is claimed by Gawande et al. [49] with value of 2.07 in comparison with other similar studies, while few studied are available on triangular flow passage SAH. In these studies, the chamfered rib roughness is classified into two basic geometries based on the location of provided chamfer, i.e., forward-facing and backward-facing chamfer [50, 51]. It is observed that the location of providing chamfer plays an important role in the SAH performance. It is noticed that the highest TPP (with value of 2.08) obtained in case of forward-facing chamfered rib roughness is approximately 4% higher than the backward-facing chamfered ribs, and this resulted due to slope occurred due to

Table 5 Different studies published on chamfered rectangular cross-sectional roughness elements

Authors	Year	Arrangement of rib element	Cross section of flow passage	Range of key parameters		
				Re ($\times 10^{-3}$)	*e*/*D*	*P*/*e*
Han and Park [46]	1988	Transverse trapezoidal	Rectangular	8–80	0.021–0.078	10–20
Karwa et al. [47]	1999	Transverse chamfered rectangular	Rectangular	3–20	0.0141–0.0328	4.5–8.5
Karwa et al. [48]	2001	Transverse chamfered	Rectangular	3.75–16.35	0.0197–0.0441	4.58–7.09
Gawande et al. [49]	2016	Chamfered rib	Rectangular	3.8–18	0.042	7.4–17.86
Kumar et al. [50]	2018	Forward chamfered rib	Triangular	4–17	0.018–0.043	12
Kumar et al. [51]	2019	Backward chamfered	Triangular	4–17.1	0.018–0.043	12
Layek et al. [52]	2009	Chamfered rib with alternative groove	Rectangular	3–21	0.03	10
Bhagoria et al. [53]	2002	Wedge	Rectangular	12.5–86.5	0.018–0.037	6.6–53.3

the chamfering over the rib-top surface. In this case, the rib-top surface allows the smooth passage of free-shear layer and reduces the eddy formation at the start of the rib element which leads to better augmentation of heat.

In the literature, two different rib patterns were studied by the researchers, i.e., chamfered rib with alternative triangular groove [52] and wedge pattern [53]. It is found that the TPP obtained in case of wedge pattern is 2.0 which is approximately 13.63% higher as that estimated in case chamfered rib with alternative triangular groove pattern.

4 Effect of Duct Cross-Section

The performance of SAH is not only affected by shape or pattern of the rib roughness over the absorber plate, but it is also shows the significant influence of cross-sectional shape of the flow passage which is considered in design of SAH. Table 6 shows the comparison of THP values for triangular and rectangular duct SAH. The comparison has been carried out using different cross-sectional-shaped rib roughnesses, and it can be seen from the table that TPP value changes noticeably with the change of flow passage of SAH. This variation in THP might be resulted due to number of available corners in the duct. There is coexistence of laminar and turbulent flow near the corner

Table 6 Comparison of THP between triangular and rectangular duct SAH

Cross-sectional of rib element	SAH with triangular passage		SAH with rectangular passage	
	Authors	TPP	Authors	TPP
Circular	Kumar et al. [25]	1.82	Verma and Prasad [19]	1.71
Square	Kumar et al. [37]	1.44	Chaube et al. [22]	1.94
Rectangular	Kumar et al. [16]	1.89	Chaube et al. [22]	2.17
Chamfered	Kumar et al. [50]	2.15	Chaube et al. [22]	1.95

region which leads to flow stagnation, and there is formation of hot spots [11]. These hot spots indicate the trapped heat which not carried away the main or primary flow, and thus, the heat transfer reduced in the duct. Beside this, the velocity profile is also an important factor which affects the heat transfer in the duct. In equilateral triangular duct, the main flow concentrated at one-third of duct height from the base surface, while it is at the center of the duct in case of rectangular duct. Due to this, the main flow is near to the base surface, and higher heat trapped may take place at the corner opposite to the base surface which leads poor thermal performance of the triangular duct in comparison with the rectangular duct.

With the use of rib roughness, the thermal performance of triangular duct can be increased to the significant level. As the main flow is near to the base surface, the turbulence created in the main flow due to secondary flow resulted better mixing and thus increases the heat transfer augmentation in SAH. It can be concluded from Table 6 that the it is better option to consider triangular duct in case of rib with circular and chamfered cross section because it resulted higher TPP, i.e., 1.82 and 2.15, in comparison with rectangular duct SAH, while rectangular duct SAH is better for square and rectangular cross-sectional rib elements.

5 Parameters Used for Predicting SAH Performance

For measuring the desired output parameters, all the properties measured during experimentation were calculated at bulk temperature, and collected data is utilized to determine thermal (as Nusselt number; Nu_{avg}) and hydraulic (as friction factor; f) performance of SAH. For the calculation of Nu and f, following equations were used:

The heat transfer through the SAH is predicted using Nusselt number (Nu) which is a non-dimensional number, and mathematically, it is calculated as:

$$Nu_{avg} = \frac{h_{avg}}{k_{air}} \times D \tag{1}$$

where

$$D = 4 \times \frac{\text{area}}{\text{perimeter}} \tag{2}$$

$$= 4 \times \left[\frac{W \times H}{2(W + H)}\right] \text{(for rectangular duct)} \tag{2.1}$$

$$= 4 \times \left[\frac{\frac{1}{2} \times WH}{2(W + H)}\right] \text{(for triangular duct)} \tag{2.2}$$

There is unavoidable effect which appeared in terms of pressure drop in the SAH, and it is calculated as:

$$f = \frac{1}{4} \times \frac{\Delta P}{L} \times \left(\frac{1}{2}\rho_{\text{air}} \times v_{\text{air}}^2\right)^{-1} \times D \tag{3}$$

where

$$\Delta P_t = \rho g \times \Delta h \tag{4}$$

It is seen that there is one desirable affect that is heat transfer which is occurred along with the unwanted affect, i.e., friction factor (f). Therefore, one more parameter needs to be considered to estimate the effectiveness of the SAH which is thermohydraulic performance parameter (TPP), and it is expressed as:

$$\text{TPP} = \frac{(\text{Nu}/\text{Nu}_2)}{(f/f_s)^{1/3}} \tag{5}$$

6 Comparative Performance Assessment of Various Roughness Used in SAH

The different proposed design of rib roughness has shown strong influence on the heat and flow characteristics of SAH, and these different designs can be compared on the basis of thermal and hydraulic performances in order the find out the best possible cross section of rib element. With the use of roughness, thermal performance (i.e., Nu) changes drastically in comparison with the smooth SAH, but there also appears an unavoidable effect, i.e., pressure drop or friction factor. In other words, it can be said that the heat transfer increases at the expense of friction factor. On the basis of cost effectives is concerned, it is important to design a SAH in such a way that the Nu should be as higher as possible with f as small as possible, such that the heat transfer dominates the f and opposite to this leads to the faulty design in which the desired outcome; i.e., heat transfer is comparatively lower than f. The external power consumed by the system is comparatively higher than the corresponding heat transfer which results in wastage of energy.

6.1 Based on Nusselt Number and Friction Factor

Figure 9 presents the comparison of highest Nu_{enh} value which is claimed by the different researchers using various cross-sectional rib elements with different patterns over the absorber plate. It is observed from the figure that the Nu_{enh} changes drastically with the variation of rib cross section and with the slight change in the cross-sectional area or pattern of the rib element resulted noticeable change in the thermal performance of SAH. By providing the rib element in the path of air flow, the dynamics of flow change considerable inside the SAH due to flow separation and its reattachment over the absorber plate. The secondary flow induced due to rib element develops local turbulence at or near the rib element. The best Nu_{enh} is observed in case of innovative arc-shaped rib pattern with value of 3.6 for circular cross-sectional rib element [25], whereas the lowest Nu_{enh} is obtained in case of transverse circular cross-sectional rib element, and it is due to the minimum rib-top area which does not effectively support the secondary flow in the SAH and resulted in poor local turbulence thus lower Nu_{enh}.

The heat transfer improvement in the ribbed SAH is also accompanied with the pressure drop or friction factor (f). For effective design of SAH, the f is required to be as small as possible. Figure 10 presents the comparison of $f_{penalty}$ obtained

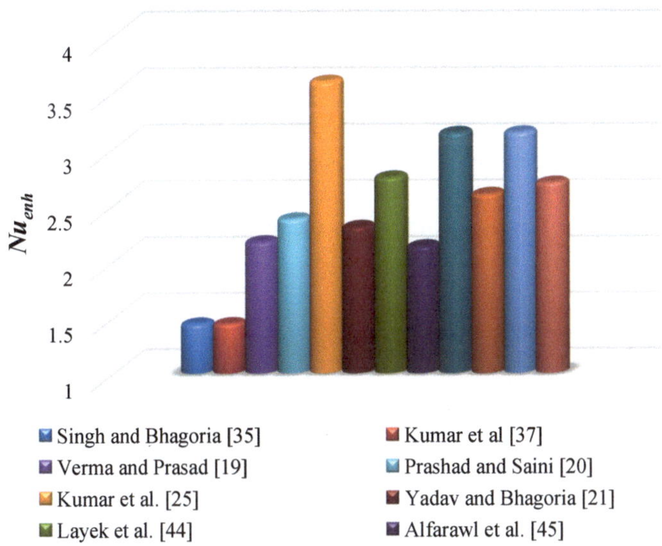

Fig. 9 Comparison of optimum Nu_{enh} claimed in various studies

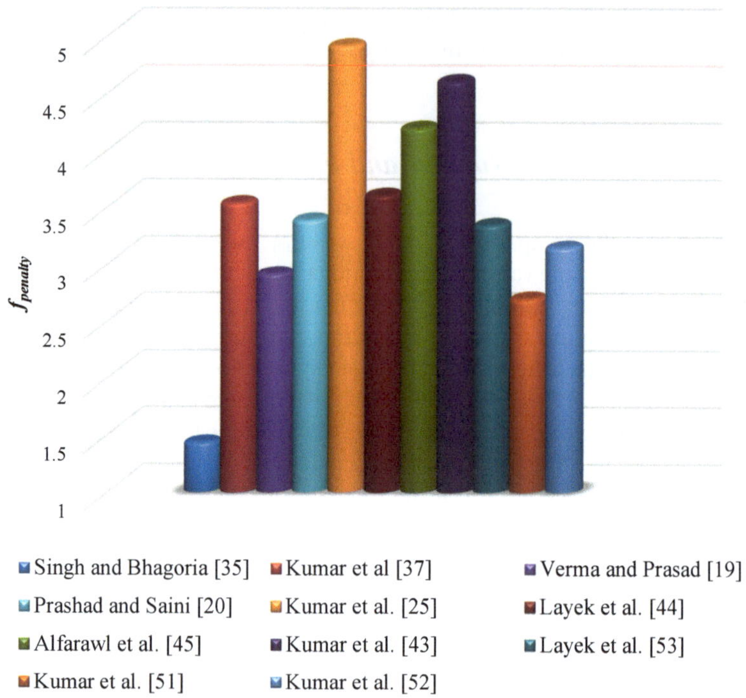

Fig. 10 Comparison of maximum f_{penalty} claimed in various studies

in SAH with different cross-sectional rib element under different rib patterns. The minimum value of f_{penalty} is claimed by Singh and Bhagoria [35], i.e., 1.44, and it is obtained in transverse circular cross-sectional rib element, while the highest f_{penalty} is obtained in innovative arc-shaped rib pattern of circular cross-sectional rib element [25]. For other combination of rib element patterns, the f_{penalty} lies in between these two values. Most importantly, it is also found that the higher f_{penalty} corresponds to lower Re, and as with the increase of Re, the f_{penalty} starts decreasing because the viscous forced in the flow is dominated by the inertia forces.

6.2 *Based on Thermohydraulic Performance Parameter*

It is noticed from Figs. 9 and 10 that both Nu_{enh} and f_{penalty} varied noticeably with the shape of rib element and it orientation over the absorber plate. Furthermore, both these values obtained maximum for forward chamfered rib pattern [50], while minimum for transverse circular rib element [37]. Thus, it is quite difficult to come to a conclusion on the basis of this observation. It is very important to introduce a parameter which includes effect of both heat transfer and flow characteristics in

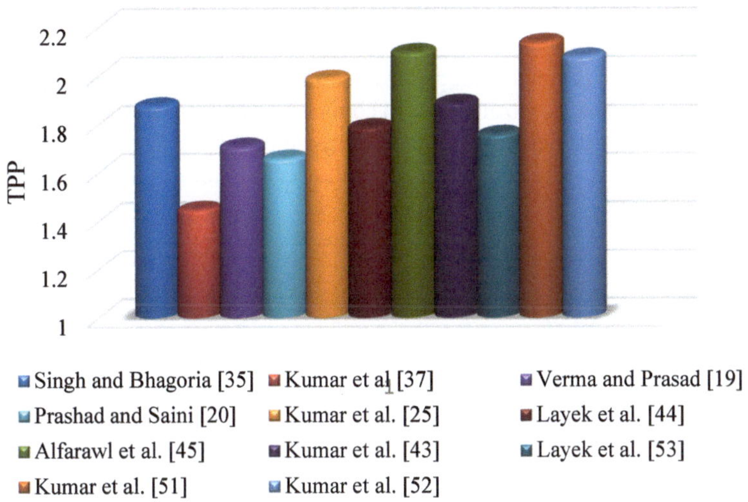

Fig. 11 Comparison of optimum TPP claimed in various studies

order to estimate the effectiveness of the rib element. Keeping this in mind, the thermohydraulic performance parameter (TPP) is calculated for the various available rib elements. According to TPP, the effectiveness of roughness can only be ensured if its value is greater than one. In other words, it can be said that if the TPP is greater than one, the heat transfer which takes place in the system is comparatively higher in comparison with the effort made to maintain the flow through the system. The comparison of TPP for various rib roughness configurations is shown in Fig. 11. The best TPP obtained is 2.14 in case of triangular duct SAH having forward chamfered rib elements of rectangular cross-sectional area [50]. In case of rectangular duct SAH, the TPP found is 2.1 for rectangular cross-sectional rib element [35].

7 Conclusions

In this chapter, the comprehensive review of literature has been carried out in order to analyze the best configuration of rib element which resulted in higher thermohydraulic performance parameter. From the current report, it is concluded that the cross-sectional area of the duct used in SAH plays an important role in the deciding thermohydraulic performance parameter in addition to rib cross-sectional area and its orientation on the absorber plate.

The heat transfer in ribbed SAH is influenced by the rib-top surface. Higher the top-surface area, more is the heat transfer through the SAH. On this fact, it is concluded that the transverse rib pattern of the highest thermohydraulic performance parameter is obtained in case of rectangular cross-sectional rib element and relatively worst thermohydraulic performance parameter is found in case of circular cross-sectional rib element.

As per rib pattern is concerned, the best thermohydraulic performance parameter corresponds to arc-shaped rib elements in comparison with other studied rib patterns such as transverse, inclined, and V-shaped. Beside this, by providing the gap in the rib element, there is further enhancement in the performance of SAH.

References

1. Halabi MA, Al-Qattan A, Al-Qtaibi A (2005) Application of solar energy in the oil industry-current status and future prospects. Renew Sustain Energy Rev 43:296–314
2. IEA (2012) World energy outlook. International Energy Agency, Head of Communication and Information Office: Paris, France. Available at www.iea.org/publications/freepublications/publication/name-49561-en.html
3. OPEC (2013) World oil outlook. Secretariat, Organisation of the Petroleum Exporting Countries: Vienna, Austria. Available at www.opec.org
4. Kleissl J (2013) Solar energy forecasting and resource assessment
5. Tiwari GN, Tiwari Shyam A (2016) Handbook of solar energy: theory, analysis and applications. Springer Nature, Berlin
6. Tiwari GN, Mishra RK (2012) Advance renewable energy sources. RSC publishing, UK
7. The Potential for Solar Energy in India Is Huge: IEA. http://www.eai.in/club/users/krupali/blogs/627
8. Solar Energy in India-Pros, Cons and the Future. http://www.mapsofindia.com/my-india/india/scope-of-solar-energy-in-india-pros-cons-and-the-future
9. Manju S, Sagar N (2017) Progressing towards the development of sustainable energy: a critical review on the current status, applications, developmental barriers and prospects of solar photovoltaic system in India. Renew Sustain Energy Rev 70:293–313
10. Lingayat AB, Chandramohan VP, Raju VRK, Meda V (2020) A review on indirect type solar dryers for agriculture crops-dryer setup, its performance, energy storage and important highlights. Appl Energy 28:11400
11. Kumar R, Varun Kumar A (2016) Thermal and fluid dynamic characteristics of flow through triangular cross-section duct: a review. Renew Sustain Energy Rev 61:123–140
12. Bharadwaj G, Varun Kumar R, Sharma A (2017) Heat transfer augmentation and flow characteristics in ribbed triangular duct solar air heater: an experimental analysis. Int J Green Energy 14(7):587–598
13. Kumar R, Goel V, Kumar A (2017) A parametric study of the 2D model of solar air heater with elliptical rib roughness using CFD. J Mech Sci Technol 31(2):959–964
14. Prasad K, Mullick SC (1983) Heat transfer characteristics of a solar air heater used for drying purposes. Appl Energy 13(2):83–93
15. Hu J, Zhang G (2019) Performance improvement of solar air collector based on airflow reorganization: a review. Appl Therm Eng 155:592–611
16. Kumar R, Kumar A, Goel V (2017) A parametric analysis of rectangular rib roughened triangular duct solar air heater using computational fluid dynamics. Sol Energy 157:1095–1107
17. Kumar R, Kumar A, Goel V (2018) Effect of rounded corners on the heat transfer and fluid flow through triangular duct. J Heat Transfer 140(12):
18. Incropera F, Dewitt PD, Bergman TL, Lavine AS, Setharamu KN, Setharama TR (2013) Fundamentals of heat and mass transfer, 5th edn. Wiley, United States
19. Verma SK, Prasad BN (2000) Investigation for the optimal thermohydraulic performance of artificially roughened solar air heaters. Renewable Energy 20(1):19–36

20. Prasad BN, Saini JS (1988) Effect of artificial roughness on heat transfer and friction factor in a solar air heater. Sol Energy 41(6):555–560
21. Yadav AS, Bhagoria JL (2013) A CFD (computational fluid dynamics) based heat transfer and fluid flow analysis of a solar air heater provided with circular transverse wire rib roughness on the absorber plate. Energy 55:1127–1142
22. Chaube A, Sahoo PK, Solanki SC (2006) Analysis of heat transfer augmentation and flow characteristics due to rib roughness over absorber plate of a solar air heater. Renewable Energy 31(3):317–331
23. Aharwal KR, Gandhi BK, Saini JS (2008) Experimental investigation on heat-transfer enhancement due to a gap in an inclined continuous rib arrangement in a rectangular duct of solar air heater. Renew Energy 33:585–596
24. Kumar R, Kumar A, Varun (2017) Computational fluid dynamics based study for analysing heat transfer and friction factor in semi-circular rib- roughened equilateral triangular duct. Int J Numer Meth Heat Fluid Flow 27(4):941–957
25. Kumar R, Goel V, Kumar A, Khurana S, Singh P, Bopche S (2018) Numerical investigation of heat transfer and friction factor in ribbed triangular duct solar air heater using computational fluid dynamics (CFD). J Mech Sci Technol 32(1):399–404
26. Aharwal KR, Gandhi BK, Saini JS (2009) Heat transfer and friction characteristics of solar air heater ducts having integral inclined discrete ribs on absorber plate. Int J Heat Mass Transf 52(25–26):5970–5977
27. Bhardwaj G, Varun, Kumar R, Sharma A (2017) Heat transfer augmentation and flow characteristics in ribbed triangular duct solar air heater: an experimental analysis. Int J Green Energy 14(7):587–598
28. Singh S, Chander S, Saini JS (2011) Heat transfer and friction factor correlations of solar air heater ducts artificially roughened with discrete V-down ribs. Energy 36(8):5053–5064
29. Singh S, Chander S, Saini JS (2015) Thermo-hydraulic performance due to relative roughness pitch in V-down rib with gap in solar air heater duct—comparison with similar rib roughness geometries. Renew Sustain Energy Rev 43:1159–1166
30. Sahu MK, Prasad RK (2016) Exergy based performance evaluation of solar air heater with arc-shaped wire roughened absorber plate. Renew Energy 96:233–243
31. Pandey NK, Bajpai VK, Varun (2016) Experimental investigation of heat transfer augmentation using multiple arcs with gap on absorber plate of solar air heater. Sol Energy 134:314–326
32. Kumar R, Goel V, Singh P, Saxena A, Kashyap AS, Rai A (2019) Performance evaluation and optimization of solar assisted air heater with discrete multiple arc shaped ribs. J Energy Storage 26:
33. Kumar R, Kashyap AS, Singh P, Goel V, Kumar K (2020) Innovatively arranged curved ribbed solar assisted air heater: performance and correlation development for heat and flow characteristics. J Solar Energy Eng 142(3)
34. Taslim D, Li ME, Kercher T (1996) Experimental heat transfer and opposite walls. J Turbomach 118:20–28
35. Singh A, Bhagoria JL (2014) A numerical investigation of square sectioned transverse rib roughened solar air heater. Int J Therm Sci 79:111–131
36. Abhay L, Chandramohan VP, Raju VRK (2018) Numerical analysis on solar air collector provided with artificial square shaped roughness for indirect type solar dryer. J Clean Prod 190:353–367
37. Kumar R, Kumar A, Goel V (2019) Performance improvement and development of correlation for friction factor and heat transfer using computational fluid dynamics for ribbed triangular duct solar air heater. Renewable Energy 131:788–799
38. Singh I, Singh S (2018) CFD analysis of solar air heater duct having square wave profiled transverse ribs as roughness elements. Sol Energy 162:442–453
39. Karwa R, Chauhan K (2009) Performance evaluation of solar air heaters having v-down discrete rib roughness on the absorber plate. Energy 35(1):398–409
40. Kumar A, Saini RP, Saini JS (2012) Experimental investigation on heat transfer and fluid flow characteristics of air flow in a rectangular duct with multi v-shaped rib with gap roughness on the heated plate. Sol Energy 86(6):1733–1749

41. Jaurker AR, Saini JS, Gandhi BK (2006) Heat transfer and friction characteristics of rectangular solar air heater duct using rib-grooved artificial roughness. Sol Energy 80(8):895–907
42. Aghaie AZ, Rahimi AB, Akbarzadeh A (2015) A general optimized geometry of angled ribs for enhancing the thermo-hydraulic behaviour of a solar air heater channel—a Taguchi approach. Renewable Energy 83:47–54
43. Layek A, Saini JS, Solanki SC (2007) Heat transfer and friction characteristics for artificially roughened ducts with compound turbulators. Int J Heat Mass Transf 50:4845–4854
44. Alfarawi S, Abdel-Moneim SA, Bodalal A (2017) Experimental investigations of heat transfer enhancement from rectangular duct roughened by hybrid ribs. Int J Therm Sci 118:123–138
45. Gupta AD, Varshney L (2017) Performance prediction for solar air heater having rectangular sectioned tapered rib roughness using CFD. Therm Sci Eng Prog 4:122–132
46. Han JC, Park JS (1988) Developing heat transfer in rectangular channels with rib tabulators. Int J Heat Mass Transf 31(1):183–195
47. Karwa R, Solanki SC, Saini JS (1999) Heat transfer coefficient and friction factor correlations for the transitional flow regime in rib-roughened rectangular ducts. Int J Heat Mass Transf 42(9):1597–1615
48. Karwa R, Solanki SC, Saini JS (2001) Thermo-hydraulic performance of solar air heaters having integral chamfered rib roughness on absorber plates. Energy 26(2):161–176
49. Gawande VB, Dhoble AS, Zodpe DB, Chamoli S (2016) Analytical approach for evaluation of thermohydraulic performance of roughened solar air heater. Case Stud Therm Eng 8:19–31
50. Kumar R, Kumar A, Goel V (2019) Simulation of flow and heat transfer in triangular cross-sectional solar-assisted air heater. J SolEnergy Eng 141(1):
51. Kumar R, Goel V, Kumar A (2018) Investigation of heat transfer augmentation and friction factor in triangular duct solar air heater due to forward facing chamfered rectangular ribs: a CFD based analysis. Renewable Energy 115:824–835
52. Layek A, Saini JS, Solanki SC (2009) Effect of chamfering on heat transfer and friction characteristics of solar air heater having absorber plate roughened with compound turbulators. Renewable Energy 34(5):1292–1298
53. Bhagoria JL, Saini JS, Solanki SC (2002) Heat transfer coefficient and friction factor correlations for rectangular solar air heater duct having transverse wedge shaped rib roughness on the absorber plate. Renew Energy 25(3):341–369

Estimation of Mechanical Behavior of Cementitious Material Using Microstructural Information

Praveen Verma and Ayushman Bajpai

Abstract Concrete is a widely used material, so estimating its mechanical properties is essential for the safety and durability of structures. It has a complex multi-component and multiscale structure that forms through a chemical reaction and then continue to change with time. Generally, cement paste is the most porous and also most chemically active part of concrete. Hence, understanding the cement microstructure and its variation over time is essential for understanding the mechanical properties. The microstructural information can be extracted computationally using modeling platform-tools. The generated microstructural information is modeled as a Representative volume element in finite element analysis software to generate some cubes of several tens of microns in dimensions subjected through prescribed deformation mode to generate effective elastic tensor of the material. The current work focusses on the estimation of effective properties of cementitious material through finite element analysis. The calculated elastic properties, such as Elastic moduli and poison's ratio, are validated through previous studies.

Keywords Microstructure · Hydration · Elastic modulus · RVE · Homogenization · Calcium silicate hydrate

Abbreviations

C_2S	Dicalcium silicate
C_3S	Tricalcium silicate
C_3A	Tricalcium aluminate
C_4AF	Tetra calcium aluminoferrite
$C\bar{S}H_2$	Gypsum

P. Verma (✉)
Department of Civil Engineering, IIT Roorkee, Roorkee, India

A. Bajpai
Department of Civil Engineering, Ram Manohar Lohia Avadh University Ayodhya, Faizabad, India

A. Gupta et al. (eds.), *Advances in Energy and Combustion*, Green Energy and Technology,
https://doi.org/10.1007/978-981-16-2648-7_24

CH	Calcium hydroxide
CSH	Calcium silicate hydrate gel
C_3AH_6	Hydrogarnet
$C_6A\overline{S_3}H_{32}$	Ettringite
$C_4A\overline{S}H_{12}$	Monosulfate AFM
FH_3	Iron hydroxide
RVE	Representative volume element
ITZ	Interfacial transition zone
MM	Multiscale modelling
FEA	Finite element analysis
VCCTL	Virtual cement and concrete testing laboratory
NIST	National Institute of Standards and Technology
E	Elastic modulus
G	Shear modulus
K	Bulk modulus
K^*	Effective bulk modulus
G^*	Effective shear modulus
μ	Poisson's ratio
ρ	Mass density
E_1	Tensile deformation (Uniaxial)
G_{12}	Shear deformation in x-y plane
E_i	Homogeneous modulus corresponding to applied axial deformation mode ($i = 1, 2, 3$)
G_{ij}	Homogeneous modulus corresponding to shear deformation mode ($i, j = 1, 2, 3$)
CV	Computational volume
α	Degree of hydration
V_f	Volume fraction of fibre material
V_m	Volume fraction of matrix material
$\overline{\sigma_{ij}}$	Average stress tensor over the entire RVE
$\overline{\varepsilon_{ij}}$	Average strain tensor over the entire RVE

1 Introduction

As we all know that the cementitious materials are highly heterogeneous and complex structure because their microstructures have random features at different length scale level such as CSH, cement paste, mortar, and concrete level [1, 2]. Every scale represents a random composite. At the nanoscale, the CSH is a random composite which exists in two different forms such as high-density CSH and low-density CSH because of different packing densities. Cement paste has a complex structure at microscale with a random arrangement of coarse and fine aggregate in concrete. However, at the meter scale generally concrete is considered to be uniform with bulk

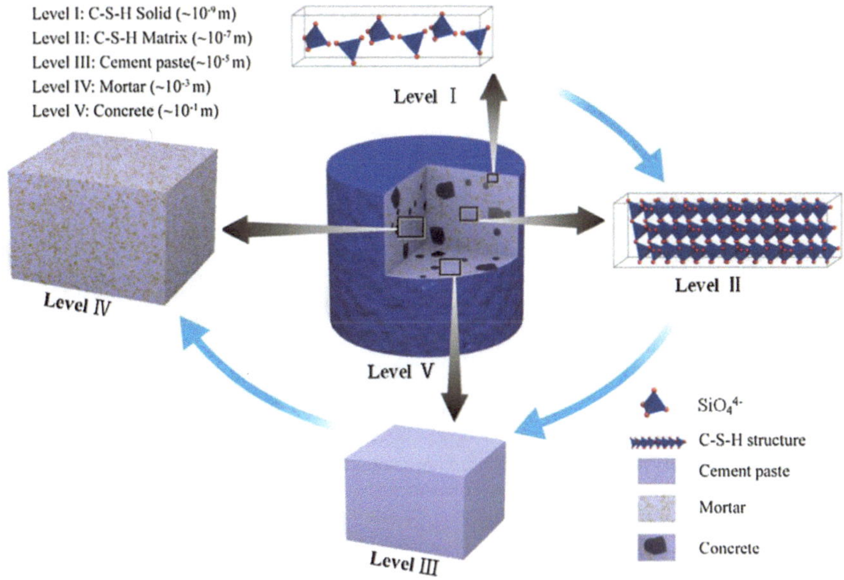

Fig. 1 Different levels of cement-based composites [6]

characteristic properties such as compressive strength, elastic modulus, and others [3]. The multiscale level for cement-based composites can be illustrated in Fig. 1. It can be seen from the figure that there are different levels of cement-based materials that effect the loading behavior and properties of concrete [4, 5]. The modeling at the level of concrete is difficult because here concrete is modeled by considering three-phase material such as cement paste, interfacial transition zone (ITZ), and aggregate phase. Therefore, finding effective properties of cementitious materials are required to help engineers during design process.

Multiscale modeling (MM) of materials is the most prominent priority of the research community for achieving material properties. Now, it is possible to model cementitious materials by selecting RVE size which governs the overall properties of material [7]. The elastic properties of individual phase and their densities such as clinkers, hydrated and unhydrated products can be estimated using molecular dynamics (MD) simulations. These properties are employed during modeling of RVE of different sizes. However, it is very difficult to analyze the behavior of the material for predicting its property using different realistic models of its microstructure However, the properties of the individual components of concrete, knowledge of its microstructure, and their relationship to each other is useful for controlling the properties of concrete [8–11].

In this paper, an attempt is made for estimating the effective properties of heterogeneous CSH-based multiphase cementitious composite, based on knowledge of individual phases. The focus of this paper is two-fold: (i) to develop a microstructural information of cement paste during hydration process. (ii) to model the cement paste

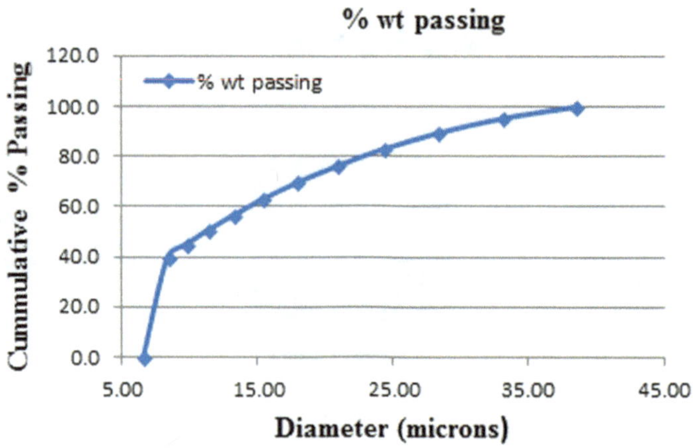

Fig. 2 Particle size distribution for simulation corresponding to Blaine Fineness 370 m^2/kg used in the hydration model

in FEA software to get homogenized elastic properties of cementitious materials. The main aim of this study is to determine the homogenized properties at macrolevel using micromechanics-based model when small strain quasi-static loading conditions are applied.

2 Methodology

2.1 Model for Cement Paste

In order to investigate the microstructure development with time and the change of its properties, in terms of its mechanical properties quantitative topological and geometrical description is needed. The most used digital microstructure model Virtual Cement and Concrete Testing Laboratory (VCCTL) modeling platform developed by National Institute of Standards and Technology (NIST) researchers which were based on CEMHYD3D developed by Bentz and Bentz [12]. There is quickly generation of microstructure using this model which is based on discretization of computational volumes into cubic voxel, each occupies by a single phase of cement constituents [13].

The modeling platform VCCTL was used to model hydration of cement paste and 3CaO·SiO or C_3S in chemistry notation are the main phase of Portland cement and resulting three-dimensional microstructures. The advantage of using microstructures from is that rather generating random spatial distribution of phases, it simulates the processes that lead to microstructural development, generating more realistic representations of microstructures.

2.2 *Simulation*

The composition of Portland cement that is simulated is shown in Fig. 3 and their Particle Size Distribution is shown in Fig. 2. There are three simulations is done at a different water-cement ratio i.e. 0.3, 0.35, and 0.4. The simulation is carried out using the VCCTL modeling tools in which user may input or choose the material characterization and the computational cubic volume is associated with one million (μm) in size and this size is sufficient for simulating hydration and porosity.

The density and elastic properties of each phases are assumed as given in Tables 1 and 2. Each reaction of cement clinkers is calculated in volumetric terms, and the hydrated products e.g. ettringite and C–S–H (Calcium Silicate Hydroxide) their deposition starts around the cement particles and the other products, i.e. Portlandite nucleates in the pores, and the particles are produced throughout the first 24 h of hydration to give a final number around 20% of the number of C_3S particles in the system.

3 Homogenization of Cementitious Composites

Homogenization simply means to find out the effective properties of heterogeneous materials. Homogenization is a method for finding the microscopic fields and properties based on the microstructure parameters and local properties of heterogeneous media and the effective properties represent the overall behavior of heterogeneous media which depends upon the phase properties and microstructure information of heterogeneous materials.

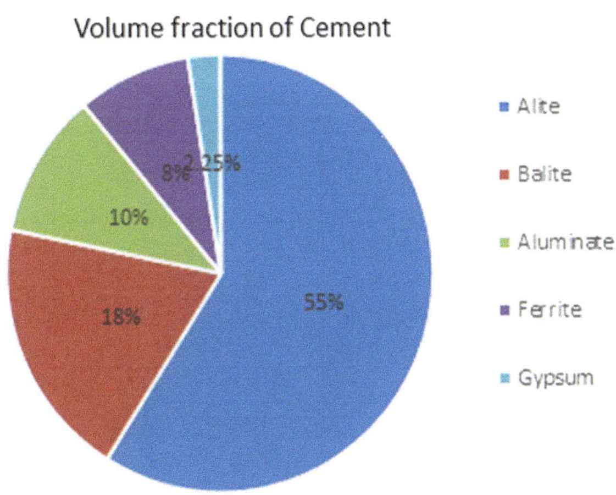

Fig. 3 Compositions of Portland cement

Table 1 Elastic properties of chemical phases measured by the mechanical tests [4, 11, 12, 14–17]

Phase	Young modulus (GPa)	Poisson's ratio (μ)
Tricalcium silicate (C_3S)	117.6	0.314
Dicalcium silicate (C_2S)	117.6	0.314
Tricalcium aluminate (C_3A)	117.6	0.314
Tetracalcium aluminoferrite (C_4AF)	117.6	0.314
Gypsum $\left(C\overline{S}H_2\right)$	45.7	0.33
Calcium hydroxide (CH)	42.3507	0.324
Calcium silicate hydrate gel (CSH)	22.4	0.25
Hydrogarnet (C_3AH_6)	22.4	0.25
Ettringite ($C_6A\overline{S_3}H_{32}$)	22.4	0.25
Monosulfate AFM ($C_4A\overline{S}H_{12}$)	0.0423	0.324
Iron hydroxide (FH_3)	22.4	0.25
Empty porosity	0.001	0.00
Water filled porosity	0.001	0.499924

Table 2 Bulk density of each phase of cement used in the simulation [18]

Phase	Chemical formula	Notation	Density (g/cm^3)
Air	–	–	0.00
Alite	$(CaO)_3(SiO_2)$	C_3S	3.21
Aluminate	$(CaO)_3(Al_2O_3)$	C_3A	3.03
Anhydrite	$CaSO_4$	CS	3.28
Belite	$(CaO)_2(SiO_2)$	C_2S	3.28
Calcium silicate hydrate	$(CaO)_3(SiO_2)$ $3H_2O$	C–S–H	2.12
Calcium silicate hydrate	$(CaO)_{1.7}(SiO_2)$ $1.8H_2O$	C–S–H	2.60
Ettringite	$(CaO)_3(Al_2O_3)(CaSO_4)_3$ $32H_2O$	C_6 A\$$_3$ H_{32}	1.78
Ferrite	$(CaO)_4(Al_2O_3)(Fe_2O_3)$	C_4AF	3.73
Free lime	CaO	C	2.52
Gypsum	$CaSO_4$ $2H_2O$	C\$$H_2$	2.32
Monosulfate	$(CaO)_3(Al_2O_3)(CaSO_4)$ $12H_2O$	C_4 A\$$H_{12}$	2.00
Periclase	MgO	M	3.58
Portlandite	$Ca(OH)_2$	CH	2.24
Quartz	SiO_2	S	2.65
Water	H_2O	H	1.00
Hydrogarnet	$(CaO)_3Al_2O_3$ $6(H_2O)$	C_3AH_6	2.52

3.1 Representative Volume Element (RVE)

During analysis, the selection of RVE size must be necessary for homogenization. The concept of representative volume element (RVE) was developed by Hill [19]. For the composite materials, the representative elementary volume (REV) (also called the representative volume element (RVE) or the unit cell) is the smallest volume over which a measurement can be made that will yield a value representative of the whole. The Representative volume element is defined as a small volume taken that is less than the actual volume of material that capable to capture the overall properties of the composite. Alternatively, Representative volume element also be defined as a Volume that shows the effective properties of chosen RVE size is same as the effective properties of overall behavior.

3.2 RVE Generation

The software package virtual cement and concrete testing laboratory (VCCTL) based on CEMHYD3D V.3 which was developed by the National Institute of Standards and technology. Researchers also use this modeling platform to simulate the hydration of cement paste and concrete and the formation of digital generated microstructures for type-1 Portland cement [12]. This software creates a microstructure information which depends particle size distribution, surface area fraction, and the volume fraction of the several constituents of cement phases. Microstructural information stored at a different degree of hydration. For obtaining the result, it is important to choose a cubic size of 4 different domains of "$10 \times 10 \times 10$" μm (1 K), "$20 \times 20 \times 20$" μm (8 K) and "$25 \times 25 \times 25$" μm (15.625 K) are considered in this study. A 3-D periodic microstructure is generated and which contains typically four cement clinkers minerals, gypsum, and other hydrated products, in the form of spherical particles [12] and these spherical particles are placed into a 3-D Computational Volume having individual voxels size is 1 μm and the particle size ranges varies from 1 to 35 μm in Diameter of material properties with respect to the "boundary condition" applied [20] (K Sab—Eur. J. Mech & 1992).

3.3 Finite Element Analysis (FEA) Based Microstructure Modeling

The microstructure voxel-based information generated by hydration model which was imported into finite element software using a python scripting, MATLAB coding, and geometry modeled using continuum C3D8 brick elements of dimension "$10 \times 10 \times 10$" μm (1 K), "$20 \times 20 \times 20$" μm (8 K) and "$25 \times 25 \times 25$" μm (15.625 K) as in Fig. 4.

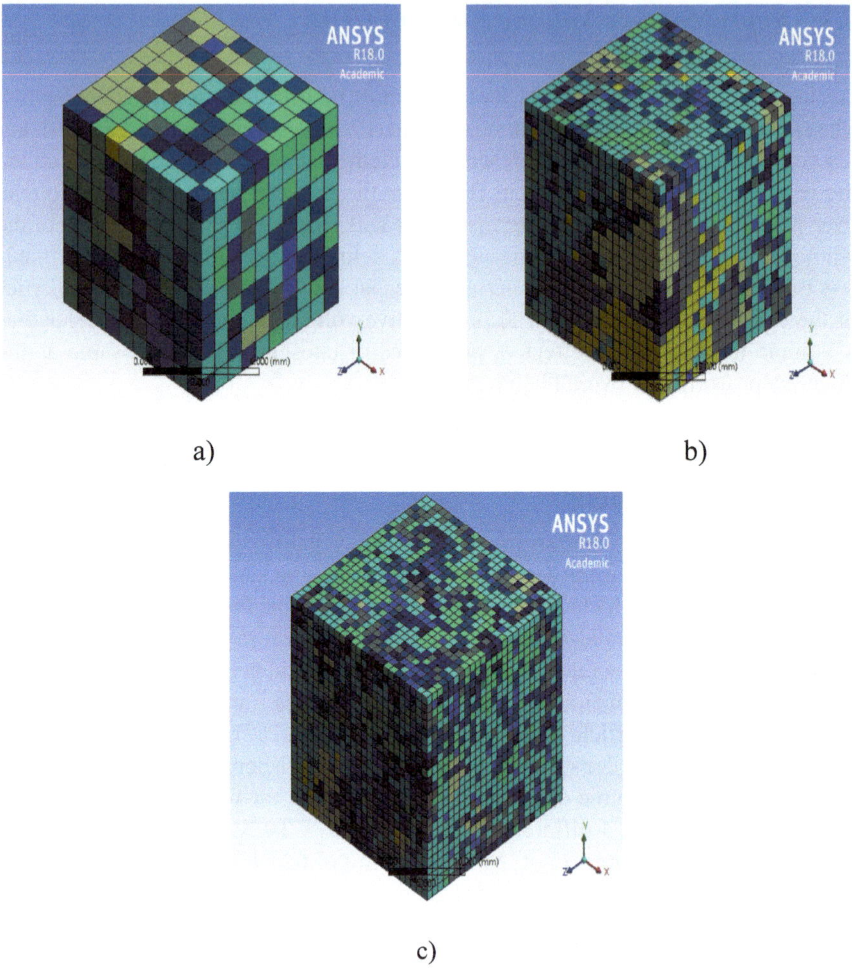

Fig. 4 **a** 1 K, **b** 8 K, **c** 15.625 K models of hydrated cement paste

Mechanical properties can find out by creating a micromechanical model based on the microstructural information available during simulation, and the appropriate constitutive equation has been created. Microstructure modeling can be divided into three parts: geometry, boundary condition, and constitutive equations. The modeled microstructure of cement paste as in Fig. 4.

3.4 Boundary Conditions

Boundary conditions are applied to the modeled geometry of various sizes which was shown in the literature [11, 12, 14, 21]. For the estimation of effective properties, of the cementitious material the modeled geometry of sizes 1, 8, and 15.625 K was subjected to various deformation modes which were shown in Fig. 4.

For the estimation of effective properties of a composite model, it may be necessary to impose uniform strain to compute elastic properties in which the pure tensile deformation and pure shear deformation are applied as shown in Fig. 5a and b. The homogenized elastic properties are calculated by assigning the elastic properties of each hydrated or unhydrated phase of cement and subjected to the RVE for imposing deformation modes. The elastic moduli are the ratio of average stress to the average strain The elastic moduli are the ratio of average stress and strain corresponding to deformation mode.

$$E_i = \frac{\sigma_i^{\text{avg}}}{\in_i^{\text{avg}}},\ G_{ij} = \frac{\sigma_i^{\text{avg}}}{\in_j^{\text{avg}}} \tag{1}$$

where E_i is a homogeneous modulus corresponding to applied deformation mode (i, $j = 1, 2, 3$) And in case of pure shear deformation the tensorial shear strain can be computed as (Fig. 6)

$$\in_{xy} = \frac{1}{2}(\theta_1 + \theta_2) = \frac{1}{2}\left(\frac{\Delta \text{x}}{\Delta y} + \frac{\Delta \text{y}}{\Delta x}\right) \tag{2}$$

Two examples related to the uniaxial deformation (E_1) along the x-axis and shear deformation (G_{12}) in the x-y plane are explained in details

Case 1 Applied strain: $\in_{11} \neq 0$ and $(\in_{22}, \in_{33}, \in_{12}, \in_{13}, \in_{23} = 0)$ for uniaxial deformation along the x-axis.

Nodes on faces: $\in_{11} \neq 0,\ i = 1$	Nodes on edges: $\in_{11} \neq 0,\ i = 1$	Nodes on vertices: $\in_{11} \neq 0,\ i = 1$
$u_1^{F2} - u_1^{F1} - L_x \in_{11} = 0$	$u_1^{E2} - u_1^{E4} - L_x \in_{11} = 0$	$u_1^{V3} - u_1^{V5} - L_x \in_{11} = 0$
$u_1^{F4} - u_1^{F3} = 0$	$u_1^{E1} - u_1^{E3} - L_x \in_{11} = 0$	$u_1^{V2} - u_1^{V8} - L_x \in_{11} = 0$
$u_1^{F6} - u_1^{F5} = 0$	$u_1^{E6} - u_1^{E8} - L_x \in_{11} = 0$	$u_1^{V7} - u_1^{V1} - L_x \in_{11} = 0$
	$u_1^{E5} - u_1^{E7} - L_x \in_{11} = 0$	$u_1^{V4} - u_1^{V6} - L_x \in_{11} = 0$
	$u_1^{E11} - u_1^{E9} = 0$	
	$u_1^{E10} - u_1^{E12} = 0$	

Case 2 Applied strain: $\in_{12}$ and $\in_{21} \neq 0$ and $(\in_{11}, \in_{22}, \in_{33}, \in_{12}, \in_{13}, \in_{23} = 0)$ for shear deformation in x-y plane.

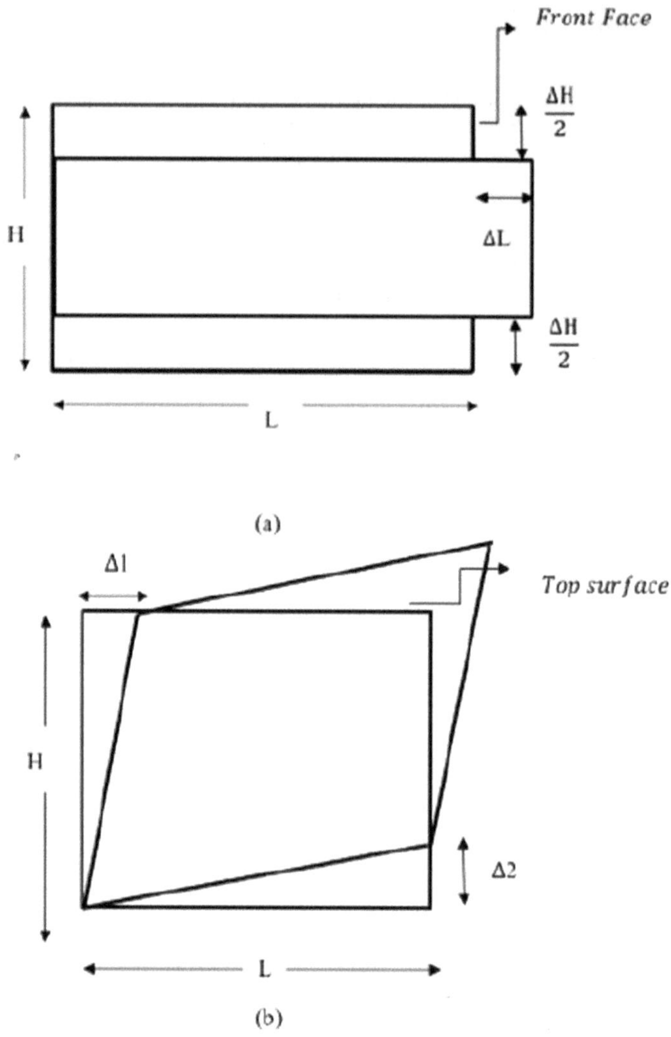

Fig. 5 Model subjected to displacements **a** tensile deformation (E_1) and **b** Pure shear (G_{12}) boundary conditions

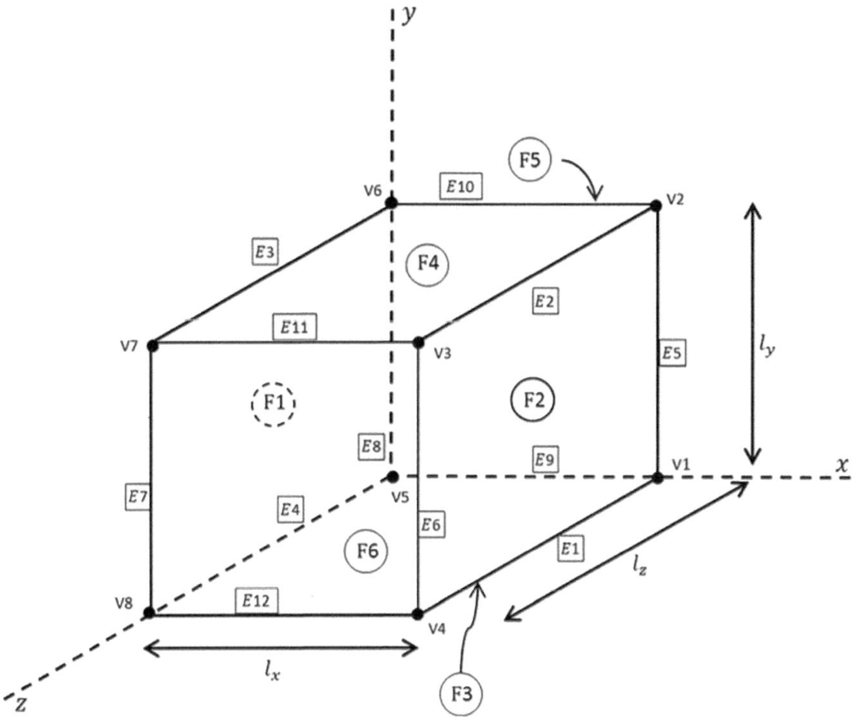

Fig. 6 Prescribed boundary condition

Nodes on faces: $\in_{12} \neq 0, \; i = 1$	Nodes on edges: $\in_{12} \neq 0, \; i = 1$	Nodes on vertices: $\in_{12} \neq 0, \; i = 1$
$u_1^{F2} - u_1^{F1} = 0$	$u_1^{E2} - u_1^{E4} - L_y \in_{12} = 0$	$u_1^{V3} - u_1^{V5} - L_y \in_{12} = 0$
$u_1^{F4} - u_1^{F3} - L_y \in_{12} = 0$	$u_1^{E1} - u_1^{E3} + L_y \in_{12} = 0$	$u_1^{V2} - u_1^{V8} - L_y \in_{12} = 0$
$u_1^{F6} - u_1^{F5} = 0$	$u_1^{E6} - u_1^{E8} = 0$	$u_1^{V7} - u_1^{V1} - L_y \in_{12} = 0$
	$u_1^{E5} - u_1^{E7} = 0$	$u_1^{V4} - u_1^{V6} + L_y \in_{12} = 0$
	$u_1^{E11} - u_1^{E9} - L_y \in_{12} = 0$	
	$u_1^{E10} - u_1^{E12} - L_y \in_{12} = 0$	

Nodes on faces: $\in_{21} \neq 0,\ i = 2$	Nodes on edges: $\in_{21} \neq 0,\ i = 2$	Nodes on vertices: $\in_{21} \neq 0,\ i = 2$
$u_2^{F2} - u_2^{F1} - L_x \in_{21} = 0$	$u_2^{E2} - u_2^{E4} - L_x \in_{21} = 0$	$u_2^{V3} - u_2^{V5} - L_x \in_{21} = 0$
$u_2^{F4} - u_2^{F3} = 0$	$u_2^{E1} - u_2^{E3} - L_x \in_{21} = 0$	$u_2^{V2} - u_2^{V8} - L_x \in_{21} = 0$
$u_2^{F6} - u_2^{F5} = 0$	$u_2^{E6} - u_2^{E8} - L_x \in_{21} = 0$	$u_2^{V7} - u_2^{V1} + L_x \in_{21} = 0$
	$u_2^{E5} - u_2^{E7} - L_x \in_{21} = 0$	$u_2^{V4} - u_2^{V6} + L_x \in_{21} = 0$
	$u_2^{E11} - u_2^{E9} = 0$	
	$u_2^{E10} - u_2^{E12} = 0$	

3.5 *Homogenization*

There are different homogenization approaches. Analytical homogenization and homogenization based on volume averages of field quantities such as stress, strain, and the effective properties can be found according to the definition of effective properties of the composites. Finite element analysis can perform the average and the calculation of local field quantities. The simplest approach of the estimation of mechanical properties, under the loading condition such as uniaxial tension test [11]. There are several studies conducted for finding the effective properties by imposing boundary conditions are briefly discussed in [14, 21].

3.5.1 Homogenization Based on FEM

FEM is a tool for micromechanical analysis that can be conveniently applied even to highly complex microstructures where a discrete mesh can be generated. Similar to the traditional FEM used in structural mechanics, here the stress and strain fields at each integration point in the CV are approximated after imposing kinematic or uniform boundary conditions. In the kinematic boundary conditions, a uniform deformation is applied to two opposite faces of the CV. In the static boundary condition, a uniform force is applied to one of the faces, and the opposite face is held fixed. The Elastic modulus and Poisson's ratio can be extracted using the averages of stress and strain fields over the whole CV. It is assumed that the average properties of a representative volume element are equal to the average properties of the heterogeneous material.

3.5.2 Average Fields

If the exact micro field σ_{ij} and $\in_{ij}$ in the RVE are Known under the applied load, the averaged stresses and strain over the RVE are given by Kurukuri and Eckardt [22].

$$\overline{\sigma_{ij}} = \frac{1}{V}\int \sigma_{ij}\mathrm{d}V \tag{3}$$

$$\overline{\varepsilon_{ij}} = \frac{1}{V}\int \epsilon_{ij}\ \mathrm{dV} \tag{4}$$

where V = Volume of RVE.

The averages are then treated as the effective stress and strain fields in the homogenized RVE. When linear elasticity is assumed, the generalized Hook's law applies

$$\overline{\sigma_{ij}} = C_{ijkl}\overline{\varepsilon_{ij}} \tag{5}$$

C_{ijkl} is a fourth-order tensor. It is also known as the constitutive matrix.

4 Results and Discussion

4.1 FEM-Based Homogenization

The microstructure of each periodic microstructure domain is captured at various degrees of hydration (DOH (α) = 0.3, 0.5, or 0.8) in order to study the evolution of elastic properties of cement paste. In case of small domain size i.e. less than 8 K, the volume fraction of various constituent's phases cannot be maintained due to low number of particles present in the computational volume. In the case of low domain size 1 K the volume fraction of individual constituents becomes unfeasible. A significant increment is observed in the volume fraction of C–S–H as domain size is increased from 1 to 8 K. This is because of more active surface of C–S–H is formed.

The elastic tensor (Voigt notation) for an isotropic material is given by

$$\begin{bmatrix} \sigma_{11} \\ \sigma_{22} \\ \sigma_{33} \\ \tau_{12} \\ \tau_{23} \\ \tau_{13} \end{bmatrix} = \begin{bmatrix} C_{11} & C_{12} & C_{13} & 0 & 0 & 0 \\ C_{21} & C_{22} & C_{23} & 0 & 0 & 0 \\ C_{31} & C_{32} & C_{33} & 0 & 0 & 0 \\ 0 & 0 & 0 & C_{44} & 0 & 0 \\ 0 & 0 & 0 & 0 & C_{55} & 0 \\ 0 & 0 & 0 & 0 & 0 & C_{66} \end{bmatrix} \begin{bmatrix} \epsilon_{11} \\ \epsilon_{22} \\ \epsilon_{33} \\ \gamma_{12} \\ \gamma_{23} \\ \gamma_{13} \end{bmatrix} \tag{6}$$

where the individual components of the stiffness tensor are given by

$$C_{11} = \frac{-E_1(1 - \nu_{23}\nu_{32})}{\Delta} \tag{7}$$

$$C_{12} = \frac{-E_1(\nu_{21} + \nu_{23}\nu_{31})}{\Delta} \tag{8}$$

$$C_{13} = \frac{-E_1(\nu_{31} + \nu_{21}\nu_{32})}{\Delta} \quad (9)$$

$$C_{21} = \frac{-E_2(\nu_{12} + \nu_{13}\nu_{32})}{\Delta} \quad (10)$$

$$C_{22} = \frac{-E_2(1 - \nu_{13}\nu_{31})}{\Delta} \quad (11)$$

$$C_{23} = \frac{-E_2(\nu_{32} + \nu_{12}\nu_{31})}{\Delta} \quad (12)$$

$$C_{31} = \frac{-E_3(\nu_{13} + \nu_{12}\nu_{23})}{\Delta} \quad (13)$$

$$C_{32} = \frac{-E_3(\nu_{23} + \nu_{13}\nu_{21})}{\Delta} \quad (14)$$

$$C_{33} = \frac{-E_3(1 - \nu_{12}\nu_{21})}{\Delta} \quad (15)$$

$$C_{44} = G_{12}, C_{55} = G_{13}, C_{66} = G_{23} \quad (16)$$

$$\Delta = \nu_{12}\nu_{21} + \nu_{13}\nu_{31}\nu_{23}\nu_{32} + \nu_{12}\nu_{23}\nu_{31} + \nu_{13}\nu_{21}\nu_{32} - 1 \quad (17)$$

To determine the anisotropy more precisely, the following parameters were described by Kanit et al. [23].

$$a = \frac{2Y_{44}}{Y_{11} - Y_{12}} \quad (18)$$

where

$$Y_{44} = \frac{C_{44} + C_{55} + C_{66}}{3}, Y_{12} = \frac{C_{12} + C_{23} + C_{13}}{3}, Y_{11} = \frac{C_{11} + C_{22} + C_{23}}{3}$$

The coefficient a is equal to 1 if the elastic behavior is perfectly isotropic. So, therefore the effective engineering constants for an isotropic material can be determined by

$$Y_{11} = C_1; Y_{12} = C_2$$
$$\lambda = C_2, G = Y_{11}, K = \lambda + \frac{2}{3}G$$
$$\nu = \frac{\lambda}{2(\lambda + \mathrm{G})}, E = \frac{G(3\lambda + 2\mathrm{G})}{\lambda + \mathrm{G}} \quad (19)$$

Table 3 Effective properties obtained from elastic matrix assuming orthotropic symmetry

Domain size (μm)	DOH	E (GPa)	G (GPa)	λ (GPa)	K (GPa)	ν	a
20 × 20 × 20	0.8	23.097	9.348567	8.312469	14.54485	0.235334	0.874179
20 × 20 × 20	0.5	18.7496	7.3485	9.034908	13.93395	0.275732	1.031
20 × 20 × 20	0.3	15.4262	5.885233	9.650537	13.57403	0.310591	0.940254
25 × 25 × 25	0.8	23.955	9.801	7.834262	14.36826	0.222119	0.854855

After applying the boundary condition, the total of six pure axial (E_i) and shear deformation (G_{ij}) cases are analyzed.

$$E_{11} = \sum_{i=1}^{n} \frac{\sigma_{11}^{i} V_i}{\epsilon_{11}^{i} V_i},\quad E_{22} = \sum_{i=1}^{n} \frac{\sigma_{22}^{i} V_i}{\epsilon_{22}^{i} V_i},\quad E_{33} = \sum_{i=1}^{n} \frac{\sigma_{33}^{i} V_i}{\epsilon_{33}^{i} V_i} \tag{20}$$

$$G_{12} = \sum_{i=1}^{n} \frac{\tau_{12}^{i} V_i}{\gamma_{12}^{i} V_i},\quad G_{23} = \sum_{i=1}^{n} \frac{\tau_{23}^{i} V_i}{\gamma_{23}^{i} V_i},\quad G_{31} = \sum_{i=1}^{n} \frac{\tau_{31}^{i} V_i}{\gamma_{33}^{i} V_i} \tag{21}$$

$$\begin{gathered}\nu_{12} = \sum_{i=1}^{n} \frac{\epsilon_{22}^{i} V_i}{\epsilon_{11}^{i} V_i},\quad \nu_{23} = \sum_{i=1}^{n} \frac{\epsilon_{33}^{i} V_i}{\epsilon_{22}^{i} V_i},\quad \nu_{31} = \sum_{i=1}^{n} \frac{\epsilon_{11}^{i} V_i}{\epsilon_{33}^{i} V_i}\\ \nu_{21} = \sum_{i=1}^{n} \frac{\epsilon_{11}^{i} V_i}{\epsilon_{22}^{i} V_i},\quad \nu_{32} = \sum_{i=1}^{n} \frac{\epsilon_{22}^{i} V_i}{\epsilon_{33}^{i} V_i},\quad \nu_{13} = \sum_{i=1}^{n} \frac{\epsilon_{33}^{i} V_i}{\epsilon_{11}^{i} V_i}\end{gathered} \tag{22}$$

Therefore all E_{11}, E_{22}, E_{33}, G_{12}, G_{23}, G_{31}, ν_{12}, ν_{21}, ν_{23}, ν_{32}, ν_{31}, ν_{13} are calculated for finding the elastic matrix obtained by applying boundary conditions to different size of RVE domain at DOH 0.8, 0.5, 0.3 using finite element software (Table 3).

$$\begin{bmatrix}\sigma_{11}\\ \sigma_{22}\\ \sigma_{33}\\ \tau_{12}\\ \tau_{23}\\ \tau_{13}\end{bmatrix} = \begin{bmatrix} 29.47 & 9.03522 & 7.6743 & 0 & 0 & 0\\ 9.03513 & 33.30632 & 8.22781 & 0 & 0 & 0\\ 7.67692 & 8.2293 & 26.3221 & 0 & 0 & 0\\ 0 & 0 & 0 & 10.28 & 0 & 0\\ 0 & 0 & 0 & 0 & 8.56 & 0\\ 0 & 0 & 0 & 0 & 0 & 9.2057\end{bmatrix}\begin{bmatrix}\epsilon_{11}\\ \epsilon_{22}\\ \epsilon_{33}\\ \gamma_{12}\\ \gamma_{23}\\ \gamma_{13}\end{bmatrix}$$

Elastic matrix for RVE size domain 8K at DOH = 0.8

The strength of cement paste is controlled by the formation of C–S–H and the refinement of pore space. As the hydration process of cement paste proceeds the formation of C–S–H results in the reduction of free water from the pores which reduces the Poisson ratio and increment in the elastic modulus as in Figs. 7, 8, and 9 respectively.

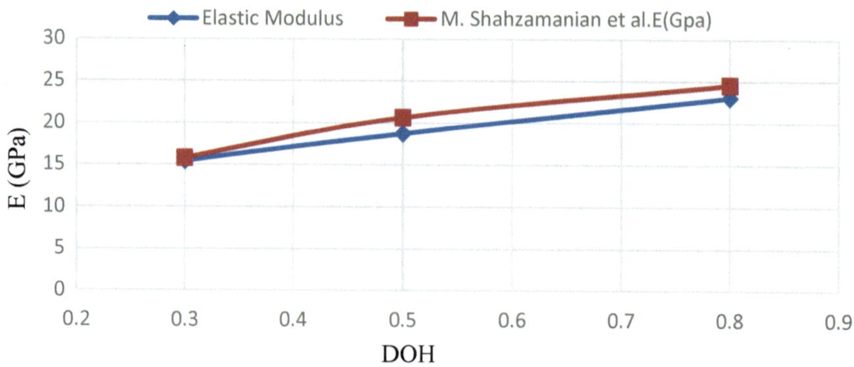

Fig. 7 Effect of degree of hydration on elastic modulus (*E*)

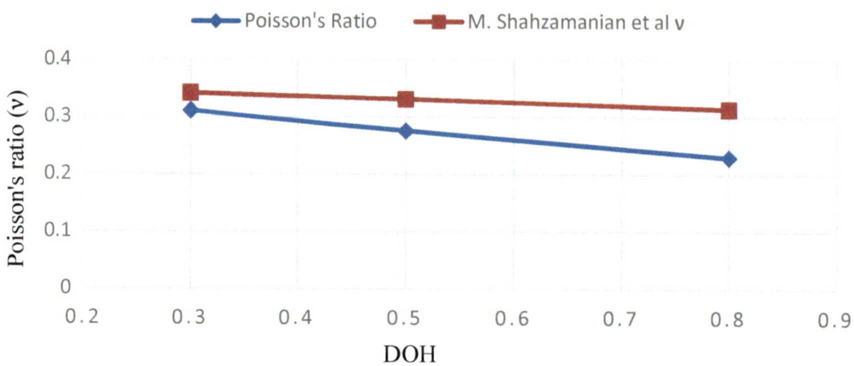

Fig. 8 Effect of degree of hydration on Poisson's ratio

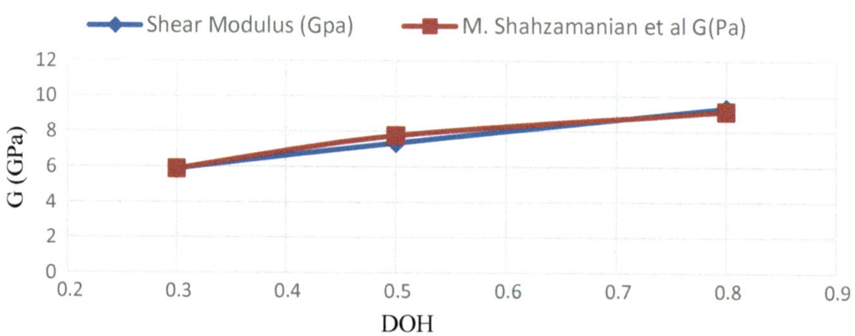

Fig. 9 Effect of degree of hydration on shear modulus

The elastic modulus (E) and shear modulus (G) of cement paste obtained by the homogenization based on FEM their variation with respect to DOH is shown in Fig. 9. The out coming results validated with the previous study by Shahzamanian et al. [24].

4.2 Analytical Homogenization

Typical representatives of homogenization methods are rule of mixtures [1], Hashin-Shtrikman bound [25], Mori-Tanaka [26] or self-consistent [19] method. This approach is also termed as analytical homogenization approach.

The assumption is taken that the analytical homogenization with the real microstructure of cementitious materials leads to significant refinements of results. The validity of results checked and compared with the homogenization based on FEM. The key term is the representative volume element (RVE) that will able to analyze and capture limited piece of the heterogeneity in the material. Then the numerical methods must be used for calculation of fluctuation of stress or starin field to bridge the length scale of the materials. This approach is referred to as a numerical approach based on FEM. Here we have to discuss the method of prediction of models and bounds for elastic predictions of composites are reviewed. The method comprise the following.

4.2.1 Rule of Mixtures Based Homogenization

A rule of mixtures approach independent of the microstructure of the material is used to compute the effective bulk properties of the cementitious material.

The theoretical extreme upper and lower bounds effective material properties of multiphase materials are the Voigt and Reuss [27] bounds. The rule of mixtures-based Voigt upper Bound on the effective bulk (K^*) and shear moduli (G) of a mixture of n phase material is given by

$$K^* = \sum_{i=1}^{n} f_i K_i \tag{23}$$

$$G^* = \sum_{i=1}^{n} f_i G_i \tag{24}$$

The inverse rule of mixtures based Reuss lower bound on effective bulk (K^*) and shear modulus (G^*) is given by

$$\frac{1}{K^*} = \sum_{i=1}^{n} \frac{f_i}{K_i} \tag{25}$$

Table 4 Volume fraction of various constituent of cement phases at w/c-0.3 and $(\alpha) = 0.8$

Phase	Phase ID	Volume fraction %	E (GPa)	Poisson's ratio (μ)	K (GPa)
Alite (C_3S)	1	5.1	117.6	0.314	105.3763
Belite (C_2S)	2	5.78	117.6	0.314	105.3763
Aluminate (C_3A)	3	0.33	117.6	0.314	105.3763
Ferrite (C_4AF)	4	3.03	117.6	0.314	105.3763
Gypsum	5	0.27	45.7	0.33	44.80392
Water-filled porosity	6	12.08	0.001	0.499924	2.192982
Calcium silicate hydrate gel (C–S–H)	7	42.4	22.4	0.25	14.93333
Calcium hydroxide	8	15.3	42.3507	0.324	40.10483
Ettringite ($C_6AS_3H_{32}$)	9	4	22.4	0.25	14.93333
Monosulfate	10	4.7	0.0423	0.324	0.040057
Iron Hydroxide	11	0.624	22.4	0.25	14.93333
Hydrogarnet	12	4.9	22.4	0.25	14.93333

$$\frac{1}{G^*} = \sum_{i=1}^{n} \frac{f_i}{G_i} \tag{26}$$

K^* = effective bulk moduli, G^* = effective shear moduli

The analytical homogenization is done at water-cement ratio 0.3, 0.35, and 0.4 corresponding to each water-cement ratio volume fraction of each hydrated and anhydrate phases are taken at the degree of hydration ($\alpha = 0.8$) as in Tables 4, 5, and 6.

The effective properties of cement paste calculated by the homogenization based on Finite element method and analytical homogenization using Voigt's and Reuss's bound which gives the upper and lower bounds of composite materials. The effective elastic properties has been calculated using voigt and reuse model as in Table 7.

So, it is estimated from the above results that the analytical homogenization overestimates the effective properties of cement paste when compared with microstructure-based homogenization.

Above Fig. 10, illustrates the variation of elastic modulus at different water-cement ratio from the graph it may be concluded that the water-cement ratio increases there is reduction of elastic properties due to more water content leads to decrease in the strength of cement paste.

Table 5 Volume fraction of various constituent of cement phases at w/c-0.35 and (α) = 0.8

Phase	Phase ID	Volume fraction %	*E* (GPa)	Poisson's ratio (μ)	*K* (GPa)
Alite (C_3S)	1	5.12	117.6	0.314	105.3763
Belite (C_2S)	2	5.81	117.6	0.314	105.3763
Aluminate (C_3A)	3	0.28	117.6	0.314	105.3763
Ferrite (C_4AF)	4	3.025	117.6	0.314	105.3763
Gypsum	5	0.25	45.7	0.33	44.80392
Water-filled porosity	6	16	0.001	0.499924	2.192982
Calcium silicate hydrate gel (C–S–H)	7	42.5	22.4	0.25	14.93333
Calcium hydroxide	8	15.25	42.3507	0.324	40.10483
Ettringite ($C_6AS_3H_{32}$)	9	4.75	22.4	0.25	14.93333
Monosulfate	10	4.12	0.0423	0.324	0.040057
Iron Hydroxide	11	0.66	22.4	0.25	14.93333
Hydrogarnet	12	5.52	22.4	0.25	14.93333

Table 6 Volume fraction of various constituent of cement phases at w/c-0.4 and (α) = 0.8

Phase	Phase ID	Volume fraction %	*E* (GPa)	Poisson's ratio (μ)	*K* (GPa)
Alite (C_3S)	1	3.18	117.6	0.314	I05.3763
Belite (C_2S)	2	4.08	117.6	0.314	I05.3763
Aluminate (C_3A)	3	0.125	117.6	0.314	I05.3763
Ferrite (C_4AF)	4	2.23	117.6	0.314	I05.3763
Gypsum	5	0.21059	45.7	0.33	44.80392
Water-filled porosity	6	21.4863	0001	0.49	2.192982
Calcium silicate hydrate gel (C–S–H)	7	37.5	22.4	0.25	14.93333
Calcium hydroxide	8	13.2	42.3507	0.324	40.10483
Ettringite ($C_6AS_3H_{32}$)	9	5.229	22.4	0.25	14.93333
Monosulfate	10	2.35	0.0423	0.324	0.040057
Iron Hydroxide	11	0.623	22.4	0.25	14.93333
Hydrogarnet	12	5.5	22.4	0.25	14.93333

Table 7 Comparison of theoretical bounds on homogenized elastic modulus for 1 M RVE (DOH = 0.8)

Effective properties	Voigt (V)	Reuss (R)
E	28.40	5.12×10^{-4}
ν	0.3145	0.29
K	25.52	0.6235
G	11.83141	2.00×10^{-4}

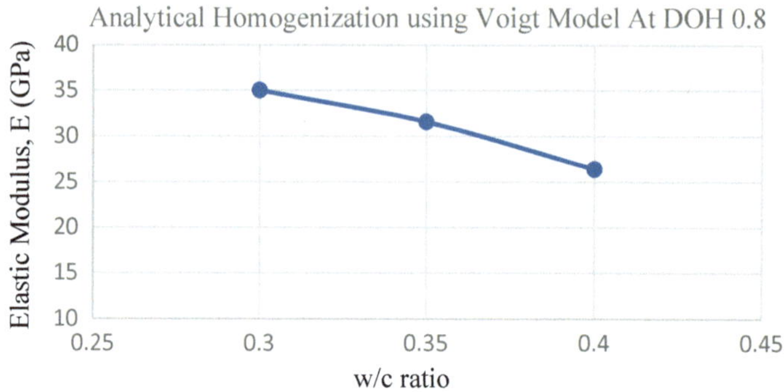

Fig. 10 Analytical homogenization using Voigt model at a different water-cement ratio at degree of hydration (α) = 0.8

5 Conclusions

Estimation of elastic properties is an important step in analysis and design of composite materials. The work aimed at the simulation of the hydration of cement paste and the overall goal is to determine the effective elastic properties of cement paste. The microstructure was produced using the modeling platforms. The study is done to determine the effective properties of type-1, Portland cement using homogenization based on FEA approach and analytical homogenization.

The elastic properties of the 3-D microstructure of hydrating cement paste were calculated at the early ages by applying finite element method. And it is found that low values of elastic modulus are found at a low degree of hydration due to not much gain of strength of cement paste at early ages. The calcium silicate hydroxide (C–S–H) gel formed between the particle and creates a percolated network that can increase the mechanical properties of the microstructure as hydration process take place. This corresponds to the increase in the elastic moduli. Due to an increase in the volume fractions of C–S–H and simultaneously decrease in volume fraction of other major constituents, which leads microstructure to become more isotropic. As the hydration of the cement paste proceeds, there is the formation of C–S–H that results in the depletion of free water content in the pores due to pore space is taken by the portlandite or calcium hydroxide which reduces the Poisson's ratio. Analytical homogenization overestimates the effective properties of cement paste as compared to homogenization based on FEM solution.

Acknowledgements We would like to thank Mr. Amardeep Dahiya, Assistant professor at National institute of technology, Uttarakhand and Dr. Siddharth Shukla, professor at Dr. Rammanohar Lohia Avadh University for providing their valuable guidance and constant encouragement throughout the work. This work is simply the reflection of their thoughts, ideas, and concepts and all their efforts. I am highly obliged to them for his kind and valuable suggestions and of course their valuable time during the period of the work. I would acknowledge my gratefulness to my parents for their support, love, patience and for being a constant source of inspiration. The financial support from Ministry of Human Resource Development (MHRD), Government of India is gratefully acknowledged.

References

1. Mehta PK, Monteiro PJM (2017) Concrete microstructure, properties and materials
2. Neville AM (1995) Properties of concrete, vol 4. Longman, London
3. Garboczi EJ, Bentz DP (1996) Multi-scale picture of concrete and its transport properties: introduction for non-cement researchers. US Department of Commerce, National Institute of Standards and Technology, Gaithersburg
4. Bernard O, Ulm F-J, Lemarchand E (2003) A multiscale micromechanics-hydration model for the early-age elastic properties of cement-based materials. Cem Concr Res 33(9):1293–1309
5. Lee J et al (2009) A multiscale model for modulus of elasticity of concrete at high temperatures. Cem Concr Res 39(9):754–762
6. Yang H et al (2017) A critical review on research progress of graphene/cement based composites. Compos A Appl Sci Manuf 102:273–296
7. Zaoui A (2002) Continuum micromechanics: survey. J Eng Mech 128(8):808–816
8. Torquato S (2002) Random heterogeneous materials. Microstructure and macroscopic properties. Interdisc Appl Math 16
9. Milton GW (2002) The theory of composites (Cambridge Monographs on Applied and Computational Mathematics). Cambridge University Press, Cambridge
10. Christensen RM (2012) Mechanics of composite materials. Courier Corporation
11. Haecker C-J et al (2005) Modeling the linear elastic properties of Portland cement paste. Cem Concr Res 35(10):1948–1960
12. Bentz DP, Bentz DP (1995) A three-dimensional cement hydration and microstructure program: I. hydration rate, heat of hydration, and chemical shrinkage. US Department of Commerce, National Institute of Standards and Technology
13. Bullard JW (2014) Virtual cement and concrete testing laboratory: version 9.5 user guide
14. Bernard F, Kamali-Bernard S, Prince W (2008) 3D multi-scale modelling of mechanical behaviour of sound and leached mortar. Cem Concr Res 38(4):449–458
15. Velez K et al (2001) Determination by nanoindentation of elastic modulus and hardness of pure constituents of Portland cement clinker. Cem Concr Res 31(4):555–561
16. Monteiro PJM, Chang CT (1995) The elastic moduli of calcium hydroxide. Cem Concr Res 25(8):1605–1609
17. Boumiz A et al (2000) Modelling the development of the elastic moduli as a function of the hydration degree of cement pastes and mortars. In: 2nd international RILEM workshop on hydration and setting: why does cement set? An interdisciplinary approach (Dijon, 11–13 June 1997)
18. Taylor HF (1997) Cement chemistry. Thomas Telford
19. Hill R (1965) Theory of mechanical properties of fibre-strengthened materials—III. Self-consistent model. J Mech Phys Solids 13(4):189–198
20. Nemat-Nasser S, Hori M (2013) Micromechanics: overall properties of heterogeneous materials, vol 37. Elsevier

21. Ren Z-Y, Zheng Q-S (2002) A quantitative study of minimum sizes of representative volume elements of cubic polycrystals—numerical experiments. J Mech Phys Solids 50(4):881–893
22. Kurukuri S, Eckardt S (2004) A review of homogenization techniques for heterogeneous materials. Weimar, Germany
23. Kanit T et al (2006) Apparent and effective physical properties of heterogeneous materials: representativity of samples of two materials from food industry. Comput Methods Appl Mech Eng 195(33–36):3960–3982
24. Shahzamanian MM et al (2014) Representative volume element based modeling of cementitious materials. J Eng Mater Technol 136(1)
25. Hashin Z, Shtrikman S (1963) A variational approach to the theory of the elastic behaviour of multiphase materials. J Mech Phys Solids 11(2):127–140
26. Mori T, Tanaka K (1973) Average stress in matrix and average elastic energy of materials with misfitting inclusions. Acta Metall 21(5):571–574
27. Zuo L, Humbert M, Esling C (1992) Elastic properties of polycrystals in the Voigt-Reuss-Hill approximation. J Appl Crystallogr 25(6):751–755